Nuclear Physics and Astrophysics VI (NPA6)

AA002684

Journal of Physics: Conference Series Volume 665

Lisbon, Portugal
19 - 24 May 2013

ISBN: 978-1-5108-1817-0
ISSN: 1742-6588

Printed from e-media with permission by:

Curran Associates, Inc.
57 Morehouse Lane
Red Hook, NY 12571

Some format issues inherent in the e-media version may also appear in this print version.

Printed by Curran Associates, Inc. (2016)

For permission requests, please contact the Institute of Physics at the address below.

Institute of Physics
Dirac House, Temple Back
Bristol BS1 6BE UK

Phone: 44 1 17 929 7481
Fax: 44 1 17 920 0979

techtracking@iop.org

Additional copies of this publication are available from:

Curran Associates, Inc.
57 Morehouse Lane
Red Hook, NY 12571 USA
Phone: 845-758-0400
Fax: 845-758-2634
Email: curran@proceedings.com
Web: www.proceedings.com

Table of contents

Volume 665

Nuclear Physics in Astrophysics VI (NPA6)
19–24 May 2013, Lisbon, Portugal

Accepted papers received: 13 November 2015
Published online: 5 January 2016

Preface

Papers

Big Bang Nucleosynthesis

Stellar Nucleosynthesis

The s-process

Explosive Nucleosynthesis (p-, rp-process)

Nuclear astrophysics with radioactive ions at FAIR OPEN ACCESS R Reifarth, S Altstadt, K Göbel, T Heftrich, M Heil, A Koloczek, C Langer, R Plag, M Pohl, K Sonnabend, M Weigand, T Adachi, F Aksouh, J Al-Khalili, M AlGarawi, S AlGhamdi, G Alkhazov, N Alkhomashi, H Alvarez-Pol, R Alvarez-Rodriguez, V Andreev, B Andrei, L Atar, T Aumann, V Avdeichikov, C Bacri, S Bagchi, C Barbieri, S Beceiro, C Beck, C Beinrucker, G Belier, D Bemmerer, M Bendel, J Benlliure, G Benzoni, R Berjillos, D Bertini, C Bertulani, S Bishop, N Blasi, T Bloch, Y Blumenfeld, A Bonaccorso, K Boretzky, A Botvina, A Boudard, P Boutachkov, I Boztosun, A Bracco, S Brambilla, J Briz Monago, M Caamano, C Caesar, F Camera, E Casarejos, W Catford, J Cederkall, B Cederwall, M Chartier, A Chatillon, M Cherciu, L Chulkov, P Coleman-Smith, D Cortina-Gil, F Crespi, R Crespo, J Cresswell, M Csatlós, F Déchery, B Davids, T Davinson, V Derya, P Detistov, P Diaz Fernandez, D DiJulio, S Dmitry, D Doré, J Dueñas, E Dupont, P Egelhof, I Egorova, Z Elekes, J Enders, J Endres, S Ershov, O Ershova, B Fernandez-Dominguez, A Fetisov, E Fiori, A Fomichev, M Fonseca, L Fraile, M Freer, J Friese, M G. Borge, D Galaviz Redondo, S Gannon, U Garg, I Gasparic, L Gasques, B Gastineau, H Geissel, R Gernhäuser, T Ghosh, M Gilbert, J Glorius, P Golubev, A Gorshkov, A Gourishetty, L Grigorenko, J Gulyas, M Haiduc, F Hammache, M Harakeh, M Hass, M Heine, A Hennig, A Henriques, R Herzberg, M Holl, A Ignatov, A Ignatyuk, S Ilieva, M Ivanov, N Iwasa, B Jakobsson, H Johansson, B Jonson, P Joshi, A Junghans, B Jurado, G Körner, N Kalantar, R Kanungo, A Kelic-Heil, K Kezzar, E Khan, A Khanzadeev, O Kiselev, M Kogimtzis, D Körper, S Kräckmann, T Kröll, R Krücken, A Krasznahorkay, J Kratz, D Kresan, T Krings, A Krumbholz, S Krupko, R Kulessa, S Kumar, N Kurz, E Kuzmin, M Labiche, K Langanke, I Lazarus, T Le Bleis, C Lederer, A Lemasson, R Lemmon, V Liberati, Y Litvinov, B Löher, J Lopez Herraiz, G Münzenberg, J Machado, E Maev, K Mahata, D Mancusi, J Marganiec, M Martinez Perez, V Marusov, D Mengoni, B Million, V Morcelle, O Moreno, A Movsesyan, E Nacher, M Najafi, T Nakamura, F Naqvi, E Nikolski, T Nilsson, C Nociforo, P Nolan, B Novatsky, G Nyman, A Ornelas, R Palit, S Pandit, V Panin, C Paradela, V Parkar, S Paschalis, P Pawłowski, A Perea, J Pereira, C Petrache, M Petri, S Pickstone, N Pietralla, S Pietri, Y Pivovarov, P Potlog, A Prokofiev, G Rastrepina, T Rauscher, G Ribeiro, M Ricciardi, A Richter, C Rigollet, K Riisager, A Rios, C Ritter, T Rodriguez Frutos, J Rodriguez Vignote, M Röder, C Romig, D Rossi, P Roussel-Chomaz, P Rout, S Roy, P Söderström, M Saha Sarkar, S Sakuta, M Salsac, J Sampson, J Sanchez, del Rio Saez, J Sanchez Rosado, S Sanjari, P Sarriguren, A Sauerwein, D Savran, C Scheidenberger, H Scheit, S Schmidt, C Schmitt, L Schnorrenberger, P Schrock, R Schwengner, D Seddon, B Sherrill, A Shrivastava, S Sidorchuk, J Silva, H Simon, E Simpson, P Singh, D Slobodan, D Sohler, M Spieker, D Stach, E Stan, M Stanoiu, S Stepantsov, P Stevenson, F Strieder, L Stuhl, T Suda, K Sümmerer, B Streicher, J Taieb, M Takechi, I Tanihata, J Taylor, O Tengblad, G Ter-Akopian, S Terashima, P Teubig, R Thies, M Thoennessen, T Thomas, J Thornhill, G Thungstrom, J Timar, Y Togano, U Tomohiro, T Tornyi, J Tostevin, C Townsley, W Trautmann, T Trivedi, S Typel, E Uberseder, J Udias, T Uesaka, L Uvarov, Z Vajta, P Velho, V Vikhrov, M Volknandt, V Volkov, P von Neumann-Cosel, M von Schmid, A Wagner, F Wamers, H Weick, D Wells, L Westerberg, O Wieland, M Wiescher, C Wimmer, K Wimmer, J S Winfield, M Winkel, P Woods, R Wyss, D Yakorev, M Yavor, J Zamora Cardona, I Zartova, T Zerguerras, M

Neutron Stars & Equation of State

Compact Stars and Supernovae

Preface

The Nuclear Physics in Astrophysics VI conference was the 6th event of the NPA biannual conference series. Previous events of this series were held at the Institute of Nuclear Research of the Hungarian Academy of Sciences (ATOMKI), Debrecen, Hungary, in 2002 and 2005; at the Forschungszentrum Dresden-Rossendorf, Dresden, Germany, in 2007; at the Laboratori Nazionali del Gran Sasso (LNGS), L'Aquila, Italy, in 2009; and in Eilat, Israel, in 2011. This edition of the NPA conference series was organized by the European Physical Society (EPS) through its Nuclear Physics Division, corresponding to the 26th edition of the Topical Conferences of the EPS.

As in previous editions, the goal of the NPA conference was to bring together the specialists in the fields of Nuclear Physics (theory and experiment) and Nuclear Astrophysics (theory and observation), providing the appropriate forum for review and discussion of the status and prospects of the field of Nuclear Astrophysics. During the discussions, special attention was given to the Nuclear Physics aspects that have an impact in Astrophysics. The list of topics discussed during the conference was:

- Big Bang Nucleosynthesis
- Stellar Nucleosynthesis and Evolution
- Experimental Facilities for Nuclear Astrophysics (Underground Laboratories)
- Experimental Facilities for Nuclear Astrophysics (Recoil Mass Separators)
- The s-process
- Explosive Nucleosynthesis I (r- and vp-process)
- Explosive Nucleosynthesis II (p and rp-process)
- Astrophysics of Compact Stars and Supernovae
- Neutron Stars and Equation of State (EoS)
- Isotope Detection Technologies for Nuclear Astrophysics
- Solar Standard Model
- Nuclear Physics of Exotic Nuclei: Experiments with RIB

The program consisted of keynote talks, providing a review on the recent advances in the field; invited talks and selected oral contributions that focused on specific topics; and poster contributions that had the opportunity to present the work to the audience via one minute oral contributions. In total 78 oral contributions and 64 posters were presented.

A total of 146 delegates participated in lively discussions during the oral presentations, the poster session scheduled during the second day of the conference, the coffee breaks and all programmed events of a social character. Regarding the participants, 41 were PhD candidates, 33 postdoctoral researchers and 72 senior scientists. The participation rate of women at the NPA-VI was 1/3. Fellowships to students and selected participants were supported by the Nuclear Astrophysics Virtual Institute (NAVI), the Joint Institute for Nuclear Astrophysics (JINA), the Network Activity ATHENA, and the ExtreMe Matter Institute (EMMI). This support was crucial to warrant the high participation of young researchers at the conference.

The excellent scientific atmosphere we experienced during the conference promoted vivid discussions, strengthened existing collaborations and motivated the continuation of the work in the interdisciplinary field of Nuclear Astrophysics, keeping the different communities on a joint working perspective.

Nuclear Physics
Astrophysics VI
LISBON · PORTUGAL · 2013

Committees

International Advisory Board (IAB)

- Marcello Baldo (Catania, Italy)
- Zsolt Fülöp (Debrecen, Hungary)
- Daniel Galaviz (Lisbon, Portugal)
- Roberto Gallino (Torino, Italy)
- Pawel Haensel (Torun, Poland)
- Michael Heil (Darmstadt, Germany)
- Karlheinz Langanke (Darmstadt, Germany)
- Ilidio Lopes (Lisbon, Portugal)
- Christopher Pethick (Nordita, Denmark)
- Claus Rolfs (Bochum, Germany)
- Friedrich Thielemann (Basel, Switzerland)
- Jim Truran (Chicago, USA)

International Program Committee (IPC)

- Nicolas Chamel (Brussels, Belgium)
- Pierre Descouvemont (Brussels, Belgium)
- Inma Dominguez Aguilera (Granada, Spain)
- Lucio Gialanella (Naples, Italy)
- Alessandra Guglielmetti (Milan, Italy)
- Adelaide P. de Jesus (Lisbon, Portugal)
- Jordi José (Barcelona, Spain)
- Franz Kaeppeler (Karlsruhe, Germany)
- Gabriel Martínez Pinedo (Darmstadt, Germany)
- Endre Somorjai (Debrecen, Hungary)
- Anton Wallner (Canberra, Australia)
- Michael Wiescher (South Bend, USA)

Journal of Physics: Conference Series **665** (2016) 011002 doi:10.1088/1742-6596/665/1/011002

Local Organizing Committee (LOC)

- Rui Agostinho (Lisbon)
- João Cruz (Lisbon)
- Daniel Galaviz (Lisbon, Chair)
- Adelaide P. de Jesus (Lisbon)
- André Moitinho (Lisbon)
- Constança Providência (Coimbra)
- Isaac Vidaña (Coimbra)

Local Organizing Staff

- Pablo Cabanelas
- Lucia Cartegni
- Ana Henriques
- Jorge Machado
- André Ornelas
- Pamela Teubig
- Paulo Velho

Journal of Physics: Conference Series **665** (2016) 011002 doi:10.1088/1742-6596/665/1/011002

Conference Organization

EPS - European Physical Society

SPF - Sociedade Portuguesa de Física

SPA - Sociedade Portuguesa de Astronomía

CFNUL - Centro de Física Nuclear da Universidade de Lisboa

FCT/UNL - Faculdade de Ciências e Tecnología da Universidade Nova de Lisboa

DF/UC - Departamento de Física - Universidade de Coimbra

OAL - Observatório Astronómico de Lisboa

SIM - Laboratório de Sistemas, Instrumentação e Modelação em Ciências e Tecnologias do Ambiente e do Espaço

Journal of Physics: Conference Series **665** (2016) 011002 doi:10.1088/1742-6596/665/1/011002

Institutional Sponsors

Corporate Sponsors

Peer review statement

All papers published in this volume of *Journal of Physics: Conference Series* have been peer reviewed through processes administered by the proceedings Editors. Reviews were conducted by expert referees to the professional and scientific standards expected of a proceedings journal published by IOP Publishing.

Nuclear Physics in Astrophysics VI (NPA6) IOP Publishing
Journal of Physics: Conference Series **665** (2016) 012001 doi:10.1088/1742-6596/665/1/012001

Primordial Nucleosynthesis

A Coc[1]

[1] Centre de Sciences Nucléaires et de Sciences de la Matière (CSNSM), CNRS/IN2P3, Université Paris Sud 11, UMR 8609, Bâtiment 104, F–91405 Orsay Campus, France

E-mail: Alain.Coc@csnsm.in2p3.fr

Abstract. Primordial nucleosynthesis, or Big Bang Nucleosynthesis (BBN), is one of the three evidences for the Big-Bang model, together with the expansion of the Universe and the Cosmic Microwave Background. There is a good global agreement over a range of nine orders of magnitude between abundances of ^{4}He, D, ^{3}He and ^{7}Li deduced from observations, and calculated in primordial nucleosynthesis. This comparison was used to determine the baryonic density of the Universe. For this purpose, it is now superseded by the analysis of the Cosmic Microwave Background (CMB) radiation anisotropies. However, there remain, a yet unexplained, discrepancy of a factor $\approx$3, between the calculated and observed lithium primordial abundances, that has not been reduced, neither by recent nuclear physics experiments, nor by new observations. Big-Bang nucleosynthesis, that has been used, to first constrain the baryonic density, and the number of neutrino families, remains, a valuable tool to probe the physics of the early Universe.

1. Introduction

There are presently three evidences for the Big-Bang Model : the universal expansion, the Cosmic Microwave Background (CMB) radiation and Primordial or Big-Bang Nucleosynthesis (BBN). The third evidence comes, indeed, from the primordial abundances of the "light elements": ^{4}He, D, ^{3}He and ^{7}Li. They were produced during the first $\approx$20 minutes of the Universe when it was dense and hot enough for nuclear reactions to take place. (Besides the ^{4}He, D, ^{3}He and ^{7}Li isotopes, only minute traces of ^{6}Li, ^{9}Be, ^{11}B and CNO are produced by BBN.) These primordial abundances are compared to astronomical observations in primitive astrophysical sites. It is worth reminding that Big Bang Nucleosynthesis has been essential in the past, to first estimate the baryonic density of the Universe, $\rho_\mathrm{B} = (1-3)\times10^{-31}$ g/cm^3 [1], and to give an upper limit on the number neutrino families $N_\nu \leq 3$ [2], both in the seventies. The number of light neutrino families is now known from the measurement of the Z^0 width by LEP experiments at CERN: N_ν = 2.9840$\pm$0.0082 [3]. The nuclear reaction rates have all been measured in nuclear physics laboratories or can be calculated from the standard theory of weak interactions. In that case, they are normalized to the experimental value for the lifetime of the neutron. Its precise value is still a matter of debate [4] τ_n = 880-884 s, but its uncertainty has only marginal effect on BBN. The last parameter to have been independently determined is the precise value of baryonic density of the Universe, which is now deduced from the observations of the anisotropies of the CMB radiation. It is usual to introduce η, the number of photons per baryon which remains constant during the expansion, and is directly related to Ω_b by $\Omega_\mathrm{b}{\cdot}h^2$=3.65$\times$10$^7\eta$ with

$$\Omega_\mathrm{b}{\cdot}h^2 = 0.02249 \pm 0.00062 \text{ and } \Omega_\mathrm{b} = 0.04455 \pm 0.0027 \tag{1}$$

("WMAP only Seven Year Mean" [5]) or

$$\Omega_b{\cdot}h^2 = 0.00207 \pm 0.00033 \tag{2}$$

according to the first release of the Planck mission collaboration [6]. The parameter h represents the Hubble constant (H_0) in units of 100 km/s/Mpc and Ω_b, the baryonic density relative to the critical density. These results are just slightly above the range of baryonic density provided of by Wagoner [1] from BBN in 1973!

Hence, the number of free parameters in Standard Big Bang Nucleosynthesis has now been reduced to zero, and the calculated primordial abundances are in principle only affected by the moderate uncertainties in some nuclear cross–sections. It may appears that Big Bang Nucleosynthesis studies are now useless, but this is certainly not the case. First, even though the agreement with observations is good or very good for ^{4}He, ^{3}He and D, there is a tantalizing discrepancy for ^{7}Li that has not yet found a consensual explanation. Second, when we look back in time, it is the ultimate process for which, *a priori*, we know all the physics involved. Hence, departure from its predictions could provide hints or constraints on new physics or astrophysics.

2. The "light elements": ^{4}He, D, ^{3}He and ^{7}Li

2.1. Abundances of the light elements

During the evolution of the Galaxy, complex nucleosynthesis takes place, mainly in massive stars which release matter enriched in heavy elements into the interstellar medium when they explode as supernovae. Accordingly, the abundance of heavy elements in the gas, at the origin of star formation, increases with time. The observed abundance of *metals* (in astrophysics, the chemical elements beyond helium) is hence an indication of its age: the oldest stars have the lowest metallicity. To derive the most primitive abundances one has to extract them from observations of astrophysical sites which are thought to be non evolved and possibly extrapolate them to zero metallicity.

Primordial lithium abundance is deduced from observations of low metallicity stars in the halo of our Galaxy where the lithium abundance is almost independent of metallicity, displaying a plateau [7]. This constant Li abundance is interpreted as corresponding to the BBN ^{7}Li yield. This interpretation assumes that lithium has not been depleted at the surface of these stars so that the presently observed abundance is supposed to be equal to the initial one. The small scatter of values around the "Spite plateau" is an indication that depletion may not have been very effective. Astronomical observations of metal poor halo stars have led to a relative primordial abundance [8] of:

$$\text{Li/H} = (1.58 \pm 0.31) \times 10^{-10}. \tag{3}$$

Note also that, observationally challenging, detections of ^{6}Li had been reported [9] to a level of $\sim 10^{-2}$ below the Spite plateau value, several orders of magnitude below the BBN predictions [10]. But the presence of a ^{6}Li plateau is now excluded after improved modeling (now including 3-dimensions and non-local thermodynamic equilibrium) of these star atmospheres [11].

Contrary to ^{7}Li which can be both produced (spallation, asymptotic giant branch (AGB) stars, novae) and destroyed (in the interior of stars), deuterium is a very fragile isotope, that can only be destroyed after BBN. Its most primitive abundance is determined from the observation of absorption lines in clouds at high redshift, on the line of sight of more distant quasars. Very few observations of these cosmological clouds are available and a weighted mean [12] (and references therein) of this data yields a D/H abundance of:

$$\text{D/H} = (3.02 \pm 0.23) \times 10^{-5}. \tag{4}$$

After BBN, ^{4}He is produced by stars. Its primitive abundance is deduced from observations in HII (ionized hydrogen) regions of compact blue galaxies. Galaxies are thought to be formed by the agglomeration of such dwarf galaxies which are hence considered as more primitive. The primordial ^{4}He abundance $Y_{\rm p}$ (^{4}He mass fraction) is given by the extrapolation to zero metallicity but is affected by systematic uncertainties such as plasma temperature or stellar absorption. Using the data compiled in Ref. [13], it was found [14] that:

$$\mathrm{Y}_p = 0.2534 \pm 0.0083. \tag{5}$$

Contrary to ^{4}He, ^{3}He is both produced and destroyed in stars so that the evolution of its abundance as a function of time is not well known, and has only been observed in our Galaxy [15]:

$$^3\mathrm{He/H} = (1.1 \pm 0.2) \times 10^{-5}. \tag{6}$$

Consequently, comparison with ^{3}He abundance from BBN is subject to caution [16].

2.2. Nuclear reactions for ^{4}He, D, ^{3}He and ^{7}Li nucleosynthesis

Unlike in other sectors of nuclear astrophysics, nuclear cross sections have usually been directly measured at BBN energies (a few 100 keV). There are ≈12 nuclear reactions responsible for the production of ^{4}He, D, ^{3}He and ^{7}Li in Standard BBN. There are many other reactions connecting these isotopes, but their cross sections are too small and/or reactants too scarce to have any significant effect. Even among these 12 reactions, a few of them (e.g. ^{3}H(d,n)^{4}He and ^{3}H(α,γ)^{7}Li) are now irrelevant at WMAP baryonic density. This can be deduced from the inspection of Table 1 of Ref. [17] which displays the sensitivity of the ^{4}He, D, ^{3}He and ^{7}Li calculated abundances w.r.t. to a change in the 12 reaction rates by a constant factor.

The weak reactions involved in n↔p equilibrium are an exception: their rates [18] come from the standard theory of the weak interaction, normalized to the experimental neutron lifetime. Until very recently, the averaged value of 885.7±0.8 s was recommended by the Particle Data Group, but new measurement lead to significantly lower values. While it has not yet been possible to solve this discrepancy [4], a reevaluation of the recommended value: 880.1±1.1 s has been proposed [19] and is now used in BBN calculations.

The influence of the ^{1}H(n,γ)^{2}H rate was unexpected: ^{7}Li final abundance depends strongly on this reaction rate while other isotopes are little affected. Its cross section is obtained from Effective Field Theory whose results are estimated to be reliable to within 1% error [20]. Indeed, the few experimental informations available for this cross section at BBN energies are in very good agreement with theory (see Fig. 1 in Ref. [21]).

For the ten remaining reactions, ^{2}H(p,γ)^{3}He, ^{2}H(d,n)^{3}He, ^{2}H(d,p)^{3}H, ^{3}H(d,n)^{4}He, ^{3}H(α,γ)^{7}Li, ^{3}He(d,p)^{4}He, ^{3}He(n,p)^{3}H, ^{3}He(α,γ)^{7}Be, ^{7}Li(p,α)^{4}He and ^{7}Be(n,p)^{7}Li, the cross sections have been measured in the laboratory at the relevant energies. We used the reaction rates from the the evaluation performed by Descouvemont *et al* [22], supplemented by results from more recent experiments and analysis that have lead to improved reaction rates for a few important reactions.

At CMB deduced baryonic density, ^{7}Li is produced indirectly by ^{3}He(α,γ)^{7}Be, that will, much later decay to ^{7}Li while it is destroyed by ^{7}Be(n,p)^{7}Li(p,α)^{4}He. This is the, second to ^{1}H(n,γ)^{2}H, most influential reaction for ^{7}Li [17] as its final abundance is directly proportional to this rate. It cross–section has long been a subject of debate because of systematic differences that were found according to the experimental technique: prompt or activation measurements. We refer to Di Leva [23] for a review of the most recent experimental data noting the relative scarcity of experimental data within the Gamow window. We use the Cyburt & Davids [24] calculated the S–factor, incorporating new data, which is significantly higher than the Descouvemont *et al* [22] R–matrix fit.

The $^2H(d,n)^3He$ reaction, also influential on ^7Li, was re-measured [together with $^2H(d,p)^3H$] by Leonard *et al* [25], after the R-matrix analysis [22] was performed. The very precisely measured cross section is in perfect agreement with the R-matrix fit (see also [26]).

2.3. BBN primordial abundances compared to observations

Figure 1 shows the up to date abundances of ^4He (mass fraction), D, ^3He and ^7Li (in number of atoms relative to H) as a function of the baryonic density. The thicknesses of the curves reflect the nuclear uncertainties, obtained by a Monte-Carlo calculation using for the nuclear rates discussed in § 2.2. The horizontal lines represent the limits on the ^4He, D and ^7Li primordial abundances deduced from spectroscopic observations § 2.1. The vertical stripes represent the baryonic density deduced from CMB observations [5, 6]. The concordance between BBN and observations is in good agreement for deuterium. Considering the large uncertainty associated with ^4He observations, the agreement with CMB+BBN is fair, but close to the observational lower limit. The calculated ^3He value is close to its galactic value showing that its abundance has little changed during galactic chemical evolution. On the contrary, the ^7Li, CMB+BBN calculated abundance is significantly higher than the spectroscopic observations. Table 1 displays the corresponding numerical values deduced from the Monte–Carlo, at CMB deduced baryonic density, compared with those deduced from spectroscopic observations (§ 2.1). It emphasize the discrepancy of a factor of ≈3 between lithium abundances deduced either from CMB+BBN or from spectroscopic observations. Table 1 shows that these results are stable when comparing the most recent result [27] with earlier work [17]. The small differences can be traced [27] to the evolutions of the baryonic density from the early WMAP results to the Planck ones, of the neutron lifetime and to the extension of the network.

The lithium discrepancy, has not yet found a satisfactory answer, and somewhat weakens the ability to use BBN as a probe of fundamental physics in the early Universe. Several ideas were addressed to try to explain this ^7Li problem. Some conceived the idea that the ^7Li deficit points toward physics beyond standard model such as decay of super-symmetric particles, mirror universe,... etc [28]. Others have suggested that the problem could due to ^7Li stellar destruction in the atmosphere of the halo stars [29] but a uniform destruction of ^7Li over the Spite-plateau region seems difficult. For a recent review of the latest Li observations and their different astrophysical aspects, see Refs. [30, 31, 32], and the proceedings of the 2012 Workshop *Lithium in the Cosmos* [1].

Table 1. Primordial abundances.

	[17]	[27]	Observations
Y_p	0.2476±0.0004	0.2463±0.0003	0.2534 ± 0.0083 [14]
D/H (×10^{-5})	2.68 ± 0.15	2.67±0.09	3.02 ± 0.23 [12]
^3He/H (×10^{-5})	1.05±0.04	1.05±0.03	1.1 ± 0.2 [15]
^7Li/H (×10^{-10})	5.14±0.50	$4.89^{+0.41}_{-0.39}$	1.58 ± 0.31 [8]

3. BBN with an extended network

3.1. Motivations and method

There are several reasons to increase the BBN network beyond the twelve or so "main reactions".

[1] `http://www.iap.fr/lithiuminthecosmos2012/index.html`, proceedings in *Mem. S.A.It.* 2012 **22**

Journal of Physics: Conference Series **665** (2016) 012001 doi:10.1088/1742-6596/665/1/012001

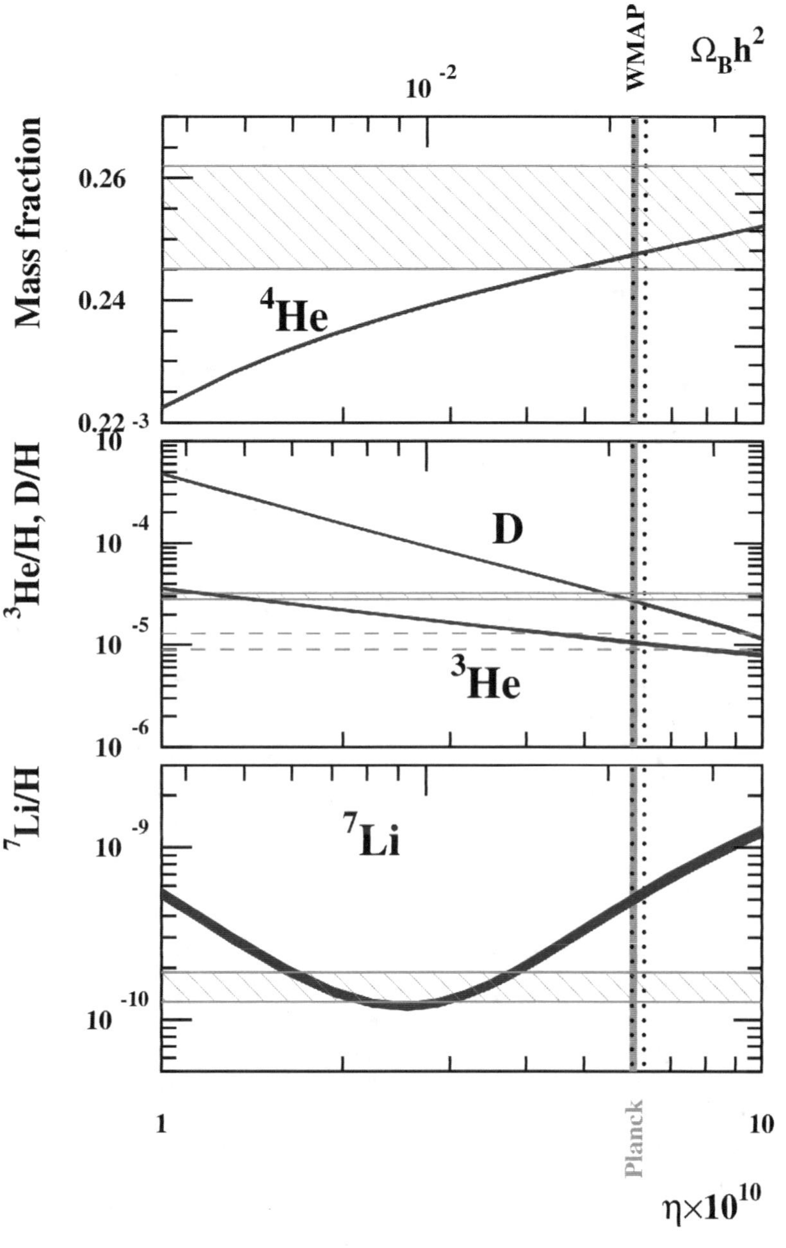

Figure 1. Abundances of ^{4}He (mass fraction), D, ^{3}He and ^{7}Li (by number relative to H) as a function of the baryon over photon ratio η (or $\Omega_b{\cdot}h^2$). The vertical stripe corresponds to the Planck baryonic density [6] (dotted lines are from WMAP [5]) while the horizontal area represent the adopted primordial abundances.

First of all one needs to definitively exclude a direct or indirect "nuclear solution" to the lithium problem. A direct solution would be the destruction of ^{7}Be (the ^{7}Li progenitor) by a reaction that was not previously included in the network. An indirect one would be, again a previously neglected reaction, but that would modify the neutron abundance near the end of BBN. This is indeed the explanation for the unexpected effect of a lower ^{1}H(n,γ)^{2}H rate (§ 2.2) that can be traced to the increased neutron abundance at ^{7}Be formation time making its destruction by neutron capture, ^{7}Be(n,p)^{7}Li(p,α)^{4}He, more efficient (see Fig. 1 in [33]). Second, it allows to calculate the tiny BBN production of ^{6}Li, ^{9}Be and ^{11}B, even though their direct detections seem highly unlikely with the present observational techniques. Third, hydrogen burning in the first generation of stars ("Population III") proceeds through the slow p–p chains until they produce enough carbon, by the triple-alpha reaction, to activate the CNO cycle so that CNO production by BBN could affect their evolution.

The main difficulty in BBN calculations up to CNO is the extensive network needed, including n-, p-, α-, but also d-, t- and ^{3}He-induced reactions on targets in the A=1 to 20 range. Most of the corresponding cross sections cannot be extracted from experimental data only. This is especially true for radioactive tritium-induced reactions, or for those involving radioactive targets like e.g. ^{10}Be. A detailed analysis of all reaction rates and associated uncertainties would be desirable but is impractical for a network of $\approx$400 reactions. In a recent analysis [34], as a first approximation, the results from the TALYS code [35] were used, for rates that are not available in the literature, followed by sensitivity tests allowing each rate to be scaled by 0.01, 0.1, 10, 100 and 1000 factors and for the most influential reactions, was followed by dedicated rate re-evaluations.

3.2. Reactions that affect ^{7}Li production

No new important reaction for ^{7}Li+^{7}Be production was found besides those previously identified [36, 37]: essentially the ^{7}Be(d,p)2α reaction. Indeed, if this reaction rate were significantly higher, ^{7}Li abundance would be brought down to the observed level [36, 38]. However, subsequent experiments and analyses [39, 40, 41] ruled out this possibility [42].

Extending this search, very recent works [43, 44, 45] suggested the possibility of overlooked resonances in nuclear reactions involving ^{7}Be. The most promising candidates were found to be in the ^{7}Be+^{3}He$\rightarrow$^{10}C [43], and ^{7}Be+^{4}He$\rightarrow$^{11}C [44, 45] channels. Indeed, the presence of a level close to the ^{7}Be+^{3}He reaction threshold (Q=15.003 MeV) in ^{10}C [43], or between 7.793 and 7.8936 MeV excitation energy in ^{11}C [45] with favorable properties would help alleviate the lithium problem. However, a recent experiment was conducted at the Tandem of the Orsay ALTO facility to improve the ^{10}C and ^{11}C spectroscopy. The ^{10}B(^{3}He,t)^{10}C and ^{11}B(^{3}He,t)^{11}C reactions were investigated at a ^{3}He beam energy of 35 MeV and the tritons analyzed by the Split-pole magnet. Only upper limits for the presence of new levels in ^{10}C and ^{11}C were obtained, too low to have an impact on ^{7}Li production [46].

3.3. Reactions that affect ^{6}Li, ^{9}Be and ^{11}B production

The most important reactions for ^{6}Li, ^{9}Be and ^{11}B productions are ^{4}He(d,γ)^{6}Li, ^{7}Li(t,n)^{9}Be, ^{7}Be(t,p)^{9}Be, ^{7}Li(d,γ)^{9}Be and ^{11}C(n,α)2α. However, some of them have been measured (^{4}He(d,γ)^{6}Li [10] and ^{7}Li(t,n)^{9}Be [47, 48]) so that the uncertainties on their cross sections are small compared to the range explored in this sensitivity study. The BBN ^{6}Li yield (^{6}Li/H$\approx$ 10^{-14}) at CMB deduced baryonic density is about two orders of magnitude below the previously reported observations in some halo stars, that nevertheless were not confirmed.

3.4. Reactions that affect CNO production

The minimum value of the initial CNO abundance that would affect Population III stellar evolution is estimated to be as low as 10^{-13} (in number of atoms relative to hydrogen, CNO/H) for the less massive ones [49]. This is only two orders of magnitude above the Standard Big Bang Nucleosynthesis CNO yield, using the current nuclear reaction rate evaluations of Iocco *et al* [50]. Among the ≈400, only a few reactions were found to have a strong impact on the CNO final abundance. The CNO production is significantly sensitive (more than by a factor of about 2) to several reaction rates. In particular, these include: ^{7}Li(d,n)2^{4}He, ^{7}Li(t,n)^{9}Be, ^{8}Li(α,n)^{11}B, ^{11}B(n,γ)^{12}C, ^{11}B(d,n)^{12}C, ^{11}B(d,p)^{12}B and ^{11}C(d,p)^{12}C. The impact of ^{7}Li(d,n)2^{4}He is unexpected and should be compared to the influence of ^{1}H(n,γ)^{2}H on ^{7}Li. Indeed, when increasing the ^{7}Li(d,n)2^{4}He reaction rate from [51] by a factor of 1000, even though the ^{4}He, D, ^{3}He and ^{7}Li *final abundances are left unchanged*, the peak ^{7}Li abundance at t ≈200 s is reduced by a factor of about 100 (see Fig. 15 in [34]), an evolution followed by ^{8}Li and CNO isotopes. Finally, the CNO Standard Big Bang Nucleosynthesis production is CNO/H $(0.5-3.)\times10^{-15}$ (with estimated uncertainties). These results are consistent with those of Iocco *et al* [50] and are too low to have an impact on Population III stellar evolution.

4. Some extensions of BBN

There are many extensions of the Standard Model of BBN and we refer to Iocco et al. [52] for an extensive review and to Fields [28] for non–Standard BBN solutions to the lithium problem. Note that these solutions quite systematically lead to an increased D production which is easier to cure than the lithium overproduction [12]. We will limit ourselves here in a few examples where nuclear physics is important.

4.1. Variation of constants

The possibility of the variation of "constants" is a key issue of modern physics (see Ref. [53] for a review). It is motivated by superstring theories and by observations of a possible [54, 55] variation of the fine structure constant in clouds at high redshift, on the line of sight of more distant quasars. Other tests of the variation of constants involve atomic clocks, the CMB and nuclear physics (Big Bang Nucleosynthesis, the triple–alpha reaction and stellar evolution, radioactivities in meteorites and the Oklo fossil reactor). They are all interesting because they probe variations on different cosmic time scale, BBN being the earliest one [56, 57, 58]. However, the impact of these variations on the (BBN) nuclear reaction rates is difficult to estimate, as in general, nuclear physics uses phenomenological models, whose parameters are not explicitly linked to fundamental constants. Key nuclear reactions, whose rate variations would be influential, in this context, are n↔p (affecting ^{4}He) and n+p→d+γ (affecting ^{4}He, D, and strongly ^{7}Li) but we will illustrate our purpose with the triple–alpha reaction: ^{4}He$(\alpha,\gamma)^8$Be$(\alpha,\gamma)^{12}$C. Here, ^{8}Be is normally slightly unbound and very short lived, acting as an intermediate state an the rate is boosted by the presence of a resonant ("Hoyle") state in ^{12}C. If constants were slightly different at BBN epoch, the rate would be much different because of the new position of the Hoyle state changes and even ^{8}Be could become stable [59]. According to textbooks, a stable ^{8}Be would allow the build–up of heavy elements, but this is not the case (Fig. 2): the C(NO) production remains ≈6 order of magnitude [34] lower than the Standard Big Bang Nucleosynthesis value (§ 3.4).

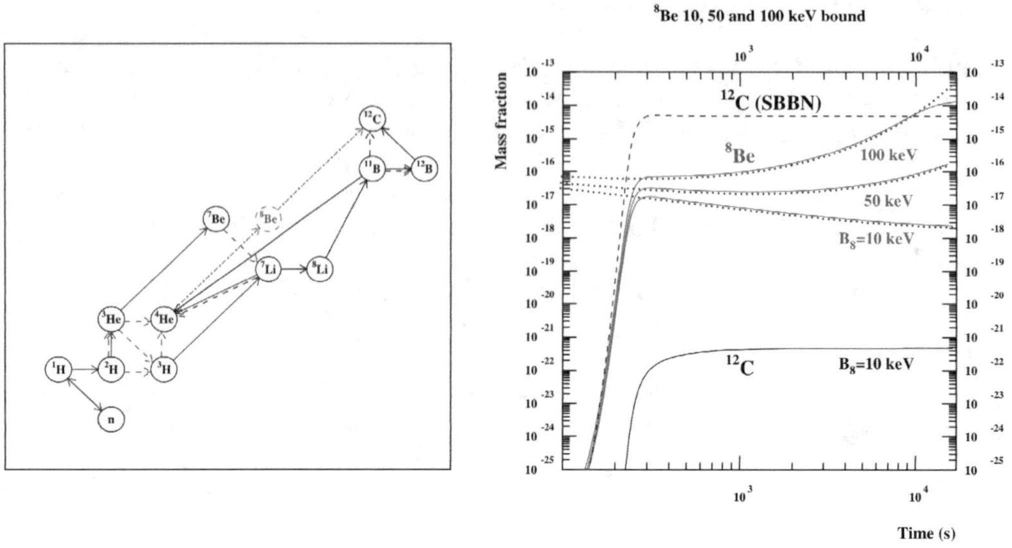

Figure 2. Left panel: the standard network up to A=7 (blue), A=12 (black) with in addition, the triple–alpha reaction (red). Right panel: ^{12}C (solid black) and ^{8}Be (red) mass fractions as a function of time, assuming ^{8}Be is bound by 100, 50 and 10 keV (dotted lines correspond to thermal equilibrium) [57] compared to Standard BBN ^{12}C production (dash) [34].

4.2. Exotic particles

There are various way, in which exotic particles can influence BBN [52]. The decay of a massive particle during or after BBN could affect the light element abundances and potentially lower the ^{7}Li abundance (see e.g. [60]). Neutrons, protons or photons produced by these decays may be thermalized but more likely have a non-thermal high energy (∼1 GeV) distributions. Interestingly, some nuclear cross–sections involved in these non-thermal processes are not known with a sufficient precision [61]. If they can be thermalized, it provides an extra source of neutrons that can alleviate the lithium problem [62, 63] as proposed in § 3. An other exotic source of thermalized neutrons could come from a "mirror world" [64] initially proposed to restore global parity symmetry. Long lived (relative to BBN time scale) negatively charged relic particles, like the supersymmetric partner of the tau lepton, could form bound states with nuclei, lowering the Coulomb barrier and hence leading to the catalysis of nuclear reactions (see e.g. [65, 66, 67]). Even though exotics, the interaction of these *electromagnetically* bound states with other nuclei can be treated by conventional nuclear physics theory.

5. Conclusions

The baryonic density of the Universe as determined by the analysis of the CMB anisotropies is in very good agreement with Standard BBN compared to D and ^{4}He primordial abundance deduced from observations. However, it disagrees with lithium observations in halo stars by a factor that has increased with the availability of improved nuclear data and astronomical observations. A possible explanation is lithium stellar depletion, but the larger needed depletion factor is hardly compatible whit the thin observed plateau. It is hence essential to determine precisely the *absolute* cross sections important for ^{7}Li nucleosynthesis.

Nevertheless, primordial nucleosynthesis remains an invaluable tool for probing the physics of the early Universe. When we look back in time, it is the ultimate process for which we, *a priori*, know all the physics involved. Hence, departure from its predictions provide hints for new physics or astrophysics.

Last but not least, we stress here the importance of sensitivity studies in nuclear astrophysics that have been done, e.g. in the context of novae [68], X–ray burst [69] or massive stars [70]. Even in the simpler context of BBN without the complexity (e.g. mixing) of stellar nucleosynthesis, it would have been very unlikely to predict the influence of the ^{1}H(n,γ)^{2}H reaction on ^{7}Li nor of the ^{7}Li(d,n)2^{4}He reaction on CNO.

Acknowledgments

I am indebted to all my collaborators on these topics: Pierre Descouvemont, Sylvia Ekström, Stéphane Goriely, Georges Meynet, Keith Olive, Jean-Philippe Uzan, and especially Elisabeth Vangioni .

References

[1] Wagoner R V 1973 *Astrophys. J.* **179** 343
[2] Yang J, Schramm D, Steigman G and Rood R T 1979 *Astrophys. J.* **227** 697
[3] The ALEPH Collaboration and The DELPHI Collaboration and the L3 Collaboration and The OPAL collaboration and The SLD Collaboration and the LEP Electroweak Working Group and the SLD Electroweak and Heavy Flavour Groups 2006 *Phys. Rep.* **427** 257
[4] Wietfeldt F E and Greene G L 2011 *Rev. Mod. Phys.* **83** 1173
[5] Komatsu E, Smith K M, Dunkley J *et al* [WMAP Collaboration] 2011 *Astrophys. J. Supp.* **192** 18
[6] Ade P A R, Aghanim N, Armitage–Caplan C *et al* [Planck Collaboration] 2013 arXiv:1303.5076
[7] Spite F and Spite M 1982 *Astron. Astrophys.* **115** 357
[8] Sbordone L, Bonifacio P, Caffau E *et al* 2010 *Astron. Astrophys.* **522** A26
[9] Asplund M, Lambert D L, Nissen P E, Primas F and Smith V V 2006 *Astrophys. J.* **644** 229
[10] Hammache F, Heil M, Typel S *et al* 2010 *Phys. Rev.* C **82** 065803
[11] Lind K, Melendez J, Asplund M, Collet R and Magic Z 2013 *Astron. Astrophys.* **554** A96
[12] Olive K A, Petitjean P, Vangioni E and Silk J 2012 *Mon. Not. R. Astron. Soc.* **426** 1427
[13] Izotov Y I, Thuan T X and Stasińska G 2007 *Astrophys. J.* **662** 15
[14] Aver E, Olive K A and Skillman E D 2012 *Journal of Cosmology and Astroparticle Physics* **04** 004
[15] Bania T, Rood R and Balser D 2002 *Nature* **415** 54
[16] Vangioni-Flam E *et al* 2003 *Astrophys. J.* **585** 611
[17] Coc A and Vangioni E 2010, *Journal of Physics Conference Series* **202** 012001
[18] Dicus D, Kolb E, Gleeson A, Sudarshan E, Teplitz V and Turner M 1982 *Phys. Rev.* D **26** 2694
[19] Beringer J *et al* [Particle Data Group] 2012 *Phys. Rev.* D **86** 010001 (URL: `http://pdg.lbl.gov`)
[20] Ando S, Cyburt R H, Hong S W and Hyun C H 2006 *Phys. Rev.* C **74** 025809
[21] Coc A 2013 *Acta Physica Polonica* **44** 521
[22] Descouvemont P, Adahchour A, Angulo C, Coc A and Vangioni–Flam E 2004 *Atomic Data and Nuclear Data Tables* **88** 203
[23] Di Leva A, *these proceedings*
[24] Cyburt R H and Davids B 2008 *Phys. Rev.* C **78** 064614
[25] Leonard D S, Karwowski H J, Brune C R, Fisher B M and Ludwig E J 2006 *Phys. Rev.* C **73** 045801
[26] Tumino A, Spitaleri C, Mukhamedzhanov A M *et al* 2011 *Phys. Lett.* B **700** 111
[27] Coc A , Uzan J P, Vangioni E *Astrophys. J. Lett. submitted*
[28] Fields B 2011 *Annual Review of Nuclear and Particle Science* **61** 47
[29] Richard O, Michaud G and Richer J 2005 *Astrophys. J.* **619** 538
[30] Spite F and Spite M *Proc. IAU Symposium No. 268, "Light Elements in the Universe" (Geneva)*, Eds. C. Charbonnel, M. Tosi, F. Primas and C. Chiappini (Cambridge University Press, 2010) p. 221.
[31] Steffen M, Cayrel R, Caffau E, Bonifacio P, Ludwig H G and Spite M 2012, *Proc. "Lithium in the Cosmos" (Paris), Mem. S.A.It.* **75** 282
[32] Iocco F 2012 *Mem. S.A.It.* **22** 19
[33] Coc A, Nunes N J, Olive K A, Uzan J P and Vangioni E 2007 *Phys. Rev.* D **76** 023511
[34] Coc A, Goriely S, Xu Y, Saimpert M and Vangioni E 2012 *Astrophys. J.* **744** 58
[35] Goriely S, Hilaire S, Koning A J 2008 *Astron. Astrophys.* **487** 767

[36] Coc A, Vangioni-Flam E, Descouvemont P, Adahchour A and Angulo C 2004 *Astrophys. J.* **600** 544
[37] Serpico P D, Esposito S, Iocco F, Mangano G, Miele G and Pisanti O 2004 *Journal of Cosmology and Astroparticle Physics* **12** 010
[38] Cyburt R H and Pospelov M 2012 *International Journal of Modern Physics* E **21** 50004
[39] Angulo C, Casarejos E, Couder M *et al* 2005 *Astrophys. J. Lett.* **630** L105
[40] O'Malley P D, Bardayan D W, Adekola A S *et al* 2011 *Phys. Rev.* C **84** 042801
[41] Scholl C, Fujita Y, Adachi T *et al* 2011 *Phys. Rev.* C **84** 014308
[42] Kirsebom O S and Davids B 2011 *Phys. Rev.* C **84** 058801
[43] Chakraborty N, Fields B D and Olive K A 2011 *Phys. Rev.* D **83** 063006
[44] Broggini C, Canton L, Fiorentini G and Villante F L 2012 *Journal of Cosmology and Astroparticle Physics* **6** 30
[45] Civitarese O and and Mosquera M E 2013 *Nucl. Phys.* A **898** 1
[46] Hammache F, Coc A, de Séréville N *et al Phys. Rev.* **C** *submitted*
[47] Brune C R, Kavanagh R W, Kellogg S E and Wang T R 1991 *Phys. Rev.* C **43** 875
[48] Barhoumi S, Bogaert G, Coc A *et al* 1991 *Nucl. Phys.* A **535** 107
[49] Ekström S, Meynet G, Chiappini C, Hirschi R and Maeder A 2008 *Astron. Astrophys.* **489** 685
[50] Iocco F, Mangano G, Miele G, Pisanti O and Serpico P D 2007 *Phys. Rev.* **D75** 7304
[51] Boyd R N, Mitchell C A and Meyer B S 1993 *Phys. Rev.* C **47** 2369
[52] Iocco F, Mangano G, Miele G, Pisanti O and Serpico P D 2009 *Phys. Rev.* **472** 1
[53] Uzan J P 2011 *Living Reviews in Relativity* **14** 2
[54] Murphy M T, Webb J K and Flambaum V V 2003 *Mon. Not. R. Astron. Soc.* **345** 609; Webb J K, King J A, Murphy M T *et al* 2011 *Phys. Rev. Lett.* **107** 191101
[55] Srianand R, Chand H, Petitjean P and Aracil B 2007 *Phys. Rev. Lett.* **99** 239002
[56] Coc A, Olive K A, Uzan J P and Vangioni E 2006 *Phys. Rev.* D **73** 083525
[57] Coc A, Descouvemont P, Olive K A, Uzan J P and Vangioni E 2012 *Phys. Rev.* D **86** 043529
[58] Berengut J C, Epelbaum E, Flambaum V V *et al* 2013 *Phys. Rev.* D **87** 085018
[59] Ekström S, Coc A, Descouvemont P, Meynet G, Olive K A, Uzan J P and E. Vangioni 2010 *Astron. Astrophys.* **514** A62
[60] Cyburt R H, Ellis J, Fields B D, Luo F, Olive K A and Spanos V C 2013 *Journal of Cosmology and Astroparticle Physics* **05** 014
[61] Cyburt R H, Ellis J, Fields B D, Luo F, Olive K A and Spanos V C 2010 *Journal of Cosmology and Astroparticle Physics* **10** 32
[62] Jedamzik K 2004 *Phys. Rev.* D **70** 083510
[63] Albornoz Vásquez D, Belikov A, Coc A, Silk J and Vangioni E, *Phys. Rev.* D **86** 063501
[64] Coc A , Uzan J P, Vangioni E 2013 *Phys. Rev.* D **87** 123530
[65] Pospelov M 2007 *Phys. Rev. Lett.* **98** 231301
[66] Kusakabe M, Kajino T, Boyd R N, Yoshida T and Mathews G J 2008 *Astrophys. J.* **680** 846
[67] Cyburt R H, Ellis J, Fields B D, Luo F, Olive K A and Spanos V C 2012 *Journal of Cosmology and Astroparticle Physics* **12** 037
[68] Iliadis C, Champagne A, José J, Starrfield S and Tupper P 2002 *Astrophys. J. Supp.* **142** 105
[69] Parikh A, José J, Moreno F and Iliadis C 2008 *Astrophys. J. Supp.* **178** 110
[70] Iliadis C, Champagne A, Chieffi A and Limongi M 2011 *Astrophys. J. Supp.* **193** 16

Experimental status of ^{7}Be production and destruction at astrophysical relevant energies

A Di Leva[1,2], **L Gialanella**[3,2], **F Strieder**[4]

[1] Dipartimento di Fisica, Università di Napoli Federico II, Naples, Italy
[2] INFN Sezione di Napoli, Naples, Italy
[3] Dipartimento di Matematica e Fisica, Seconda Università di Napoli, Caserta, Italy
[4] Institut für Experimentalphysik, Ruhr-Universität Bochum, Bochum, Germany

E-mail: antonino.dileva@unina.it

Abstract. The production and destruction of ^{7}Be plays a significant role in the Big Bang Nucleosynthesis as well as in the framework of the solar neutrino. The ^{3}He$(\alpha,\gamma)^{7}$Be reaction cross sections has been measured several times in the last decades, but the precision achieved on reaction rate determinations at the relevant astrophysical energies is not yet satisfactory. The experimental status of this reaction will be critically reviewed, and the theoretical descriptions available will be discussed.

The production and destruction of ^{7}Be plays a significant role in the Big Bang Nucleosynthesis (BBN) as well as in the framework of the solar neutrino physics. The production of ^{7}Be proceeds essentially through the ^{3}He$(\alpha,\gamma)^{7}$Be reaction, while the main destruction channels in the Sun are ^{7}Be$(p,\gamma)^{8}$B and ^{7}Be$(e^{-},\nu)^{7}$Li. On the contrary in BBN several other channels are present ^{7}Be$(n,p)^{7}$Li, , ^{7}Be$(n,\alpha)\alpha$, ^{7}Be$(d,p\alpha)\alpha$, ^{7}Be$(\alpha,\gamma)^{11}$C.

In this paper the experimental status of the ^{3}He$(\alpha,\gamma)^{7}$Be cross section is reviewed and among the available theoretical models the one that best describes simultaneously the larger number of the existing datasets is determined. This will be done using a statistically consistent analysis, avoiding an inflation of the errors and trying to use as much experimental information as possible. As regards the solar neutrino physics, for the ^{3}He$(\alpha,\gamma)^{7}$Be reaction the relevant energy range is 22 keV that is the energy of the Gamow peak at the Sun's central temperature of 15 MK. For BBN, considering that the Gamow peak has a width of $\sim$ 300 keV, the relevant energy range spans from $\sim$ 100 to $\sim$ 900 keV.

The S-factor of the ^{3}He$(\alpha,\gamma)^{7}$Be has been measured many times either by detecting the prompt γ-rays emitted in the reaction [1, 2, 3, 4, 5], or by detecting the γ-ray that follows the electron capture decay of ^{7}Be (*activation method*) [6, 7, 8, 9] and [10]. In some cases both methods were used concurrently: in [11], in [12, 13, 14] (the LUNA measurements), and in the work of Brown *et al.* [15]. Additionally the direct detection of the ^{7}Be nuclei produced in the reaction was exploited with the recoil separator ERNA [16], whereas activation and prompt γ-ray measurement were also performed. This measurement spans the largest energy range 0.65 to 3.15 MeV, and shows a significant discrepancy, on both the absolute scale and the energy dependence, with the results of [1], i.e. the data set with the largest energy range overlap. This questioned the fact that the simple picture of a pure external capture is not sufficient to describe this process. The results of ERNA have been recently confirmed by [9] and [10]. The results of a new measurement [17] became available after the present analysis was completed, and are therefore not considered in the following.

An excellent agreement is found among the different datasets around $E_{\rm cm} = 1\,\text{MeV}$, see figure 1. The comparison of data sets with overlapping energy range is straight forward, for the others the use of theoretical models is mandatory. The quantitative estimation on the agreement of the various experiments with each other needs a proper statistical analysis, taking into account the statistical and systematic uncertainty for each measurement.

Several models predicting the S-factor of the $^3\text{He}(\alpha,\gamma)^7\text{Be}$ are available in literature. Since the $^3\text{He}(\alpha,\gamma)^7\text{Be}$ is considered an external capture, the cross section is calculated from the electromagnetic operator expanded as electric and magnetic multipole operators, connecting the initial and final state wave functions. Depending on the way the wave functions are determined the models can be classified in two different groups: potential models and microscopic models. In the potential models the initial wave function is expressed as a function of the experimentally determined elastic scattering phase shifts. The final wave function depends on the reduced widths of the bound state, that can be derived from the measurement of $\sigma_{\gamma 429}/\sigma_{\gamma\rm gs}$. Thus the only free parameter in the model is normalisation constant that is obtained from a fit to the cross section data.

The microscopic models, instead, try to describe the 7 nucleon system, in a more or less detailed way, as clusters of nucleons interacting through a nucleon-nucleon (NN) potential. The parameters of the NN potential are adjusted to match the properties of target, projectile, and compound system, like scattering phase shifts and energy levels. The cross section calculated in this approach contains no absolute intensity parameter.

In the following the models given in [18, 19, 20, 21, 22, 23, 24] will be considered. Although the calculation of [23] is based on R-matrix, since no resonances are present in the energy range considered, it is similar to a potential model calculation, indeed very similar to the predictions of [25]. The models are usually claimed to be valid for $E_{\rm cm} \lesssim 2\,\text{MeV}$, therefore comparison with data will be limited to this energy range.

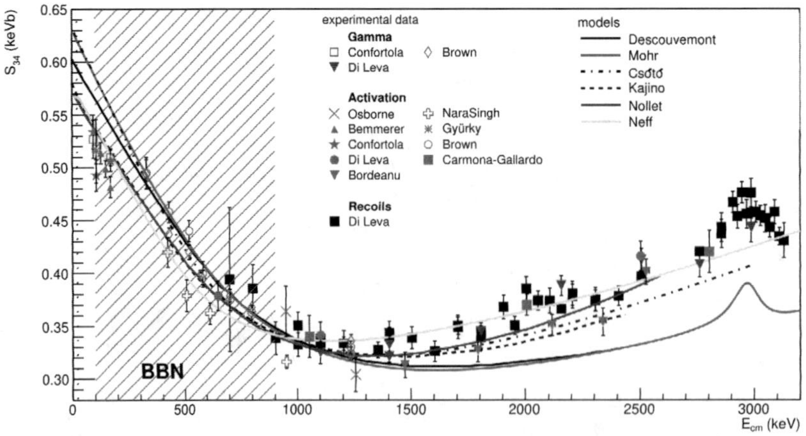

Figure 1. Comparison of the models with respect to experimental data. In this plot the curves are arbitrarily scaled in order to match the experimental data a $E_{\rm cm} \simeq 1\,\text{MeV}$.

Unfortunately a global analysis of all the experimental data available is not possible, since in several works [1, 2, 3, 5] the quoted uncertainties include systematic errors, and a deconvolution of the errors is not possible on the basis of the published information. Thus the results obtained by least squares fits to these data sets would be not statistically correct. The works of [6, 4, 7] measured each a single data point, so they provide no test of the model energy dependence, even though they may be used to estimate the absolute intensity. Anyhow the results of [7] are model dependent, and therefore will not be considered.

Hence to establish which calculation better describes the experimental data a least squares fit of each model is performed separately to: [11] prompt γ, [11] activation, [8], LUNA prompt γ [14], LUNA activation [12, 13, 14]), [15] prompt γ, [15] activation, ERNA recoil and activation [16], [9] and [10]. The fits of the model to the data are obtained by scaling each calculation by a constant factor c. This procedure is somewhat questionable for microscopic models, but the possible inaccuracy resulting from the scaling stays small if $c \approx 1$.

The minima of the least squares function L are given in table 1. If L follows a χ^2 probability

Table 1. Least square function minima, $\mathrm{LSF}_{\mathrm{min}}$, of the fits of the different models to the data sets in the energy range $E_{\mathrm{cm}} \leq 2.0\,\mathrm{MeV}$. The number of degree of freedom ν is indicated for each data set. The colours indicate different confidence level: fits with are marked green if $P(\chi^2 > \mathrm{LSF}_{\mathrm{min}}) > 5\%$ (good fit), yellow if $1\% < P(\chi^2 > \mathrm{LSF}_{\mathrm{min}}) < 5\%$, red if $P(\chi^2 > \mathrm{LSF}_{\mathrm{min}}) < 1\%$ (poor fit).

	[11] prompt	[11] activation	[8]	LUNA prompt	LUNA activation	[15] prompt	[15] activation	ERNA recoils	ERNA activation	[9]	[10]
ν	18	1	3	2	6	7	7	21	1	1	1
Descouvemont [23]	86	1.6	1.1	0.8	6.6	20	5.3	90	1.8	4.4	0.7
Kajino *et al.* [19]	83	2.3	1.4	0.7	6.6	9.6	12.3	38	0.4	1.4	0.2
Csoto [21]	95	2.5	0.9	1.4	6.7	19	3.4	25	1.1	0.5	0.1
Liu *et al.* [18]	79	2.4	2.8	0.1	7.3	17	21.5	41	0.3	1.1	0.3
Nollett [22]	83	2.5	0.9	0.7	6.6	11	10.0	23	0.7	0.3	0.02
Mohr [20]	91	1.5	2.0	1.2	6.6	31	6.6	93	2.7	4.5	0.7
Neff [24]	87	3.9	4.0	1.2	6.6	55	46	25	0.1	0.01	0.03

distribution its expectation value is the number of degrees of freedom ν. The fits are assumed consistent with the data considering a conservative confidence level of 99%, i.e. the probability that the model is discarded although it is correct is 1% or less.

The prompt γ-ray data of [11] are not fitted by any model, this is due to the fact that the external error is larger than the quoted internal error, therefore little information on the model can be obtained from this data set. The LUNA data sets are fitted equally well by all models. This is due to the fact that the energy range covered by the experiments is too narrow to give any information on energy behaviour of the S-factor. Although the data of [8], [9] and [10] cover a relatively wide energy interval they also do not constrain the model since they are well fitted by all calculations, because of the relatively low precision. Thus the only measurements that actually discriminate among available models are [15] and ERNA. The only models that are compatible with both data sets are [19] and [22], although one should be aware that the normalisation factor c is significantly different from 1 in both cases, ~ 1.15 and ~ 1.4, respectively. We could not quantify the uncertainty introduced by the normalisation.

It is worth noting that the recent microscopic calculation of [24] requires a normalisation close to 1, but it cannot describe sufficiently well the data of [15]. In fact the low energy points in [15] suggest a significantly steeper slope towards solar energies.

Journal of Physics: Conference Series **665** (2016) 012002 doi:10.1088/1742-6596/665/1/012002

A quantitative analysis of the agreement between the different data sets is performed comparing the S-factor at zero energy $S_{34}(0)$ from the best fit of the models from [19] and [22], respectively. In figure 2 top panel the results according to the model of Kajino *et al.* [19] are shown. The total error σ_i affecting the extrapolation is obtained adding quadratically the systematic error of each individual measurement and the statistical error obtained by the fitting procedure. In those cases where two experimental techniques were used, the results are combined, taking into account the common systematic uncertainty according to the prescription of [26].
Although the measurements of [8] and [9] were carried out with the same experimental setup, if a common systematic uncertainty of 1% is assumed, the $S_{34}(0)$ values obtained in the present

Figure 2. Top panel: extrapolated $S_{34}(0)$ values according to the model of [19], the continuous line represents the weighted average of the compatible values, while the dashed lines the uncertainty. Very similar results are obtained with the model of [22]. Bottom panel: solid curve calculation of [22], dashed [19]. The curves are scaled to match the average value of $S_{34}(0)$ according to the model.

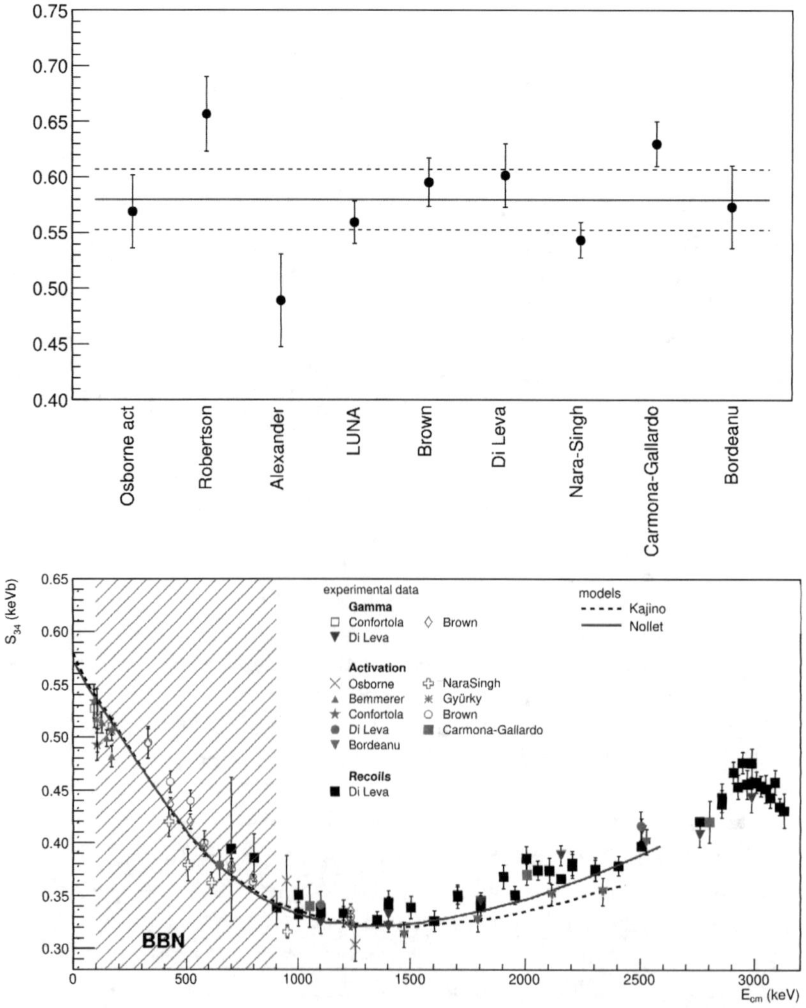

Table 2. Probability P of datasets to be compatible to each other, see text for details. The colours indicate different confidence level: $P \geq 32\%$ (agreement within 1σ) is highlighted in green, $5\% \leq P < 32\%$ (agreement within 2σ) is highlighted in yellow, $P < 5\%$ (no agreement) is highlighted in red. Top: extrapolation according to the [19] model. Bottom: extrapolation according to the [22] model.

	[6]	[4]	LUNA	[15]	ERNA	[8]	[9]	[10]
[11] activation	0.01	0.03	0.84	0.68	0.78	0.53	0.28	0.94
[6]		0.00	0.01	0.03	0.07	0.00	0.36	0.02
[4]			0.03	0.00	0.00	0.10	0.00	0.03
LUNA				0.26	0.42	0.64	0.15	0.77
[15]					0.93	0.27	0.59	0.73
ERNA						0.46	0.77	0.81
[8]							0.07	0.49
[9]								0.33

	[6]	[4]	LUNA	[15]	ERNA	[8]	[9]	[10]
[11] activation	0.01	0.02	0.82	0.71	0.82	0.51	0.42	0.86
[6]		0.00	0.00	0.02	0.04	0.00	0.11	0.01
[4]			0.02	0.00	0.00	0.08	0.00	0.05
LUNA				0.28	0.47	0.64	0.28	0.98
[15]					0.99	0.28	0.79	0.61
ERNA						0.49	0.87	0.75
[8]							0.12	0.68
[9]								0.36

analysis, with both the model of [22] or [19], are only marginally in agreement, therefore they are treated as completely independent experiments.

The obtained $S_{34i}(0)$ values, where i spans over the datasets, and relative uncertainty, are very similar for both calculations since their energy dependence differ significantly only for $E_{\rm cm} \gtrsim 1.5\,\mathrm{MeV}$, figure 2 bottom panel.

Given the model, the compatibility of the different $S_{34i}(0)$ values is evaluated with a Student's t-test with ν_i, ν_j degrees of freedom, for each pair i, j. In case $\nu = 1$ for one, or both, data sets the compatibility with zero of the difference is tested with a normal distribution with variance $(\sigma_i^2 + \sigma_j^2)$. The probability for the results to be compatible are reported in table 2.

The compatibility of the extrapolations is tested with a confidence limit of 95%. This test shows that the $S_{34}(0)$ extrapolated from the data of [6] and [4] are not statistically compatible with the determinations of some other experiments, and therefore are not used in averaging the values.

The best estimate for $S_{34}(0)$ is the weighted average of the results of the compatible experiments. The mean values for $S_{34}(0)$ are $0.579 \pm 0.027\,\mathrm{keV\,b}$ and $0.571 \pm 0.022\,\mathrm{keV\,b}$ for the [19] and [22] model, respectively. Since the internal error is smaller than the external error, the latter is quoted as the error on the mean value. Thus on the basis of this analysis the best estimate of

$$S_{34}(0) = 0.574 \pm 0.027\,\mathrm{keV\,b}$$

is suggested.

The quoted error represents the uncertainty to the determination of $S_{34}(0)$ due to statistical fluctuations of the experimental information used. This analysis assumes that the energy dependence of the S-factor at energies below 1 MeV is correctly predicted by the [19] and [22] calculations. The contribution to the error that may come from the rather large normalisation of the models is difficult to evaluate. Recently the Solar Fusion II compilation [27] proposed a 0.02 keV b error, that has to be added to the statistical uncertainty quoted above, as a result of the uncertainty on the slope.

The result of the present analysis is similar to the one recommended by [27]. However it has to be stressed that the analysis described in this paper requires no inflation of the errors, thus preserving rigorously the probabilistic information of the uncertainty.

The authors thank F Terrasi for fruitful discussions about the manuscript.

[1] Parker P and Kavanagh R 1963 *Phys. Rev.* **131** 2578
[2] Nagatani K, Dwarakanath M and Ashery D 1969 *Nucl. Phys. A* **128** 325
[3] Kräwinkel H *et al.* 1982 *Z. Phys. A* **304** 307
[4] Alexander T *et al.* 1984 *Nucl. Phys.* **A427** 526
[5] Hilgemeier M *et al.* 1988 *Z. Phys.* **A329** 243
[6] Robertson R *et al.* 1983 *Phys. Rev.* **C27** 11
[7] Volk H *et al.* 1983 *Z. Phys.* **A310** 91
[8] NaraSingh B *et al.* 2004 *Phys. Rev. Lett.* **93** 262503
[9] Carmona-Gallardo M, Nara Singh B S, Borge M J G *et al.* 2012 *Phys. Rev. C* **86** 032801
[10] Bordeanu C, Gyürky G, Halász Z, Szücs T, Kiss G, Elekes Z, Farkas J, Fülöp Z and Somorjai E 2013 *Nucl. Phys. A* **908** 1 – 11 ISSN 0375-9474
[11] Osborne J *et al.* 1984 *Nucl. Phys.* **A419** 115
[12] Bemmerer D, Confortola F, Costantini H *et al.* 2006 *Phys. Rev. Lett.* **97** 122502 (pages 5)
[13] Gyürky G, Confortola F, Costantini H *et al.* 2007 *Phys. Rev. C* **75** 035805–+
[14] Confortola F, Bemmerer D, Costantini H *et al.* 2007 *Phys. Rev. C* **75** 065803–+ (*Preprint* arXiv:0705.2151)
[15] Brown T A D, Bordeanu C, Snover K A, Storm D W, Melconian D, Sallaska A L, Sjue S K L and Triambak S 2007 *Phys. Rev. C* **76** 055801–+ (*Preprint* arXiv:0710.1279)
[16] Di Leva A, Gialanella L, Kunz R, Rogalla D, Schürmann D, Strieder F, De Cesare M, De Cesare N, D'Onofrio A, Fülöp Z, Gyürky G, Imbriani G, Mangano G, Ordine A, Roca V, Rolfs C, Romano M, Somorjai E and Terrasi F 2009 *Phys. Rev. Lett.* **102** 232502 (pages 4) erratum: [Phys. Rev. Lett. 103, 159903 (2009)]
[17] Kontos A, Uberseder E, deBoer R, Görres J, Akers C, Best A, Couder M and Wiescher M 2013 *Phys. Rev. C* **87**(6) 065804
[18] Liu Q, Kanada H and Tang Y 1981 *Phys. Rev. C* **23** 645
[19] Kajino T, Austin S M and Toki H 1987 *ApJ* **319** 531–540
[20] Mohr P, Abele H, Zwiebel R, Staudt G, Krauss H, Oberhummer H, Denker A, Hammer J and Wolf G 1993 *Phys. Rev. C* **48** 1420–1427
[21] Csótó A and Langanke K 2000 *Few-Body Systems* **29** 121–130 (*Preprint* arXiv:nucl-th/9906053)
[22] Nollett K M 2001 *Phys. Rev. C* **63** 054002–+ (*Preprint* arXiv:nucl-th/0102022)
[23] Descouvemont P, Adahchour A, Angulo C, Coc A and Vangioni-Flam E 2004 *ADNDT* **88** 203–236
[24] Neff T 2011 *Phys. Rev. Lett.* **106**(4) 042502
[25] Tombrello T and Parker P 1963 *Phys. Rev.* **131** 2582
[26] D'Agostini G 1994 *Nucl. Instr. Meth. A* **346** 306–311
[27] Adelberger E G *et al.* 2011 *Rev. Mod. Phys.* **83**(1) 195–245

Measurement of the photodissociation of the deuteron at energies relevant to Big Bang nucleosynthesis

R Hannaske[1,2], D Bemmerer[1], R Beyer[1], E Birgersson[1], A Ferrari[1], E Grosse[1,2], A R Junghans[1], M Kempe[1], T Kögler[1,2], K Kosev[1], M Marta[1], R Massarczyk[1,2], A Matic[1], K D Schilling[1], G Schramm[1,2], R Schwengner[1], A Wagner[1], D Yakorev[1,2]

[1] Helmholtz-Zentrum Dresden-Rossendorf, Bautzner Landstr. 400, 01328 Dresden, Germany
[2] Technische Universität Dresden, 01062 Dresden, Germany

E-mail: r.hannaske@hzdr.de

Abstract. The photodissociation of the deuteron is a key reaction in Big Bang nucleosynthesis, but is only sparsely measured in the relevant energy range. To determine the cross section of the d(γ,n)p reaction we used pulsed bremsstrahlung and measured the time-of-flight of the neutrons. In this article, we describe how the efficiency of the neutron detectors was experimentally determined and how the modification of the neutron spectrum by parts of the experimental setup was simulated and corrected.

1. Introduction, motivation, experimental setup and data analysis

A brief introduction to Big Bang nucleosynthesis, the motivation to measure the photodissociation of the deuteron at energies relevant to it, the experimental setup at the ELBE accelerator at Helmholtz-Zentrum Dresden-Rossendorf and the data analysis are described in ref. [1, 2].

In this article, we discuss two important systematic effects: the neutron detector efficiency and the modification of the neutron spectrum from the d(γ,n)p reaction due to interactions with the target and other parts of the experimental setup.

2. Efficiency calibration of the neutron detectors

Neutrons from the d(γ,n)p reaction have been detected using a time-of-flight method with 1000 mm long, 42 mm and 11 mm thick plastic scintillators (EJ 200) coupled to high-gain photomultiplier tubes (PMTs) at both ends. To suppress γ-ray induced events the detectors are surrounded by a 1 cm thick lead shield. The efficiency of unshielded detectors of this type had been determined before [3], but later we observed discrepancies in measurements of shielded detectors relative to a ^{235}U fission chamber at the neutron time-of-flight facility nELBE [4] as well as in simulations using the NEFF7 code [5], see figure 1.

In 2011 we did a new efficiency calibration of the shielded neutron detectors at Physikalisch-Technische Bundesanstalt (PTB) Braunschweig. A pulsed Van-de-Graaff generator [6] provided charged ions to produce quasi-monoenergetic neutrons in the energy ranges 25 – 565 keV, 1 – 2.5 MeV and around 5 MeV via the reactions ^{7}Li(p,n), T(p,n) and D(d,n), respectively. The

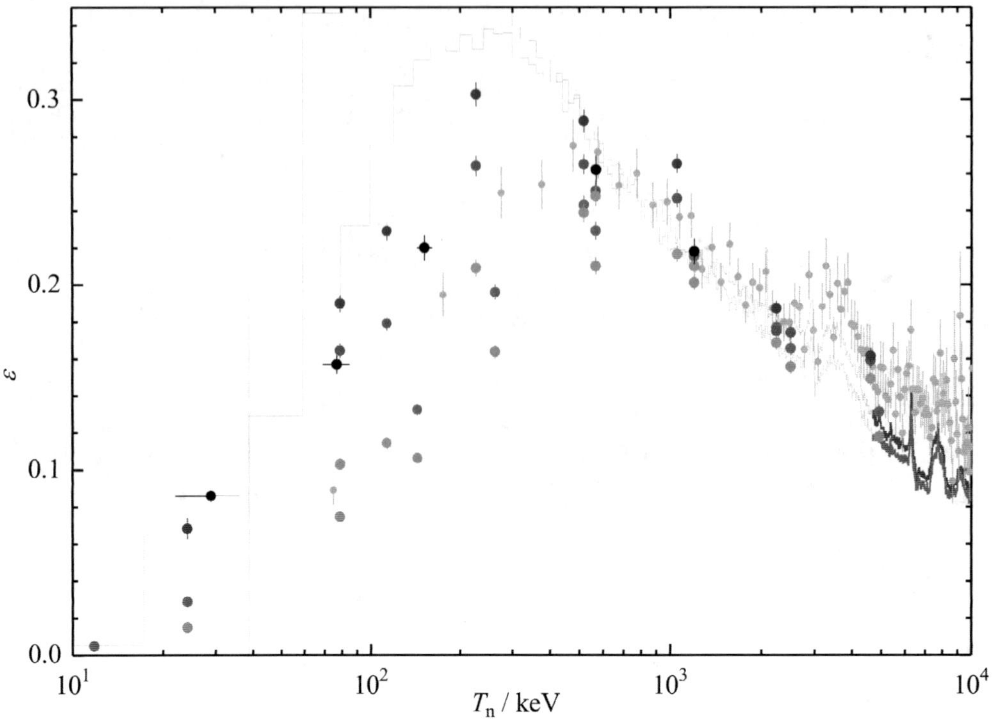

Figure 1. Efficiency ε of plastic scintillation detectors as a function of kinetic neutron energy T_{n}. Older absolute measurements at PTB Braunschweig (black symbols) [3] and measurements relative to a ^{235}U fission chamber at nELBE [4] (grey symbols) are compared to NEFF7 simulations [5] with its original, cylindrical geometry (blue line) and with a modified, rectangular geometry (magenta line). The symbols coloured blue, dark green, light green, red and orange are the results of the efficiency calibration of the five neutron detectors at PTB Braunschweig described in this article.

low-scattering environment reduced the in-scattering of neutrons. We simulated the contribution of neutrons scattered in the target with the TARGET code [7] and found it to be negligible. We used shadow bars made of borated polyethylene, which absorb neutrons on the direct source-detector path, to measure the background of scattered neutrons. Figure 2 shows typical time-of-flight spectra with and without shadow bars. Before the background is subtracted, a time-of-flight dependent dead-time correction as described in ref. [8] is applied, which can differ by up to 1.5 % from the integral dead-time correction at high count rates. Finally, the number of detected neutrons n_{det} is obtained by setting a gate on the neutron distribution in the dead-time corrected, background-subtracted time-of-flight spectra.

Two long counters, a ^{3}He Geiger counter and a charge integrator at the target monitored the intensity of the neutron production. To obtain the neutron flux, these monitors were calibrated using a recoil-proton proportional counter, a recoil-proton telescope or a NE213 detector as reference instruments [9]. After taking into account the neutron attenuation in air, the solid angle of the detector and the angular distribution of the neutron-producing reaction, the number of incoming neutrons n_{inc} can be determined.

Figure 1 shows the efficiency $\varepsilon = n_{\mathrm{det}}/n_{\mathrm{inc}}$ of the five calibrated detectors. While the energy dependence is quite similar among the detectors, the absolute values differ by a factor of two. Scintillator and PMTs are connected optically with silicone grease and mechanically with opaque

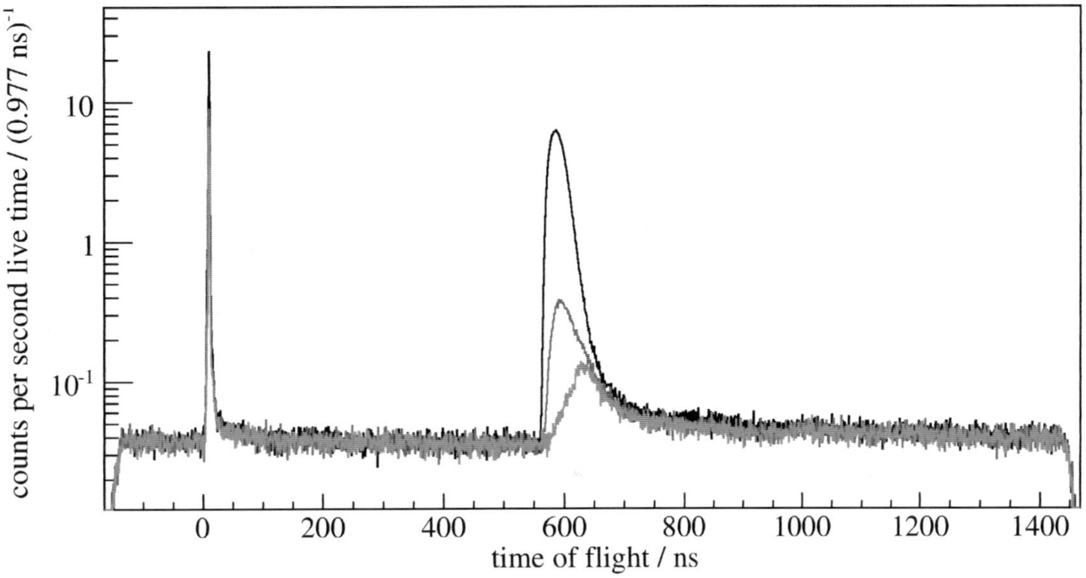

Figure 2. Live-time normalized time-of-flight spectra of a detector at an angle of 0° and a distance of 3 m exposed to 140 keV neutrons without shadow bar (black line), with shadow bar covering only the solid angle of the scintillator (red line) and with shadow bar covering the solid angle of the scintillator and its lead shield (green line). The peak at 10 ns stems from photons created in the target. The neutron distribution starts at 550 ns.

heat shrink tube. We assume that the optical coupling between scintillator and PMTs partly degraded due to ageing or handling, because a comparison of the charge spectra of all PMTs from the calibration measurements with quasi-monoenergetic neutrons shows large differences. The coincident read-out using high-gain PMTs results in a detection threshold of about 10 keV neutron energy, confirmed by the efficiency of 0.004 measured at 12 keV with one detector.

3. Modification of the neutron spectrum

The neutron spectrum from the photodissociation of the deuteron measured at the ELBE accelerator contained more low-energy neutrons than expected. To understand this discrepancy, we used the FLUKA Monte Carlo code [10, 11], version 2011.2, to simulate, how neutron scattering on different parts of the experimental setup modifies the neutron spectrum. The simulation included walls, floor and ceiling of the experimental area, the target consisting of alternating layers of aluminium and deuterated polyethylene (PE), all six neutron detectors and their lead shield, lead walls shielding the photon collimator and the beam dump, the lead shield of γ-ray detectors and air. It was not possible to use the incoming photons as primary particles, because the angular distribution of the d(γ,n)p reaction is not properly implemented in the FLUKA code. We calculated a neutron source distribution using the shape of the bremsstrahlung spectrum [12], theoretical values of the E1 and M1 contributions to the total cross section [13, 14] and a parametrisation of the angular distribution (M1: isotropic, E1: dipole). Each neutron started at a random position in the PE layers of the target and its energy and polar angle was randomly sampled from the calculated distribution. It should be noted that the normalisation of the distribution is arbitrary and the results of the simulation do not depend on the absolute value of the cross section, but only on its energy dependence.

Journal of Physics: Conference Series **665** (2016) 012003 doi:10.1088/1742-6596/665/1/012003

Table 1. Relative abundance of events hitting the detector located at 90° depending on their previous interaction place. The relative abundance of events, in which the arriving particles are not neutrons, is given in parentheses, if it is larger than zero.

prev. interaction place	rel. abundance	
no interaction	36.4 %	
walls, floor and ceiling	35.8 %	(7.5 %)
lead shield γ-ray det.	9.8 %	
target PE	4.9 %	
lead shield other det.	4.4 %	(0.2 %)
air	3.8 %	(0.3 %)
target aluminium	1.6 %	
lead shield beam dump	1.6 %	
scintillator other det.	0.9 %	
lead shield collimator	0.8 %	

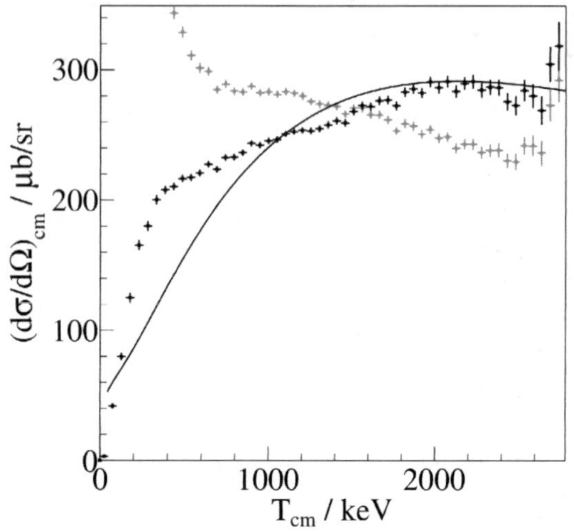

Figure 3. Differential cross section of the d(γ,n)p reaction as a function of total kinetic energy, both in the center-of-momentum (cm) frame. The symbols are preliminary results from the measurement at a neutron angle of 90° in the laboratory system with (black) and without (grey) correction of the modified neutron spectrum. The uncertainties of the data points printed here are due to the bin width (50 keV) and neutron statistics (3 % at 300 keV, 1 % at 1 MeV). For comparison the black line shows a theoretical calculation [15].

Table 1 shows which parts of the setup are the main scatterers for neutrons. Of all simulated particles reaching the detector located at 90°, one third are neutrons from the d(γ,n)p reaction which had no interactions with the setup. Another third of the detector hits had their previous interaction in the walls, the floor or the ceiling. These events are often neutrons slowed down below the detection threshold or capture γ-rays without time-of-flight correlation, which can be subtracted from the experimental time-of-flight spectrum as a constant background. Other major sources of scattered neutrons are the lead shields of the γ-ray detectors, which are close to the target, the deuterated PE in the target, the lead shield of the other detectors and the air. In the analysis of the experiment, the neutron energy is calculated from the time-of-flight assuming they took the direct flight path from the target to the detector. Thus, the energy assigned to scattered neutrons will be wrong and the efficiency correction cannot be applied. To avoid this problem, we apply the efficiency correction to the simulated neutron events, of which the true energy and the time-of-flight are known. By comparing the simulated efficiency-corrected neutron spectrum to the spectrum calculated from the neutron source distribution, we get a correction factor for the experimental neutron spectrum, that includes the experimentally determined detection efficiency and the simulated modification of the neutron spectrum from interactions with the setup. In figure 3 the cross section is plotted with and without the correction of the modified neutron spectrum. At $T_{\rm cm} > 800$ keV the correction results in a realistic description of the cross section whereas at lower energies discrepancies remain.

4. Conclusions

The efficiency of the neutron detectors has been calibrated at PTB Braunschweig in the energy range 25 keV to 5 MeV. The low detection threshold of about 10 keV neutron energy was confirmed experimentally. The large efficiency differences between the detectors require a review of the optical and mechanical coupling of scintillators and PMTs. The successful calibration is very important for the absolute d(γ,n)p cross section measurement at the ELBE accelerator at Helmholtz-Zentrum Dresden-Rossendorf, in which the detectors have been used several months before.
The d(γ,n)p experiment at ELBE required an investigation of the modification of the neutron spectrum due to interactions of the neutrons with the experimental setup. First results of our FLUKA simulation demonstrated the importance of this correction. Although the correction results in a realistic description of the cross section at $T_{\mathrm{cm}} > 800$ keV, the discrepancies at lower energies are to be investigated. In the next step our simulation will include more parts of the setup, namely further lead shields, the full photon beam dump, the target holder and the γ-ray detectors (high-purity germanium crystals surrounded by a anti-Compton detectors).

Acknowledgments

Technical assistance by Andreas Hartmann, Manfred Sobiella and Daniel Stach is greatly acknowledged. We thank Ralf Nolte for his continuous support. We also thank the accelerator crews of ELBE and PTB Braunschweig for providing stable beams during the experiments. This work is supported by the Deutsche Forschungsgemeinschaft under contract no. JU 2705/1-1 and by the German Federal Ministry for Education and Research under contract no. 02NUK013A (TRAKULA project).

References

[1] Hannaske R *et al.* 2010 *Proc. of the 11th Symp. on Nuclei in the Cosmos (19.-23.07.2010, Heidelberg, Germany) PoS* **(NIC XI)** 090
[2] Hannaske R 2014 Ph.D. thesis Technische Universität Dresden to be submitted
[3] Beyer R *et al.* 2007 *Nucl. Instrum. Methods* A **575** 449–455
[4] Beyer R *et al.* 2013 *Nucl. Instrum. Methods* A **723** 151–162
[5] Dietze G and Klein H 1982 Tech. Rep. PTB-ND-22
[6] Brede H *et al.* 1980 *Nucl. Instrum. Methods* **169** 349–358
[7] Schlegel D 2005 Tech. Rep. PTB-6.42-05-2
[8] Hannaske R *et al.* 2013 *Eur. Phys. J.* A **49** 137
[9] Schlegel D and Guldbakke S 2002 Tech. Rep. PTB-6.41-2002-03
[10] Battistoni G, Cerutti F, Fassò A, Ferrari A, Muraro S, Ranft J, Roesler S and Sala P R 2007 *Proc. of the Hadronic Shower Simulation Workshop 2006 (Fermilab 6–8 September 2006)* ed M Albrow and R Raja *AIP Conf. Proc.* **896** 31–49
[11] Ferrari A, Sala P, Fassò A and Ranft J 2005 Tech. Rep. CERN-2005-10, INFN/TC_05/11, SLAC-R-773
[12] Schiff L I 1951 *Phys. Rev.* **83** 252–253
[13] Chen J W and Savage M J 1999 *Phys. Rev.* C **60** 065205
[14] Rupak G 2000 *Nucl. Phys.* A **678** 405–423
[15] Arenhövel H 2005 Private communication

The Underground Nuclear Astrophysics in the Precision Era of BBN: Present Results and Future Perspectives

C Gustavino
Istituto Nazionale di Fisica Nucleare (INFN), Sezione di Roma, Rome (Italy)
E-mail: carlo.gustavino@roma1.infn.it

Abstract. The abundance of light isotopes such as D, ^3He, ^4He, ^6Li and ^7Li produced during Big Bang Nucleosynthesis (BBN) only depends on particle physics, baryon density and relevant nuclear processes. At BBN energies ($0.01 \div 1\ MeV$) the cross section of many BBN processes is very low because of the Coulomb repulsion between the interacting nuclei. As low-energy measurements on earth's surface are predominantly hampered by the effects of cosmic rays in the detectors, it is convenient to study the relevant reactions with facilities operating deep underground. Starting from the present uncertainty of the relevant parameters in BBN (i.e. baryon density, observed abundance of isotopes and nuclear cross-sections), it will be shown that the study of several reactions of the BBN chain, with existing or proposed underground accelerator facilities, can improve the accuracy of BBN calculations, providing a powerful tool to constrain astrophysics, cosmology and particle physics. In particular, a precise measurement of $D(p,\gamma)^3He$ reaction at BBN energies is of primary importance to calculate the baryon density of universe with an accuracy similar to the one obtained by Cosmic Microwave Background (CMB) experiments, and to constrain the number of active neutrino species. For what concern the so called "Lithium problems", i.e. the disagreement between computed and observed abundances of the ^7Li and ^6Li isotopes, it will be also shown the importance of a renewed study of the $D(\alpha,\gamma)^6Li$ reaction.

1. Introduction

In its standard picture, the Big Bang nucleosynthesis occurs during the first minutes of universe, with the formation of light isotopes such as D, ^3He, ^4He, ^6Li and ^7Li through the reaction network shown in figure 1. Their abundance essentially depends on the competition between the relevant nuclear processes and the expansion rate of the early universe. Therefore, their abundance depends on the standard model physics, on the baryon-to-photon ratio η and on the cross-sections of nuclear processes. Indeed, the BBN theory makes definite predictions for the abundances of the light elements as far as the knowledge of relevant parameters is accurate enough. Presently, we are in the exciting era of "high-precision" cosmology, with most of the cosmological parameters known to within a few percent. In particular, Cosmic Microwave Background (CMB) experiments provide a precise measurement of baryon density Ω_b and the derived η parameter [1, 2]. The universal expansion rate (i.e. Hubble parameter H) is determined through the Friedman equation by the total energy density, which is dominated by the number of relativistic species at these epochs. In the standard model, CMB photons, $e^{\pm}$ pairs and neutrinos. In addition, through their charged-current weak interactions, the electron-type neutrinos contribute to control the fraction of free neutrons available, which effectively determine the primordial abundance of ^4He. The nucleosynthesis chain begins with the formation of deuterium in the process $p(n,\gamma)D$. However, the density of photons is huge relative to the baryon

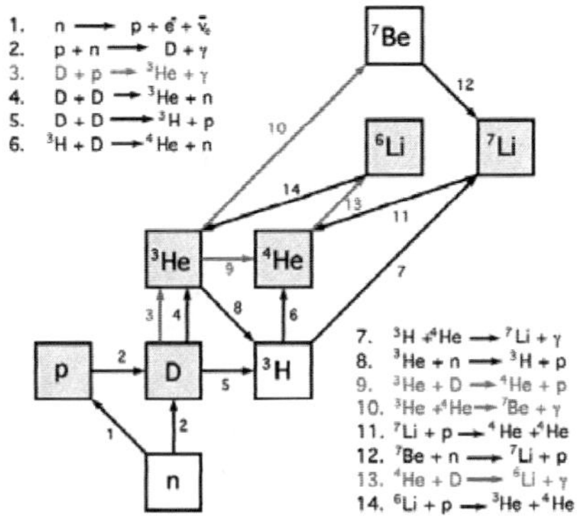

Figure 1. Leading processes of Big Bang Nucleosynthesis. Yellow boxes mark stable isotopes. The red arrows show the reactions that can be studied by exploiting existing or proposed underground accelerators.

density ($\eta \simeq 6 \cdot 10^{-10}$). Because of the large number of photons per baryon, photo-dissociation delays the production of deuterium well after T drops below the binding energy of deuterium ($Q = 2.23\ MeV$, the "deuterium bottleneck"). Nearly all the free neutrons end up bound in the most stable light element ^4He, while heavier nuclei do not form in any significant quantity both because of the absence of stable nuclei with mass number 5 or 8 (which impedes nucleosynthesis via $^4He + n$, $^4He + p$ or $^4He +^4 He$ reactions) and of the large Coulomb barriers for reactions such as the $^3H(^4He,\gamma)^7Li$ and $^3He(^4He,\gamma)^7Be$. The uncertainty of ^4He abundance weakly depends on details of the reaction network, and it is almost entirely due to neutron lifetime error. The error of computed D abundance is presently dominated by the $D(p,\gamma)^3He$ reaction. In fact, the $D(^2H,p)^3H$ and the $D(^2H,n)^3He$ reactions have recently been measured with high accuracy [4]. the ^3He error is mainly due to the $D(^3He,p)^4He$ process and, in a lower extent, to the $D(p,\gamma)^3He$ reaction [3]. For what concern the ^7Li isotope, there are several reactions contributing to the total uncertainty for this nuclide. For $\eta \simeq 6 \cdot 10^{-10}$, leading contributions to the theoretical estimate come from $^7Be(n,\alpha)^4He,^3He(^4He,\gamma)^7Be$ and $^7Be(^2H,p)2\alpha$ reactions. Finally, the theoretical error of ^6Li is very large, and it is almost entirely due to the poor knowledge of the $D(\alpha,\gamma)^6Li$ cross-section.

The computed abundances of D and ^4He are in agreement with observations within errors, confirming the overall validity of BBN theory. The ^3He observations are presently affected by large systematics errors, making this isotope not a powerful probe for the BBN theory. For what concern the "heavy" isotopes of Lithium, the observed abundance of ^7Li is a factor 2-4 lower than the predicted one ("The lithium problem") [5]. Recently, some observation indicates that the amount of ^6Li observed in metal poor stars is unexpectedly large compared to BBN predictions (see figure 2). Even though many of the claimed ^6Li detections are controversial, for a very few metal-poor stars there still seems to be a significant amount of ^6Li ("The second Lithium problem") [5].

As stated above, the knowledge of several nuclear processes is affected by large uncertainties because of their low cross-section at BBN energies. Therefore, to reduce the background induced by cosmic rays, their study must be performed with facilities operating underground. Presently, the only underground accelerator in the world is the LUNA 400 kV accelerator, located inside the Gran Sasso laboratory, Italy [6], while many new facilities operating in the MV range have recently been proposed [7, 8, 9]. In the following, it will be discussed the feasibility of accurate measurements of the $D(\alpha,\gamma)^6Li$ and $D(p,\gamma)^3He$ reactions, and their impact in cosmology and particle physics.

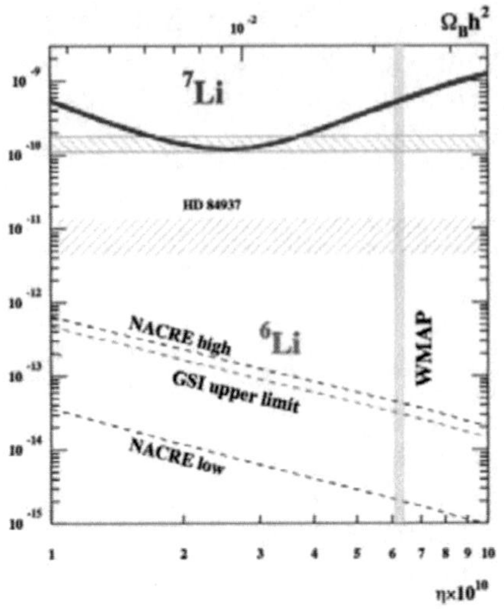

Figure 2. Predicted and observed BBN production ratios for ^6Li and ^7Li as a function of η. The blue band denotes the range of predicted ^7Li yields, The dashed black lines indicate the calculated ^6Li uncertainty and the dashed violet line is the upper limit obtained with the recent measurement by Hammache et al. [19]. Observational data are indicated by horizontal green-hatched areas. The yellow vertical band shows the WMAP η value [1]. The large uncertainty in the predicted ^6Li abundance is due to the lack of direct measurements of the $D(\alpha,\gamma)^6Li$ reaction in the BBN energy region.

2. The Lithium problem(s)

As shown in figure 2, the abundance of ^7Li observed in metal poor stars is several σs lower with respect to BBN predictions, while the amount ^6Li is order of magnitudes higher with respect to calculations [5]. The difference between observed and calculated abundances of ^7Li and ^6Li may reflect systematics in the observed values, post-primordial processes or physics beyond the Standard Model (see for example [15] and references therein). However, as shown in figure 2, the theoretical error of ^6Li abundance is very large because the cross-section of the leading process to synthesize ^6Li is poorly known. As a matter of fact, the $^2H(\alpha,\gamma)^6Li$ cross-section is very small at BBN energies ($30 \div 400$ keV), because of the Coulomb barrier and because electric dipole transition is suppressed for the iso-scalar particles ^2H and ^4He. Therefore, the $^2H(\alpha,\gamma)^6Li$ cross-section has been measured only for energies greater than 1 MeV and around the 711 keV resonance [16, 17]. As shown in figure 3, there are two indirect attempts to determine the $^2H(\alpha,\gamma)^6Li$ cross-section at BBN energies, using the Coulomb dissociation technique [18, 19]. A direct attempt to directly measure the $^2H(\alpha,\gamma)^6Li$ cross-section in the BBN energy range by in-beam γ spectroscopy resulted only in an upper limit [11]. Very recently this cross-section have been measured at BBN energies by the LUNA collaboration [12, 13, 14]. The measurement has been done with the underground 400 kV accelerator, by using an α beam impinging a deuterium gas target and detecting the prompt γs with a lithium-drifted germanium detector. In this measurement the advantage of operating underground is reduced by the existence of a weak but inevitable beam induced background. In fact, the $^2H(\alpha,\alpha)^2H$ Rutherford scattering induces a small amount of $^2H(^2H,n)^3He$ reactions, with the production of few neutrons/second. The neutrons produced by the $^2H(^2H,n)^3He$ reaction induce $(n,n'\gamma)$ reactions in the germanium detector and in the surrounding materials, generating a neutron-induced background in the spectrum. Therefore, the LUNA measurement is limited to a narrow energy range around $E_{cm} = 130$ kV. In fact, the maximal energy of the α particle beam in LUNA is only 400 keV and the Coulomb barrier makes unfavorable the signal-to-noise ratio at much lower beam energies. In any case, this first positive measurement of the $^2H(\alpha,\gamma)^6Li$ cross-section establishes that the study of this reaction at BBN energy is feasible, if it is carried out underground. Therefore, the proposed underground accelerators operating in the MeV region [7, 8, 9] allow a detailed study of this reaction in a wide energy range. In fact, it will be possible to cover the full BBN energy range and also overlap the Mohr's direct measurement around the 711 keV resonance ($E_{lab} = 2.1$ MeV), making possible the calculation of ^6Li abundance with unprecedented precision and without any theoretical assumptions.

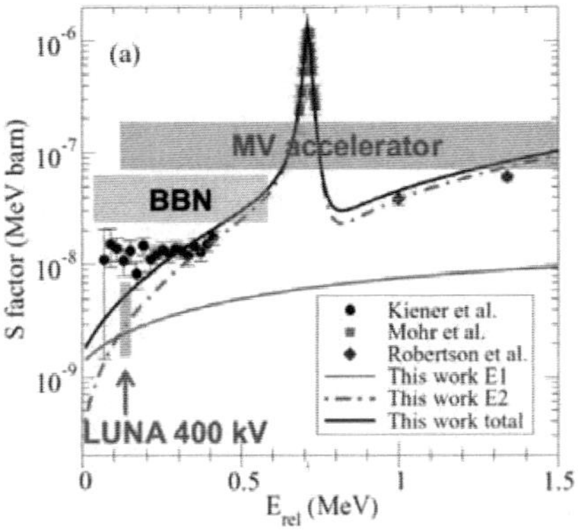

Figure 3. S-factor data for the reaction $D(\alpha,\gamma)^6Li$. Direct measurements [16, 17] are reported together with the indirect Coulomb Dissociation (CD) measurement by [18]. The Theoretical E1, E2, and total S factors curves are inferred from the CD data of Hammache et al. [19]. The energy range of the direct measurement performed with the LUNA 400 kV accelerator is also shown [12, 13, 14], as well as the energy region that can be investigated with an underground MegaVolt accelerator.

3. The deuterium abundance and $D(p,\gamma)^3He$ reaction.

Deuterium is the most sensitive nuclide to η (or equivalently to the related baryon density $\Omega_b h^2$). Therefore it provides an independent measure of this parameter with respect to CMB experiments, which give the η value as it was after the recombination epoch, when the universe was some 400,000 years old [1, 2]. The most accurate measurement of baryon density is presently the one released by the PLANCK collaboration, $\Omega_{b,0}(CMB) = (2.205 \pm 0.028)/h^2$, where $\Omega_{b,0}$ is the present day baryon density of universe and h is the Hubble constant in units of 100 $km\ s^{-1}Mpc^{-1}$. On the other hand, the abundance of primordial deuterium has recently been measured with unprecedented precision, $10^5(D/H) = 2.53 \pm 0.05$ [2], improving the BBN-derived baryon density to $\Omega_{b,0}(BBN) = (2.22 \pm 0.03 \pm 0.08)/h^2$ [20]. The error terms reflect the uncertainties in, respectively, observed deuterium abundance and BBN calculation. The last is due to the uncertainties in the nuclear reaction rates used in BBN codes, and in particular the one related to the $D(p,\gamma)^3He$ cross-section. Therefore, even a modest reduction of the uncertainty of the $D(p,\gamma)^3He$ cross-section would make the precision of $\Omega_{b,0}(BBN)$ comparable to that of $\Omega_{b,0}(CMB)$ achieved by the PLANCK mission [21, 22]. For what concern the particle physics sector, D and ^4He abundances depends on the expansion rate of universe, therefore allow to bound the number of active neutrino families (and any other relativistic species) much better than with CMB alone [21, 22, 23]. Furthermore, unlike the CMB, The abundance of D and ^4He has the potential to probe a non-zero lepton asymmetry (neutrino degeneracy).

As discussed above, the accuracy of BBN constraints in cosmology and particle physics depends on the knowledge of the $D(p,\gamma)^3He$ cross-section. Figure 4 shows the S-factor data for this reaction. It is worth to point out that the precise low-energy data comes from underground measurements, by using a 50 kV accelerator [24]. The data at higher energies are much less accurate and show some tension with respect to theoretical models. In this concern, a measurement in the BBN energy range is mandatory, possibly with an underground accelerator. It is worth to point out that the existing 400 kV accelerator of LUNA can be used to measure the $D(p,\gamma)^3He$ S-factor up to $E_{cm} = 260\ keV$, with a substantial improvement of the calculated abundance of deuterium.

4. Conclusions

BBN theory represent a powerful tool to probe cosmology and standard model physics. In the age of "precision cosmology" we currently are, it is mandatory to improve the accuracy of BBN reactions, because they severely affects the uncertainty of calculations. This program must be be performed with facilities operating deep underground, to reduce the background induced by cosmic rays. To give an example, the muon and neutron fluxes inside the Gran Sasso Laboratory are reduced by a factor 10^6 and 10^3 with respect to the Earth's surface typical values, and the suppression of the γ-ray activity is a factor $10^2 - 10^5$, depending on the photon energy [25]. The $D(p,\gamma)^3He$ and $^2H(\alpha,\gamma)^6Li$ are of crucial importance to calculate the D and ^6Li isotopes and therefore to probe cosmology and particle physics. In this concern, a renewed study of $D(^3He,p)^4He$ and $^3H(^4He,\gamma)^7Li$ reactions, to respectively improve the calculation ^3He and ^7Li abundances, is also desirable.

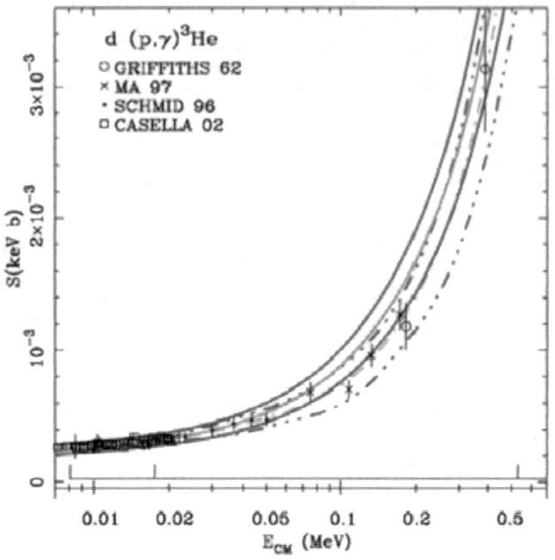

Figure 4. S-factor data for the reaction $D(p,\gamma)^3He$. the best-fit curve (dash-dot curves) and theoretical calculation (solid) are shown. All errors are shown as 2 σs.

References

[1] WMAP collaboration: arXiv:1212.5225v3 [astro-ph.CO] 4 Jun 2013.
[2] PLANCK collaboration: arXiv:1303.5076v1 [astro-ph.CO] 20 Mar 2013.
[3] P.D. Serpico et al., J. Cosmol. Astropart. Phys. 2004, 010 (2004).
[4] D.S. Leonard et al., Phys. Rev. C 73 045801 (2006).
[5] See proceedings of "Lithium in the Cosmos", 27-29 February 2012, Paris.
[6] A. Formicola et al., Nucl. Instr. and Meth. A 507 (2003) 609.
[7] M. Junker: "Experiences and Prospects of Nuclear Astrophysics in Underground Laboratories", proceedings of Nuclear Physics in Astrophysics VI, 19-24 May 2013, Lisbon.
[8] D. Bemmerer: "5 MeV underground accelerator in the Dresden Felsenkeller" , proceedings of Nuclear Physics in Astrophysics VI, 19-24 May 2013, Lisbon.
[9] A. Lemut: "The DIANA Underground Accelerator Facility", proceedings of Nuclear Physics in Astrophysics VI, 19-24 May 2013, Lisbon.
[10] F. Hammache et al., Phys. Rev. C 82, 065803 (2010), 1011.6179.
[11] F.E. Cecil, J. Yan, and C. S. Galovich, Phys. Rev. C 53, 1967 (1996).
[12] "Results of the first direct measurement of the $^2H(\alpha,\gamma)^6Li$ S-factor at Big Bang Energies", proceedings of Nuclear Physics in Astrophysics VI, 19-24 May 2013, Lisbon.
[13] C. Gustavino: "LUNA data on the $^2H(\alpha,\gamma)^6Li$ reaction and the Big Bang Nucleosynthesis", PoS(NIC XII)074 (2012).
[14] C. Gustavino: "The $^2H(\alpha,\gamma)^6Li$ experiment at LUNA", Proceedings of Int. Nucl. Phys. Conf., 2-7 june 2013, Florence, Italy.
[15] B.D. Fields, Annu. Rev. Nucl. Part. Sci. 61, 47 (2011).
[16] R.G.H. Robertson et al., Phys. Rev. Lett. 47, 1867 (1981).
[17] P. Mohr et al., Phys. Rev. C 50, 1543 (1994).
[18] J. Kiener et al., Phys. Rev. C 44, 2195 (1991).
[19] F. Hammache et al., Phys. Rev. C 82, 065803 (2010), 1011.6179.
[20] M. Pettini and R. Cooke: Mon. Not. R. Astron. Soc. 425, 24772486 (2012).
[21] K.M. Nollett and G.P. Holder: arXiv:1112.2683v1 [astro-ph.CO] 12 Dec 2011.
[22] R.J. Cooke et al.: arXiv:1308.3240v1 [astro-ph.CO] 14 Aug 2013.
[23] G. Steigman: arXiv:1208.0032v1 [hep-ph] 31 Jul 2012.
[24] C. Casella et al.: Nuclear Physics A 706 (2002) 203216.
[25] A. Caciolli et al., Eur. Phys. J. A 39, 179 - 186 (2009).

Astrophysical Probes of Varying Constants and Unification

C J A P Martins

Centro de Astrofísica, Universidade do Porto, Rua das Estrelas, 4150-762 Porto, Portugal

E-mail: `Carlos.Martins@astro.up.pt`

Abstract. The observational evidence for the acceleration of the universe demonstrates that canonical theories of gravitation and particle physics are incomplete, if not incorrect. A new generation of astronomical facilities will soon carry out precision consistency tests of the standard cosmological model and search for evidence of new physics beyond it. I describe recent work of CAUP's Dark Side team on some of these tests, focusing on the stability of nature's fundamental couplings and tests of unification scenarios.

1. Varying couplings & Unification

Nature is characterized by a set of physical laws and fundamental dimensionless couplings, which historically we have assumed to be spacetime-invariant. For the former this is a cornerstone of the scientific method, but for the latter it is only a simplifying assumption without further justification. Fundamental couplings are expected to vary in many extensions of the current standard model, including in theories with additional spacetime dimensions, such as string theory. A detection of varying fundamental couplings will be revolutionary: it will immediately prove that the Einstein Equivalence Principle is violated (and therefore that gravity can't be purely geometry), that there is a fifth force of nature, and so on.

Recent astrophysical evidence from quasar absorption systems [1] observed with HIRES/Keck and UVES/VLT suggests a parts-per-million spatial variation of the fine-structure constant α; although no known model can explain such a result without considerable fine-tuning, there is also no identified systematic effect that can explain it.

In theories where a dilaton-like dynamical scalar field yields varying α, the other gauge and Yukawa couplings are also expected to vary. In Grand Unified Theories, the variations of α are related to those of the proton-to-electron mass ratio μ and the proton gyromagnetic ration g_p by model-dependent parameters. Specifically [2]

$$\frac{\Delta\mu}{\mu} = [0.8R - 0.3(1+S)]\frac{\Delta\alpha}{\alpha}\,, \qquad \frac{\Delta g_p}{g_p} = [0.10R - 0.04(1+S)]\frac{\Delta\alpha}{\alpha}\,, \tag{1}$$

where R is related to QCD and S is related to electroweak physics. This parametrization is valid for any model where gauge coupling unification occurs at some (unspecified) high energy scale. Thus joint measurements of these couplings in various systems can provide key consistency tests.

This approach has been used in [2] for BBN, [3] for Population III stars and in [4] for the Sun. Here we report on analogous work for neutron stars [5] and quasar absorption systems [6], as well as for laboratory measurements using atomic clocks [7].

2. Neutron Stars

We can exploit the complementarity provided by the extreme conditions of matter described by the nuclear equation of state (EoS) in heavy ion and neutron star physics. Although the precision of inferred nuclear observables is much lower than for BBN, one can assess the constraining power of these high density environments and scan the (α, R, S) parameter space.

For this we choose a physically motivated set of equations of state for dense matter [5] and, for the purposes of this poster, assume a variation of $\Delta\alpha/\alpha = 5 \times 10^{-3}$. This is larger than suggested by [1], but that value holds for very low density systems (absorption lines along the line of sight of quasars) and in the context of environment-dependent variations will not apply to high-density environments. We note that our analysis differs from that of [2] where some unification models were studied, by choosing fixed values of α, R and S. Here our goal is to *scan* the phenomenological (R, S) space for fixed choices of α.

The EoS of nuclear matter is crucial for describing neutron star structure and, in particular, to obtain its mass-radius relationship. This same EoS is also relevant for understanding the properties of nuclei and yields of heavy ion collisions (at higher temperatures) experimentally accessible on Earth. The possible variation of masses of different population species under beta equilibrium, namely nucleons and electrons, can be modified by the non vanishing values of a variation in α. There is a vast number of model EoSs in the literature.

In [5] we focused on parametrizations such as TM1, and we have verified that more advanced parameterizations as PK1 yield same trend results. In order to partially account for the spread of the nuclear EoS we introduce a phenomenological parameter δ to parameterize the softness of the EoS: for $0 \leq \delta \leq 0.5$ the pressure is modified as $P(1-\delta)$. Since our reference EoS describes better the stiff side of the constraints, $\delta \geq 0$. The unchanged TM1 EoS describes the neutron rich systems and as $\delta > 0$ more isospin symmetric systems can be accessed. With this procedure we partially size the combined effect of the uncertainty in the isospin content and the softness of the EoS and the robustness of our findings.

In Fig. 1 we plot values of R and S yielding pressure values compatible with EoS from neutron star constraints (using the TM1 EoS) as well as from heavy ion collision data (both for symmetric nuclear matter and pure neutron matter). We see that, globally, allowed values lie on a variable-width strip. Without being fully comprehensive on the scan of the allowed R and S values we can see that equally stiff or soft EoS have constraining power in their phase space.

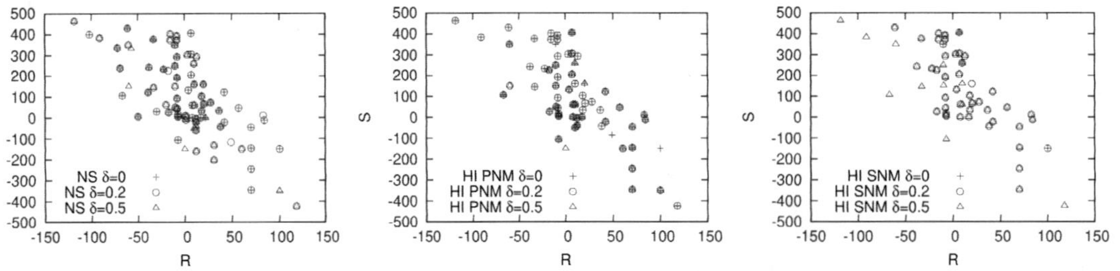

Figure 1. R, S phase space exclusion region. Crosses, circles and triangles denote values of pressure compatible with neutron star (left) and heavy ion symmetric nuclear matter (middle) and pure neutron matter (right) constraints when applying a spread of $\delta = 0, 0.2, 0.5$ for $\Delta\alpha/\alpha = 5 \times 10^{-3}$.

One must note that the experimental uncertainties here are much larger that in the low density BBN case. Nevertheless in all cases we studied the tendency is the same, showing robustness, and we find that the allowed region in the (R, S) plane partially overlaps the values allowed by BBN and in particular the 'typical' values $R \sim 30$ and $S \sim 160$ [2].

3. Atomic Clocks and QSOs

By measuring the rate of two different atomic clocks one obtains a constraint on the relative shift of the corresponding characteristic frequencies. These are proportional to certain products of fundamental couplings, and thus the measurement can be translated into a constraint of the drift of that combination. Different clock comparisons are sensitive to different products of these couplings, and therefore a combined analysis of all existing measurements, summarized in [7], leads to constraints on each of them.

From this analysis we obtain the following 68% confidence intervals

$$\frac{\dot{\mu}}{\mu} = (6.8 \pm 26.3) \times 10^{-17}\,\mathrm{yr}^{-1}\,, \qquad \frac{\dot{g}_p}{g_p} = (-7.2 \pm 4.4) \times 10^{-17}\,\mathrm{yr}^{-1}\,. \tag{2}$$

These should be compared to the result of [8] for the fine-structure constant (also at the 68% confidence level)

$$\frac{\dot{\alpha}}{\alpha} = (-1.7 \pm 2.3) \times 10^{-17}\,\mathrm{yr}^{-1}\,. \tag{3}$$

This highlights the importance of improved experimental bounds using pairs of clocks with different sensitivities to α, μ and g_p.

The above formalism can be used to obtain constraints on the $R-S$ parameter space, shown the left panel on Fig. 2. There is a degeneracy between the two, so that only a combination of them is well constrained. The degeneracy direction is $(S+1)-2.7R = -5 \pm 15$, and the naively expected values are allowed. By separately fixing each of them we find

$$R = 61 \pm 71\,, \quad \text{assuming}\, S = 160 \tag{4}$$

$$S = 76 \pm 197\,, \quad \text{assuming}\, R = 30\,. \tag{5}$$

An analogous study can be made using existing astrophysical measurements of combinations of the fine-structure constant α, the proton-to-electron mass ratio μ and the proton gyromagnetic ratio g_p towards the radio source PKS1413+135, at redshift $z \sim 0.247$ [6]. In this case one finds (at the one-sigma confidence level)

$$\frac{\Delta\alpha}{\alpha} = (-5.1 \pm 4.3) \times 10^{-5} \tag{6}$$

$$\frac{\Delta\mu}{\mu} = (4.1 \pm 3.9) \times 10^{-5} \tag{7}$$

$$\frac{\Delta g_p}{g_p} = (9.9 \pm 8.6) \times 10^{-5}\,; \tag{8}$$

these constraints are still relatively weak, and in particular do not yet provide a test of the dipole result [1]. However, improvements of one order of magnitude in each of the combined measurements (which are well within the reach of forthcoming facilities) should turn this into a stringent test. Moreover, the current results are intriguing when translated into the $R-S$ plane, as shown in the right panel of Fig. 2.

From this analysis we obtain the best-fit values for R and S at the one-sigma confidence level

$$R = 277 \pm 24\,, \qquad S = 742 \pm 65\,. \tag{9}$$

Although the notion that there is a 'standard' model for unification is debatable, it is worth noticing that these values are significantly different from the suggested $R \sim 30$ and $S \sim 160$. A discussion of the causes and implications of this result is beyond the scope of the present work.

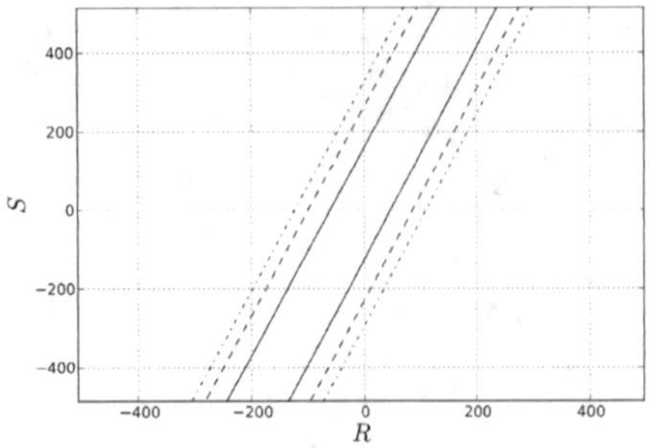

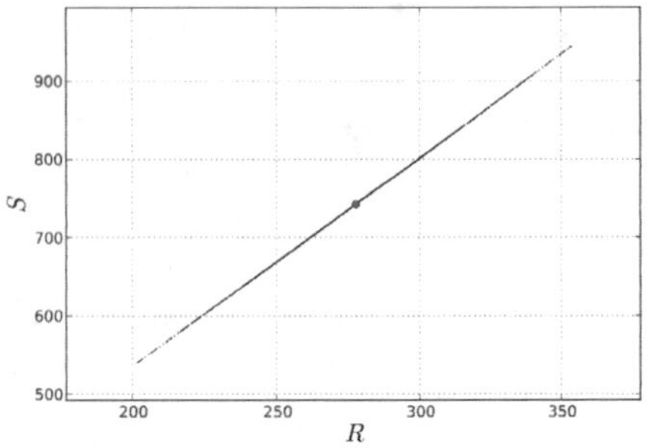

Figure 2. Regions in the R, S parameter space in agreement with current tests of the stability of fundamental constants, from local measurements with atomic clocks (left) and from astrophysical measurements towards the radio source PKS1413+135 (right). Solid, dashed and dotted lines correspond to 68.3%, 95.4% and 99.97% likelihood contours, respectively.

4. Conclusions

Our results show the potential for using the high density region of the density phase space of matter to constrain fundamental physics *in addition* to low density tests using BBN and the CMB. Both our high-density and our atomic clock results are compatible with theoretical expectations on unification scenarios, but much freedom exists (especially in the latter case) due to the presence of a degeneracy direction in the relevant parameter space.

This degeneracy persists for astrophysical measurements towards the radio source PKS1413+135, but in this case the preferred values shift significantly, and are no longer compatible with (arguably naive) expectations on unification. Our analysis motivate the interest of further, more precise astrophysical measurements of fundamental couplings in this and other astrophysical systems.

Finally, this ongoing project highlights the point that the early universe is an ideal laboratory in which to carry out precision consistency tests of our standard cosmological paradigm and search for and constrain new physics. Future facilities such as ALMA, the E-ELT, the SKA and other will play a key role in this endeavour. Future work to explore the relevant parameter space and break parameter degeneracies by combining measurements from systems with different $R-S$ sensitivities is needed.

Acknowledgments

Many interesting discussions with my collaborators in this work (Miguel Ferreira, Mariana Julião, Mafalda Monteiro, Mário Monteiro, Ángeles Pérez-García and José Pedro Vieira) have shaped my views on this subject, and are gratefully acknowledged. This work was done in the context of the project PTDC/FIS/111725/2009 from FCT, with additional funding from Grant No. PP-IJUP2011-212 (funded by U. Porto and Santander-Totta).

References

[1] Webb J K *et al.* 2011 *Phys. Rev. Lett.* **107** 191101
[2] Coc A *et al.* 2007 *Phys. Rev.* **D76** 023511
[3] Ekstrom S *et al.* 2010 *Astron. & Astrophys.* **514** A62
[4] Vieira J P P, Martins C J A P and Monteiro M J P F G 2012 *Phys. Rev.* **D86** 043013
[5] Perez-Garcia M A and Martins C J A P 2012 *Phys. Lett.* **B718** 241
[6] Ferreira M C, Julião M D, Martins C J A P and Monteiro A M R V L 2013 *Phys. Lett.* **B724** 1
[7] Ferreira M C, Julião M D, Martins C J A P and Monteiro A M R V L 2012 *Phys. Rev.* **D86**125025
[8] Rosenband T *et al.* 2008 *Science* **319** 1808

R-matrix analysis of reactions in the ^{9}B compound system applied to the ^{7}Li problem in BBN

M Paris, G Hale, A Hayes-Sterbenz, & G Jungman
T-2 Nuclear & Particle Physics, Astrophysics & Cosmology,
Los Alamos National Laboratory, Los Alamos, New Mexico 87545
E-mail: `mparis@lanl.gov`

Abstract. Recent activity in solving the 'lithium problem' in big bang nucleosynthesis has focused on the role that putative resonances may play in resonance-enhanced destruction of ^{7}Li. Particular attention has been paid to the reactions involving the ^{9}B compound nuclear system, d+^{7}Be $\rightarrow^9$B. These reactions are analyzed via the multichannel, two-body unitary R-matrix method using the code *EDA* developed by Hale and collaborators. We employ much of the known elastic and reaction data, in a four-channel treatment. The data include elastic ^{3}He +^{6}Li differential cross sections from 0.7 to 2.0 MeV, integrated reaction cross sections for energies from 0.7 to 5.0 MeV for ^{6}Li(^{3}He,p)^{8}Be* and from 0.4 to 5.0 MeV for the ^{6}Li(^{3}He,d)^{7}Be reaction. Capture data have been added to an earlier analysis with integrated cross section measurements from 0.7 to 0.825 MeV for ^{6}Li(^{3}He,γ)^{9}B. The resulting resonance parameters are compared with tabulated values, and previously unidentified resonances are noted. Our results show that there are no near d+^{7}Be threshold resonances with widths that are 10's of keV and reduce the likelihood that a resonance-enhanced mass-7 destruction mechanism, as suggested in recently published work, can explain the ^{7}Li problem.

1. Introduction

Calculations of the abundance of ^{7}Li [1] overestimate the value extracted from observations of low-metallicity halo dwarf stars [2], where the stellar dynamics are supposed to be sufficiently understood to isolate the primordial ^{7}Li component. The discrepancy with this (and others, including [3]) observation by a factor of $2.2 \rightarrow 4.2$ corresponds to a deviation of $4.5\sigma \rightarrow 5.5\sigma$, a result that has only become more severe with time. It is essential to determine the nature of this discrepancy as big-bang nucleosynthesis (BBN) probes conditions of the very early universe and our understanding of physical laws relevant in an extreme environment.

Recent attention has focused on the role of reactions that destroy $A = 7$ nuclei at early times $\lesssim 1$ s in the big-bang environment [4, 5]. The authors of Ref. [4], citing the TUNL-Nuclear Data Group (NDG) evaluation tables [6], (See Table 1.) conjecture that the putative $\frac{5}{2}^+$ resonance near 16.7 MeV may enhance the destruction of ^{7}Be through reactions like ^{7}Be (d,p)$\alpha\alpha$ and ^{7}Be (d,γ)^{9}B if the resonance parameters are within given ranges. These studies employ approximations to determine an upper bound on the contribution of resonances, particularly ^{9}B, to a resonant enhancement in reactions that destroy mass-7 nuclides, ^{7}Li in particular. Because there is a paucity of data in the region near the d+^{7}Be threshold where the assumed $\frac{5}{2}^+$ ^{9}B resonance inhabits, we wondered if the existing data may indicate the presence of such a resonance if a multichannel, unitary R-matrix evaluation is pursued.

E_x(MeV ± keV)	$J^\pi; T$	$\Gamma_{\rm cm}$(keV)	Decay
16.024 ± 25	$T = (\frac{1}{2})$	180 ± 16	
16.71 ± 100	$(\frac{5}{2}^+); (\frac{1}{2})$		
17.076 ± 4	$\frac{1}{2}^-; \frac{3}{2}$	22 ± 5	$(\gamma, {}^3\mathrm{He})$
17.190 ± 25		120 ± 40	p, d, ^{3}He
17.54 ± 100	$(\frac{7}{2}^+);(\frac{1}{2})$		
17.637 ± 10		71 ± 8	p, d, ^{3}He, α

Table 1. The TUNL-NDG/ENSDF resonances in the ^{9}B compound nuclear system [6] for resonances that are low-lying with respect to the d-^{7}Be threshold, which occurs at 16.4901 MeV.

We are interested in updating the evaluation of the ^{9}B compound system to address the key question outlined above for BBN: does a resonance near the d+^{7}Be threshold cause an increase in the destruction of mass-7 nuclides in the early universe and possibly explain the ^{7}Li overprediciton problem?

2. The R-matrix formalism and EDA code

The R-matrix approach [7, 8, 9] is a unitary, multichannel parametrization that has proven useful for an array of nuclear reaction phenomenology, particularly for light nuclei. The reader is referred to the literature for details[10]. It is implemented in the EDA (Energy Dependent Analysis) code developed by Hale and collaborators [11]. The available two-body scattering and reaction data is described by minimization of the χ^2 function with respect to variation of the R-matrix parameters E_λ and $\gamma_{\lambda c}$.

3. Analysis and results

The R-matrix configuration, constructed for input into the EDA code, is given in terms of the included channel partitions (pairs), the LS terms for each partition, and the channel radii and boundary conditions, B_c for each channel.

We have included in the analysis the hadronic channels: d+^{7}Be partition with threshold of 16.5 MeV with up to D-waves, ^{3}He +^{6}Li at 16.6 MeV up to P-waves, and p+^{8}Be* at 16.7 MeV up to P-waves. The channel radii were constrained to lie in the range between 5.5 fm and 7.5 fm for these. The electromagnetic γ+^{9}B channels included were $E_1^{3/2}$, $M_1^{5/2}$, $M_1^{3/2}$, $M_1^{1/2}$, $E_1^{5/2}$, and $E_1^{1/2}$ with a channel radius of 50.0 fm.

The ^{9}B analysis is based upon data gathered from the literature. (See Ref.[12] for details.)

Using about 40 parameters, the results of the χ^2 minimization result in a T matrix which gives the solid curves appearing in Figs.1–4, plotted along with the data obtained from references cited in the paragraphs above. The fit quality is fair with χ^2/datum of 1.91, 0.55, 2.38, and 0.37 for Figs.1–4, respectively. The fit to the capture data, Fig.4 has been folded with a Gaussian acceptance function whose width is 5 keV to match the quoted energy resolution in Ref. [16].

The present R-matrix parametrization gives a resonance structure as presented in Table 2. The resonance poles of the T matrix are determined by diagonalization of the complex "energy-level" matrix

$$\mathcal{E}_{\lambda'\lambda} = E_\lambda \delta_{\lambda'\lambda} - \sum_c \gamma_{c\lambda'}[L_c(E) - B_c]\gamma_{c\lambda}, \qquad (1)$$

where $L_c(E) = r_c(\partial\mathcal{O}/\partial r_c)\mathcal{O}^{-1}\big|_{r_c=a_c}$, $\mathcal{O}$ is the outgoing Coulomb wave function, and B_c is the boundary condition given at the channel radius, a_c. Details are given in Ref. [17].

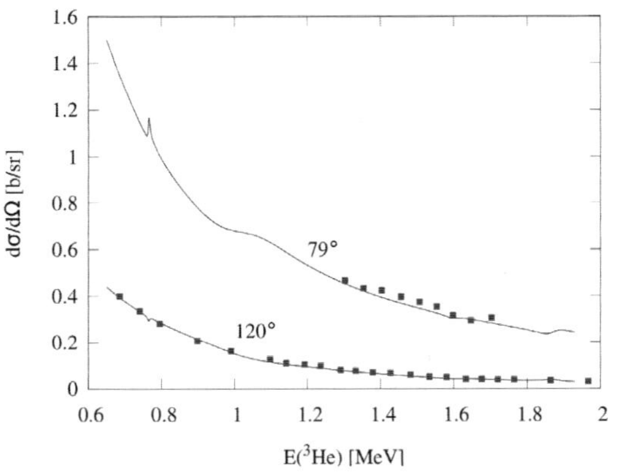

Figure 1. (a)Elastic scattering data from [13] plotted against the R-matrix fit (solid curve) for center-of-mass differential cross section vs. ^{3}He lab energy.

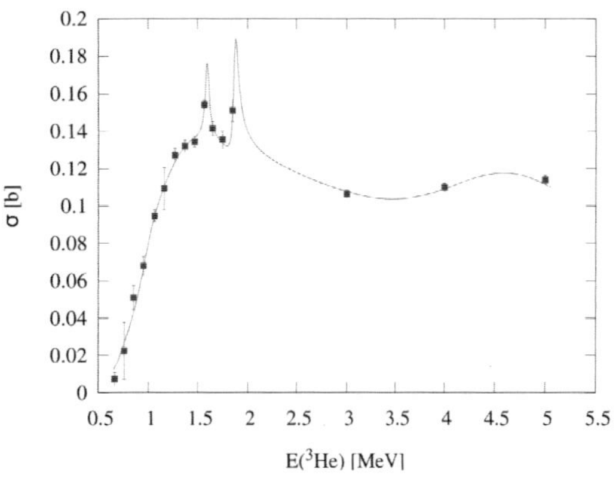

Figure 2. Reaction data from [14], integrated cross section vs. ^{3}He lab energy.

E_x(MeV)	J^π	Γ(keV)	ReE_0(MeV)	ImE_0(keV)	$E(^3\text{He})$(MeV)	Strength	
16.4754	$1/2^-$	768.46	−0.1369	−384.2	−0.2054	0.06	weak
17.1132	$1/2^-$	0.14	0.5109	−0.07	0.7664	1.00	strong
17.2012	$5/2^-$	871.63	0.5989	−435.8	0.8984	0.40	weak
17.2809	$3/2^-$	147.78	0.6785	−73.9	1.0178	0.77	strong
17.6754	$5/2^+$	33.33	1.0631	−16.7	1.5947	0.98	strong
17.8462	$7/2^+$	2036.21	1.2339	−1018.1	1.8509	0.15	weak
17.8577	$3/2^-$	42.52	1.2454	−21.3	1.8681	0.97	strong
18.0582	$3/2^+$	767.11	1.4459	−383.6	2.1689	0.54	weak
18.4229	$1/2^+$	5446.32	1.8206	−2723.2	2.7309	0.03	weak
18.6872	$1/2^-$	10278.41	2.0749	−5139.2	3.1124	0.15	weak
19.6192	$3/2^-$	1478.22	3.0069	−739.1	4.5104	0.52	weak

Table 2. The resonance structure determined in the present 4-channel fit to data as described in the text. The table displays the pole location along with J^π and pole-strength information, as described in the text.

The first column of Table 2 gives the real part of the pole position, $E_0 = E_r - i\Gamma/2$, where E_0 is one of the eigenvalues of the energy-level matrix, Eq.(1) relative the ground state of ^{9}B. The spin-parity is given in the second column. The width Γ is the center-of-mass width in keV. The following column restates the real part of the resonance pole position relative the ^{3}He +^{6}Li threshold in the center-of-mass. The column labeled $E(^3\text{He})$ is the corresponding lab energy. The 'Strength' function is the ratio of the sum of the channel widths (defined in Ref. [17]) divided by the total width, $\Gamma^{-1}\sum_c \Gamma_c$. Resonances labeled 'strong' are clearly seen in at least one of the figures. The resonance structure shown in Table 2 differs significantly from that in Table 1.

Returning the problem of the overprediction of ^{7}Li in current treatments of BBN, we see that the requirements for a near-threshold resonance of Refs. [4] and [5] are difficult to arrange given the resonance structure of Table 2. Both of these works require a narrow resonance, a few 10's of keV in width with a 100 keV of the ^{3}He +^{6}Li (that is, 200 keV within the d+^{7}Be) threshold in order to explain the overproduction of ^{7}Li in BBN reaction network codes [1].

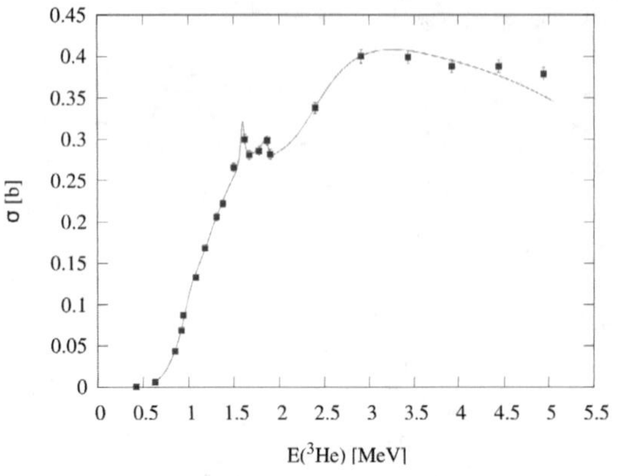

Figure 3. Reaction data from [15], integrated cross section vs. ^{3}He lab energy.

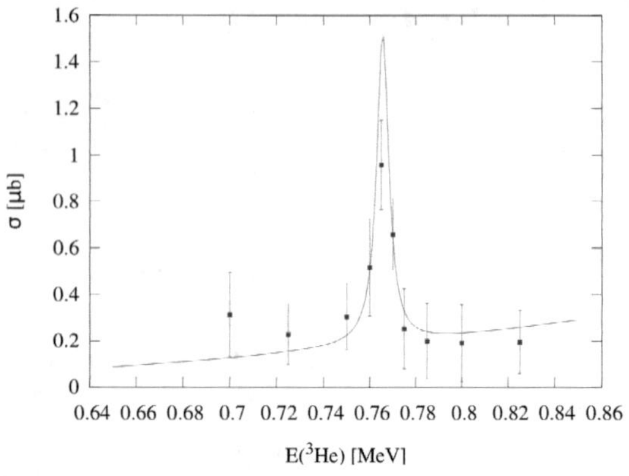

Figure 4. Capture data from [16], integrated cross section vs. ^{3}He lab energy.

4. Summary and findings

We've studied a possible resonant enhancement of the destruction of mass-7 (^{7}Be, in particular) in BBN scenarios. The near threshold, narrow state anticipated in Refs. [4] and [5] appear not to be supported by our multichannel, two-body unitary R-matrix analysis.

Our analysis determines a resonance structure significantly different from that published in the TUNL-NDG/ENSDF compilation [6], as can be seen by comparing the results from the present analysis in Table 2 with the table, Table 1 for the TUNL-NDG/ENSDF analysis.

Our findings for the role of a putative resonance in ^{9}B near the d+^{7}Be threshold as envisioned in Refs. [4] and [5] is that their particular mechanism of resonant enhancement of mass-7 destruction is an unlikely explanation to the ^{7}Li problem in BBN, though low-energy data would allow a more conclusive statement of this finding or its converse.

This work was carried out under the auspices of the National Nuclear Security Administration.

References

[1] B.D. Fields, Ann. Rev. Nucl. Part. Sci. **61**, 47 (2011), and references therein.
[2] F. Spite and M. Spite, Astron. Astrophys. **115**, 357 (1982).
[3] González Hernández, *et al.*Astron. Astrophys. **505**, L13 (2009).
[4] R.H. Cyburt and M. Pospelov, Int. J. Mod. Phys. E **21**, 1250004 (2012).
[5] N. Chakraborty, B.D. Fields and K.A. Olive, Phys. Rev. D **83**, 063006 (2011)
[6] D.R. Tilley, J.H. Kelley, J.L. Godwin, D.J. Millener, J.E. Purcell, C.G. Sheu and H.R. Weller, Nucl. Phys. A **745**, 155 (2004).
[7] P.L. Kapur and R. Peierls, Proc. R. Soc. Lond. A **166** 277 (1938).
[8] E.P. Wigner, Phys. Rev. **70**, 15 (1946).
[9] E.P. Wigner and L. Eisenbud, Phys. Rev. **72**, 29 (1947).
[10] A.M. Lane and R.G. Thomas, Rev. Mod. Phys. **30**, 257 (1958).
[11] G.M. Hale, Nucl. Data Sheets, **109**, 2812 (2008).
[12] M. Paris, *et al.*, Nucl. Data. Sheets, Proc. of the International Conference on Nuclear Data for Science & Technology, 2013, *accepted*
[13] S. Buzhin'ski, *et al.*, Izv. Rossiiskoi Akademii Nauk, Ser.Fiz. **43**, 158 (1979).
[14] A.J. Elwyn, R.E. Holland, C.N. Davids, J.E. Monahan, F.P. Mooring and W. Ray, Phys. Rev. C **22**, 1406 (1980).
[15] Cited in Holland *et al.*, IEEE Trans. Nucl. Sci. **NS-28**, 1344 (1981).
[16] M.R. Aleksic and R.V. Popic, Fizika **10**, 273 (1978).
[17] G.M. Hale, R.E. Brown and N. Jarmie, Phys. Rev. Lett. **59**, 763 (1987).

Indirect study of $^{12}C(\alpha,\gamma)^{16}O$ reaction

F Hammache[1], N Oulebsir[1,2], P Roussel[1], MG Pellegriti[1,6], L Audouin[1], D Beaumel[1], A Bouda[2], P Descouvemont[3], S Fortier[1], L Gaudefroy[4,7], J Kiener[5], A Lefebvre-Schuhl[5] and V Tatischeff[5]

[1] Institut de Physique Nucléaire d'Orsay, UMR8608, IN2P3-CNRS, Université Paris sud 11, 91406 Orsay, France
[2] Université Abderrahmane Mira, 06000 Bjaa ALGERIE
[3] Physique théorique et Mathématique, ULB CP229, B-1050 Brussels, Belgium
[4] GANIL,CEA/DSM-CNRS/IN2P3,Bd Henri Becquerel,BP 55027,F-14076 Caen Cedex 5, France
[5] CSNSM, UMR 8609, CNRS/IN2P3 and Université Paris Sud 11, F-91405 Orsay, France

E-mail: hammache@ipno.in2p3.fr

Abstract. The radiative capture reaction $^{12}C(\alpha,\gamma)^{16}O$ plays an important role in helium burning in massive stars and their subsequent evolution [1]. However, despite various experimental studies, the cross section of this reaction at stellar energies remains highly uncertain. The extrapolation down to stellar energy (E_{cm} ~300 keV) of the measured cross sections at higher energies is made difficult by the overlap of various contributions of which some are badly known such as that of the 2^+ (E_x=6.92 MeV) and 1^- (E_x=7.12 MeV) sub-threshold states of ^{16}O. Hence, to further investigate the contribution of these two-subthreshold resonances to the $^{12}C(\alpha,\gamma)^{16}O$ cross section, a new determination of their α-reduced widths and so their α-spectroscopic-factors was performed using $^{12}C(^7Li,t)^{16}O$ transfer reaction measurements at two incident energies and a detailed DWBA analysis of the data [2]. The measured and calculated differential cross sections are presented as well as the obtained spectroscopic factors and the α-reduced widths as well as the assymptotic normalization constants (ANC) for the 2^+ and 1^- sub-threshold states. Finally, the results obtained from the R-matrix calculations of the $^{12}C(\alpha,\gamma)^{16}O$ cross section using our obtained α-reduced widths for the two sub-threshold resonances are presented and discussed.

1. Introduction

During He burning phase in massive stars, the two main and important reactions through which He is consumed are the triple-α reaction and $^{12}C(\alpha,\gamma)^{16}O$. The ratio of these two reaction rates determine the $^{12}C/^{16}O$ abundance ratio in stars at the end of their helium burning phase. This ratio has important consequences on further hydrostatic burning stages and so on nucleosynthsis in massive stars and consequently has an effect on the final fate of the stars [1]. The rate of the triple-α reaction is well determined but it is not the case of $^{12}C(\alpha,\gamma)^{16}O$. The low-energy cross section of the latter remains highly uncertain despite the various experiments

[6] Present address: Dipartimento di Fisica e Astronomia, Universitá di Catania and Laboratori Nazionali del Sud - INFN, Catania, Italy
[7] Present address: CEA, DAM, DIF, F-91297 Arpajon, France

Journal of Physics: Conference Series **665** (2016) 012007 doi:10.1088/1742-6596/665/1/012007

performed this last four decades. At the Gamow peak of 300 keV where this reaction occurs during the He burning stage, the expected cross section is about 10^{-8} nbarn which makes it impossible to measure directly. Hence, direct measurements were performed down to 900 keV in center-of-mass system and then extrapolated to the energy of interest. The extrapolation in case of this reaction is made difficult by the presence of several contributions, the most important ones being the E1 and the E2 transitions to the ground state via the low energy tail of the 1^- broad resonant state at 9.58 MeV of ^{16}O and the high energy tails of the 2^+ and 1^- sub-threshold resonant states at 6.92 and 7.1 MeV of ^{16}O. Unfortunately, the effect of these two sub-threshold states is badly known because their measured alpha spectroscopic factors S_α and so their corresponding reduced alpha width are spread over a large range of values [2]. So, in view of the importance of $^{12}C(\alpha,\gamma)^{16}O$ reaction and the large uncertainties surrounding the S_α and the γ_α^2 of the two sub-threshold states, we perfomed a new measurement of these quantities via the transfer reaction $^{12}C(^7Li,t)^{16}O$ [3].

2. Experiment description

The experiment was carried out at two incident energies, 28 and 34 MeV, using a $^7Li^{3+}$ beam provided by the Orsay TANDEM. The used target consisted of a self-supporting enriched ^{12}C which thickness was of 80 ± 4 $\mu g/cm^2$. The reaction products were detected with a position-sensitive gas chamber and a ΔE proportional gas-counter located in the focal plane of the Enge Split-pole magnetic spectrometer.

The tritons were detected at angles ranging from 0 to 31° corresponding to angles up to 43° in the center of mass frame. The beam and ^{12}C amount were continuously monitored with a telescope of silicon detectors mounted inside the scattering chamber at θ_{lab}=35°.

A typical excitation energy spectrum of ^{16}O obtained at 11.5° is displayed in Figure 1.

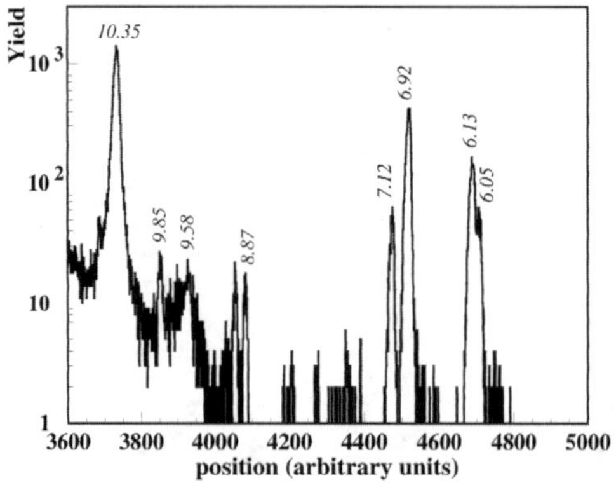

Figure 1. Triton spectrum obtained at 11.5° [3] with the 34 MeV ^{7}Li beam on ^{12}C target in the excitation energy region from 6 to 11 MeV . The excitation energy (MeV) of ^{16}O levels are indicated.

One can notice in the spectrum the strong population of the α-cluster states, the 6.92 and 10.35 MeV states which indicates that the data are consistent with a direct α-transfer mechanism. The other feature one can also notice is the weak population of the non-natural parity state 2^-, the 8.87 MeV state of ^{16}O which can not be populated by direct transfer mechanism. It is probablly populated by the compound nucleus mechanism. This is used to evaluate the contribution of the compound nucleus mechanism in this transfer reaction [3].

3. Results

3.1. DWBA analysis and results

The experimental $^{12}C(^{7}Li,t)^{16}O$ differential cross sections of the 6.05, 6.13, 6.92, 7.12, 8.87, 9.58, 9.85 and 10.35 were measured at the two incident energies of 28 and 34 MeV [3] but only those of 6.13, 6.92, 7.12 and 8.87 states are displayed in Figure 2 together with the Finite-range DWBA and HF calculations.

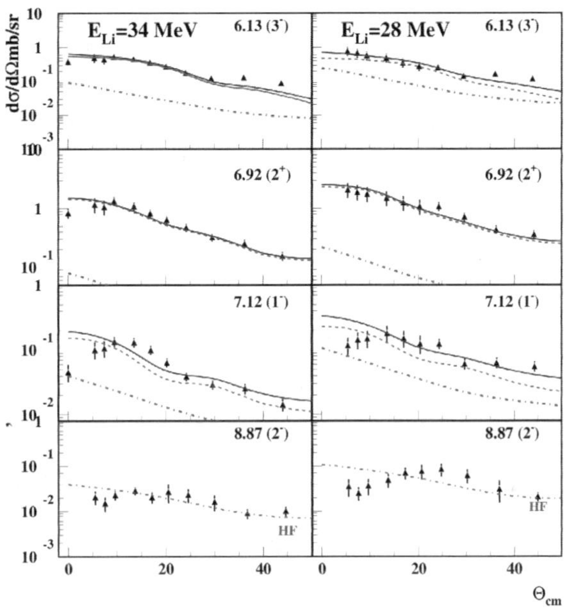

Figure 2. Experimental differential cross sections of the $^{12}C(^{7}Li,t)^{16}O$ reaction obtained at 34 MeV (left column) and 28 MeV (right column) for the 6.13, 6.92, 7.12 and 8.87MeV states [3], compared with FRDWBA calculations (dashed red curve) normalized to the data, Hauser-Feshbach (HF) calculations (dashed-dotted pink line) and the sum HF+FRDWBA (blue solid line).

The Finite-range DWBA (FRDWBA) calculations were performed using the FRESCO code [4]. For the entrance channel, we used the optical potential parameters of Schumacher et al. [5] who performed ^{7}Li elastic scattering measurements on ^{12}C at the energy of 34 MeV. Concerning the exit channel, the optical model potentials used were taken from Garrett et al. [6]. The optical potential parameters finally selected are those giving the best fit for all the studied transitions in the $(^{7}Li,t)$ reaction.

For the α wave function in ^{16}O, an $\alpha+^{12}C$ Wood-Saxon potential was used. A range of radius (3.5 fm $\leq$ R $\leq$ 4.5 fm) and diffuseness (0.53 fm $\leq$ a $\leq$ 0.93 fm) was selected by using the maximum likelihood function (set at 3σ level) on the angular distributions of all measured levels except the non-natural parity 8.87 MeV state and the 9.85 MeV state which displays a quasi-flat distribution [3]. Within this radius and diffuseness range, the boundary values R=4.5 fm and a=0.73 fm provide the best fit for the angular distributions of all the studied states at both incident energies (fig.2) except the 8.87 and the 9.85 MeV states. Details in the DWBA analysis procedure is given in [3]. The displayed FRDWBA calculations in Figure 2 are normalized to the data.

Except for the 8.87 MeV state, the good agreement observed between the DWBA calculations and the measured differential cross sections of the different populated states of ^{16}O at the two bombarding energies of 28 MeV and 34 MeV respectively, gives strong evidence of the direct nature of the (^{7}Li,t) reaction populating these levels and confidence in our DWBA analysis. However, as one can see in Figure 2, a disagreement at angles smaller than 10° is observed for the 7.12 MeV state and for both incident energies. This discrepancy is not understood and it was also observed in ^{12}C(^{7}Li,t)^{16}O experiment of Becchetti et al. [7] at 34 MeV. To try to understand if the decrease of the cross section at angles smaller than 10° is due to a multi-step effect mechanism, coupled channel calculations are needed and they are in progress.

From a χ^2 minimization of the DWBA differential cross sections to the measured ones, S_α mean values of 0.15±0.05 and 0.07±0.03 are deduced for the states of interest at 6.92 MeV and 7.12 MeV of ^{16}O respectively The uncertainty on the extracted α spectroscopic factors for the states of interest was evaluated from the dispersion of the deduced S_α values at the two incident energies and using different sets of optical potentials in the entrance [5] and exit channels [6] and different α-^{12}C well geometry parameters selected above.

Once the S_α of the states of interest are determined, we deduced their α-reduced widths using the expression, $\gamma_\alpha^2 = \frac{\hbar^2 R}{2\mu} S_\alpha |\varphi(R)|^2$ [8] where μ is the reduced mass and φ(R) is the radial part of the α-^{12}C wave function. The calculations of γ_α^2 were performed at the radius R=6.5 fm where φ(R) reaches its asymptotic behavior. The obtained values are 26.7±10.3 keV and 7.8±2.7 keV for the 6.92 MeV and 7.12 states respectively

The asymptotic normalisation constants (ANC) [9] of the two sub-threshold states were also deduced. The obtained values $\tilde{C}^2$=(2.07±0.80)10^{10} fm^{-1} and $\tilde{C}^2$=(4.00±1.38)10^{28} fm^{-1} for the 6.92 and 7.12 MeV states respectively were found in excellent agreement with those obtained by Brune et al. [10] from a sub-coulomb ANC measurement.

3.2. R-matrix calculations and results

R-matrix calculation of the E1 and E2 astrophysical S-factor of $^{12}C(\alpha,\gamma)^{16}O$ reaction, using P. Descouvemont code, were performed using our deduced α-reduced widths for the two sub-threshold states. In the R-matrix calculations, both the $^{12}C(\alpha,\gamma)^{16}O$ astrophysical S-factors obtained by direct measurements at higher energies and the phase shifts data from elastic scattering $^{12}C(\alpha,\alpha)$ measurements were fitted and the two components were fitted separately.

For the E2 component, four 2^+ states were considered in the calculation; the 6.92 MeV state for which we determined the γ_α^2, the 9.85 MeV, the 11.52 MeV and a background equivalent state which takes into account the tails of other higher-lying 2^+ states. In the R-matrix fitting procedure, the resonance parameters of all states except the background state are kept fixed [3]. From the best fits displayed in figure 3, we deduced an E2 S-factor at 300 keV of 50±19 keV-b.

The same fitting procedure was applied for the E1 component. The 1^- states considered are the 7.12 MeV state, the broad resonant state at 9.58 MeV and a background equivalent state which takes into account the tails of other higher-lying 1^- states and the direct component. The only free parameters are those of the 1^- background equivalent state. From the best fits shown in Figure 4, an E1 S-factor at 300 keV of 100±28 keV-b was deduced.

To validate furthermore our R-matrix fits and results, especially for the E1 component, we performed a p-wave calculation of the β-delayed α-spectrum of ^{16}N. For the calculation, we used equation 3 of ref [20] and our R-matrix parameters while we considered the β-feeding amplitudes, $A_{\lambda l}$ (see equation 3 of ref. [20]), as free parameters. As one can see in Figure 5, our calculation describes well the measured data of Tang et al. [20] and this gives strong confidence in our R-matrix calculations. The disagreement between the calculation and the data in the energy

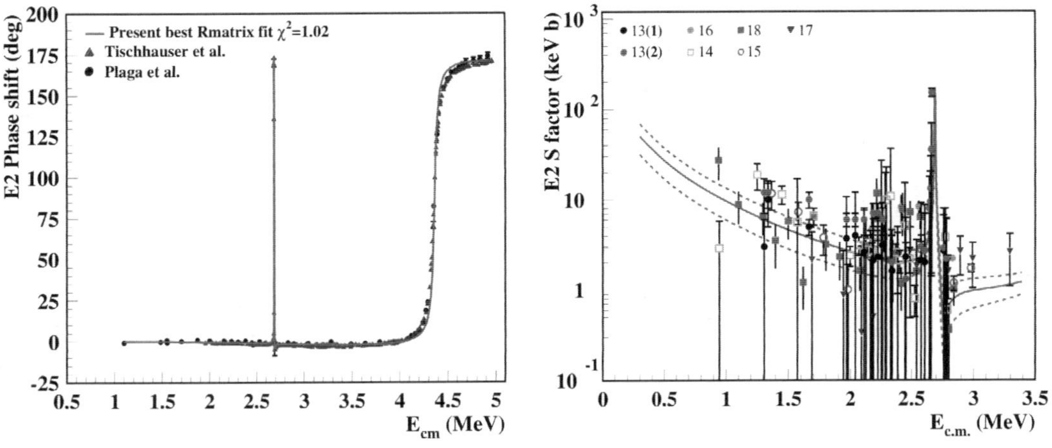

Figure 3. left: Phase shifts for $^{12}C(\alpha,\alpha)^{12}C$ elastic diffusion reaction with R-matrix calculations of the E2 component [3]. Data points are from [11] (black points) and [12] (blue triangles). The solid line correspond to our best R-matrix fit with χ^2=1.02. right:Astrophysical S-factor for the $^{12}C(\alpha,\gamma)^{16}O$ reaction with R-matrix calculations of the E2 component [3]. Experimental data are from [13, 14, 15, 16, 17, 18]. The solid line is our best R-matrix fit using our deduced γ_α^2 for the 6.92 MeV state and the dashed lines when using our upper and lower values for γ_α^2.

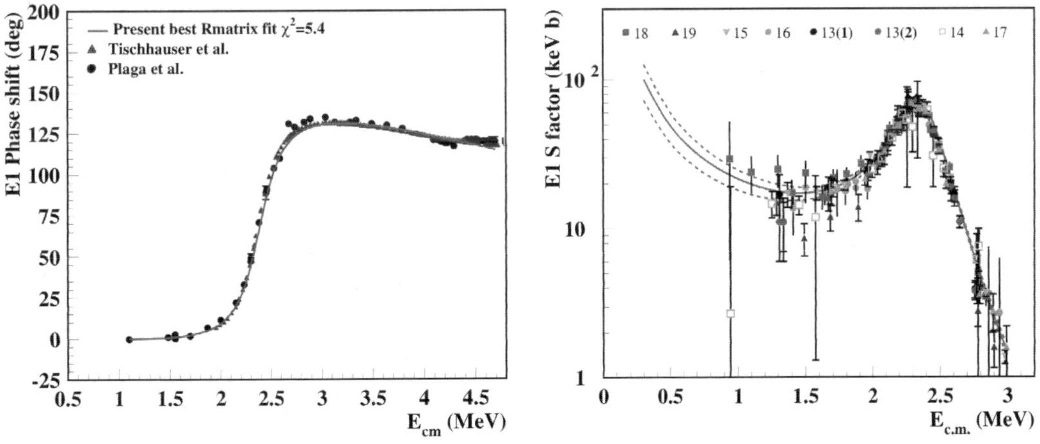

Figure 4. Left:Phase shifts for $^{12}C(\alpha,\alpha)^{12}C$ elastic diffusion reaction with R-matrix calculations of the E1 component [3]. Data points are from [11] (black points) and [12] (blue triangles). The solid line correspond to our best R-matrix fit with χ^2=5.4. Right:Astrophysical S-factor for the $^{12}C(\alpha,\gamma)^{16}O$ reaction with R-matrix calculations of the E1 component [3]. Experimental data are from [13, 14, 15, 16, 17, 18, 19]. The solid line is our best R-matrix fit using our deduced γ_α^2 for the 7.12 MeV state and the dashed lines when using our upper and lower values for γ_α^2

region between 1.3 and 1.5 MeV is due to the f-wave contribution which was not considered in the calculation as it is not contributing to the E1 component.

A comparison of our results for the E1 and E2 S-factors at 300 keV with those of some previous works are given in table 1.

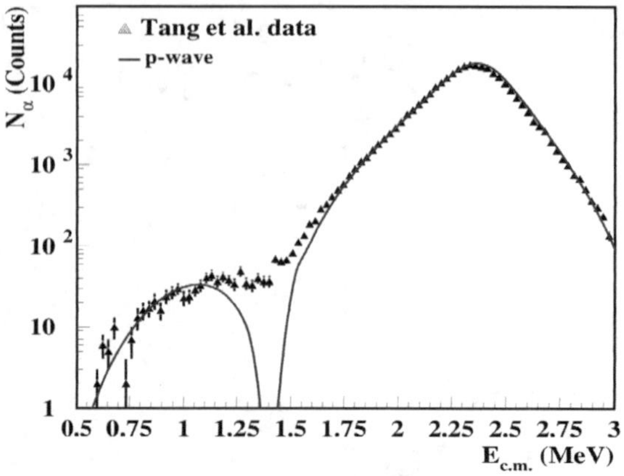

Figure 5. R-matrix calculation [3] (see text) of the β-delayed α-spectrum of ^{16}N together with data obtained in [20]. Only the p-wave was considered in the calculation.

Table 1. Comparison of the astrophysical S-factor at 300 keV obtained in various experiments including this work for the E1 and E2 component as well as the total.

Experiment	S_{E1}(0.3 MeV) (keV-barn)	S_{E2}(0.3 MeV) (keV-barn)	S_{total}(0.3 MeV) (keV-barn)
This work [3]	100±28	50±19	175^{+63}_{-62}
Brune et al. [10]	101±17	44^{+19}_{-23}	170^{+52}_{-55}
Tischhauser et al. [12]	-	53±13	-
Tang et al. [20]	84±21	-	-
Kunz et al. [14]	76±20	85±30	186^{+66}_{-65}
NACRE [21]	79±21	120±60	224^{+97}_{-96}

Our deduced values for the E1 and E2 S-factors are in excellent agreement with those of Brune et al. [10] and Tischhauser et al [12] and in good agreement with those of Tang et al. [20] and Kunz et al. [14] within the error bars. Concerning NACRE recommended values, our E1 S-factor is in agreement with NACRE result within the error bars while for the E2 component, our central value is two times smaller than NACRE recommended value and our error bar is much smaller.

If we consider for the cascade transition, the value of 25±16 keV-b obtained by Matei et al. [22] which is in total disagreement with the recent value (< 1 keV-b) obtained in Schürmann et al. work [23], we obtain a total S-factor of 175±16 keV-b. This results is in good agreement with Brune's et al result (see table 1) and some previous works [3].

Note that in our work as well as in Brune's et al one, the alpha-reduced widths of the two sub-threshold states were fixed in the R-matrix calculations at the measured values which is not the case of all other works where they are considered as free parameters. It is interesting to notice that the two works where the γ_α^2 of the 6.92 and 7.12 MeV states are kept fixed obtain very close results. This is likely due to the additional constraint provided by this fixed parameters in the R-matrix calculations.

4. Conclusion

The α-spectroscopic factors S_α and the reduced α-widths γ_α^2 of the two sub-threshold 2^+ (E_x=6.92 MeV) and 1^- (E_x=7.12 MeV) states of ^{16}O were determined from the analysis of the transfer reaction ^{12}C(^{7}Li,t)^{16}O measurement performed recently at two incident energies. The uncertainties on the S_α and γ_α^2 of the 6.92 and 7.12 MeV states of interest are now well and carefully determined thanks to the detailed finite range DWBA analysis of the measured data. The obtained γ_α^2 for the 2^+ and 1^- sub-threshold resonances were introduced in R-matrix calculations in order to determine the E2 and E1 S-factor at the energy of 300 keV. The result for the E1 S-factor at 300 keV confirms the values obtained in various direct and indirect measurements as well NACRE compilation [3] while for the E2 component, the central value of our result is found to be nearly two times smaller than NACRE recommended value. Our results are in excellent agreement with Brune's et al. [10] ones and in both works, the γ_α^2 or the ANC's of the two sub-threshold states were fixed in the R-matrix calculation at the measured values leading to a bigger constraint in the fitting procedure.

Nevertheless, more precise cascade transitions measurements as well as more precise E2 direct data at higher energies are needed to have a better determination of the total S-factor of $^{12}C(\alpha,\gamma)^{16}O$ at 300 keV.

References

[1] T.A. Weaver and S.E. Woosley, Phys. Reports 227 (1993) 65-96
[2] A. Belhout et al., Nucl. Phys. A793, 178 (2007) and references therein
[3] N. Oulebsir, F. Hammache et al., Phys. Rev. C 85, 035804 (2012)
[4] I. J. Thomson et al., Comp. Phys. Rep. 7, 167 (1988)
[5] P. Schumacher et al., Nucl. Phys. A212, 573 (1973)
[6] J. D. Garrett et al., Nucl. Phys. A212, 600 (1973)
[7] F. D. Becchetti and J. Jänecke, Nucl. Phys. A 305, 293 (1978)
[8] F. D. Becchetti and J. Jänecke, Nucl. Phys. A305, 313 (1978)
[9] A. M. Mukhamedzhanov and R. E. Tribble, Phys. Rev. C 59, 3418 (1999)
[10] C. R. Brune, W. H. Geist, R. W. Kavanagh and K. D. Veal, Phys. Rev. Lett. 83, 4025 (1999) and references therein
[11] R. Plaga et al., Nucl. Phys. A465, 291 (1987)
[12] P. Tischhauser et al., Phys. Rev. C 79, 055803 (2009)
[13] M. Assunção et al., Phys. Rev. C 73, 055801 (2006) and references therein
[14] R. Kunz et al., Phys. Rev. Lett. 86, 3244 (2001)
[15] J. M. Ouellet et al. Phys. Rev. C 54, 1982 (1996)
[16] P. Dyer et al., Nucl. Phys. A233, 495 (1974)
[17] G. Rotters et al., Eur. Phys. J A6, 451 (1999)
[18] A. Redder et al, Nucl. Phys. A462, 385 (1987)
[19] R. M. Kremer et al., Phys. Rev. Lett. 60, 1475 (1988)
[20] X. D. Tang et al. Phys. Rev. C 81, 045809 (2010)
[21] C. Angulo et al., Nucl. Phys. A656, 3 (1999)
[22] C. Matei et al., Phys. Rev. Lett. 97, 242503 (2006)
[23] D. Schürmann et al. Phys. lett. B 703, 551 (2011)

Do we really know M_{up} (i.e. the transition mass between Type Ia and core-collapse supernova progenitors)?

O Straniero, L Piersanti and S Cristallo

INAF – Osservatorio Astronomico di Teramo, 64100 Teramo, Italy and INFN, sezione Napoli, Italy

straniero@oa-teramo.inaf.it

Abstract. M_{up} is the minimum stellar mass that, after the core-helium burning, develops temperature and density conditions for the occurrence of a hydrostatic carbon burning. Stars whose mass is lower than this limit are the progenitors of C-O white dwarfs and, when belong to a close binary system, may give rise to explosive phenomena, such as novae or type Ia supernovae. Stars whose mass is only slightly larger than M_{up} ignite C in a degenerate core and, in turn, experience a thermonuclear runaway. Their final fate may be a massive O-Ne WDs or, if the core mass approaches the Chandrasekhar limit, an e-capture SNe. More massive objects ignite C in non-degenerate conditions. These "massive " stars are the progenitors of various kind of core-collapse supernovae (type IIp. IIL, IIN, Ib, Ic). It goes without saying that M_{up} is a fundamental astrophysical parameter. From its knowledge depends our understanding of the SNe progenitors, of their rates, of the chemical evolution, of the WD luminosity functions and much more. A precise evaluation of M_{up} relies on our knowledge of various input physics used in stellar modeling, such as the plasma neutrino rate, responsible of the cooling of the core, the equation of state of high density plasma, which affects the heating of the contracting core and its compressibility, and some key nuclear reaction rates, such as, in particular, the $^{12}C+^{12}C$ and the $^{12}C+\alpha$. In this paper we review the efforts made to determine this important parameter and we provide an up-to-date evaluation of the uncertainties due to the relevant nuclear physics inputs.

1. Introduction

At the end of the core-He burning, stars with $M \leq 10\ M_{\odot}$ develop a degenerate C-O core surrounded by a He-burning shell. While the He burning moves progressively outward, the core becomes more massive, contracts and heats up. Owing to the progressive increase of the pressure of the degenerate electrons, the contraction rate slows down. Meanwhile, plasma neutrinos are efficiently produced near the centre, so that the central temperature reaches a maximum and, then, decreases. As a consequence, the highest temperature within the star moves outward, toward the external border of the C-O core. In intermediate mass stars the growth of the C-O core is stopped by the occurrence of the second dredge up, when the convective envelope penetrates into the H-exhausted core attaining layers located immediately above the shell He-burning. In summary, owing to the simultaneous actions of i) electron degeneracy, ii) neutrino cooling and iii) the second dredge up, only the more massive stars, those with a sufficiently large C-O core, may attain the physical conditions for the C ignition. In practice, the C ignition must occur before that the second dredge up is completed, otherwise temperature and density remain below the critical values.

The minimum initial mass of a single star able to experience a C-burning phase is called M_{up} [1]. If the initial mass is lower than M_{up}, after the second dredge up a star enters the AGB phase, during which the H-rich envelope is almost completely eroded by a huge stellar wind. These stars end their life as C-O white dwarfs (WDs). If they belong to a close binary system, the interaction with the companion stars may lead to explosive phenomena like Novae, cataclysm variables, low-mass X-ray binaries or type Ia supernovae. On the contrary, if the mass is slightly larger than M_{up}, an off-centre C ignition takes place in degenerate conditions, giving rise to a thermonuclear runaway followed by a quiescent C burning [2]. As a consequence, the star develops a degenerate O-Ne core and enters a super-AGB phase. The final fate of super-AGB stars is not very well known. Depending on the mass loss rate, they may lose the whole envelope ending their life as massive O-Ne WDs [3]. Alternatively, if the core attains the Chandrasekhar mass limit before the end of the super-AGB phase, a supernova triggered by electron captures may occur. Finally, more massive objects ignite carbon in the centre of a non-degenerate core. After the C burning, they experience the well-known sequence of central burnings (O, Ne, Si) until the formation of an iron core. These stars are the progenitors of "normal" core-collapse supernovae (type IIp, IIL, IIN, Ib, Ic). The scenario described so far is summarized in Figure 1.

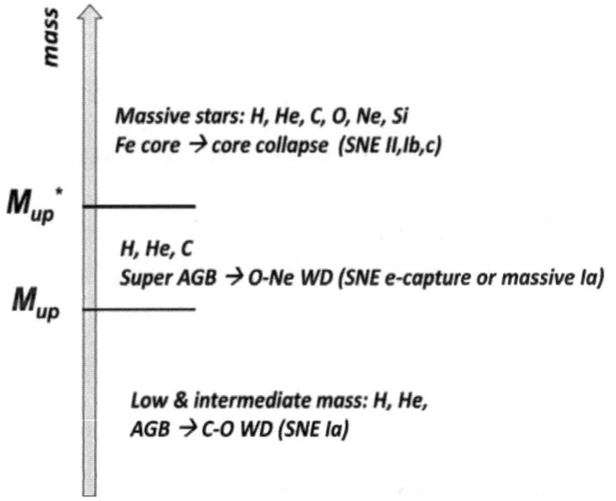

Figure 1. A scheme showing how the evolution of a star and its final fate change with the initial mass. M_{up} is the minimum initial mass of the stars that ignited C in a degenerate C-O core, while M_{up}^* is the minimum mass for a non-degenerate C ignition.

The first systematic attempt to determine the value of M_{up} was done by [1]. They provided an analytical relation between M_{up}, the metallicity (Z) and the initial helium content (Y). For a nearly solar composition, i.e. Z=0.02 and Y=0.28, this relation gives $M_{up} = 9\ M_{\odot}$. In the last 20 years many improvements in the physics inputs needed to derive this relations have been obtained. In particular, we have now a much better knowledge of the equation of state [4, 5], of the radiative opacity [6] or of the neutrinos production rate [7, 8, 9]. An up-to-date version of this relation is plotted in Figure 2. The corresponding value of M_{up} for the revised solar composition is 7.8 $M_{\odot}$.

In this paper we analyse the uncertainties affecting such a relation. First of all, M_{up} depends on the relation between the initial mass and the mass of the H-exhausted core at the end of the core-He burning (M_H). The larger this core mass, the larger the maximum temperature attained within the core during the post core-He burning phase. M_{up} also depends on the amount of C left in the core at the end of the He burning, a quantity that directly affect the carbon burning rate. Both this quantities are the

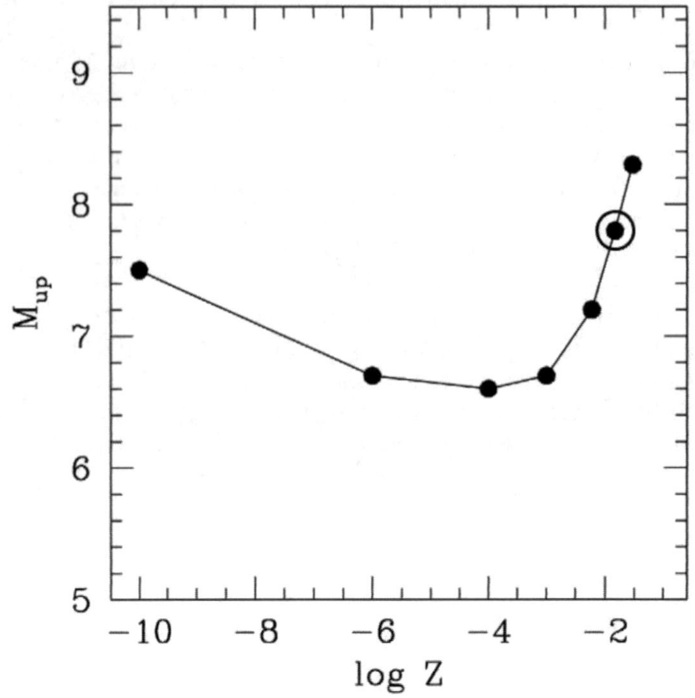

Figure 2. Variation of M_{up} with Z. The circled point corresponds to the solar metallicity

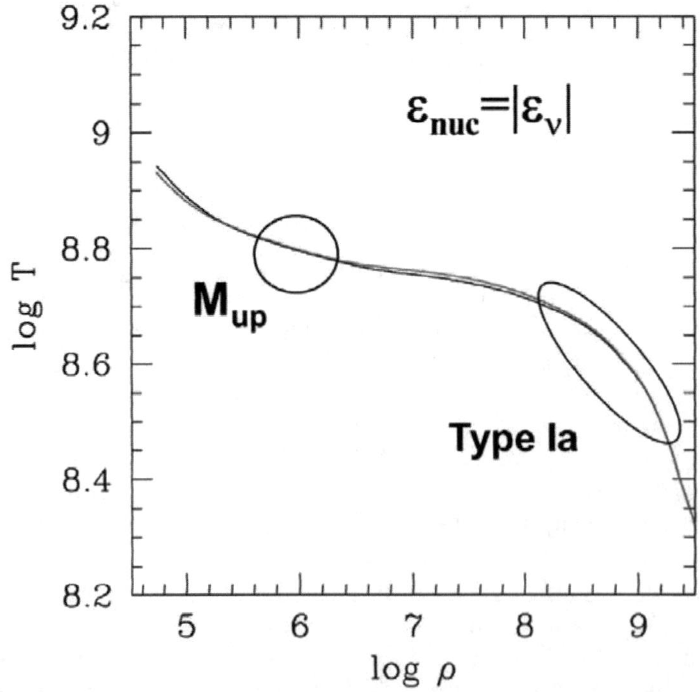

Figure 3. The carbon ignition curve (see text for the definition). The red and the black curves have been obtained under different assumptions for the neutrinos production rate, namely: [8] (red) and [9] (black). The two circled regions correspond to the typical ignition conditions for core-C burning in single massive stars and type Ia supernovae, respectively.

legacy of the evolutionary phases preceding the C burning. The core mass developed by a star with a given initial mass, for example, is affected by uncertainties in the determination of the extension of the convective core during both the core-H and the core-He burnings. On the other hand, the amount of carbon left in the core by the He burning depends on the adopted rate of the $^{12}C(\alpha,\gamma)^{16}O$ reaction.

Other uncertainties are directly connected to the physical description of the structure of the degenerate C-O core when the star approaches the C ignition. Figure 3 show the so-called ignition curve, i.e a line in the log T – log ρ plane on which the nuclear energy production rate is equal to the neutrinos energy loss rate. In stars with M< M_{up} the core peak temperature always remain below this curve so that they skip the C burning. The locations of the C ignition in stars with M>M_{up} as well as in progenitors of type Ia SNe are also shown. The following sections are mainly focused on the possible variations of the estimated value of M_{up} due to uncertainties affecting some key reaction rates, namely the $^{12}C(\alpha,\gamma)^{16}O$, the $^{12}C(^{12}C,\alpha)^{20}Ne$ and the $^{12}C(^{12}C,p)^{23}Na$.

2. The $^{12}C(\alpha,\gamma)^{16}O$ reaction rate and its impact on M_{up}

This reaction plays a fundamental role during the late part of the He-burning phase [10, 11]. In particular, its competition with the 3α reaction determines the amount of C (and O) left in to the core. Since the carbon-burning rate depends by the square of the C abundance, this quantity directly affects the value of M_{up}. However, a change of the $^{12}C(\alpha,\gamma)^{16}O$ rate also affect the core-He-burning lifetime and, in turn the final value of M_H.

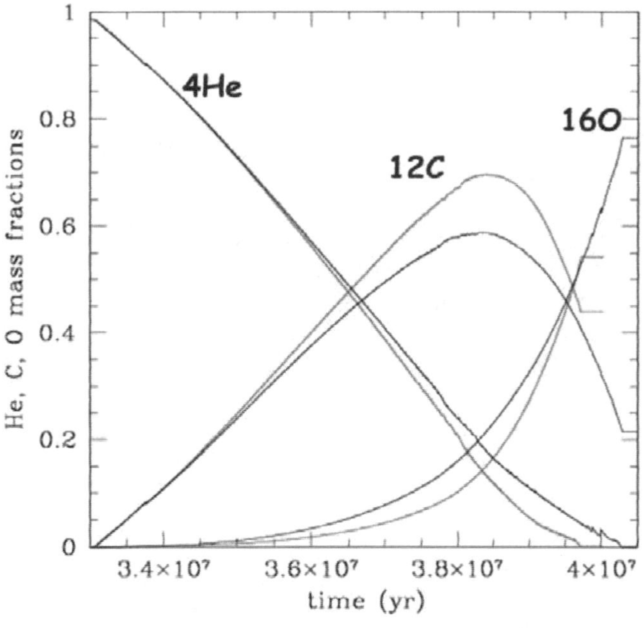

Figure 4. Core-He burning phase in a 7.6 $M_{\odot}$ model (solar composition). Evolution of the central mass fractions of ^4He, ^{12}C and ^{16}O. The black lines correspond to the model computed with S(300)=160 keVb for the $^{12}C(\alpha,\gamma)^{16}O$, while the red lines correspond to the model with S(300)=80 keVb.

Journal of Physics: Conference Series **665** (2016) 012008 doi:10.1088/1742-6596/665/1/012008

The He burning in intermediate mass stars occurs at a temperature of about 100-200 MK corresponding to a Gamow's peak energy of about 300 keV. The latest experimental investigations on this reaction rate (see [12, 13] and references therein) converge on an S(300 keV) ~160 keVb. However, the available measurements con not definitely exclude a destructive resonant configuration for this reaction. In that case, the E1 contribution, which account for about half of the total astrophysical factor at 300 keV would be fully suppressed.

According to such a consideration, we have compared stellar models obtained under opposite assumptions for the S(300) , namely 160 and 80 keVb. The result is illustrated in Figure 4.

The final central C mass fraction scales inversely with S(300), namely $X(^{12}C)=0.42$, in the S(300)=80 keVb model (red lines), and $X(^{12}C)=0.21$, in the S(300)=160 keVb model (black lines). On the contrary, the higher the $^{12}C(\alpha,\gamma)^{16}O$ reaction rate, the longer the He-burning lifetime, and, in turn, the larger the final core mass. As a result, M_H at the end of the core-He burning is 1.66 and 1.77 $M_\odot$ for the 80 and 160 keVb model, respectively. In practice, the increase of the core mass compensates the decrease of the C mass fraction within the core, so that M_{up} remains almost unaffected by the change of the $^{12}C(\alpha,\gamma)^{16}O$ reaction rate.

3. The $^{12}C+^{12}C$ reaction rate and its impact on M_{up}

A change of this reaction directly affects the C ignition curve (see section 1). The Gamow's peak energy for the C burning in massive stars is about 1.5 MeV, while the available experimental investigations barely attain a lower limit of about 2.5 MeV (see [14, 15] and references therein). In the last years it has been speculated about the occurrence of a narrow resonance between 1.3 and 1.7 MeV [16, 17]. The effect of such a resonance on the estimated value of M_{up} would be dramatic.

Table 1. Variations of M_{up} with the $^{12}C+^{12}C$ rate at different Z

Z	M_{up} ($M_\odot$)	
	CF88	CF88 + 1.4 MeV res.
0.0001	6.6	4.5
0.0010	6.7	4.7
0.0060	7.2	5.3
0.0149	7.8	5.8
0.0300	8.3	6.1

Figure 5 illustrate the resulting variation of the ignition curve. The black curve has been obtained by using the rate published in [14] (CF88), whereas an additional contribution due to a narrow resonance at 1.4 MeV has been added in the red curve calculation (for the details see section 2 in [17]). The corresponding values of M_{up} are reported in Table 1. If such a resonance really exists, M_{up} would be up to 2 $M_{\odot}$ smaller.

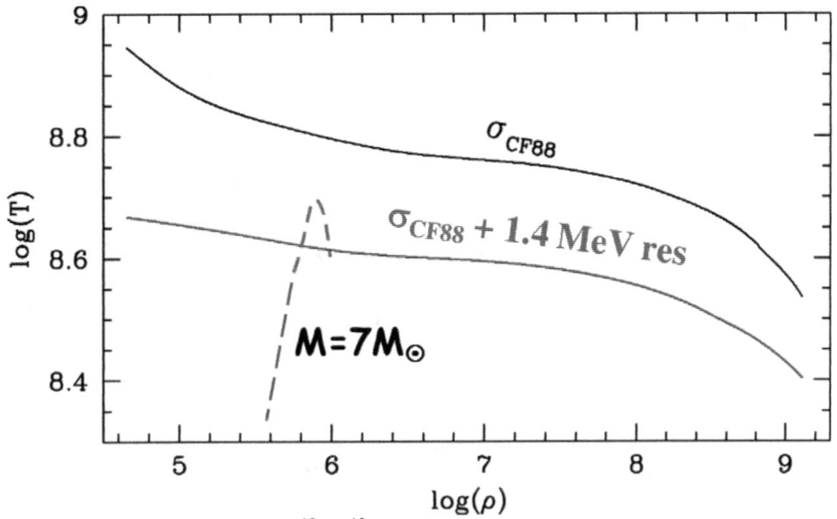

Figure 5. Carbon ignition curve and the ^{12}C+^{12}C reaction. Effects of a hypothetical narrow resonance at E=1.4 MeV. The dashed line represents the evolution of the maximum temperature within the core of a 7 $M_{\odot}$ model, when the additional resonant contribution is 0.

4. Conclusions

In spite of the many progresses made in our understanding of stellar physics, a precise evaluation of M_{up} is still missing. In this paper we have shown how a change of the ^{12}C+^{12}C reaction rate may substantially affect M_{up}.

The astrophysical consequences would be extremely important. Moving up or down the M_{up} value, a change of the mass range of supernova progenitors is expected. Depending on the steepness of the mass function, e.g. a power law MF with exponent 2.35 is adequate for the galactic disk [18], the supernova rate would drastically change. For example, we have calculated that a reduction of about 2 $M_{\odot}$ of M_{up} would reduce the type Ia frequency up to a factor of 4, either in single or double degenerate scenarios. Since in this case the promptest SNe are suppressed, i.e. those with more massive progenitors, the onset of the SNe Ia chemical pollution would be significantly delayed.

Concerning core-collapse SNe (e-capture SNe included), the minimum mass of the progenitors would be significantly reduced. For example, if M_{up}=6 $M_{\odot}$, the smallest "normal" core-collapse progenitor, i.e. the supernovae powered by the collapse of an iron core, is about 8 $M_{\odot}$, in good agreement with recent surveys devoted to the search of SNe IIp progenitors which found 8.5 ($_{-1.5}^{+1}$) $M_{\odot}$ [19]. By adopting a Salpeter's MF, the ratio between super-AGB stars, possible progenitors of e-capture SNe, and the more massive stars progenitors of "normal" core collapse SNe is 0.57, 0.45 or <0.45 when M_{up} is 6, 8 or >8 $M_{\odot}$, respectively[1].

A reduction of M_{up} of about 2 $M_{\odot}$, as due to a substantial increase of the ^{12}C+^{12}C rate, would reduce the maximum C-O white dwarf mass from 1.1 down to 0.95 $M_{\odot}$. Accordingly, the minimum mass for a O-Ne WD would be significantly reduced, while their fraction with respect to the total number of WDs would increase.

[1]These calculations have been made by assuming that the mass range of super–AGB stars is 6-8 or 8-10 $M_{\odot}$, while the mass range of core-collapse SNe progenitors is 8-30 or 10-30 $M_{\odot}$. In both cases dN/dm is proportional to $m^{-2.35}$.

Finally, the maximum mass of the stars that experience an AGB phase follows the variations of M_{up}. This occurrence have relevant effects on the contribution of massive AGB to the galactic chemical evolution. Indeed, only the more massive AGB stars develop the high temperature at the base of the convective envelope required for the activation of proton capture reactions (the so called *hot bottom burning*, HBB). Therefore, a reduction of the maximum AGB mass would imply a reduction of the number of stars undergoing hot bottom burning.

In summary, a precise evaluation of the $^{12}C+^{12}C$ rate at E $\approx$1.5 MeV is a necessary condition to improve our knowledge of M_{up} and of many related astrophysical open issues. On the contrary, we have shown that M_{up} is almost insensitive to a variation of the $^{12}C+\alpha$ reaction rate.

On the other hand, a reliable evaluation of M_{up} also requires a better understanding of the initial-mass to final-core-mass relation. This task implies an adequate evaluation of the effects of rotation on the evolution of intermediate mass stars and, in particular, an improved comprehension of the mechanisms controlling angular momentum transport and rotational induced mixings. As a matter of fact, larger stellar lifetime might result when rotation is included in stellar models, thus leading to larger core masses. In addition, as shown by [20], the onset of the second dredge up is delayed in fast rotating models. These occurrences may obviously affect the estimated value of M_{up}. Finally, deviations from the *standard model* of the particle physics may also modify the derived value of M_{up}. For example, axions are elementary particles introduced by Peccei and Quinn to solve the strong CP violation problem and are considered possible dark matter candidates. They may be produced within the degenerate C-O core of a star approaching the C burning [21]. Like neutrinos, axions cool down the core, thus preventing the C ignition.

From the discussion presented in this paper, we may conclude that M_{up} is the astrophysical *Saint Graal*[2]. Its search would imply a synergy between theoretical and observational astrophysics, nuclear physics and theoretical physics.

References

[1] Becker S and Iben I. 1980 *ApJ* 237 111
[2] Ritossa C, Garcia-Berro E and Iben I 1996 *ApJ* 460 489
[3] Siess L 2007 *A&A* 476 893
[4] Straniero O 1988 *A&AS* 76 157
[5] Rogers F, Swenson, F and Iglesias C 1996 *ApJ* 456, 902
[6] Iglesias C and Rogers F 1996 *ApJ* 464, 943
[7] Haft M, Raffelt G and Weiss A 1994 *ApJ* 225 222
[8] Esposito S, Mangano G, Miele G, Picardi I and Pisanti O 2003 *Nuclear Physics B* 658, 217
[9] Itoh N, Nishikawa A and Kohyama Y 1996 *ApJ* 470 1015
[10] Imbriani G, Limongi M, Gialanella L, et al. 2001 *ApJ* 558 903
[11] Straniero O., Dominguez I, Imbriani G and Piersanti L 2003 *ApJ* 583, 878
[12] Kunz R, Fey M., Jaeger M et al. 2002 *ApJ* 567, 643
[13] Schurmann D, Gialanella L, Kunz R. and Strieder F 2012 *Physics Letters B* 711, 35
[14] Caughlan G. and Fowler W. 1988 *At. Data Nucl. Data Tab.* 40, 283
[15] Spillane T, Raiola F, Rolfs C, et al. 2007 *Physical Review Letters* 98 122501
[16] Cooper R, Steiner A and Brown E 2009 *ApJ* 702 660
[17] Bravo E, Piersanti L, Dominguez I, et al. 2011 *A&A* 535 114
[18] Salpeter E 1955 *ApJ* 121 161
[19] Smartt S, Eldridge J, Crockett R and Maund J 2009 *MNRAS* 395 1409
[20] Dominguez I, Straniero O, Tornambe A and Isern, J 1996 *ApJ* 472 783

[2] Holy Grail

[21] Dominguez I, Straniero O and Isern J 1999 *MNRAS* 306 L1

Nuclear Astrophysics with the Trojan Horse Method

A Tumino[1,2], C Spitaleri[2,3], L Lamia[3], RG Pizzone[2], S Cherubini[2,3], M Gulino[1,2], M La Cognata[2], SMR Puglia[2,3], GG Rapisarda[2,3], S Romano[2,3], ML Sergi[2,3], R Spartá[2,3]

[1]Facoltá di Ingegneria e Architettura, Universitá degli Studi di Enna "Kore", Enna, Italy
[2]Laboratori Nazionali del Sud - INFN, Catania, Italy
[3]Dipartimento di Fisica e Astronomia - Università degli Studi di Catania, Catania, Italy

E-mail: tumino@lns.infn.it

Abstract. The Trojan Horse Method (THM) represents the indirect path to determine the bare nucleus astrophysical S(E) factor for reactions between charged particles at astrophysical energies. This is done by measuring the quasi free cross section of a suitable three body process. The basic features of the THM will be presented together with some applications to demonstrate its practical use.

1. Introduction

All nuclei have positive charge and any fusion process requires considerable energy to take place, due to the Coulomb repulsion between like charges. This is a critical issue in nuclear astrophysics that makes the study of all nucleosynthesis processes at the relevant temperatures extremely challenging. The Coulomb repulsion is indeed responsible for the exponential decrease of the cross section $\sigma(E)$ at those temperatures. For this reason, the behavior of $\sigma(E)$ at low energy is usually extrapolated from higher energies (usually E>100 keV) by means of the astrophysical S(E)-factor

$$S(E) = E\sigma(E)\exp(2\pi\eta), \tag{1}$$

with η the Coulomb parameter of the colliding nuclei, and $\exp(2\pi\eta)$ the inverse of the Gamow factor that removes the energy dependence of $\sigma(E)$ due to the barrier tunneling. However, even an easier extrapolation can be source of additional uncertainties for $\sigma(E)$ due, for instance, to the presence of unexpected resonances.

Another critical issue in the laboratory measurement of nucleosynthesis processes is represented by the electron screening effect. Indeed both target and projectile are usually embedded in neutral/ionized atoms or molecules and their electron clouds are responsible for a reduction of the Coulomb barrier. This, in turn, leads to an increased cross section for screened nuclei, $\sigma_s(E)$, compared to the cross section for bare nuclei $\sigma_b(E)$ [1, 2]. Therefore, the so called screening factor, defined as

$$f_{\rm lab}(E) = \sigma_s(E)/\sigma_b(E) \approx \exp(\pi\eta U_e/E)\,, \tag{2}$$

where U_e is the so-called "electron screening potential" [1, 2], has to be taken into account to determine the bare nucleus cross section.

In the stellar plasma, the cross section σ_{pl}(E) is related to the bare nucleus cross section by a similar enhancement factor:

$$f_{pl}(E) = \sigma_{pl}(E)/\sigma_b(E) \approx \exp(\pi\eta U_{pl}/E) \tag{3}$$

that can be calculated once the plasma screening potential U_{pl} is known, depending on important properties of the plasma such as the Debye-Hückel radius. A measurement of U_e, which is needed to calculate σ_s(E) from Eq. 2, would also help to better understand U_{pl}.
Low-energy fusion reactions measured so far have indeed shown the exponential enhancement according to equation (2)[2]. However, the deduced U_e values are larger than the adiabatic limit, provided by the atomic models as the difference between the electron binding energies of the separate atoms in the entrance channel and that of the composite atom [2, 3]. This disagreement in laboratory experiments is yet to be justified, and prevents the effects under astrophysical conditions to be fully understood.
A week point in the laboratory approach - and thus in the deduced U_e value - is the necessity to make an assumption for the energy dependence of σ_b(E) at ultra-low energies. Thus, indirect techniques ([4] and references therein) such as the Trojan Horse Method (THM) ([5] and references therein) , have been introduced to overcome all these difficulties. In particular, the THM provides a successful alternative path to determine σ_b(E) for reactions between charged particles. Here we recall the basic ideas of the THM and we show some recent results.

2. Features of the Trojan Horse Method

The THM ([5, 6] and references therein) makes use of a suitable $A + a \rightarrow c + C + s$ two-to-three body process to measure the astrophysical $A + x \rightarrow c + C$ two-body reaction of interest by establishing a relation between the two reactions based on the nuclear reaction theories. The three-body process is chosen in such a way that target a (or equivalently the projectile) has a wave function with a large amplitude for a $x - s$ cluster configuration, x being the target (or equivalently the projectile) of the two-body reaction. The selected part of the three-body phase space is the one where the quasi free kinematics applies: the other cluster s remains spectator to the process, and $A + x \rightarrow c + C$ can be regarded as an half-off-energy-shell (HOES) two-body reaction, usually referred to as a quasi-free reaction. Since the three-body process occurs at an energy above the Coulomb barrier, the main feature is the real suppression of both Coulomb barrier and screening effects in the HOES two-body cross section. Nevertheless, the quasi-free $A+x$ process can occur even at very low sub-Coulomb energies because the $A+a$ relative motion is compensated for by the $x - s$ binding energy. This is indeed a different approach to the THM [5] compared to the original idea of Baur [7], where the initial velocity of the projectile A is compensated for by the $x - s$ intercluster motion. In that framework a quite large momentum of the order of 200 MeV/c or more is needed. But the relative yield of the experimental momentum distribution at such momenta can be very small, in particular for a l=0 inter-cluster motion (for example p-n motion inside ^{2}H or α-d motion inside ^{6}Li). This would make very critical the separation from other competitive reaction mechanisms. Moreover, the theoretical description of the tails of the momentum distribution is a hard task, their shape being very sensitive to it. In our approach to the THM, the intercluster motion is only needed to determine the accessible astrophysical energy region. It corresponds to a cutoff in momentum distribution of s of few tens of MeV/c. In this framework, the so called "quasi-free two-body energy" is given by:

$$E_{QF} = \frac{m_x}{m_x + m_A} E_A - B_{x-s}. \tag{4}$$

where E_A represents the beam energy, m_x and m_A are the masses of x and A particles respectively, and B_{x-s} is the binding energy for the $x - s$ system.

In the Impulse Approximation, based essentially on the assumption that the interaction of the spectator with particles C and c is neglected, the three body-cross cross section can be factorized as:

$$\frac{d^3\sigma}{dE_c d\Omega_c d\Omega_C} \propto [KF\,|\varphi_a(\mathbf{p}_{sx})|^2] \left(\frac{d\sigma}{d\Omega_{c.m.}}\right)^{\text{HOES}} \tag{5}$$

where KF is a kinematical factor containing the final state phase-space factor. It is a function of the masses, momenta and angles of the outgoing particles [29]; $\varphi_a(\mathbf{p}_{sx})$ is proportional to the Fourier transform of the radial wave function $\chi(\mathbf{r})$ for the $x-s$ inter-cluster relative motion; $(d\sigma/d\Omega_{c.m.})^{\text{HOES}}$ is the half-off-energy-shell (HOES) differential cross section for the binary reaction at the center of mass energy $E_{c.m.}$ given in post-collision prescription by

$$E_{c.m.} = E_{cC} - Q_{2b}\,. \tag{6}$$

Here, Q_{2b} is the Q-value of the binary reaction and E_{cC} is the relative energy of the outgoing particles c and C, which spans the accessible astrophysical region defined above.

The factorization of Eq.5 is strictly valid in Plane Wave Impulse Approximation, which changes only the absolute magnitude but not the energy dependence of the two-body cross section.

3. Experimental ingredients

In THM experiments, the decay products (c and C) of the virtual two-body reaction of interest are usually detected and identified by means of telescopes (silicon detector or ionization chamber as ΔE detector and PSD as E detector) placed in a phase space region where quasi free kinematics is dominant. After the selection of the reaction channel, a critical point is the separation of the quasi free mechanism from other reaction mechanisms feeding the same particles in the final state, e.g. sequential decay and direct break-up. An observable which is very sensitive to the reaction mechanism is the shape of the experimental momentum distribution $\Phi(p_s)$ of the spectator. To reconstruct the experimental $\Phi(p_s)$ distribution, the energy sharing method [8] can be applied for each pair of coincidence QF angles: the quasi-free coincidence yield with a cutoff in relative energy of few hundreds of keV at most is divided by the kinematic factor, providing a quantity which is proportional to the product of the momentum distribution for the spectator with the differential two-body cross section (see Eq.5). The window is chosen in such a way that the differential two-body cross section in this range can be considered almost constant. Thus, the quantity defined above represents essentially the momentum distribution for the spectator that in PWIA can be compared with the Fourier transform of the radial $x-s$ bound state wave function.

Data analysis is limited to the region where the agreement between the two distributions exists. Usually a window of few tens of MeV/c is chosen not to move too far from E_{QF}, according to the prescriptions reported in [14].

Therefore, $((d\sigma/d\Omega_{cm})^{\text{HOES}}$ can be extracted from the three-body coincidence yield by simply inverting Eq.5. The Coulomb barrier in $((d\sigma/d\Omega_{cm})^{\text{HOES}}$ is suppressed [9] and this is due to the virtuality of particle x. This seems to be the only consequence of off-energy-shell effects as suggested by the agreement between HOES and OES cross-sections for the ^{6}Li(n,α)^{3}H reaction [10].

The suppression of the Coulomb barrier has been tested also in scattering processes. In particular, we have address the study of the $p+p$ elastic scattering through the $^2\text{H}(p,pp)n$ reaction [11, 12]. The $p+p$ system is the simplest system where Coulomb and nuclear forces co-exist and are coherent. Coherence means interference, which is destructive due to the opposite sign of the two forces. Interference is responsible for the deep minimum in the free $p+p$ cross section shown in Fig.1 as solid line as a function of the $p-p$ relative energy E_{pp}. The

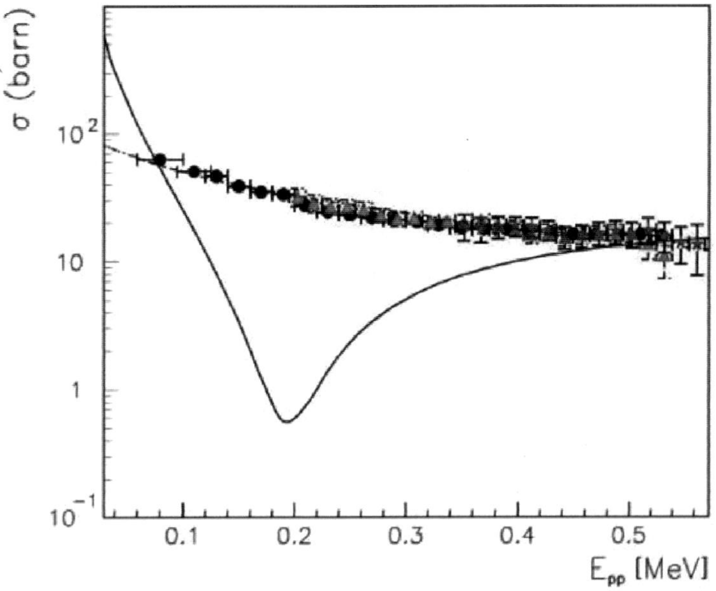

Figure 1. THM two-body cross section (black and colored symbols) vs.*pp* relative energy [11]. Solid line represents the theoretical OES p-p cross section [13]. The dashed-dotted line is the calculated HOES cross section [11].

extracted $p-p$ HOES cross section is shown as black and colored symbols. The dashed-dotted line represents the calculated HOES $p-p$ cross section as reported in [11] where the Coulomb amplitude is strongly suppressed compared to the nuclear one. We observe a striking disagreement between the THM (HOES) and the free $p-p$ (OES) cross sections throughout the region of the interference minimum, which is missing in the THM data. Instead, the calculated HOES $p-p$ nicely fits the THM data. A full account of the work is reported in [11, 12]. The good agreement provides conclusive evidence of the suppression of Coulomb effects in the THM method.

Thus, in a final step, to relate the HOES excitation function to the relevant on-energy-shell (OES) one, Coulomb corrections have to be considered. If one looks at the angular distributions no correction is needed because once the energy is fixed, it would mean to introduce simply a scaling factor. Thus, the OES data are directly comparable with HOES ones projected onto the emission angle of C (or c) in the $C-c$ center of mass system, $\theta_{c.m.}$, as given by [15]:

$$\theta_{c.m.} = \arccos \frac{(\mathbf{v}_{\mathrm{A}} - \mathbf{v}_{\mathbf{x}}) \cdot (\mathbf{v}_{\mathbf{c}} - \mathbf{v}_{\mathrm{C}})}{|\mathbf{v}_{\mathrm{A}} - \mathbf{v}_{\mathbf{x}}||\mathbf{v}_{\mathbf{c}} - \mathbf{v}_{\mathrm{C}}|} \tag{7}$$

where the vectors $\mathbf{v}_{\mathrm{A}}$, $\mathbf{v}_{\mathbf{x}}$, $\mathbf{v}_{\mathbf{c}}$, $\mathbf{v}_{\mathrm{C}}$ are the velocities of projectile, transferred particle and outgoing nuclei respectively. These quantities are calculated from their corresponding momenta in the lab-system, where the momentum of the transferred particle is equal and opposite to that of s when the quasi-free break-up takes place in the target, otherwise a little different formula has to be used [15]. If HOES data are projected onto the E_{Cc} axis, Coulomb suppression has to be introduced before comparison with OES data. In a simple approach this is done by means of

Table 1. Two-body reactions studied via the THM with measured two-to-three TH reaction and relevant references for each reaction

Direct reaction	TH reaction	ref
^{7}Li(p,α)^{4}He	^{7}Li(d,α α)n	[18]
^{7}Li(p,α)^{4}He	^{7}Li(^{3}He,α α)^{2}H	[19]
^{6}Li(d,α)^{4}He	^{6}Li(^{6}Li,α α)^{4}He	[20]
^{6}Li(p,α)^{3}He	^{6}Li(d,α ^{3}He)n	[21]
^{11}B(p,α)^{8}Be	^{11}B(d ,^{8}Be α)n	[29]
^{10}B(p,α)^{7}Be	^{10}B(d,^{7}Be α)n	[30, 31, 32]
^{9}Be(p,α)^{6}Li	^{9}Be(d,^{6}Li α)n	[33, 34, 35]
^{2}H(^{3}He,p)^{4}He	^{6}Li(^{3}He,p α)^{4}He	[36]
^{2}H(d,p)^{3}H	^{2}H(^{6}Li,t p)^{4}He	[37]
^{15}N(p,α)^{12}C	^{15}N(p,α^{12}C)n	[38]
^{18}O(p,α)^{15}N	^{18}O(p,α^{15}N)n	[39, 40]
^{1}H(p,p)^{1}H	^{2}H(p,pp)n	[11, 12]
^{2}H(d,p)^{3}H	^{2}H(^{3}He,t p)^{1}H	[41, 42, 43, 44, 45]
^{2}H(d,n)^{3}He	^{2}H(^{3}He,^{3}He n)^{1}H	[41, 42, 44]
^{19}F(p,α)^{16}O	^{2}H(^{19}F,α^{16}O)n	[46]
^{17}O(p,α)^{14}N	^{2}H(^{17}O,α^{14}N)n	[47]
^{4}He(^{12}C,^{12}C)^{4}He	^{6}Li(^{12}C,α^{12}C)^{2}H	[48]
n(^{6}Li,t)^{4}He	^{2}H(^{6}Li,t^{4}He)^{1}H	[10, 49]
^{11}B(p,α)^{8}Be	^{2}H(^{11}B,α^{8}Be)n	[51]
^{13}C(α,n)^{16}O	^{6}Li(^{13}C,n^{16}O)^{2}H	[52]

the penetrability factor:

$$P_l(k_{Ax}R) = \frac{1}{G_l^2(k_{Ax}R) + F_l^2(k_{Ax}R)} \qquad (8)$$

with F_l and G_l regular and irregular Coulomb wave functions and R the so called cutoff radius [21], which is usually chosen as the sum of the radii of nuclei A and x. This procedure does not allow us to extract the absolute value of the two-body cross section. However, this is not a real problem since the absolute magnitude can be derived from a scaling to the direct data available at higher energies.

4. Applications

The THM was applied already many times to charged particle reactions connected with fundamental astrophysical problems [16], from Big-Bang to stellar nucleosynthesis. A list of reactions studied by means of the THM is given in Table 1 together with the relevant references. In particular, low-energy cross sections for reactions producing or destroying Lithium isotopes are fundamental for several astrophysical problems yet to be solved, e.g. understanding Big Bang nucleosynthesis and the "Lithium depletion" in the Sun or in other galactic stars. As for the solar "Lithium problem", the present Lithium abundance on the solar surface corresponds to a mass fraction of about 7.0x10^{-11} [53]; this value needs to be compared with the meteoritic value of about 9.9 x 10^{-9}. Since the proto-solar nebula is assumed to be homogeneous at

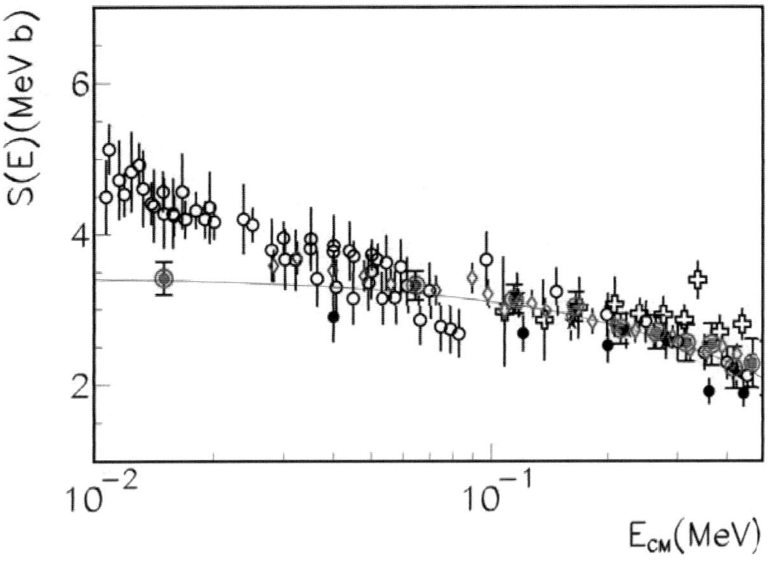

Figure 2. THM bare nucleus S(E) factor for the $^6Li(p,\alpha)^3He$ reaction (red solid dots from [28] and black solid dots for [21]. Direct data represented with open symbols as described in [28]. The blue and red ones refer to the most recent direct experiment reported in [55, 56]. The solid line represents the polynomial fit to the THM bare nucleus S(E) factor as reported in [28].

the beginning of the solar system, the Sun has clearly depleted Lithium by a factor 140. An experimental program was undertaken to study p-capture reactions on $^{6,7}Li$, main responsible for their destruction [17, 18, 21, 22, 23, 24, 25, 26, 27, 28]. The extracted astrophysical S(E) factors were compared with those from direct measurements and found in fair agreement in the energy range where screening effects are negligible. Recently, an updated reaction rate for the $^6Li(p, \alpha)^3He$ reaction has been determined using direct data of [55, 56] for a more accurate normalization in the energy region below 300 keV (see [28] for the detailed description). The new normalization returns a S(0)=3.44 $\pm$ 0.39 MeV b, with an S(E) factor shown in Fig.2 as red solid dots from [22, 28] and black solid dots from [21]. Direct data are given as open symbols as described in [28]. The blue and red ones refer to direct data of [55, 56]. The solid line represents the polynomial fit to the THM bare nucleus S(E) factor as reported in [28]. Fitting the low-energy data below E_{cm}=70 keV with the screening function (Eq.2) and leaving U_e as free parameter, this provides U_e=350$\pm$100 eV. The new reaction rate deviates from 5% to 15% with respect to the value reported in the Nacre compilation [54] as the temperature decreases from 1 down to 10^{-3} T9 (see [28] for details). Astrophysical implications of these results have been investigated performing calculations on PMS models for ^6Li and on the solar "Lithium problem" for ^7Li. A full account is reported in [28] and [24] respectively. In particular for the $^7Li(p,\alpha)^3He$ case, it appears that the THM cross section does not significantly change the surface Lithium abundance for the present Sun, with respect to previous results reported in the NACRE compilation [54] because of the small discrepancy between the two rates. The Lithium problem for the Sun is not solved and the observed Lithium surface abundance is not reproduced by the model.

However, uncertainties on these nuclear reaction rate are now so small to move the "Lithium problem" to astrophysics. Other input parameters such as original metallicity, helium abundance, convection efficiency and stellar mass have much larger uncertainties that seriously affect any comparison between theory and observations and need to be better constrained.

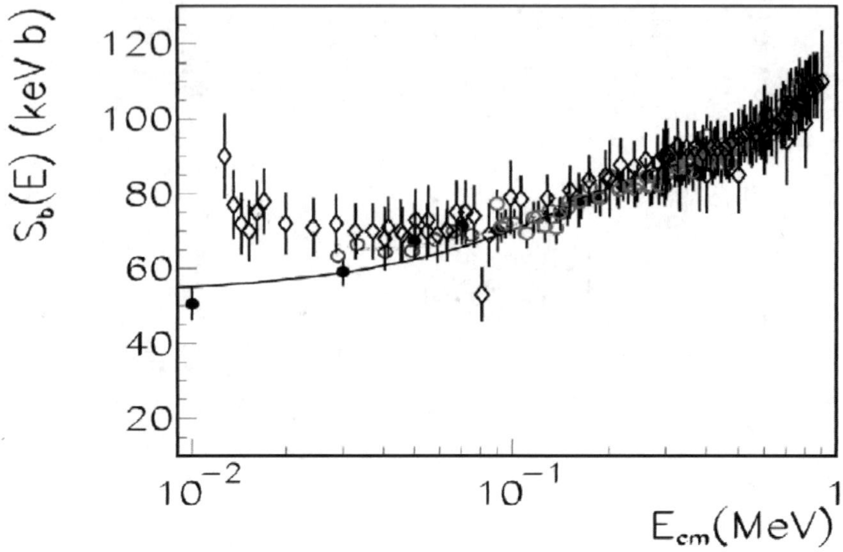

Figure 3. THM bare nucleus S(E) factor for the ^{7}Li(p,α)^{4}He reaction (black solid dots from [18]. Direct data represented with open symbols as described in [26]. The red and blue open circles refer to the most recent direct experiment reported in [55] and [56] respectively. The solid line represents the polynomial fit to the THM bare nucleus S(E) factor as reported in [26].

Recently, the new direct data by [55, 56] have been used for a better normalization also in the ^{7}Li(p,α)^{4}He channel [26]. The new normalization of the THM S(E) factor is shown in Fig.3 as black solid dots. Red and blue open circles represent data from [55] and [56] respectively. The other open symbols refer to direct data of [59]. The solid line represents the polynomial fit to the THM bare nucleus S(E) factor as reported in [26]. The new S(0) turns out to be S(0)=53 $\pm$ 5 keV b, while the fit of the low energy region including the screening function (Eq.2) returns a U_e value of U_e=425$\pm$60eV. The udated S(0) value deviates by 4% when compared with the previous normalization in [18], while Ue deviates by about 23%. The new reaction rate experiences a variation with respect to the Nacre one [54] from about 13% at around T9=10^{-3} to about 5% at higher temperatures and astrophysical calculations need to be readdressed for several scenarios.

5. Bibliography

[1] HJ Assenbaum *et al.* 1987, *Z. Phys.* A **327** 461
[2] F Strieder *et al.* 2001, *Naturwissenschaften* **88** 461
[3] G Fiorentini *et al.* 1995 *Z. Phys.* A **350** 289
[4] A Tumino *et al.* 2013, *Few Body Syst.*, **54**, Issue 7-10, 869
[5] C Spitaleri *et al.* 2011, *Phys. At. Nucl.*, **74**, 1763
[6] A Tumino *et al.* 2013, *Few Body Syst.*, **54**, Issue 5-6 745
[7] G Baur 1986, *Phys. Lett.* B **178** 135
[8] J Kasagi *et al.* 1975, *Nucl. Phys.* A **239** 233
[9] AM Mukhamedzhanov et al. 2006, *Eur. Phys. J.* A **27** 205
[10] A Tumino *et al.* 2005, *Eur. Phys. J.* A **25** 649
[11] A Tumino *et al.* 2007, *Phys. Rev. Lett.* **98** 252502
[12] A Tumino *et al.* 2008, *Phys. Rev.* C **78** 064001
[13] JD Jackson and J.M. Blatt 1950, Rev. Mod. Phys. **22**, 77
[14] IS Shapiro *et al.* 1965, *Nucl. Phys.* A, **61**, 353
[15] M Jain *et al.* 1970, *Nucl. Phys.* A **153** 49
[16] CJ Copi, DN Schramm, MS Turner 1995, *Science* **627**, 192
[17] C Spitaleri *et al.* 1999, *Phys. Rev.* C **60** 055802

[18] M Lattuada *et al.* 2001, *Astrophysical J.* **562**, 1076
[19] A Tumino *et al.* 2006, *Eur. Phys. J. A* direct, 1 DOI: 10.1140/epja/i2006-08-038-1
[20] C Spitaleri et al. 2001 *Phys. Rev.* C **63** 005801
[21] A Tumino et al. 2003, *Phys. Rev.* C **67** 065803
[22] A Tumino *et al.* 2004, *Prog. Theor. Phys. Suppl.* **154** 341
[23] M Aliotta *et al.* 2000 *Eur. Phys. J. A* **9**, 435
[24] RG Pizzone *et al.* 2003, *A&A* **398**, 423
[25] RG Pizzone *et al.* 2011, *Phys. Rev.* C **83**, 045801
[26] L Lamia *et al.* 2012, *A&A* **A185** 541
[27] RG Pizzone *et al.* 2005, *A&A* **438**, 779
[28] L Lamia *et al.* 2013, *Astrophysical J.* **768**, 65
[29] C Spitaleri *et al.* 2004, *Phys. Rev.* C **63** 055806
[30] L Lamia *et al.* 2007, *Nucl. Phys. A* **787** 309
[31] L Lamia *et al.* 2008, *Il Nuovo Cimento* **31** N.4 423
[32] C Spitaleri *et al.* 2014, *Phys. Rev.* C **90** 035801
[33] S Romano *et al.* 2006, *Eur. Phys. J. A* **27** 221
[34] Qun-Gang Wen *et al.* 2008, *Phys. Rev.* C **78** 035805
[35] Q Wen *et al.* 2011, *J.Phys. G: Nucl. Part. Phys.*, **38**, 085103
[36] M La Cognata *et al.* 2006, *Eur. Phys. J.* A **27** 249
[37] A Rinollo *et al.* 2005, *Nucl. Phys.* A **758** 146
[38] M La Cognata *et al.* 2007, *Phys. Rev.* C **76** 065804
[39] M La Cognata *et al.* 2008, *Phys. Rev. Lett.* **101** 152501
[40] M La Cognata *et al.* 2010, *Astrophysical J.* **708** 796
[41] A Tumino *et al.* 2011, *Phys. Lett.* B **700**, 111
[42] A Tumino *et al.* 2011, *Phys. Lett.* B **705**, 546
[43] RG Pizzone *et al.* 2013, *Phys. Rev.* C **87**, 025805
[44] A Tumino *et al.* 2014, *Astrophysical J.* **785**, 96
[45] Li Chengbo *et al.* 2015, *Phys. Rev.* C **92**, 025805
[46] M La Cognata *et al.* 2011, *Astrophysical J. Lett.* **739**, L54
[47] ML Sergi *et al.* 2010, *Phys.Rev.* C **R82**, 032801
[48] C Spitaleri *et al.* 2000, *Eur. Phys. J. A* **7**,181
[49] M Gulino *et al.* 2010, *J.Phys. G: Nucl. Part. Phys.*, **37**, 125105
[50] L Lamia *et al.* 2009, *Il Nuovo Cimento*, **31**, 423
[51] L Lamia *et al.* 2012, *Journ. of Phys. G: Nucl. Part. Phys.* **39** 015106
[52] M La Cognata *et al.* 2013, *Phys. Rev. Lett.* **49**, 015106
[53] N Grevesse & AJ Sauval 1998, *Space Sci. Rev.*, **85** 161
[54] C Angulo et al. 1999, *Nucl. Phys.* A **656** 3
[55] J Cruz *et al.* 2005, *Phys. Lett. B* **624** 181
[56] J Cruz *et al.* 2008, *Journ. of Phys. G: Nucl. Part. Phys.* **635** 192
[57] S Degl'Innocenti *et al.* 2008, *Ap&SS* **316** 25
[58] F Ciacio, S Degl'Innocenti, B Ricci 1997, *A&A* **123**, 449
[59] S Engstler et al. 1992, *Z. Phys.* A **342 471**

Journal of Physics: Conference Series **665** (2016) 012010 doi:10.1088/1742-6596/665/1/012010

^{12}C+^{12}C fusion in a multichannel folding model

M Assunção[1] and P Descouvemont[2]

[1]Departamento de Ciências Exatas e da Terra, Universidade Federal de São Paulo, 09972-270, Campus Diadema, São Paulo, Brazil
[2]Physique Nucléaire Théorique et Physique Mathématique, C.P. 229, Université Libre de Bruxelles (ULB), B 1050 Brussels, Belgium

Abstract.
The ^{12}C+^{12}C fusion reaction is investigated using a folding potential in a multichannel approach involving the ^{12}C($0_1^+, 2^+, 0_2^+, 3^-$) states. The ^{12}C densities (including transition densities) are taken from the RGM calculation of Kamimura. For the nucleon-nucleon interaction, we use the DDM3Y density-dependent interaction. Owing to the explicit presence of inelastic channels, the imaginary part of the optical potential only contains a short-range fusion contribution. The S-factor is then virtually insensitive to the precise value, and the model is free of any fitting parameter. From the coupled-channel system, we determine the elastic and fusion cross sections simultaneously. As elastic data are available around the Coulomb barrier, this simultaneous treatment offers a good test for the reliability of the model. In the fusion cross section, the role of the inelastic channels and, in particular of the ^{12}C(0_1^+)+^{12}C(0_2^+) channel involving the Hoyle state, is discussed.

1. Introduction

The ^{12}C+^{12}C reaction plays an important role in stellar nucleosynthesis [1], and in particular in the evolution of massive stars [2]. The extrapolation of the S-factor at low energies is made difficult by the presence of resonances whose interpretation is not clear (see a recent review in [2]). A recent work [3] suggests that excitations of ^{12}C are important for a reliable determination of the S factor. Most fusion calculations to date are performed in a single-channel model, i.e. involving the ^{12}C ground-state only, while the absorption is simulated by a phenomenological imaginary potential [4]. In light systems, however, it is known that inelastic channels may be important and require to be explicitly included in the calculation. In Ref. [3], the authors suggest that mutual excitations play an important role even at low energies, where excited channels are closed. At first sight, this effect may seem surprising since only a single channel is open. It is explained by distortion effects in the wave functions: the cross section is mostly sensitive to the inner part of the wave functions, where closed channels may have a significant amplitude.

The aim of our calculation is to investigate the ^{12}C+^{12}C fusion in a multichannel folding method [5]. We include the ^{12}C($0_1^+, 2^+, 0_2^+, 3^-$) states and also the corresponding mutual excitations. A folding method is performed using the density-dependent M3Y (DDM3Y) interaction [7] to describe the nucleon-nucleon (NN) interaction, and ^{12}C densities are taken from the RGM values of Kamimura [6]. These densities (elastic and inelastic) are obtained from a microscopic triple-alpha model, and are known to provide a precise description of many scattering data. In particular, the 0_2^+ state of ^{12}C is well described by the three-α microscopic calculation of Kamimura, and is expected to play a significant role in the ^{12}C+^{12}C system [8].

Another aspect that makes this work even more interesting is that the calculation is free of parameter, except for a weak dependence on the absorption potential. It provides a simultaneous description of elastic scattering and of fusion.

2. Theoretical framework

In a coupled-channel formalism, the ^{12}C+^{12}C potentials are defined as

$$V_{\alpha_1\alpha_2,\alpha_1'\alpha_2'}(\boldsymbol{r}) = \int\int d\boldsymbol{r}_1 d\boldsymbol{r}_2\, v_{NN}(\boldsymbol{r}-\boldsymbol{r}_1+\boldsymbol{r}_2)\,\rho_1^{\alpha_1\alpha_1'}(\boldsymbol{r}_1)\,\rho_2^{\alpha_2\alpha_2'}(\boldsymbol{r}_2), \tag{1}$$

where $v_{NN}(\boldsymbol{r})$ represents the nucleon-nucleon nuclear interaction, $\boldsymbol{r} = (r, \Omega_r)$ is the relative coordinate, $\rho_k^{\alpha\alpha'}(\boldsymbol{r}_k)$ are the ^{12}C nuclear densities, and labels α_k refer to different ^{12}C states. The same formalism is applied to the Coulomb interaction. In the present work, we include ^{12}C$(0_1^+, 2^+, 0_2^+, 3^-)$ states, which means that ten ^{12}C+^{12}C channels are introduced in the coupled-channel system.

In practice, the evaluation of the double integral (1) is performed by using Fourier transforms for the nuclear as well as for the Coulomb interactions [5]. The densities $\rho_k^{\alpha\alpha'}(\boldsymbol{r})$ are taken from the 3α microscopic calculation of Kamimura [6]. These densities, elastic ($\alpha = \alpha'$) as well as inelastic ($\alpha \neq \alpha'$), are known to reproduce many experimental data.

As usual, the densities $\rho_k^{\alpha\alpha'}(\boldsymbol{r})$ are expanded in multipoles [6], and the radial wave functions $g_c^{J\pi}(r)$ are obtained from the coupled-channel system

$$-\frac{\hbar^2}{2\mu}\left[\frac{d^2}{dr^2} - \frac{\ell(\ell+1)}{r^2}\right] g_c^{J\pi}(r) + \sum_{c'} V_{cc'}^{J\pi}(r) g_{c'}^{J\pi}(r) = (E - E_1^c - E_2^c) g_c^{J\pi}(r), \tag{2}$$

where μ is the reduced mass of the system, and E_i^c are the ^{12}C energies. In this equation, label c stands for $c = (\alpha_1, \alpha_2, I, \ell)$, where I is the channel spin and ℓ the relative orbital momentum.

In this approach, the absorption to other channels is simulated by adding an imaginary component to the potentials as (see Ref.[12] for details)

$$V_{cc'}^{J\pi}(r) \longrightarrow V_{cc'}^{J\pi}(r) + iW_{cc'}^{J\pi}(r). \tag{3}$$

In the present calculation, the coupled-channel system (2) explicitly includes inelastic channels in a wide energy range. To define the fusion component of the potential, a short-range absorption potential [11] is included as

$$W_{cc'}(r) = -\frac{W_0}{1+\exp((r-R_0)/a)}\delta_{cc'}. \tag{4}$$

The range R_0 is chosen smaller than the barrier radius, and this potential acts at short distances only. In this multichannel calculation, the values were taken as $W_0 = 10$ MeV, $R_0 = 3$ fm, and $a = 0.1$ fm, and the same conditions were employed to investigate the elastic-scattering and fusion processes. The calculation are stable within $1-2\%$ when these parameters are modified. An important consequence is that the model is free of parameters, and that all cross sections are obtained without any adjustment.

The coupled-channel system (2) is solved with the R-matrix method [9]. In the internal region ($r \leq a$), the radial functions $g_c^{J\pi}(r)$ are expanded over a Lagrange basis [10]. In the external region, they are given by linear combinations of Coulomb functions. The matching provides the collision matrix $\boldsymbol{U}^{J\pi}$. Notice that, at low energies, most of ^{12}C+^{12}C excited channels are closed (the first open channel is ^{12}C(0_1^+)+^{12}C(2^+) which opens at 4.44 MeV).

The elastic cross sections are computed from the collision matrices by using standard formulas [4]. The fusion cross section is defined as [13]

$$\sigma_F(E) = \frac{2\pi}{k^2} \sum_{J\ \mathrm{even}} (2J+1) P_J(E), \tag{5}$$

where k is the wave number, and where the fusion probability $P_J(E)$ is obtained from

$$P_J(E) = -\frac{2}{\hbar v} \sum_c \int |g_c^{J\pi}(r)|^2\, W_{cc}(r)\, dr, \tag{6}$$

where v is the relative velocity [4].

At low energies, the fusion and reaction cross sections are identical, and $P_J(E)$ can be expressed as

$$P_J(E) = 1 - |U_{11}^J|^2, \tag{7}$$

where U_{11}^J is the elastic element of the collision matrix, associated with the ^{12}C+^{12}C ground-state channel.

For the ^{12}C+^{12}C reaction, the fusion cross section is traditionally converted in a modified S factor as

$$\tilde{S}(E) = \sigma_F(E) E \exp(2\pi\eta + 0.46\, E), \tag{8}$$

where η is the Sommerfeld parameter. The linear term in the exponential accounts for an additional energy dependence (E is expressed in MeV).

3. Results and Discussions

The present folding model was first applied to ^{12}C+^{12}C elastic scattering at energies around the Coulomb barrier, where experimental data are available [14]. Thus, the idea was to assess the accuracy of the model, which is well known experimentally at energies close to the Coulomb barrier [14], and then apply it to the calculation of fusion cross sections. The procedure was started from a single-channel approximation, and we progressively included additional channels.

The comparison between theory and experiment is presented in Fig.1. When the energy increases, and in particular at $E = 10$ MeV, inelastic channels significantly improve the theoretical cross section. At 6 MeV, the physics of the problem is essentially determined by the Coulomb interaction, and the role of the inelastic channels is hardly visible. The most sensitive angular range is beyond $\theta = 70°$, where the single-channel approximation provides a poor fit of the data. Including the 2^+ state improves the overall agreement, but adding further the 0_2^+ Hoyle state provides an excellent agreement with the data. Note that good fits can be obtained even in the single-channel approximation [15], but after fitting the imaginary potential to optimize the agreement with experiment. The same behavior is observed at higher energies, i. e., the inclusion of the Hoyle state in the calculation significantly improves the agreement with experiment (see Figure in Ref. [16]). Most likely, breakup channels start playing a role at these energies, and the imaginary potential should be adapted.

The modified S factor is displayed in Fig.2, where the experimental data have been corrected as suggested by Aguilera *et al.* [25]. Above the Coulomb barrier (≈ 6.5 MeV) the data are well reproduced by the calculation, and the role of inelastic channels is minor. When the energy decreases, the sensitivity with respect to the number of excited channels is more and more important, in agreement with Ref.[3]. At $E = 1$ MeV, the multichannel calculation provides an enhancement of about a factor of three, in comparison with the single-channel approach.

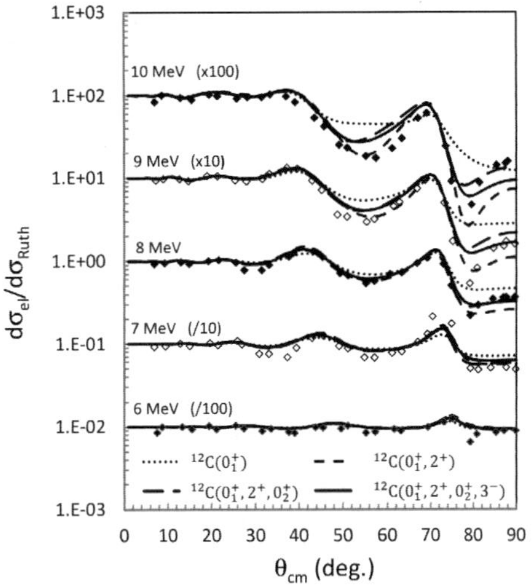

Figure 1. Ratios of the elastic and Rutherford cross sections around the Coulomb barrier, for increasing numbers of ^{12}C+^{12}C inelastic channels. Labels correspond to c. m. energies. Experimental data are from Ref.[14].

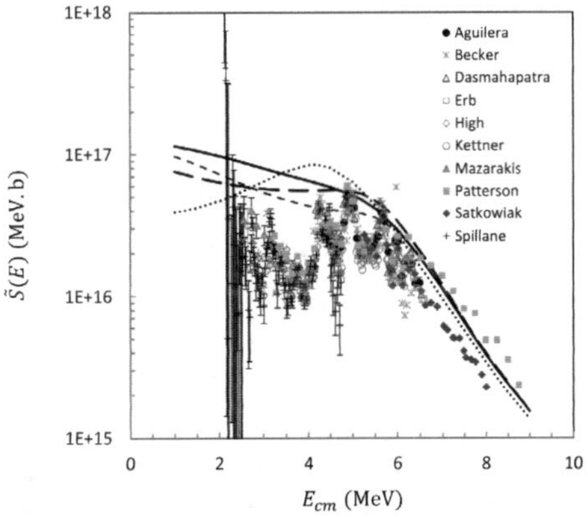

Figure 2. (Color online). Modified S factor (8) for increasing numbers of ^{12}C+^{12}C inelastic channels (the curves are as in Fig.1). Experimental data are taken from Refs.[17–26].

Of course, fluctuations are absent from the present theory. Although molecular resonances are predicted by the calculation with a real potential, they are strongly hindered by the absorption part of the potential.

The interpretation of the theoretical S factor can be seen in Fig.3 through a decomposition in angular momenta J (upper panel) and in the various channels (lower panel). The fusion cross section is essentially given by the contribution of $J = 0^+$ and $J = 2^+$; $J = 4^+$ provides less than 10 %, and other partial waves are negligible. The contributions of the different channels confirm that the fusion cross sections are strongly affected by inelastic channels. These channels are closed at low energies, but the corresponding wave functions $g_c^{J\pi}(r)$ have a significant amplitude in the inner region. Even if they tend to zero at large distances, the short-range potential $W(r)$ makes integrals (6) sensitive to the inner part of the wave function only. Consequently the contribution of inelastic channels to the fusion cross section (5) may be important, and even larger than the ground-state contribution. The role of the Hoyle state is supported by the importance of the $^{12}C(0_1^+)$+$^{12}C(0_2^+)$ channel, which is even dominant above 3.5 MeV.

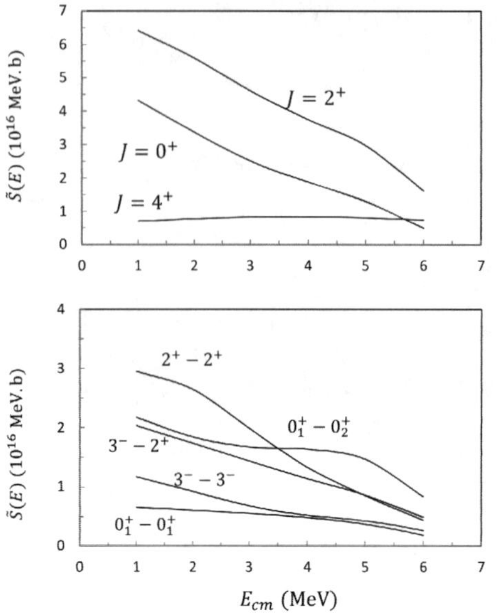

Figure 3. Decompositions of the modified S factor (8) in partial waves (upper panel), and in the main channel contributions (lower panel).

4. Conclusion

In the present work, the ^{12}C+^{12}C fusion process was investigated in a multichannel model. The coupling potentials are generated from ^{12}C densities obtained in a microscopic cluster model. The calculation does not contain any fitting parameter, and provides simultaneously the fusion and elastic cross sections. Around the Coulomb barrier the elastic data are well reproduced by the model provided that all inelastic channels, and in particular those involving the 0_2^+ state, are included. The results confirm the conclusion of Ref.[3], i.e. that inelastic channels play an important role, and must be taken into account for a precise description of the fusion cross section.

Acknowledgments

We are grateful to Prof. M.S. Hussein for useful discussions, and to Profs. M. Kamimura and F. Strieder for providing us with the ^{12}C densities, and with the data of Ref.[26]. We acknowledge FAPESP for financial support. This text presents research results of the IAP programme P7/12 initiated by the Belgian State Federal Services for Scientific, Technical and Cultural Affairs.

References

[1] Wiescher M, Käppeler F, Langanke K 2012 *Annu. Rev. Astron. Astrophys* **50** 165
[2] Pignatari M, *et al.* 2013 *The Astrophysical Journal* **762** 31
[3] Esbensen H, Tang X and Jiang C L 2011 *Phys. Rev. C* **84** 064613
[4] Canto F and M. S. Hussein M S 2013 *Scattering Theory of Molecules, Atoms and Nuclei, World Scientific Publishing*
[5] Satchler G R and Love W G 1979 *Phys. Rep.* **55C** 183
[6] Kamimura M 1981 *Nucl. Phys. A* **351** 456
[7] Kobos A, *et al.* 1982 *Nucl. Phys. A* **384** 65
[8] Ito M, Y.Sakuragi Y, Hirabayashi Y 2001 *Phys. Rev. C* **63** 064303
[9] Descouvemont P and Baye D 2010 *Rep. Prog. Phys.* **73** 036301

[10] Baye D 2006 *Phys. Stat. Sol. (b)* **243** 1095
[11] Rhoades-Brown M, Braun-Munzinger P 1984 *Phys. Lett. B.* **136** 19
[12] T. Khoa D T and Satchler G 2000 *Nucl. Phys. A* **668** 3
[13] F. Canto L F,*et al.* 2006 *Phys. Rep.* **424** 1
[14] Treu W, *et al.* 1980 *Phys. Rev. C* **22** 2462
[15] Gasques L R, *et al.* 2005 *Phys. Rev. C* **72** 025806
[16] Assunção M and Descouvemont P 2013 *Phys. Lett. B* **723** 355
[17] Patterson J R and Zaidinis S 1969 *Astrophys. J.* **157** 367
[18] Mazarakis M G and W. E. Stephens W E 1973 *Phys. Rev. C* **7** 1280
[19] High M and Cujec B 1977 *Nucl. Phys. A* **282** 181
[20] Erb K A, *et al.* 1980 *Phys. Rev. C* **22** 507
[21] Kettner K, Lorenz-Wirzba H and Rolfs 1980 *Z. Phys. A* **298** 65
[22] Becker H, *et al.* 1981 *Z. Phys. A* **303** 305
[23] Dasmahapatra B, Cujec B and Lahlou F 1982 *Nucl. Phys. A* **384** 257
[24] Satkowiak L J, *et al.* 1982 *Phys. Rev. C* **26** 2027
[25] Aguilera E F, *et al.* 2006 *Phys. Rev. C* **73** 064601
[26] Spillane T, *et al.* 2007 *Phys. Rev. Lett.* **98** 122501

Gamma-Rays from Nucleosynthesis Ejecta

R Diehl[1,2]

[1] *Max Planck Institut für extraterrestrische Physik, D-85748 Garching, Germany*
[2] *Excellence Cluster Universe, D-85748 Garching, Germany*

E-mail: rod@mpe.mpg.de

Abstract. Gamma-ray lines from radioactive decay of unstable isotopes produced in massive-star and supernova nucleosynthesis have been measured with INTEGRAL over the past ten years, complementing the earlier COMPTEL survey. ^{26}Al has become a tool to study specific source regions, such as massive-star groups and associations in nearby regions which can be discriminated from the galactic-plane background, and the inner Galaxy where Doppler shifted lines add to the astronomical information. Recent findings are that superbubbles show a remarkable asymmetry, on average, in the spiral arms of our galaxy. ^{60}Fe is co-produced by the sources of ^{26}Al, and the isotopic ratio from their nucleosynthesis encodes stellar-structure information. Annihilation gamma-rays from positrons in interstellar space show a puzzling bright and extended source region central to our Galaxy, but also may be partly related to nucleosynthesis. ^{56}Ni and ^{44}Ti isotope gamma-rays have been used to constrain supernova explosion mechanisms. Here we summarize latest results using the accumulated multi-year database of observations, and discuss their astrophysical interpretations. We also add a comparison of isotopic ratios between the ISM of the current Galaxy and the solar vicinity at solar-system formation time.

1. Introduction

Cosmic nucleosynthesis is expected to be ongoing in star-forming galaxies such as our Milky Way. Co-produced unstable nuclides, if ejected into interstellar space and sufficiently long-lived to not decay within the source already, can be observed through characteristic gamma-ray line emission with current gamma-ray telescopes. This makes nuclides ^{7}Be, ^{22}Na, ^{56}Ni, ^{44}Ti, ^{26}Al, and ^{60}Fe prime candidates for such gamma-ray line studies of ongoing nucleosynthesis in our galaxy, through massive stars, supernovae, and novae as candidate sources [1; 2].

After ^{26}Al observations with HEAO-C [3] had confirmed this connection, gamma-ray line surveys have been undertaken by NASA with the Compton Gamma-Ray Observatory (1991-2000) [4], and by ESA with the International Gamma-Ray Astrophysics Laboratory observatory INTEGRAL (since 2002 and currently ongoing) [5]. Instruments and missions up to now are characterized by sensitivities of a few 10^{-5} photons cm^{-2}s^{-1} that limit studies essentially to sources in our own galaxy[1]. Advances by up to two orders of magnitude in sensitivity would be feasible [6], but are unlikely to happen[2].

[1] except for ^{56}Ni from rare supernovae in nearby galaxies, and the ^{44}Ti measurements from SN1987A in the nearby LMC galaxy.
[2] because of high cost and competing astrophysics goals with broader community support than nuclear astrophysics has at present.

Observations from these missions and their results and implications have been reviewed before [2; 7–9]. Here we therefore only summarize the status and results (Section 2), and add some details to stimulate a broader view, e.g. connecting gamma-ray data with presolar grain and early solar system data from meteorites. We finish this short contribution with comments on the open issues and prospects for this field of astronomy and astrophysics.

2. Selected Observational Results

2.1. Supernovae: ^{56}Ni *and* ^{44}Ti

Shortlived radioactivity in the ^{56}Ni and ^{44}Ti decay chains is expected to be bright enough to be detectable from supernova explosions even beyond our Galaxy. For ^{56}Ni telescopes such as INTEGRAL should reach out to a few Mpc distance for SNIa which eject up to $\sim$1 $M_\odot$ of 56Ni [10]. Even for core-collapse supernovae, nearby galaxies could be reached, in particular for hypernovae and pair instability supernovae. ^{44}Ti as the next nuclide in ranking gamma-ray brightness[3] can only be seen from Galactic young supernova remnants, and for the SN1987A supernova in the nearby LMC at 50 kpc distance.

INTEGRAL spent substantial observing time (4Ms), therefore, when a SNIa exploded on 23 Aug 2011 in M101 at about 6 Mpc distance [11]. It was clear that this was a challenging case for gamma-ray observers at the outer limb of the usable distance range, but an excellent one for astronomers in many other fields - the prospect of multi-wavelength measurements provides for another unique learning experience. The disappointment of not seeing ^{56}Ni decay gamma-rays [12] was not surprising after all, but we needed to take our chance of contributing to SNIa brightness calibrations through measurement of the gamma-rays directly from ^{56}Ni decay.

^{44}Ti had been detected from one young supernova remnant only, up to now: the 360-year old Cas A remnant at 3.4 kpc distance [13; 14]. For this object, the ^{44}Ti gamma-ray flux is rather well determined at 2.5 10^{-5} ph $cm^{-2}s^{-1}$. This corresponds to an amount of ^{44}Ti ejected by the explosion which is about a factor of three above expectations from models [15]. Spectroscopy with SPI, together with the IBIS flux measurement, sets a limit on the ejecta kinematics with INTEGRAL observations, requiring ^{44}Ti ejecta to coast at velocities above 500 km s^{-1} [16], compatible with expectations from supernova simulations [17; 18]. The imaging hard X-ray mission NuStar launched in 2012 [19] has recently reported first images of ^{44}Ti X-rays from Cas A (Harrison, priv.comm.). This will allow to test for the extent of explosion asymmetries in ^{44}Ti versus other observables; highly non-spherical explosions are suspected to be the reason of high ejected ^{44}Ti amounts in core-collapse supernovae [7; 15].

Surveys have been performed to search for more ^{44}Ti emitting young supernova remnants [20–22], and several candidates have been discussed, most intensely the Vela Junior/GROJ0852-4642 [23; 24] and G1.9+0.3 [25] remnants. But no other source has been assessed as detected so far in ^{44}Ti decay lines in our Galaxy apart from Cas A.

In 2011, IBIS data then showed a signature from SN1987A in the LMC [26] that plausibly is attributed to ^{44}Ti line emission, though IBIS cannot resolve the X-ray lines from ^{44}Sc decay (which follows ^{44}Ti decay). Although once more the ^{44}Ti amount inferred from the ^{44}Ti gamma-ray intensity lies well above model expectations, recent NuStar observations have confirmed this result (Harrison, priv.comm.).

2.2. Diffuse Radioactivity: ^{26}Al, ^{60}Fe, *and* e^+

The long-lived radio-isotopes of ^{26}Al and ^{60}Fe accumulate in interstellar medium around their sources, and result in a diffuse afterglow of nucleosynthesis with a My decay time. This is still shorter than the evolutionary time scale for star clusters (tens of My), but much longer

[3] This involves amounts produced in the event, multiplied by decay rate. Therefore short-lived nuclides up to the iron group are best candidates. See table/list in [2]

Journal of Physics: Conference Series **665** (2016) 012011 doi:10.1088/1742-6596/665/1/012011

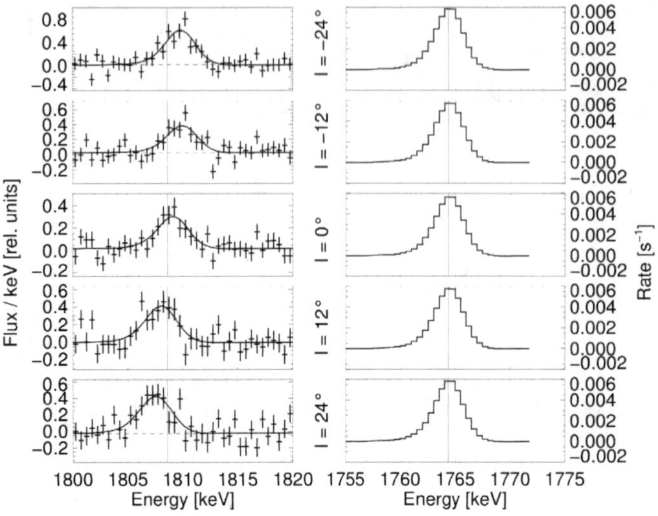

Figure 1. ^{26}Al line measurements with SPI along the plane or the Galaxy, for different lines of sight, show the systematic Doppler shift from large-scale rotation in the Galaxy (*left*). A nearby instrumental line at 1764 keV (*right*) does not show differences in line position, from the same data.

than the time scale of other observables related to nucleosynthesis events such as supernova remnants (tens of ky). The sky survey from COMPTEL had shown ^{26}Al emission correlating to known massive star activity in our Galaxy [27]. ^{60}Fe was an expected gamma-ray line emitter ejected from supernovae, but only detected in 2004/2005 with Ge spectrometers on RHESSI and INTEGRAL[28–30]. Its gamma-ray brightness is only about 1/7 of ^{26}Al, and therefore the measurement of its spatial distribution in the Galaxy and its gamma-ray line shape characteristics will be challenging even with many more years of INTEGRAL/SPI measurements.

Galactic ^{26}Al has been studied in more detail, and provides an independent observable for astrophysics questions relating to the Galaxy's massive-star population and the interstellar medium. From the total Galactic ^{26}Al amount, one can infer the Galactic core-collapse supernova rate (or the rate of star formation) through a measurement of penetrating gamma-rays. The result is consistent with expectations within uncertainties of the various star formation / supernova rate observables: A supernova rate of 1.x $\pm$0.x per 100 years corresponds to a star formation rate of x $M_\odot$ y^{-1} in the Galaxy as a whole. While that confirms our Galaxy forming stars at a rate that requires gas replenishing from outside the Galaxy (e.g. through infalling clouds from intergalactic space or from collisions with nearby galaxies), the question of where stars are being formed exactly, and why, is a key issue for evolution of galaxies. Studying the multi-messenger picture of massive-star feedback in specific regions hosting well-observed groups of massive stars is therefore a rich field for such studies, e.g. the Cygnus, Carina, Scorpius-Centaurus, and Orion regions.

With the exquisite spectral resolution of INTEGRAL's spectrometer SPI, a new measurement (Fig. 1) of the ^{26}Al line properties as it is Doppler shifted by large-scale Galactic rotation may shed additional and new light on this issue. Observations show Doppler shifts in the ^{26}Al line which exceeds values seen in other Galactic objects such molecular clouds. This surprising finding suggests that there may be a global asymmetry in outflows from massive-star groups, i.e. the ejecta show a tendancy of less-inhibited free streaming away from ^{26}Al sources in the direction of global galactic rotation. Studies are underway to investigate if massive-star groups may not be populated symmetrically on leading and training sides of spiral arms, and if that could lead to the observed effect. Superbubbles play a role here, as ejecta flows are directed by the density structures that have ben pre-shaped around the sources. It will be interesting to see how ^{26}Al sources relate to the drivers of the structure and dynamics of the interstellar medium in general, and the different phases of the interstellar medium.

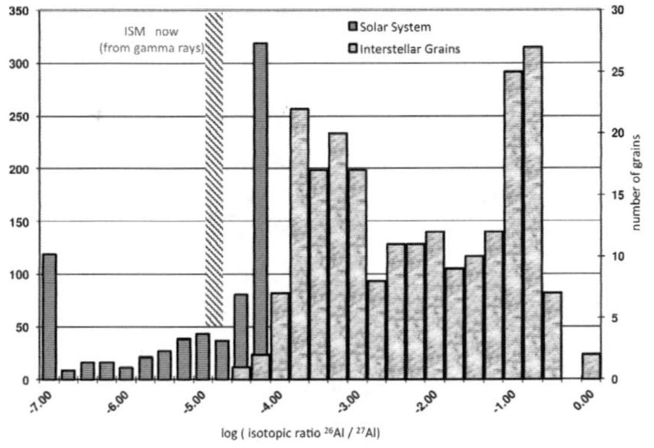

Figure 2. The $^{26}Al/^{27}Al$ isotopic ratio from meteoritic measurements of different types of presolar-grain inclusions (grey/shaded; A. Davis, priv.comm.), and CAI inclusions (green narrow columns, from [31])and from the gamma-ray result for the current interstellar medium average in the Galaxy (hatched).

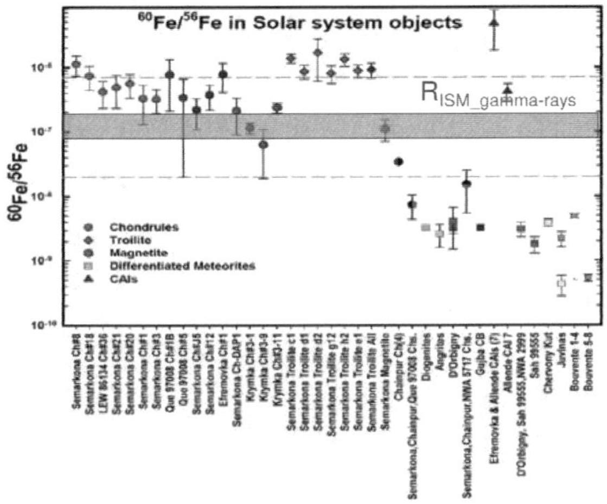

Figure 3. $^{60}Fe/^{56}Fe$ Isotopic ratios from different measurements relating to the early solar system [32], compared to the gamma-ray result for the current interstellar medium average in the Galaxy (shaded). (Figure adapted from [32])

Many nucleosynthesis sources involve H burning and the production of proton rich nuclei in general; examples are ^{26}Al from massive stars, but also ^{13}N and ^{18}F from novae, and even p-n symmetric nuclides ^{56}Ni and ^{44}Ti fall on the p-rich side of the valley of nuclide stability, as neutron number excesses compensate for increasing Coulomb repellence in heavier nuclei. Incurred β^+-decays release positrons, which eventually annihilate with ambient electrons to produce characteristic gamma-rays as the particle-antiparticle mass is radiated away. Characteristic e^+ annihilation gamma-rays at 511 keV have been measured since decades, and finally been mapped across the sky with INTEGRAL/SPI. This map presents a major puzzle, however: The expected e^+ sources are distributed throughout the Galaxy and in particular its plane, where massive stars are being formed and eject nucleosynthesis products plus leave behind compact stars that again can turn into e^+ sources (see for a detailed review and discussion). Again the structure and morphology of interstellar medium may hold the answer to this puzzle: positrons are typically produced with relativistic energies, and need to slow down before annihilating with interstellar electrons. Propagation of e^+ in the different phases of ISM and the impacts of interstellar magnetic fields are currently being studied. Interplanetary positron measurements have recently added excitement and may also teach us about propagation aspects. Once this is understood, nucleosynthesis e^+ and their role in observed annihilation will become clear, and next steps could then be exciting to possibly reveal new sources of e^+.

Ongoing nucleosynthesis activity ever increases the heavy-element content of interstellar gas. Our solar system samples ambient interstellar gas from 4.568 Gy ago, and solar-system matter composition is measured in great detail through solar radiation absorption-line spectroscopy and also through mass spectrometry of meteoritic inclusions [33; 34]. Solar abundances constitute our best-known reference for the study of cosmic abundances, and chemical evolution prescriptions are required to properly connect our models of cosmic nucleosynthesis sources to the various abundance measurements. We therefore compare the ^{26}Al and ^{60}Fe gamma-ray measurements of the current interstellar gas in our Galaxy to the record of those same isotopic ratios with respect to the stable reference isotope in the early solar system (ESS) (Fig's 2 and 3). The total Galactic ^{26}Al mass of 2.25 $M_\odot$ translates into an isotope ratio of 1.0 10^{-5}, assuming that the ^{26}Al mass reflects star forming activity and is in steady state. Here, we use a total Galactic gas mass of 4.95 $10^9 M_\odot$ [35] and solar abundances from [33]. The gamma-ray flux ratio for ^{60}Fe versus ^{26}Al of 15% yields a steady-state mass of ^{60}Fe of 1.53 $M_\odot$ for the Galaxy, or an isotopic ratio ^{60}Fe/^{56}Fe of 1.5 10^{-7}. The fractions of the unstable isotopes apparently decrease with time, consistent with proceeding enrichment of heavy nuclei in general and the decay of the unstable nuclei. We may ask if the early solar system isotopic ratios are consistent with large-scale averaged ISM composition as inferred from the current-ISM gamma-ray data and solar stable-isotope abundances, or if special enrichments may be suggested by these meteoritic data. There is a rich literature on special nearby events injecting short-lived radioactivities into the forming solar system. In recent years, increased precision of isotopic abundance measurements have reduced the significance of required enrichments for several of those radioisotopes, however. Within present uncertainties, it appears that ^{26}Al enrichments still persist, while this is less clear for ^{60}Fe. Simulations of abundance evolutions in a star-forming gas disk [36] also support these conclusions.

3. Open Issues and Prospects

Gamma-ray survey missions have shown the brightest sources, and confirm the approach of testing cosmic nucleosynthesis by measuring radioactive decays of trace isotopes co-produced herein. ^{56}Ni is an exception, as its decay is the dominating power source for supernova light Major open issues still are the variety of models discussed for all types of supernova explosions, the poor constraints for massive-star interior structure in late phases of the pre supernova evolution, and the inadequacy of nuclear explosion models on compact stars in general. Continued and deeper observations in gamma-rays are our best chance to learn about these nucleosynthesis sources as they happen in the present nearby universe, so that improved models can be developed and tested on the variety of nuclide abundance measurements with the help of chemical-evolution prescriptions and source population models.

Acknowledgments

This work was supported by the DFG-funded Cluster of Excellence "Origins and evolution of the universe". The INTEGRAL/SPI project has been completed under the responsibility and leadership of CNES; we are grateful to ASI, CEA, CNES, DLR (grants 50 OG 1101 and 50 OR 0901), ESA, INTA, NASA and OSTC for support of this ESA space science mission and its science analysis.

References

[1] Lingenfelter R E and Ramaty R 1978 *Physics Today* **31** 40–47

[2] Diehl R, Hartmann D H and Prantzos N 2011 *Astronomy with Radioactivities* (*Lecture Notes in Physics, Berlin Springer Verlag* vol 812) (Springer: Berlin, Heidelberg)

[3] Mahoney W A, Ling J C, Wheaton W A and Jacobson A S 1984 *ApJ* **286** 578–585

[4] Gehrels N, Fichtel C E, Fishman G J, Kurfess J D and Schonfelder V 1993 *Scientific American* **269** 68–+

[5] Winkler C, Courvoisier T J L, Di Cocco G, Gehrels N, Giménez A, Grebenev S, Hermsen W, Mas-Hesse J M, Lebrun F, Lund N, Palumbo G G C, Paul J, Roques J P, Schnopper H, Schönfelder V, Sunyaev R, Teegarden B, Ubertini P, Vedrenne G and Dean A J 2003 *A&A* **411** L1–L6

[6] Greiner J, Mannheim K, Aharonian F, Ajello M, Balasz L G, Barbiellini G, Bellazzini R, Bishop S, Bisnovatij-Kogan G S, Boggs S, Bykov A, DiCocco G, Diehl R, Elsässer D, Foley S, Fransson C, Gehrels N, Hanlon L, Hartmann D, Hermsen W, Hillebrandt W, Hudec R, Iyudin A, Jose J, Kadler M, Kanbach G, Klamra W, Kiener J, Klose S, Kreykenbohm I, Kuiper L M, Kylafis N, Labanti C, Langanke K, Langer N, Larsson S, Leibundgut B, Laux U, Longo F, Maeda K, Marcinkowski R, Marisaldi M, McBreen B, McBreen S, Meszaros A, Nomoto K, Pearce M, Peer A, Pian E, Prantzos N, Raffelt G, Reimer O, Rhode W, Ryde F, Schmidt C, Silk J, Shustov B M, Strong A, Tanvir N, Thielemann F K, Tibolla O, Tierney D, Trümper J, Varshalovich D A, Wilms J, Wrochna G, Zdziarski A and Zoglauer A 2012 *Experimental Astronomy* **34** 551–582 (*Preprint* 1105.1265)

[7] Prantzos N 1996 *A&AS* **120** C303+

[8] Diehl R, Prantzos N and von Ballmoos P 2006 *Nuclear Physics A* **777** 70–97 (*Preprint* arXiv:astro-ph/0502324)

[9] Diehl R 2013 *Reports on Progress in Physics* **76** 026301 (*Preprint* 1302.3441)

[10] Isern J, Bravo E and Hirschmann A 2008 *New Astronomy Review* **52** 377–380

[11] Nugent P E, Sullivan M, Cenko S B, Thomas R C, Kasen D, Howell D A, Bersier D, Bloom J S, Kulkarni S R, Kandrashoff M T, Filippenko A V, Silverman J M, Marcy G W, Howard A W, Isaacson H T, Maguire K, Suzuki N, Tarlton J E, Pan Y C, Bildsten L, Fulton B J, Parrent J T, Sand D, Podsiadlowski P, Bianco F B, Dilday B, Graham M L, Lyman J, James P, Kasliwal M M, Law N M, Quimby R M, Hook I M, Walker E S, Mazzali P, Pian E, Ofek E O, Gal-Yam A and Poznanski D 2011 *Nature* **480** 344–347 (*Preprint* 1110.6201)

[12] Isern J, Jean P, Bravo E, Diehl R, Knödlseder J, Domingo A, Hirschmann A, Hoeflich P, Lebrun F, Renaud M, Soldi S, Elias-Rosa N, Hernanz M, Kulebi B, Zhang X, Badenes C, Domínguez I, Garcia-Senz D, Jordi C, Lichti G, Vedrenne G and Von Ballmoos P 2013 *A&A* **552** A97 (*Preprint* 1302.3381)

[13] Iyudin A F, Diehl R, Bloemen H, Hermsen W, Lichti G G, Morris D, Ryan J, Schoenfelder V, Steinle H, Varendorff M, de Vries C and Winkler C 1994 *A&A* **284** L1–L4

[14] Renaud M, Vink J, Decourchelle A, Lebrun F, den Hartog P R, Terrier R, Couvreur C, Knödlseder J, Martin P, Prantzos N, Bykov A M and Bloemen H 2006 *ApJ* **647** L41–L44 (*Preprint* arXiv:astro-ph/0606736)

[15] The L S, Clayton D D, Diehl R, Hartmann D H, Iyudin A F, Leising M D, Meyer B S, Motizuki Y and Schönfelder V 2006 *A&A* **450** 1037–1050 (*Preprint* arXiv:astro-ph/0601039)

[16] Martin P, Knödlseder J, Vink J, Decourchelle A and Renaud M 2009 *A&A* **502** 131–137

[17] Kifonidis K, Plewa T, Janka H and Müller E 2003 *A&A* **408** 621–649 (*Preprint* arXiv:astro-ph/0302239)

[18] Wongwathanarat A, Janka H T and Müller E 2013 *A&A* **552** A126 (*Preprint* 1210.8148)

[19] Harrison F A, Craig W W, Christensen F E, Hailey C J, Zhang W W, Boggs S E, Stern D, Cook W R, Forster K, Giommi P, Grefenstette B W, Kim Y, Kitaguchi T, Koglin J E, Madsen K K, Mao P H, Miyasaka H, Mori K, Perri M, Pivovaroff M J, Puccetti S, Rana V R, Westergaard N J, Willis J, Zoglauer A, An H, Bachetti M, Barrière N M, Bellm E C, Bhalerao V, Brejnholt N F, Fuerst F, Liebe C C, Markwardt C B, Nynka M, Vogel J K, Walton D J, Wik D R, Alexander D M, Cominsky L R, Hornschemeier A E, Hornstrup

A, Kaspi V M, Madejski G M, Matt G, Molendi S, Smith D M, Tomsick J A, Ajello M, Ballantyne D R, Baloković M, Barret D, Bauer F E, Blandford R D, Niel Brandt W, Brenneman L W, Chiang J, Chakrabarty D, Chenevez J, Comastri A, Dufour F, Elvis M, Fabian A C, Farrah D, Fryer C L, Gotthelf E V, Grindlay J E, Helfand D J, Krivonos R, Meier D L, Miller J M, Natalucci L, Ogle P, Ofek E O, Ptak A, Reynolds S P, Rigby J R, Tagliaferri G, Thorsett S E, Treister E and Urry C M 2013 *ApJ* **770** 103 (*Preprint* 1301.7307)

[20] Dupraz C, Bloemen H, Bennett K, Diehl R, Hermsen W, Iyudin A F, Ryan J and Schoenfelder V 1997 *A&A* **324** 683–689

[21] Iyudin A F, Schönfelder V, Bennett K, Bloemen H, Diehl R, Hermsen W, Knödlseder J, Lichti G G, Oberlack U, Ryan J, Strong A W and Winkler C 1999 *Astrophysical Letters Communications* **38** 383–+

[22] Renaud M, Vink J, Decourchelle A, Lebrun F, Terrier R and Ballet J 2006 *New Astronomy Review* **50** 540–543 (*Preprint* arXiv:astro-ph/0602304)

[23] Aschenbach B 1998 *Nature* **396** 141–142

[24] Schönfelder V, Bloemen H, Collmar W, Diehl R, Hermsen W, Knödlseder J, Lichti G G, Plüschke S, Ryan J, Strong A and Winkler C 2000 *American Institute of Physics Conference Series* (*American Institute of Physics Conference Series* vol 510) ed McConnell M L and Ryan J M pp 54–+

[25] Borkowski K J, Reynolds S P, Hwang U, Green D A, Petre R, Krishnamurthy K and Willett R 2013 *ApJ* **771** L9 (*Preprint* 1305.7399)

[26] Grebenev S A, Lutovinov A A, Tsygankov S and Winkler C 2012 *Nature* **(tbd)** (accepted for publication)

[27] Diehl R, Dupraz C, Bennett K, Bloemen H, Hermsen W, Knoedlseder J, Lichti G, Morris D, Ryan J, Schoenfelder V, Steinle H, Strong A, Swanenburg B, Varendorff M and Winkler C 1995 *A&A* **298** 445–+

[28] Smith D M 2004 *5th INTEGRAL Workshop on the INTEGRAL Universe* (*ESA Special Publication* vol 552) ed Schoenfelder V, Lichti G and Winkler C pp 45–+

[29] Harris M J, Knödlseder J, Jean P, Cisana E, Diehl R, Lichti G G, Roques J P, Schanne S and Weidenspointner G 2005 *A&A* **433** L49–L52 (*Preprint* arXiv:astro-ph/0502219)

[30] Wang W, Harris M J, Diehl R, Halloin H, Cordier B, Strong A W, Kretschmer K, Knödlseder J, Jean P, Lichti G G, Roques J P, Schanne S, von Kienlin A, Weidenspointner G and Wunderer C 2007 *A&A* **469** 1005–1012 (*Preprint* 0704.3895)

[31] MacPherson G J, Davis A M and Zinner E K 1995 *Meteoritics* **30** 365–+

[32] Mishra R, Chaussidon M and Marhas K 2012 *Nuclei in the Cosmos (NIC XII)*

[33] Asplund M, Grevesse N, Sauval A J and Scott P 2009 *ARA&A* **47** 481–522 (*Preprint* 0909.0948)

[34] Lodders K 2003 *ApJ* **591** 1220–1247

[35] Robin A C, Reylé C, Derrière S and Picaud S 2003 *A&A* **409** 523–540 (*Preprint* arXiv:astro-ph/0401052)

[36] Vasileiadis A, Nordlund Å and Bizzarro M 2013 *ApJ* **769** L8 (*Preprint* 1302.0843)

Abundance analysis of extremely metal-poor stars

T Hansen[1], C J Hansen[1], N Christlieb[1], and J Andersen[2]

[1] Landessternwarte, Heidelberg University
[2] Niels Bohr Institute, University of Copenhagen

E-mail: thansen@lsw.uni-heidelberg.de

Abstract. The outer atmosphere of the first generations of low-mass ($M \leq 0.8\ M_{\odot}$) stars retain to a great extent the original chemical composition of the interstellar medium (ISM) at the time and place of their birth. The composition of this pristine gas represents the nucleosynthesis of the very first massive stars, that produced and ejected the first heavy elements into the early ISM. Thus a detailed abundance analysis of low-mass, metal-poor stars can help us track these gasses and provide insight into the formation processes that took place in the very early stages of our Galaxy. Preliminary result of a 25-star homogeneously analysed sample of metal-poor candidates from the Hamburg/ESO survey is presented. The main focus is on the most metal-poor stars of the sample; stars with [Fe/H] < -4. The abundance pattern of these ultra metal-poor (UMP) stars is used to extract key information of the earliest ongoing formation processes (ranging from hydrostatic burning to neutron-capture processes).

1. Introduction

Metal-poor stars are the key to unlock the first chapter in the history of the chemical enrichment of our Galaxy. The first stars to form in the universe were presumably massive and short lived ending their lives in a supernova (SN) explosions, and thereby enriching the inter stellar medium (ISM) with heavy elements. The metal-poor stars we observe today are the first generation of stars formed from this freshly enriched ISM. The ESO large program First Stars resulted in the first abundances trends for metal-poor stars. They found a low scatter in the abundances of α and iron-peak elements, and a small over abundance of the α-elements, consistent with a well mixed early inter stellar medium [1]. But they also found a large scatter in the abundances of neutron-capture elements [2]. Large surveys have also shown that a large fraction of the metal-poor stars especially at very low metallicity are carbon-enhanced [3], but the origin of this of carbon in the early universe is still not completely understood. To better constrain the nucleosynthesis processes taking place in the early universe, for both the light and heavy elements, more studies of low metallicity stars are needed. Here abundance analysis of four newly discovered ultra metal-poor (UMP) stars is presented.

2. Observation and Analysis

The stars have been observed with the VLT/UVES, covering a wavelength range from 3100 Å to 5000 Å in the blue and 5800 Å to 9500 Å in the red and have a resolving power of R$\sim$45000. The data was reduced with the UVES reduction pipeline version 2.4.0. In Figure 1 reduced and normalized spectra around the Sr II line at 4077 Å are displayed for three different stars, with decreasing metallicity going from the top to the bottom of the figure. The spectrum in

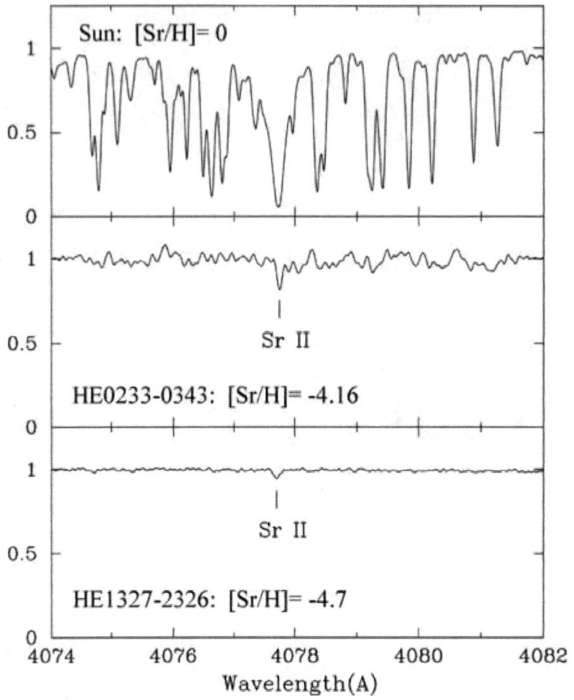

Figure 1. Spectra of the sun (top), HE 0233−0343 (middle) and HE 1327−2326 (bottom) around the Sr II line at 4077 Å.

the top panel is that of the Sun showing multiple lines, the middle panel shows the spectrum of HE 0233−0343, where the Sr line is still visible and at the bottom HE 1327−2326 is shown, here the Sr II line is barely visible. This figure clearly displays the increasing difficulty in abundance determination with decreasing metallicity. Thus highlighting the need for high resolution high signal to noise spectra to facilitate this work.

The abundance analysis was performed using the LTE (local thermal equilibrium) stellar line analysis program MOOG [4]. The model atmospheres are α-enhanced ($[\alpha/\mathrm{Fe}]^1$=+0.4) models from the NEWODF grid of ATLAS9 models by Castelli & Kurucz 2003[5]. Temperatures where deduced from the infrared flux method, by fitting model atmospheres to spectrophotometric observations, for this α-enhanced ($[\alpha/\mathrm{Fe}] = 0.5$) MARCS models [6] where used. Gravities are from Y^2 isochrones [7] assuming age=10 Gyr and $[\alpha/\mathrm{Fe}] = +0.3$ and the metallicity was deduced from Fe I lines. The stars with their parameters and derived abundances of carbon, barium and strontium are listed in Table 1.

3. Results

3.1. α and iron-peak elements

In Figure 2 the derived abundances for the α-elements magnesium, calcium and titanium (left) and the iron-peak elements manganese and nickel (right) are plotted as a function of metallicity. For comparison abundances for the most metal-poor stars from the Yong et al 2013 [8] and Frebel 2010 [9] samples plus abundances for SDSS J102915+172927 [10] are added to the plot. The four newly discovered UMP stars clearly follow the trend of the other samples with a small

[1] $[\mathrm{X/Y}]=\log\left(\frac{N_X}{N_Y}\right)_* - \log\left(\frac{N_X}{N_Y}\right)_\odot$

Table 1. UMP stars in this sample

Stellar ID	T_{eff}	Logg	[Fe/H]	[C/Fe]	[Ba/Fe]	[Sr/Fe]
HE 0134−1519	5525	3.22	−4.0	+1.1	< −0.8	−0.3
HE 0233−0343	6075	3.49	−4.7	+3.5	< +0.9	+0.5
HE 1310−0536	4975	1.88	−4.2	+2.5	−0.5	−1.1
HE 2239−5019	6125	3.57	−4.2	< +1.8	<0.0	< −0.6

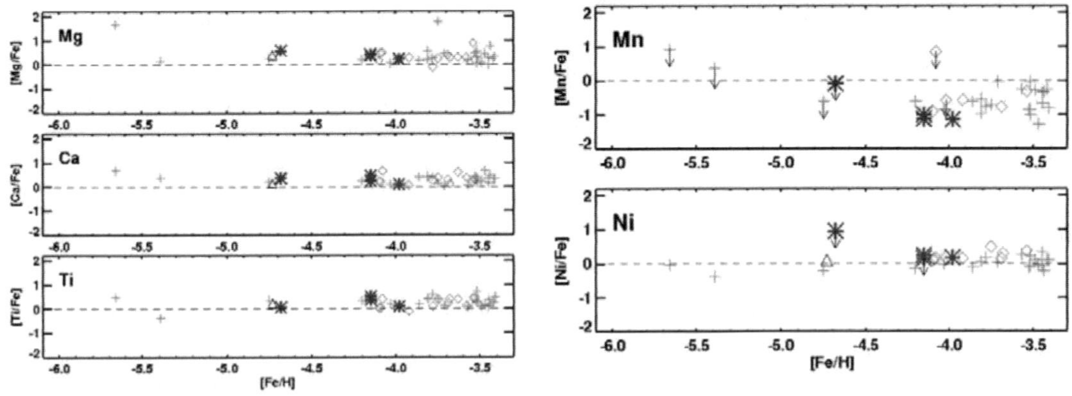

Figure 2. Derived abundances for the α-elements Mg, Ca and Ti (left) and the iron-peak elements Mn and Ni (right). Blue asterisks: Hansen et al. in prep., Green diamonds: Yong et al. 2013 [8], red crosses: Frebel 2010 [9] grey triangle: Caffau et al. 2011[10]. Dashed line is solar value.

over abundance in the α elements. Consistent with the existing picture of α abundances in metal-poor stars reflecting the enrichment from core-collapse SNe in the early universe. The abundances for the iron-peak elements also follow the trends seen in the abundances for the larger samples, displaying a very low scatter in the abundances of these elements. Hence these four new UMP stars behave normally compared to other metal-poor stars.

3.2. neutron-capture elements

The abundances derived for the neutron-capture elements show a very different picture from that of the α and iron-peak elements. In Figure 3 the strontium (left) and barium (right) abundances are plotted as a function of metallicity. A large star-to-stars scatter is seen for both elements. Recent work by Hansen et al. 2013 [11] show that this scatter at least for Sr persists even with correction for NLTE and with using 3D model atmospheres, i.e. the scatter is real and not just an artifact of a simplified analysis procedure. This large scatter points to more than one formation process for these elements. There are several candidates for the formation sites of neutron-capture elements in the early universe such as the r-process in the neutrino driven wind following supernova explosions [12] or in low mass faint electron capture supernovae [13] or the s-process in massive fast rotating stars [14]. To put further constraints on the production of the neutron-capture elements key information can be derived from the correlation or anti-correlation of one element with another. In Figure 4 the Sr abundances are plotted as a function of the Ba abundances. A clear correlation between the two elements is seen (dashed line is a 1:1

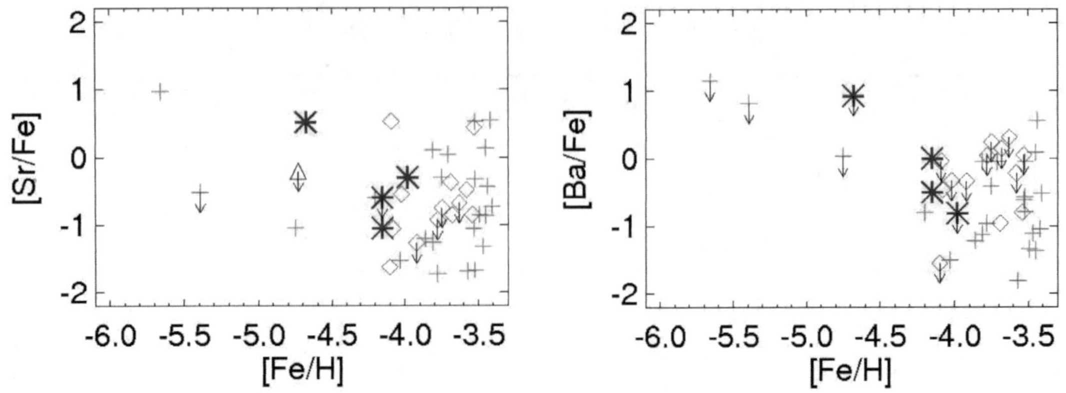

Figure 3. Abundances of Sr (left) and Ba (right) as function of metallicity. Symbols as in Figure 2

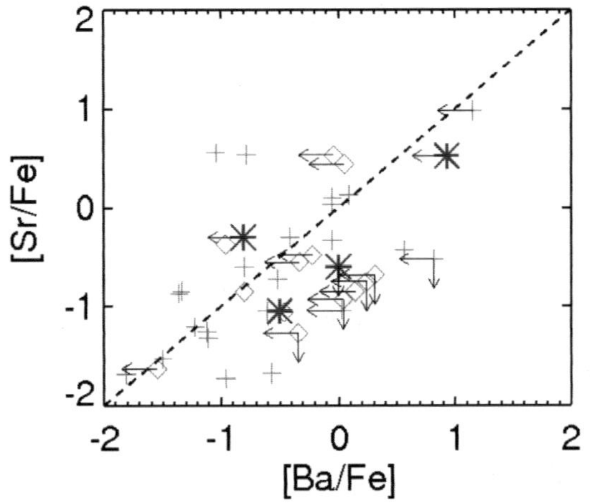

Figure 4. Sr abundances as function of Ba abundances. Symbols as in Figure 2, the dashed line corresponds to a 1:1 correlation.

correlation). The same result have been seen in Andrievsky et al. 2011 [15] for a smaller but NLTE corrected sample and in Roederer 2013 [16] for a much larger sample. The correlation of the two elements show that their abundances grow simultaneously over time. So regardless of the production processes or sites, roughly equal amount of Sr and Ba have been produced at a given time in the history of the Galaxy.

3.3. Carbon enhanced stars

Three of the four stars in our sample are carbon-enhanced metal-poor stars (CEMP), but none of these three show any enhancement in neutron-capture elements (see Table 1), hence they belong to the CEMP-no class [3]. The origin of the large carbon enhancements but low neutron-capture element abundances of these stars are unknown. One possible explanation is that they are part of a binary system, so the star we observe have received mass from a former AGB companion, the companion having now evolved into a white dwarf. To test this hypothesis a radial velocity monitoring program have been under way at the Nordic Optical Telescope for the last of 6 years. Preliminary results from this program show that nine out of the eleven CEMP-s program stars are definite long period binaries where as only two of the eight CEMP-no program stars are binaries.

4. Summery

Four new ultra metal-poor stars have been presented. The derived abundances show very low star-to-star scatter in α and iron-peak elements abundances and hence follow the picture seen for abundances of these elements in larger samples of metal-poor stars. A large star-to-star scatter is observed for the neutron-capture element abundances, pointing to different formation processes for these elements. But with Ba and Sr correlating, making the formation of these two co-dependent. Among the four stars, three are carbon-enhanced, but with no enhancement in neutron-capture elements, i.e. CEMP-no stars. The nature of these stars is unknown but preliminary results from radial velocity monitoring show that they are not binaries and hence points to an alternative explanation than mass transfer from an AGB companion.

References

[1] Cayrel R, Depagne E, Spite M, Hill V, Spite F, François P, Plez B, Beers T, Primas F, Andersen J, Barbuy B, Bonifacio P, Molaro P and Nordström B 2004 *Astronomy and Astrophysics* **416** 1117–1138 (*Preprint* arXiv:astro-ph/0311082)

[2] François P, Depagne E, Hill V, Spite M, Spite F, Plez B, Beers T C, Andersen J, James G, Barbuy B, Cayrel R, Bonifacio P, Molaro P, Nordström B and Primas F 2007 *Astronomy and Astrophysics* **476** 935–950 (*Preprint* 0709.3454)

[3] Beers T C and Christlieb N 2005 *Annual Review of Astronomy & Astrophysics* **43** 531–580

[4] Sneden C 1973 *The Astrophysical Journal* **184** 839–849

[5] Castelli F and Kurucz R L 2003 *Modelling of Stellar Atmospheres* (*IAU Symposium* vol 210) ed Piskunov N, Weiss W W and Gray D F p 20P

[6] Gustafsson B, Edvardsson B, Eriksson K, Jørgensen U G, Nordlund Å and Plez B 2008 *Astronomy and Astrophysics* **486** 951–970 (*Preprint* 0805.0554)

[7] Demarque P, Woo J H, Kim Y C and Yi S K 2004 *The Astrophysical Journals* **155** 667–674

[8] Yong D, Norris J E, Bessell M S, Christlieb N, Asplund M, Beers T C, Barklem P S, Frebel A and Ryan S G 2013 *The Astrophysical Journal* **762** 26 (*Preprint* 1208.3003)

[9] Frebel A 2010 *Astronomische Nachrichten* **331** 474–488 (*Preprint* 1006.2419)

[10] Caffau E, Bonifacio P, François P, Sbordone L, Monaco L, Spite M, Spite F, Ludwig H G, Cayrel R, Zaggia S, Hammer F, Randich S, Molaro P and Hill V 2011 *Nature* **477** 67–69 (*Preprint* 1203.2612)

[11] Hansen C J, Bergemann M, Cescutti G, François P, Arcones A, Karakas A I, Lind K and Chiappini C 2013 *Astronomy Astrophysics* **551** A57

[12] Arcones A and Montes F 2011 *The Astrophysical Journal* **731** 5 (*Preprint* 1007.1275)

[13] Wanajo S, Janka H T and Müller B 2011 *The Astrophysical Journal Letters* **726** L15 (*Preprint* 1009.1000)

[14] Pignatari M, Gallino R, Meynet G, Hirschi R, Herwig F and Wiescher M 2008 *The Astrophysical Journal* **687** L95–L98 (*Preprint* 0810.0182)

[15] Andrievsky S M, Spite F, Korotin S A, François P, Spite M, Bonifacio P, Cayrel R and Hill V 2011 *Astronomy and Astrophysics* **530** A105 (*Preprint* 1104.0476)

[16] Roederer I U 2013 *The Astronomical Journal* **145** 26 (*Preprint* 1211.3427)

Reaction rate of the ^{13}C$(\alpha, n)^{16}$O neutron source using the ANC of the -3 keV resonance measured with the THM

M La Cognata[1], C Spitaleri[1,2], O Trippella[1,3], GG Kiss[1,4], GV Rogachev[5], AM Mukhamedzhanov[6], M Avila[5], GL Guardo[1,2], E Koshchiy[5], A Kuchera[5], L Lamia[2], SMR Puglia[1,2], S Romano[1,2], D Santiago[5], R Spartà[1,2]

[1]Istituto Nazionale di Fisica Nucleare, Laboratori Nazionali del Sud, Catania, Italy
[2]Dipartimento di Fisica e Astronomia, Università di Catania, Catania, Italy
[3]Istituto Nazionale di Fisica Nucleare, Sezione di Perugia, & Dipartimento di Fisica, Università di Perugia, Perugia, Italy
[4]Institute of Nuclear Research (ATOMKI), Debrecen, Hungary
[5]Department of Physics, Florida State University, Tallahassee, Florida, USA
[6]Cyclotron Institute, Texas A&M University, College Station, Texas, USA

E-mail: lacognata@lns.infn.it

Abstract. The *s*-process is responsible of the synthesis of most of the nuclei in the mass range $90 \leq A \leq 208$. It consists in a series of neutron capture reactions on seed nuclei followed by β-decays, since the neutron accretion rate is slower than the β-decay rate. Such small neutron flux is supplied by the ^{13}C$(\alpha, n)^{16}$O reaction. It is active inside the helium-burning shell of asymptotic giant branch stars, at temperatures $\leq 10^8$ K, corresponding to an energy interval of $140 - 230$ keV. In this region, the astrophysical $S(E)$-factor is dominated by the -3 keV sub-threshold resonance due to the 6.356 MeV level in ^{17}O. In this work, we have applied the Trojan Horse Method (THM) to the ^{13}C$(^6$Li$, n^{16}$O$)d$ quasi-free reaction to extract the 6.356 MeV level resonance parameters, in particular the asymptotic normalization coefficient (ANC) $(\tilde{C}_{\alpha^{13}C}^{^{17}O(1/2^+)})^2$. A preliminary analysis of a partial data set has lead to $(\tilde{C}_{\alpha^{13}C}^{^{17}O(1/2^+)})^2 = 6.7^{+0.9}_{-0.6}$ fm^{-1}, slightly larger than the values in the literature. However, the deduced ^{13}C$(\alpha, n)^{16}$O reaction rate is in agreement with most results in the literature at $\sim 10^8$ K, with enhanced accuracy thanks to our innovative approach merging together ANC and THM.

1. Physics case

Nuclear reaction cross sections are among the most important input data for nucleosynthesis calculations and stellar modeling, as nuclear processes power stars supplying the thermal pressure that contrasts the inward force of gravity. Typical stellar temperatures for hydrostatic nuclear burning range between $10^7 - 10^8$ K, corresponding to kinetic energies smaller than 1 MeV [1, 2]. These energies are usually well below Coulomb barriers for interacting nuclei, thus typical cross sections can be as small as few 10^{-12} barn. Indeed, the Coulomb barrier exponentially reduces the cross section σ, rapidly pushing the signal-to-noise to zero. Under these conditions, experimental measurements become exceedingly challenging and experimental uncertainties grow up owing to statistical and systematic errors.

As a consequence, at astrophysical energies (the so-called Gamow window [1, 2]), direct measurements of the cross section are often impossible and extrapolation, supported by nuclear theory such as the R-matrix [3], has been used to determine the cross section. Since it drops exponentially with decreasing energy, large uncertainties might be introduced by the extrapolation procedure. To this purpose, the astrophysical $S(E)$-factor has been introduced [1] to improve the accuracy of the extrapolation procedure:

$$S(E) = E\,\sigma(E)\exp(2\pi\eta)\,, \qquad (1)$$

where $\exp(2\pi\eta)$ is the reciprocal of the Coulomb barrier penetration factor for s-wave and center-of-mass energies much smaller than the Coulomb barrier and η the Sommerfeld parameter. The astrophysical factor has a weaker dependence on the energy than the cross section at astrophysical energies, as Coulomb effects are partially compensated for by the $\exp(2\pi\eta)$ factor, making the uncertainty introduced by extrapolation significantly smaller.

Even in those few cases where reaction cross sections have been measured down to astrophysical energies (check, for instance, [4]), the electron screening effect [1] prevented one to assess the nuclear cross section or, equivalently, the astrophysical factor. In the laboratory, projectile and target are in the form of ions and atoms or molecules, respectively, thus electrons shield the nuclear charges and determine an exponential increase of S(E) for center-of-mass energies $E_{c.m.} \to 0$ [1, 2]. Atomic electron screening is very different from electron screening in stars, where matter is in the form of plasma, making it necessary to divide out the exponential increase of the S-factor to attain the nuclear cross section. However, our current understanding of the electron screening effect is rather incomplete as experimental screening potentials exceed the theoretical upper limits in many cases [5, 4, 6, 7, 8, 9], though this is still a debated topic [10]. Therefore, potential systematic errors might be introduced in the evaluation of the bare-nucleus astrophysical factor, making extrapolation from higher energies, where the electron screening enhancement is negligible, necessary for astrophysical applications

However, nuclear models are not able yet to supply reliable predictions for the trend of the astrophysical factor at low energies. Unaccounted resonances might significantly alter the behavior of the S-factor and spoil present-day astrophysics predictions. In particular, broad sub-threshold resonances can influence the shape of the low-energy astrophysical factor by interfering with other resonances as in the case of the $^{12}C(\alpha,\gamma)^{16}O$ reaction ([2] and references therein), or through their high energy tails. In this last case, the increase of the S(E)-factor as $E \to 0$ competes with the electron screening enhancement, and disentangling the two effects might further increase the total error budget. This is the case, for instance, of the $^{13}C(\alpha,n)^{16}O$ reaction. At low energies, the S-factor is dominated by the contribution of the sub threshold resonance at -3 keV due to the 6.356 MeV level of ^{17}O, having $\Gamma_n = 124 \pm 12$ keV [11]. At the lowest energies, direct data, ending up at ~ 280 keV [12], had to be corrected for atomic electron screening determining an enhancement of less than 20% for the lowest-energy data point.

The $^{13}C(\alpha,n)^{16}O$ cross section has been extensively measured since it is considered as the main neutron source necessary to build up heavy elements from iron-peak seed nuclei through the s-process. In low-mass ($\leq 3M_\odot$) asymptotic giant branch (AGB) stars [13], where the presence of a ^{13}C pocket [14] allows for neutron production through the $^{13}C(\alpha,n)^{16}O$ reaction, $90 \leq A \leq 208$ nuclei are synthesized through a sequence of neutron capture reactions. Because of the relatively low neutron fluxes generated by the $^{13}C(\alpha,n)^{16}O$ reaction (on the order of 10^5 to 10^{11} neutrons per cm^2 per second), the neutron accretion rate is slower than the β-decay rate, thus only heavy isotopes along the stability valley can be produced (s-process, s for slow) [15]. At $0.9\,10^8$ K, a typical temperature characterizing ^{13}C-burning [16], the energy range where the $^{13}C(\alpha,n)^{16}O$ reaction is most effective is $\sim 100 - 270$ keV.

Table 1. Summary of the S-factor values evaluated at $E_{cm} = 100$ keV.

Ref.	$S(100$ keV$)$ $(10^6$ MeVb$)$	Approach
[17]	3.3	R-matrix
[18]	2.7	microscopic cluster approach
[19]	5.3	microscopic cluster approach
[20]	6.3	R-matrix
[21]	1.2	ANC
[22]	3.4	Spectroscopic factor
[23]	2.5	Spectroscopic factor

Measurements of the $^{13}C(\alpha,n)^{16}O$ astrophysical factor include direct and indirect studies, focused on the extraction of the 6.356 MeV ^{17}O level resonance parameters using transfer reactions. Indirect measurements deduced the spectroscopic factor or the asymptotic normalization coefficient (ANC) of this level, as these parameters fix the resonance top value, allowing for the calculation of the $S(E)$-factor beyond the energy region explored by means of direct measurements. Table 1 summarizes the astrophysical S-factors evaluated at 100 keV by different authors as well as the procedures adopted to obtain them. Pellegriti et al. [22], Guo et al. [23] and Johnson et al. [21] established $S(100$ keV$)$ using the spectroscopic factor and the ANC, respectively. The R-matrix calculations reported in [17, 20] are extrapolations of direct measurements to astrophysical energies, the microscopic cluster approach uses nucleon-nucleon interaction to calculate the S-factor. Table 1 clearly shows a large scatter of the $S(E)$-factor values at low energies, suggesting the possible existence of systematic errors causing such indetermination in the $S(100$ keV$)$ parameter.

Therefore, new and improved measurements are necessary to pin down the $^{13}C(\alpha,n)^{16}O$ $S(E)$-factor at astrophysical energies and calculate a reliable reaction rate for astrophysical applications.

2. The THM measurement of the $^{13}C(\alpha,n)^{16}O$ $S(E)$-factor

The Trojan Horse Method (THM) (see [24] for a review of the method) is very suited to investigate the $^{13}C(\alpha,n)^{16}O$ reaction as it allows us to determine the resonance parameters even for sub threshold energies, as in the case of the 6.356 MeV ^{17}O state. In the THM approach, the astrophysical factor of the $^{13}C(\alpha,n)^{16}O$ reaction is obtained by extracting the quasi-free (QF) contribution to the $^{13}C(^6Li,n^{16}O)^2H$ process. In QF kinematics, ^6Li, characterized by a prominent $\alpha \oplus d$ cluster structure, is used to transfer the participant cluster α and feed the excited states of ^{17}O, while the other constituent cluster d is emitted without interacting with ^{17}O, thus behaving as a spectator to the $^{13}C(\alpha,n)^{16}O$ sub-process. Then, the modified R-matrix formula is used to fit the THM data and deduce the resonance parameters [25, 26]. Indeed, the same reduced widths appear in the THM cross section and in the direct data; therefore, those extracted from THM data can be introduced into a standard R-matrix code to establish the trend of the $^{13}C(\alpha,n)^{16}O$ S-factor [27, 28, 29, 30]. In the case of sub threshold resonances, the THM approach can be used to deduce the ANC of such resonances, as in the case of the 6.356 MeV ^{17}O state. This possibility discloses the deep connection between the two indirect methods, which is exhaustively addressed in Refs. [25, 26].

The $^{13}C(^6Li,n^{16}O)^2H$ reaction was measured at he Florida State University Tandem-LINAC facility, which delivered a $E_b = 7.82$ MeV, 1 mm spot ^6Li beam impinging onto 99% ^{13}C enriched foils, whose thicknesses were 53 $\mu g/cm^2$ and 107 $\mu g/cm^2$. A preliminary analysis of the

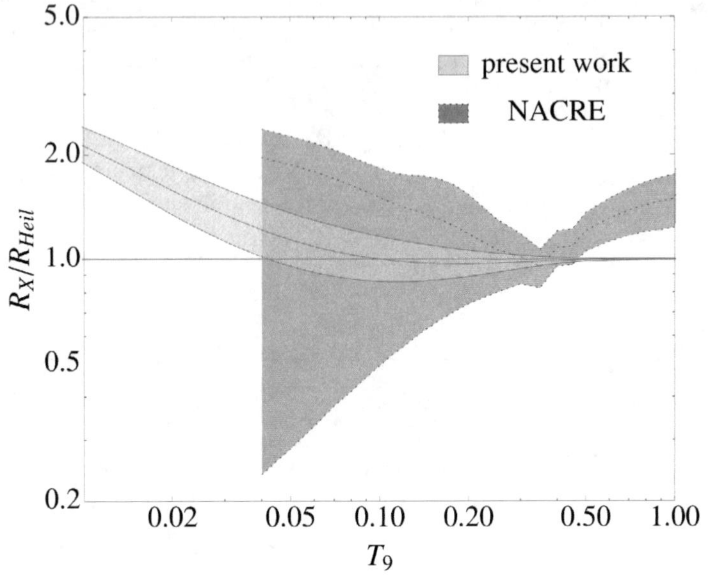

Figure 1. Comparison of the THM reaction rate [25, 26] with other prominent rates in the literature, as a function of the temperature $T_9 = T/10^9$ K. Since reaction rates span several orders of magnitude, their ratio to the Heil et al. one [17] is reported. In detail, in this plot the Heil et al. rate [17] is obviously represented by the horizontal blue line ($R_X/R_{Heil} = 1$). The light red band highlights the R_{THM}/R_{Heil} range allowed by statistical, normalization and analysis uncertainties. Finally, the gray band is used to emphasize the R_{NACRE}/R_{Heil} range [31].

THM experiment, based on the data taken with the thinner target, has lead to an ANC for the 6.356 MeV ^{17}O state $(\tilde{C}^{^{17}O(1/2^+)}_{\alpha^{13}C})^2 = 6.7^{+0.9}_{-0.6}$ fm^{-1} slightly larger than the values in the literature [21, 22, 23], determining a ^{13}C$(\alpha, n)^{16}$O S-factor at 100 keV of $4.0 \pm 0.7 \times 10^6$ MeVb. This result agrees quite well with the largest $S(100\,\text{keV})$ listed in table 1, with an improved accuracy due to a reduced systematic uncertainty (check Refs. [25, 26] for more details). Since the THM cross section is given in arbitrary units, normalization to direct data was accomplished by spanning the $0.5 - 1.2$ MeV energy window in the indirect measurement, an energy region covered by direct data as well. Then, the THM cross section was scaled to match the direct one, which is given in absolute units. Therefore, the resonance parameters deduced from the THM cross section are normalized to those extracted from direct data [28, 25, 26]. The choice of this energy region, where the high energy tail of the -3 keV resonance has a vanishingly small contribution, allows us to reduce the systematic uncertainty connected with the normalization procedure for two reasons, namely, the presence of four resonances in this interval, whose parameters are well known from previous studies [17] and the negligible effect of electron screening.

3. The ^{13}C$(\alpha, n)^{16}$O reaction rate

The reduced widths γ, deduced from the THM data, were used to calculate the astrophysical factor of the ^{13}C$(\alpha, n)^{16}$O reaction [25, 26]. It is worth noting that in this case the R-matrix formula is not used to extrapolate the S-factor but to parameterize it, as the resonance parameters of the 6.356 MeV ^{17}O state were measured in the THM experiment. The reaction rate was then calculated by means of the standard equation [2]:

$$R_{THM} = N_A \sqrt{\frac{8}{\pi\mu(k_BT)^3}} \int_0^\infty e^{-2\pi\eta} S(E) e^{\frac{-E}{k_BT}} dE \tag{2}$$

where N_A is the Avogadro number, μ the reduced mass of the projectile-target system, k_B the Boltzmann constant and T the temperature, and compared with the most recent one [17]. For ease of comparison, the ratio of the THM rate to the Heil et al. one [17] (R_{Heil}) is given in figure 1; in this representation, the Heil et al. rate [17] is displayed by the horizontal blue line ($R_X/R_{Heil} = 1$). The THM recommended rate R_{THM}/R_{Heil} is shown by the red middle line, while the upper and lower red lines are used to highlight the uncertainty range fixed by experimental errors (including statistical, normalization and analysis uncertainties). Similarly, the NACRE [31] recommended rate, upper and lower limits are marked by black dotted lines. While at 0.9×10^8 K a difference of only 1% is found, at 10^7 K the THM reaction rate is two times larger than in Ref. [17], up to a factor of 3 for lower temperatures (figure 1). Moreover, the present reaction rate is affected by a much smaller error (18%) than the NACRE one, namely +17% and -69% [31] and slightly smaller than the Heil et al. one, about 22% [17].

In the light of the recent work by Kimura and Bonasera [32], the results shortly summarized in the present work (more details are given in Refs. [25, 26]) deserve further investigation to understand their astrophysical consequences. As discussed in [32], when a resonance is present in the astrophysical factor the definition of Gamow window has to be modified, as this is deduced by assuming a slowly varying S-factor. In particular, they found that the energy range of astrophysical interest approaches the resonance energy, since the astrophysical factor is larger right at this energy. Therefore, in the $^{13}C(\alpha, n)^{16}O$ reaction it is expected that the energy region where this reaction is most effective in the intershell region of AGB stars draws closer to zero energy, since the $^{13}C(\alpha, n)^{16}O$ astrophysical factor is characterized by a sub threshold resonance. In this region the difference between the THM $S(E)$ and those in the literature is maximum, making it necessary to study the effects on astrophysics. Such work is currently in progress.

References

[1] Rolfs C E and Rodney W S 1988 *Cauldrons in the cosmos* (Chicago, IL: University of Chicago Press)
[2] Iliadis C 2007 *Nuclear Physics of Stars* (Weinheim: Wiley-VCH Verlag)
[3] Lane A M and Thomas R G 1958 *Rev. Mod. Phys.* **30** 257
[4] Aliotta M et al. 2001 *Nuclear Physics* A **690** 790
[5] La Cognata M et al. 2005 *Phys. Rev. C* **72** 065802
[6] Engstler S et al. 1992 *Z. Phys.* A **342** 471
[7] Zahnow D et al. 1997 *Z. Phys.* A **359** 211
[8] Angulo C et al. 1993 *Z. Phys.* A **345** 231
[9] Greife U et al. 1995 *Z. Phys.* A **351** 107
[10] Barker F C 2002 *Nuclear Physics* A **707** 277
[11] Tilley D R, Weller H R and Cheves C M 1993 *Nucl. Phys.* A **564** 1
[12] Drotleff H W et al. 1993 *Astrophys. J.* **414** 735
[13] Meyer B 1994 *Annu. Rev. Astron. Astrophys.* **32** 153
[14] Gallino R et al. 1998 *Astrophys. J.* **497** 388
[15] Käppeler F et al. 2011 *Reviews of Modern Physics* **83** 157
[16] Straniero O et al. 1995 *Astrophys. J.* **440** L85
[17] Heil M et al. 2008 *Phys. Rev.* C **78** 025803
[18] Descouvemont P 1987 *Phys. Rev.* C **36** 2206
[19] Dufour M and Descouvemont P 2005 *Phys. Rev.* C **72** 015801
[20] Hale G M 1997 *Nucl. Phys.* A **621** 177
[21] Johnson E D et al. 2006 *Phys. Rev. Lett.* **97** 192701
[22] Pellegriti M G et al. 2008 *Phys. Rev.* C **77** 042801
[23] Guo B et al. 2012 *Astrophys. J.* **756** 193
[24] Spitaleri C et al. 2011 *Physics of Atomic Nuclei* **74** 1725

[25] La Cognata M et al. 2012 *Phys. Rev. Lett.* **109** 232701
[26] La Cognata M et al. 2013 submitted to the *Astrophys. J.*
[27] La Cognata M et al. 2010 *Astrophys. J.* **723** 1512
[28] La Cognata M et al. 2011 *Astrophys. J.* **739** L54
[29] Mukhamedzhanov A M et al. 2008 *J. Phys. G Nucl. Part. Phys.* **35** 014016
[30] Mukhamedzhanov A M 2011 *Phys. Rev.* C **84** 044616
[31] Angulo C et al. 1999 *Nucl. Phys.* A **656** 3
[32] Kimura S and Bonasera A 2013 *Phys. Rev.* C **87** 058801

Journal of Physics: Conference Series **665** (2016) 012014 doi:10.1088/1742-6596/665/1/012014

Lithium abundances in AGB stars and a new estimate for the ^{7}Be life-time

S Palmerini[1], M Busso[2,6], S Simonucci[3,6], and S Taioli[4,5,6]

[1]Departamento de Fisica Teórica y del Cosmos, Universidad de Granada, Spain
[2]Dipartimento di Fisica, Università degli Studi di Perugia, Italy
[3]Dipartimento di Fisica, Università degli Studi di Camerino, Italy
[4]Interdisciplinary Lab for Computational Science, FBK/CMM , Trento, Italy
[5]Dipartimento di Chimica, Università degli Studi di Bologna, Bologna, Italy
[6]INFN sezione di Perugia, Italy

Abstract. In most cases RGB and AGB stars with M$\leq$ 2M$_\odot$ destroy Li (which is instead synthesized trough electron-captures on ^{7}Be). This occurs through the combined operation of mixing processes and proton captures, when H-burning operates close to the envelope. Observed Li abundances are however difficult to explain, as they cover a wide spread. Various uncertainties affect model attempts, but so far the largest one concerns the processes of bound and free e-captures on ^{7}Be, hence its life-time, whose known estimates are valid only for solar conditions. RGB and AGB stages have temperatures and densities below the envelope covering a wide range and differing from solar by up to a factor of five for T and up to five orders of magnitudes for ρ, hence extrapolations are unreliable. Recently, we presented an estimate of the ^{7}Be half-life based on a fully quantistic method that goes beyond the Debye–Hückel approximation. Here we discuss its consequences on Li nucleosynthesis in low mass AGB stars.

1. Introduction

The cosmological Lithium abundance remains one of the most important challenges for the Nuclear Astrophysics. Studies of Big Bang nucleosynthesis have not yet alleviated the discrepancy between predictions and observations of ^{7}Li abundances [1] and the stellar contributions to Li evolution in Galaxies have not been completely clarified so far.

^{7}Li, which is the most abundant stable isotope of Li, is produced by stellar H-burning via the p-p chain. In this reaction network the ^{3}He(^{4}He, γ)^{7}Be channel provides the synthesis of ^{7}Be that, undergoing electron captures, decays into ^{7}Li. However, the ^{3}He $+^4$ He reaction is efficient at temperatures higher than 4$\cdot$10^{7}K, while proton captures on ^{7}Be and ^{7}Li take place down to much lower temperatures (2$\cdot$10^{7}K and 2-3$\cdot$10^{6}K, respectively). Therefore, both ^{7}Be and ^{7}Li turn out to be easily burnt into two α-particles. Because of its fragility ^{7}Li is destroyed already in pre-Main Sequence stars, where convection carries it at temperatures of few million Kelvins [2], and observations disagree with theoretical expectations also in further evolutionary stages.

In fact, the spectra of low mass Main Sequence stars reveal that Li is depleted during central hydrogen burning, in solar-like objects as well as in hotter Main Sequence stars of the Galactic disc (producing the so-called Li-dip) [3]. Furthermore, evolved stars, which climb the Red Giant Branch (RGB) and the Asymptotic Giant Branch (AGB), show an even more complicate

Journal of Physics: Conference Series **665** (2016) 012014 doi:10.1088/1742-6596/665/1/012014

scenario: the observed Li abundances actually lie in an large spread of values ($-2 \leq$ A(Li) ≤ 3)[1]. In principle, massive red giants (M$> 5-6M_{\odot}$) are recognized to be Li-rich objects (A(Li) ≈ 3[4], because of the Hot Bottom Burning phenomenon [5]. Nevertheless, low mass evolved stars (M$\leq 2M_{\odot}$, where envelope temperatures are too low to modify the surface composition) show Li abundances $-2 \leq$ A(Li) ≤ 2, pretty difficult to be accounted for by canonical stellar models; the known Li-depletion phenomena, which occur during the previous evolutionary stages, do not seem to justify such a wide distribution of Li abundances.

The unexpected abundances of light elements in low mass stars have been interpreted in terms of mixing episodes of a nature different from pure convection [6], often attributed to rotationally-induced effects [7], atomic diffusion, or the action of stellar magnetic field [8–10].

The energy balance of RGB and AGB stars is mainly ensured by the radiative burning of a thin H-shell below the base of the envelope. Since non-convective mixing processes (often called extra-mixing) should link the convection interface with regions where thermonuclear reactions take place, these phenomena can modify the surface abundances of nuclei involved in H-burning (e.g. C, N, O and Al isotopes) [11]; they are therefore expected to further deplete Li as compared to Main Sequence stages [12].

Despite the uncertainties related to the physical mechanisms driving the phenomenon, parametric models of extra-mixing can reproduce the observational constrains, which are instead inexplicable by standard stellar models [13]. Successful examples are the reproduction of C and N isotopic ratios in low mass RGB stars [14] and of the oxygen isotopic mix in meteorite grains of AGB origins ([15] and reference therein). As already mentioned, extra-mixing episodes might account for the observed trend of Li depletion in RGB stars (with M ≤ 2.3M$_{\odot}$), but a limited number (2%) of giants show Li enhancement at the photosphere and Li observations in most AGB stars do not reveal special trends. In general the observational evidence is believed to be the outcome of the coupling between fast mixing and nucleosynthesis. Indeed, in extra-mixing phenomena Li is destroyed by downward streams of materials in competition with upward transport of ^{7}Be from burning regions; therefore, if the mixing is fast enough ^{7}Be can survive proton captures and undergoes only electron captures in the stellar envelope, so that Li is produced.

Such a complicate scenario is affected by an important uncertainty from nuclear physics: the rate of electron capture on ^{7}Be employed in calculations is often used outside its range of validity. It has been actually estimated by [16] (and further updates e.g.[17]) for the physical conditions of the solar core, while in H-burning regions of RGB and AGB stars, and in those crossed by extra-mixing as well, densities (ρ) are lower than in the solar center (from one to five orders of magnitudes) and temperatures (T) span a range from 70-80 MK down to a few MK.

2. The new estimate for the $^7\mathrm{Be} + \mathrm{e}^-$ rate in stellar plasmas

To correctly estimate ^{7}Be life-time in a plasma it is crucial to determine the contributions of electrons from both bound and continuum states and the total-to-continuum capture ratio. In works devoted to study the $^7\mathrm{Be}(\mathrm{e}^-, \nu_\mathrm{e})^7\mathrm{Li}$ reaction, a Maxwellian distribution of free electron around a Be nucleus is usually assumed (Debye–Hückel approach [18]). This approximation is suitable to describe systems with a large number of electrons and high temperatures, as it is based on the statistical consideration of smooth changes of the potential over a characteristic distance (λ_D), which is larger than the thermal De-Broglie wave-length of the electrons (λ_{DB}).

However, the conditions are at the limits of validity of the Maxwell-Boltzmann gas approximation, already for our Sun. At T$= 1.6 \cdot 10^7$K and $\rho = 150$g/cm^3, $\lambda_D = 0.407$ and $\lambda_{DB} = 0.352$ a.u.. The situation worsens at physical conditions of the H-burning shell in Red Giants, where λ_D might be smaller than λ_{DB}.

[1] Spectroscopic notation $\mathrm{A(Li)} \equiv \log \epsilon(\mathrm{Li}) = \log \frac{\mathrm{N(Li)}}{\mathrm{N(H)}} + 12$

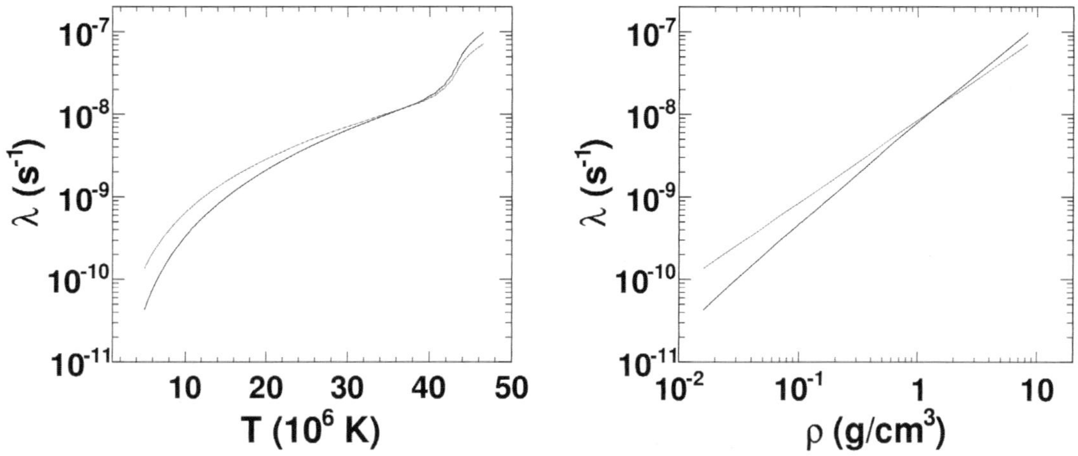

Figure 1. A comparison between the decay rate (λ) of ^{7}Be in the layers between the convective envelope and the H-burning shell during the fourth inter-pulse period of a $2M_\odot$ AGB star, adopting data by [19] (red line) and the one extrapolated by [17] (black line). λ is plotted as a function of T and ρ (left and right panel, respectively). Notice that although λ is shown as a function of T and ρ separately, it actually depends on both these variables in each layer of the stellar structure.

Simonucci et al. 2013 [19] explore the problem of electron captures on ^{7}Be in stellar environments introducing a new treatment, in which electrons are considered as a Fermi gas and a mean-field adiabatic approximation to the scattering process is used. This is justified by the fact that the Debye–Hückel approach can be derived as a two-step approximation of the more general Thomas-Fermi model, by using the linearized Maxwell-Boltzmann distribution. Furthermore, the two-body (electron-^{7}Be) system is extended by the authors to a many-body interaction scheme by including the Hartree–Focks approximation. In this way electron captures might be computed over a much wider range of T and ρ, with a more precise estimate of the contributions from both bound and free electronic states to the capture. We refer to the paper by [19] for further details about the electron density modeling and the calculation of the ^{7}Be half-life. Here we want to underline that the results of the quoted paper agree well with the previous findings for the Sun (so that no changes to the solar neutrino problem are introduced).

Figure 1 reports a comparison between the ^{7}Be(e$^-$, ν_e)^{7}Li rate computed by [19] and that obtaing using the analytical expression by [17] in the radiative region of a $2M_\odot$ AGB star of solar composition (as a function of ρ and T). In Fig.1, it is evident that the difference between the two rates of electron capture is larger at lower T and ρ. This portion of the parameter space is very important for Li nucleosynthesis, as only in these conditions does the efficiency of proton captures on ^{7}Be and ^{7}Li decrease sufficiently, so that extra-mixing phenomena can affect the Li abundance in the stellar envelope: this will be a production or a dilution, depending on the mixing rate [12].

3. Lithium abundances in AGB stars

We now analyze the Li nucleosynthesis in AGB stars employing the new estimate for the ^{7}Be + e$^-$ reaction rate. Our reference models for this scope are the ones presented by [12] to discuss the effects of extra-mixing phenomena on Li nucleosynthesis. In the quoted paper, the mixing velocity is related to the mixing rate trough the mass conservation law, thus the extra-mixing

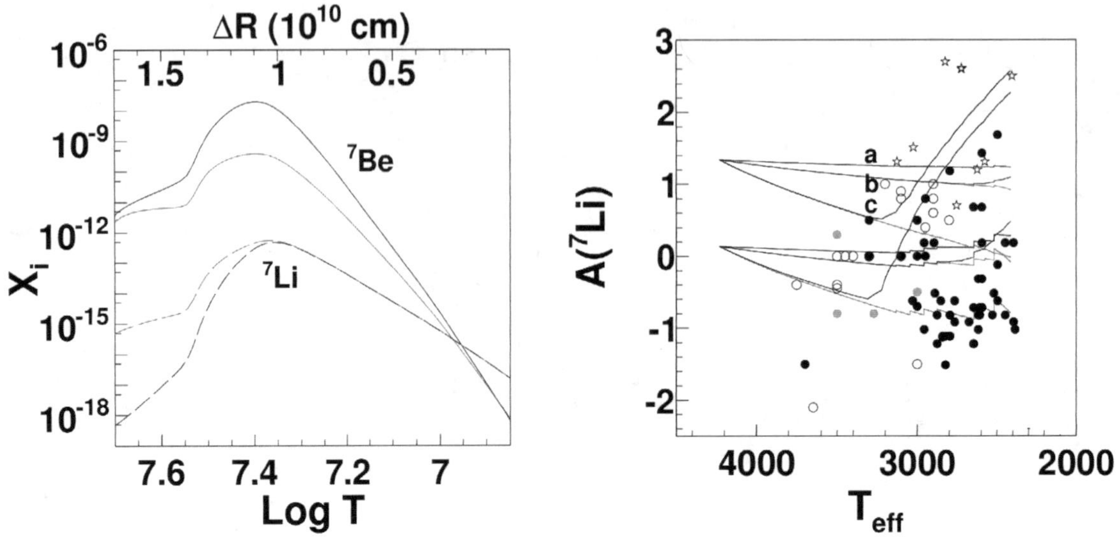

Figure 2. Left panel: comparison between the equilibrium abundances of ^{7}Be and ^{7}Li achieved in the layers above the H-burning shell, adopting the ^{7}Be lifetime by [19] (red line) and that extrapolated by [17] (black line), in the stellar model considered in Fig. 1. On the right side the position $\Delta R = 0$ represents the base of the envelope, while on the left there is the region where the maximum energy is released from H-burning. The matter density is also shown (blue line) and refers to the scale on the right axis. Right panel: curves refer to calculations of Li abundance evolution in the envelope of a $2M_\odot$ AGB stars of solar compositions: the black line refers to the ^{7}Be life-time from [17] and the red the one to that from [19]. Model curves are characterized by the following extra-mixing parameters: $\Delta T = 0.22$, $\dot{M}_6 = 0.1$(a), 0.3(b), 1(c). The Li abundances observed in O- and C-rich stars on the AGB are also reported. Open, grey and black dots refer to S, SC and C(N) stars, respectively, while open stars show the CJ stars (see [12] and references therein).

models that reach the highest ^{7}Li abundances are those with the largest mixing efficiency $\dot{M}$[2] (black line in the right panel of figure 2). However, another phenomenon contributes to modify the Li abundance in the envelopes of AGB stars: this is the third dredge-up (TDU), which is the deep convective mixing occurring at the end of each recurrent thermal instability, after the enrgy release accompanying the onset of the He burning. In red giant stars, a sort of ^{7}Be-pocket (left panel of figure 2) forms in the H rich zone between the envelope and the H-burning shell. During the TDU such a reservoir is brought by the convection to the surface where it undergoes electron captures enriching the outermost stellar layers in ^{7}Li The ensuing abundance can afterwards be destroyed or further increased by extra-mixing at different rates. The competition between TDU and extra-mixing can determine the large spread of Li abundances observed in low mass AGB stars. The left panel of figure 2 shows the abundances of ^{7}Be and ^{7}Li in the radiative layers below the convective envelope, during the H-burning phase of an AGB star (the model star is the same $2M_\odot$, solar composition case discussed in [12]). The red lines illustrate the amounts of Li and Be obtained by using the new rate by [19] in calculations, while the black line shows the results with the rate by [17]. The variations shown in the abundances of ^{7}Be and ^{7}Li are entirely due to the different choices adopted for the rate of electron capture on ^{7}Be.

[2] Extra-mixing mechanisms are generally parameterized and depend on two main parameters; in the approach by [12, 15] these parameters are the mass circulation rate (in units of $10^{-6}M_\odot$/yr) induced by the transport processes ($\dot{M}_6$) and the maximum depth ΔT they reach.

The same nuclear physics inputs and stellar structure have been adopted to calculate the evolution of surface Li abundances along the whole AGB evolutionary stage (from the early phase up to the last thermal pulse) due to the interplay of extra-mixing and TDU. Results, presented in the right panel of figure 2, are achieved from 2 different initial values of the Li abundance, chosen inside the observed spread left by RGB stages. The black curves show the evolution of Li abundances in the stellar envelope induced by 3 cases of extra-mixing (with parameters $\Delta T = 0.22$ and $\dot{M}_6 = 0.1, 0.3, 1$) and using the estimate of ^{7}Be half-life extrapolated by [17]. Results of the same calculation adopting the new Hartree–Focks estimate are instead represented by red curves. The curves showing the Li depletion during the early-AGB are almost insensitive to the rate for e-capture adopted. On the contrary, large changes emerge at the on-set of the thermal pulses (and thus of TDU).

The reason for the above dichotomy can be understood with reference to the left panel of figure 2, where one can see how the ^{7}Be reservoir in the stellar radiative region computed with the new decay rate is smaller than in the previous scenario (at least by a factor of 10). Moreover, the reduction of the ^{7}Be lifetime is strong enough to allow its decay into Li before it is saved to the envelope, even when a fast extra-mixing is at play. As a consequence, Li production becomes impossible in low mass stars with the new $^7\mathrm{Be} + \mathrm{e}^-$ rate, being the destruction of Li in the envelope not compensated. As a consequence, the highest observed Li abundances (typical of CJ stars) cannot be reached.

This finding is now in line with other indications provided by various chemical anomalies of the very peculiar CJ stars, which jointly suggest that these last are not actually normal C-stars affected by deep mixing, but rather the outcome of a more complex evolution, perhaps deriving originally from a binary system. On the other hand, the extra-mixing calculations using the new rate for electron captures on ^{7}Be better reproduce the abundances shown by normal C(N)-stars(solid symbols), whose chemical composition and photometry reveal instead that they do represent a normal family of C-rich AGB stars.

S.P. is grateful to Spanish Grant AYA2011-22460 for the post-doc contract during which this research was carried out.

References

[1] Coc A 2012 *ApJ* **10** 87
[2] Sestito P, Degl'Innocenti S, Prada Moroni P G and Randich S 2006 *A&A* **454** 311
[3] Boesgaard A M, Deliyannis C P, Stephens A and Lambert D L 1998 *ApJ* **492** 727
[4] D'Antona F and Ventura P 2010 *in Light elements in the Universe, IAU Symposyum 268* 395
[5] Lattanzio J C, Frost C A, Cannon R C and Wood P R 1997 *Nuclear Physics A* **621** 435
[6] Michaud G 1986 *ApJ* **302** 650
[7] Charbonnel C and Lagarde N 2010 *A&A* **522** 10
[8] Eggenberger P, Meynet G, Maeder A and et al 2010 *A&A* **519** 116
[9] Busso M, Wasserburg G, Nollett K and Calandra A 2007 *ApJ* **671** 802
[10] Denissenkov P, Pinsonneault M and MacGregor K 2009 *ApJ* **696** 1823
[11] Sackmann I J and Boothroyd A I 1999 *ApJ* **510** 217
[12] Palmerini S, Cristallo S, Busso M, Abia C, Uttenthaler S, Gialanella L and Maiorca E 2011 *ApJ* **741** 26
[13] Nollett K M, Busso M and Wasserburg G J 2003 *ApJ* **582** 1036
[14] Abia C, Palmerini S, Busso M and Cristallo S 2012 *A&A* **548** 55
[15] Palmerini S, La Cognata M, Cristallo S and Busso M 2011 *ApJ* **729** 3
[16] Bahcall J 1962 *Phys. Rev.* **126** 1143
[17] Adelberger E and et al 2011 *Rev. Mod. Phys.* **83** 195
[18] Debye P and Hückel E 1923 *Phys. Z.* **24** 185
[19] Simonucci S, Taioli S, Palmerini S and Busso M 2013 *Apj* **764** 118

Dark matter constraints from stellar evolution

A Ayala [1], **I Domínguez** [1], **O Straniero** [2]

[1] Departamento de Física teórica y del cosmos, Granada University, Fuentenueva Campus sn, 18071, Granada, Spain

[2] Instituto Nazionale di Astrofisica, Osservatorio Astronomico di Collurania, Via Mentore Maggini s.n.c. 64100, Teramo (Italy)

E-mail: aayala@ugr.es

Abstract. The study of dark matter constraints from its effect on star evolution has been discussed in recent years. We propose a star evolution simulation approach to determine those costraints from properties related to star evolutionary stages and propose globular cluster observables in order to check those constraints. My work in progress (my PhD project research) employs FRANEC code to simulate complete star evolution from pre-main sequence to AGB phase, and regards several DM candidates like axions or WIMPs, motivated by different unsolved physical problems. Detailed energy production or energy loss due to DM particles are included, taking into account the expected interaction between dark matter particles and stellar plasma within different models.

1. Introduction

Dark matter (DM) remains as one of the most intriguing problems of modern physics. We actually don't know 80% of the mass in the universe. Different standard model or "beyond standard model" particles have been proposed as contituents of DM. Direct and indirect experiments are looking for evidences of these particles. We propose a complementary approach to the field, addressed, among other, by Raffelt [1]: the study of dark matter production, trapping or annhilation inside stars by means of computer simulations. Constraints to dark matter particles within different theoretical frames can be obtained from accurate enough simulations, which might even detect new effects, and star cluster population surveys.

We focus on axions and WIMPs. Theoretical motivations and features for both particles as DM particles are well established. We summarize them briefly.

1.1. Axions

- Proposed by Wilczek in the late seventies, axions are needed to explain the small value of the CP-violating terms in QCD lagrangian, $\theta\,(\alpha_s/8\pi)\,G\tilde{G}$. They are actually the quantum of a new field called Pecci-Quinn field, which practically makes the CP-violating term vanish and precludes its rising to higher values. So, axions appear by means of a spontaneous symmetry breaking mechanism. The additional term in the lagrangian includes axion scalar field, a, referred to a certain energy scale, a/f_a.

$$\mathcal{L}_\theta = (\bar{\theta} - a/f_a)\,(\alpha_s/8\pi)\,G\tilde{G} \tag{1}$$

- They are scalar zero-spin massive particles. Good properties to consider them cold dark matter candidates.
- They interact with photons, hadrons and charged leptons in some models like DFSZ and only with hadrons in hadronic models [6]. Compton effect and bremsstrahlung processes are important in first model. The axion coupling to two photons, "Primakoff" effect is present in both models.

1.2. WIMPs

- Weakly Interacting Massive Particles in the range of GeV. Related in some cases to supersymmetry theory.
- They can be trapped inside stars close to halo region, and depending on the density, modify normal stellar evolution. So they can annhilate and produce energy, becoming an additional energy source for the star [2].

2. Expected effect on stellar evolution

The effect of axion and WIMPs particles on stellar evolution depends on the energy carried away or introduced by them. Some of the expected effects are the following.

2.1. Axion effects

- In stars in the main sequence, it is expected that nuclear reactions will be enhanced by axion energy losses, because hydrostatic equilibrium implies that additional energy loss by axion must be balanced with a higher energy production inside stars [1].
- Inside red giant degenerate cores, energy losses due to axions produce a more efficient cooling that can increase the He core mass at the He ignition. Therefore red giant luminosity at tip phase should be greater [3] and, in principle, it could modify globular cluster isochrones.
- AGB thermal pulse period could be increased if axion emission is taken into account [3].
- Time of neutrino emission in supernovae should behave as a function of axion-nucleon coupling constant [1]. It decreases up to a certain coupling constant value and then raises again with coupling constant values.
- White dwarf cooling decreases, so number of WD in a bolometric magnitude range is reduced [4].

2.2. WIMPS effects

- WIMPs might release energy inside stars, because of annihilation process, and increase the expected time a star remains in the main sequence. Globular cluster isochrones are brighter in some regions when WIMPs processes are considered [2] [5].
- The smaller a star mass is, the greater the aforementioned effect becomes [2].

3. Approach and method

Our idea is to implement an energy loss code in FRANEC (Frascati Newton Raphson Evolutionary Code) program. Different DM particles will be studied. We begin with axions. In that case star additional energy loss is due mainly to axion bremsstrahlung, Compton and Primakoff processes. These processes depend strongly on stellar plasma properties, like temperature and density. Energy carried away by dark matter particles should be able to modify different stages as explained in previous section. The FRANEC code solves (employing Henyey method) a coupled differential equation system, which drives star evolution:

$$\frac{dP}{dr} = -\frac{GM(r)}{4\pi r^4} \tag{2}$$

$$\frac{dr}{dM(r)} = \frac{1}{4\pi r^2 \rho} \tag{3}$$

$$\frac{dL}{dM(r)} = \epsilon_g + \epsilon_n + \epsilon_\nu \tag{4}$$

$$\frac{dT}{dM(r)} = -\nabla \frac{GM(r)}{4\pi r^4} \frac{T}{P} \tag{5}$$

$$\frac{\partial X_i}{\partial t} = \frac{m_i}{\rho} \left(\sum_j r_{ji} - \sum_k r_{ik} \right) \tag{6}$$

Energy loss is driven by 4, where energy released by gravity and nuclear reactions are ϵ_g and ϵ_n and energy carried away by neutrinos is ϵ_ν. This equation can be modified including energy loss due to axions. $\frac{dL}{dM(r)} = \epsilon_g + \epsilon_n + \epsilon_\nu + \epsilon_{axion}$. That means energy loss carried by axion has an influence on stellar evolution and should be detected in our simulations, accelerating, or delaying evolutionary phases. Moreover, we can take into account WIMPS and modelize a star evolution affected by both, energy released by WIMPs annihilation or WIMPs transport of energy inside a star, by adding suitable terms in 4.

4. Observables

The observables we are looking for in order to check our simulations and finally constrain DM properties are mainly related to globular clusters (GC) and high precision spectroscopy, this last focused on nucleosynthesis in certain kind of stars, like AGB. We plan review several astrophysical catalogues and, perhaps, take advantage of new data acquired by different missions like GAIA. On the other hand, it's important to consider all uncertainties (experimental and due to theoretical bias) in order to achieve reliable constraints.

5. References

[1] Raffelt G 1990 *Physics Reports* **198** 1–113
[2] Casanellas J and Lopes I 2011 *ApJ Letters* **733** L51
[3] Domínguez I, Straniero O and Isern J 1999 *Monthly Notices of the Royal Astronomical Society* **306** L1–L7
[4] Corsico AH, Althaus LG, Miller-Bertolami MM, Romero AD, García-Berro E, Isern J and Kepler SO 2012 *Monthly Notices of the Royal Astronomical Society* **424** 2792–99
[5] Scott P, Fairbairn M and Edsjö J 2008 *Identification of Dark Matter 2008* (*Preprint* 0810.5560)
[6] Raffelt, G. G. and Dearborn, D. S. P. 1987 *Physical Review D* **36** 2211-2225

Constraints on asymmetric dark matter from asteroseismology

J Casanellas[1], **I Lopes**[2]

[1] Max Planck Institute for Gravitational Physics, Albert Einstein Institute, Germany
[2] CENTRA, Instituto Superior Técnico, Universidade de Lisboa, Portugal

E-mail: jordicasanellas@ist.utl.pt ; ilidio.lopes@ist.utl.pt

Abstract. We report recent results on the impact of asymmetric dark matter (DM) particles on low-mass stars. First, we found that the small convective core expected in stars with masses between 1.1 and 1.3 $M_\odot$ is suppressed due to DM cooling. Moreover, stars with masses below 1 $M_\odot$ have their central temperatures and densities more strongly influenced by DM than in the solar case. We were able to put limits to the DM mass and spin-dependent DM-proton scattering cross section by comparing the modelling of the nearby star α Cen B with photometric, spectroscopic and asteroseismic observations.

1. Introduction

If the dark matter (DM) of the Universe is made of weakly interacting massive particles (WIMPs) their accumulation inside stars may lead to observable effects that depend on the unknown characteristics of the dark particles [1–16]. Thus, the DM parameter space can be constrained using this approach, in particular in the case of models of DM particles that do not annihilate in stellar interiors. A good example is asymmetric DM [17–28], a DM model motivated to explain the similar cosmic abundance of dark and baryonic matter and that naturally predicts a DM mass of the order of the few GeV (see for instance the recent reviews [29, 30]).

A fraction of the WIMPs in the galactic halo may scatter with nucleons of the stellar plasma and become gravitationally captured by stars [31–33]. The efficiency of this process increases with the stellar density (among other factors), so compact stars can be in principle strongly influenced [34–38]. The destruction of neutron stars by black hole formation in their interior has been predicted, and limits on the properties of asymmetric DM have been derived using this approach [39–43].

In the case of the Sun and solar-type stars, the conduction of DM particles in the stellar interior acts as a new energy transport mechanism, cooling the center of the stars [44–50]. The impact on the properties of these stars tends to be less significant than for compact stars, but on the other hand the physics of the interior of solar-type stars is much better known and observations are much more precise, in particular with the complementary diagnostic of the solar interior provided by solar neutrino measurements [51] and helioseismology [52].

Table 1. Summary of observational constraints on the modelling and selected results.

Star	M ($M_\odot$)	L ($L_\odot$)	T_{eff} (K)	$(Z/X)_s$	$\langle\Delta\nu_{n,0}\rangle^a$ (μHz)	$\langle\delta\nu_{02}\rangle^a$ (μHz)
KIC 8006161						
Observations [b]	0.92-1.10	0.61 ± 0.02	5340 ± 70	0.043 ± 0.007	148.94 ± 0.13	10.10 ± 0.16
Stand. model.	0.92	0.63	5379	0.039	149.03	10.12
DM model.[c]	0.92	0.63	5379	0.039	149.08	9.13
HD 52265						
Observations [b]	1.18-1.25	2.09 ± 0.24	6100 ± 60	0.028 ± 0.003	98.07 ± 0.19	8.18 ± 0.28
Stand. model.	1.18	2.22	6170	0.028	97.92	8.16
DM model.[c]	1.18	2.22	6170	0.028	98.05	7.65
α Cen B						
Observations [b]	0.934 ± 0.006	0.50 ± 0.02	5260 ± 50	0.032 ± 0.002	161.85 ± 0.74	10.94 ± 0.84
Stand. model.	0.934	0.51	5245	0.031	162.56	10.23
DM model.[c]	0.934	0.51	5230	0.031	162.45	8.95

[a] The averages are calculated for the intervals $2750 < \nu(\mu\text{Hz}) < 3900$ for KIC 8006161, $1600 < \nu(\mu\text{Hz}) < 2600$ for HD 52265, and $3300 < \nu(\mu\text{Hz}) < 5500$ for α Cen B.
[b]Data from [55] and [56] for KIC 8006161, [57] for HD 52265, and [58] for α Cen B.
[c] $m_\chi = 5$ GeV, $\sigma_{\chi,SD} = 3 \cdot 10^{-36}$ cm^2, $\rho_\chi = 0.4$ GeV cm^{-3}.

In these proceedings we report the study of the impact of asymmetric DM on stars with masses similar to that of the Sun [53], stars embedded in halos of DM with a density equal to that estimated for the solar neighbourhood, $\rho_\chi = 0.4$ GeV cm^{-3} [54]. We focus on stars whose acoustic oscillations have already been precisely identified. The characteristics of these stars are shown in Table 1. Our approach is motivated mainly by the following reasons: i) stars with lower masses are more strongly influenced by DM, ii) stars with a small convective core (1.1-1.3 $M_\odot$ stars) can have their dominant energy transport mechanisms altered due to DM, and iii) asteroseismology (the study of the stellar oscillations) is a technique sensitive to properties of the stellar cores, where the DM impact occurs.

As we will show, we were able to put the first limits to the nature of DM using an asteroseismic analysis of α Cen B. The reader is addressed to reference [53] for a more thorough description of the methods used and the results obtained.

2. Results

2.1. Impact on the central temperatures and densities. Suppression of core convection

We modelled the star KIC 8006161, reproducing all its observed characteristics, including the large and small frequency separations as measured by the *Kepler* mission. We found a $\sim 10\%$ decline in its central temperature when the influence of the accumulation of asymmetric DM particles with $m_\chi = 5$ GeV and $\sigma_{\chi,SD} = 3 \cdot 10^{-36}$ cm^2 was taken into account.

Furthermore, we also studied the star HD 52265, a 1.2 $M_\odot$ star whose oscillation frequencies have been measured by the *CoRoT* mission. We found that, whereas in the standard modelling this star is expected to have a convective core, the inclusion of the DM energy transport removes this core and the star is left with a radiative interior. Specific combinations of frequencies of low-degree modes have been shown to be sensitive to the presence of small convective cores (see Ref. [59]). We leave this interesting diagnostic approach for a further work.

Journal of Physics: Conference Series **665** (2016) 012016 doi:10.1088/1742-6596/665/1/012016

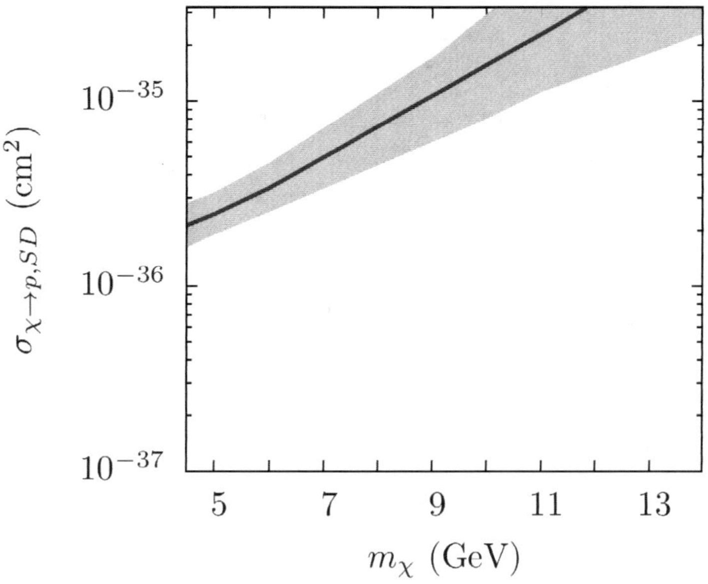

Figure 1. Upper limits for the WIMP-proton spin-dependent scattering cross section as a function of the WIMP mass from an asteroseismic analysis of the star α Cen B. Asymmetric DM particles with properties above the blue line produce a strong impact on the core of the star, leading to a mean small frequency separation more than 2 σ away from the observations. The filled region shows the uncertainty in the modelling when the observational errors are taken into account. A density of $\rho_\chi = 0.4$ GeV cm^{-3} was assumed. Figure adapted from Ref. [53].

2.2. Constraints on the DM properties

We have chosen to study the star α Cen B because it is close to the Sun and belongs to a binary system, which has allowed a very accurate measurement of its mass and other characteristics (see Table 1). This fact strongly reduces the uncertainties in the modelling of the star. We used the stellar oscillations, in particular the mean small separation between the modes of low degree $\langle \delta\nu_{02} \rangle$, to identify the modifications introduced by asymmetric DM in the core of the star.

We found that the stellar models strongly influenced by the DM cooling mechanism, while reproducing all the global properties of α Cen B within observational errors, cannot account for the observed small frequency separations. The existence of asymmetric DM particles with m_χ and $\sigma_{\chi,SD}$ above the blue line in Figure 1 can be excluded because they lead to models of α Cen B with a $\langle \delta\nu_{02} \rangle$ more than 2 σ away from the observed value. The 1 σ variation on $\langle \delta\nu_{02} \rangle$ when the classical stellar parameters (M, L, T_{eff} and $(Z/X)_s$) vary within the observational errors is plotted as a filled region around the blue line. Further uncertainties not included in this analysis are those in other parameters of stellar modelling, such as nuclear reaction rates and chemical abundances, with an expected minor impact on our diagnostic and, more importantly, the error in the determination of the local DM density, which will introduce an estimated uncertainty of a factor smaller than 2 in $\sigma_{\chi,SD}$ [54, 60].

Our method provides a competitive exclusion plot on $\sigma_{\chi,SD}$ for low-mass DM particles. These constraints are particularly valuable taking into account the present controversy among different direct detection experiments in the same region of the DM parameter space [61].

References

[1] Spergel D N and Press W H 1985 *ApJ* **294** 663–673
[2] Lopes I P, Bertone G and Silk J 2002 *Mon.Not.Roy.Astron.Soc.* **337** 1179–1184 (*Preprint* astro-ph/0205066)
[3] Spolyar D, Freese K and Gondolo P 2008 *Phys.Rev.Lett.* **100** 051101 (*Preprint* 0705.0521)
[4] Iocco F 2008 *Astrophys.J.* **677** L1–L4 (*Preprint* 0802.0941)
[5] Scott P, Fairbairn M and Edsjo J 2009 *Mon.Not.Roy.Astron.Soc.* **394** 82 (*Preprint* 0809.1871)
[6] Casanellas J and Lopes I 2009 *Astrophys.J.* **705** 135–143 (*Preprint* 0909.1971)
[7] Gondolo P, Huh J H, Kim H D and Scopel S 2010 *JCAP* **1007** 026 (*Preprint* 1004.1258)
[8] Zackrisson E, Scott P, Rydberg C E *et al.* 2010 *Astrophys.J.* **717** 257–267 (*Preprint* 1002.3368)
[9] Sivertsson S and Gondolo P 2011 *Astrophys.J.* **729** 51 (*Preprint* 1006.0025)
[10] Casanellas J and Lopes I 2011 *Mon.Not.Roy.Astron.Soc.* **410** 535–540 (*Preprint* 1008.0646)
[11] Scott P, Venkatesan A, Roebber E *et al.* 2011 *Astrophys.J.* **742** 129 (*Preprint* 1107.1714)
[12] Casanellas J and Lopes I 2011 *Astrophys.J.* **733** L51 (*Preprint* 1104.5465)
[13] Zentner A R and Hearin A P 2011 *Phys.Rev.* **D84** 101302 (*Preprint* 1110.5919)
[14] Ilie C, Freese K, Valluri M, Iliev I T and Shapiro P R 2012 *Mon.Not.Roy.Astron.Soc.* **422** 2164–2186
[15] Li X, Wang F and Cheng K 2012 *JCAP* **1210** 031 (*Preprint* 1210.1748)
[16] Leung S C, Chu M C, Lin L M and Wong K W 2013 *Phys. Rev. D* **87** 123506 (*Preprint* 1305.6142)
[17] Griest K and Seckel D 1987 *Nuclear Physics B* **283** 681–705
[18] Khlopov M Y and Kouvaris C 2008 *Phys. Rev. D* **78** 065040 (*Preprint* 0806.1191)
[19] Davoudiasl H, Morrissey D E, Sigurdson K and Tulin S 2011 *Phys.Rev.* **D84** 096008 (*Preprint* 1106.4320)
[20] Davoudiasl H and Mohapatra R N 2012 *New Journal of Physics* **14** 095011 (*Preprint* 1203.1247)
[21] Cirelli M, Panci P, Servant G and Zaharijas G 2012 *JCAP* **1203** 015 (*Preprint* 1110.3809)
[22] Tulin S, Yu H B and Zurek K M 2012 *JCAP* **1205** 013 (*Preprint* 1202.0283)
[23] Blennow M, Fernandez Martinez E, Mena O, Redondo J and Serra e P 2012 *JCAP* **7** 022 (*Preprint* 1203.5803)
[24] March-Russell J, Unwin J and West S M 2012 (*Preprint* 1203.4854)
[25] Gu P H 2013 *Nucl.Phys.* **B872** 38–61 (*Preprint* 1209.4579)
[26] Pearce L and Kusenko A 2013 *Phys. Rev. D 87,* **123531** (*Preprint* 1303.7294)
[27] Bhattacherjee B, Matsumoto S, Mukhopadhyay S and Nojiri M M 2013 *ArXiv e-prints* (*Preprint* 1306.5878)
[28] McCullough M and Randall L 2013 (*Preprint* 1307.4095)
[29] Petraki K and Volkas R R 2013 (*Preprint* 1305.4939)
[30] Zurek K M 2013 (*Preprint* 1308.0338)
[31] Gould A 1987 *ApJ* **321** 571–585 URL http://dx.doi.org/10.1086/165653
[32] Gondolo P, Edsjö J, Ullio P *et al.* 2004 *JCAP* **7** 008 (*Preprint* arXiv:astro-ph/0406204)
[33] Lopes I, Casanellas J and Eugenio D 2011 *Phys.Rev.* **D83** 063521 (*Preprint* 1102.2907)
[34] Moskalenko I V and Wai L L 2007 *Astrophys.J.* **659** L29–L32 (*Preprint* astro-ph/0702654)
[35] Bertone G and Fairbairn M 2008 *Phys.Rev.* **D77** 043515 (*Preprint* 0709.1485)
[36] McCullough M and Fairbairn M 2010 *Phys.Rev.* **D81** 083520 (*Preprint* 1001.2737)
[37] Fan Y Z, Yang R Z and Chang J 2011 *Phys. Rev. D* **84** 103510 (*Preprint* 1110.2819)
[38] Bramante J, Fukushima K and Kumar J 2013 *Phys.Rev.* **D87** 055012 (*Preprint* 1301.0036)
[39] Kouvaris C and Tinyakov P 2011 *Phys.Rev.* **D83** 083512 (*Preprint* 1012.2039)
[40] Kouvaris C and Tinyakov P 2011 *Phys.Rev.Lett.* **107** 091301 (*Preprint* 1104.0382)
[41] Goldman I, Mohapatra R, Nussinov S, Rosenbaum D and Teplitz V 2013 (*Preprint* 1305.6908)
[42] Kouvaris C and Tinyakov P 2013 *Phys. Rev. D* **87** 123537 (*Preprint* 1212.4075)
[43] Bell N F, Melatos A and Petraki K 2013 *Phys. Rev. D* **87** 123507 (*Preprint* 1301.6811)
[44] Bottino A, Fiorentini G, Fornengo N *et al.* 2002 *Phys.Rev.* **D66** 053005 (*Preprint* hep-ph/0206211)
[45] Taoso M, Iocco F, Meynet G, Bertone G and Eggenberger P 2010 *Phys.Rev.D* **82** (*Preprint* 1005.5711)
[46] Cumberbatch D T, Guzik J, Silk J *et al.* 2010 *Phys.Rev.* **D82** 103503 (*Preprint* 1005.5102)
[47] Lopes I and Silk J 2012 *Astrophys.J.* **757** 130 (*Preprint* 1209.3631)
[48] Turck-Chièze S, García R A, Lopes I, Ballot J, Couvidat S *et al.* 2012 *Astrophys. J. Lett.* **746** L12
[49] Lopes I and Silk J 2012 *Astrophys.J.* **752** 129
[50] Iocco F, Taoso M, Leclercq F and Meynet G 2012 *Phys.Rev.Lett.* **108** 061301 (*Preprint* 1201.5387)
[51] Lopes I and Silk J 2010 *Science* **330** 462–
[52] Lopes I and Silk J 2010 *Astrophys.J.* **722** L95 (*Preprint* 1009.5122)
[53] Casanellas J and Lopes I 2013 *Astrophys.J.* **L21** (*Preprint* 1212.2985)
[54] Garbari S, Liu C *et al.* 2012 *Mon. Not. Roy. Astron. Soc.* **425** 1445–1458 (*Preprint* 1206.0015)
[55] Mathur S, Metcalfe T, Woitaszek M *et al.* 2012 *Astrophys.J.* **749** 152 (*Preprint* 1202.2844)
[56] Bruntt H, Basu S, Smalley B *et al.* 2012 *Mon. Not. Roy. Astron. Soc.* **423** 122–131 (*Preprint* 1203.0611)
[57] Ballot J, Gizon L, Samadi R *et al.* 2011 *Astron.Astrophys.* **530** A97 (*Preprint* 1105.3551)
[58] Kjeldsen H, Bedding T R, Butler R P *et al.* 2005 *Astrophys.J.* **635** 1281–1290 (*Preprint* astro-ph/0508609)
[59] Cunha M S and Brandão I M 2011 *Astronomy and Astrophysics* **529** A10 (*Preprint* 1008.1299)
[60] Pato M, Agertz O, Bertone G, Moore B and Teyssier R 2010 *Phys.Rev.* **D82** 023531 (*Preprint* 1006.1322)
[61] Buckley M R and Lippincott W H 2013 (*Preprint* 1306.2349)

Towards a study of ^{22}Ne(p,γ)^{23}Na at LUNA

R Depalo for the LUNA collaboration
Università degli Studi di Padova and INFN Padova, Italy
E-mail: rdepalo@pd.infn.it

Abstract. The ^{22}Ne(p,γ)^{23}Na reaction is involved in the NeNa cycle of hydrogen burning. This cycle plays an important role for nucleosynthesis in the Red Giant Branch and Asymptotic Giant Branch phases of stellar evolution, as well as in classical novae and type Ia supernovae explosions.
The ^{22}Ne(p,γ)^{23}Na reaction rate is highly uncertain because of a large number of resonances lying in the energy region of the Gamow peak. Several of these resonances have never been studied in either direct or indirect experiments, and only upper limits exist for their strengths.
A measurement of the ^{22}Ne(p,γ)^{23}Na cross section is on-going at the Laboratory for Underground Nuclear Astrophysics (LUNA) in Gran Sasso. With the LUNA setup, it will be possible to study the ^{22}Ne+p reaction inside the Gamow window.
The results of a feasibility test, as well as the measurement strategy and the setup for the first experimental campaign are discussed.

1. Introduction

The ^{22}Ne(p,γ)^{23}Na reaction takes part in the neon - sodium cycle of hydrogen burning (fig. 1). This cycle is active in the Red Giant and Asymptotic Giant Branch phases of stellar evolution [1], as well as in classical novae [2] and type Ia supernovae explosions [3]. In these scenarios, the NeNa cycle is particularly important for the synthesis of the elements between ^{20}Ne and ^{27}Al.

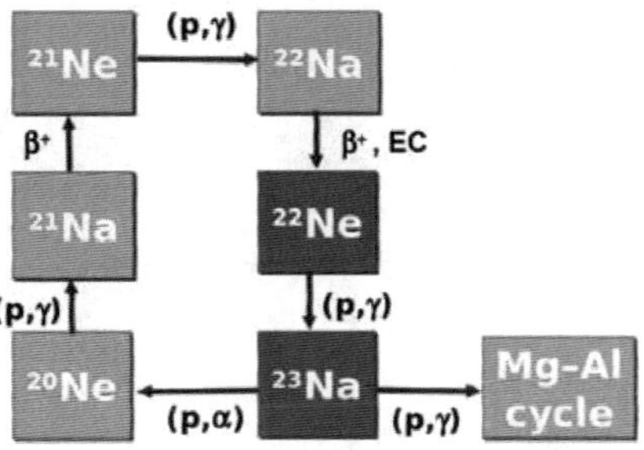

Figure 1. Schematic view of the neon - sodium cycle.

Among the reactions of the NeNa cycle, the ^{22}Ne(p,γ)^{23}Na has the most uncertain reaction rate.
Inside the Gamow window, the reaction cross section is dominated by the contribution of a

large number of resonances (fig. 2). So far, none of the resonances between 50 and 400 keV has been directly observed, and only upper limits exist for the resonance strengths. Moreover, three resonances at 68, 100 and 206 keV, respectively, have been reported as tentative by Powers et al. [4], but not observed in a similar experiment by Hale et al. [5].

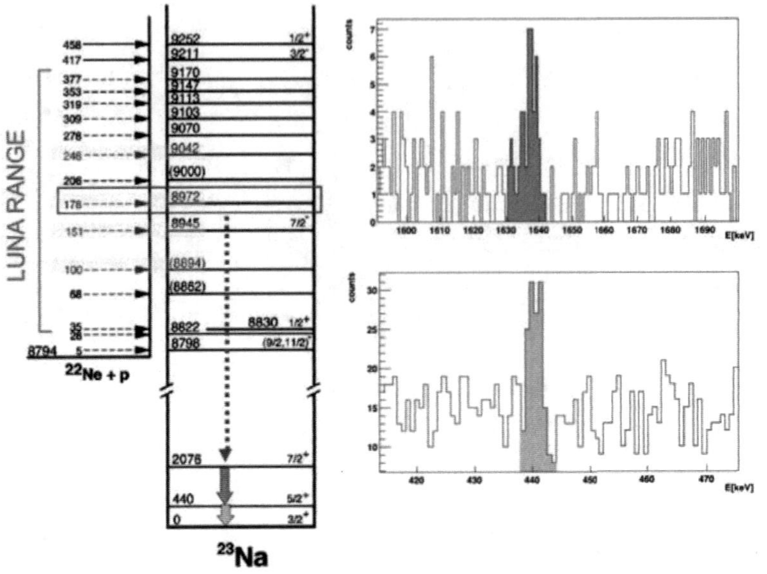

Figure 2. Left: partial level scheme of ^{23}Na. For each excited state, the corresponding resonance enrgy is reported. Right: 178 keV resonance peaks observed in the gamma - ray spectrum. The peaks correspond to the decay of the first (bottom) and second (top) excited states of ^{23}Na.

Because of this lack of information, the reaction rate reported in the NACRE compilation [6] and the one adopted by Iliadis and collaborators [7] differ by a factor of 2000.
In order to improve the experimental knowledge on the ^{22}Ne(p,γ)^{23}Na reaction, a new direct measurement will be performed at the Laboratory for Underground Nuclear Astrophysics (LUNA).

2. Experimental setup

The LUNA facility is located at Gran Sasso National Laboratories (Italy) [8]. The laboratory is shielded against cosmic radiation by 1400 m of rocks, reducing the cosmic muon and neutron flux by six and three orders of magnitude, respectively.
The LUNA setup is composed of a 400 kV electrostatic accelerator providing an high intensity ($\sim$200 μA) proton or alpha beam. The beam can be delivered to a solid or gas target system. Different gamma-ray or particle detectors can be used, depending on the characteristics of the nuclear reaction to be studied.
For the ^{22}Ne(p,γ)^{23}Na cross section measurement, a proton beam will be delivered to a windowless gas target filled with 99.9% enriched ^{22}Ne.
The gamma-rays emitted in the de-excitation of ^{23}Na will be detected by two HPGe detectors placed at 55° and 90° angle with respect to the beam direction.
In order to reduce the environmental background, the two detectors will be surrounded with a 25 cm thick lead and copper shielding (fig. 3). GEANT 4 simulations of the setup indicate that this shielding will ensure about four orders of magnitude background reduction in the γ - ray energy range below 3 MeV.

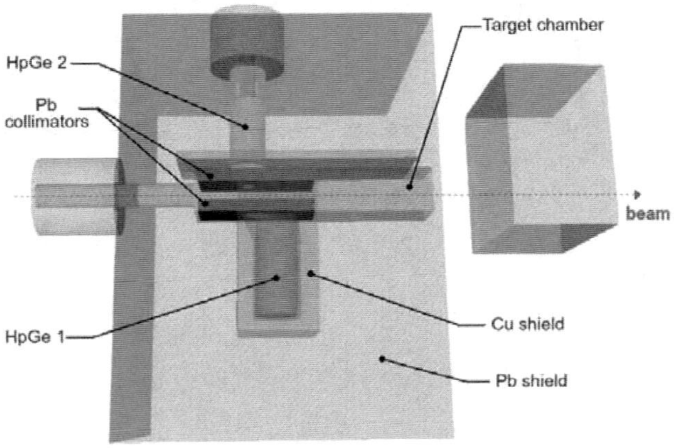

Figure 3. GEANT 4 simulation of the setup for the ^{22}Ne(p,γ)^{23}Na measurement.

3. Measurement plan

As a first step, a feasibility test was performed using the setup of the previous ^{2}H(α,γ)^{6}Li experiment [9].
For this test, neon gas with natural isotopic composition was used (90.48% ^{20}Ne, 0.27% ^{21}Ne and 9.25% ^{22}Ne). The aim of the test was to study the possible sources of beam induced background, and to have some hints on the sensitivity to the ^{22}Ne+p resonant cross section.
During the test, runs on the ^{22}Ne(p,γ)^{23}Na resonances at 100, 151, 178 and 248 keV have been performed. Despite the use of non enriched gas (only 9.25% ^{22}Ne) and a setup which was not optimized for this measurement, gamma-rays from the 178 keV resonance have been observed in a 12 h run (fig. 2). This resonance was never observed in previous experiments, and the literature upper limit for the resonance strength is $\omega\gamma < 2.6 \cdot 10^{-6}$ eV.
Following the feasibility test, the characterization of the setup for the first experimental campaign on the ^{22}Ne(p,γ)^{23}Na resonances was started.
In particular, the pressure and temperature profile of the gas inside the target chamber have been studied using a dedicated setup. Then the beam heating effect in natural neon gas has been meaured with the resonance scan technique, using the intense ^{21}Ne(p,γ)^{22}Na resonance at 271 keV beam energy.
The preparation of the final setup is currently on-going, and the data taking is planned to start in fall 2013.

4. Conclusions

An experimental study of the ^{22}Ne(p,γ)^{23}Na reaction rate is needed to understand the nucleosynthesis of the elements with mass number around 20 in many astrophysical scenarios.
A new measurement of the ^{22}Ne(p,γ)^{23}Na cross section is now starting at LUNA.
The first test run demonstrated that the sensitivity to the weak resonances lying in the Gamow window can be significantly improved thanks to the low background. This will allow to observe some resonances for the first time, or significantly lower the existing upper limits on the resonance strengths.

References

[1] N. Prantzos et al. 2007 *Astronomy and Astrophysics* **470** 179-190
[2] C. Iliadis et al. 2002 *Astrophys. Jour. Suppl. Series* **142** 105-137
[3] A. Parikh et al. 2013 *Preprint* astro-ph/1306.6007
[4] J. R. Powers et al. 1971 *Phys. Rev. C* **4** 2030-2046
[5] S. E. Hale et al. 1971 *Phys. Rev. C* **65** 015801
[6] C. Angulo et al. 1999 *Nuclear Physics A* **656**, 3-187
[7] C. Iliadis et al. 2010 *Nucl. Phys. A* **841** 1-322
[8] C. Broggini et al. 2010 *Ann. Rev. Nucl. Part. Sci.* **60** 53-73
[9] M. Anders et al. 2013 *Eur. Phys. J. A* **49** 28

Journal of Physics: Conference Series **665** (2016) 012018 doi:10.1088/1742-6596/665/1/012018

Study of the ^{26}Al(n,p)^{26}Mg and ^{26}Al(n,α)^{23}Na reactions using the ^{27}Al(p,p')^{27}Al inelastic scattering reaction

S Benamara[1,2], N de Séréville[1], P Adsley[3], A M Laird[3], F Hammache[1], I Stefan[1], P Roussel[1], S Ancelin[1], M Assié[1], C Barton[3], A Coc[4], C Diget[3], I Deloncle[4], S Fox[3], J Guillot[1], C Hamadache[4], J Kiener[4], B Le Crom[1], L Lefebvre[1], A Lefebfre-Schuhl[4], G Marquinez Duran[5], G Mavilla[1], P Morfouace[1], A Mutschler[1], C T Nsangu[3], L Perrot[1], N Oulebsir[6], Á-M Sánchez-Benítez[5], D Suzuki[1] and V Tatischeff[4] and M Vandebrouck[1],

[1] Institut de Physique Nucléaire d'Orsay, UMR8608, IN2P3-CNRS, Université Paris Sud 11, 91406 Orsay, France

[2] Laboratoire de Physique et Chimie Quantique, Université Mouloud Mammeri, Tizi-Ouzou, Algeria

[3] Department of Physics, University of York, York YO10 5DD, United Kingdom

[4] CSNSM, IN2P3-CNRS et Université Paris Sud, 91405 Orsay campus, France

[5] Departamento de Física Aplicada, Universidad de Huelva, E-21071 Huelva, Spain

[6] Laboratoire de Physique Théorique, Université Abderahmane Mira, 06000 Béjaia, Algeria

E-mail: deserevi@ipno.in2p3.fr

Abstract. ^{26}Al was the first cosmic radioactivity ever detected in the galaxy as well as one of the first extinct radioactivity observed in refractory phases of meteorites. Its nucleosynthesis in massive stars is still uncertain mainly due to the lack of nuclear information concerning the ^{26}Al(n,p)^{26}Mg and ^{26}Al(n,α)^{23}Na reactions. We report on a single and coincidence measurement of the ^{27}Al(p,p')^{27}Al(p)^{26}Mg and ^{27}Al(p,p')^{27}Al(α)^{23}Na reactions performed at the Orsay TANDEM facility aiming at the spectroscopy study of ^{27}Al above the neutron threshold. Fourteen states are observed for the first time within 350 keV above the ^{26}Al+n threshold.

1. Introduction

^{26}Al is a radioisotope produced in a variety of stellar sites such as massive stars (Wolf-Rayet phase and core-collapse supernovae), asymptotic giant branch (AGB) stars and nova outbursts. It β^+ decays ($T_{1/2} = 7.2 \times 10^5$ yr) to the first excited state of ^{26}Mg which then de-excites to its ground state emitting a characteristic gamma-ray line at $E_\gamma = 1.809$ MeV. The induced gamma-ray emission and ^{26}Mg excesses are observed since the late 70's thanks to gamma-ray [1] and meteoritic [2] observations, respectively. For example, these observations can give insight of the massive star forming conditions as well as of the astrophysical context of the Solar System formation.

Production of ^{26}Al in massive stars occurs at different stages of their evolution which includes core hydrogen burning, neon/carbon convective shell burning as well as explosive neon burning. For a typical 20 $M_\odot$ massive star the largest contribution to the ^{26}Al yield comes from the explosive phase [3]. Sensitivity studies [3, 4] have shown that this yield depends strongly on the ^{26}Al(n,p)^{26}Mg and ^{26}Al(n,α)^{23}Na reactions.

For typical temperatures achieved during explosive burning ($\approx$ 2.3 GK) the relevant energy range above the ^{26}Al+n threshold goes up to $\approx$ 500 keV. Unfortunately very little experimental data are available in this energy range. Indeed neutron induced measurements on ^{26}Al above the thermal energy are scarce [5, 6, 7]. In this work we investigate ^{27}Al spectroscopy up to 350 keV above the ^{26}Al+n threshold by means of the ^{27}Al(p,p')^{27}Al* reaction.

2. Experimental method and set-up

The ^{27}Al spectroscopy above ^{26}Al+n threshold (S_n = 13.057 MeV) has been investigated with a high-precision study of the proton inelastic scattering off ^{27}Al. High-resolution particle spectroscopy was used to derive the energy of the levels of interest. Their decay modes were studied performing the coincidence measurement of the ^{27}Al(p,p')^{27}Al(p)^{26}Mg and ^{27}Al(p,p')^{27}Al(α)^{23}Na reactions.

The ^{27}Al(p,p')^{27}Al* reaction was studied at the Tandem-Alto facility in Orsay, using a 18-MeV proton beam of about 300 nA. The beam impinged a self-supporting ^{27}Al target with a thickness of 89 μg/cm^2 and the light reaction products were momentum analysed with an Enge Split-Pole magnetic spectrometer. The focal-plane detection system includes a position sensitive gas chamber, a ΔE proportional gas counter and an plastic scintillator. The energy loss and position informations were used to identify the inelastic protons from other light particles. A resolution (FWHM) in the laboratory frame of 22 keV and 12 keV was obtained for excitation energies of 4.5 MeV and 13 MeV, respectively.

In case of the Split-Pole inclusive measurement ^{27}Al levels from the ground state up to about 14 MeV were studied by changing the magnetic field four times. Measurements were performed at spectrometer angles of 10°, 40° and 45°. In case of the coincidence measurement the inelastic protons were detected with the magnetic spectrometer positioned at an angle of 40°. The decaying protons and alpha-particles were detected using three Double-Sided Silicon Strip Detectors (DSSSD) of 5×5 cm^2 with 16 channels on each side (W1 model from Micron Semiconductor Ltd.). While two DSSSDs had a thickness of 300 μm the third one had a thickness of 1 mm. The DSSSDs were placed at about 10 cm from the target at backward angles in the laboratory to avoid high counting rate from elastic scattering. They covered about 6% of the 4π steradians. The beam intensity was reduced down to 100 nA for this measurement.

3. Split-Pole inclusive measurement

A careful focal-plane calibration was performed using the low-excitation energy exposure where well-known isolated states are strongly populated. Eleven ^{27}Al states across the whole focal plane were used to obtain a relation between the radius of curvature ρ and the focal-plane position. The one-degree polynomial function describing the previous relation was used for all other magnetic field settings. Proton energy losses in the target were taken into account.

Proton focal-plane spectrum obtained with the ^{27}Al target is shown in Fig. 1 for a spectrometer angle of 40°. The spectrum covers an energy region about 350 keV above the ^{26}Al+n threshold. The main proton peak observed at $B\rho \approx 0.308$ T.m. corresponds to the 12.710-MeV ^{12}C state. The proton focal-plane spectrum obtained at 45° (not shown here) was used to check that the proton peaks were following the mass-kinematic displacement associated to the ^{27}Al(p,p')^{27}Al* reaction. This confirmed that all observed proton peaks in Fig. 1 correspond to ^{27}Al states.

Deconvolution of the proton spectrum after background subtraction is superimposed to the data in Fig. 1. While our results are in excellent agreement with the ^{23}Na(α,γ)^{27}Al measurement

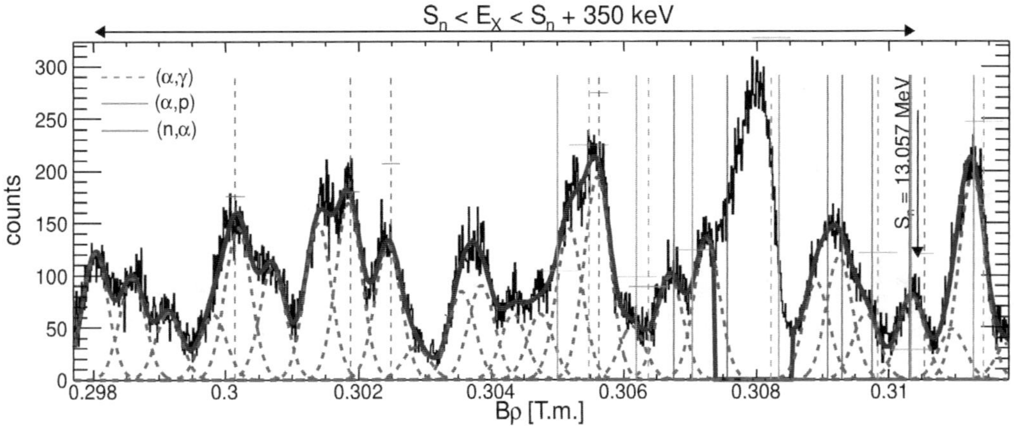

Figure 1. (Color on-line) Proton rigidity spectrum at spectrometer angle of 40°. The spectrum deconvolution is shown after background subtraction. Vertical lines represent the levels that have been observed in previous experiments (see text). The energy uncertainty of each level is represented by an horizontal line.

Table 1. Energies for new ^{27}Al states above the ^{26}Al+n threhsold. Uncertainty is about 4-5 keV. Experimental rigidity associated to each state is also given.

E_X (keV)	$B\rho$ (T.m.)	E_X (keV)	$B\rho$ (T.m.)	E_X (keV)	$B\rho$ (T.m.)
13104	0.3089	13249	0.3038	13363	0.2998
13152	0.3072	13258	0.3035	13381	0.2992
13211	0.3052	13277	0.3029	13397	0.2986
13224	0.3047	13319	0.3014	13412	0.2980
13235	0.3043	13338	0.3007		

of de Voigt et al. [8], they agree marginaly with the ^{23}Na(α,p$_{0,1}$)^{26}Mg measurements of Whitmire et al. [9]. Recently, neutron resonances have been populated using the ^{26}Al(n,α)^{26}Mg reaction [7], confirming the energies of neutron resonances obtained from direct ^{26}Al+n measurements [5] and time-reverse measurements [10, 11]. All these neutron resonances were observed in the present data. Excitation energies were extracted with an uncertainty of about 4-5 keV. The excitation energies for the new observed levels are listed in Table 1.

4. Coincidence measurement

For the coincidence measurement the Split-Pole spectrometer was set to an angle of 40° and covered excitation energies in ^{27}Al in the vicinity of the ^{26}Al+n threshold. The DSSSDs placed in the reaction chamber were used to record the energy and time information of the decaying protons and alpha-particles. Their energy resolution was about 20 keV. The DSSSDs signal was used as a stop for the time-of-flight measurement while the start signal was given by the Split-Pole plastic scintillator. After time-of-flight selection and after ensuring that the two sides of the DSSSDs measure the same energy deposit, the Split-Pole position information is plotted as a function of the DSSSD energy (see Fig. 2). Proton decays to the ground state and first excited state of ^{26}Mg are clearly observed. At lower DSSSD energies two loci are observed which could correspond to alpha-decay to ground state and first excited state of ^{23}Na as well as proton decay to the second excited state of ^{26}Mg. This measurement confirms that the ^{26}Al(n,p)^{26}Mg is dominated by the (n,p$_1$) component. Extraction of the proton and alpha branching ratio is in progress.

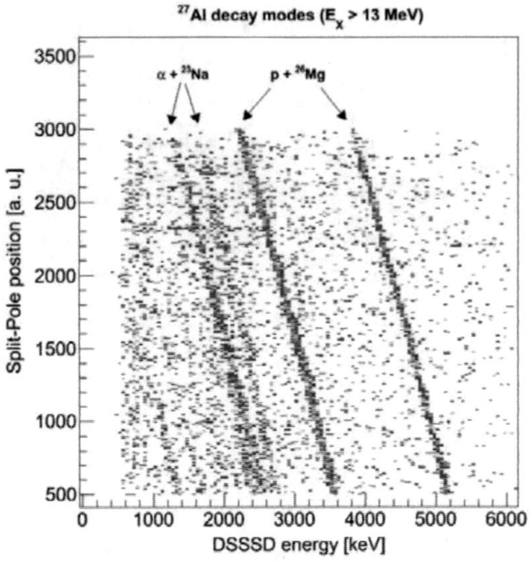

Figure 2. Split-Pole position information as a function of the energy deposit of one DSSSD. The Split-Pole spectrometer covers excitation energies in ^{27}Al in the vicinity of the ^{26}Al+n threshold.

5. Conclusions

A high-resolution coincidence measurement of the ^{27}Al(p,p')^{27}Al(p)^{26}Mg and ^{27}Al(p,p')^{27}Al(α)^{23}Na reactions has been performed at the Tandem-Alto facility using the Enge Split-Pole spectrometer coupled to a DSSSDs array. The energy range up to 350 keV above the ^{26}Al+n threshold was covered and 14 new states were observed. The coincidence measurement confirms that the ^{26}Al(n,p)^{26}Mg reaction rate is dominated by the (n,p$_1$) component. The branching ratio extraction is in progress.

Acknowledgments

The continued support of the staff of the Tandem-Alto facility as well as the target laboratory staff is gratefully acknowledged. UK personnel acknowledge the support of the Science and Technology Funding Council. This work has been supported by the European Community FP7 - Capacities-Integrated Infrastructure Initiative- contract ENSAR n° 262010 and the french-spanish AIC-D-2011-0820 project.

References

[1] Mahoney W A, Ling J C, Wheaton W A and Jacobson A S 1984 *Astrophys. J.* **286** 578–585
[2] Lee T, Papanastassiou D A and Wasserburg G J 1976 *Geophysical Research Letters* **3** 41–44
[3] Limongi M and Chieffi A 2006 *Astrophys. J.* **647** 483
[4] Iliadis C, Champagne A, Chieffi A and Limongi M 2011 *Astrophys. J. Supl.* **193** 16
[5] Koehler P E, Kavanagh R W, Vogelaar R B and Gledenov Y M e a 1997 *Phys. Rev.* **C56**(2) 1138–1143
[6] Trautvetter H P, Becker H W, Heinemann U, Buchmann L, Rolfs C, Käppeler F, Baumann M, Freiesleben H, Lütke-Stetzkamp H J, Geltenbort P and Gönnenwein F 1986 *Z Phys A - Hadron Nuc* **323** 1–11
[7] de Smet L, Wagemans C, Wagemans J, Heyse J and van Gils J 2007 *Phys. Rev.* **C76** 045804
[8] De Voigt M, Maas J, Veenhof D and Van Der Leun C 1971 *Nucl. Phys.* **A170** 449 – 466
[9] Whitmire D P and Davids C N 1974 *Phys. Rev.* **C9** 996–1001
[10] Doukellis G and Rapaport J 1987 *Nucl. Phys.* **A467** 511 – 527
[11] Skelton R T, Kavanagh R W and Sargood D G 1987 *Phys. Rev.* **C35**(1) 45–54

The FRUITY database on AGB stars: past, present and future

S Cristallo, L Piersanti, O Straniero

E-mail: cristallo@oa-teramo.inaf.it
INAF Osservatorio Astronomico di Teramo, via Maggini sn 64100 - Teramo (Italy)
INFN Sezione Napoli, Napoli (Italy)

Abstract. We present and show the features of the FRUITY database, an interactive web-based interface devoted to the nucleosynthesis in AGB stars. We describe the current available set of AGB models (largely expanded with respect to the original one) with masses in the range $1.3 \leq M/M_\odot \leq 3.0$ and metallicities $-2.15 \leq$[Fe/H]$\leq +0.15$. We illustrate the details of our s-process surface distributions and we compare our results to observations. Moreover, we introduce a new set of models where the effects of rotation are taken into account. Finally, we shortly describe next planned upgrades.

1. Introduction

Asymptotic Giant Branch stars (hereafter AGBs) are exceptional laboratories to test the robustness of stellar models. During this evolutionary phase, both light (C, N, F, Na) and heavy elements can be produced. The latter are synthesized via the slow neutron capture process (the s-process). A detail description of AGBs evolution and nucleosynthesis can be found in [1].
The structure of an AGB consists of a partially degenerate CO core, surrounded by an He-shell and an H-shell separated by a thin layer (He-intershell) and by a cool and expanded convective envelope. The surface luminosity is mainly sustained by the H-burning shell, active for most of the time. This situation is recurrently interrupted by the growing up of thermonuclear runaways driven by violent ignitions of 3α reactions at the base of the He-intershell (Thermal Pulses, hereafter TPs). The energy suddenly released by TPs can not be transported outward radiatively and, thus, convective episodes develop. These convective shells efficiently mix the He-intershell, enriching this layer in carbon (produced by 3α reactions) and in s-process elements. Moreover, the energy boosted by the TP forces the overlying layers to expand, possibly switching off the H-shell. This allows the convective envelope to penetrate downward, carrying to the surface the isotopes freshly synthesized in the He-intershell. This phenomenon is called Third Dredge Up (hereafter TDU) and it is know to work in AGB stars since the early fifties [2].
The major neutron source in AGB stars is the ^{13}C(α,n)^{16}O reaction. Neutrons are released within the He-intershell during the interpulse phase (thus in radiative conditions) when T$\sim 1.0 \times 10^8$ K [3]. Actually, the amount of ^{13}C requested to fit observations is one of the major sources of uncertainty in AGB models ([4]; see also Section 2). A marginal contribution comes from the ^{22}Ne(α,n)^{25}Mg reaction, which is activated at higher temperatures (T$> 2.5 \times 10^8$ K) during TPs.

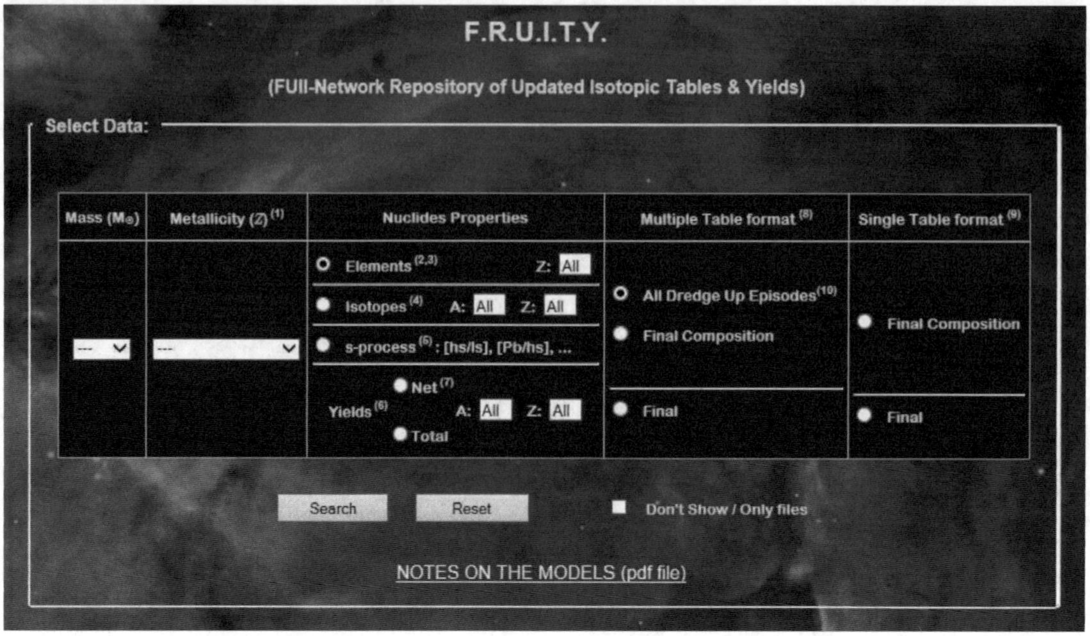

Figure 1. The FRUITY database homepage.

The great amount of available spectroscopic data such as the need of yields for Chemical Evolution Models require the computation of a large number of detailed AGB evolutionary models. We fit this request by making available our theoretical results on the on-line web pages of the FRUITY (FUNs Repository of Updated Isotopic Tables & Yields) database, entirely dedicated to the nucleosynthesis in AGB stars.

In Section 2 we briefly present our evolutionary code (FUNs), while in Section 3 we describe the set of models currently available on FRUITY. In Section 5 we present a recently published set of rotating AGB stars. Finally, in Section 6 we list our planned upgrades.

2. The FUNs evolutionary code

The FUNs (FUll Network stellar) evolutionary code is a one-dimension hydrostatic code (see [1] and references therein). The adopted mass-loss rate has been calibrated on the Period-Luminosity and Period- Mass loss relations observed in Long Period Variable Stars [1]. In the envelope, atomic and molecular opacities are calculated according to the changes in the chemical composition due to the occurrence of TDU episodes [4]. The radiation/convection interface at the inner border of the convective envelope is treated by applying an exponential decay of the convective velocities. As a by-product, we obtain the self-consistent formation of the ^{13}C pocket after each TP followed by TDU. The extension of the that pocket varies from TP to TP following the shrinking of the He-intershell [5]. In order to avoid the possible loss of accuracy inherent to post-process techniques, we directly coupled our models to a full nuclear network, which includes all isotopes from hydrogen to bismuth (at the ending-point of the s-process path).

3. The FRUITY database

The original set of FRUITY has been presented in [6]. It consists of 28 evolutionary models, with different combinations of masses (1.5, 2.0, 2.5, and 3.0 $M_\odot$) and metallicities ([Fe/H]= -1.15, -0.67, -0.37, -0.24, -0.15, 0.00, +0.15). Different He contents and scaled solar compositions [7]

are used (see [6] for details).
Recently, new models have been uploaded. In particular, the mass range has been expanded down to 1.3 $M_\odot$ models. Note that, at large metallicities, these masses do not experience TDU episodes.
In order to cover the metallicity range of Galactic Globular Clusters, we compute two additional metallicities: [Fe/H]=-2.15 and [Fe/H]=-1.67, both of them with α-enhanced elements ([α/Fe]=0.5).
For each model, users can freely download pulse by pulse surface isotopic compositions and elemental overabundances[1]. Tables are available in two different formats (see Figure 1). In the Multiple Table Format, the query returns multiple table, depending on the number of selected models. In the Single Table Format, the query returns a single table containing all the selected models; in this case, only single elements (or isotopes) can be visualized. Among other quantities available on FRUITY there are the [ls/Fe][2], the [hs/Fe][3] and the s-process indexes [hs/ls], [Pb/hs] and [Pb/ls]. Finally, stellar yields can be selected. With respect to the original version of the FRUITY database, beside net yields we also made available total stellar yields.
Users interested in upgrades of the FRUITY database can register to the dedicated mailing list.

4. Comparison to observations

In order to verify the robustness of our models, we compare them to their observational counterparts. In particular, we concentrate on the Luminosity Function of Carbon Stars (hereafter LFCS) and on *s*-process spectroscopic indexes at different metallicities.

The LFCS links the luminosity of these objects (and thus the core mass; see [8]) with their carbon surface abundance. Thus, by studying this quantity, we can test our prescriptions on convection and mass-loss law. In 2006, [9] analyzed a sample of Galactic Carbon Stars and show how the radiation emitted by these objects at infrared wavelengths strongly weights the LFCS toward large lambdas. Recently, [10] proposed a re-analysis of the same sample and found that the high luminosity tail of the LFCS described by [9] disappears, in good agreement with theoretical expectations [6].

Another useful litmus test for theoretical models is the study of s-process spectroscopic indexes, in particular the [hs/ls] and the [Pb/hs]. Before introducing them, a distinction between intrinsic and extrinsic s-process rich stars has to be done.
TP-AGB stars are intrinsic s-rich stars, since they are presently undergoing thermal pulses and TDU episodes. Post-AGB stars belong to the same class.
Extrinsic s-rich stars are less-evolved stars (dwarfs or giants) belonging to binary systems, whose s-enhancement is likely due to the pollution caused by the intense wind of an AGB companion. Ba-stars and, at lower metallicities, CH stars and Carbon Enhanced Metal Poor s-process rich stars (hereafter CEMP-s stars) belong to this class of objects.
The only possibility to group together all s-rich stars is to study relative surface abundances. In such a way, any problem correlated to possible further dilution process or to their evolutionary status is avoided.
We have already shown that our theoretical models match observations, even if they are not able to explain the observed spread for a fixed metallicity (see Fig. 12 and 13 of [6]). Although it is possible that (at least part of) the observed spread can be ascribed to the large observational uncertainties, theoretical scenarios able to explain such a spread has to be explored. Among them, we verified is the mixing induces by rotation can lead to a certain spread of the spectroscopic indexes.

[1] In the usual spectroscopic notation: [El/Fe] = log(N(El)/N(Fe))$_*$ - log(N(El)/N(Fe))$_\odot$
[2] [ls/Fe] = ([Sr/Fe]+[Y/Fe]+[Zr/Fe])/3
[3] [hs/Fe] = ([Ba/Fe]+[La/Fe]+[Nd/Fe]+[Sm/Fe])/4

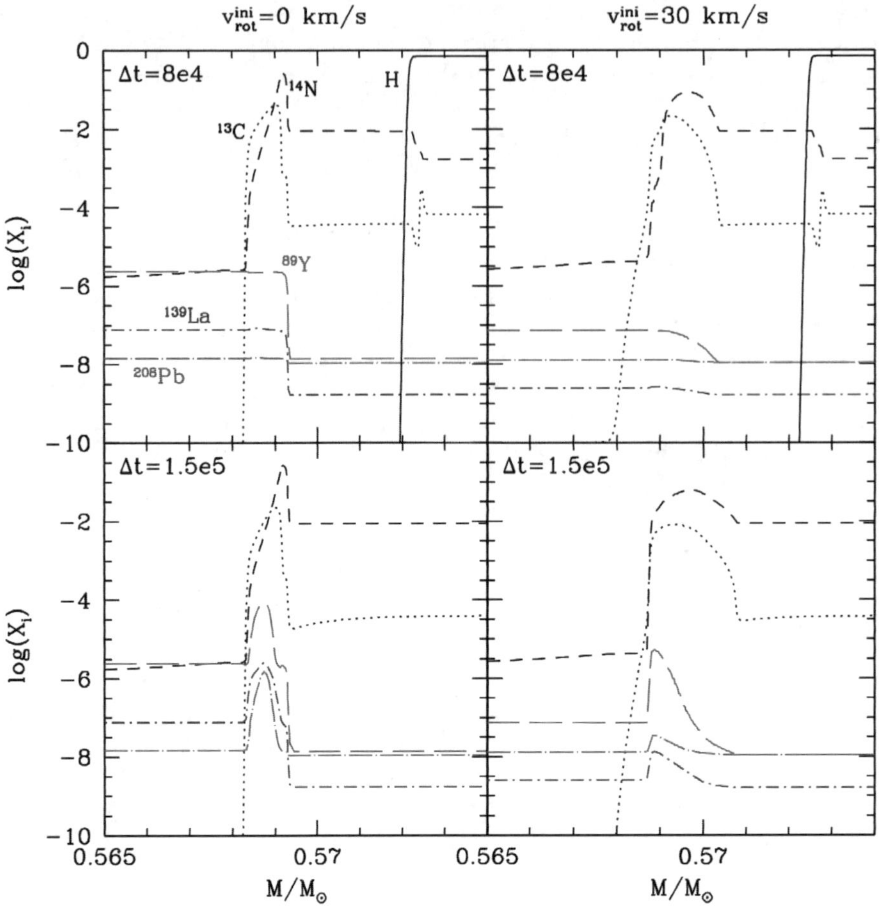

Figure 2. Temporal evolution of key elements during the radiative burning of the 2nd ^{13}C pocket of a 2 $M_\odot$ star with Z=10^{-2} and different initial rotation velocities. See text for details.

5. Low mass rotating AGB models

Even if low mass stars are generally slow rotators, the lifting due to the centrifugal force and the mixing induced by dynamical and secular instabilities can modify their physical structure and chemical composition. In order to verify the effect that rotation can have on the nucleosynthesis during the AGB phase, we implement our models with rotation. In particular, we add rotation-induced mixing to the other mixing already considered in our previous works, those due to convection in particular. We found that a variation in the initial velocity can lead to a consistent spread in the final surface s-process enhancements and spectroscopic indexes for stars with the same initial mass and metallicity [11].

Rotation does not determine the formation of the ^{13}C pocket. Notwithstanding, rotation-induced mixing modify the mass extension of both the ^{13}C and the ^{14}N pockets and their overlap, thus reducing the average neutrons-to-seeds ratio. In Figure 2 we report the ^{13}C and the ^{14}N abundances in the region where the 2nd ^{13}C pocket of a 2 $M_\odot$ star with Z=10^{-2} forms. Left panels refer to a non-rotating model, while right panels to a model with initial rotation

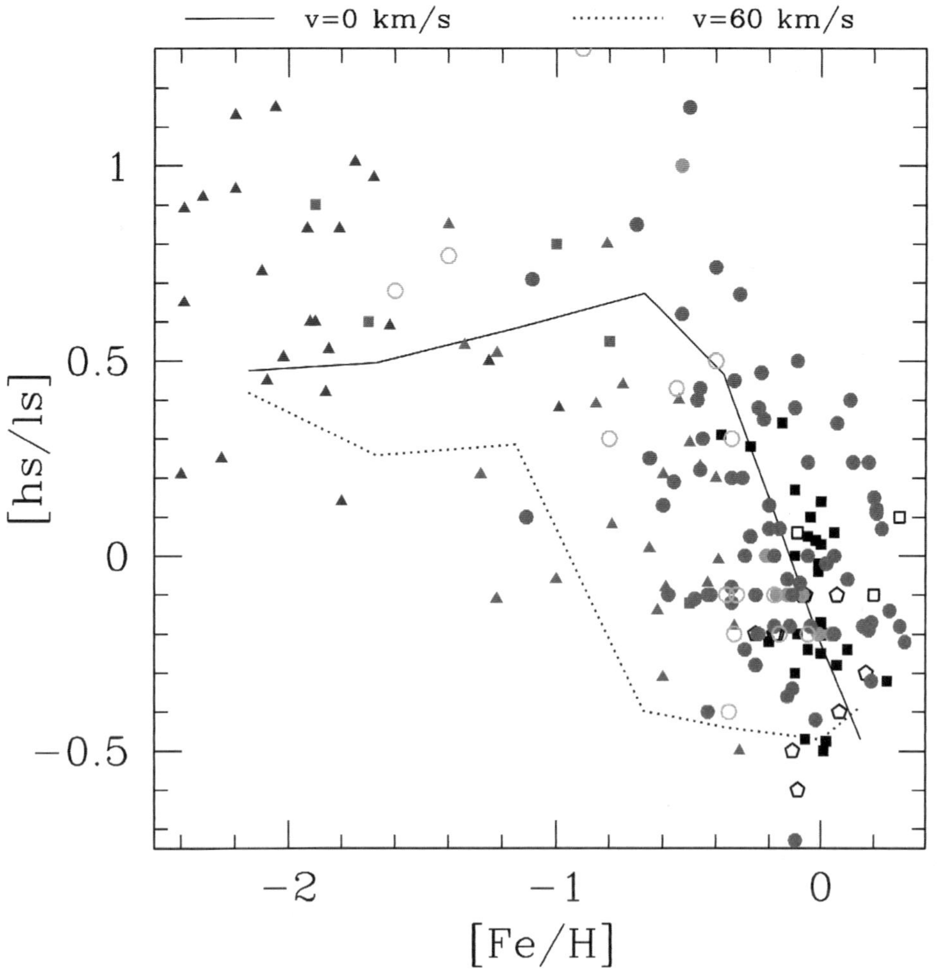

Figure 3. [hs/ls] s-process index as a function of metallicity for models with and without rotation. See text for details.

velocity v_{rot}^{ini}=30 km/s . In the same plot we also report the local abundances of ^{89}Y, ^{139}La and ^{208}Pb, assumed as representative of the three s-process peaks. We find that the Goldreich-Schubert-Fricke instability, active at the interface between the convective envelope and the rapid rotating core, contaminates the ^{13}C-pocket with ^{14}N. Thus,the mixing induced by rotation locally decreases the neutron-to-seed ratio, leading to a reduction of the total amount of heavy elements produced by the s process and favoring the light-s elements with respect to the heavier ones. As a matter of fact, both the [hs/ls] and the [Pb/hs] spectroscopic indexes decrease as the initial rotation velocity increases (see Figures 6 and 9 of [11]). At low metallicity, the combined effect of Goldreich-Schubert-Fricke instability and meridional circulations determines an increase of light-s and, to a less extent, heavy-s elements, while lead is strongly reduced.
In Figures 3 and 4 we report, as a function of the metallicity, the [hs/ls] and [Pb/hs] s-process indexes for models with and without rotation. Two values of initial rotation velocities (v_{rot}^{ini}=60 km/s and v_{rot}^{ini}=120 km/s) have been considered. With respect to the observational data reported in [6], we add recent measurements of post-AGB stars [12,13,14,15], Ba-stars [16,17], CH stars

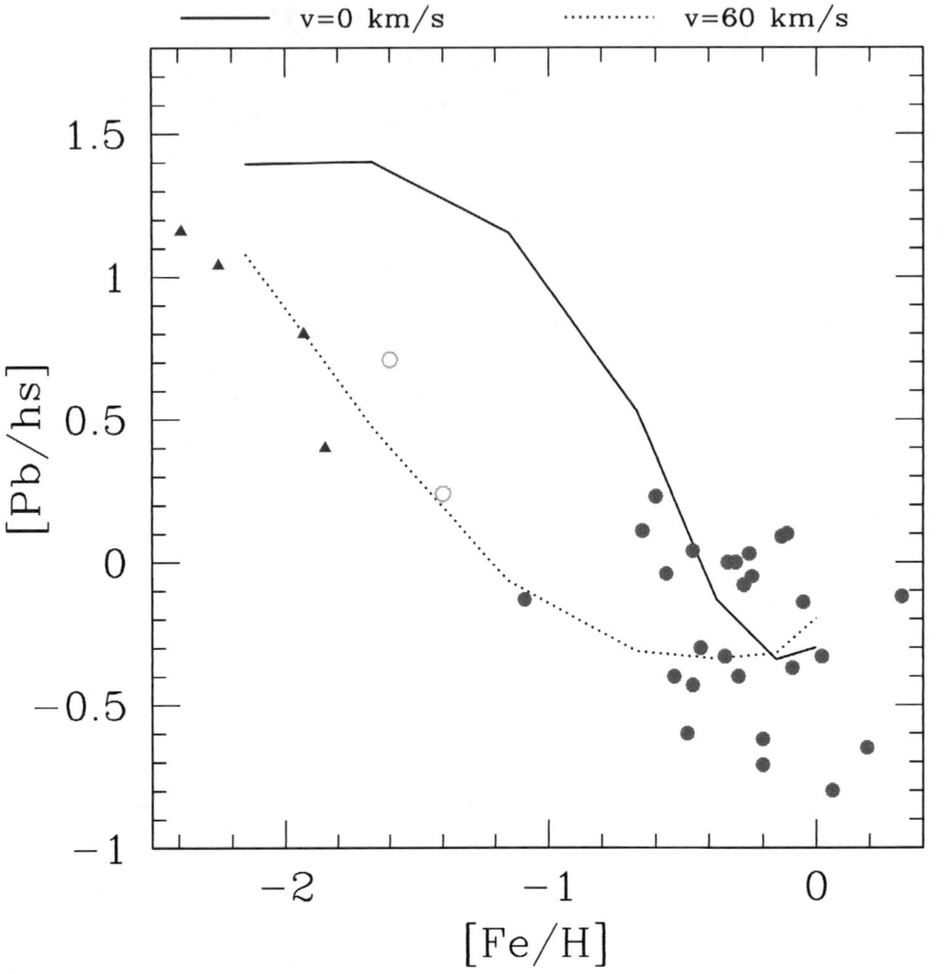

Figure 4. As in Figure 3, but for the [Pb/hs] s-process index.

[18,19,20] and a selection of well measured CEMP-s stars (see [21] and references therein).

Our results suggest that rotation can be regarded as a possible physical mechanism responsible for the observed spread of s-process spectroscopic indexes. In particular, the large spread in the [hs/ls] at intermediate metallicities (-1.0<[Fe/H]<-0.6) can be reproduced by hypothesizing different initial rotation velocities. Notwithstanding, our models cannot match the large [hs/ls] (up to 1) measured at low metallicities. Interestingly, the lower [Pb/hs] values characterizing rotating models are in agreement with observations. In fact, as shown in Figure 4, models with v_{rot}^{ini}=60 km/s better reproduce the sequence, starting from high Z, of Ba-stars (full dots), CH stars (open dots) and CEMP-s stars (full triangles). Unfortunately, firm conclusions cannot be drawn due to the paucity of lead measurements and to the large errors affecting observations.

6. Planned Upgrades

The current version of FRUITY spans on a reasonable range of metallicities and well covers the low mass range. Intermediate mass AGB stars (4< $M/M_\odot$ < 7; hereafter IM-AGBs) are currently missing. These stars are particulary important for the chemical evolution of Globular

Clusters (see e.g. [22] and references therein). We already conducted an explorative study of IM-AGBs [23], but without publishing detailed nucleosynthetic predictions. Therefore, we intend to explore the evolution and nucleosynthesis of IM-AGBs, starting from the metallicities of interest for the study of Galactic Globular Clusters [24]. Later, we aim to extend these calculations to all the metallicities considered in FRUITY.
Finally, as a long-term project, we are planning to extend the mass range of FRUITY to massive stars, in order to study the weak component of the s-process, at work during the core-He burning and the C-burning shell phases particularly efficient when rotation is explicitly taken into account. (see [25] and references therein).

6.1. Acknowledgments

The authors warmly thank Prof. Roberto Gallino for continuous and fruitful scientific discussions.

7. References

[1] Straniero O, Gallino R and Cristallo S 2006 *Nucl. Phys.* A **777** 311
[2] Merrill P W 1952 *Astrophys. J.* **116** 21
[3] Straniero O, Gallino R, Busso M, Chieffi A, Limongi M & Salaris M 1995 *Astrophys. J. Lett.* **440** 85
[4] Gallino R, Arlandini C, Busso M, Lugaro M, Travaglio C, Straniero O, Chieffi A and Limongi M 1998 *Astrophys. J.* **497** 388
[5] Cristallo S, Straniero O, Lederer M T and Aringer B 2007 *Astrophys. J.* **667** 489
[6] Cristallo S, Piersanti L, Straniero O, Gallino R, Dominguez I, Abia C, Di Rico G, Quintini M and Bisterzo S 2011 *Astrophys. J. Suppl.* **197** 17
[7] Lodders K 2003 *Astrophys. J.* **591** 1220
[8] Paczynski B 1970 *Acta Astron.* **20** 47
[9] Guandalini R, Busso M, Ciprini S and Silvestro G 2006 *Astron. & Astrophys.* **445** 1069
[10] Guandalini R and Cristallo S 2013 *Astron. & Astrophys.* **555** 120
[11] Piersanti L, Cristallo S and Straniero O 2013 *Astrophys. J. accepted* (arXiv1307.2017)
[12] Sumangala Rao S, Pandey G, Lambert D L and Giridhar S 2011 *Astrophys. J. Lett.* **737** 7
[13] Pereira C B, Gallino R and Bisterzo S 2012 *Astron. & Astrophys.* **538** 48
[14] De Smedt K, Van Winckel H, Karakas A I, Siess L, Goriely S and Wood P R 2012 *Astron. & Astrophys.* **541** 67
[15] van Aarle E, Van Winckel H, De Smedt K, Kamath D and Wood P R 2013 *Astron. & Astrophys.* **554** 106
[16] Pereira C B, Sales Silva J V, Chavero C, Roig F and Jilinski E 2011 *Astron. & Astrophys.* **533** 51
[17] Lebzelter T, Uttenthaler S, Straniero O and Aringer B 2013 *Astron. & Astrophys.* **554** 30
[18] Pereira C B and Drake N A 2011 *Astron. Journ.* **141** 79
[19] Pereira C B, Jilinski E, Drake N A, de Castro D B, Ortega V G, Chavero C and Roig F 2012 *Astron. & Astrophys.* **543** 58
[20] Liu S, Nissen P E, Schuster W J, Zhao G, Chen Y Q and Liang Y C 2012 *Astron. & Astrophys.* **541** 48
[21] Bisterzo S, Gallino R, Straniero O, Cristallo S and Kappeler F 2012 *MNRAS* **422** 849
[22] Ventura P and D'Antona F 2009 *Astron. & Astrophys.* **402** 72
[23] Straniero O, Limongi M, Chieffi A, Dominguez I, Busso M and Gallino R 2000 *MemSAIt* **71** 719
[24] Straniero O, Cristallo S and Piersanti L 2013 *MemSAIt* **84** 105
[25] Pignatari M, Gallino R, Meynet G, Hirschi R, Herwig F and Wiescher M 2008 *Astrophys. J. Lett.* **687** 95

Experiments with neutron beams for the astrophysical *s* process

C Lederer[1,2,*] S Altstadt[1], J Andrzejewski[3], L Audouin[4], M Barbagallo[5], V Bécares[6], F Bečvář[7], F Belloni[8], E Berthoumieux[8,9], J Billowes[10], V Boccone[9], D Bosnar[11], M Brugger[9], M Calviani[9], F Calviño[12], D Cano-Ott[6], C Carrapiço[13], F Cerutti[9], E Chiaveri[8,9], M Chin[9], N Colonna[5], G Cortés[12], MA Cortés-Giraldo[14], M Diakaki[15], C Domingo-Pardo[16], I Duran[17], R Dressler[18], N Dzysiuk[19], C Eleftheriadis[20], A Ferrari[9], K Fraval[8], S Ganesan[21], AR García[6], G Giubrone[16], MB Gómez-Hornillos[12], IF Gonçalves[13], E González-Romero[6], E Griesmayer[22], C Guerrero[9], F Gunsing[8], P Gurusamy[21], A Hernández-Prieto[9,12], DG Jenkins[23], E Jericha[22], Y Kadi[9], F Käppeler[24], D Karadimos[15], N Kivel[18], P Koehler[25], M Kokkoris[15], G Korschinek[26], M Krtička[7], J Kroll[7], C Lampoudis[8], C Langer[1], E Leal-Cidoncha[17], H Leeb[22], LS Leong[4], R Losito[9], A Mallick[21], A Manousos[20], J Marganiec[3], T Martínez[6], C Massimi[27], PF Mastinu[19], M Mastromarco[5], M Meaze[5], E Mendoza[6], A Mengoni[28], PM Milazzo[29], F Mingrone[27], M Mirea[30], W Mondalaers[31], C Paradela[17], A Pavlik[2], J Perkowski[3], M Pignatari[32], A Plompen[31], J Praena[14], JM Quesada[14], T Rauscher[32], R Reifarth[1], A Riego[12], MS Robles[17], F Roman[9,30], C Rubbia[9,33], M Sabaté-Gilarte[14], R Sarmento[13], A Saxena[21], P Schillebeeckx[31], S Schmidt[1], D Schumann[18], G Tagliente[5], JL Tain[16], D Tarrío[17], L Tassan-Got[4], A Tsinganis[9], S Valenta[7], G Vannini[27], V Variale[5], P Vaz[13], A Ventura[28], R Versaci[9], MJ Vermeulen[23], V Vlachoudis[9], R Vlastou[15], A Wallner[2], T Ware[10], M Weigand[1], C Weiß[22], T Wright[10], P Žugec[11]

[1]Johann-Wolfgang-Goethe Universität, Frankfurt, Germany
[2]University of Vienna, Faculty of Physics, Austria
[3]Uniwersytet Łódzki, Lodz, Poland
[4]Centre National de la Recherche Scientifique/IN2P3 - IPN, Orsay, France
[5]Istituto Nazionale di Fisica Nucleare, Bari, Italy
[6]Centro de Investigaciones Energeticas Medioambientales y Tecnológicas (CIEMAT), Madrid, Spain
[7]Charles University, Prague, Czech Republic
[8]Commissariat à l'Énergie Atomique (CEA) Saclay - Irfu, Gif-sur-Yvette, France
[9]European Organization for Nuclear Research (CERN), Geneva, Switzerland
[10]University of Manchester, Oxford Road, Manchester, UK
[11]Department of Physics, Faculty of Science, University of Zagreb, Croatia
[12]Universitat Politecnica de Catalunya, Barcelona, Spain
[13]Instituto Tecnológico e Nuclear, Instituto Superior Técnico, Universidade Técnica de Lisboa, Lisboa, Portugal
[14]Universidad de Sevilla, Spain

[15]National Technical University of Athens (NTUA), Greece
[16]Instituto de Física Corpuscular, CSIC-Universidad de Valencia, Spain
[17]Universidade de Santiago de Compostela, Spain
[18]Paul Scherrer Institut, Villigen PSI, Switzerland
[19]Istituto Nazionale di Fisica Nucleare, Laboratori Nazionali di Legnaro, Italy
[20]Aristotle University of Thessaloniki, Thessaloniki, Greece
[21]Bhabha Atomic Research Centre (BARC), Mumbai, India
[22]Atominstitut, Technische Universität Wien, Austria
[23]University of York, Heslington, York, UK
[24]Karlsruhe Institute of Technology, Campus Nord, Institut für Kernphysik, Karlsruhe, Germany
[25]Department of Physics, University of Oslo, N-0316 Oslo, Norway
[26]Physik Department E12 and Excellence Cluster Universe, Technische Universität München, Munich, Germany
[27]Dipartimento di Fisica, Università di Bologna, and Sezione INFN di Bologna, Italy
[28]Agenzia nazionale per le nuove tecnologie, l'energia e lo sviluppo economico sostenibile (ENEA), Bologna, Italy
[29]Istituto Nazionale di Fisica Nucleare, Trieste, Italy
[30]Horia Hulubei National Institute of Physics and Nuclear Engineering - IFIN HH, Bucharest - Magurele, Romania
[31]European Commission JRC, Institute for Reference Materials and Measurements, Retieseweg 111, B-2440 Geel, Belgium
[32]Department of Physics and Astronomy - University of Basel, Basel, Switzerland
[33]Laboratori Nazionali del Gran Sasso dell'INFN, Assergi (AQ),Italy
*Present Address: School of Physics and Astronomy, University of Edinburgh, UK.

Abstract. Neutron capture cross sections are the key nuclear physics input to study the slow neutron capture process, which is responsible for forming about half of the elemental abundances above Fe. Stellar neutron capture cross section can be measured by the time-of-flight technique, or by activation. Both techniques will be discussed and recent experiments in the Fe/Ni mass region will be presented.

1. Introduction

Neutron capture reactions are the main mechanisms for synthesizing elements heavier than Fe in stars. Two different processes contribute about equally to the overall elemental abundance pattern, the slow neutron capture process (s process) and the rapid neutron capture process (r process). The r process is associated to explosive scenarios with high neutron densities and nuclear reactions involve nuclei far from the stability valley. In the s process, heavy elements are produced by subsequent neutron captures on seed nuclei (mainly Fe). Neutron capture timescales are of the order of years, therefore, if an unstable element is produced, β-decays towards the stability valley are faster than subsequent neutron captures. An exception are long lived radionuclides, where β-decay and neutron capture may compete. Such nuclei are called branching points and their study can provide important information such as temperatures and neutron densities in s process environments. The s process can be further divided in two different components, the main component taking place in thermally pulsing Asymptotic Giant Branch stars, where mainly elements between Zr and Bi are formed, and the weak s process which takes place in massive stars ($M > 8M_{\odot}$), where mainly elements between Fe and Zr are produced [1]. The nucleosynthesis path for the s process from Fe to Se is illustrated in Figure 1.

The main nuclear physics quantities that are required to study the s process are neutron capture cross sections and β-decay half lives. The effective stellar neutron capture cross section, called Maxwellian Averaged Cross Section (MACS), is the energy dependent cross section

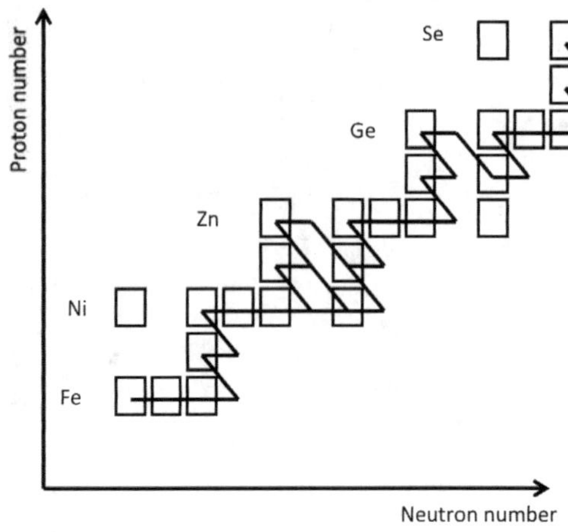

Figure 1. The *s* process nucleosynthesis path starting from Fe, indicated by the black solid line.

averaged over the stellar neutron spectrum and defined as:

$$<\sigma>=\frac{2}{\sqrt{\pi}}\frac{1}{(k_BT)^2}\int_0^{\infty}\sigma(E)E\exp(-\frac{E}{k_BT})dE \quad (1)$$

Maxwellian Averaged Cross Sections need to be known over a wide range of kT values (8-100 keV) since s process environments exhibit temperatures from 0.1-1 Gigakelvin.
This paper will present neutron capture measurements at the neutron time-of-flight facility n_TOF/CERN in the Fe/Ni mass region.

2. Measuring Neutron Capture Cross sections

There are two complementary techniques to measure stellar neutron cross sections, the time-of-flight technique, and the activation technique.
The time-of-flight technique allows measurement of the energy dependence of the cross section, since the neutron energy is determined via the time-of-flight of the neutron for each capture event. A pulsed neutron beam is produced by reactions of a charged particle beam with a neutron production target. This can be achieved, for example, by spallation reactions (e.g. n_TOF/CERN [2] and LANSCE/Los Alamos [3]), or by photon-induced reactions by bremsstrahlung produced by an electron beam hitting a high Z target (e.g. GELINA/Geel [4]), or by (p, n) reactions using low-energy accelerators (e.g. the Karlsruhe Van de Graaff [1]). Capture reactions are measured by detecting the prompt γ radiation emitted after each capture event. Background effects can be minimized by using isotopically enriched samples, and structure materials with low neutron cross sections. The time-of-flight technique allows measurements of neutron cross sections over a wide energy range, enabling calculation of Maxwellian Averaged Cross Sections for several values of kT from a single experiment.

The activation technique comprises of irradiating a sample with neutrons, and afterwards counting the reaction product. This can be done by decay counting or by direct atom counting

Journal of Physics: Conference Series **665** (2016) 012020 doi:10.1088/1742-6596/665/1/012020

techniques, e.g. Accelerator Mass Spectrometry [5]. The latter is especially suited if the reaction product is long lived and/or the decay is not easily measurable by its radiation. To measure stellar cross sections via activation, a neutron spectrum resembling the stellar distribution is needed. A quasi-stellar spectrum around $kT = 25$ keV can be generated using the ^{7}Li(p,n) reaction with a proton beam of about 2 MeV energy [6, 7, 8]. This possibility was developed by Beer and Käppeler [9] at the Forschungszentrum Karlsruhe and extensively used for measuring stellar (n,γ) cross sections over the entire periodic table (e.g. [6, 10]). Quasi-stellar spectra have been obtained for $kT = 5$ keV [11] using the ^{18}O(p,n) reaction and for $kT = 52$ keV [12] using the ^{3}H(p,n) reaction, but both reactions yield significantly reduced intensities. To convert the measured spectrum averaged cross section to a MACS, some assumptions on the energy dependence of the cross section have to be made, but the respective conversion factors are usually close to 1. Besides the limitation that the activation technique can only be applied to reactions producing radioactive nuclei, there are also several advantages compared to the time-of-flight technique. Sample material can be used in natural composition since the reaction is detected either via characteristic γ-ray emission or by directly counting the reaction product. Additionally, small amounts of sample material are sufficient (of the order of mg vs. hundreds of mg to g for the time-of-flight method), since the neutron beam is continuous and the sample can be put close to the neutron source.

3. (n,γ) Measurements at the n_TOF Facility and Astrophysical Implications

In recent years, several neutron capture measurements of interest for the s process have been performed at the neutron-time-of-flight facility n_TOF, CERN. This section will concentrate on measurements in the Fe/Ni mass region. As mentioned in the introduction, masses from Fe to Zr are produced by the weak s process component. It was found that especially in this mass region, neutron capture cross sections of all involved isotopes have to be known with high accuracy because a single cross section can significantly influence the abundances of a number of heavier isotopes [13]. This need has further been underlined by a discrepancy that was found between elemental abundances in ultra-metal-poor stars and calculated abundances in the Sr-Y mass region [14, 15], which led to the suggestion of an additional nucleosynthesis process [16]. Since neutron capture cross sections are a crucial input for these calculated abundances, cross sections should be known with high accuracy for a better study of this discrepancy.

A campaign was started at n_TOF to measure cross sections of all stable isotopes of Fe and Ni. Measurements of (n,γ) reactions on ^{54,56,57}Fe [17] and ^{58,62}Ni [18, 19] have been completed.

At n_TOF, an intense neutron beam is produced by spallation reactions of a 20 GeV proton bunch of 7 ns width from the CERN-PS with a massive lead target. Around the target is a water layer for cooling and for moderating the initially very energetic neutrons. The resulting neutron spectrum at n_TOF ranges from thermal neutron energies (0.025 eV) to few GeV. The experimental area is located at a distance of 185 m from the neutron target, which ensures a high energy resolution of 3×10^{-4} at 1 eV to 5×10^{-3} at 1 MeV neutron energy [2]. Prompt capture γ rays were detected using a pair of scintillation detectors, filled with deuterated benzene (C_6D_6), which were optimized to a very low sensitivity towards reactions with neutrons [20].

For the ^{62}Ni(n,γ) reaction, we obtain a MACS at $kT = 25$ keV which is in good agreement with a time-of-flight measurement by Alpizar-Vicente et al. [21] and activation measurements of Nassar et al. [13] and Dillmann et al. [22]. For $kT > 40$ keV, however, our stellar cross sections are systematically smaller than the results of Alpizar-Vicente et al. [21]. The final MACSs has total uncertainties of about 5% at $kT = 30$ keV and 10% at $kT = 100$ keV.

Additionally to the stable Ni isotopes, the neutron capture cross section of the long lived radionuclide ^{63}Ni, with a half life of 101.2 ± 1.5 years [23] has been measured for the first time [24]. The neutron capture yield measured at n_TOF compared to backgrounds due to presence of ^{62}Ni in the sample and due to reactions of neutrons with the empty sample holder is shown in Figure 2.

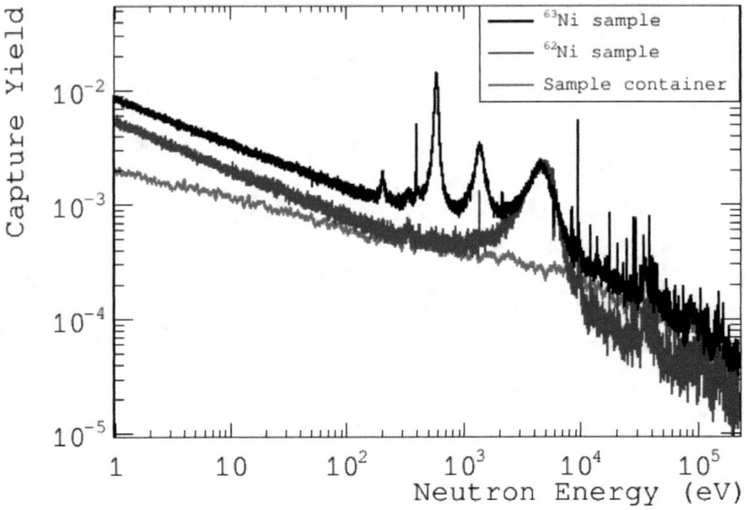

Figure 2. Capture yield of the ^{63}Ni(n,γ) reaction measured at n_TOF. The yield is compared to backgrounds coming from reactions on ^{62}Ni, which was present in the sample, and reactions of neutrons with the sample container [24].

^{63}Ni is a branching point in the s process. The weak s process takes place in two different burning stages in massive stars. First, neutrons are produced at the end of helium core burning via the ^{22}Ne(α,n) reaction. The neutron densities around 10^6 cm^{-3} are not sufficient for neutron capture to compete with the decay of ^{63}Ni, which means that the reaction flow is carried only by the decay channel. In the later carbon shell burning phase, the ^{22}Ne(α,n) neutron source is reactivated, leading to several magnitudes larger neutron densities. In this scenario, ^{63}Ni acts like a stable isotope and is mainly depleted via subsequent neutron capture towards ^{64}Ni, by-passing completely the production of ^{63}Cu, which is only produced by β-decay of the ^{63}Ni that is left after the neutron irradiation [25]. We determined stellar cross sections, which were about a factor of 2 higher than previous theoretical estimates listed in the KADoNiS compilation [26]. With the new cross section, we calculated s process abundances of a 25 $M_\odot$ star and obtained an increase of 20% in the ^{64}Ni abundance and a 15 and 30% decrease of the ^{63}Cu and ^{64}Zn abundances, respectively [24].

4. Outlook

Maxwellian Averaged Cross Sections are an important quantity to study the s process. Most challenging are (n,γ) measurements on radioactive species, since usually only small amounts of sample material are available, as well as measurements of very small (n,γ) cross sections (typically cross sections of light nuclei which may act as neutron poison in the s process). Some of those reactions require higher neutron fluxes than presently available. Several new facilities and upgrades are currently under construction or have been recently completed which are designed to obtain neutron fluxes 10-1000 times higher than existing facilities, e.g. the FRANZ [27] facility at the University of Frankfurt (Germany), SARAF [28] at the Soreq research centre (Israel), and a second experimental area at a shorter flight path at n_TOF [29]. These new facilities will enable a number of new cross section measurements, e.g. ^{85}Kr(n,γ) or ^{204}Tl(n,γ), and will provide crucial information for understanding the formation of the heavy elements.

References

[1] Reifarth R, Lederer C, Käppeler F, *J. Phys. G: Nucl. Part. Phys.* **41**, 053101 (2014).
[2] Guerrero C, and the n_TOF collaboration, *Eur. Phys. J.* A **49**, 27 (2013), www.cern.ch/ntof
[3] http://lansce.lanl.gov/
[4] http://irmm.jrc.ec.europa.eu/about_IRMM/laboratories/Pages/gelina_neutron_time_of_flight_facility.aspx
[5] Wallner A, *Nucl. Instrum. Meth.* B **268**, 1277 (2010).
[6] Ratynski W and Käppeler F, *Phys. Rev.* C **37**, 595 (1988).
[7] Lederer C, and the n_TOF Collaboration, *Phys. Rev.* C **85**, 055809 (2012).
[8] Feinberg G, et al., *Phys. Rev.* C **85**, 055810 (2012).
[9] Beer H and Käppeler F, *Phys. Rev.* C **21**, 534 (1980).
[10] Bao Z Y, et al., *At. Data Nucl. Data Tab.* **76**, 70 (2000).
[11] Heil M, et al., *Phys. Rev.* C **71**, 025803 (2005).
[12] Käppeler F, Naqvi A, Al-Ohali M, *Phys. Rev.* C **35**, 936 (1987).
[13] Nassar H, et al., *Phys. Rev. Lett.* **94**, 092504 (2005).
[14] Sneden C, et al., *Astroph. J.* **467**, 819 (1996).
[15] Sneden C, et al., *Astroph. J. Lett.* **533**, L139 (2000).
[16] Travaglio C, et al., *Astroph. J.* **601**, 864 (2004).
[17] Giubrone G, and the n_TOF Collaboration, in preparation (2014).
[18] Žugec P, and the n_TOF Collaboration, *Phys. Rev.* C **89**, 014605 (2014).
[19] Lederer C, and the n_TOF Collaboration, *Phys. Rev.* C **89**, 025810 (2014).
[20] Plag R, et al., *Nucl. Instrum. Meth.* A **496**, 425 (2003).
[21] Alpizar-Vicente A M, *et al.*, *Phys. Rev.* C **77**, 015806 (2008).
[22] Dillmann I, et al., *Nucl. Instrum. Meth.* B **268**, 1283 (2010).
[23] Colle R, et al., *Appl. Radiat. Isotopes* **66**, 60 (2008).
[24] Lederer C, and the n_TOF Collaboration, *Phys. Rev. Lett.* **110**, 022501 (2013).
[25] Pignatari M, et al., *Astroph. J.* **710**, 1557-1577 (2010).
[26] Dillmann I, et al., *AIP Conf. Proc.* **819**, 123 (2005); online at http://www.kadonis.org"
[27] Reifarth R, et al., *Publ. Astr. Soc. Aus.* **26**, 255 (2009).
[28] Feinberg G, et al., *Nuc. Phys.* A **827**, 590 (2009).
[29] Chiaveri E, and the n_TOF Collaboration, "Proposal for n_TOF Experimental Area 2 (EAR-2)", CERN-INTC-2012-029 / INTC-O-015 (2012).

Current hot questions on the *s* process in AGB stars

M Lugaro[1,2], S W Campbell[3,2], V D'Orazi[4,2,5], A I Karakas[6], D A Garcia-Hernandez[7], R J Stancliffe[8], G Tagliente[9], C Iliadis[10], and T Rauscher[11,12]

[1]Konkoly Observatory, Research Centre for Astronomy and Earth Sciences, Hungarian Academy of Sciences, Konkoly Thege Miklós út 15-17, H-1121 Budapest, Hungary
[2]Monash Centre for Astrophysics (MoCA), School of Mathematical Sciences, Building 28, Monash University, Clayton 3800, Victoria, Australia
[3]Max-Planck-Institut für Astrophysik, Karl-Schwarzschild-Str. 1, D-85741 Garching bei München, Germany
[4]INAF - Osservatorio Astronomico di Padova, vicolo dell'Osservatorio 5, I-35122 Padova, Italy
[5]Department of Physics and Astronomy, Macquarie University, Balaclava Rd, North Ryde, NSW 2109, Australia
[6]Research School of Astronomy & Astrophysics, Mount Stromlo Observatory, Weston Creek ACT 2611, Australia
[7]Instituto de Astrofisica de Canarias, C/ Via Lactea s/n, 38200 La Laguna (Tenerife), Spain
[8]Argelander-Institut fur Astronomie, Auf dem Hugel 71, D-53121 Bonn Germany
[9]Istituto Nazionale di Fisica Nucleare (INFN), Bari, Italy
[10]Department of Physics & Astronomy, University of North Carolina at Chapel Hill, Chapel Hill, North Carolina 27695-8202, USA
[11]Centre for Astrophysics Research, School of Physics, Astronomy, and Mathematics, University of Hertfordshire, Hatfield AL10 9AB, UK
[12]Department of Physics, University of Basel, 4056 Basel, Switzerland

E-mail: `maria.lugaro@csfk.mta.hu`

Abstract.
Asymptotic giant branch (AGB) stars are a main site of production of nuclei heavier than iron via the *s* process. In massive (>4 $M_\odot$) AGB stars the operation of the ^{22}Ne neutron source appears to be confirmed by observations of high Rb enhancements, while the lack of Tc in these stars rules out ^{13}C as a main source of neutrons. The problem is that the Rb enhancements are not accompanied by Zr enhancements, as expected by *s*-process models. This discrepancy may be solved via a better understanding of the complex atmospheres of AGB stars. Second-generation stars in globular clusters (GCs), on the other hand, do not show enhancements in any *s*-process elements, not even Rb. If massive AGB stars are responsible for the composition of these GC stars, they may have evolved differently in GCs than in the field. In AGB stars of lower masses, ^{13}C is the main source of neutrons and we can potentially constrain the effects of rotation and proton-ingestion episodes using the observed composition of post-AGB stars and of stardust SiC grains. Furthermore, independent asteroseismology observations of the rotational velocities of the cores of red giants and of white dwarves will play a fundamental role in helping us to better constrain the effect of rotation. Observations of carbon-enhanced metal-poor stars enriched in both Ba and Eu may require a neutron flux in-between the *s* and the *r* process, while the puzzling increase of Ba as function of the age in open clusters, not accompanied by increase in any other element heavier than iron, require further observational efforts. Finally, stardust SiC provides us high-precision constraints to test nuclear inputs such as neutron-capture cross sections of stable and unstable isotopes and the impact of excited nuclear states in stellar environments.

Journal of Physics: Conference Series **665** (2016) 012021 doi:10.1088/1742-6596/665/1/012021

1. Introduction

Since the 1950s the products of *slow* neutron captures (the *s* process) have been observed at the surface of asymptotic giant branch (AGB). These include the presence of the radioactive element Tc, which represented the first evidence that nuclear reactions produce heavy elements in stars [1, 2]. Observationally, AGB stars are red giants characterised by strong stellar winds, which drive most of the stellar envelope into the surroundings. Theoretically, they represent the final phase of the lives of stars with initial mass between roughly 1 and 10 $M_\odot$, before their degenerate C-O, or Ne-O, cores are left as cooling white dwarves. During the AGB phase, *thermal pulses* (TP) occur in the He-rich *intershell* region located in-between the H- and the He-burning shells. During a TP, He burning releases a large amount of energy ($\sim 10^7$ $L_\odot$), which drives a convective zone in the whole intershell. The outer layers of the star expand and H burning shuts off. Eventually this convective zone extinguishes, He burning also shuts off, and the convective envelope may sink in mass, penetrate into the intershell, and carry the products of partial He burning to the stellar surface (the *third dredge-up*, TDU). These products include C, F, and the elements heavier than Fe produced by the *s* process. In AGB stars of initial mass greater than ~ 4 $M_\odot$, the base of the convective envelope can become hot enough to trigger proton-capture reactions, whose products are carried to the stellar surface directly by the envelope convection (*hot bottom burning*). See [3] for a detailed review on AGB stars.

The sources of free neutrons for the *s* process in the He-rich intershell of AGB stars are the $^{13}C(\alpha,n)^{16}O$ and the $^{22}Ne(\alpha,n)^{25}Mg$ reactions. The $^{22}Ne(\alpha,n)^{25}Mg$ reaction is activated inside the convective TPs when the temperature reaches above 300 MK, as it is happens in AGB stars of initial mass > 3 $M_\odot$ [4, 5, 6, 7, 8, 9]. In these conditions there is a significant impact of the still uncertain rate of the neutron source reaction on the final *s*-process abundances [8].

The $^{13}C(\alpha,n)^{16}O$ reaction is activated at lower temperatures, from $\sim$90 MK. While required by the observations that show that low-mass AGB stars are *s*-process enhanced, the formation of this neutron source is still a matter of debate. It is usually accounted for in the models by means of more or less artificial mixing of protons from the envelope into the intershell. These protons react with the abundant ^{12}C to produce ^{13}C. In the most common scenario partial mixing of protons is included at the end of each TDU episode, under the assumption that the sharp discontinuity at the border between the convective envelope and the radiative intershell should favour the occurrence of such mixing. In practice, this mixing has been modelled via direct inclusion of ^{13}C [10, 11, 12, 13], direct inclusion of the protons leading to the formation of ^{13}C [14, 15, 7], or inclusion of an exponential decreasing profile of the diffusion coefficient [16] or of the convective velocity [17]. In all cases, free parameters allow us to adjust the extent in mass of the region affected by the mixing in order to match the observations. Usually, this extent represents a small fraction of the intershell ($1/10^{th}$-$1/20^{th}$) and the resulting thin ^{13}C-rich layer is refereed to as the ^{13}C *pocket*. The bottom line is that we still do not know the actual mechanism by which the ^{13}C pocket forms. It could be overshoot of convective border beyond the standard Schwarzschild criterion [16, 17], gravity waves [18], semiconvection [19], rotational mixing [20], or other processes not yet investigated. In any case, all the mechanisms proposed so far give us a pretty much exponentially decreasing proton profile, which is why this is the choice made in the parametric models of, e.g., [7]. Clearly, a fully self-consistent 3D hydro-dynamical model of the formation of the ^{13}C pocket is needed but not available yet.

In AGB stars of mass between ~ 1.75 $M_\odot$ and 3 $M_\odot$, once formed, the ^{13}C pocket burns releasing neutrons in radiative conditions, before the onset of the following TP [10, 11, 14, 12, 15, 17, 21, 13, 7]. In this conditions, the total number of free neutrons at any given metallicity is determined uniquely by the number of ^{13}C nuclei minus the number of the ^{14}N neutron poison, whose neutron-capture reaction $^{14}N(n,p)^{14}C$ is relatively efficient. In this situation the impact of the uncertainties related to the $^{13}C(\alpha,n)^{16}O$ reaction is minimal [22]. On the other hand, stellar rotation may have a large effect on the final *s*-process distribution: a possible difference in the angular velocity between the contracting core and the expanding envelope generate mixing in the ^{13}C pocket. As a consequecnce, ^{14}N produced in the top layers of the pocket (for initial proton numbers > 0.01) is mixed into the underlying ^{13}C-rich region, lowering the number of free neutrons. In this situation the *s* process can be completely inhibited, or modulated to lower efficiencies [23, 24, 25]. (See also S. Cristallo et al., this conference.) For stellar masses < 1.75 $M_\odot$, some ^{13}C can be left in the pocket to be ingested and burn in the following TP [17, 21, 7]. In this case, the efficiency of the *s*-process is lower than in the (non-rotating) radiative ^{13}C-pocket scenario as the ^{14}N abundance is higher because this nuclues is ingested in the TP from the pocket and from the H-burning ashes.

Another way to produce the ^{13}C neutron source is via ingestion of a small number of protons directly inside the TP [26, 27, 7]. Also in this case, the efficiency of the *s*-process is lower than in the non-rotating radiative ^{13}C-pocket scenario. Such *proton-ingestion episodes* are well known to occur in AGB stars of low mass (~ 1 $M_\odot$) and low metallicity (< 0.0001) [28, 29, 30, 31, 27, 32, 33, 7], as well as in post-AGB stars [34]. The details of PIE events and the mass and metallicity range for which they occur are very uncertain because, as in the case of the formation of the ^{13}C pocket, they rely on our incomplete understanding of convective boundaries in stars. First hydro-dynamical 3D models find many more proton ingested than 1D models [35] and preliminary 2D and 3D models of 2 $M_\odot$ stars of solar metallicity show that there is a finite mixing of material [36, 37]. When ^{13}C burns convectively inside the TPs, the uncertainties related to the $^{13}C(\alpha,n)^{16}O$ reaction have a more significant impact [22] as they determine the time scale at which ^{13}C burns, as compared to the time scale against which ^{14}N is destroyed via α captures.

Results for AGB stars of initial masses between 0.9 $M_\odot$ and 6 $M_\odot$ at metallicity 0.0001 [7] have shown that the the final *s*-process abundance distributions for different stellar masses depend on the interplay of the different regimes described above. In general, when rotation is not included, ^{13}C burning in radiative conditions produces higher total number of neutrons than ^{13}C burning in convective conditions, due to the effect of ^{14}N described above, but lower neutron densities because the burning time scale is longer.

2. Key Questions

2.1. Is the operation of the ^{22}Ne neutron source confirmed in massive (>4 $M_\odot$) AGB stars? Is the ^{13}C source also at work in these stars? Can massive AGB stars be responsible for the composition of the second stellar generation in globular clusters (GCs)?

Massive AGB stars at the end of the AGB phase show [Rb/Fe] ratios from $\sim$1 to $\sim$5 dex [38, 39]. This can be considered as the signature of the high neutron density produced by the ^{22}Ne source because ^{87}Rb, a magic nucleus with a low neutron-capture cross section, is produced via the branching points at ^{85}Kr and ^{86}Rb. While this qualitative argument is probably correct (it also predicts the observed increases of Rb with increasing the stellar mass and decreasing the metallicity [6]) it has been shown that only models with a delayed mass loss have enough TP to reach close to the data [8]. Another main issue is that in the same stars that show high Rb enhancements [Zr/Fe]$\sim$0 [40] while *s*-process models can at most produce [Rb/Zr] up to 0.5 dex [6, 8]. This main problem is currently under investigation by means of updated models of the complex atmospheres of AGB stars, where pulsation and dust formation may also play a role in defining the stellar spectra.

While the observed massive AGB Rb-rich stars, being enshrouded by dust, are believed to represent the end of the AGB phase, massive AGB stars observed at the start of the AGB phase show solar abundances of both Zr and Rb together with no sign of the presence of Tc [41]. This constraint can be matched only by models where the ^{13}C pocket is not included. The lack of neutrons from the ^{13}C source in massive AGB stars was predicted theoretically due to the effect of "hot dredge-up" [42, 43].

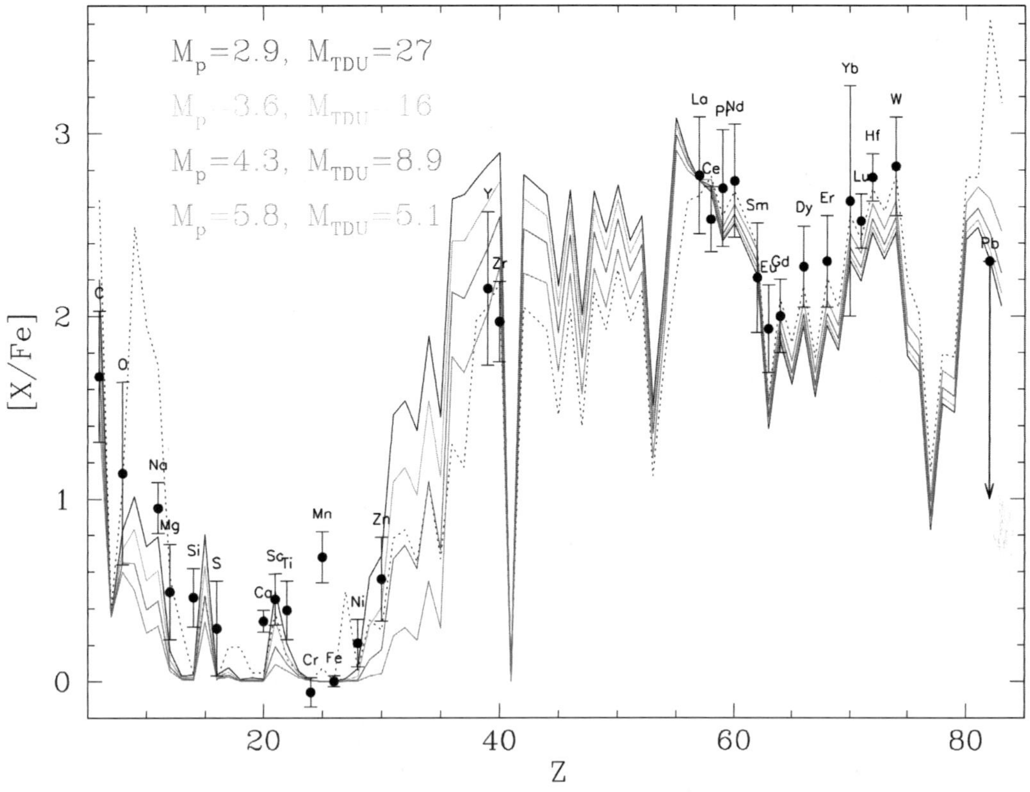

Figure 1. Comparison of AGB model predictions, computed on the basis of a stellar stucture with initial 1.3 $M_\odot$ and [Fe/H]=−1.3, to the composition of the post-AGB star J004441.04-732136.4 [46]. The dotted black line represents the results obtained introducing a ^{13}C pocket resulting from the mixing an exponentially decreasing proton profile over a mass of 0.002 $M_\odot$ and with a parametric TDU of 0.0096 $M_\odot$. The TDU is fixed to match the observed [La/Fe] ratio. The colored lines represent the results obtained by artificially ingesting in the third-last TP a mass of protons between 2.9 and 5.8 (in units of 10^{-6} $M_\odot$), and with a parametric TDU between 5.1 and 27 (in units of 10^{-4} $M_\odot$).

Due to hot bottom burning, massive AGB stars represent one of the most popular candidate to explain the O, Na, Mg, and Al composition of the different populations in GC stars [44]. However, variations in these elements are not accompanied by any variations in *s*-process elements, not even Rb. This *s*-process constraint can be matched only if massive AGB models are evolved using a strong mass loss [9]. However, as discussed above, direct observations of Rb appear to require a weaker mass loss [8]. This may indicate that massive AGB stars evolved differently in GCs than in the field, perhaps due to different binary properties of the stellar population, affecting the stellar lifetime [45]. This needs to be investigated via stellar population synthesis models.

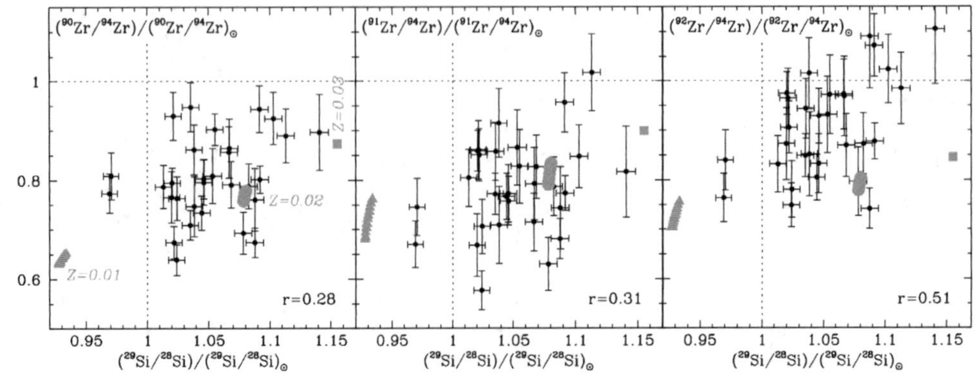

Figure 2. Comparison of Zr and Si data from SiC grains (black dots with 2σerror bars) with AGB models of 3 $M_\odot$ (green symbols for different metallicities Z, where solar = 0.014). The observed range of Zr ratios can be explained by changing the metallicity (as in the figure) or by the effect of rotation. If metallicity is the primary effect, we expect correlations between the Zr and the Si isotopic ratios, since the latter depend on the initial composition of the parent star. These correlations are hinted at in the small data sample currently available and it will be possible to confirm their existence (or lack of) via forth-coming studies. Also note that the SiC ^{92}Zr/^{94}Zr ratios are on average significantly higher than the models, and new experiments at GELINA (Belgium) and n_TOF are aimed at re-evaluating the neutron-capture cross section of ^{92}Zr. (Figure adapted from Lugaro et al. [48].)

2.2. How does the ^{13}C pocket operate in low-mass (<4 $M_\odot$) AGB stars? Can we constrain the effects of rotation and proton-ingestion episodes?

Recent observations of post-AGB stars in the Magellanic Clouds have provided us *s*-process abundances more reliable than those derived from observations of AGB star together with the opportunity to estimate the initial mass of the star to typical values ~ 1 1.5 $M_\odot$ [46, 47]. These post-AGB stars are characterised by *s*-process patterns that point to *s*-process efficiencies lower than those resulting by the non-rotating ^{13}C-pocket models, as well as lower C abundances than predicted. As discussed above, both rotation and proton-ingestion episodes can produce lower *s*-process efficiencies than the non-rotating ^{13}C pocket and we need to identify observational discriminants that can allow us to understand which of the two processes is responsible for the observed abundance patterns. We have started a parametric study to check for differences between the two scenarios. The first results of parametric models of proton-ingestion episodes are shown in Figure 1 and compared to J004441.04-732136.4. We confirm the results of de Smedt et al. [46] that the standard ^{13}C-pocket scenario produces too much Pb and too much C to reproduce the observations. The proton-ingestion models can better reproduce the observed abundance pattern, including C, however, it is not possible to find an *s*-process efficiency (as determined by the number of protons ingested) that can reproduce the observed abundance of Zr, as well as of all the elements between La and W and at the same time does not, even if slightly, overproduce Pb above the given upper limit. This problem needs to be further investigated.

More constraints on the *s*-process in low-mass AGB stars come from the interpretation of the composition of the elements heavier than Fe in silicon carbide (SiC) grains recovered from primitive meteorites. The isotopic composition of these grains have been analysed to very high precision using resonance or secondary ion mass spectrometry (RIMS and SIMS, respectively) and show the clear signature of an origin in low-mass AGB stars, including strong signatures of the *s* process [49]. These measurements are extremely powerful especially when data are available for the same grain on a number of different elements. Lugaro et al. [48] suggested that an observed correlation (or lack of) between the Zr and Si isotopic ratios of SiC grains can be used to evaluate if rotation or metallicity variations are responsible for the range of Zr isotopic ratios measured in the grains (Figure 2). So far, only roughly 30 data points are available to study this effect, but more data will become available soon also thanks to the upcoming CHILI RIMS instrument [50].

Another way to independently constrain the effect of rotation is use asteroseismology observations. As described above, the efficiency of the *s* process in the ^{13}C pocket depends on how fast the core rotates, which in turn depends on the initial velocity and the evolution of the angular momentum in the star. The latter can be modified by effects such as magnetic fields and gravity waves, which have not been considered in rotating *s*-process models so far. While it is difficult to infer rotational velocities for the cores of AGB stars from asteroseismology, it will be possible to derive them from models aimed at matching asteroseismic observations of the rotational velocities of the cores of red giants and of white dwarves, the stellar evolutionary phases just before and just after the AGB. Currently, the rotational velocities of white dwarves call for some braking effect due to, e.g., magnetic fields [51]. Furthermore, Mosser et al. [52] have observed a spin down of the core rotation in red giants, which requires a transfer of angular momentum in the star to spin down the core. Tayar & Pinsonneault [53] have shown that these observations can be explained only by complete coupling between the core and the envelope. The consequences of such studies on the *s* process needs to be investigated.

2.3. Is the standard s process enough to understand all the observations?

Some of the most interesting objects in the halo of our Galaxy are the carbon-enhanced metal-poor (CEMP) stars. The majority of them is believed to have gained their C and *s*-process enhancement via mass transfer from a more massive binary companion while it was evolving through the AGB. About half of CEMP stars, the CEMP-s/r stars, have enhancements in Ba, as well as in Eu, which cannot be explained by standard *s*-process models [13, 7]. Moreover, the Ba and Eu enhancements present a correlation ([Ba/Eu]$\sim$0.6, while the *s* process always produces [Ba/Eu]$\sim$0.9), which cannot be recovered by simply assuming high initial [Eu/Fe] ratio. As suggested by Lugaro et al. [7] the composition of these stars needs to be investigated in the light of a possible s/r process with neutron fluxes in-between the *s* and the *r* processes, and possibly linked to proton-ingestion episodes.

Recent observations of elements heavier than Fe in open clusters also present us with a puzzle: they show Ba abundances increasing with decreasing the age of the cluster, however, all the other observed neutron-capture elements, e.g., Zr, La, and Eu are constant [54, 55, 56, 57, 58]. (See also T.V. Mishenina et al., this conference.) Is this an observational problem or we need another process that produces only Ba? Attempting an answer to this question requires first to confirm which trends are real. To this aim large homogeneous data samples are mandatory.

2.4. Given the important uncertainties in the stellar models can we still learn something on the nuclear physics of the s process?

In spite of the large stellar model uncertainties, laboratory analysis of stardust SiC grains provide us with the isotopic ratios and the high precision needed to address nuclear physics issues. For example, the ^{92}Zr/^{94}Zr ratios in SiC grains are on average still higher than models predictions [48] even when computed using the latest ^{92}Zr(n,γ)^{93}Zr cross section measured at n_TOF [59]. Analysis of new experiments is underway to resolve this issue.

A new indirect estimate of the neutron-capture cross section of the unstable ^{85}Kr, via ^{86}Kr(γ,n)^{85}Kr at TUNL has allowed AGB *s*-process models to predict the ^{86}Kr/^{82}Kr ratio with the precision required to analyse this ratio in stardust SiC grains and derive that AGB models of low mass <1.5 $M_\odot$, where a large fraction of the ^{13}C neutron source is ingested in the TPs, provide a possible match to the high ratios observed in SiC grains of large size (a few μm) [60].

Finally, new reliable data on Eu isotopic ratios in SiC, obtained after careful investigation of molecular interferences in SIMS [61] have pointed out the need of a revision of the ^{151}Sm(n,γ)^{152}Sm reaction rate from the rate measured at n_TOF with very high precision [62], in line with the analysis of the effect of population of higher nuclear energy levels at stellar temperatures presented by Rauscher [63] (see T. Rauscher, this conference).

Acknowledgments

ML is a Momentum Project leader of the Hungarian Academy of Sciences. AIK is an ARC Future Fellow (FT10100475).

References

[1] Merrill S P W 1952 *Astrophy. J.* **116** 21–26
[2] Burbidge E M, Burbidge G R, Fowler W A and Hoyle F 1957 *Rev. Mod. Phys.* **29** 547–650
[3] Herwig F 2005 *Ann. Rev. Astron. Astrophys.* **43** 435–479
[4] Truran J W and Iben Jr I 1977 *Astrophy. J.* **216** 797–810
[5] Abia C, Busso M, Gallino R, Domínguez I, Straniero O and Isern J 2001 *Astrophy. J.* **559** 1117–1134 (*Preprint* arXiv:astro-ph/0105486)
[6] van Raai M A, Lugaro M, Karakas A I, García-Hernández D A and Yong D 2012 *Astron. Astrophys.* **540** A44 (*Preprint* 1202.2620)
[7] Lugaro M, Karakas A I, Stancliffe R J and Rijs C 2012 *Astrophy. J.* **747** 2 (*Preprint* 1112.2757)
[8] Karakas A I, García-Hernández D A and Lugaro M 2012 *Astrophy. J.* **751** 8 (*Preprint* 1203.2931)
[9] D'Orazi V, Campbell S W, Lugaro M, Lattanzio J C, Pignatari M and Carretta E 2013 *Mon. Not. Royal Astron. Soc.* **433** 366–381 (*Preprint* 1304.7009)
[10] Straniero O, Gallino R, Busso M, Chiefei A, Raiteri C M, Limongi M and Salaris M 1995 *Astrophy. J. Lett.* **440** L85–L87
[11] Gallino R, Arlandini C, Busso M, Lugaro M, Travaglio C, Straniero O, Chieffi A and Limongi M 1998 *Astrophy. J.* **497** 388
[12] Busso M, Gallino R, Lambert D L, Travaglio C and Smith V V 2001 *Astrophy. J.* **557** 802–821
[13] Bisterzo S, Gallino R, Straniero O, Cristallo S and Käppeler F 2011 *Mon. Not. Royal Astron. Soc.* **418** 284–319 (*Preprint* 1108.0500)
[14] Goriely S and Mowlavi N 2000 *Astron. Astrophys.* **362** 599–614
[15] Lugaro M, Herwig F, Lattanzio J C, Gallino R and Straniero O 2003 *Astrophy. J.* **586** 1305–1319
[16] Herwig F 2000 *Astron. Astrophys.* **360** 952–968
[17] Cristallo S, Straniero O, Gallino R, Piersanti L, Domínguez I and Lederer M T 2009 *Astrophy. J.* **696** 797–820 (*Preprint* 0902.0243)
[18] Denissenkov P A and Tout C A 2003 *Mon. Not. Royal Astron. Soc.* **340** 722–732
[19] Hollowell D and Iben Jr I 1988 *Astrophy. J. Lett.* **333** L25–L28
[20] Langer N, Heger A, Wellstein S and Herwig F 1999 *Astron. Astrophys.* **346** L37–L40
[21] Cristallo S, Piersanti L, Straniero O, Gallino R, Domínguez I, Abia C, Di Rico G, Quintini M and Bisterzo S 2011 *Astrophy. J. Supp.* **197** 17 (*Preprint* 1109.1176)
[22] Guo B and et al 2012 *Astrophy. J.* **756** 193 (*Preprint* 1208.0714)
[23] Herwig F, Langer N and Lugaro M 2003 *Astrophy. J.* **593** 1056–1073
[24] Siess L, Goriely S and Langer N 2004 *Astron. Astrophys.* **415** 1089–1097
[25] Piersanti L, Cristallo S and Straniero O 2013 *Astrophy. J.* **774** 98 (*Preprint* 1307.2017)
[26] Ulrich R K 1974 *Astrophy. J.* **192** 507–516
[27] Cristallo S, Piersanti L, Straniero O, Gallino R, Domínguez I and Käppeler F 2009 *Pub. Astron. Soc. Aust.* **26** 139–144 (*Preprint* 0904.4173)
[28] Cassisi S, Castellani V and Tornambe A 1996 *Astrophy. J.* **459** 298
[29] Fujimoto M Y, Ikeda Y and Iben I J 2000 *Astrophy. J.* **529** L25–L28
[30] Campbell S W and Lattanzio J C 2008 *Astron. Astrophys.* **490** 769–776 (*Preprint* 0901.0799)

[31] Lau H H B, Stancliffe R J and Tout C A 2009 *Mon. Not. Royal Astron. Soc.* **396** 1046–1057 (*Preprint* 0903.2324)
[32] Iwamoto N 2009 *Pub. Astron. Soc. Aust.* **26** 145–152
[33] Suda T and Fujimoto M Y 2010 *Mon. Not. Royal Astron. Soc.* **405** 177–193 (*Preprint* 1002.0863)
[34] Herwig F, Pignatari M, Woodward P R, Porter D H, Rockefeller G, Fryer C L, Bennett M and Hirschi R 2011 *Astrophy. J.* **727** 89 (*Preprint* 1002.2241)
[35] Stancliffe R J, Dearborn D S P, Lattanzio J C, Heap S A and Campbell S W 2011 *Astrophy. J.* **742** 121 (*Preprint* 1109.1289)
[36] Herwig F, Freytag B, Hueckstaedt R M and Timmes F X 2006 *Astrophy. J.* **642** 1057–1074 (*Preprint* arXiv:astro-ph/0601164)
[37] Herwig F, Freytag B, Fuchs T, Hansen J P, Hueckstaedt R M, Porter D H, Timmes F X and Woodward P R 2007 *Why Galaxies Care About AGB Stars: Their Importance as Actors and Probes* (*Astronomical Society of the Pacific Conference Series* vol 378) ed Kerschbaum F, Charbonnel C and Wing R F p 43 (*Preprint* 0709.0197)
[38] Garcia-Hernandez D A, Garcia-Lario P, Plez B, D'Antona F, Manchado A and Trigo-Rodriguez J M 2006 *Science* **314** 1751 (*Preprint* astro-ph/0611319)
[39] García-Hernández D A, Manchado A, Lambert D L, Plez B, García-Lario P, D'Antona F, Lugaro M, Karakas A I and van Raai M A 2009 *Astrophy. J. Lett.* **705** L31–L35 (*Preprint* 0909.4391)
[40] Garcia-Hernandez D A, Garcia-Lario P, Plez B, Manchado A, D'Antona F, Lub J and Habing H 2007 *Astron. Astrophys.* **462** 711–730 (*Preprint* astro-ph/0609106)
[41] García-Hernández D A, Zamora O, Yagüe A, Uttenthaler S, Karakas A I, Lugaro M, Ventura P and Lambert D L 2013 *Astron. Astrophys.* **555** L3 (*Preprint* 1306.2134)
[42] Goriely S and Siess L 2004 *Astron. Astrophys.* **421** L25–L28
[43] Herwig F 2004 *Astrophy. J.* **605** 425–435 (*Preprint* arXiv:astro-ph/0312616)
[44] Ventura P, D'Antona F, Mazzitelli I and Gratton R 2001 *Astrophy. J. Lett.* **550** L65–L69 (*Preprint* arXiv:astro-ph/0103337)
[45] D'Orazi V, Lugaro M, Campbell S W, Bragaglia A, Carretta E, Gratton R G, Lucatello S and D'Antona F 2013 *ArXiv e-prints* (*Preprint* 1308.4977)
[46] De Smedt K, Van Winckel H, Karakas A I, Siess L, Goriely S and Wood P R 2012 *Astron. Astrophys.* **541** A67 (*Preprint* 1203.4413)
[47] van Aarle E, Van Winckel H, De Smedt K, Kamath D and Wood P R 2013 *Astron. Astrophys.* **554** A106 (*Preprint* 1304.7103)
[48] Lugaro M, Tagliente G, Karakas A I, Milazzo P M, Käppeler F, Davis A M and Savina M R 2014 *Astrophy. J.* **780** 95 (*Preprint* 1311.2660)
[49] Lugaro M, Davis A M, Gallino R, Pellin M J, Straniero O and Käppeler F 2003 *Astrophy. J.* **593** 486–508
[50] Stephan T, Pellin M J, Rost D, Davis A M, Savina M R, Trappitsch R and Liu N 2013 *Lunar and Planetary Institute Science Conference Abstracts* (*Lunar and Planetary Inst. Technical Report* vol 44) p 2536
[51] Suijs M P L, Langer N, Poelarends A J, Yoon S C, Heger A and Herwig F 2008 *Astron. Astrophys.* **481** L87–L90 (*Preprint* 0802.3286)
[52] Mosser B, Goupil M J, Belkacem K, Marques J P, Beck P G, Bloemen S, De Ridder J, Barban C, Deheuvels S, Elsworth Y, Hekker S, Kallinger T, Ouazzani R M, Pinsonneault M, Samadi R, Stello D, García R A, Klaus T C, Li J, Mathur S and Morris R L 2012 *Astron. Astrophys.* **548** A10 (*Preprint* 1209.3336)
[53] Tayar J and Pinsonneault M H 2013 *ArXiv e-prints* (*Preprint* 1306.3986)
[54] D'Orazi V, Magrini L, Randich S, Galli D, Busso M and Sestito P 2009 *Astrophy. J. Lett.* **693** L31–L34 (*Preprint* 0901.2743)
[55] Maiorca E, Randich S, Busso M, Magrini L and Palmerini S 2011 *Astrophy. J.* **736** 120 (*Preprint* 1105.2208)
[56] Yong D, Carney B W and Friel E D 2012 *Astronom. J.* **144** 95 (*Preprint* 1206.6931)
[57] Jacobson H R and Friel E D 2013 *Astronom. J.* **145** 107 (*Preprint* 1303.4283)
[58] Mishenina T, Korotin S, Carraro G, Kovtyukh V V and Yegorova I A 2013 *Mon. Not. Royal Astron. Soc.* **433** 1436–1443 (*Preprint* 1305.1909)
[59] Tagliente G and et al 2010 *Phys. Rev. C* **81** 055801
[60] Raut R, Tonchev A P, Rusev G, Tornow W, Iliadis C, Lugaro M, Buntain J, Goriely S, Kelley J H, Schwengner R, Banu A and Tsoneva N 2013 *Physical Review Letters* **111** 112501 (*Preprint* 1309.4159)
[61] Ávila J N, Ireland T R, Lugaro M, Gyngard F, Zinner E, Cristallo S, Holden P and Rauscher T 2013 *Astrophy. J. Lett.* **768** L18 (*Preprint* 1303.5932)
[62] Marrone S and et al 2006 *Phys. Rev. C* **73** 034604
[63] Rauscher T 2012 *Astrophy. J. Lett.* **755** L10 (*Preprint* 1207.1664)

Nuclear Physics in Astrophysics VI (NPA6) IOP Publishing
Journal of Physics: Conference Series **665** (2016) 012022 doi:10.1088/1742-6596/665/1/012022

Experiments with radioactive target samples at FRANZ

K Sonnabend, S Altstadt, C Beinrucker, M Berger, A Endres, S Fiebiger, J Gerbig, J Glorius, K Göbel, T Heftrich, O Hinrichs, A Koloczek, A Lazarus, C Lederer, A Lier, B Mei, O Meusel, E Mevius, J Ostermöller, R Plag, M Pohl, R Reifarth, S Schmidt, Z Slavkovská, B Thomas, T Thomas, M Weigand, and C Wolf

Institut für Angewandte Physik, Goethe Universität Frankfurt a.M., Max-von-Laue-Str. 1, D-60438 Frankfurt a.M., Germany

E-mail: sonnabend@iap.uni-frankfurt.de

Abstract. The FRANZ facility is currently under construction at Goethe Universität Frankfurt a.M., Germany. It is designed to produce the world's highest neutron intensities in the astrophysically relevant energy range between 10 keV and 1 MeV and consists of a high-intensity proton linac providing energies close to the threshold of the ^{7}Li(p,n) reaction at $E_p = 1880$ keV. The high intensities of both the proton and the neutron beam allow the investigation of reactions of unstable target isotopes since the needed amount of target material is significantly reduced. We will present two examplary reactions relevant for the *s* process and the nucleosynthesis of *p* nuclei, respectively.

1. Introduction

The study of nuclear reactions relevant for stellar evolution or the nucleosynthesis of elements is often hampered by very low reaction yields due to low reaction cross-sections and/or small amounts of target material especially in case of radioactive samples [1]. To overcome these limits, it is mandatory to develop high-intensity beams and enable their usage by sophisticated cooling methods of the targets. The FRANZ facility is currently under construction at Goethe Universität Frankfurt a.M., Germany, and aims to produce highest neutron intensities in the astrophysically relevant energy range of about 10 keV to 1 MeV. Therefore, it will provide a high-intensity proton beam with energies close to the reaction threshold of $E_p = 1880$ keV of the reaction ^{7}Li(p,n) [2]. In addition to the neutron production, the beam can be used directly to study proton-induced reactions.

A schematic layout of the FRANZ facility is shown in Fig. 1 [3]. After the production in a volume-type ion source, the protons are transported via the Low Energy Beam Transport (LEBT) section which ends in a chopper device to impose a suitable time structure on the beam for the further processing. The acceleration to an energy of $E_p = 2$ MeV is realized with a combination of RFQ and IH structure. This solution is very space saving and can handle the aspired high intensities up to beam currents of 20 mA. After this acceleration, a CH rebuncher is foreseen to tune the beam energy without losses in intensity within ±10% to allow a specific modulation of the energy spectrum of the produced neutron beam.

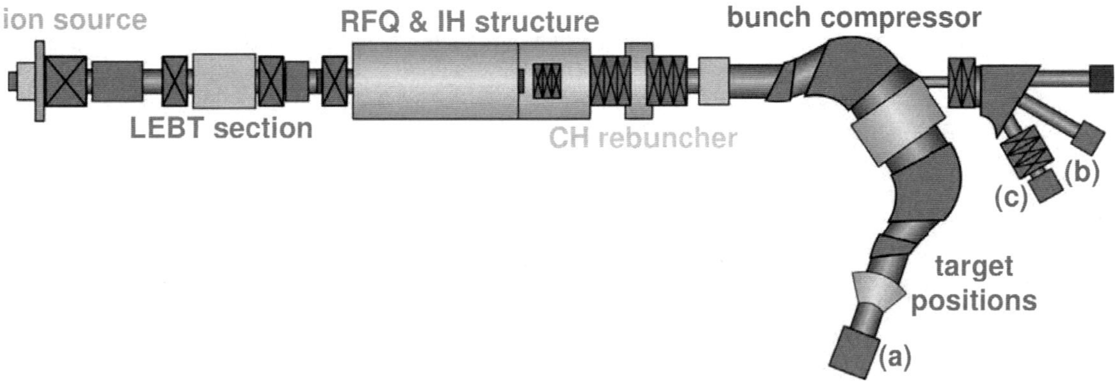

Figure 1. Schematic layout of the FRANZ facility at Goethe Universität Frankfurt a.M., Germany. A high-intensity low-energy proton beam will produce neutrons with the world's highest intensities in the keV energy-range. For details, see text.

At the FRANZ facility, different intensities of the neutron beam will be available. Using a fast calorimeter [4] at the experimental position labeled (a) in Fig. 1, the time-of-flight (TOF) technique provides the energy of the neutrons in an in-beam experiment. The bunch compressor samples nine of the proton bunches provided by the accelerating structures at the neutron-production target. Therefore, the lithium target deals with a proton beam of a mean current of 2 mA but with a very pronounced time structure: bunches with a length of 1 ns are repeated with a rate of 250 kHz and contain a current of 8 A. The resulting neutron beam has the world's highest intensity of 10^7 neutrons $s^{-1}cm^{-2}$.

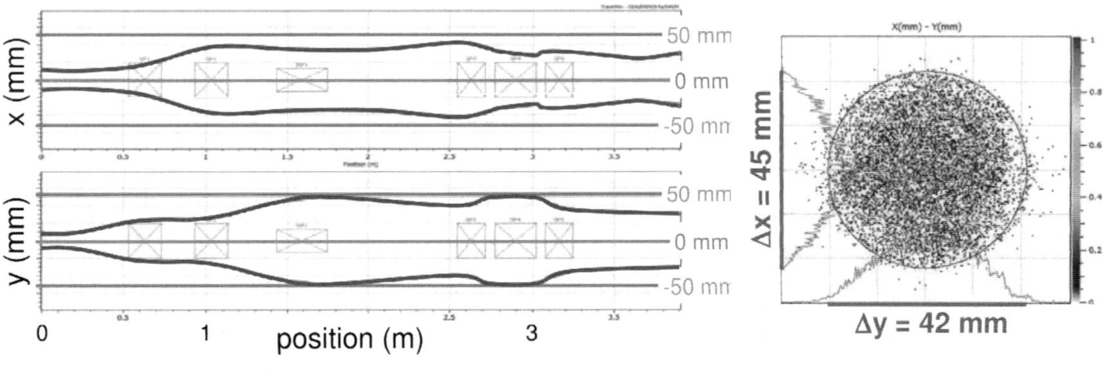

Figure 2. Simulation of beam transport at FRANZ. The transport from the CH rebuncher to experimental position (c) in Fig. 1 is shown in the left panel. The envelope of the beam passing a quadrupole doublet, a deflecting dipole magnet, and a quadrupole triplet is depicted. The right panel illustrates the dimension of the beam at target position (c) in Fig. 1. For details, see text.

For the activation technique realized at position (b) in Fig. 1, the proton beam need not to be shaped by the bunch compressor. Hence, the beam is used in the quasi-continuous wave mode as provided by the accelerating structures. Mean currents of up to 20 mA provide a neutron beam in the keV energy-range with intensities of up to 10^{12} neutrons s^{-1}. Position (c) in Fig. 1 is dedicated to the investigation of proton-induced reactions. Again, the proton beam is used in the quasi-continuous wave mode in combination with a 4π BaF_2 detector for γ calorimetry [4].

Fig. 2 shows the results of a simulation of the beam transport behind the bunch compressor towards position (c). A focussing quadrupole doublet, a deflecting dipole magnet, and a focussing and aligning quadrupole triplet are implemented in the simulation software TraceWin [5]. The left panels indicate the envelope of the beam in x and y direction, respectively. The right panel depicts the dimension of the beam at position (c). The quadrupole triplet yields an almost circular and homogeneous distribution with a diameter of about 4 cm. While a broad distribution is desirable to reach a reduced areal power density and prevent the samples from damage, further optimization of the components of the beam line is needed as the envelope is partly to close to the limiting diameter of the beam pipe [6].

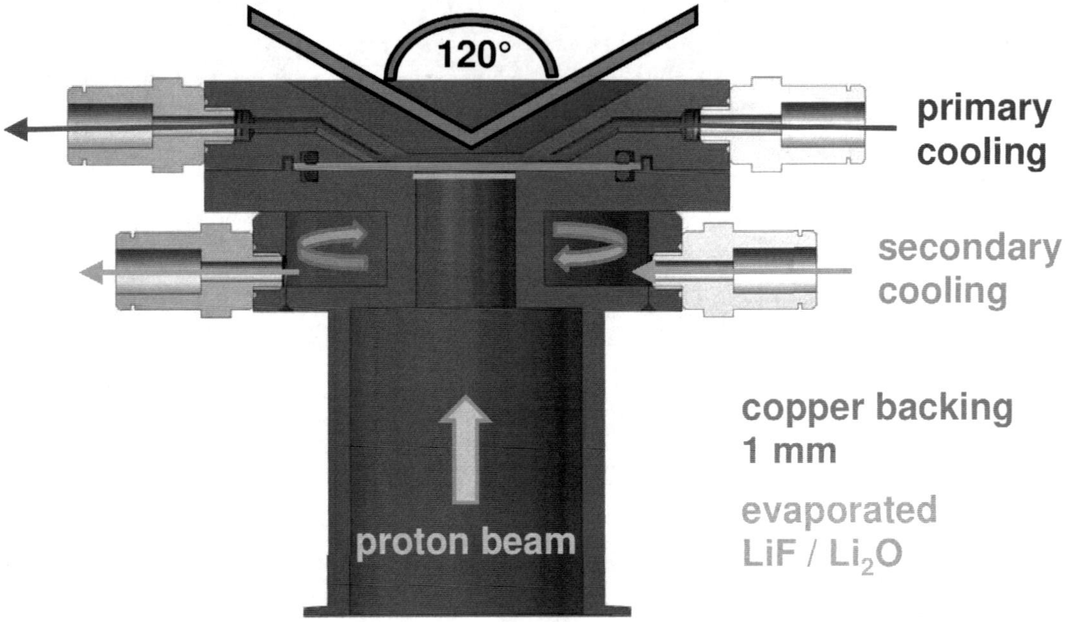

Figure 3. Sketch of a prototype of a neutron converter target. The protons are converted to neutrons by the ^{7}Li(p,n) reaction. The deposited power is carried away by a high-pressure water-cooling. Neutrons emitted in a cone with opening angle 120° are used for activation experiments. For details, see text and [7].

The high proton current provided by the linear accelerator at FRANZ is transformed to a high-intensity neutron beam by the ^{7}Li(p,n) reaction. The lithium is evaporated as a thin layer on a backing material with high thermal conductivity, *e.g.*, copper. In a prototype design (compare Fig. 3 and [7]), the copper backing is 1 mm thick and a μm layer of lithium fluoride (LiF) or lithium oxide (Li_2O) serves as neutron converter target. The melting temperatures of LiF and Li_2O are $T_M = 1118.15$ K and $T_M = 1843.15$ K, respectively, exceeding the one of metallic lithium ($T_M = 453.69$ K) by far [8]. As the loss of neutron conversion due to the stoichiometry remains small and the spectral composition of the neutrons remains shapeable by the thickness of the lithium layer, these chemical compounds might be a good alternative compared to a liquid lithium target [9]. A primary cooling of high-pressure water-flow carries away the heat from the power deposited by the proton beam. To avoid modulation of the neutron spectrum towards lower energies, the water layer is restricted to a thickness of 0.2 mm. A secondary cooling circle prevents the mounting structure from heating. First tests at IRMM,

Geel, Belgium, proved the vacuum-tightness and water-proofness of the prototype. In addition, the influence of the water cooling on the neutron spectrum was examined [7].

The high intensities of both the neutron and proton beams allow the investigation of reactions with very low yields as in the case of a radioactive probes [1]. The FRANZ facility will be one of the scarce accelerators worldwide where this is possible in standard kinematics and provides a remarkable alternative to experiments in inverse kinematics at large-scale facilities providing radioactive ion beams.

2. Investigation of the ^{85}Kr(n,γ) reaction

The cross section of the ^{85}Kr(n,γ) reaction is one of the important ingredients to understand the wide range of isotopic ratios $\delta(^{86}\mathrm{Kr}/^{82}\mathrm{Kr})$ as found in silicon carbide grains [10]. While the observed spread might be caused in a temperature dependent half-life of ^{85}Kr due to transitions between the ground-state and the isomeric state at E = 304.87 keV [11] the absolute value depends critically on the cross section of the ^{85}Kr(n,γ) reaction. There are attempts to derive this cross section from the inverse ^{86}Kr(γ,n) reaction [12]. However, the transition matrix elements connect very different levels in both cases and the uncertainty of the extracted neutron-capture cross section is still very large. Therefore, it is desired to attempt the direct measurement of the neutron-capture cross section at FRANZ [13] to clarify the nuclear physics of the branching point isotope ^{85}Kr (compare Fig. 4).

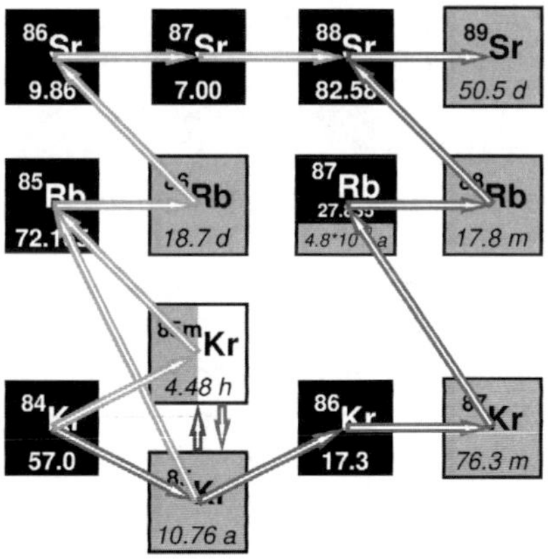

Figure 4. The s-process branching at ^{85}Kr. The situation is complicated by an isomeric state at E = 304.87 keV with a very short half-life of 4.48 h compared to the ground-state of ^{85}Kr with $t_{1/2}$ = 10.76 a. The red arrows indicate the main reaction path connecting the isotopes ^{84}Kr, ^{85}Kr, ^{86}Kr, ^{87}Kr, ^{87}Rb, ^{88}Rb, ^{88}Sr, and ^{89}Sr. The blue arrows between ^{85m}Kr and ^{85}Kr indicate the temperature dependent excitation and de-excitation reactions of the two nuclear levels. The orange arrows show alternative reactions paths due to the branching at ^{85}Kr.

An important step towards such an experiment is the production of a sufficient amount of target material in a suitable chemical form. The unstable isotope ^{85}Kr is produced by the ^{82}Se(α,n) reaction as well as by ^{82}Se(α,p) reaction and the subsequent β decay of ^{85}Br. This allows the almost isotopically pure production of ^{85}Kr by the irradiation of an isotopically enriched metallic ^{82}Se with α particles of energies between 6 MeV and 10 MeV (compare Fig. 5). If the temperature during the irradiation is below 50°C the krypton gas remains trapped in the selenium [14].

At a sample preparation site, the ^{85}Kr can be released from the selenium if the sample is heated beyond its melting point. A spherical container cooled by liquid-nitrogen can be used to condensate the krypton gas. An experiment with neutron intensities as provided by FRANZ requires an amount of at least 10^{17} atoms of ^{85}Kr. This number can be reached if the total amount of α particles impinging on the selenium target is about $2.5 \cdot 10^{18}$ projectiles [13].

Since ^{85}Kr emits a non-negligible amount of γ radiation after the β decay of both the isomeric and the ground-state, the amount of sample material to be placed inside a high-efficiency γ calorimeter without causing non-manageable background is limited. Therefore, an increase of the neutron intensity beyond the current design value of FRANZ is favourable for the measurement

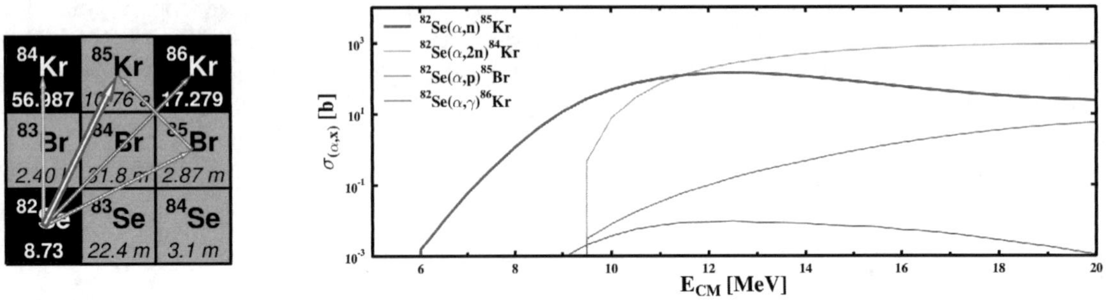

Figure 5. Production of target sample of ^{85}Kr. The irradiation of an enriched ^{82}Se target with α particles between 6 MeV and 10 MeV leads to the almost isotopically pure production of ^{85}Kr. For details, see [13].

of the ^{85}Kr(n,γ) reaction. This can be achieved by a reduced flight path of the neutrons provided by an optimized γ calorimeter as described in [13].

3. Investigation of the ^{91}Nb(p,γ) reaction

The most abundant p nucleus ^{92}Mo ($I_{\%}$ =14.84%) can be produced by the so-called γ process [15] or a series of subsequent proton capture reactions on stable or long-lived isotopes of the N = 50 closed neutron shell starting from ^{86}Kr (compare Fig. 6 and [16, 17]). The ^{91}Nb(p,γ) reaction is not experimentally studied so far since a direct measurement requires the availability of a radioactive target sample of ^{91}Nb isotopes. Due to the increasing Coulomb barrier, the proton-capture cross sections on ^{90}Zr and ^{91}Nb decrease compared to the other semi-magic isotopes at $N = 50$. Thus, these cross sections determine how much ^{92}Mo can be produced in the explosive scenario of a type Ia supernova. As an unstable isotope, ^{91}Nb is not available in the seed distribution. Thus, every ^{91}Nb atom has to be produced before it can capture a proton to become ^{92}Mo. Therefore, the cross sections of the ^{90}Zr(p,γ) and ^{91}Nb(p,γ) reactions are the most important ones to understand the production of ^{92}Mo. Besides the production of ^{92}Mo via proton-capture reactions its destruction by the same reaction type has to be taken into account, too. However, the produced ^{92}Mo will not be destroyed directly because of the low cross section of the reaction ^{92}Mo(p,γ) [18].

The fundamental difficulty of the measurement of the ^{91}Nb(p,γ) reaction is the instability of the ^{91}Nb isotope. At the FRANZ facility, a small amount of ^{91}Nb is sufficient to determine the cross section of interest. The ^{91}Nb sample will be produced by irradiating a molybdenum target enriched in ^{92}Mo with protons of energies of 25 MeV energy. Due to the closed neutron shell, the dominating reaction channels in this energy range are ^{92}Mo(p,2p), ^{92}Mo(p,pn), and ^{92}Mo(p,2n). Since ^{91}Nb is either produced directly or via subsequent electron capture decays, the produced niobium is almost isotopically pure. An amount of $1.2{\cdot}10^{16}$ ^{91}Nb nuclei yields after the irradiation of 300 mg sample material enriched to 95% in ^{92}Mo with a current of about 80 μA for about 24 hrs. This can be realized for example at the Oslo Cyclotron Laboratory, Norway. Prior to that production run, the cross sections shown in Fig. 7 will be experimentally determined at the same facility [20].

As can be seen from Fig. 7, the production of ^{93}Nb by the ^{92}Mo(p,γ) reaction is hampered by the much lower reaction cross section as well as by the very large half-life of ^{93}Mo ($t_{1/2}$ = 4000 a). The chemical purification of the sample material has to be optimized for the reduction of molybdenum in the sample. However, even if some ^{92}Mo remains in the sample material the contribution from the ^{92}Mo(p,γ) reaction to the measured spectra can be easily separated because of its different Q value (Q = 4087 keV) compared to the ^{91}Nb(p,γ) reaction

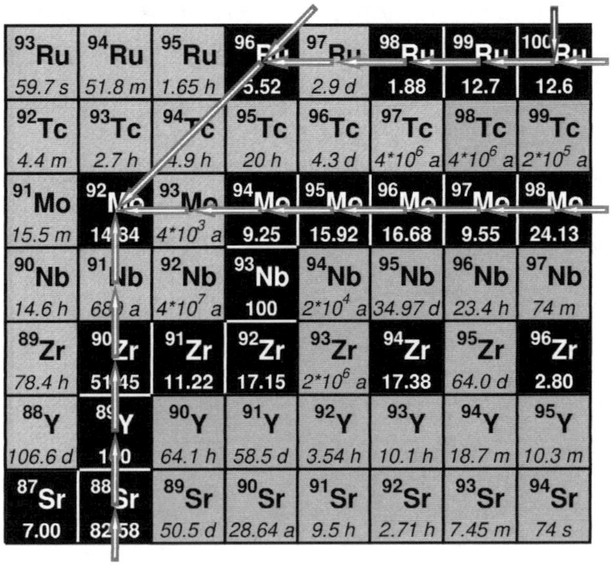

Figure 6. Nucleosynthesis of the most abundant p nucleus ^{92}Mo. The blue arrows indicate a part of the reaction network for the so-called γ process at a temperature of $T = 2.5 \cdot 10^9$ K (calculated by [19]). A series of γ induced reactions on a seeds distribution in an explosive scenario leads to the production of proton-rich isotopes. The red arrows show an alternative production scenario by a series of radiative proton-capture reactions as proposed by [16, 17]. Due to clearness, the reaction paths are halted at ^{92}Mo.

($Q = 7462$ keV).

The cross section of the ^{91}Nb(p,γ) reaction is predicted to amount about 10 μb at $E_p = 2$ MeV. An amount of 10^{16} ^{91}Nb nuclei yields a count rate of approximately 1.5 kHz using the 4π BaF_2 calorimeter available at FRANZ. This rate is clearly above the expected background and can be processed by the data acquisition system. Therefore, a sample of suitable size can be produced by the proposed procedure.

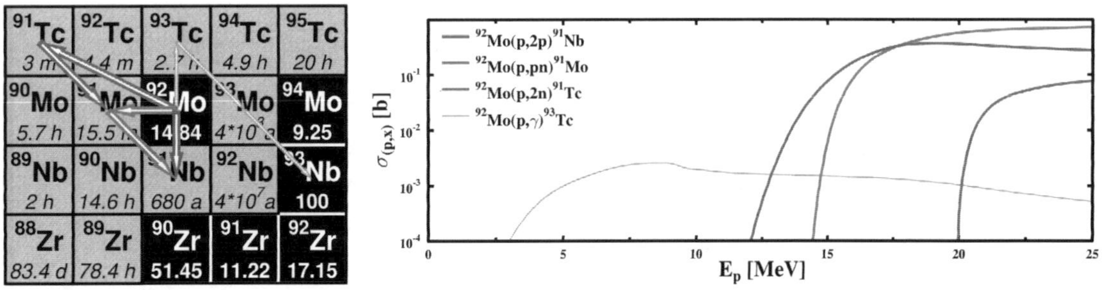

Figure 7. Production of target sample of ^{91}Nb. The irradiation of an enriched ^{92}Mo target with protons of 25 MeV leads to the almost isotopically pure production of ^{91}Nb. For details, see text.

4. Summary and outlook

The Frankfurt neutron-source FRANZ will provide the world's highest neutron intensities for both activation and TOF experiments in the keV energy-range. The study of key reactions for s-process nucleosynthesis is foreseen at this facility. In addition, the high-intensity low-energy proton beam will serve as a source to study proton-induced reactions related to the production of the p nuclei. The study of reactions on unstable isotopes necessitates the production of radioactive samples. For each case, as shown exemplarily for ^{85}Kr and ^{91}Nb, a special production mechanism leading to isotopically pure samples has to be developed.

Short half-lives or high intrinsic activities prevent a measurement in standard kinematics. Then, experiments in inverse kinematics can be performed as explained, *e.g.*, in the contributions of A.M. Laird and R. Reifarth to this volume.

Acknowledgements

This work is supported by Deutsche Forschungsgemeinschaft (contract SO907/2-1), German-Israeli Foundation for Scientific Research and Development (GIF, contract G-1051-103.7/2009), and by HIC for FAIR within the framework of LOEWE launched by the State of Hesse, Germany.

References

[1] Couture A and Reifarth R 2007 *At. Data Nucl. Data Tables* **93** 807
[2] Reifarth R, Lederer C and Käppeler F 2014 *J. Phys. G* **41** 053101
[3] Reifarth R, Chau L P, Heil M, Käppeler F, Meusel O, Plag R, Ratzinger U, Schempp A and Volk K 2009 *Publ. Astron. Soc. Aust.* **26** 255
[4] Wisshak K, Guber K, Käppeler F, Krisch J, Müller H, Rupp G and Voss F 1990 *Nucl. Instr. Meth. A* **292** 595
[5] Uriot D TraceWin online description (http://irfu.cea.fr/Sacm/logiciels/index3.php, July 29, 2014)
[6] Hinrichs O *Design und Aufbau der Hochenergie-Strahltransportführung an FRANZ* Ph.D. thesis Goethe University Frankfurt a.M., in progress
[7] Fiebiger S 2012 *Entwicklung eines Prototypen zur Neutronenproduktion via* ^{7}*Li(p,n)* bachelor thesis, Goethe University Frankfurt a.M., unpublished
[8] Berger M 2012 *Untersuchung verschiedener Li-Verbindungen zur Verwendung als Neutronenquelle* bachelor thesis, Goethe University Frankfurt a.M., unpublished
[9] Friedman M, Cohen D, Paul M, Berkovits D, Eisen Y, Feinberg G, Giorginis G, Halfon S, Krása A, Plompen A and Shor A 2013 *Nucl. Instr. and Meth. Phys. Res. A* **698** 117
[10] Pignatari M, Gallino R, Amari S and Davis A M 2006 *Memorie della Societa Astronomica Italiana* **77** 897
[11] Thomas B 2013 *Isomere im s-Prozess am Beispiel von* ^{85}*Kr* master thesis, Goethe University Frankfurt a.M., unpublished
[12] Raut R, Tonchev A P, Rusev G, Tornow W, Iliadis C, Lugaro M, Buntain J, Goriely S, Kelley J H, Schwengner R, Banu A and Tsoneva N 2013 *Phys. Rev. Lett.* **111** 112501
[13] Reifarth R, Glorius J, Kräckmann S, Langer C, Meusel O, Plage R, Pohl M, Schmitt S, Sonnabend K, Heil M and O'Donnell J M 2013 ed Garrett P E and Hadinia B (World Scientific Publishing) pp 337–346
[14] Suzuki K, Blessing G, Qaim S and Stöcklin G 1982 *Int. J. Appl. Rad. and Iso.* **33** 1445
[15] Arnould M and Goriely S 2003 *Phys. Rep.* **384** 1
[16] Kusakabe M, Iwamoto N and Nomoto K 2011 *Astrophys. J.* **726** 25
[17] Travaglio C, Röpke F, Gallino R and Hillebrandt W 2011 *Astrophys. J.* **739** 93
[18] Gy Gyürky, Vakulenko M, Zs Fülöp, Halász Z, Kiss G G, Somorjai E and Szücs T 2014 *Nucl. Phys.* **A922** 112
[19] Rauscher T 2006 *Phys. Rev. C* **73** 015804
[20] Thomas B *Produktion radioaktiver Proben zum Einsatz an FRANZ* Ph.D. thesis Goethe University Frankfurt a.M., in progress

AGB yields and Galactic Chemical Evolution: last updated

S Bisterzo[1,2], C Travaglio[2], M Wiescher[3], R Gallino[1], F Käppeler[4], O Straniero[5], S Cristallo[5], G Imbriani[6], J Görres[3] and R J deBoer[3]

[1] Department of Physics, University of Turin, Italy
[2] INAF - Astrophysical Observatory of Turin, Italy
[3] Department of Physics, University of Notre Dame, Notre Dame, IN
[4] Karlsruhe Institute of Technology, Karlsruhe, Germany
[5] Osservatorio Astronomico di Collurania, Teramo, Italy
[6] Dipartimento di Scienze Fisiche, Università di Napoli Federico II, Italy

E-mail: bisterzo@to.infn.it,sarabisterzo@gmail.com

Abstract. We study the s-process abundances at the epoch of the Solar-system formation as the outcome of nucleosynthesis occurring in AGB stars of various masses and metallicities. The calculations have been performed with the Galactic chemical evolution (GCE) model presented by [1, 2]. With respect to previous works, we used updated solar meteoritic abundances, a neutron capture cross section network that includes the most recent measurements, and we implemented the s-process yields with an extended range of AGB initial masses. The new set of AGB yields includes a new evaluation of the ^{22}Ne(α, n)^{25}Mg rate, which takes into account the most recent experimental information.

1. Introduction

The understanding of the s-process contribution to the isotopic abundances of heavy elements in the Solar System is fundamental to disentangle between several nucleosynthesis processes that have competed during the evolution of the Milky Way. The s-process abundances observed in the Solar System are the result of a complex Galactic chemical evolution process, which mainly accounts for the pollution of several AGB generations with different initial masses and metallicities.

It was shown that AGB stars with low initial masses (M = 1.5 and 3 $M_\odot$), half solar metallicity, and a specific ^{13}C-pocket choice (called case ST) reproduce the *main* component of the s-process [3], which synthesized about half of the elements from Sr to Pb. This approximation still provides strong information about the s-process contribution to isotopes in the region between ^{134}Ba and ^{204}Pb (see discussion in [4]).

Two additional s-process components should be considered to reproduce the solar abundances of light neutron capture isotopes up to Sr (the $weak-s$ component) and the stable isotopes at the termination point of the s-path, ^{208}Pb and ^{209}Bi (the $strong-s$ component). The $weak-s$ process occurs in massive stars during core He and shell C burning and partly produces neutron capture isotopes lighter than A $\sim$ 90 (^{86,87}Sr $\sim$10%; lower contribution to Y and Zr isotopes; see e.g., [5]). AGB stars with low metallicity and low initial mass synthesize about half of solar ^{208}Pb, the so called $strong-s$ component (see [6, 7]). The ^{13}C in the pocket is primary, which means it is directly synthesized in the star independently of the initial composition. Therefore, by decreasing the metallicity, the number of free neutrons per iron seed becomes so large to overcome the first and the second *s*-peaks (Sr-Y-Zr and Ba-La-Ce, respectively), to feed directly ^{208}Pb (and ^{209}Bi).

Actually, the understanding of the origin of light *s*-process isotopes Sr, Y and Zr is more enigmatic. GCE models by [2] found that AGB yields underestimate the solar abundance of Sr, Y, Zr, and the solar composition of isotopes from ^{86}Sr to ^{130}Xe by about 20–30% (including the *s*-only ^{96}Mo, ^{100}Ru, ^{104}Pd, ^{110}Cd, ^{116}Sn, ^{122,123,124}Te and ^{130}Xe). The $weak - s$ process can not compensate the missing solar abundance, because it mainly produces isotopes up to Sr, with a negligible contribution up to ^{130}Xe. Spectroscopic observations of peculiar metal-poor stars showing large enhancement in *r*-process elements[1] suggest that ~10% of solar Sr-Y-Zr is due to the *r*-process. However, both *s*- and *r*-processes are not sufficient to explain the solar observations of light neutron capture elements. [2] hypothesized the existence of an additional process of unknown origin, called by the authors LEPP (light element primary process since it was supposed to be of primary origin). Several scenarios have been recently explored, both involving the secondary *s*-process in massive stars (e.g., *cs*-component by [9], which may explain the missing *s*-process component) or primary *r*-process during the advanced phases of explosive nucleosynthesis (see review [10], which may account of complementary *r*-contributions). Therefore, even if promising theoretical improvements related to the explosive phases of massive stars and core collapse Supernovae, as well as recent spectroscopic investigations [11, 12] have been made, a fully understanding about the origin of the neutron capture elements from Sr up to Xe is still lacking.

We aim to investigate the effects of a new set of AGB yields (with updated nuclear cross section network and solar abundances) on the solar *s*-process composition (Section 2).

2. Results

We focus the analysis on the variations of the Solar-system *s*-process GCE predictions by adopting updated AGB yields, and by testing the effects of different prescriptions on nuclear cross sections. The Galactic evolution is computed as function of time up to the present epoch ($t_{Today} = 13.73 \pm 0.12$ by WMAP), following the three zones of the Galaxy, halo, thick and thin disk. We adopted the yields by [13] and [14] for SNe II and Ia, respectively.

Major revisions involve the solar system meteoritic abundances by [15], a neutron capture cross section network with the most recent published measurements, as well as a larger set of AGB yields that extend toward lower initial mass (down to $M = 1.3\ M_\odot$). We started from the AGB models presented by [16], which were based on the FRANEC code by [17, 18, 19].
A series of thousands of new AGB models have been run, for a total of 28 metallicities from [Fe/H] $= +0.2$ down to -3.6 (most of them focused on the metallicity range between solar and [Fe/H] $= -1.6$, where the isotopes of the three *s*-peaks are largely produced, see [1, 7, 2, 20]. Yields of a set of five AGB models with low initial masses (M = 1.3, 1.4, 1.5, 2, and 3 $M_\odot$) and two AGB models with intermediate initial masses (M = 5 and 7 $M_\odot$) have been interpolated/extrapolated over the whole metallicity and mass range (hereafter LMS refers to the mass range between $1.3 \leq M/M_\odot < 4$, and IMS to $4 \leq M/M_\odot < 8$). This assures a sufficiently high accuracy in the AGB mass and metallicity ranges.

Particularly challenging for the *s*-process is the understanding of the formation of the ^{13}C-pocket, specifically the mass fraction of ^{13}C and ^{14}N in the pocket and the mass involved. Both uncertainties largely affect the *s*-yields (see e.g., [21, 22]). Current full evolutionary AGB models still adopt a free parametrization to reproduce the ^{13}C-pocket, by means of overshooting

[1] E.g., CS 22892–052 and CS 31081–001 (see review by [8]), where the *rapid* neutron capture process is ascribed to explosive nucleosynthesis phases of massive stars.

Journal of Physics: Conference Series **665** (2016) 012023 doi:10.1088/1742-6596/665/1/012023

(e.g., [23, 24, 25]) or other better physically justified prescriptions (e.g., new FRUITY models by [26, 27], and references therein; or mixing produced by magnetic fields, [28]). As suggested by the *s*-process spread observed at a given metallicity in different stellar populations (e.g., post-AGB, Ba, CH and CEMP-s stars; [8, 29]), a range of the *s*-process efficiency strengths is needed. The theoretical reason of this spread observed among *s*-elements is still under investigation (AGB initial mass, magnetic fields, gravitational waves, or rotation, see [30]).

As discussed by [16], we artificially introduce the ^{13}C-pocket in our post-process AGB calculations, and we treated it as a free parameter kept constant pulse by pulse. Starting from the ^{13}C-pocket ST case, similar to that adopted by [6], we multiply or divide the ^{13}C (and ^{14}N) abundances in the pocket by different factors. We considered an accurate weighted average of the ^{13}C-pocket efficiencies in order to reproduce ~100% of solar ^{150}Sm and the other *s*-only isotopes heavier than $A \sim 90$. Note that ^{150}Sm has well defined solar abundance, and it is the *s*-only isotope less affected by branchings and nuclear cross section uncertainties as well. A second reaction, the ^{22}Ne(α, n)^{25}Mg, starts to partially burn at $T_8 = 3$ during thermal pulses and produces an efficient neutron burst mainly affecting isotopes close to the branching points.

Results are shown in Figure 1. The *s*-only isotopes are indicated by solid circles. Different symbols have been used for isotopes that receive additional contributions: ^{128}Xe, ^{152}Gd, and ^{164}Er (open squares), which have a non-negligible p contribution; ^{176}Lu (open triangle), a long-lived isotope (3.8 x 10^{10} yr), which decays into ^{176}Hf; ^{187}Os (open triangle), which is affected by the long-lived decay of ^{187}Re (4.1 x 10^{10} yr); ^{180}Ta (open circle), which also receives contributions from the p process and from ν-nucleus interactions in massive stars; ^{208}Pb (filled big square), half of which is produced by the strong-*s* component. We compare the *s*-process predictions, calculated with the GCE model at the epoch of the Solar System formation ($t_\odot$ = 9.17 Gyr from the beginning of the birth of the Universe), with the meteoritic abundances by [15]. As shown in the *top panel*, LMS (black symbols) reproduce almost all the Solar System *s*-only isotopes between A = 140 and 210. An additional small contribution (<10%) comes from IMS for isotopes with $A < 140$. The total *s*-percentages (LMS + IMS) is represented by blue symbols. IMS AGBs play a minor role in the Galactic enrichment, because their He-intershell is smaller than in LMS by one order of magnitude, with an uncertain formation of the ^{13}C-pocket and a less efficient third dredge-up. The ^{22}Ne(α, n)^{25}Mg neutron source is efficiently activated in IMS due to the high temperature reached at the bottom of the thermal pulses (T_8 = 3.5). The peak neutron density reached in IMS easily allows a strong overproduction of ^{86}Kr, ^{87}Rb, and ^{96}Zr, three neutron rich isotopes affected by the branchings at ^{85}Kr and ^{95}Zr, which are very sensitive to the neutron density.

Particularly large is the abundance of ^{96}Zr (big asterisk). Note that ^{96}Zr is strongly sensitive to the number of thermal pulses experienced by the AGB models, as well as to the ^{22}Ne(α, n)^{25}Mg rate adopted. In general, an over-prediction of ^{96}Zr may suggest a low number of thermal pulses for IMS. As discussed by [16], we assume a strong mass loss for IMS, which allows a total of 24 thermal pulses. Under this hypothesis, IMS produce about 30% of solar ^{96}Zr, while an additional 70% comes from LMS (Figure 1, *top panel*). ^{96}Zr is the most neutron rich stable Zr isotope, and a contribution of 100% from AGB is surely overestimated. Moreover, it disagrees with recent *p*-process predictions by [32], which estimate an additional non-negligible contribution to ^{96}Zr by SNIa (up to 30%).

As a further improvement, we computed a new set of AGB yields that includes a new evaluation of the ^{22}Ne(α, n)^{25}Mg rate, which takes into account the most recent experimental information. The recommended value we are suggesting is very close to [34] and agrees with the recent determination of [35]. At AGB temperature (T_8 ~ 2.5 to 3.5) the new ^{22}Ne(α, n)^{25}Mg rate is about a factor of 2 lower than our rate used so far (corresponding to the lower limit suggested by [36][2]. The related new Solar-system abundances are displayed in Figure 1, *bottom*

[2] Note that we modified accordingly the rate of the ^{22}Ne(α, γ)^{26}Mg. Specifically, at the temperature of interest

panel, red symbols. As expected, major differences are shown close to the branchings points. In particular, ∼50% of solar ^{96}Zr is produced by AGBs, in better agreement with expectations. Updated GCE calculations plausibly reproduce within the uncertainties all *s*-only isotopes with $A > 130$, and confirm the missing 20% Solar *s*-contribution of *s*-only isotopic abundances between $A = 96 - 130$ found by [2]. Variations with respect to the results presented by [29], their Fig. 15, bottom panel) are mainly due to new solar abundances, recent neutron capture cross section measurements, and the new evaluation of the ^{22}Ne(α, n)^{25}Mg rate.
Note that ^{192}Pt and ^{198}Hg are affected by large uncertainties: concerning ^{192}Pt, the neutron capture cross section of ^{191}Os and ^{192}Ir evaluated theoretically at 22%, the extrapolation of the ^{192}Ir measurement in stellar conditions (see discussion by [37], as well as the old measurement of the ^{192}Pt(n, γ) reaction, with 20% of uncertainty at 30 keV [38]; moreover, Hg is too volatile for a reliable experimental determination of the solar abundance ([15] estimated an uncertainty of 20%). ^{204}Pb (and all Pb isotopes) have well determined neutron capture cross sections (see [39]), but it is strongly affected by the branching at ^{204}Tl, with variations of ∼10%.

Updated Solar *s*-process abundances of some selected elements are compared with previous GCE computations [7, 2] in Table 1. Marginal differences (<5%) are seen in general. La and Ce are among the few exceptions: the larger *s*-contribution obtained by this work (+12% and +6%) is the consequence of the new ^{139}La(n, γ)^{140}La rate measured by [40]. A more detailed comparison and a discussion concerning the most relevant updated information will be provided in a future work.

Table 1. The *s*-process contribution from LMS and IMS AGBs at the epoch of the Solar System formation (in percentages). Results by [7, 2] in column 2 are compared with updated results (this work, column 3), computed with all recently updated information, including the new ^{22}Ne(α, n)^{25}Mg rate. The *r*-process contribution (evaluated with the residual method $N_r = N_s - N_\odot$) is given in column 4 (in percentages).

Element	s(2001-2004)	s(This work)	r(This work)
Sr	71	67	—
Y	69	70	—
Zr	65	64	—
Ba	80	83	17
La	61	73	27
Ce	75	81	19
Pr	47	49	51
Nd	54	56	44
Sm	30	31	69
Eu	6	6	94
Pb	91	88	12

Acknowledgments

S.B. wishing to acknowledge JINA for financial support under ND Fund #201387 and 305387.

References

[1] Travaglio, C., Galli, D., Gallino, R., Busso, M., Ferrini, F., Straniero, O. 1999, ApJ, 521, 691
[2] Travaglio, C., Gallino, R., Arnone, E., Cowan, J., Jordan, F., Sneden, C. 2004, ApJ, 601, 864

of AGBs, between T_8 = 2.5 to 3.5, this rate is almost unchanged.

[3] Arlandini, C., Käppeler, F., Wisshak, K., Gallino, R., Lugaro, M., Busso, M., Straniero, O. 1999, ApJ, 525, 886
[4] Bisterzo, S., Gallino, R., Straniero, O., Cristallo, S., Käppeler F. 2011, MNRAS, 418, 284
[5] Pignatari, M., Gallino, R., Heil, M., Wiescher, M., Käppeler, F., Herwig, F., Bisterzo, S. 2010, ApJ, 710, 1557
[6] Gallino, R., Arlandini, C., Busso, M., Lugaro, M., Travaglio, C., Straniero, O., Chieffi, A., Limongi, M. 1998, ApJ, 497, 388
[7] Travaglio, C., Gallino, R., Busso, M., Gratton, R. 2001, ApJ, 549, 346
[8] Sneden, C., Cowan, J. J., Gallino, R. 2008, ARA&A, 46, 241
[9] Pignatari, M., Hirschi, R., Wiescher, M., et al. 2013, ApJ, 762, 31
[10] Thielemann, F.-K., Arcones, A., Käppeli, R., et al. 2011, Progress in Particle and Nuclear Physics 66, 346
[11] Roederer I. U. 2012, ApJ, 756, 36
[12] Hansen, C. J., Primas, F., Hartman, H., et al. 2012, A&A, 545, 31
[13] Rauscher, T., Heger, A., Hoffman, R.D., & Woosley, S.E. 2002, ApJ, 576, 323
[14] Travaglio, C., Hillebrandt, W., Reinecke, M., & Thielemann, F.-K. 2004b, A&A, 425, 1029
[15] Lodders, K., Palme, H., Gail, H.-P. 2009, in Trümper J. E., ed., Landolt- Börnstein Group VI Astronomy and Astrophysics Numerical Data and Functional Relationships in Science and Technology, 4B: Solar System, 4.4. Springer-Verlag, Berlin (doi:10.1007/978-3-540-88055-4-34)
[16] Bisterzo, S., Gallino, R., Straniero, O., Cristallo, S., Käppeler F. 2010, MNRAS, 404, 1529
[17] Straniero, O., Gallino, R., Busso, M., Chieffi, A., Raiteri, C. M., Limongi, M., Salaris, M. 1995, ApJ, 440, L85
[18] Straniero, O., Limongi, M., Chieffi, A., Domínguez, I., Busso, M., Gallino, R. 2000, Mem. Soc. Astron. Ital., 71, 719
[19] Straniero, O., Domínguez, I., Cristallo, S., Gallino, R. 2003, PASA, 20, 389
[20] Serminato A., Gallino R., Travaglio C., Bisterzo S., Straniero O., PASA, 26, 153
[21] Straniero, O., Gallino, R., Cristallo, S. 2006, Nucl. Phys. A, 777, 311
[22] Herwig, F. 2005, Annu. Rev. Astron. Astrophys., 43, 435
[23] Herwig, F. 2000, Astron. Astrophys. 360, 952968
[24] Herwig, F., Pignatari, M., Woodward, P. R., et al. 2011, ApJ, 727, 89
[25] Karakas, A. I. 2010, MNRAS, 403, 1413
[26] Cristallo, S., Straniero, O., Gallino, R., Piersanti, L., Domínguez, I., Lederer, M. T. 2009, ApJ, 696, 797
[27] Cristallo, S., Piersanti, L., Straniero, O., et al. 2011, ApJS, 197, 17
[28] Busso, M., Palmerini, S., Maiorca, E., Trippella, O., Magrini, L., Randich, S. 2012, 12th Symposium on Nuclei in the Cosmos, August 5 - 12, Cairns, Australia, 20
[29] Käppeler, F., Gallino, R., Bisterzo, S., Aoki, W., 2011, Rev. Mod. Phys., 83, 157
[30] Piersanti, L., Cristallo, S., & Straniero, O. 2013, ApJ, 774, 98
[31] Bisterzo, S., Gallino, R., Käppeler, F., et al. 2015, MNRAS, 449, 506
[32] Travaglio, C., Röpke, F. K., Gallino, R., Hillebrandt, W., 2011, ApJ, 739, 93
[33] Winckler, N., Dababneh, S., Heil, M., et al. 2006, ApJ, 647, 685
[34] Jaeger, M., Kunz, R., Mayer, A., Hammer, J. W., Staudt, G., Kratz, K. L., Pfeiffer, B. 2001, Phys. Rev. Lett., 87, 202501,
[35] Longland, R., Iliadis, C., & Karakas, A. I. 2012, Phys. Rev. C, 85, 065809
[36] Käppeler, F., Wiescher, M., Giesen, U., et al. 1994, ApJ, 437, 396
[37] Rauscher, T. 2012, ApJL, 755,10
[38] Bao, Z. Y., Beer, H., Käppeler F., Voss, F., Wisshak, K. 2000, Atomic Data & Nuclear Data Tables, 76, 70
[39] KADoNiS v0.3, Karlsruhe Astrophysical Data Base of Nucleosynthesis in Stars, web site http://www.kadonis.org/
[40] Winckler, N., Dababneh, S., Heil, M., et al. 2006, ApJ, 647, 685

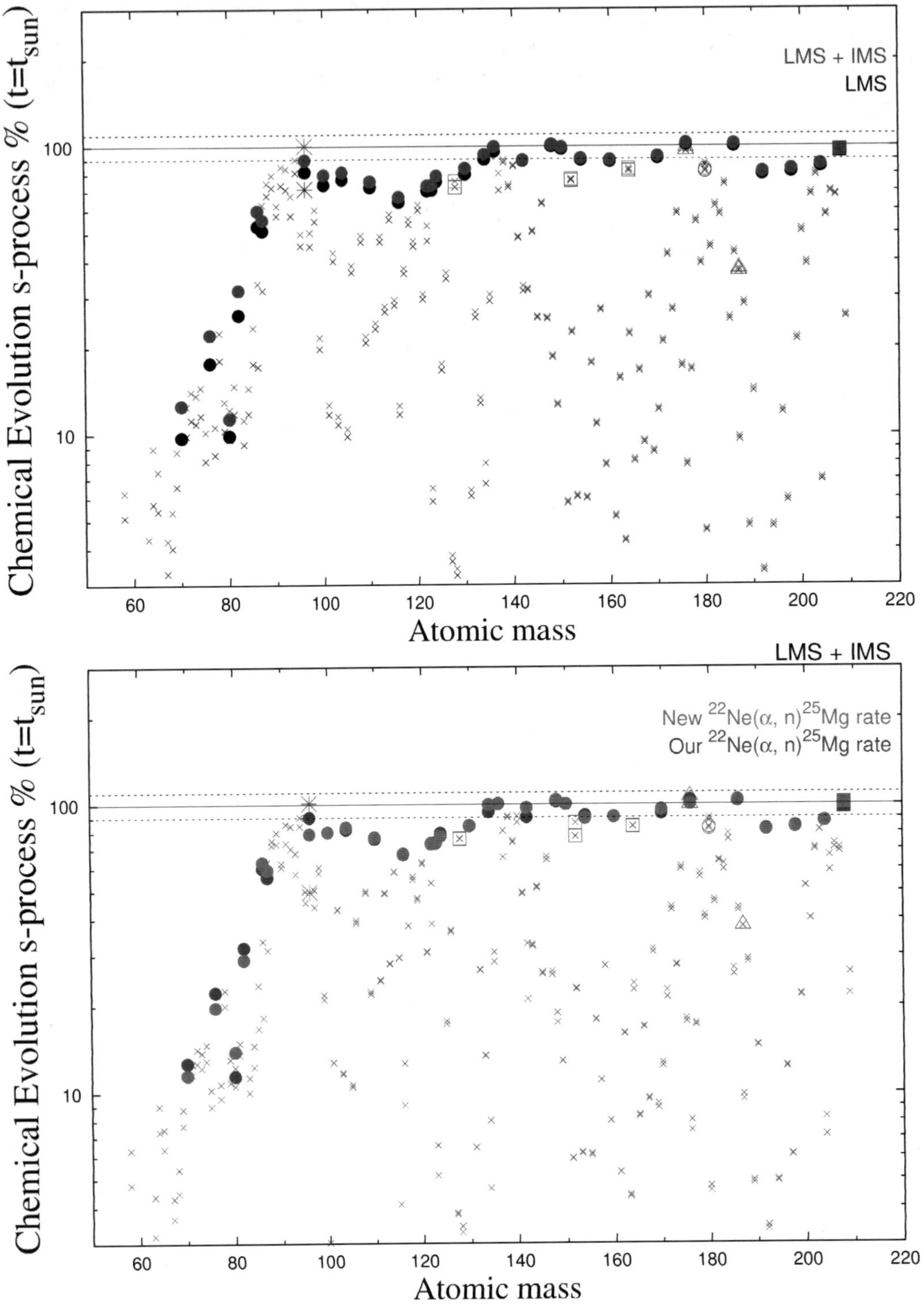

Figure 1. Reproduction of the Solar *s*-process abundances (in %) obtained at the epoch of the Solar System formation with GCE model. We used a neutron capture cross section network that includes the most recent measurements (see [31]) and we implemented the s-process yields with an extended range of AGB initial masses. Updated solar meteoritic abundances by [15] are adopted. The *s*-only isotopes are indicated by solid circles. ^{96}Zr is represented by a big asterisk. Different symbols have been used for isotopes that receive additional contributions (see text). *Top panel*: we distinguish between the contribution of LMS alone (black symbols) and the total *s*-contribution of all AGB masses (LMS + IMS; blue symbols). *Bottom panel*: we display the total *s*-contribution shown in top panel (blue) in comparison to the results obtained with a new evaluation of the ^{22}Ne(α, n)^{25}Mg reaction (see text).

Nuclear Physics in Astrophysics VI (NPA6) IOP Publishing
Journal of Physics: Conference Series **665** (2016) 012024 doi:10.1088/1742-6596/665/1/012024

Revision of the derivation of stellar rates from experiment and impact on Eu s-process contributions

T Rauscher[1,2,3]

[1] Centre for Astrophysics Research, School of Physics, Astronomy and Mathematics, University of Hertfordshire, Hatfield AL10 9AB, United Kingdom
[2] Institute of Nuclear Research (ATOMKI), H-4001 Debrecen, POB 51, Hungary
[3] Department of Physics, University of Basel, 4056 Basel, Switzerland

E-mail: t.rauscher@herts.ac.uk

Abstract. A new, general formalism to include experimental data in revised stellar rates is discussed, containing revised uncertainties. Application to the s-process shows that the actual uncertainties in the neutron capture rates can be larger than would be expected from the experimental errors alone. As a specific example for how astrophysical conclusions can depend on the approach selected to derive stellar rates, the ^{151}Eu/(^{151}Eu+^{153}Eu) abundance ratio from AGB star models is presented. Finally, a recommended workflow for the derivation of stellar rates from experiment is laid out.

1. Introduction

The estimate of uncertainties in the astrophysical reaction rates included in extended reaction networks is essential for nucleosynthesis studies to disentangle uncertainties stemming from the astrophysical model from those of the nuclear input. Most of the rates included in complete rate libraries are not experimentally constrained and thus prone to theoretical uncertainties which are difficult to assess [1]. On the other hand, experimental efforts over decades have provided reaction cross sections (CS) along the line of stability. Measurements of neutron captures for the s-process have been especially successful because there is no Coulomb barrier in such reactions which would prevent data to be taken in the astrophysically important energy range. Experiments studying only reactions on the ground state (g.s.) of a nucleus cannot always constrain the stellar rate well, even at s-process temperatures [2, 3]. In a stellar environment, nuclei are also present in excited states and thus a measurement of the g.s. reaction only constrains a fraction of the actual stellar rate. Excited state contributions to stellar rates are especially pronounced for nuclei with low-lying excited states or a high inherent nuclear level density [4], as found in many nuclei participating in s-process nucleosynthesis. This requires a detailed re-evaluation of the stellar rates and their uncertainties derived from experiment, as it is not possible by default to assign the same uncertainties as the ones derived for the experiment.

2. Ground- and excited state contributions to the stellar rate

The astrophysical reaction rate r^* for an interaction between two particles or nuclei in a stellar environment is obtained by folding the Maxwell-Boltzmann energy distribution Φ, describing the thermal c.m. motion of the interacting nuclei in a plasma of temperature T, with a measure

σ^* of the probability that the reaction occurs and by multiplying the result with the number densities n_a, n_A, i.e., number of interacting particles in a unit volume,

$$r^* = \frac{n_a n_A}{1+\delta_{aA}} \int_0^\infty \sigma^*(E)\Phi(E,T)\,dE = \frac{n_a n_A}{1+\delta_{aA}} R^* \quad . \tag{1}$$

The stellar reactivity (or rate per particle pair) is denoted by R^*. To avoid double counting of pairs, the Kronecker symbol δ_{aA} is used. Depending on temperature and nuclear level structure, a fraction of nuclei is present in an excited state in the plasma, instead of being in the ground state (g.s.). The relative contribution X_i of a specific level i to the total stellar rate r^* is given by [2]

$$X_i(T) = \frac{2J_i+1}{2J_0+1} e^{-E_i/(kT)} \frac{\int \sigma_i(E)\Phi(E,T)dE}{\int \sigma^{\mathrm{eff}}(E)\Phi(E,T)dE} \quad . \tag{2}$$

For the g.s. ($i=0$), this simplifies to [5]

$$X_0(T) = \frac{\int \sigma_0(E)\Phi(E,T)dE}{\int \sigma^{\mathrm{eff}}(E)\Phi(E,T)dE} \quad . \tag{3}$$

The *effective cross section* appearing above is defined as [4, 6]

$$\sigma^{\mathrm{eff}}(E) = \sum_i \sum_j \frac{2J_i+1}{2J_0+1} \frac{E-E_i}{E} \sigma^{i\to j}(E-E_i) \quad , \tag{4}$$

with $\sigma_i = \sum_j \sigma^{i\to j}$ being the sum of partial CS leading to final states j. As usual, CS for individual transitions $\sigma^{i\to j}$ are zero for negative energies.

Although not discussed in further detail here, it is worth mentioning that also weak interactions and decays are affected by thermal population of excited states and similar quantities can be defined for them in complete analogy.

3. Implications for the experimental determination of astrophysical reaction rates

Laboratory measurements can only determine the g.s. CS $\sigma_0^{\mathrm{exp}} = \sigma_0$ (or the CS of a long-lived isomeric state). This is only sufficient for the determination of the stellar reactivity when the contribution X_0 is close to unity. Designing an experiment, it should be taken care to measure a reaction with the largest possible X_0. This also implies that the reaction should be measured in the direction of largest X_0 [3, 7].

When $X_i < 1$, only a combination of experimental data and theory can yield a stellar rate. It is not trivial to find the correct combination to derive the stellar rate as well as its new uncertainty after σ_0 has been measured. Two extreme cases can be found, depending on whether any errors in the prediction of the g.s. CS are correlated with the theory errors in the $\sigma_{i>0}$ or not. The first approach is to assume no such correlation. It leads to the largest uncertainties. It should be used whenever a detailed theoretical investigation of the deviations between experiment and theory is missing. Therefore it is the approach of choice for experimentalists or stellar modelers without further support in reaction theory.

Strictly speaking, the experimental CS can only replace one of the contributions to the stellar rate while the others remain unconstrained by the data. Approach A starts from the notion that only measured quantities should be included in the new stellar rate and its uncertainty estimate, while all other (theoretical) ingredients remain unchanged. Since the stellar reactivity is given by a weighted sum of reactivities

$$R^*_{\mathrm{th}} = \frac{w_0 R_0^{\mathrm{th}} + w_1 R_1^{\mathrm{th}} + w_2 R_2^{\mathrm{th}} + \ldots}{w_0 + w_1 + w_2 + \ldots} \quad , \tag{5}$$

where $w_i = (2J_i+1)\exp(-E_i/(kT))$ and J_i, E_i are spin and excitation energy of the i-th excited state, respectively, this implies

$$R^*_{\mathrm{new,A}} = \frac{w_0 R_0^{\mathrm{exp}} + w_1 R_1^{\mathrm{th}} + w_2 R_2^{\mathrm{th}} + \ldots}{w_0 + w_1 + w_2 + \ldots} \quad . \tag{6}$$

Therein, R_0^{exp} is the reactivity obtained from folding the experimentally obtained CS on a g.s. (or isomeric state) with a Maxwell-Boltzmann distribution, similarly as it is done with the stellar CS σ^* in equation (1). With the experimentally determined reactivity R_0^{exp}, equation (6) yields the (temperature dependent) correction factor $f^*_{\mathrm{A}}(T)$, which has to be applied to the old stellar reactivity to provide the new $R^*_{\mathrm{new}}(T) = f^*_{\mathrm{A}}(T) R^*_{\mathrm{th}}(T)$,

$$f^*_{\mathrm{A}}(T) = 1 + X_0(T)\left(\frac{R_0^{\mathrm{exp}}(T)}{R_0^{\mathrm{th}}(T)} - 1\right) \quad , \tag{7}$$

containing the ratio between experimental and theoretical g.s. reactivity.

It is important to note that the determination of the rate is closely entwined with its associated uncertainty or "error" and should never be quoted and used without it. Frequently used in astrophysics are uncertainty factors U, which assume that the "true" value R^*_{true} of a (semi-)theoretical reactivity R^* is in the range $UR^* \geq R^*_{\mathrm{true}} \geq R^*/U$. It is neither trivial to estimate the theory uncertainty nor the experimental one, especially when it has to be in a consistent, intercomparable manner. The fundamental differences between error determinations in experiment and theory are discussed in [1] and appropriate choices are suggested in [2]. Here, it should just be mentioned that, unlike experimental measurements, theory results cannot usually be assumed to be randomly drawn from a defined statistical distribution. A theoretical model or parametrization can simply be inadequate and it is impossible to define a set of statistically distributed models. Thus, the theory uncertainty is more closely related to a systematic error and equally difficult to estimate.

Once having estimated the uncertainties in some way, nevertheless, in the above approach A it is simple to calculate the new uncertainty factor of the stellar rate from a combination of the uncertainty factor $U^*_{\mathrm{th}} = U_0^{\mathrm{th}}$ of the theoretical stellar reactivity R^* and an experimental uncertainty $U_{\mathrm{exp}} = U_0^{\mathrm{exp}} \leq U^*_{\mathrm{th}}$ of the measured g.s. contribution[1]. This yields the uncertainty factor of the new stellar rate

$$U^*_{\mathrm{new,A}}(T) = U_{\mathrm{exp}}(T) + (U^*_{\mathrm{th}}(T) - U_{\mathrm{exp}}(T))(1 - X_0(T)) \quad . \tag{8}$$

As pointed out in [2], the other extreme approach would be to assume that all theoretical uncertainties, i.e., also those of the reactions proceeding on excited states, are removed once the g.s. reactions and their deviation from the predictions have been determined experimentally. This implies two strong assumptions on theory: (a) the cause of any discrepancy between R_0^{th} and R_0^{exp} also causes a similar deviation of the same magnitude in all $R^{\mathrm{th}}_{i>0}$, and (b) there are no further uncertainties in predicted excited state CS. Only when both assumptions are valid, the same renormalization can be applied to g.s. and excited state transitions, i.e.,

$$\begin{aligned} R^*_{\mathrm{new,B}} &= \frac{w_0 R_0^{\mathrm{exp}} + w_1 R_1^{\mathrm{th}} \frac{R_0^{\mathrm{exp}}}{R_0^{\mathrm{th}}} + w_2 R_2^{\mathrm{th}} \frac{R_0^{\mathrm{exp}}}{R_0^{\mathrm{th}}} + \ldots}{w_0 + w_1 + w_2 + \ldots} = \\ &= \frac{R_0^{\mathrm{exp}}}{R_0^{\mathrm{th}}} \times \frac{w_0 R_0^{\mathrm{th}} + w_1 R_1^{\mathrm{th}} + w_2 R_2^{\mathrm{th}} + \ldots}{w_0 + w_1 + w_2 + \ldots} \quad . \end{aligned} \tag{9}$$

[1] For simplicity, $U^*_{\mathrm{th}} = U_0^{\mathrm{th}}$ is assumed here but a new uncertainty can also be derived rigorously when using different uncertainties for each R_i^{th}. It is further possible to consider the uncertainty of X_0 in U^*_{new}. Its impact, however, is small with respect to the other experimental and theoretical uncertainties [5].

In this approach B, therefore, the renormalization factor f^*_{B} for the stellar reactivity (and rate) is $f^*_{\mathrm{B}} = R_0^{\mathrm{exp}}/R_0^{\mathrm{th}}$, so that $R^*_{\mathrm{new,B}} = f^*_{\mathrm{B}} R^*_{\mathrm{th}}$, as before. It has to be noted that this is equivalent to multiplying the measured g.s. reactivity R_0^{exp} by the so-called *stellar enhancement factor* (SEF) $f_{\mathrm{SEF}} = R^*_{\mathrm{th}}/R_0^{\mathrm{exp}}$ because of assumption (a). Until recently this was the favored approach, even when nothing was known about the causes of the deviations between prediction and experiment and also not whether these also apply similarly to reactions from the excited states [8,9].

The uncertainty for the new stellar rate obtained in approach B is difficult to quantify. Only if both assumptions (a) and (b) really apply, and only then, it is just the uncertainty of the measurement. If this is not the case, the uncertainty will be larger. In order to be on the safe side, it may be advisable to use the error estimate $U^*_{\mathrm{new,A}}$ from equation (8) even when using approach B to derive $R^*_{\mathrm{new,B}}$.

The above two approaches are the two extreme cases. Realistically, the actual stellar rate and its uncertainty may be anywhere between these two cases. Detailed theoretical investigations have to be performed separately for each nucleus and each reaction before conclusions can be drawn from the comparison of the experimental results and the prediction. In the absence of such a detailed theoretical investigation, it is preferrable to apply the "pessimistic view" of approach A, contrary to the historical custom of using the seemingly simpler approach B. Although approach A will lead to larger uncertainties, it encompasses the values of approach B and thus covers all possibilities.

There are a number of reasons why the uncertainties in the predictions of the transitions originating on excited states can be different than those of the ones commencing on the g.s. of the target nucleus. First, it is clearly seen in equation (4) that σ^{eff} includes transitions at a range of relative interaction energies $0 \leq E - E_i \leq E_{\mathrm{G}} - E_i$, where E_{G} is the upper end of the relevant energy window at each temperature. It is known that the sensitivity of the $\sigma^{i \to j}$ are strongly energy-dependent (see, e.g., [1]) and therefore transitions from excited states (occurring at lower relative energy) may be sensitive to different nuclear properties than those from the g.s. For example, for neutron captures in the *s*-process this can be important mainly for nuclei with large capture Q-value and high nuclear level densities, for which the neutron widths may become comparable to or smaller than the γ-widths within the covered energy range. Second, even if the sensitivities to different widths are not changing, different spins and parities of the excited states imply different angular momenta and therefore different angular momentum barriers in particle transitions, and also may give rise to a different selection of electromagnetic multipolarities in γ-transitions. This may be more important in nuclei with low level densities. The actual circumstances and sensitivities will be different in each application and have to be thoroughly investigated for each reaction separately.

4. Impact on the Eu production in the s-process

An illustrative example for the importance of choosing the right approach to derive the stellar rate from experiment is the reproduction of the abundance ratio $\mathcal{R} = Y_{^{151}\mathrm{Eu}}/Y_{\mathrm{Eu}}$ found in meteoritic material [10] and in CEMP stars [11]. Recently, individual mainstream stardust silicon carbide (SiC) grains and a SiC-enriched bulk sample (KJB) from the Murchison carbonaceous meteorite have been analyzed for Eu isotopes [10]. The mainstream grains are believed to have condensed in the outflows of $\approx 1.5 - 3\ M_\odot$ carbon-rich asymptotic giant branch (AGB) stars with close-to-solar metallicity and $\mathcal{R}$ is determined by the values of the ^{151}Sm(n,γ) rate during *s*-processing. There are also abundances available from previous astronomical observations of carbon-enhanced metal-poor stars enriched in *s*-process elements [11]. Despite the difference in metallicity between the parent stars of the grains and the metal-poor stars, the $\mathcal{R}$ values derived from the meteoritic data agree well with those derived from astronomical observations. They appear systematically higher than the solar value of 0.48 but show large error bars. The new data from the SiC aggregate KJB, however, provides a more precise result ($\mathcal{R} = 0.54 \pm 0.03$,

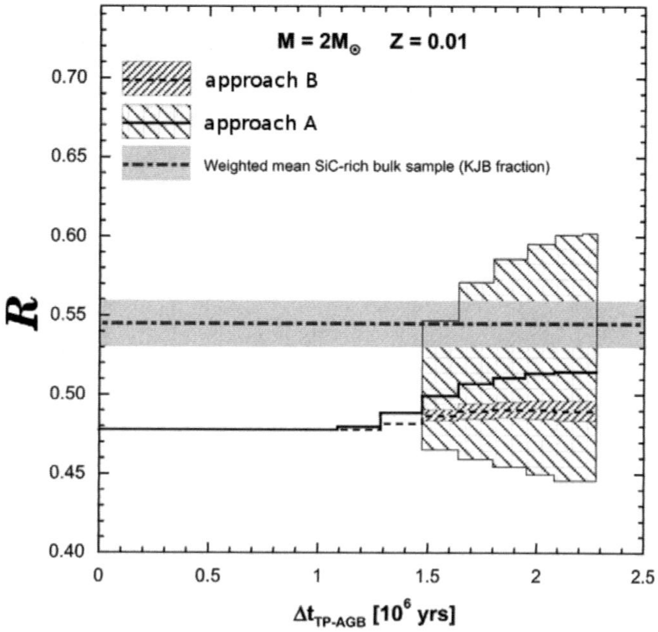

Figure 1. Evolution of $\mathcal{R}$ in the AGB envelope (time from the beginning of the TP-AGB phase) for two models using the same experimental values [12] for ^{151}Sm(n,γ), but the two approaches discussed in the text. The shaded area represents the range of results obtained when varying the rate with the derived uncertainty factors. [This is a modified version of Fig. 4 in [10].]

95% conf.), which is compatible with both the mainstream and the stellar data. It clearly shows an enhancement with respect to the solar value.

Sensitivity studies find that ^{151}Sm(n,γ)^{152}Sm is the crucial reaction determining $\mathcal{R}$ [10]. Although the stellar β-decay rate of ^{151}Sm is only theoretically known, this neutron capture reaction dominates the uncertainties [10]. A recent measurement of ^{151}Sm(n,γ) constrained R_0^{exp} within a few percent [12]. Current s-process predictions using modern AGB models and including these (n,γ) data, however, cannot reproduce the enhancement in $\mathcal{R}$ when choosing the traditional approach B to derive the stellar rate. Figure 1 shows that, on the other hand, the results obtained with a rate derived with approach A are consistent with the meteoritic ratio.

It is sometimes stated that the s-process is the nucleosynthesis process with the best constrained nuclear input (see, e.g., [13, 14]). This statement, however, is based on the uncertainties in the measured (n,γ) CS. The resulting uncertainties U^*_{new} in the stellar reactivities and rates may be larger in many cases [2].

5. Conclusion

The revised approach to improve stellar rates with experimental CS presented here allows for the first time to also consistently discuss how rate uncertainties are affected by the inclusion of experimental data. In intermediate and heavy nuclei, excited state contributions to the stellar rate are not negligible even at the comparatively low s-process temperatures and limit the ability of reaction measurements to constrain the stellar rates. This is even more pronounced in charged-particle reactions as well as in nucleosynthesis processes at higher temperature.

If nothing else is known, approach A as described above is the recommended method to derive the stellar rate und its uncertainty. Nevertheless, the best way to derive stellar rates from experiment is to perform a detailed theoretical study of all transitions and their sensitivities to nuclear input which contribute in the reaction in question. Only such a study allows to judge how well the excited state CS are constrained by a measurement of the g.s. CS.

The preferred workflow after having measured σ_0^{exp} across the range of astrophysically relevant energies, therefore, would be the following:

(i) Check whether $X_0(T)$ deviate from unity within the astrophysical temperature range; if $X_0 \approx 1$, the stellar rate can be directly calculated from the measured CS and the uncertainty is given by the experimental one. (X_0 can be found in [1, 2, 15].)

(ii) For $X_0 < 1$, if no further theory work is available, use approach A, i.e., determine the new stellar rate using f_{A}^* from equation (7) and its uncertainty from equation (8). The required R_0^{th} can be, e.g., found in [6, 15] but have to be consistent with the renormalized R_{th}^*.

(iii) For $X_0 < 1$, if theory work is desired, the transitions on the g.s. and excited states and their sensitivities have to be studied in detail:

  (a) If discrepancies between g.s. prediction and experiment are found to apply similarly to the excited state transition, use f_{B}^* to derive the stellar rate.

    1. If there are no remaining uncertainties (from other sources) in the predicted transitions from excited states, the final uncertainty of the stellar rate is the experimental one.
    2. If there are further uncertainties but no theory insights on their scaling, use equation (8) for the uncertainty of the stellar rate.
    3. Otherwise, theory should attempt to find an "intermediate" uncertainty factor, i.e., with a value between those obtained with approaches A and B.

  (b) If the theoretical excited state CS cannot be scaled in the same way as the g.s. CS:

    1. If there are no further theory insights, use approach A for both stellar rate (from equation 7) and its uncertainty (equation 8).
    2. Otherwise, theory should attempt to find "intermediate" renormalization and uncertainty factors, i.e., with values between those obtained with approaches A and B.

Acknowledgments

This work was supported in part by the Swiss NSF, the European Commission within the FP7 ENSAR/THEXO project, and the EuroGENESIS Collaborative Research Programme. TR also acknowledges support through a "Distinguished Guest Scientist Fellowship" from the Hungarian Academy of Sciences.

References

[1] Rauscher T 2012 *Ap. J. Suppl.* **201** 26
[2] Rauscher T 2012 *Ap. J. Lett.* **755** L10
[3] Rauscher T 2013 *J. Phys. Conf. Ser.* **420** 012138
[4] Rauscher T 2011 *Int. J. Mod. Phys.* E **20** 1071
[5] Rauscher T, Mohr P, Dillmann I and Plag R 2011 *Ap. J.* **738** 143
[6] Holmes J A, Woosley S E, Fowler W A and Zimmerman B A 1976 *At. Data Nucl. Data Tables* **18** 305
[7] Rauscher T, Kiss G G, Gyürky Gy, Simon A, Fülöp Zs and Somorjai E 2009 *Phys. Rev.* C **80** 035801
[8] Bao Z Y, Beer H, Käppeler F, Voss F, Wisshak K and Rauscher T 2000 *At. Data Nucl. Data Tables* **76** 70
[9] Sallaska A L, Iliadis C, Champagne A E, Goriely S, Starrfield S and Timmes F X 2013 *Ap. J. Suppl.* **207** 18
[10] Ávila J N, et al 2013 *Ap. J. Lett.* **768** L18
[11] Aoki W, et al 2003 *Ap. J. Lett.* **592** L67
[12] Marrone S, et al 2006 *Phys. Rev.* C **73** 034604
[13] Käppeler F, Gallino R, Bisterzo S and Aoki W 2011 *Rev. Mod. Phys.* **83** 157
[14] Wiescher M, Käppeler F and Langanke K 2012 *Ann. Rev. Astron. Astrophys.* **50** 165
[15] Online interface to the NON-SMOKER cross section and rate library, http://nucastro.org/nonsmoker.html

On the subject of the Ba overabundance in the open clusters stars

T V Mishenina[1], S A Korotin[1], G Carraro[2,3], V V Kovtyukh[1], and I A Yegorova[2]

[1] Astronomical Observatory of Odessa National University, Ukraine
[2] European Southern Observatory, Chile
[3] Dipartimento di Fisica e Astronomia, Universita di Padova, Italy

E-mail: `tmishenina@ukr.net`

Abstract. For eight distant open clusters, namely Ruprecht 4, Ruprecht 7, Berkeley 25, Berkeley 73, Berkeley 75, NGC 6192, NGC 6404, and NGC 6583, we determined the yttrium and barium abundances using the UVES, VLT spectra (ESO, Chile). The stars of one young cluster (Ruprecht 7) demonstrate significant barium overabundance($\sim$0.55 dex) that can not be due to the determination error. We have considered the Ba abundance determination errors due to LTE approach, saturation of the lines, synthetic and observed barium line fitting, and the causes of the Ba overabundance associated with the Galactic disc enrichment or the origin of open clusters. Possible explanation for this overabundance can be the origin of n-capture elements enrichment of the clusters (galactic or extragalactic) or additional sources of the Ba production.

1. Introduction

Yttrium and barium are the neutron capture elements, and at the solar metallicity, the main sources of the n-capture element enrichment with interstellar and pre-stellar medium are the AGB stars of the small and moderate masses. The production of the n-capture elements and their presence in the stars of different ages and location in the Galaxy play important role in the modeling of the Galactic chemical evolution. Among them barium is a particular neutron-capture element, it shows a remarkable overabundance with respect to iron in some open clusters and disc stars of the Galaxy. This fact still has no clear interpretation, but there are some possible reasons: 1) determination of reliable barium abundance (e.g. under the NLTE approximation, taking the HFS into account, the saturation of the Ba lines); 2) dependence of the barium abundances on the age in the galactic disc; 3) additional sources of production of barium; 4) the origin of the open clusters. In this study we try to clarify the questions of the Ba overabundance. To do that we measured the abundance of yttrium and barium in eight distant open clusters, namely Ruprecht 4, Ruprecht 7, Berkeley 25, Berkeley 73, Berkeley 75, NGC 6192, NGC 6404, and NGC 6583.

2. Observation and atmospheric parameters

For our study, we used the spectra obtained with the echelle spectrograph UVES on board VLT with resolving power R = 40,000 for the wavelength range 4750–6800 Å by [1] and [2]. In order

Journal of Physics: Conference Series **665** (2016) 012025 doi:10.1088/1742-6596/665/1/012025

to check the atmospheric parameters obtained in [1], we compared the equivalent widths EWs of the lines and the temperatures derived by [1] with our data for two stars Ruprecht 4_3 and Ruprecht 7_2. We measured the EWs using the DECH20 software package [3]. The effective temperatures were computed based on the calibration of the central depths ratio of the lines with different potentials for the lower excitation level [4]. We have good agreement for both the equivalent widths and effective temperatures. And then we used the atmospheric parameters obtained by [1] and [2] to determine the yttrium and barium abundances in the studied stars.

3. Abundance determination

The models by [5] were used to determine the Y and Ba abundances. The LTE yttrium abundance was obtained with a new version of the STARSP software package [6] and the VALD atomic data [7] using three lines of Y II (4854.873, 4883.690 and 5087.426 Å). The adopted solar yttrium abundance is $(Y/H)_\odot = 2.24$, where $\log A(H) = 12$. The barium abundance was computed under the non-local thermodynamic equilibrium (NLTE) approach with a version of MULTI code [8], modified by S.A. Korotin ([9]). To determine the barium abundance, we used three lines of Ba II (5853, 6141 and 6496 Å). For line 6496 Å we took the HFS into account. The HFS for other lines (5853 and 6141 Å) is not significant. Departures from the LTE do not significantly affect Ba abundances (the NLTE correction value is about 0.1 dex).

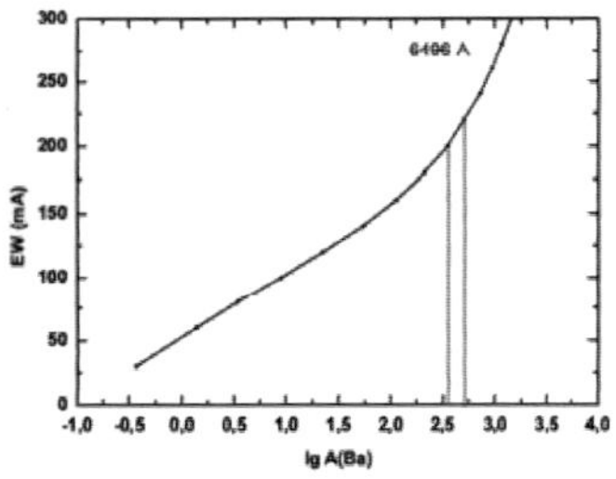

Figure 1. The growth curve for the barium line 6496 Å.

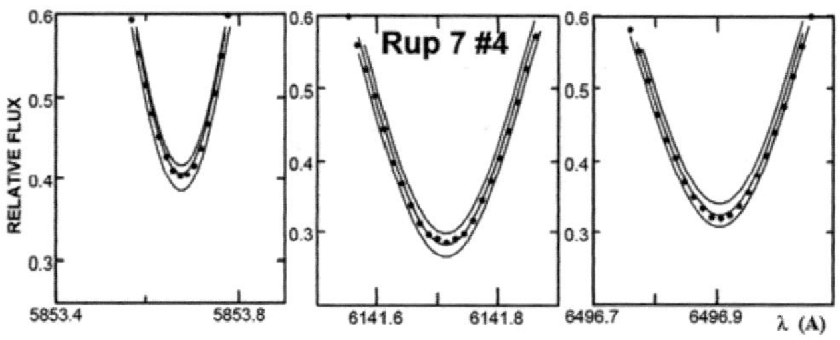

Figure 2. The observed and synthetic spectra fitting with different barium abundances ± 0.1 dex.

3.1. Determination errors

Since the equivalent widths of barium are strong enough in the spectra of the studied stars, the line saturation can provoke a low accuracy of the barium content determination. Figs. 1 and 2 show the growth curve for the barium line 6496 Å and the observed and synthetic spectra fitting. We can see from these Figures that the equivalent widths and profiles are rather sensitive to the barium abundance. The lines with EW up to 200 mÅ are very sensitive to the element abundance, while the stronger lines, obtained by the computed synthetic spectra, allow us to obtain abundance values with an accuracy of ±0.1 dex. In this case, the total error of determination of the Y and Ba abundances does not exceed 0.2 dex accounting for the uncertainty in the stellar parameters and the accuracy of the synthetic spectrums fitting.

4. Results

The dependence of [Y/Fe] and [Ba/Fe] from the age is shown in Figs. 3, 4. The Y abundances by [10] and Ba abundances by [11] are marked as black circles; those by [12] and [13] are marked as magenta triangles; those by [14] are marked as blue triangles; those by [15] are marked as green triangles; the Ba abundances by [16] as yellow triangles and by [17] as asterisks; by [18] - the thin disc (marked as black dots); for the Cepheids, the average values <[Y/Fe]> by [19] and <[Ba/Fe]> by [20] are marked as empty circles, and those by the present study are marked as red circles.

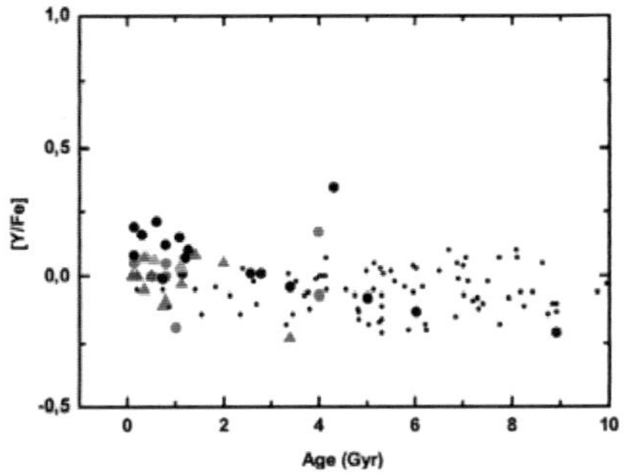

Figure 3. [Y/Fe] vs. Age.

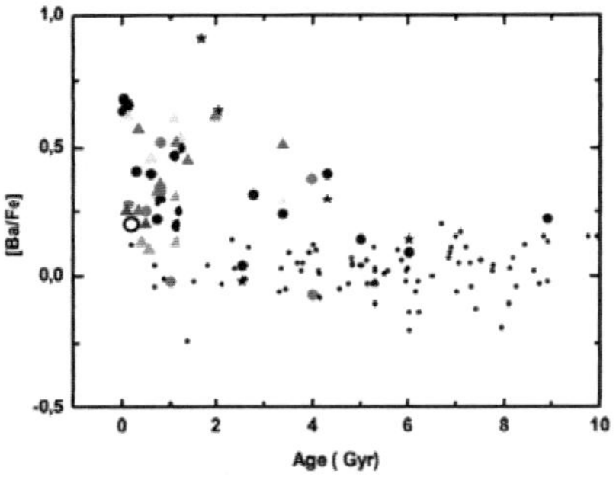

Figure 4. [Ba/Fe] vs. Age.

As it is seen in Fig. 3, there is a slight trend for yttrium (from -0.2 to 0.2 dex) with increasing age. As one can see, this trend is similar for the clusters and the thin disc that supports the idea of a similarity of origin of the stars of the thin disc and open clusters. And also it can be indicative of the increase in contributions into the enrichment of the interstellar medium from the least massive AGB stars ($M < 1.5 M_{\odot}$). For barium (Fig. 4), we observe a significant trend and a large spread of the values. The obtained average values for two youngest clusters Ruprecht 4 and Ruprecht 7 in our example (with ages about 0.8 Gyr) are <[Ba/Fe]> = 0.33 and 0.56 dex, respectively. To assess the extent to which those values are consistent with the values of barium content in young stars, we compared the data of the yttrium and barium abundances for younger clusters with data of Galactic Cepheids (the disc young stars with ages around 0.2–0.4 Gyr). Good agreement is observed for the yttrium, but the average values of barium in the Cepheids (≈ 0.2 dex, [20]) are significantly lower than 0.6 dex. In (Figs. 5, 6) we show the behaviour of the yttrium and barium abundances in young clusters and Cepheids with galaxy-centric distances. Cepheids confirms the existence of the yttrium abundance gradient in the Galactic disc. Behavior of the barium differs from that of yttrium - the barium abundance doesn't show any trend with metallicity for the Cepheids, and four clusters exhibit significant barium overabundances that exceed the spread of values ([Ba/H] > 0.5 dex) in the Cepheids. Those are IC 2391, NGC 2324, IC 2602 [11] and Ruprecht 7 (our determinations). The overabundance, especially for Ruprecht 7, remains an open question. It is not due to deviations from the LTE, since in our calculations we took those effects into account. Moreover, as shown by [14], this overabundance cannot be related to the chromospheric activity. The assumption that all these stars are the stars of barium class stars is also doubtful. We note that Ruprecht 7 has an abnormal orbit according to [21], see Fig. 7. However, the accuracy in the orbit determination (large proper motion errors for distant clusters and the adopted fixed Galactic potential) does not allow us to consider the result as reliable and it requires further investigation.

5. Conclusions

We found that the barium overabundance is real and it is not the result of determination errors. The origin of open clusters or the presence of an alternative way of enrichment (not AGB stars) could be possible reasons of that overabundance. New accurate determinations of barium and neutron-capture elements are needed.

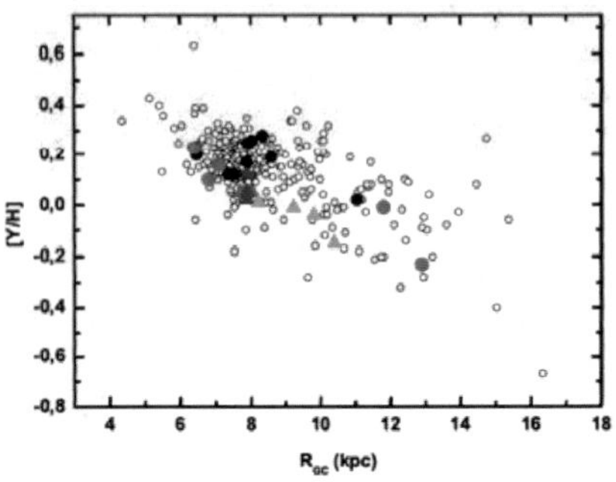

Figure 5. [Y/H] vs. R_{GC}

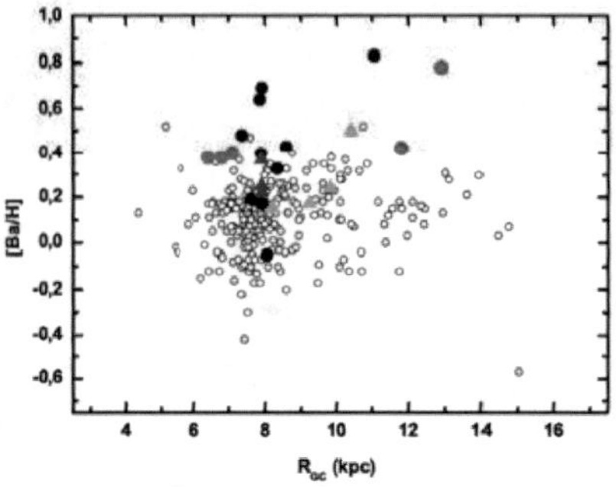

Figure 6. [Ba/H] vs. R_{GC}

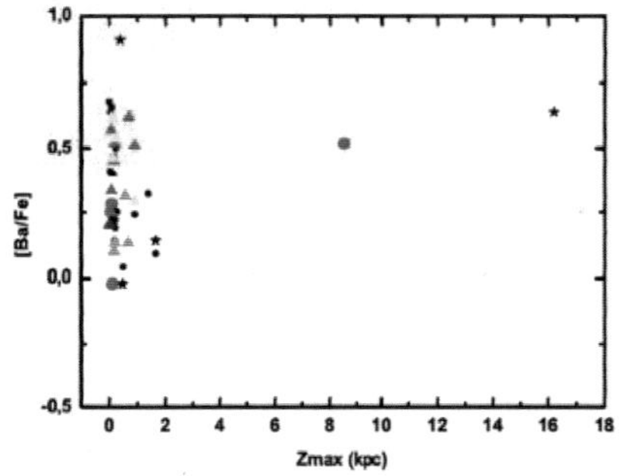

Figure 7. [Ba/Fe] vs. Zmax

Acknowledgments
T.M. and S.K. thank for the support from the Swiss National Science Foundation, project SCOPES No. IZ73Z0-12

References
[1] Carraro G Geisler D Villanova S *et al* 2007 *A&A* **476**, 217
[2] Magrini L Randich S and Zoccali M *et al* 2010, *A&A*, **523**, id.A11
[3] Galazutdinov G A 1992 *Preprint SAO RAS.*, **92**, 96
[4] Kovtyukh V V Soubiran C Belik S I and Gorlova N I 2003, *A&A* **411**, 559
[5] Castelli F and Kurucz R L 2004 *arXiv:astro-ph/0405087*
[6] Tsymbal V V 1996 *ASP Conf. Ser.*, **108**, 198
[7] Kupka F Piskunov N E Ryabchikova T A *et al* 1999 *A&AS*, **138**, 119
[8] Carlsson M 1986 *Uppsala Obs. Rep.*, **33**
[9] Korotin S A Mishenina T V Gorbaneva T I and Soubiran C 2011 *MNRAS*, **415**, 2093
[10] Maiorca E Randich S Busso M *et al* 2011 ApJ, **736**, 120
[11] D' Orazi V Magrini L Randich S *et al* 2009 *ApJ*, **693**, 31
[12] Pancino E Carrera R Rossetti E and Gallart C 2010 *A&A*, **511**, id.A56
[13] Carrera R and Pancino E 2011 *A&A* **535**, 30
[14] D' Orazi V Biazzo K Desidera S *et al* 2012 *MNRAS*, **423**, 2789
[15] Reddy A B S Giridhar S and Lambert D L 2012 *MNRAS*, **419**, 1350
[16] Bragaglia A Sestito P Villanova S *et al* 2008 *A&A*, **480**, 79
[17] Yong D Carney B W and Teixera de Almeida M-L 2005 *AJ*, **130**, 597
[18] Mishenina T V Pignatari M Korotin S A *et al* 2013, *A&A*, **552**, 128
[19] Luck R E and Lambert D L 2011 *AJ*, **142**, 136
[20] Andrievsky S M Lépine J R D Korotin S A *et al* 2013 *MNRAS*, **428**,
[21] Gozha M L Borkova T V and Marsakov V A 2012 *Astron. Letters*, **38**, 506

The ^{20}Ne(d,p)^{21}Ne transfer reaction in relation to the s-process abundances

C T Nsangu[1,*], A M Laird[1], A Parikh[2,3], P Adsley[1], M D Birch[4], A A Chen[4], T Faestermann[5], S P Fox[1], B R Fulton[1], R Hertenberger[5], D Irvine[4], B P Kay[1], R Longland[2], S Manwell[4], A St J Murphy[6], K Schmitt[7], N de Séréville[8], J R Tomlinson[1], H -F Wirth[5]

[1] Department of Physics, University of York, Heslington, YO10 5DD, York, UK
[2] Departament de Física i Enginyeria Nuclear, Universitat Politècnica de Catalunya, E-08036 Barcelona, Spain
[3] Institut d'Estudis Espacials de Catalunya, E-08034 Barcelona, Spain
[4] Department of Physics and Astronomy, McMaster University, Hamilton, Ontario L8S 4M1, Canada
[5] Maier-Leibnitz Laboratorium der Münchner Universitäten, D-85748 Garching, Germany
[6] SUPA, School of Physics and Astronomy, University of Edinburgh, EH9 3JZ Edinburgh, UK
[7] Oak Ridge National Laboratory (ORNL), TN 37831 Oak Ridge, USA
[8] Institut de Physique Nucléaire d'Orsay, UMR8608, IN2P3-CNRS, Université Paris Sud 11, 91406 Orsay, France

E-mail: *tn507@york.ac.uk

Abstract.
A study of the ^{20}Ne(d,p)^{21}Ne transfer reaction was performed using the Quadrupole Dipole Dipole Dipole (Q3D) magnetic spectrograph in Garching, Germany. The experiment probed excitation energies in ^{21}Ne ranging from 6.9 MeV to 8.5 MeV. The aim was to investigate the spectroscopic information of ^{21}Ne within the Gamow window of core helium burning in massive stars. Further information in this region will help reduce the uncertainties on the extrapolation down to Gamow window cross sections of the ^{17}O(α,γ)^{21}Ne reaction. In low metallicity stars, this reaction has a direct impact on s-process abundances by determining the fate of ^{16}O as either a neutron poison or a neutron absorber.

The experiment used a 22-MeV deuteron beam, with intensities varying from 0.5-1 μA, and an implanted target of ^{20}Ne of 7 μg/cm^2 in 40 μg/cm^2 carbon foils. Sixteen ^{21}Ne peaks have been identified in the E_x = 6.9-8.5 MeV range, of which only thirteen peaks correspond to known states. Only the previously-known E_x = 7.960 MeV state was observed within the Gamow window.

1. Introduction

Most nuclei heavier than iron are thought to be formed by exposing iron-peak seed nuclei to a source of neutrons, such that neutron-capture reactions can be initiated [1]. Neutron capture occurs either at slow rates, through a process known as the slow or s-process, or at high rates via the rapid or r-process.

Neutron capture is said to be slow when the time between two successive neutron captures is longer than the average β-decay timescale. This leads the s-process path to be located along the line of β stability. The s-process proceeds from iron-seed nuclei through most stable nuclei up to ^{209}Bi where it terminates. The s-process abundances are highly dependent on the magnitude of neutron-capture cross sections and the total neutron availability. Therefore lighter nuclei may affect the s-process abundances by either adding (neutron source), permanently removing (neutron poison) or only temporarily removing (neutron absorber) the neutrons available for the s-process.

Massive stars are responsible for one component of the s-process, known as the weak s-process, which produces nuclei with mass number between $A = 70$ and $A = 90$. The weak s-process operates in the helium core or carbon burning shell where ^{22}Ne is the main source of neutrons. ^{16}O, ^{20}Ne, ^{24}Mg and ^{25}Mg are, however, known to act as neutron poisons [2].

In low-metallicity massive stars, due to the lower abundance of heavier elements, and depending on the neutron flux, the high neutron-to-seed ratio can allow the production of isotopes beyond the Sr-Zr peak and, possibly to ^{137}Ba [3]. In this case, ^{16}O is thought to be the most dominant neutron absorber through the ^{16}O(n,γ)^{17}O reaction. Neutrons can be made available again for the s-process through the ^{17}O(α,n)^{20}Ne reaction, depending on the strength of the competing reaction channel ^{17}O(α,γ)^{21}Ne. The neutron recycling process in this instance is therefore highly dependent on the ratio of the ^{17}O(α,γ)^{21}Ne to ^{17}O(α,n)^{20}Ne reaction rate. However, despite its importance, this ratio is highly uncertain and the fate of ^{16}O as either neutron poison or neutron absorber is yet to be established. Only two theoretical predictions exist, from Caughlan and Fowler [4] and from Descouvement [5], and differ by three orders of magnitude at relevant energies.

The ^{17}O(α,n)^{20}Ne reaction has already been measured at relatively low energies and extrapolations to lower energies are available through the NACRE compilation [6] and through recent measurements by Best *et al.* in [7], for example. However, there is a lack of experimental data for the ^{17}O(α,γ)^{21}Ne reaction, for which only two recent direct measurements exist. The two measurements by Taggart *et al.* [3] and Best *et al.* [8] do not, however, cover the Gamow window which in the core helium burning is in the region of 0.3 GK which corresponds in the centre of mass to an energy range of 0.3-0.7 MeV. The uncertainties on the extrapolation of the cross sections in the Gamow window were linked to the poor knowledge of the level scheme of ^{21}Ne within the region of interest, raising the need for further spectroscopic information.

2. Experimental Method

The ^{20}Ne(d,p)^{21}Ne reaction experiment aimed to investigate and extract the ^{21}Ne level information within Gamow window located between E_x = 7.65-8.05 MeV.
The experiment took place at the Q3D magnetic spectrograph at the Maier-Leibnitz Laboratorium (MLL) of the Technische Universität München and the Ludwig-Maximilians-Universität (LMU). A 22-MeV deuteron beam between 500-1000 nA impinged on a target of ^{20}Ne implanted in ^{12}C. The nominal target area density was 7 μg/cm^2 of ^{20}Ne in 40 μg/cm^2 of ^{12}C. The experiment populated excited states of ^{21}Ne ranging from 6.9 MeV to 8.5 MeV which enabled the states around the Gamow window to be investigated. This experiment was the first ^{20}Ne(d,p)^{21}Ne measurement to be performed beyond 6-MeV in excitation energy.

The magnetic field of the three dipole magnets of the Q3D spectrograph separated the trajectories of the particles according to their magnetic rigidity. The detector used has the same features as the one described by J. Ott *et al* in [9] with the only difference being an increased active length. It is composed of two proportional gas-filled detectors followed by a scintillator. The first proportional detector provides energy loss and the second the position information. Excellent particle identification is obtained by the combination of both energy loss signals with energy collected from the scintillator.

Measurements were taken at various angles ranging from 8° to 30°. For calibration purposes ^{28}Si(d,p)^{29}Si was also measured, and for background estimation, data from the ^{12}C(d,p)^{13}C reaction was used.

3. Results and Discussion

In total, sixteen ^{21}Ne peaks were identified of which only thirteen are at energies that correspond to states reported in the literature. Table 1 summarises these results.

Table 1: Summary of identified ^{21}Ne peak energies with existing values from literature [10]. Both measured energies and uncertainties are preliminary.

Meaured Energies (keV)	Literature Energies (keV)
6960(2)	-
6977(17)	-
7032(13)	7022.6(10)
7055(13)	7042.1(5)
7239(11)	7226(5)
7330(9)	7320(5)
7485(7)	7465(10)
7558(10)	7547(10)
7955(2)	7960.3(10)
8039(12)	8062(10)
8167(15)	8154.9(6)
8301(9)	8303(10)
8328(6)	-
8349(3)	8360(10)
8437(3)	8430(10)
8520(4)	8522(3)

The observed peaks at 8328(6) keV, 6977(17) keV and 6960(2) keV are not found in the literature. All peaks were observed at most angles. For the measured energies presented in this work, it is important to underline that uncertainties due to energy determinations at different angles have been considered, energy loss effects due to the different ^{20}Ne targets employed have not been carefully considered yet. In the literature, 32 different levels exist between the 7022.6 keV and 8522 keV states and 13 could be associated to the proton peaks observed in the data. Figure 1 shows the outgoing proton spectra at 8°, 11° and 25°. As can be seen, within the Gamow window itself, only the 7960.3(10) keV peak has been observed. For most levels around that region, very little spectroscopic information is known. During the course of the experiment, although the original target thickness was known, the implanted ^{20}Ne was gradually released from the carbon foil as was seen from the ratio of ^{13}C to ^{21}Ne peak intensities. The boiling-off rate of ^{20}Ne was not monitored precisely which therefore contributes an additional uncertainty to the target thickness. Due to this fact, only the relative differential cross section at different angles may be used to infer the angular momentum and therefore the spin-parity of different states.

The results presented in this work are preliminary and analysis is ongoing.

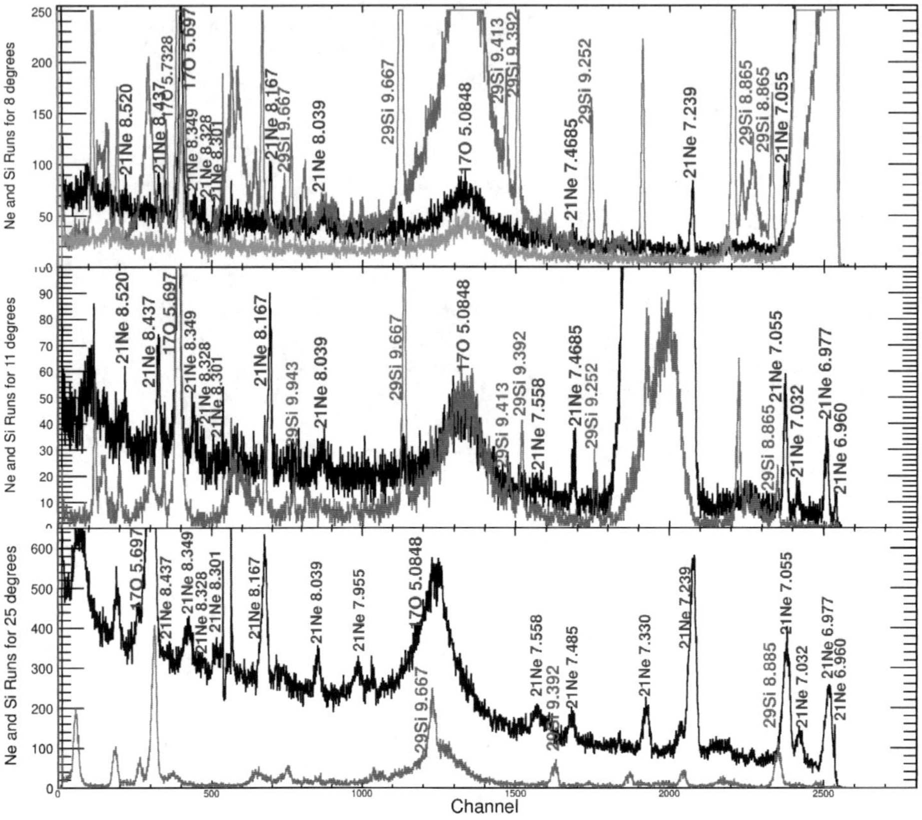

Figure 1: Excitation energy spectra at 8°, 11° and 25° respectively. This figure shows Ne runs (black), Si runs used during calibration (red) and C run at 8° only (green). Different states identified at those angles are shown. Principal contaminants, mainly ^{28}Si and ^{16}O (leading to states in ^{29}Si and ^{17}O), are also shown.

Acknowledgements

The authors would like to thank the beam operators at MLL and colleagues from Dresden-Rossendorf, Seattle, ORNL and Yale who provided the targets, as well as Christian Aa. Diget for useful discussions. Thanks should also be given to the FfWG commitee for the grant provided during this academic year. York authors especially would like to thank the University of York for providing funding of this work.

References

[1] Jose J and Iliadis C 2011 *Rep. Prog. Phys.* **74** 096901
[2] Frischknecht U, Hirschi R and Thielemann F K 2012 *Astronomy & Astrophysics* **538**
[3] Taggart M *et al.* 2010 *POS(NIC XI)* 045
[4] Caughlan and Fowler 1998 *Atomic Data and Nuclear Data Tables* **40** 2, 283–334
[5] Descouvement P 1993 *Phys. Rev.* C **48** 2746
[6] NACRE Collaboration 1999 [http://pntpm3.ulb.ac.be/Nacre/nacre.htm]
[7] Best A *et al.* 2013 *Phys. Rev.* C **87** 045805
[8] Best A *et al.* 2011 *Phys. Rev.* C **83** 052802(R)
[9] Ott J *et al.* 1995 *Nuclear Instruments and Methods in Physics Research* A **367** 280–28
[10] Brookhaven National Laboratory [http://www.nndc.bnl.gov/ensdf]

Neutron Energy Spectra and Yields from the ^{7}Li(p,n) Reaction for Nuclear Astrophysics

M Tessler[1], M Friedman[1], S Schmidt[2], A Shor[3], D Berkovits[3], D Cohen[1], G Feinberg[1,3], S Fiebiger[2], A Krása[4], M Paul[1*], R Plag[2], A Plompen[4] and R Reifarth[2]

[1] Racah Institute of Physics, Hebrew University, Jerusalem 91904, Israel
[2] Goethe University, Max-von-Laue-Straße 1, 60438 Frankfurt, Germany
[3] Nuclear Physics and Engineering Division, Soreq Nuclear Research Center, Yavne 81800, Israel
[4] European Commission, Joint Research Centre, Institute for Reference Materials and Measurements, Retieseweg 111, 2440 Geel, Belgium

E-mail: *paul@vms.huji.ac.il

Abstract. Neutrons produced by the ^{7}Li$(p,n)^7$Be reaction close to threshold are widely used to measure the cross section of s-process nucleosynthesis reactions. While experiments have been performed so far with Van de Graaff accelerators, the use of RF accelerators with higher intensities is planned to enable investigations on radioactive isotopes. In parallel, high-power Li targets for the production of high-intensity neutrons at stellar energies are developed at Goethe University (Frankfurt, Germany) and SARAF (Soreq NRC, Israel). However, such setups pose severe challenges for the measurement of the proton beam intensity or the neutron fluence. In order to develop appropriate methods, we studied in detail the neutron energy distribution and intensity produced by the thick-target ^{7}Li$(p,n)^7$Be reaction and compared them to state-of-the-art simulation codes. Measurements were performed with the bunched and chopped proton beam at the Van de Graaff facility of the Institute for Reference Materials and Measurements (IRMM) using the time-of-flight (TOF) technique with thin (1/8″) and thick (1″) detectors. The importance of detailed simulations of the detector structure and geometry for the conversion of TOF to a neutron energy is stressed. The measured neutron spectra are consistent with those previously reported and agree well with Monte Carlo simulations that include experimentally determined ^{7}Li(p,n) cross sections, two-body kinematics and proton energy loss in the Li-target.

1. Introduction

Nucleosynthesis of heavy elements (A $>$ 60) in stars involves neutron-capture processes [1, 2]. Abundances of these elements are determined by the stellar rates of the slow (s-) process and rapid (r-) process and the half-life of relevant nuclides. Hence neutron capture cross sections on stable and unstable nuclides in the stellar regimes of energy are essential quantities in nuclear astrophysics. Due to the need for an appropriate intense neutron source, these values have not been measured yet for a large number of nuclides at or near important branching points of the s-process. The neutrons produced by the ^{7}Li$(p,n)^7$Be reaction for an incident proton energy around 30 keV above the reaction threshold (1880.4 keV) on a thick Li target (thick enough that the proton energy is reduced to below the threshold energy while still in the Li) are emitted in a cone of $\sim 120^\circ$ angular opening with an energy distribution close to that of a Maxwellian flux

Journal of Physics: Conference Series **665** (2016) 012027 doi:10.1088/1742-6596/665/1/012027

with $kT \approx 25$ keV which is close to the temperature of some of the s-process sites [3]. Intense $^7\mathrm{Li}(p,n)^7\mathrm{Be}$ neutron sources are being developed for the measurements of such cross-sections at the Soreq Applied Research Accelerator Facility (SARAF) and Goethe University Frankfurt, based on high-intensity RF accelerators.

2. SARAF

The Soreq Applied Research Accelerator Facility (SARAF) [4] is based on a continuous wave (CW), proton/deuteron RF superconducting linear accelerator capable of delivering currents up to 2 mA. Phase I of SARAF (see figure 1) will produce proton and deuteron beams with energies up to $\sim$ 4 and 5 MeV respectively.

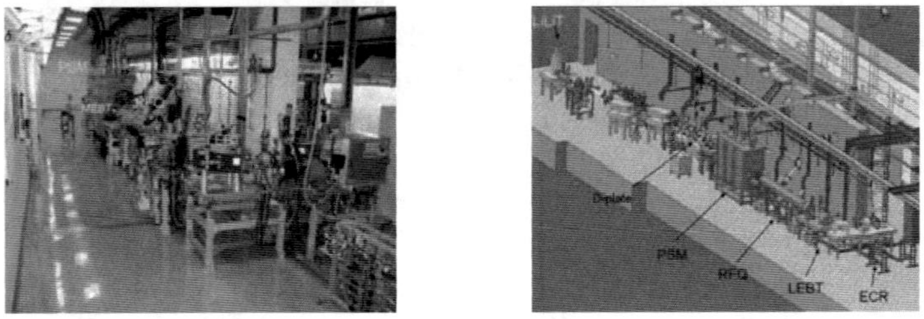

Figure 1. SARAF Phase I: Photograph (left) and sketch (right) showing ion source (ECR), low energy beam transport (LEBT), radio-frequency quadrupole (RFQ), and prototype superconducting module (PSM) consisting of six half-wave resonator accelerating cavities.

3. FRANZ

The Frankfurt neutron source at the Stern-Gerlach-Zentrum (FRANZ) [5] is currently under construction at the Institute for Applied Physics at the Goethe University in Frankfurt (see figure 2). It will be capable of delivering a proton beam with a current up to 20 mA in CW mode, resulting in beam powers of $\sim$ 40 kW, with energies up to 2 MeV. There will be several target stations, for activation and time-of-flight measurements.

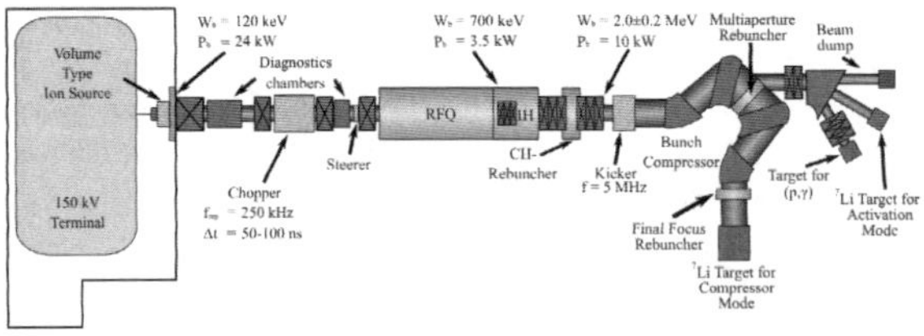

Figure 2. FRANZ: Frankfurt neutron source at the Stern-Gerlach-Zentrum.

4. Li targets

4.1. LiLiT target at SARAF

The SARAF high intensity beam requires a Li target that can withstand its beam power, incompatible with solid Li or Li-compounds. The Liquid-Lithium Target (LiLiT) [7] is a windowless forced-flow closed loop of liquid Li, producing a jet with a velocity of 3-7 m/s, acting as the target (figure 3 (left)). The target was successfully tested with an electron beam (1.6 kW) showing dissipation of $\sim$ 2 MW/cm^3, equivalent to a proton beam of $\sim$ 3 mA. The

Journal of Physics: Conference Series **665** (2016) 012027 doi:10.1088/1742-6596/665/1/012027

combination of SARAF and LiLiT will produce an intense quasi-stellar neutron source peaked at $E_n \approx$ 25-30 keV with a neutron flux of about 10^{10}-10^{11} neutrons per second (a factor 10-100 larger than presently available).

4.2. The FRANZ target

The current design of the FRANZ setup is based on a solid Li target (with a Cu or Ag backing) cooled by water flowing in two water channels (figure 3 (right)). A large diameter cooling channel is expected to take a major part of the heat from the copper target assembly. A small second channel, which cools the target backing directly, is an attempt to move as close to the heat source as possible while trying to keep neutron distribution aberrations minimal.

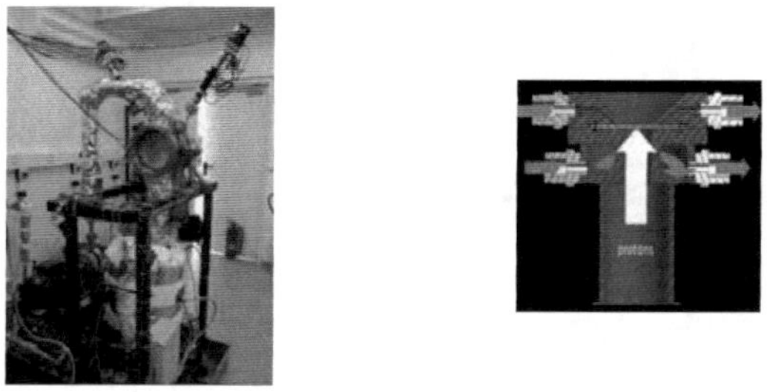

Figure 3. (left) Photograph of the Liquid Lithium Target - LiLiT; (right) FRANZ Li Target design. Blue arrows indicate water flow.

5. ^{7}Li(p,n) Simulated Neutron Spectra

A simulation code, SimLiT [6], that uses experimentally determined ^{7}Li(p,n) cross sections, two-body kinematics and proton energy loss in the Li-target, was developed to calculate neutron spectra, intensities and angular distribution. In the LiLiT setup there is a significant amount of material surrounding the target, which may affect the neutron energy spectrum and intensity. We therefore need to simulate this environment for actual experiments. Detailed simulations using the codes SimLiT as the neutron source and GEANT4 for neutron transport, with a particular emphasis on the detector response, were carried out (see figures 4 and 5).

6. ^{7}Li(p,n) Experimental Neutron Spectra

A series of experiments were conducted at the IRMM Van de Graaff accelerator. The 1912 keV proton beam, with energy spread of $\sim$ 1.5 or 15 keV, irradiated a LiF target. We used ^{6}Li glass detectors (1/8″ and 1″ thick) to measure the neutron time-of-flight (TOF) in the relevant angular range. If we denote the neutron flightpath by L, and the neutron TOF by t, the nominal neutron energy is given by: $E_n = \frac{1}{2}m_n(\frac{L}{t})^2$. However, the effect of the neutron detector thickness is important for a reliable extraction of the neutron energy spectrum. We present below experimental spectra (figure 4) obtained with each detector compared with the detailed simulations done with the SimLiT and GEANT codes. The simulations reproduce correctly the TOF spectra. In figure 5, we show that a simple approach taking the mean geometrical distance between target and detector results in consistent extracted energy spectra for the two detector thicknesses, in reasonable agreement also with the neutron spectrum emitted by the lithium target as calculated by the code SimLiT. Figure 6 shows the experimental spectrum integrated over all angles compared with the simulation, as reported in [8].

Figure 4. Time-of-flight (TOF) spectrum at 0° detected with (left) thick (1″) and (right) thin (1/8″) detectors. We can see good agreement between the experimentally measured spectrum (black) and the simulated (SimLiT+GEANT4) spectrum (red).

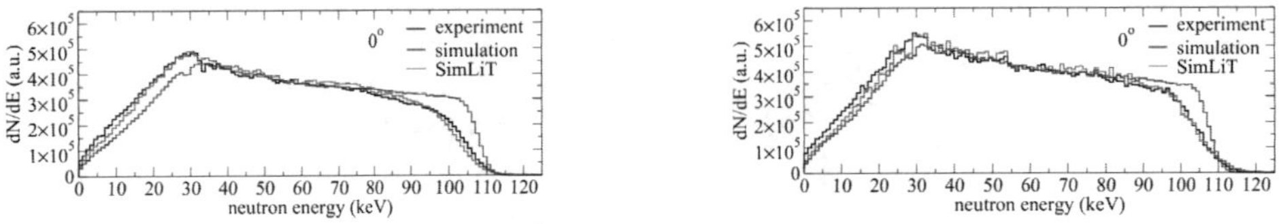

Figure 5. The energy distribution at 0° converted from the detected TOF spectrum with (left) thick (1″) and (right) thin (1/8″) detectors. The blue spectrum is calculated by SimLiT for neutrons emitted by the Li target.

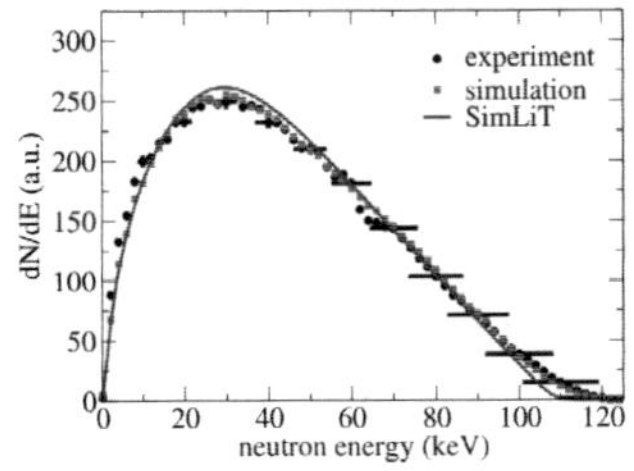

Figure 6. The energy distribution, after integrating over all angles, for the thick (1″) detector. We can see good agreement between the experimentally measured spectrum (black), the simulated (SimLiT+GEANT4) spectrum (red) and the spectrum calculated directly by SimLiT (blue). The horizontal bars represent uncertanties in neutron energy determination. Typical statistical counting errors are shown.

Acknowledgments
M.P. and R.R. wish to gratefully acknowledge the support of the German-Israeli Foundation (GIF Research Grant No. G-1051-103.7/2009). M.P. gratefully acknowledges the Pazi Foundation for its support in the design and construction of the LiLiT setup.

References
[1] Burbidge E M *et al.* 1957 *Rev. Mod. Phys.* **29** 547
[2] Käeppeler F *et al.* 2011 *Rev. Mod. Phys.* **83** 157
[3] Ratynski W and Käeppeler F 1988 *Phys. Rev.* C **37** 595
[4] Kreisel A *et al.* 2014 Proceedings of Linac 2014, Geneva (Switzerland), WEIOB02 770; and references therein; available online: `http://accelconf.web.cern.ch/AccelConf/LINAC2014/papers/weiob02.pdf`
[5] Meusel O *et al.* 2012 Proceedings of LINAC 2012, Tel Aviv (Israel), MO3A03 130; available online: `http://accelconf.web.cern.ch/accelconf/LINAC2012/papers/mo3a03.pdf`
[6] Friedman M *et al.* 2013 *Nucl. Instr. Meth. Phys. Res.* A **698** 117
[7] Feinberg G *et al.* 2009 *Nucl. Phys.* A **827** 590c
[8] Feinberg G *et al.* 2012 *Phys. Rev.* C **85** 055810

A sensitivity study of s-process: the impact of uncertainties from nuclear reaction rates

N Vinyoles and A Serenelli

Institut de Ciències de l'Espai (CSIC-IEEC), campus UAB, Facultat de Ciències, 08193, Bellaterra, Spain

E-mail: vinyoles@ice.cat

Abstract.
The slow neutron capture process (s-process) is responsible for the production of about half the elements beyond the Fe-peak. The production sites and the conditions under which the different components of s-process occur are relatively well established. A detailed quantitative understanding of s-process nucleosynthesis may yield light in physical processes, e.g. convection and mixing, taking place in the production sites. For this, it is important that the impact of uncertainties in the nuclear physics is well understood. In this work we perform a study of the sensitivity of s-process nucleosynthesis, with particular emphasis in the main component, on the nuclear reaction rates. Our aims are: to quantify the current uncertainties in the production factors of s-process elements originating from nuclear physics and, to identify key nuclear reactions that require more precise experimental determinations.

In this work we studied two different production sites in which s-process occurs with very different neutron exposures: 1) a low-mass extremely metal-poor star during the He-core flash (n_n reaching up to values of $\sim 10^{14}\mathrm{cm}^{-3}$); 2) the TP-AGB phase of a $M_\odot$, Z=0.01 model, the typical site of the main s-process component (n_n up to $10^8 - 10^9\mathrm{cm}^{-3}$). In the first case, the main variation in the production of s-process elements comes from the neutron poisons and with relative variations around 30%-50%. In the second, the neutron poison are not as important because of the higher metallicity of the star that actually acts as a seed and therefore, the final error of the abundances are much lower around 10%-25%.

1. Introduction

S-process production sites are well identified although the s-process mechanisms given in the stars are still not really understood as the creation of the ^{13}C pocket, the exact temperatures or mixing and convection processes. In order to improve the models and understand those physical processes, is necessary to have a good understanding of the nuclear physics, identify how the uncertainties of the nuclear rates affect the final results of the models and try to minimize this effect [1].

In this work, the impact of the nuclear uncertainties, concretely the uncertainties of the neutron capture rates, have been studied in two different s-process production sites.

The first site occurs in extremely metal poor stars (EMP). In our case, the mass of the star is $M = 1M_\odot$ and the metallicity $Z = 10^{-7}$. Models of low-mass EMP stars predict that, during the He-core flash, a proton ingestion episode (PIE) takes place in which protons are ingested into the He-burning core [2]. Ingested protons are swiftly captured by ^{12}C at typical temperatures of $T \sim 2 \cdot 10^8$ K, triggering the production of neutrons by the sequence

^{12}C(p, γ)^{13}N(β+)^{13}C(α, n)^{16}O. The free neutrons lead to the production of s-process elements. Because of the high temperatures, alpha-captures on ^{13}C are very fast, leading to large neutron densities, with typical values of the order of 10^{14}cm^{-3} [3, 4].

The second case studied is a more common place for the production of s-process, the thermal pulses on the AGB phase of low- and intermediate-mass stars. Specifically, we have studied a $M = 3M_\odot$ and $Z = 0.01$ model. Low-mass AGB stars are characterized by the nuclear burning of hydrogen and helium shells on top of the electron degenerate core. After each thermal pulse there is a mixing of protons from the envelope with the inter-shell rich in ^{12}C. Then, this ^{12}C captures the protons forming a ^{13}C pocket in the top of the inter-shell region. The advance in mass of the H-burning shell compresses and heats the pockets and the ^{13}C nuclides start capturing α particles and releasing neutrons by ^{13}C(α, n)^{16}O. This free neutrons allow a very efficient s-process nucleosynthesis in a small radiative zone. Another marginal neutron source is the ^{22}Ne(α, n)^{25}Mg that is activated at the bottom of the He-convective shell if temperature reaches values of $T \sim 3 \cdot 10^8$ K [5, 6].

2. Method

The calculations presented here are based on the GARSTEC code [7]. The rates used are taken from KADoNiS database [8], NACRE database [9] and JINA REACLIB library [10]. To do the uncertainty study, a post-processing code is used. This code contains 2383 different reactions and 804 isotopes. The first step of our work was to associate an error or uncertainty to each neutron capture rate. To do so, we used the error factors for each neutron capture recommended in [11]. In that work, the experimental errors of the radiative neutron capture rates are taken from KADoNiS database [8] corrected with the stellar enhancement factor (SEF) and presented as error factors (x_{var}) with 2σ. In addition to the radiative neutron capture rates two extra neutron capture rates are taken into account: ^{14}N(n, p)^{13}C and ^{17}O(n, α)^{14}C. The experimental errors are taken at 30 keV from [12] and [13] and converted into factors following [11]. For the neutron capture rates without experimental error, a factor of 2 uncertainty is assumed.

Two different methods have been used in order to study the impact of the nuclear uncertainties on the final abundances.

2.1. Monte-Carlo

The Monte-Carlo method consists in varying all the rates simultaneously, where each rate is drawn from a random distribution associated with its error. Following Rauscher's work [11] and in order to avoid negative rates we use $r_j = f \cdot r_{0j}$ where $f = e^{x_{rand}}$ and x_{rand} is uniformly distributed between $-log(x_{var}) < x_{rand} < log(x_{var})$. This method consistently quantifies the effect of the uncertainties of all the neutron capture rates over the final abundances of the different elements. It does not allow, however, to identify which reactions are the dominant source of errors for each elements. This is an important limitation of the method. Therefore, we have also developed an individual rate study, as described below.

2.2. Power law expansion

In order to find out which reactions contribute most to the final error of the abundance of each element is necessary to do an individual study of the rate variation effects.

First, for each reaction j several s-process calculations are performed, where the rate of the reaction, $< \sigma v >_j$ is varied systematically between a factor 2 up and down. The variation of element abundances obtained are well described by a logarithmic expansion:

$$\alpha_{ij} = \frac{\partial log(X_i)}{\partial log(< \sigma \cdot v >_j)}. \quad (1)$$

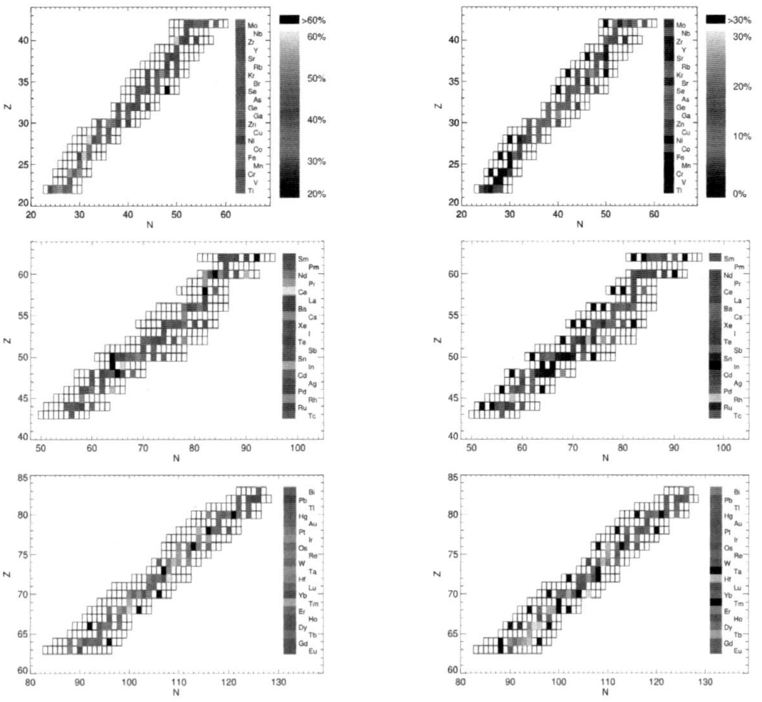

(a) EMP (b) AGB

Figure 1: Total error on the isotopes final abundances. The column on the right is the total error of the element calculated using the isotopes errors. Left: EMP stars. Right: Low mass AGB stars

X_i is the mass fraction of the i isotope and X_{0i} is the mass fraction of the isotope without any rate variation. To derive the final abundance for an element taking into account all the s-reactions we used a power-law expansion [14],

$$\frac{X_i}{X_{0i}} = \prod_j \left(f^{\alpha_{ij}}\right). \tag{2}$$

This method is very useful for studying the individual rate uncertainties contributions to the final abundances and to determine which reactions significantly affect the final results. Comparing the results with the Monte-Carlo, we found differences between methods around $-2.5\% \pm 5\%$ for the EMP case and $2.5\% \pm 10\%$ for the AGB one. These values are lower than the errors and thus, we can consider that the methods give equivalent results and the second one is used to do the study.

3. Results and Summary

In figure 1 the errors of the final abundances for the different elements and isotopes resulting from the contribution of all the neutron capture rates uncertainties are plotted. For the EMP case (left) we find that the average error is around 30%-50% and for the AGB case (right), the error is much lower, between 10% and 25%. From figure 2 it is possible to identify the reactions whose uncertainties contribute most to the final error. For the EMP, the main contributions to the error come from the neutron poison, light isotopes (as ^{16}O,^{17}O or ^{14}N) of primary origin synthesized by helium burning that capture neutrons. As the star has very low metallicity, the ratio between the neutron captures of the light isotopes of primary origin and heavy isotopes is

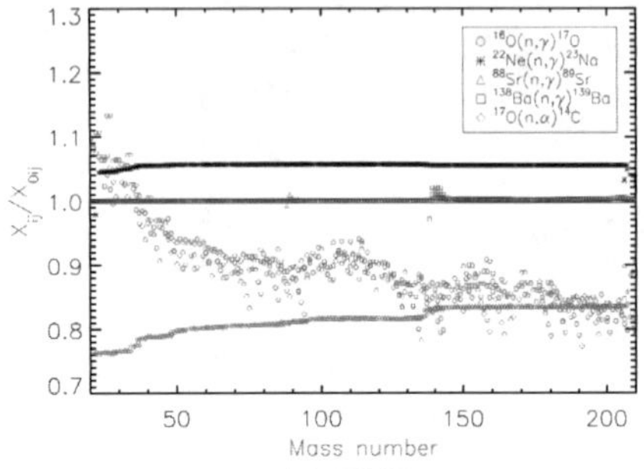

(a) EMP

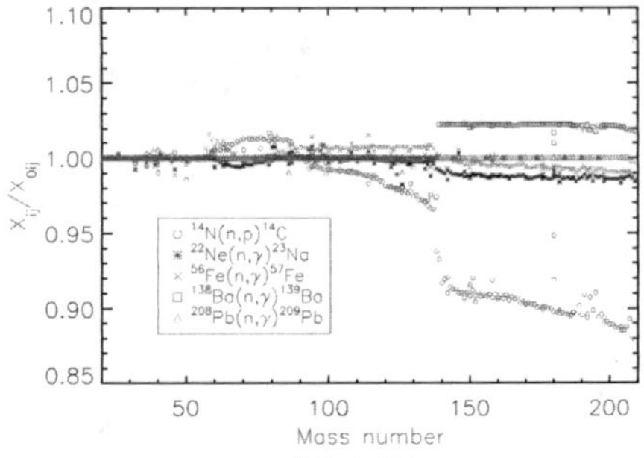

(b) AGB

Figure 2: Effect of the variation of certain rates within their uncertainties to the isotopes final abundances. This errors are plotted against the mass number. Left: EMP stars. Right: Low mass AGB stars

very high ($\sigma_{light}X_{light}/\sigma_{heavy}X_{heavy} \sim 1500$ [4]). Then, the uncertainties on the nuclear capture rates of these light isotopes will be the main source of the errors due to the nuclear rates.

For the AGB case, as the metallicity is higher, the ratio $\sigma_{light}X_{light}/\sigma_{heavy}X_{heavy}$ will be much lower (~ 100) and the impact of the neutron capture rates of the heavy isotopes more significant. The average error of the final abundances for this case is around 10%-25%. In figure 2 is possible to identify the reactions that significantly affect the final abundances. The major contribution comes from the uncertainty on the neutron capture rate of ^{14}N, a neutron poison that captures the neutrons that are released. Other contributions to the error come from the uncertainties on the neutron captures rates of the heavy isotopes that affect the final abundances of the local isotopes (see for example the effect of the neutron capture rate of the ^{138}Ba). The diminution of the final errors with comparison with the EMP case is due to the lower uncertainties associated to the significant neutron captures rates of heavy isotopes compared with the uncertainties of the neutron poison.

Further details and complementary results will be published in a paper on preparation.

Acknowledgements. This work is partially supported by the European Union International Reintegration Grant PIRG-GA-2009-247732 and the MICINN grant AYA2011- 24704.

References

[1] Käppeler F, Gallino R, Bisterzo S and Aoki W 2011 *Reviews of Modern Physics* **83** 157–194
[2] Schlattl H, Cassisi S, Salaris M and Weiss A 2001 *ApJ* **559** 1082–1093
[3] Campbell S W, Lugaro M and Karakas A I 2010 *A&A* **522** L6
[4] Cruz M, Serenelli A and Weiss A 2013 S-process in low-mass extremely metal-poor stars accepted in A&A
[5] Gallino R, Arlandini C, Busso M, Lugaro M, Travaglio C, Straniero O, Chieffi A and Limongi M 1998 *ApJ* **497** 388
[6] Herwig F 2005 *ARA&A* **43** 435–479
[7] Weiss A and Schlattl H 2008 *Ap&SS* **316** 99–106
[8] Dillmann I, Heil M, Käppeler F, Plag R, Rauscher T and Thielemann F K 2009 Kadonis - the karlsruhe astrophysical database of nucleosynthesis in stars URL `www.kadonis.org`
[9] Angulo et al 1999 *Nuclear Physics A* **656** 3–183
[10] Cyburt R H, Amthor A M, Ferguson R, Meisel Z, Smith K, Warren S, Heger A, Hoffman R D, Rauscher T, Sakharuk A, Schatz H, Thielemann F K and Wiescher M 2010 *ApJS* **189** 240–252
[11] Rauscher T 2012 *ApJ* **755** L10
[12] Koehler P E and O'brien H A 1989 *Phys. Rev. C* **39** 1655–1657
[13] Koehler P E and Graff S M 1991 *Phys. Rev. C* **44** 2788–2793
[14] Bahcall J N, Huebner W F, Lubow S H, Parker P D and Ulrich R K 1982 *Reviews of Modern Physics* **54** 767–799

Nuclear Physics in Astrophysics VI (NPA6) IOP Publishing
Journal of Physics: Conference Series **665** (2016) 012029 doi:10.1088/1742-6596/665/1/012029

Experiences and Prospects of Nuclear Astrophysics in Underground Laboratories

M Junker
INFN – Laboratori Nazionali del Gran Sasso, Località Assergi, 67100 L'Aquila, Italy

E-mail: matthias.junker@lngs.infn.it

Abstract. Impressive progress has been made in the course the last decades in understanding astrophysical objects. Increasing precision of nuclear physics data has contributed significantly to this success, but now a better understanding of several important findings is frequently limited by uncertainties related to the available nuclear physics data. Consequently it is desirable to improve significantly the quality of these data. An important step towards higher precision is an excellent signal to background ratio of the data. Placing an accelerator facility inside an underground laboratory reducing the cosmic ray induced background by six orders of magnitude is a powerful method to reach this goal, even though careful reduction of environmental and beam induced background must still be considered. Experience in the field of underground nuclear astrophysics has been gained since 20 years due to the pioneering work of the LUNA Collaboration (Laboratory for Underground Nuclear Astrophysics) operating inside the underground laboratories of the Laboratori Nazionali del Gran Sasso (LNGS) in Italy. Based on the success of this work presently also several other projects for underground laboratories dedicated to nuclear astrophysics are being pursued worldwide. This contribution will give a survey of the past experience in underground nuclear astrophysics as well as an outlook on future developments.

1. Introduction

Accurate knowledge of thermonuclear reaction rates is important in understanding the energy generation, the neutrino luminosity and the element synthesis in stars. Due to the Coulomb barrier (height E_c) of the entrance channel, the reaction cross section $\sigma(E)$ drops nearly exponentially with decreasing energy E. Thus it becomes increasingly difficult to measure $\sigma(E)$ and to deduce the astrophysical $S(E)$ factor defined by the equation $\sigma(E)=S(E)/E\cdot\exp(-2\pi\eta)$ with the Sommerfeld parameter given by $2\pi\eta=31.29\,Z_1 Z_2\sqrt{(\mu/E)}$. Z_1 and Z_2 are the nuclear charges of the interacting particles in the entrance channel, μ is the reduced mass in amu, and E is the center-of-mass energy in keV [1].

Experimental techniques have improved significantly to push the limit of measurements close to or even inside the thermal energy region in stars, which is determined by the Gamow peak $E_0 \pm \delta E_0$ and frequently lies far below the height of the Coulomb barrier, approximately at $E_0/E_c=0.01$. But even in those cases in which direct measurements prove to be feasible, uncertainties in low energy stopping

power data or the insufficient understanding of the electron screening effect introduce significant systematic uncertainties [2].
As a result extrapolations of data from higher energies using e.g. R-matrix fits [3] are often required. In order to minimize the risks related to the extrapolation, the data sets used should cover a wide range of beam energies. It is important to aim for direct measurements at energies as close as possible to the Gamow Peak maintaining small statistical and systematical errors. Scattering experiments or transfer reactions can provide additional information. Indirect methods like the Troian Horse technique can be exploited as an alternative tool for $\sigma(E)$ determination [2].
In the following we will focus on direct experiments at underground accelerators as a powerful tool to obtain nuclear cross sections inside the Gamow peak.

2. Background in detectors and its reduction

Due to their small reaction rate nuclear astrophysics experiments have to compete with background originating from cosmic rays (CR) and environmental radioactivity. At earth surface CR are mainly composed of muons [4]. Due to their high energy CR can penetrate even massive shielding. The energy deposition in a specific detector is related to its material and shape. Also secondary particles caused by CR interaction in surrounding materials interact with the detector.
While the direct interactions of CR can be efficiently rejected by active shielding (e.g. with plastic scintillators), events related to secondary particles (e.g. neutrons) are difficult to suppress and limit the efficiency of active shielding. Only the very massive shielding available in underground laboratories can provide a significant reduction of the CR flux of several orders of magnitude. This approach is commonly exploited by low event rate experiments in the field of Dark Matter research and Neutrino physics while presently only the LUNA[1] (Laboratory for Underground Nuclear Astrophysics) experiment performs accelerator-based nuclear astrophysics experiments in an environment deep underground.
Environmental radioactivity is another source of background which reaches gamma energies of up to 3.5 MeV. It is caused by radionuclides belonging to natural radioactive series (^{214}Bi, ^{226}Ra, ^{214}Pb, ^{208}Tl…), long-lived natural radionuclides (^{40}K, ^{87}Rb ..), radionuclides of cosmogenic origin (^{3}H, ^{14}C, ^{22}Na, ^{7}Be..) and radionuclides of artificial origin (^{95}Nd, ^{137}Cs, ^{90}Sr, ^{144}Ce..). These nuclides can be found outside but also inside the detectors themselves. The related background can be reduced by careful material selection and cleaning and by massive passive shielding with high purity lead and copper. As CR induce secondary particles when interacting with passive shielding, the reduction of the CR flux inside an underground laboratory makes massive Pb and Cu shielding more efficient underground than above ground. Furthermore, the higher the rock coverage of a site the higher the reduction of the underground CR flux and the higher the suppression of background achieved by lead and copper shieldings in the gamma energy range up to 3.5 MeV [5].
The interaction of the ion beam with traces of material on the target can give rise to beam induced background which is specific for each reaction. In particular materials with atomic numbers lower than or similar to the target material are candidates for beam induced backgrounds. Consequently each reaction must be considered separately taking into account beam energy and particle spectrum involved as well as Q-value and resonance structure of the background reactions. The type of the target (gaseous or solid) and the target production process are of fundamental importance. In some particular cases background can be induced also by target nuclei scattered by the impinging projectiles [6].

---

[1]http://luna.lngs.infn.it

The excellent signal to background ratio in the high resolution spectra of HPGe detectors favors the identification of the background, but they provide a limited efficiency with respect to high density materials like BGO or suffer consistent intrinsic background (e.g. $LaBr_3$). On the other hand detectors with low resolution disfavor a reliable background recognition. In any case, the spectra obtained when operating underground improve significantly the identification and subsequent suppression of beam induced background.

3. The LUNA experiment at LNGS

Presently the LUNA Collaboration operates the worldwide only underground accelerator inside the Laboratori Nazionali del Gran Sasso (Italy) run by the Italian Istituto Nazionale di Fisica Nuclear (INFN)[2]. The underground laboratories consist in three halls (100m long, 20m wide, 18m high). They are covered by 1400 m of rock corresponding to 3800 meter of water equivalent (m.w.e.) which reduce the cosmic muon flux be six orders of magnitude[3]. Located long side the motorway Teramo – Rome the underground site is accessible even for trucks of extra width and length. Inside the laboratories lifting equipment up to 40t, ventilation and 3 MW electrical power provided by redundant sources are available. Safety relevant plants like ventilation and fire detection are monitored by a laboratory wide supervision system together with selected experiment specific information. Firefighters and guards are available permanently in the underground control room.

LNGS hosts clean rooms usable for material cleaning, assembling and decontamination, a chemistry laboratory equipped, among others, with an ICPMS facility, electronic and mechanical workshops and an experienced engineering division. The LNGS facility STELLA (SubTErranean Low Level Assay) is available for measurements of low radioactivity samples. LNGS is a reference point for a vivid international scientific community involving more than 900 scientist.

The LUNA collaboration started its activities at LNGS in 1992 installing a home made 50kV accelerator [7]. A concept of an MV scale underground accelerator facility was published in 1993 [8]. Since year 2000 LUNA is operating a 400kV Singletron machine [9]. Reviews of the scientific activities of LUNA can be found in [10] [11].

4. Experience after 20 Years of LUNA

As in all fields of science it is important to identify the scientific goal of an experiment beforehand. As an example the $^3He(^3He,2p)^4He$ experiment was set up to exclude a resonance proposed by W. Fowler as a solution to the so called "Solar Neutrino Problem"[12]. To rule out the "desperate explanation of the solar neutrino puzzle" great efforts were taken to push the limit of the measurements to the lowest possible energy accepting data points with large errors (up to 100%). Finally, LUNA showed that there is not any evidence for a narrow resonance at low energies and that the cross section $\sigma(E)$ increases at solar energies as expected from the electron screening effect. This finding ruled out the astrophysical solution of the solar neutrino problem [13] [14].

Also the LUNA measurements on $^{25}Mg(p,\gamma)^{26}Al$ have been performed to directly determine the resonance strength $\omega\gamma$ of the important low energy resonances at 198 keV, 130 keV and 92 keV of this reaction which is the galactic source of ^{26}Al. As this isotope is frequently observed in astronomy via the well known 1.809 MeV gamma line associated to its radioactive decay, an understanding of these nucleosynthesis processes is important [15]. The levels in the compound nuclei ^{26}Al corresponding to the

[2]http://www.lngs.infn.it

[3]https://www.lngs.infn.it/lngs_infn/contents/lngs_en/research/europe/web_ILIAS_A1/JRA1/ilias-last/workpack1.htm

resonances of interest were already well known and the $\omega\gamma$s of the resonances at 198 keV and 92 keV had been measured by direct and indirect experiments.

A nuclear reaction emitting gamma rays can be studied by detecting all gammas of a cascade using a BGO summing detector in 4π geometry. This yields a single peak (called summing peak) in the spectrum located at $Q+E_{beam}$ where Q is the reaction's Q-value and E_{beam} is the beam energy[16]. In this case the summing peak is far above the energy region effected by room background as the Q-value of the reaction is 6.3 MeV . The efficiency of a summing detector depends on the branching ratio of the reaction [16]. Thus the resonances at 304 and 198 keV were studied at LUNA beforehand with a standard HPGe detector. The resulting level information and $\omega\gamma$s have been used in a Monte Carlo simulation to determine the detector's efficiency. It should be pointed out that in the measurements of the 98 keV resonance the intrinsic detector background as well as the beam induced background posed severe problems due to the low resolution of BGO detector. A Monte Carlo code had to be used to disentangle the spectra exploiting the segmentation of the detector. Finally LUNA was able to redetermine the $\omega\gamma$ of the resonance at 198 keV reducing the related uncertainty by a factor of two. The $\omega\gamma$ of the resonance at 92 keV has been measured for the first time by a direct experiment [17].

The goal for investigating the reaction of $^3He(^4He,\gamma)^7Be$ by LUNA was different from the cases discussed previously. This reaction had be studied already by direct gamma spectroscopy and by detection the ^7Be ejectiles. As the data available obtained with the two different techniques appeared to show a systematical discrepancy [18] LUNA aimed to measure $\sigma(E)$ with both techniques with low statistic and systematic uncertainties.

The cross section in the solar Gamow Peak of $^3He(^4He,\gamma)^7Be$ is about 6 orders of magnitude smaller than $\sigma(E)$ of $^3He(^3He,2p)^4He$. At the same time the prompt gammas emitted by $^3He(^4He,\gamma)^7Be$ have energies of about 1.6 MeV, within the energy range dominated by environmental background. Thus energy resolution as well as environmental and intrinsic radioactivity were important issues. LUNA studied the prompt gammas and the beta-delayed gammas emitted in the decay of ^7Be with HPGe detectors [19]. Thus, the efficiency reached is significantly lower than in the experiments described previously. Massive lead and copper shielding including a radon box were installed to suppress environmental radioactivity. Detailed studies of the detector geometry were performed to avoid uncertainties as effect of the angle distribution. In several runs prompt gammas and ^7Be were detected simultaneously in collaboration with the LNGS facility STELLA excluding several systematic uncertainties.

Due to the extremely low background and the expertise of low event counting gained at LNGS it was possible to measure few data points with small systematic and statistical errors. However, even though LUNA has published the lowest energy data points available in literature they have been taken at beam energies above the solar Gamow Peak [19]. Consequently the LUNA data must be extrapolated together with data available at higher energies. The small uncertainties of the LUNA results put strong constrains on the extrapolation.

Overall, the precision of the new data stimulated renewed efforts to develop improved nuclear models ([20][21][22] and references therein). To further reduce the uncertainty of the extrapolation a measurement should be performed which covers a wide energy range from the lowest energies investigated by LUNA to the highest energy data by ERNA [23] overcoming the need to match several data sets with different systematic uncertainties.

Another example for the impact of the underground data at LUNA are the measurements of the reaction $^{14}N(p,\gamma)^{15}O$. This reaction is of great importance for stellar evolution, solar neutrinos and for the age determination of Globular Clusters [24]. It had been measured by Schröder et al. [25] via gamma spectroscopy using different accelerators covering a beam energy range from 3.3 to 0.24 MeV. As the data did not enter inside the solar Gamow Peak (E_g = 26 keV) an extrapolation had been performed based on a Breit Wigner formalism indicating $S(0) = 3.20 \pm 0.5$ keVb. In particular the authors assumed an effect of a sub-threshold state at -507 keV for the direct transition to the ground state. Later Angulo et al. [26] re-analyzed the data set of [25] and published a result of 1.77 ± 0.20 keV almost a factor of two lower than the previously accepted result. This difference is due to the fact that Angulo et al. could not reproduce the effect of the sub-threshold state indicated by Schröder et al.

LUNA investigated this reaction between 120 keV and 370 keV with a HPGe detector and solid targets [27][28] and between 70 keV and 370 keV with a summing BGO detector and a gas target system [29]. The good signal to background ratio in the HPGe data of LUNA allowed to identify the branching ratios of the different gamma cascades also at the lowest energies. In this context it became obvious that the ground state transition data published in [25] had not been corrected for the "Summing in" effect. This effect describes the coincident detection of two gamma rays belonging to the same cascade which simulates a single gamma ray related to a direct ground state transition. If this correction is omitted the extrapolation of the direct ground state transition and, to a minor extend, of the other transitions are falsified. Based on the LUNA data it was possible to correct the Schröder data a posteriori (see [28] for details) and a reliable extrapolation has be performed using the corrected Schröder and the LUNA data.

The measurements performed by LUNA with the BGO detector extended to even lower energies. Still the solar Gamow Peak has not been reached. Due to the summing of all different gammas of one cascade the BGO data do not contain nuclear physics information usable to support an extrapolation of high energy data. Still these data are of great value to verify independently the extrapolations made on the basis of the available data [19][29].

Also in the measurements of $^{15}N(p,\gamma)^{16}O$ high energy data taken above ground and low energy data obtained underground have been used to support an R-matrix fit down to the stellar Gamow peak. In a joint effort of the University of Notre Dame (USA) and the LUNA collaboration high energy data between 1.8 MeV and 285 keV were taken at the University of Notre Dame and data between 400 and 130 keV where taken at LUNA. In this context the limited energy range of the present LUNA 400kV accelerator posed some problems as the resonance at E = 440 keV in $^{15}N(p,\gamma)^{16}O$, commonly used for normalization and target checks, is not accessible at LUNA [30].

Similar issues had to be faced in the present work at LUNA for $^{17}O(p,\gamma)^{18}F$. This reaction has been studied in a wide energy region (E_{cm} = 160 to 370 keV) appropriate to explosive hydrogen burning in classical novae. The reaction cross section was measured using both prompt γ-ray detection and activation approaches taking advandage of the knowledge gained in during the measurements of $^3He(^4He,\gamma)^7Be$. Results from the non-resonant reaction contributions have been analyzed in a global fit and lead to a total astrophysical S-factor $S(0)$ = 4.81 keV barn. The strength of the E_{cm} = 183 keV resonance, $\omega\gamma = 1.67 \pm 0.12$ μeV, has been determined with the highest precision to date and was found to dominate the astrophysical reaction rate at temperatures $0.1 < T < 0.4$ GK relevant to novae explosions. A global fit to the experimental data and other data from the literature has led to an improved recommended reaction rate. The abundances of key isotopes such as ^{18}F, ^{18}O, ^{19}F and ^{15}N have been evaluated through nova models calculations and have been obtained with a precision of 10%, i.e. sufficient to put firmer constraints on observational features of novae nucleosynthesis. The results presented in this study are now sufficiently precise for nova model calculations [31].

5. Upcoming possibilities in underground nuclear astrophysics

The examples above show that underground nuclear astrophysics is a powerful tool not only in pushing measurements to lower energies but also, and even more, in providing high quality data which can be used to reduce the uncertainties of unavoidable extrapolations.

Stimulated by the success of the LUNA experiment several next generation facilities for underground nuclear astrophysics have been proposed in the last years. A series of workshops has been organized in Europe (Barcelona 2009, Dresden 2010, Gran Sasso 2011, Canfranc 2012) [32][4]. In its Long Range Plan 2010 [33] NUPECC recommends to "fully exploit the currently existing large-scale research infrastructures (...) and (to) perform limited-size upgrades to ensure the best use of the large investments made in the past: (…)" and in particular "the nuclear astrophysics underground accelerator LUNA at INFN Gran Sasso, and the exploration of advanced new facilities."

Already in 2007 LUNA submitted a Letter of Intend (LOI) to the LNGS scientific committee proposing to measure $^{12}C(\alpha,\gamma)^{16}O$, and (α,γ) reactions on $^{14,15}N$ and ^{18}O, which are key reactions of the He burning. In addition the LOI suggested investigations on $^{13}C(\alpha,n)^{16}O$, $^{22}Ne(\alpha,n)^{25}Mg$, the neutron sources for the s-process. A renewed study of the reaction $^{3}He(\alpha,\gamma)^{7}Be$ over a wide energy range has been included later into the scientific program.

As these reactions are relevant at higher temperatures (larger energies) than reactions belonging to the hydrogen-burning studied so far at LUNA, the collaboration proposed to install a single ended machine with a terminal voltage of approximately 3.5 MV in the underground laboratories of LNGS. The physics program of the LOI has been approved by the LNGS Scientific Committee, but as the impact of the neutrons produced by the proposed measurements was of concern, this issue has been evaluated by an external committee of experts. Implementing a proper shielding of the neutron background of the proposed experiments can be reduced to a level which is a factor 100 lower than the natural neutron flux inside LNGS. The installation of the MV scale accelerator at LNGS has been approved by the Scientific Committee in 2010. In summer 2012 the Italian Ministry of Research has granted financial support to INFN which allows for site preparation and accelerator installation.

In 2009 CUNA (Canfranc Underground Nuclear Astrophysics) has been proposed at the Laboratorio Subterráneo de Canfranc (LSC) located under the Pyrenees (2500 m.w.e)[5]. The accelerator-based nuclear astrophysics project foresees the installation of a multi MV accelerator coupled with a RF source delivering proton and alpha beams. An engineering design has recently been completed for an independent experimental hall, distant from the rest of the caves at the LSC. A viable alternative presently under consideration is the refurbishing and extension of one of the existing caves. Neutron background measurements and simulations have recently been completed [34]. Funds for the accelerator are being sought at present. A Letter of Intent, extending the case presented in the Expression of Interest from 2009, will be presented to the LSC Scientific Committee by the collaboration in the next few months [35].

The Felsenkeller Underground Accelerator[6] has been proposed by a group of scientists from the Helmholtz Zentrum Dresden Rossendorf (HZDR) and the University of Dresden several of them involved also in LUNA and in other research activities of LNGS. They foresee the installation of a used MV scale accelerator in a site shielded by 112 m.w.e., called Felsenkeller, an old brewery cellar close to Dresden

---

[4]Web pages of the workshops on next generation underground nuclear astrophysics installations:
http://www.fnuc.es/workshop/LSC_Program.htm http://www.hzdr.de/db/Cms?pNid=2088
http://luna.lngs.infn.it/luna-mv, http://nuclear.fis.ucm.es/nucastro/

[5]http://www.lsc-canfranc.es/

[6]http://www.hzdr.de/db/Cms?pNid=2088

(Germany). Given the shallow shielding the background is significantly higher than at LNGS. The facility will be shared with other research groups and will thus be available for nuclear astrophysics only for limited times. Recently, a used 5 MV NEC Pelletron tandem accelerator system has been delivered to HZDR and is now undergoing a refurbishing at HZDR [36].
In the USA the DIANA project (Dakota Ion Accelerators for Nuclear Astrophysics)[7], a collaboration of University of Notre Dame, University of North Carolina, Western Michigan University, Colorado School of Mines and Lawrence Berkeley National Laboratory aims to build a nuclear astrophysics underground accelerator facility. DIANA is designed to achieve large laboratory reaction rates by delivering ion beam currents up to 100 mA to a high density super-sonic jet-gas target as well as to a solid target. It will consist of two accelerators, 50-400 kV and 0.4-3 MV that will cover a wide range of ion beam intensities, with sufficient energy overlap to consistently connect the results to measurements above-ground. DIANA will address scientific issues in the field of solar neutrino sources and the metallicity of the sun, carbon-based nucleosynthesis and neutron sources for the production of trans-Fe elements in stars .

6. References

[1] Iliadis C 2007 Nuclear *Physics in Stars,* Wiley-VCH
[2] Tumino A et al 2013 *Few-Body Sys.,* **54** 745
[3] Lane A M and Thomas R G 1958 *Rev. Mod. Phys.* **30** 257
[4] Beringer J et al. (Particle Data Group) 2012 *Phys. Rev.* D **86**, 010001
[5] Heusser G 1995 *Ann. Rev. Nucl. Part. Sci.* **45**, 543
[6] Anders M, Trezzi D, Bellini A et al 2013, *EPJ* A **49**, 28
[7] Greife U et al 1994 *Nucl. Inst. Meth.* A **350**, 327
[8] Fiorentini G, Kavanagh R W and Rolfs C, 2005, *Z. Phys.* A **350,** 289
[9] Formicola A, Imbriani G, Junker M et al 2003 *Nucl. Inst. Meth.* A **507**, 609
[10] Costantini H, Formicola A, Imbriani G, Junker M, Rolfs C and Strieder F 2009 *Rep. Prog. Phys.* **72** 086301
[11] Broggini C, Bemmerer D, Guglielmetti A and Menegazzo R 2010 *Ann. Rev. Nucl. Part. Sci.* **60**, 53
[12] Fowler W 1972 *Nature* **238**, 24
[13] Junker M, D'Alessandro A, Zavatarelli S, et al 1998 *Phys. Rev.* C **57**, 2700
[14] Bonetti R et al 1999 *Phys Rev Lett* **82**, 5205
[15] Straniero O, Imbriani G, Strieder F et al 2013 *Astrophys. J.* **763**,100
[16] Casella C, Costantini H, Lemut A, Limata B et al 2002 *Nucl. Inst. Meth.* A **489**, 160
[17] Strieder F, Limata B, Formicola A, Imbriani G, Junker M et al 2012 *Phys. Lett.* B **707**, 60
[18] Adelberger E et al 1998 *Rev. Mod. Phys.* **70**, 055502
[19] Costantini H, Bemmerer D, Confortola F, Formicola A, Gyurky Gy et al 2008 *Nucl. Phys* A **814**, 144
[20] Neff T 2011 *Phys. Rev. Lett.* **106**, 042502
[21] Di Leva A et al 2009, *Phys. Rev. Lett.* **102**, 232502
[22] Kontos A ,Uberseder E, Görres J, Akers C, Best A, Couder M, Wiescher M et al 2013 *Phys. Rev.* C **87**, 065804.
[23] Di Leva A, Gialanella L, Kunz R, Rogalla D, Schürmann D, Strieder F et al 2009 *Phys. Rev. Lett.* **102**, 232502

[7] http://www.jinaweb.org/underground/DIANA/index.html

[24] Imbriani G, Costantini H, Formicola A, et al 2003 *A & A* **420**, 625
[25] Schröder U et al. 1987 *Nucl. Phys.* A **467**, 240
[26] Angulo C and Descouvemont P 2001 *Nucl. Phys.* A **690**, 755
[27] Formicola A, Imbriani G, Costantini H et al 2004 *Phys. Lett.* B **591**, 61
[28] Imbriani G, Costantini H, Formicola A, Vomiero A et al 2005 *Eur. Phys. J.* A **25**, 455
[29] Lemut A, Bemmerer D, Comfortola F et al 2006 *Phys. Lett.* B **634**, 483
[30] LeBlanc P J, Imbriani G, Görres J, Junker M et al. 2010 *Phys. Rev. C* **82**, 055804
[31] Scott D A, Caciolli A, Di Leva A, Formicola A et al 2013 Phys Rev Lett **109**, 208001
[32] Proceedings of the round table workshop "LUNA-MV at Gran Sasso" Feb. 10-11, 2011, *http://agenda.infn.it/getFile.py/access?resId=0&materialId=3&confId=3083*
[33] European Science Foundation 2010, Perspectives of Nuclear Physics in Europe, NuPECC Long Range Plan 2010, *http://www.nupecc.org/pub/hb12/hb2012.pdf*
[34] Jordán D et al 2013 *Astrop. Phys.* **42**, *1*
[35] Fraile L M 2012 *private communication*
[36] Bemmerer D 2013, Proceedings of the conference "Nuclear Physics in Astrophysics VI", *J. Phys.: Conf Series*

Journal of Physics: Conference Series **665** (2016) 012030 doi:10.1088/1742-6596/665/1/012030

Program and status for the planned underground accelerator in the Dresden Felsenkeller

D Bemmerer[1], S Akhmadaliev[1], T Al-Abdullah[1,3], M Anders[1,2], T E Cowan[1,2], Z Elekes[1], A R Junghans[1], St Gohl[1,2], J Krause[1,2], T P Reinhardt[2,1], S Reinicke[1,2], B Rimarzig[1], M Röder[1,2], K Schmidt[1,2], R Schwengner[1], T Szücs[1], M Takács[1,2], A Wagner[1], L Wagner[1,2], J Wielicki[1,2], and K Zuber[2]

[1] Helmholtz-Zentrum Dresden-Rossendorf (HZDR), Dresden, Germany
[2] Technische Universität Dresden (TU Dresden), Dresden, Germany
[3] Physics Department, The Hashemite University, Zarqa, Jordan

E-mail: d.bemmerer@hzdr.de (Daniel Bemmerer)

E-mail: zuber@physik.tu-dresden.de (Kai Zuber)

Abstract. The scientific program and curent status of the planned accelerator laboratory in the Felsenkeller shallow-underground facility in Dresden, Germany, are reviewed.

1. Introduction

The increasing precision of nuclear reaction cross section data is one of the main drivers of progress in nuclear astrophysics [1]. For reactions induced by light, stable ions, the preferred experimental technique to obtain such data is to perform ultra-low background experiments at an underground ion accelerator [2]. The only such device worldwide is the 0.4 MV LUNA (Laboratory for Underground Nuclear Astrophysics) accelerator in the Gran Sasso underground facility, Italy.

Based on the great success of LUNA [2, and references therein], the nuclear physics communities in Europe and in North America have recognized the need for one or several new, higher-energy underground accelerators, stating e.g. that

> "*Providing an underground multi-MV accelerator facility is a high priority. There are a number of proposals being developed in Europe and it is vital that construction of one or more facilities starts as soon as possible.*" [3]

As a consequence, a number of related projects are being developed, amongst others, in Europe [4] and North America [5]. The planned underground accelerator in the Dresden Felsenkeller, Germany, is described in the present contribution.

2. Site and background

The Felsenkeller tunnel system in Plauenscher Grund, Dresden/Germany, consists of nine mutually connected tunnels that have been dug in the 1850s to serve a nearby brewery (fig. 1).

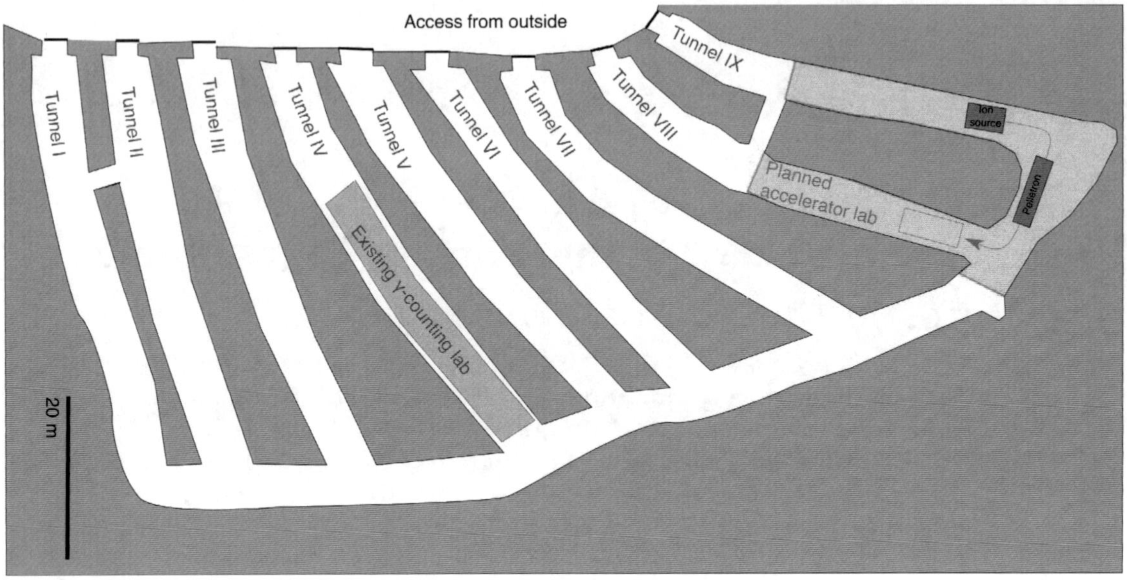

Figure 1. Schematic view of the Felsenkeller underground site. The existing γ-counting laboratory [6] is shown in blue, the planned accelerator laboratory in red.

Since 1982, an underground low-background activity counting laboratory is hosted in one of the tunnels, currently including about ten well-shielded germanium γ-ray detectors [6].

Inside the Felsenkeller tunnels, the cosmic ray muon flux is suppressed overall by a factor of 20-40 when compared to the Earth's surface. A detailed measurement with the REGARD muon tomograph has recently been performed showing the flux and angular distribution of incoming muons [7].

It has been shown previously that an actively vetoed high-purity germanium (HPGe) detector, when used in a shallow underground site like Felsenkeller, has a very low background level in the crucial 6-8 MeV γ-ray energy range, comparably even to the ultra-low background at LUNA [8]. To give an example, in a 44 keV wide region of interest near 6 MeV, the background level in one and the same actively vetoed HPGe detector was found to be $1.39\pm0.05\,h^{-1}$ at the Earth's surface, $0.039\pm0.007\,h^{-1}$ in Felsenkeller, and $0.021\pm0.006\,h^{-1}$ at LUNA [8]. It should be noted that the deep-underground (LUNA) background level can in principle be further improved by a passive neutron shield [9]. However, even without such a shield the background at LUNA is already low enough to enable highly sensitive nuclear astrophysics measurements [2].

The Felsenkeller site offers horizontal access, by lorry, to all parts of the tunnel system and is located inside the Dresden city limits at a distance of 4 km to Dresden central railway station and the TU Dresden campus, and at 25 km distance to the HZDR Rossendorf campus. This urban location greatly facilitates scientific and educational use and the maintenance of the planned facility by qualified specialists.

3. Accelerator status and planned upgrades

In order to speed up progress, an immediately available high-energy accelerator system was acquired in June 2012 and transported to Dresden (fig. 2). It consists of a National Electrostatics (NEC) 5 MV pelletron tandem, model 15SDH-2, external 134 MC-SNICS multi-cathode cesium sputter ion source [10], electromagnets for the low and high energy ends of the tandem, and complete beam lines. This system has previously served as an accelerator mass spectrometer for ^{14}C analyses in an industrial setting [11].

Figure 2. Main accelerator tank of the 5 MV pelletron system being unloaded at HZDR.

The cesium sputter ion source supplies a $^{12}C^-$ current of $50\,\mu A$, as measured behind the injector magnet, enabling high-current carbon beam experiments for astrophysics. The accelerator is equipped with double pelletron charging chains, thus permitting $250\,\mu A$ of upcharge current.

However, an external ion source such as the existing one is not optimized for noble-gas beam implantations, as are routinely done at the HZDR ion beam center for industrial applications. Therefore, it is presently under study whether the existing pelletron can be upgraded by placing an internal radio-frequency (RF) ion source on the terminal (fig. 3). Such an ion source has the advantage that easily obtainable positive ions can be directly accelerated, obviating the need for inefficient charge-exchange channels.

One possibility currently under study is to adapt and use the design for an RF ion source that has been used for more than 30 years now on the high voltage terminal of the HZDR 2 MV Van de Graaf accelerator. The source would be placed at an angle of 30-45° with respect to the beam axis, either below or above the beam axis. In the latter case, the foil stripper would need to be sacrificed, maintaining the gas stripper intact. In order to couple in the beam emitted from such a source, one option being studied is to use an electrostatic deflector with a hole in it, as has been suggested previously [12]. In any case, the tandem capability will be maintained also with the additional RF ion source.

4. Windowless gas target

For the low background, low counting rate experiments of underground nuclear astrophysics, special care must be taken to ensure that the targets being irradiated with intensive ion beams are stable and sufficiently pure. In particular, any remaining impurity may serve as target for the light-ion beam and give rise to parasitic γ-ray emissions.

High purity can be most easily achieved when the target is in a gas phase. Windowless gas targets, either of the static or of the jet gas type, give the additional benefit that they are continuosly flushed with either commercially available fresh or chemically purified gas, so that any impurities transported in with the ion beam or diffusing out of the vacuum chamber walls

Figure 3. Left: Photograph of the high voltage terminal, with the dimensions of the space that may be gained by sacrificing the stripper foil changer. Right: Drawing of a radio-frequency ion source based on the one currently used at the HZDR 2 MV Van de Graaf accelerator.

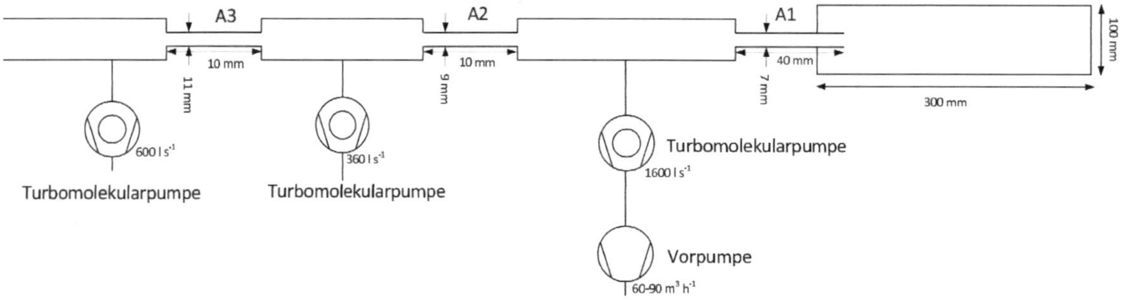

Figure 4. Schematic view of the planned static windowless gas target system for Felsenkeller.

are kept on a manageable level. What is more, windowless gas targets are inherently stable and do not suffer from disturbing effects in target entrance foils, as e.g. energy and radial straggling of the incoming beam, foil heating, and accidental breaches of the foil.

The main disadvantage of windowless gas target systems are the high demands placed on the pumping system. In order to limit geometrical footprint, investment, and running cost all at the same time, a windowless gas target system is being developed for Felsenkeller that relies purely on turbomolecular pumps and one powerful backing pump of 60-90 m^3/h (fig. 4).

Ion optical calculations using the SIMION code show that this gas target system has satisfactory ion optical properties when its entrance is placed at a distance of 1.7 m downstream of the analyzing magnet, both for carbon beam (1-5 MeV) and even for an extreme test case with low-energy hydrogen beam. Pumping speed calculations indicate that it can be operated purely with turbomolecular pumps (appropriately equipped with backing pumps) for hydrogen, helium, carbon dioxide, and nitrogen gases at target pressures up to 2 mbar. Such a target pressure value should be sufficient for most low-energy nuclear astrophysics experiments.

5. Scientific program

The existing, 0.4 MV LUNA underground accelerator has made landmark contributions to the study of solar fusion reactions [2, 13]. However, in order to fully address key cases such as the ^{14}N(p,γ)^{15}O and ^{12}C(p,γ)^{13}N reactions that are necessary to precisely understand the Bethe-Weizsäcker or CNO cycle of hydrogen burning, data over a wide energy range including higher energies than those available at LUNA are needed. Therefore, solar fusion reactions are part of the scientific motivation for the planned Felsenkeller accelerator. They can be studied either in direct kinematics, using the intense hydrogen beam by the RF ion source, or in inverse kinematics using carbon and nitrogen beams provided by the cesium sputter source (table 1). The planned new data on CNO cycle nuclear reactions are a necessary ingredient in order to address the so-called solar abundance problem [14]. With such new nuclear cross section data, the expected measurement of the flux of CNO neutrinos e.g. by the BOREXINO [15] detector for low-energy solar neutrinos may then be used to constrain the carbon and nitrogen elemental abundances in the solar core [16].

A second science field to be addressed concerns the nuclear reactions producing radionuclides that are observable in space-based observatories [17]. Such radionuclides offer a unique view on nucleosynthesis in action, either integrated over the galaxis [18] or for one particular object [19]. Of particular interest is the α-rich freezeout in supernovae, which is believed to dominate the synthesis of ^{44}Ti (half-life 60 years). This radionuclide has recently been proven to be the energy source in the present-day emissions of supernova remnant 1987A [19]. Experiments addressing this case require an intense ^{4}He beam at energies of a few MeV, as in a recent related study at HZDR [20].

A third science field is given by the reactions of carbon burning [21]. Here, some disputed claims of low-energy resonances affecting the stellar reaction rate need to be clarified [22], and overall the uncertainty in the reaction rates can be significantly reduced by new experimental data. As is the case for several hydrogen burning reactions, also the carbon burning reactions may affect the nucleosynthesis in supernovae of type Ia, which are not only cosmological standard candles but also important contributors to the chemical enrichment of the universe.

A fourth topic is the applied physics potential of a high-current, low background accelerator. One relevant example is ion beam based analytics, where in particular the proton-induced γ-ray emission (PIGE) technique [23] may benefit from the drastically reduced laboratory background when very clean samples must be analyzed. In an underground low-background environment such as Felsenkeller, hydrogen depth profiling by ^{15}N beam induced reactions can be performed at much lower beam intensity, conserving sample integrity, and for much lower hydrogen concentrations, increasing sensitivity. Finally, the planned RF ion source will enable intensive noble-gas beams that may serve a useful purpose for industrial implantations e.g. in the production of power transistors.

When sorting the possible reactions by the required beams and, by extension, ion source, it is apparent that several topics can already be addressed by the existing MC-SNICS ion source, while other topics will require the use of the RF ion source that is under development (table 1). Detailed counting rate estimates and feasibility studies for selected cases have been presented elsewhere [8].

In addition to work on the research topics outlined here, the planned accelerator will be open to the worldwide scientific community as a facility. Proposals from all fields of science will be

Table 1. Scientific program of the planned accelerator in Felsenkeller, arranged by ion beams.

Ion	Ion source	Scientific topic
^{1}H	MC-SNICS (Tandem) or	Solar fusion reactions
	RF (single-ended)	Proton-induced γ-ray emission (PIGE)
^{4}He	RF (single-ended)	Reactions of the α-rich freezeout in supernovae
		High-energy ion implantations
$^{12/13}$C	MC-SNICS (Tandem)	Solar fusion reactions
		Stellar carbon burning and type Ia supernovae
$^{14/15}$N	MC-SNICS (Tandem)	Solar fusion reactions
		Hydrogen depth profiling

welcomed.

6. Status, permissions, and operations

The features of the Felsenkeller site from a nuclear physics point of view are by now known: The background in a typical detector for nuclear physics experiments has already been studied experimentally [8]. Recently, also the muon flux and angular distribution have been measured, indicating a muon flux at the site of the proposed accelerator that is even slightly lower than at the nearby site of the existing γ-counting lab [7].

The work planned here will require a construction permit and an operating permit from the radiation safety point of view. Test measurements carried out during experiments at the HZDR 3 MV Tandetron, which can be extrapolated to the ones planned here, have shown that no special shielding will be required for the typical, low-background experiments. It is envisaged that the high voltage will be removed by an automatic system if the measured radiation dose rate exceeds a predetermined threshold.

The site is privately owned, and a long-term rental contract similar to the existing one for the low-activity counting laboratory is envisaged. The tunnels have been scanned with a three-dimensional laser-based system. Based on these data and the known dimensions of the pelletron, a full feasibility study and detailed cost estimate for the preparation of the site are now complete. As of the writing of this report, a third of the necessary construction cost has been pledged, but construction presupposes full funding.

For the day to day operations, a scientist and an engineer, both HZDR staff, will be available on a full time basis. In addition to this, students (currently four PhD students and three Diploma students) and postdocs (currently one) will take part in the work on site, based on projects. This team will work on the above described scientific program (sec. 5) and also support external users.

A large share of the beam time will be allocated to external users from all fields of science, based on the recommendations by an independent panel of outside scientific advisors judging the scientific merits of proposals brought before them. For the external users selected by this panel, the accelerator and laboratory will function like a user facility providing all necessary support.

7. Summary and outlook

A project for a new, 5 MV underground accelerator is under development for the Dresden Felsenkeller, Germany. This shallow-underground site offers good background conditions and excellent access capabilities. A used, high-current 5 MV pelletron accelerator was already bought for this laboratory and recently transported to Dresden. Areas of ongoing development work include a radio frequency ion source to be placed on the high voltage terminal and a windowless gas target system.

The highly promising scientific program that can be addressed with such a laboratory has been outlined. A full feasibility study has been performed for the site preparation. This civil construction work is at present only partially funded. As key components are already available, the laboratory is expected to be running less than one year after the point in time when all of the construction funding is available.

Acknowledgments

The authors are grateful to the organizers of the "Nuclear Physics in Astrophysics VI" conference in Lisbon, Portugal, for a stimulating workshop. Support by the Helmholtz Association Nuclear Astrophysics Virtual Institute (NAVI, VH-VI-417) is gratefully acknowledged.

References

[1] José J and Iliadis C 2011 *Rep. Prog. Phys.* **74** 096901 (*Preprint* 1107.2234)
[2] Broggini C, Bemmerer D, Guglielmetti A, and Menegazzo R 2010 *Annu. Rev. Nucl. Part. Sci.* **60** 53–73
[3] Nuclear Physics European Collaboration Committee 2010 *Long Range Plan: Perspectives of Nuclear Physics in Europe* (NuPECC) URL `http://www.nupecc.org/index.php?display=lrp2010/main`
[4] Junker M *these proceedings*
[5] Lemut A *these proceedings*
[6] Köhler M, *et al.* 2009 *Appl. Radiat. Isot.* **67** 736–740
[7] Oláh L, *et al.* 2013 *Proceedings of Nuclear Physics in Astrophysics VI, Lisbon, Portugal*
[8] Szücs T, *et al.* 2012 *Eur. Phys. J. A* **48** 8
[9] Wulandari H, Jochum J, Rau W, and von Feilitzsch F 2004 *Astropart. Phys.* **22** 313–322
[10] Loger R L, Raatz J E, Sundquist M L, Garner R C, and Barker J 1999 *AIP Conf. Proc.* **475** 640–643 URL `http://link.aip.org/link/?APC/475/640/1`
[11] Young G C, Corless S, Felgate C C, and Colthup P V 2008 *Rapid Commun. Mass Spectrom.* **22** 4035–4042 ISSN 1097-0231 URL `http://dx.doi.org/10.1002/rcm.3829`
[12] Kreckel H, *et al.* 2010 *Rev. Sci. Instrum.* **81** 063304
[13] Adelberger E, *et al.* 2011 *Rev. Mod. Phys.* **83** 195–246
[14] Serenelli A M, Basu S, Ferguson J W, and Asplund M 2009 *Astrophys. J. Lett.* **705** L123–L127 (*Preprint* 0909.2668)
[15] Bellini G, *et al.* 2012 *Phys. Rev. Lett.* **108** 051302 (*Preprint* 1110.3230)
[16] Haxton W C and Serenelli A M 2008 *Astrophys. J.* **687** 678–691
[17] Diehl R, Hartmann D, and Prantzos N (eds) 2011 *Astronomy with Radioactivities* (*Lecture Notes in Physics* vol 812) (Berlin and Heidelberg: Springer)
[18] Diehl R *et al.* 2006 *Nature* **439** 45–47
[19] Grebenev S A, Lutovinov A A, Tsygankov S S, and Winkler C 2012 *Nature* **490** 373–375 (*Preprint* 1211.2656)
[20] Schmidt K, *et al.* 2013 *Phys. Rev. C* **88** 025803 (*Preprint* 1307.6516)
[21] Pignatari M, *et al.* 2013 *Astrophys. J.* **762** 31 (*Preprint* 1212.3962)
[22] Spillane T, *et al.* 2007 *Phys. Rev. Lett.* **98** 122501
[23] Abriola D and Pedro de Jesus A 2010 Assessment of Nuclear Data Needs for Particle Induced Gamma Ray Emission (PIGE) Tech. Rep. INDC(NDS)-0568 International Atomic Energy Agency, Nuclear Data Section

Journal of Physics: Conference Series **665** (2016) 012031 doi:10.1088/1742-6596/665/1/012031

Measurement of very low (α,n) cross sections of astrophysical interest

J L Tain[1], D Jordán[1], J Agramunt[1], A Algora[1], I Bandac[2], A Bettini [2,3] R Caballero-Folch[4], F Calviño[4], D Cano-Ott[5], G Cortés[4], C Domingo-Pardo[1], L M Fraile[6], A R García[5], J J Gomez-Cadenas[1], J José[4], R Longland[4], E Mendoza[5], A Parikh[4]

[1] Instituto de Física Corpuscular, CSIC-Univ. Valencia, Valencia, Spain
[2] Laboratorio Subterraneo de Canfranc, Canfranc Estación, Spain
[3] Università di Padova and INFN, Padova, Italy
[4] Universitat Politècnica de Catalunya, Barcelona, Spain
[5] Centro de Investigaciones Energéticas, Medioambientales y Tecnológicas, Madrid, Spain
[6] Universidad Complutense de Madrid, Madrid, Spain

E-mail: jose.luis.tain@ific.uv.es

Abstract. The reactions $^{13}C(\alpha,n)^{16}O$ and $^{22}Ne(\alpha,n)^{25}Mg$ are the primary sources of neutrons for the astrophysical *s*-process. The feasibility of cross section measurements within the respective Gamow windows is discussed in quantitative terms for a 4π neutron counter, based on ^{3}He tubes and a neutron moderator, placed in an underground lab.

1. Introduction

The astrophysical *s*-process is responsible for the production of about half of the abundance of the elements beyond iron in the Universe. The process is described as a succession of neutron captures followed by β-decay taking place along the valley of nuclear stability. Starting with seed nuclei in the iron group, elements up to bismuth can be synthesized. Actually two processes are distinguished [1]: the main *s*-process, reaching Pb-Bi, for which the abundance distribution shows the characteristic pattern of formation through a steady flow of neutrons; and the weak s-process, reaching A $\sim$ 90, for which the neutron fluence is much lower. The latter is associated with the Red Giant phase of massive stars ($M > 10M_{\odot}$), where the temperature can reach values high enough to activate the reaction $^{22}Ne(\alpha,n)^{25}Mg$ during the He burning ($T \sim 3 \times 10^{8}$K) and C-shell burning ($T \sim 10^{9}$K) stages. This reaction becomes the source of neutrons for the weak *s*-process. The main *s*-process takes place in thermally pulsing low mass asymptotic giant branch (AGB) stars. Elements are synthesized at the He inter-shell, in the time interval between recurrent ignition events (thermal pulse, TP). The reaction $^{13}C(\alpha,n)^{16}O$ at $T \sim 10^{8}$K provides the source of neutrons. The so-called ^{13}C pocket is formed by proton capture reactions on the ^{12}C formed during TP. In addition to this, during TP when the temperature reaches $T \sim 3 \times 10^{8}$K, the reaction $^{22}Ne(\alpha,n)^{25}Mg$ is activated but contributes little to the final abundances. Although the general picture is well established a detailed description that can be correlated with the different stellar scenarios requires precise knowledge of both (α,n) reaction rates at the relevant stellar temperatures. This question is one of the current key issues in nuclear astrophysics.

Experimentally the cross sections at the relevant energies (Gamow window) have not been measured and currently they are estimated from extrapolations. The limiting factor in the measurements has been the relatively high level of neutron background in comparison with the weak reaction rates expected at these energies. All measurements up to now have been performed at surface level, where the natural neutron background flux is about $10^{-2}cm^{-2}s^{-1}$. The major part of the background has cosmic ray origin, mostly the final product of muon interactions. Since the level of muon flux is reduced by several orders of magnitude underground it has been proposed that future experiments should be pursued at an underground laboratory.

The measurement of these important (α, n) cross sections is the main objective of the project to build a nuclear astrophysics facility at the Canfranc Underground Laboratory (LSC) in Spain [2]. The project foresees the installation of a MV electrostatic accelerator in a dedicated hall, and the development of the specific instrumentation required for the measurements. As part of this project we are studying the feasibility of the proposed measurements. Apart from the generic argument related to the reduction in background the sensitivity that can be reached with specific setups has not been discussed up to now, to our knowledge. In this contribution we present the results of the work done so far and outline how the work will continue.

2. Measurement of very low neutron rates

Over the years a variety of different detection techniques have been applied to the measurement of the $^{13}C(\alpha,n)^{16}O$ and $^{22}Ne(\alpha,n)^{25}Mg$ cross sections. They included: organic scintillation detectors [3], plastic neutron time-of-flight detectors in conjunction with γ-ray detectors (NaI) [4], BF_3 neutron counters embedded in a graphite neutron moderator [5], ^3He neutron spectrometers [6] germanium γ-ray detectors, ^3He neutron counters embedded in a polyethylene (PE) neutron moderator [7, 8] and a 4π BaF_2 γ-ray calorimeter with a cadmium loaded paraffin neutron-gamma converter [9].

The lowest energy at which a cross section was measured in the case of $^{13}C(\alpha,n)^{16}O$ [7] is $E_\alpha \sim 360$ keV with a value $\sigma \sim 60$ pb. In the case of $^{22}Ne(\alpha,n)^{25}Mg$ [8] the lowest point is at about 800 keV with a cross section of about 130 pb. In order to make some rate estimates one can fix the goal reference cross section value for the future measurements at 1 pb, a reduction of two orders of magnitude in the sensitivity limit. Using a realistic value of target thickness: $n = 10^{-6}b^{-1}$ (equivalent to about 20 $\mu g/cm^2$ or an energy loss of 50 keV in the target) and alpha beam intensity: $i = 5 \times 10^{14}s^{-1}$ (80 pμA), one obtains a reaction rate of $5 \times 10^{-4}s^{-1}$ or 1.8 per hour. In addition one has to consider the detection efficiency of the neutron detector. Since in general the efficiency is quite sensitive to neutron energy, it is important to note the energy range of the emitted neutrons. The $^{13}C(\alpha,n)^{16}O$ reaction has a positive Q-value of 2.216 MeV. Assuming the beam energy reaches up to about $E_\alpha = 1.2$ MeV, the range of neutrons to be detected will be around 2.2-3.5 MeV. In the case of $^{22}Ne(\alpha,n)^{25}Mg$ the reaction has a negative Q-value of -0.478 MeV and the neutrons will have an energy of less than 1 MeV.

The combination of ^3He proportional counters with a neutron moderator has been the most frequently used setup until now. In our opinion it is also the best candidate to reach the lowest sensitivity limit because: a) The detection of the thermalized neutrons through the reaction $^3He + n \rightarrow ^3H + n$ (Q=764 keV) provides a very distinctive signature of neutron detection, b) Helium filled proportional tubes are rather insensitive to gamma radiation, c) Large detection efficiencies ($\geq 50\%$) can be achieved using a sufficient number of long ^3He tubes surrounding the source, and d) The detection efficiency can be made rather independent of neutron energy in the range of interest by an appropriate spatial distribution of tubes. Such devices are commonly used also to measure β-delayed neutron emission probabilities and in fact we have built and used for this purpose one such detector (see Fig 1), named BELEN [10]. In the following we investigate the suitability of the BELEN detector (or a similar one) for the measurement of very low rate (α, n) reactions. The rate of natural neutron background detected in BELEN measured

Figure 1. View of the BELEN 4π neutron counter.

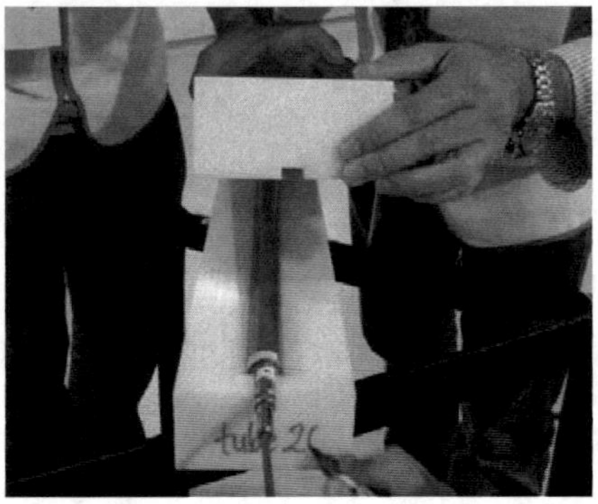

Figure 2. View of one of the PE blocks with ^{3}He tube used in the neutron background measurement at LSC.

at several surface labs is around 0.6-1.2 s^{-1}. For comparison the detector of Ref. [8], which is the most sensitive setup used until now, had a background rate of $2.8 \times 10^{-2}s^{-1}$. This very low background rate was achieved thanks to the use of massive passive shielding and also an active shield to veto cosmic rays. In order to evaluate the expected rate of detected background neutrons for any experiment in the underground lab it is necessary to know accurately the neutron background flux at the site. More importantly it is necessary to know the energy distribution of the neutron background field, since the cross section for neutron interactions is strongly energy dependent. Accordingly the first step in our work has been the measurement of the neutron flux energy distribution at LSC. The results have been published already [11] and are described briefly here.

The measurement was performed using six ^{3}He tubes from the BELEN detector, each enclosed in a PE block of different size (see Fig. 2). The size of the blocks varied between 4.5cm $\times$ 4.5cm $\times$ 70cm and 27cm $\times$ 27cm $\times$ 70cm. The setup was placed in the middle of Hall A at LSC and the measurement lasted for four weeks. The set of neutron counters is treated as a Bonner Sphere Spectrometer (BSS) [12]. The working principle of the BSS is based on the dependence of the number of thermalized neutrons giving a signal in the ^{3}He counter on both the initial neutron energy and the thickness of the moderator. The tubes from BELEN have an active volume of $\oslash$2.5cm $\times$ 60cm and a pressure of 20 atmosphere. This provides a gain of two orders of magnitude in sensitivity when compared with conventional BSS systems. The response of the spectrometer was calculated using the MCNPX Monte Carlo simulation code [13] and verified with calibration measurements using a ^{252}Cf neutron source. The deconvolution of the neutron rates measured in the six detectors with the spectrometer response allows the extraction of the distribution of the flux as a function of the energy. Several codes were used in order to check for systematic deviations. Figure 3 shows the solution obtained with two different codes and corresponds to Fig. 6 of Ref. [11]. The distribution is relatively well determined between 1 eV and 10 MeV but it is rather uncertain at thermal energies and above 10 MeV. As explained in Ref. [11] this reflects the lack of sensitivity of our setup to those energies and is a feature of these kinds of ill-posed inverse problems. The energy integrated flux however is rather insensitive and was determined to be $\Phi = 3.44(35) \times 10^{-6}$cm^{-2}s^{-1} which is about four orders of magnitude smaller than at earth's surface. Having the flux versus energy distribution, it is now possible to make a realistic estimation of the background rate in a neutron sensitive detector using Monte Carlo simulations. We chose the MCNPX code for this work based on our positive experience when comparing experiment and simulation. The geometry of the BELEN detector consisted of a total of twenty ^{3}He tubes distributed in two rings around a central longitudinal hole of

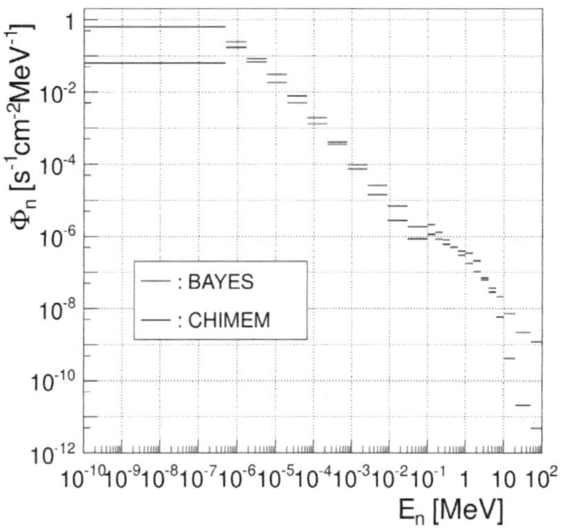

Figure 3. Measured neutron flux at LSC obtained with two different deconvolution codes.

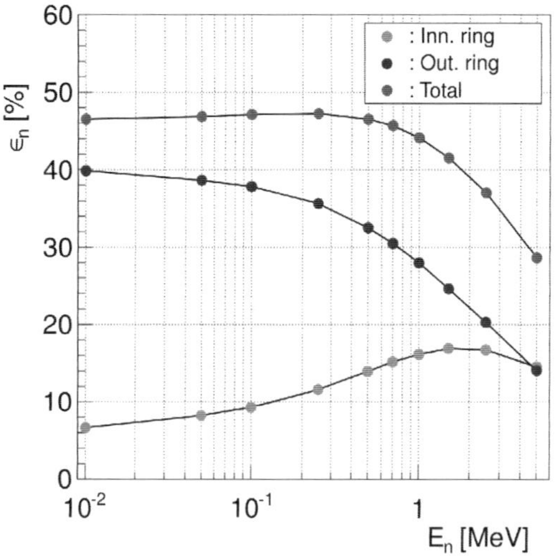

Figure 4. Efficiency of the BELEN detector as a function of neutron energy obtained with MCNPX.

5.5 cm radius in the PE block. Eight tubes were placed at a radius of 9.5 cm and twelve tubes at 14.5 cm. The PE block dimensions were varied to study the shielding effect on the incoming background neutrons. The minimum size was 50cm × 50cm × 80cm (the actual PE size for the BELEN detector is 90cm × 90cm × 80cm.) With this arrangement the efficiency for detecting neutrons emitted randomly from the centre of the detector was calculated as a function of energy as shown in Fig. 4. It can be observed that for this configuration the efficiency is rather constant up to 1 MeV, with a maximum value of 47%, and drops to 37% at 2.5 MeV. To calculate the background rate in this detector, events are simulated with energy sampled from the flux versus energy distribution and a random direction. The neutron initial position is randomly distributed over all outer surfaces of the PE block. The result of the simulation was sensitive to the different flux distributions obtained in Ref. [11]. We found that the rate for the minimal PE block size varied between $6 \times 10^{-4}\mathrm{s}^{-1}$ and $9 \times 10^{-4}\mathrm{s}^{-1}$. Increasing the size of the PE block, which acts as shielding for the background neutrons, has a significant impact. For example an extra layer of 20 cm surrounding the initial PE block reduces the detected rate to a range between $4 \times 10^{-5}\mathrm{s}^{-1}$ and $1 \times 10^{-4}\mathrm{s}^{-1}$ depending on the flux distribution used. For a 40 cm layer the rate is in the range $8 \times 10^{-6}\mathrm{s}^{-1}$ to $6 \times 10^{-5}\mathrm{s}^{-1}$. Comparing these numbers with the expected rates for the (α, n) reactions given above one can conclude that from the point of view of neutron background a PE moderated ^{3}He neutron counter array is capable of measuring cross sections of a fraction of a picobarn.

One of the advantages of using the ^{3}He tubes from the BELEN detector in this measurement is that we have gained essential information about background signals with an origin different from ambient neutrons. We show in Fig. 5 the count versus amplitude histogram recorded during the whole measuring time for one of the detectors. We should mention that detector operation was very stable during the four week period, except for sporadic sudden increases of noise in some detectors. These noise signals are easily removed in the off-line analysis. Superposed on the spectrum of Fig. 5 is the characteristic spectrum of signals due to thermal neutron capture in ^{3}He. The shape of this spectrum, measured with a neutron source, shows the full absorption peak and the tail due to partial deposition of energy by either the triton or proton when the capture happens very close to the walls (wall effect). Fitting this shape to the measured spectrum provides the number of true neutron counts. As can be observed there are many counts in the spectrum besides those attributable to neutron capture. It is more instructive to look at the spectrum after subtraction of the neutron signals. This is done in Fig. 6 which shows the superposition of the rate spectra belonging to the six neutron detectors. The peak visible in the spectra in the range between 1.6-2.0 MeV corresponds to a fixed frequency (10 Hz) pulse generator signal which was used to measure the lifetime of the data acquisition system (DACQ). It is to be noticed that the six spectra are very similar even if the PE blocks

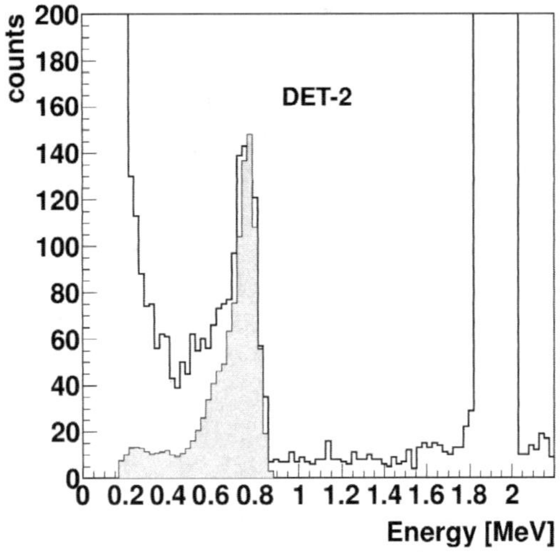

Figure 5. Spectrum registered with one of the neutron detectors. Superposed is the pure neutron response (shaded area).

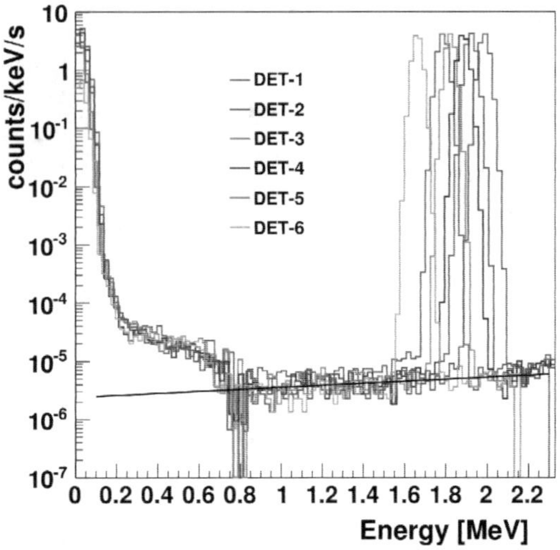

Figure 6. Superposition of the spectra registered with the six neutron detectors after subtraction of the neutron response.

differ greatly in size. This points to a similar origin for the signals which could be the tube itself or the electronics. The counts below 100 keV are associated with electronic noise. We should mention that the data was taken with the self-triggered digital DACQ developed for BELEN. The signals from each counter (conveniently shaped) are sent to one channel of a 100 MHz sampling digitizer. Trapezoidal filters implemented in the firmware are applied to derive time and amplitude information for any signal above a given threshold. Typically the threshold is adjusted to let a certain fraction of noise signals to pass. The counts observed above 800 keV (a flat slightly increasing component) are related to α-radioactivity present in the walls and other materials of the tube. If the emitter is sufficiently close to the surface, the α-particle can enter the gas volume after losing part of its initial energy. This produces a continuum component (indicated by the line in Fig. 6) which actually extends down to zero energy. The steep component in the spectra between 100 and 200 keV could be related to the interaction of γ-rays in the tube. The component extending between 200 keV and 800 keV above the flat component is not readily explained and is of unknown origin. This region coincides with the region of neutron signals. The total rate in this energy interval varies between $5 \times 10^{-4}\text{s}^{-1}$ and $9 \times 10^{-4}\text{s}^{-1}$ depending on the counter. From this number about 20% could be attributed to the flat α-component. Such a rate is too high to allow the measurement of the proposed (α, n) reaction cross sections of interest. For a twenty tube neutron counter such as the one described here the total background rate will be of the order of 10^{-2}s^{-1}. The origin of this intrinsic background will be investigated further. However, it should be noted that it has already been demonstrated [14] that ^{3}He tubes with a very low intrinsic background can be built. The tubes were fabricated using very low activity materials and special manufacturing processes, achieving an intrinsic background rate about 1000 times smaller than the present tubes.

3. Summary and outlook

We have investigated the issue of the feasibility of the measurement of the important $^{13}C(\alpha, n)^{16}O$ and $^{22}Ne(\alpha, n)^{25}Mg$ reaction cross sections at energies in the region of the Gamow window in an underground lab. For this we have measured the energy distribution of the neutron background flux at LSC and used this information to make a realistically simulation of the rate of background events in an existing neutron counter when placed underground. The neutron detector consists of an array of ^{3}He tubes embedded in a PE neutron moderator which we believe is the best candidate for this type of measurement. The results show that this type of detector when placed underground can provide two to three orders of magnitude better sensitivity when compared with measurements at earth's surface, thus allowing the measurement of cross sections of a fraction of picobarn. We have also investigated the intrinsic background of the ^{3}He tubes from this detector and found that it must be strongly reduced if this application is to be successful.

The work towards the development of a dedicated neutron detector will continue. We plan to perform a new more accurate measurement of the neutron flux energy distribution at LSC. The measurement will use a larger set of neutron detectors than before, including bare tubes to enhance the sensitivity at thermal energies and neutron converter-multipliers to enhance the sensitivity at high energies. A complementary measurement using low background liquid scintillation detectors will be carried out. The issue of the intrinsic background in ^{3}He tubes will be further investigated. We plan to use detectors with different characteristics (size, pressure, gas composition and manufacturer). In particular we will study the performance of a new ^{3}He detector specially designed to have low intrinsic background rate. The walls of the tube are internally covered with a thin carbon layer to prevent the penetration of charged particle activity from the walls into the gas volume. In addition to this, the accelerator induced neutron background will be investigated.

3.1. Acknowledgments

This work was supported by the Spanish Ministerio de Economía y Competitividad under Grants CONSOLIDER-Ingenio 2010 CSD2008-0037 (CUP), PRI-PIMNUP-2011-1348, FPA2011-28770-C03-03, FPA2011-24553, FPA2010-17142 and CSD-2007-00042 (CPAN).

References

[1] Kaeppeler F *et al* 2011 *Rev. Mod. Phys.* **83** 157
[2] Fraile L M *et al* 2012 *Letter of Intent to LSC: A nuclear astrophysics facility for LSC: the sources of neutrons in the stars and other reactions of astrophysical interest*
[3] Davids C N 1968 *Nucl. Phys. A* **110** 619
[4] Ashery D 1969 *Nucl. Phys. A* **136** 481
[5] Haas F X and Bair J K 1973 *Phys. Rev. C* **7** 2432
[6] Harms F X *et al* 1991 *Phys. Rev. C* **43** 2849
[7] Drotleff H W *et al* 1993 *Astroph. J.* **414** 735
[8] Jaeger M *et al* 2001 *Phys. Rev. Lett.* **87** 202501
[9] Heil M *et al* 2008 *Phys. Rev. C* **78** 025803
[10] Gomez-Hornillos M B *et al* 2011 *J. Phys. Conf. Ser.* **312** 052008
[11] Jordan D *et al* 2013 *Astrop. Phys.* **42** 1
[12] Bramblett R L *et al* 1960 *Nucl. Instrum. Methods* **9** 1.
[13] D. Pelowitz (Ed.), MCNPX Users Manual Version 2.5.0, Los Alamos National Laboratory, Report LA-CP-05-0369, April 2005.
[14] McDonald A B 2009 *J. Phys. Conf. Ser.* **173** 012002

Journal of Physics: Conference Series **665** (2016) 012032 doi:10.1088/1742-6596/665/1/012032

Cosmic Background Measurements at a Proposed Underground Laboratory by the REGARD Muontomograph

L Oláh[1,2], G Surányi[3], G G Barnaföldi[2], D Bemmerer[4], G Hamar[2], H G Melegh[2], D Varga[1]

[1] Eötvös Loránd University, 1/A Pázmány P. s., H-1117, Budapest, Hungary
[2] Wigner Research Centre for Physics of HAS, PO Box 49, H-1525, Budapest, Hungary
[3] Geological, Geophysical and Space Science Research Group of the HAS, 1/A Pázmány P. s., H-1117, Budapest, Hungary
[4] Helmholtz-Zentrum Dresden Rossendorf, Bautzner Landstraße 400, 01328 Dresden, Germany

E-mail: olah.laszlo@wigner.mta.hu

Abstract. A portable cosmic particle tracking detector has been developed by the REGARD group with angular resolution of 10 mrad. The Close Cathode Chamber-based tracking system is optimized for environmental and geophysical applications with its weight of 15 kg and size of 51 cm × 43 cm × 32 cm. Our aim was to determine the cosmic background at the site of the proposed accelerator and experimental facilities at an approximate 50 meter depth in Felsenkeller, Dresden, Germany. Here, we present our high-precision muon flux measurements, which have been performed during 44 days in one of the tunnels. Angular acceptance of our mapping covered full 2π solid angle of the upper hemisphere. The maximum flux value is found to be below 2.5 $m^{-2}sr^{-1}s^{-1}$.

1. Introduction

The next generation of low energy ion beam experiments will provide more details about the reactions of stable nuclei. For better understanding of these processes, more precise cross-section data are needed at low energies which require low-radiation background [1]. Due to this, experimental facilities of this kind are usually located underground, since the muonic component of a cosmic ray induced background is attenuated by the rock overburden. Therefore, the knowledge on the muon flux is a key element to propose and design underground accelerator experiments. Our aim is to perform a high-precision measurement of the cosmic muon background to cover the full 2π of the upper hemisphere at a proposed laboratory at shallow depth underground in Felsenkeller, Dresden, Germany [2]. In this contribution we briefly describe the detector and present the analysed data.

2. Portable tracking system for charged particle detection

A portable cosmic particle tracking detector has been developed by the REGARD group (Collaboration of the Wigner RCP of HAS and the Eötvös Loránd University on Gaseous Detector R&D). The Close Cathode Chamber based tracking system [3, 4] is optimized for geophysical applications [5, 6] with its weight of 15 kg, size of 51 × 46 × 32 cm^3, sensitive area of 0.1 m^2 and angular resolution of 10 mrad. Our muon telescope is housed in a plexiglass box, which besides giving mechanical support, provides environmental isolation as well. In addition the Muontomograph was upgraded with stand including a swivel joint in order to perform tilted measurements in different angles at a fixed location.

The CCC-chambers of the detector require continuous $1 - 5$ l/h gas flow during data taking. In our case the gas is a standard, non-flammable mixture of Ar and CO_2 in 82:18 proportion. A PIC32 microcontroller based data acquisition (DAQ) system was developed and its small size units fit well between the middle tracking layers. All functions (high/low-voltage units, trigger system, data handling, environmental control, human-machine interface, etc.) are integrated into the common system plan including a simple, user-friendly control system.

The measured track coordinates are read out by the chains of the Front-End Electronics (FEE) and written to a standard Secure Digital (SD) card. Such a memory card with e.g a 2 GB capacity enables a full year of run time recording in 50 meter-rock-equivalent depth. The total power consumption of the complete detector is less than 5 W, thus the Muontomograph can operate for a week supplied by a standard 50 Ah battery. Using a 10 l gas bottle is enough for about a month of data taking.

3. Cosmic background measurements in Felsenkeller, Dresden, Germany

Our aim was to map the muon flux at the proposed place of the future underground laboratory facilities, which is covering the full 2π solid angle of the upper hemisphere with high-enough statistics. The cosmic muon flux measurements have been performed between 12th February and 27th March 2013 in Felsenkeller. Our detector was at a fixed position during the whole period, where we performed multiple-directional, partially overlapping measurements. Besides the vertical setting (highest statistics) the detector was oriented to 350° relative to the magnetic North, and 5 further directions were also measured, tilted by 45° to the azimuth. Data taking was without degradation in detector performance, which was above 97% efficient – as it used to be in earlier measurements as well [5, 6].

Table 1 summarizes all the parameters of the runs with all angular settings: detector directions given in spherical coordinates (the azimuth angle to the magnetic North and the zenith angle, both are in degree units), the duration of measurements are in days. The number of detected cosmic muon tracks is listed too.

Table 1. The summary table of our measurements in Felsenkeller: the detector positions in azimuth to the magnetic North and in zenith, the duration of the measurements as well as the number of penetrating muon tracks.

Azimuth (deg)	Zenith (deg)	Time (day)	Number of tracks
350°	0°	17.5	215,000
350°	45°	7.5	60,000
80°	45°	4.5	110,000
170°	45°	3.5	26,000
260°	45°	4.5	45,000
305°	45°	1.5	21,000

Standard high-energy physics tracking procedures have been applied during the offline data analysis. The first step was the track reconstruction based on the recorded clusters on the CCC planes. The determined muon flux have been corrected for all systematic effects such as detector acceptance, time of measurement, trigger efficiency and angular-dependent tracking efficiency as well. See more details in Refs. [5, 6].

4. Results

The muon telescope reliably operated during the 44 days meanwhile, about 477k events were collected all together. This statistics were satisfactory for our analysis. The first step was to compare the measured vertical flux to earlier data in the units of $m^{-2}sr^{-1}s^{-1}$. As shown in Fig. 1 data are in good agreement with the earlier measurements and empirically parametrized curve [7].

In Fig. 1, the *blue circles* are for data and *green dashed lines* both taken from Ref. [7]. *Red* and *black dots* are taken by the Muontomograph from earlier and this measurement respectively. As *black triangle* denotes, the vertical flux is 1.78 ± 0.23 in the tunnel of Felsenkeller under 50 meter-rock-equivalent depth with the rock density of 2.40 ± 0.2 g cm^{-3}.

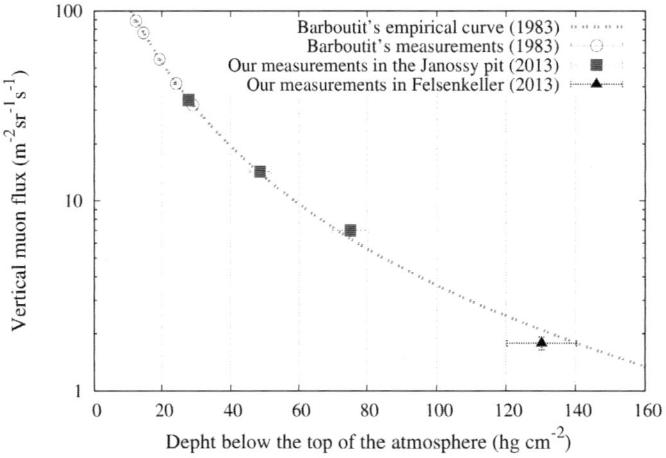

Figure 1. The vertical absolute cosmic muon flux ($m^{-2}sr^{-1}s^{-1}$) versus density depth (hg cm^{-2}). Our measurements are compared to the empirical formula, which based on an earlier work [7].

Using the 6 data sets listed in Table. 1 all overlapping zenith and azimuth directions were merged and binned using the SURFER 9.0 [8] software. The obtained flux-map of the cosmic muon background is shown in Fig. 2, where the flux of cosmic muons appears with *color-scale contours* in units, $m^{-2}sr^{-1}s^{-1}$ as a function of zenith and azimuth angles. The *red contour lines* show the overburden rock thickness above the detector in meter-rock-equivalent, which has been calculated based on our laser scanning total station shots — in parallel with the data taking.

As expected, the measured muon flux correlates well with the overburden rock thickness: the colour scaling and the red contours are mainly parallel to each other in Fig. 2. The maximum muon flux is found to be below 2.5 $m^{-2}sr^{-1}s^{-1}$. The highest flux were measured in the direction of the zenith and the entrance of the tunnel to West.

The obtained vertical absolute flux in Fig. 1 and the flux map in Fig. 2 provide well defined baselines for the design of the proposed accelerator-based experiments in the Felsenkeller site. However, we note, other natural background sources might also exist, which should be targets of forthcoming checks.

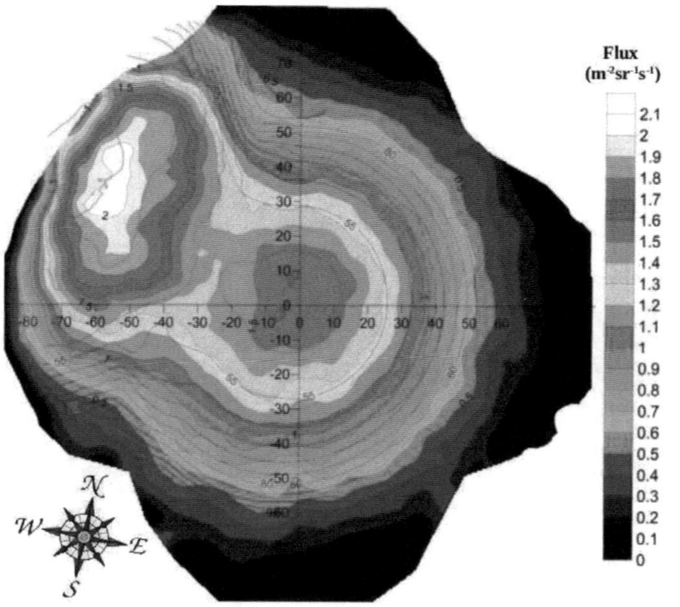

Figure 2. The measured 2π cosmic background in $m^{-2}sr^{-1}s^{-1}$ units from the detector point of view at the place of the proposed accelerator-based experiment. The detector was oriented to 350° to the magnetic North. The measurement error is typically 3-10 % statistical and 5 % systematic.

5. Summary and Conclusions

The cosmic background have been measured in the full 2π solid angle of the upper hemisphere by the newly developed portable tracking detector at an underground laboratory in Felsenkeller, Dresden, Germany. As we expected, the maximum muon flux value is found to be below 2.5 $m^{-2}sr^{-1}s^{-1}$. The results quantify the shielding of Felsenkeller tunnel system for the proposed radioactive ion beam experiments. The portability, reliable tracking performance, low power consumption and the good angular resolution make the presented Muontomograph to be a useful tool to perform reference measurements of cosmic background.

Acknowledgement

Technical support by D. Degering (VKTA Dresden) is gratefully acknowledged. This work was supported by Hungarian OTKA grants NK77816, NK106119 and K104260, the NIH TET_10-1_2011-0061, ZA-15/2009, and Eötvös University. Authors (GGB,DV) also thank the János Bolyai Research Scholarship of the HAS. The project is also supported by the Momentum ("Lendület") grant of the HAS.

References

[1] T. Szűcs *et al.*: 2012 *Eur. Phys. J.* A **48:** 8
[2] D. Bemmerer *et al.*: 2013 NPA6 conference proceedings
[3] D. Varga *et al.*: 2011 *Nucl. Inst. and Meth.* A **648** 163167
[4] D. Varga *et al.*: 2013 *Nucl. Inst. and Meth.* A **698** 1118
[5] G. G. Barnaföldi *et al.*: 2012 *Nucl. Inst. and Meth.* A **689** 60-69
[6] L. Oláh *et al.*: 2013 *Advances in High Energy Physics* Vol **2013** Article ID 560192, 7
[7] A. I. Barboutit and B. C. Rastin: 1983 J. Phys. G: Nucl. Phys. **9** 1577-1595
[8] http://www.ssg-surfer.com/

Nuclear Physics in Astrophysics VI (NPA6) IOP Publishing
Journal of Physics: Conference Series **665** (2016) 012033 doi:10.1088/1742-6596/665/1/012033

An ERC Starting Grant project on p-process nucleosynthesis concluded

Gy Gyürky, Z Halász, T Szücs, G G Kiss, Zs Fülöp

Institute for Nuclear Research (MTA Atomki), H-4001 Debrecen, POB.51., Hungary

E-mail: gyurky@atomki.mta.hu (Gy. Gyürky)

Abstract.
In 2008 a Starting Grant project supported by the European Research Council titled "*Nuclear reaction studies relevant to the astrophysical p-process nucleosynthesis*" was launched. After five years of successful research related to the experimental investigation of proton- and alpha-induced nuclear reaction for the astrophysical p-process, the project came to an end. In this paper a summary of the research and the most important achievements is given.

1. Introduction

Nucleosynthesis, the production of chemical elements that build up our universe is an extremely complicated scientific problem. Not surprisingly, a whole new interdisciplinary science called nuclear astrophysics was born at around the middle of the 20th century to devise and study the processes leading to the production of various groups of chemical elements [1]. After several decades of progress, we have now a reasonably good general understanding of the astrophysical sites and processes contributing to the nucleosynthesis. Our models are, however, still very far away from being able to describe well the different astrophysical phenomena and the composition of the universe, so the quest for the origin of the elements is still unfinished [2].

The chemical elements heavier than iron represent a special group in nuclear astrophysics. Since their synthesis requires rather than produces energy, their production cannot be the source of energy that powers a star, their existence in nature must be attributed to some kind of a subsidiary process. Since the capture of neutrons is not hindered by the Coulomb-barrier, the heavy elements are thought to be produced mainly through neutron capture reactions. Two distinct processes of neutron capture, the s- and r-processes play a role in the heavy element synthesis which takes place in different stellar environments [3, 4].

About three dozens of heavy stable isotopes located on the proton rich side of the valley of nuclear stability, on the other hand, cannot be produced by any of these two neutron capture processes. These are the so-called p-isotopes and their stellar production mechanism is generally referred to as the astrophysical p-process [5]. The general feature of the p-isotopes is their typically very low elemental abundance (with a few exceptions below 1 %). As opposed to the s- and r-isotopes, no heavy element has a dominant p-isotope. Therefore, abundance information of the p-isotopes is available only from the Solar System (from meteorites or from the bulk), the performance of telescopes does not allow the identification of the p-isotopes in stellar spectra.

Therefore, p-process models try to reproduce the Solar System p-isotope abundances. However, in spite of the significant progress huge differences between the observed and calculated p-isotope abundances persist [6]. In order to cure the discrepancies, several different processes are considered which might contribute to the p-isotope production. Indeed, the p-process is actually thought to comprise several sub-processes such as the ν-, νp-, rp-, pn-processes [7, 8, 9, 10, e.g.]. None of these processes can, however, be responsible for the production of the p-isotopes in the whole mass range. Only the so-called γ-process is able to produce proton-rich isotopes in the whole mass range and thus the γ-process seems to be by far the most important sub-process of the p-process.

The γ-process proceeds through γ-induced reactions initiated on pre-existing heavy isotopes. In a high temperature stellar environment the high energy wing of the Planck distribution is energetic enough to remove neutrons from heavy isotopes. Consecutive (γ,n) reactions drive thus the material towards the proton-rich region producing this way the p-isotopes or their radioactive progenitors. As the isotopes become more and more neutron-deficient, charged particle emitting (γ,p) and (γ,α) reactions start to play a role and influence strongly the resulting p-isotope abundances. As the high temperature is a necessary prerequisite of the γ-induced reactions, explosive scenarios are thought to host the γ-process. Perhaps the most likely and most intensively studied site is the O/Ne layer of core collapse supernovae [11], but type Ia supernovae are also considered and being studied [12].

2. Data needs for the γ-process modelling

For a successful γ-process modelling the astrophysical conditions must of course be known. This involves the knowledge about the seed isotope abundances and the conditions of the explosion such as the spatial and temporal variation of density and temperature. Equally important is the nuclear physics input. A full γ-process network calculation involves thousands of reactions, the rate of which must be known for a reliable calculation. Most of the reactions take place on radioactive nuclei, therefore it is not surprising that experimental data are very scarce and the models have to rely on theoretical reaction rates obtained typically from the Hauser-Feshbach statistical model. The fact that the γ-process models fail to reproduce the observed p-isotope abundances can be related to the possibly incorrect reaction rates from theory. Therefore, the experimental study of γ-process related reactions is of high importance in order to provide direct data for the γ-process networks and to check the reliability of statistical model calculations.

The γ-induced reactions which play the key role in the γ-process can be studied directly if a suitable photon beam is available. However, it is technically much easier to study the inverse capture reactions and to obtain the γ-induced reaction rate by using detailed balance. Moreover, the study of capture reactions gives a more direct information for the astrophysical process, since the ground state contribution (which is always measured in the laboratory) is much larger in the case of capture reactions than for γ-induced ones. The measurement of capture cross sections in the relevant mass and energy range can thus provide important input for the γ-process models and can contribute to the better understanding of the origin of the p-isotopes.

Owing to their high importance in several different fields of science and technology, neutron capture cross sections of stable isotopes are extensively studied and relatively well known. In the case of charged particle capture reactions, however, the experimental database relevant for the γ-process is very limited [13]. The reason for this is that in the astrophysically relevant energy range the cross sections are typically tiny and thus practically no experimental effort was devoted to these reactions before the urgent need from the γ-process modelling arose. Therefore, the statistical models in the case of charged particle induced reactions are largely untested. On the other hand, dedicated sensitivity studies indicate that the resulting p-isotope abundances from a network calculation depend strongly on the reaction rates of the (γ,p) and (γ,α) reactions [14, 15]. It is therefore necessary to extend the available experimental database for charged particle induced reactions providing this way a stringent test of the statistical model calculations and also the direct input for the γ-process networks.

Table 1. List of α-induced reactions studied in this project. The quoted energies are given in the center-of-mass frame. The study of several other reactions not listed in this table is in progress.

Reaction	Energy range [MeV]	Ref.	Reaction	Energy range [MeV]	Ref.
^{64}Zn(α,γ)^{68}Ge	8.0 – 12.4	[17, 18]	^{130}Ba(α,γ)^{134}Ce	11.6 – 16.0	[23, 24]
^{64}Zn(α,n)^{67}Ge	9.0 – 12.4	[17, 18]	^{130}Ba(α,n)^{133}Ce	12.1 – 16.0	[23, 24]
^{64}Zn(α,p)^{67}Ga	5.8 – 12.4	[17, 18]	^{132}Ba(α,n)^{135}Ce	12.1 – 16.0	[23, 24]
^{113}In(α,γ)^{117}Sb	8.7 – 13.6	[19, 20, 21]	^{151}Eu(α,γ)^{155}Tb	12.3 – 15.6	[25, 26]
^{113}In(α,n)^{116}Sb	9.7 – 13.6	[19, 20, 21]	^{151}Eu(α,n)^{154}Tb	11.3 – 15.6	[25, 26]
^{127}I(α,γ)^{131}Cs	9.5 – 15.2	[22]	^{169}Tm(α,γ)^{173}Lu	12.3 – 17.1	[27, 28, 29]
^{127}I(α,n)^{130}Cs	9.6 – 15.2	[22]	^{169}Tm(α,n)^{172}Lu	11.2 – 17.1	[27, 28, 29]

Answering the call from astrophysics modelers, a systematic study of charged particle induced reactions relevant for the γ-process was started about 15 years ago [16]. Significant part of the experimental database was created by the nuclear astrophysics group of Atomki. The Starting Grant of the European Research Council in the last five years (2008-2013) made a very important contribution to the success of the experiments. Some details of the project and the obtained results are presented in the next sections.

3. Alpha-induced reaction cross section measurements

The above mentioned sensitivity studies indicate that the calculated p-isotope abundances are especially sensitive to the rates of the (γ,α) reactions in the heavier mass range of the p-isotopes. This mass range was unfortunately completely unexplored experimentally. With the exception of ^{144}Sm no (α,γ) cross section measurement above A $\approx$ 120 was available. Therefore, one of the main aims of the project was to extend the available experimental database towards the heavy mass region.

Table 1 shows the list of α-induced reactions which were studied in the framework of the ERC project. The experiments were always carried out at the cyclotron accelerator of Atomki using the activation method. The reader is referred to the original publications for further experimental details.

The obtained cross sections were always compared with the predictions of statistical model calculations. For this purpose the NON-SMOKER code [30] was typically used since reactions rates provided by this code are often used in astrophysical network calculations. Other codes like TALYS [31] or the recently developed SMARAGD [32] were also used occasionally in order to assess the influence of statistical model implementation and the default input parameters of the codes on the resulting cross sections.

The general observation is that the (α,γ) cross sections calculated using the default settings of the NON-SMOKER code always overestimate significantly the measured data. The deviation is between a factor of two and almost one order of magnitude. In order to draw an astrophysical conclusion from this observation, however, two things have to be considered. First, the results of the statistical model calculations depend on a few input parameters, such as e.g. the optical potentials, level densities, γ-ray strength functions, etc. We do not have a priori information about which of the several possible choices of these input parameters are correct. The comparison with experiment can help choose the best parameters. Second, the studied energy range of the (α,γ) reactions is unfortunately higher than the astrophysically relevant one. The experimental energy range is limited by the falling cross section towards low energies. Therefore, the comparison of theory and experiment can be done only at energies without direct astrophysical

Table 2. List of α-elastic scattering reactions studied in this project. The quoted energies are given in the center-of-mass frame. The study of several Sm and Nd isotopes is in progress.

Reaction	Energy [MeV]	Ref.
^{64}Zn$(\alpha,\alpha)^{64}$Zn	11.3, 15.2	*
^{89}Y$(\alpha,\alpha)^{89}$Y	15.5, 18.6	[34]
^{110}Cd$(\alpha,\alpha)^{110}$Cd	15.6, 18.9	[35, 36]
^{116}Cd$(\alpha,\alpha)^{116}$Cd	15.6, 18.9	[35, 36]
^{113}In$(\alpha,\alpha)^{113}$In	15.6, 18.8	[37]
^{115}In$(\alpha,\alpha)^{115}$In	15.6, 18.8	[38]*

* Analysis in progress

relevance.

Unfortunately the calculated cross sections show a different sensitivity to various input parameters at different energy regions. For a detailed study of these sensitivities see Ref. [33]. At the experimentally accessible energy range the calculations show high sensitivity to more than one input parameter, the cross sections are typically sensitive to the α-, γ- and neutron widths. Therefore it is not possible to constrain one of these input parameters based on the data. Studying a different reaction channel can, however, help. Although not directly relevant astrophysically, the (α,n) reaction channel can be used to constrain the α-width since the calculated (α,n) cross section is typically sensitive only to this width. Fortunately, in those cases where the (α,γ) reaction leads to a radioactive isotope, the final nucleus of the (α,n) channel is often also radioactive. Using the activation method the simultaneous determination of the (α,γ) and (α,n) cross sections is thus possible. Therefore, along with the (α,γ) cross section measurement, the (α,n) channel was also studied, as indicated in table 1.

The α-width which is obtained from the α-nucleus optical potential, turns out to be perhaps the most important input parameter of the calculations for reactions involving α-particles. Different global α-nucleus optical potentials lead to largely different calculated cross section. See e.g. figure 3. in ref. [29]. This introduces a high uncertainty in γ-process network calculations and therefore the study of the low energy α-nucleus optical potential is of high importance. The presently available experimental dataset of (α,γ) reactions does not show a clear picture yet about which global α-nucleus optical potential is the best suited for the γ-process network calculations. Further α-induced cross section data is therefore clearly needed.

4. Alpha elastic scattering experiments

The important α-nucleus optical potential can be directly studied with high precision α-elastic scattering experiments. Such experiments were the other main objective of the present ERC project. Table 2 shows those scattering reactions which were studied in the framework of the present project.

The angular distribution of elastically scattered α-particles were measured in a full angular range between typically 20 and 175 degrees. The chosen energies represent a compromise between the astrophysically relevant energies and energies where the scattering cross section deviates significantly from the Rutherford cross section. The experiments were thus carried out at around the Coulomb-barrier. The angular distributions were measured at a high precision scattering chamber mounted on one of the beamlines of the cyclotron accelerator of Atomki.

The measured angular distributions were compared with the predictions using several different global α-nucleus optical potentials. The general observation is that none of the optical potentials can give a good description of the measured data and they fail to reproduce the evolution of the elastic scattering cross section along isotopic and isotonic chains. The best optical potentials can still be selected based on the scattering data, but unfortunately it is found that those potentials which give the best description of e.g. the (α, γ) cross sections, fail completely to reproduce the scattering data. This also points to the need of an improved low energy α-nucleus optical potential parametrization for astrophysical purposes.

Based on the scattering data obtained in this project and previously at Atomki, we have developed a new, global optical potential [39]. The aim of this new potential is to provide input for astrophysical applications, i.e. the potential focuses on low energies. The potential has only five parameters, so the extrapolation towards unstable isotopes is hoped to be more reliable than in the case of a many parameter global potential. The new potential predicts well the total cross section of α-induced reactions and provides a relatively good reproduction of the available scattering data. Further development of this potential is in progress with the inclusion of additional scattering data and fine tuning based on α-induced reaction data.

5. Proton-induced reactions and auxiliary measurements

Realizing their importance and the lack of experimental data, the main focus of the ERC project was the study of α-induced reactions and α-elastic scattering. Nevertheless, some proton-induced reactions relevant for the γ-process were also studied. (p,γ) and (p,n) cross sections of several stable Se isotopes was measured in the Gamow window relevant for the γ-process [40]. The standard NON-SMOKER calculations give a reasonably good description of the measured data. Through the study of the ^{85}Rb(p,n)^{85}Sr reaction we demonstrated the Coulomb suppression of the stellar enhancement factor, i.e. that the stellar effects can be minimized in the charged particle channel, even when the reaction Q value is negative [41, 42]. For the ^{92}Mo(p,γ)^{93}Tc reaction ambiguous cross section data can be found in the literature. In order to study the discrepancy, this reaction was studied with a method never applied for any γ-process related reaction so far. The cross section is derived from thick target yield measurement with activation [43].

The cross sections of charged particle induced reactions were determined with the activation method in this project. The uncertainty of the half-life of the reaction products influences therefore directly the accuracy of the measured cross sections. In some cases where the half-lives in question were not known with sufficient precision, dedicated precise half-life measurement was carried out. The half-life of an isomer in ^{154}Tb was determined to be 9.994 ± 0.039 h, a value which is one order of magnitude more precise than the literature value [44]. The half-life of ^{133m}Ce was measured to be 5.326 ± 0.011 h, a factor of 30 more precise value than the literature one [45]. The half-life of ^{66}Ga, an isotope important for high energy efficiency calibration of γ-detectors, was measured to be 9.312 ± 0.032 h, supporting the validity of one of the two recent, strongly contradicting measurements [46].

One of the most important methodological development of the project was the extension of the activation method by the detection of characteristic X-ray radiation. In many cases, especially in the higher mass region of p-isotopes the decay of the radioactive reaction products is not followed by a high intensity γ-radiation. The conventional γ-counting based activation is therefore not applicable. These isotopes, on the other hand, typically decay by electron capture which is followed by characteristic X-ray emission. The detection of this radiation can be used for the cross section determination. The applicability of the method using a LEPS (Low Energy Photon Spectrometer) detector was demonstrated with the ^{169}Tm$(\alpha, \gamma)^{173}$Lu reaction [28] and is used and will be used for several other reactions.

6. Conclusions and outlook

During the five years of the project charged particle induced reactions and elastic scattering reactions relevant for the astrophysical γ-process were extensively studied. The experimental database relevant for this poorly known nucleosynthetic process was substantially increased.

Our present knowledge about the origin of p-isotopes has been summarized recently in a review paper [47], where the experimental situation of the nuclear physics side of γ-process is also surveyed.

The number of studied reactions, however, still remains very low compared to the thousands of reactions involved in a γ-process reaction network. And although great progress was made in astrophysical modeling recently, the models are still far from being able to reproduce the observed p-isotope abundances. Therefore, the experimental study of γ-process relevant reactions cannot stop.

The study of several other reactions not listed in tables 1 and 2 is in progress at Atomki[1]. These include among others: ^{121}Sb(α,γ)^{125}I, ^{121}Sb(α,n)^{124}I and ^{123}Sb(α,n)^{126}I [48]; ^{107}Ag(α,γ)^{111}In and ^{107}Ag(α,n)^{110}In [49]; ^{115}In(α,γ)^{119}Sb, ^{162}Er(α,γ)^{166}Yb, ^{162}Er(α,n)^{165}Yb [50], ^{164}Er(α,n)^{167}Yb, ^{166}Er(α,n)^{169}Yb [51], ^{191}Ir(α,γ)^{195}Au, ^{191}Ir(α,n)^{194}Au and ^{193}Ir(α,n)^{196}Au [52]; and elastic scattering on Sm and Nd isotopes.

The experimental methods have to be developed, however, as more and more difficult-to-study reactions are faced. The so far very fruitful activation method can still be extended and further reactions can be investigated with this technique. For example, the amount of produced long-lived reaction products can be determined by the AMS (Accelerator Mass Spectrometry) technique. Pilot measurements with this technique is in progress for the ^{64}Zn(α,γ)^{68}Ge and ^{142}Nd(α,γ)^{146}Sm reactions.

In those cases where the reaction leads to a stable product, the activation method cannot be applied and the cross section must be determined with the technically much more challenging in-beam method. Such measurements are in progress in several institutions with different experimental approaches [53, 54, 55, 56, e.g.].

The actual γ-process path is located on the proton rich side of the valley of stability mainly in the region of radioactive isotopes. Thus most of the reactions relevant for a γ-process network cannot be studied with a stable target experiment. In order to be able to measure capture cross sections on radioactive isotopes, the rapidly developing field of radioactive ion beams needs to be exploited. A pilot experiment at the Experimental Storage Ring of GSI on the ^{96}Ru(p,γ)^{97}Rh reaction demonstrated the feasibility of such a measurement [57] and we hope that the measurements will be extended towards the radioactive proton-rich isotopes in the near future.

Acknowledgments

This work would not have been possible without the valuable support from the European Research Council Starting Grant No. 203175. Part of the research was also supported by OTKA grants K101328, K108459, PD104664 and NN83261(EuroGENESIS). G G Kiss is a Bolyai fellow. Besides the present nuclear astrophysics group of Atomki signing this paper, many other people contributed to the success of the project. Therefore, we thank J. Farkas, D. Galaviz, Z. Korkulu, P. Mohr, A. Ornelas, E. Somorjai, P.M. Takács, Zs. Török, C. Yalçın and many others for their work. The theoretical analysis and interpretation of the measured data was always done by T. Rauscher.

References

[1] Burbidge E M *et al.* 1957 *Rev. Mod. Phys.* **29** 547
[2] Jose J and Iliadis C 2011 *Rep. Prog. Phys.* **74** 096901
[3] Käppeler F *et al.* 2011 *Rev. Mod. Phys.* **83** 157
[4] Arnould M *et al.* 2007 *Phys. Rep.* **450** 97
[5] Woosley S E and Howard W M 1978 *Astrophys. J. Suppl. Ser.* **36** 285

[1] Since the submission of this manuscript the results of some of these measurements have been published. The list below is updated with these references.

[6] Dillmann I *et al.* 2008 *J. Phys G* **35** 014029
[7] Woosley S E *et al.* 1990 *Astrophys. J.* **356** 272
[8] Fröhlich C *et al.* 2006 *Phys. Rev. Lett.* **96** 142502
[9] Schatz H *et al.* 1998 *Phys. Rep.* **294** 167
[10] Goriely S *et al.* 2002 *Astron. Astrophys.* **383** L27
[11] Rayet M *et al.* 1995 *Astron. Astrophys.* **298** 517
[12] Travaglio C *et al.* 2011 *Astrophys. J.* **739** 93
[13] Szücs T, Dillmann I, Plag R, Fülöp Zs 2012 *J. Phys.: Conf. Series* **337**, 012033
[14] Rauscher T 2006 *Phys. Rev. C* **73** 015804
[15] Rapp W *et al.* 2006 *Astrophys. J.* **653** 474
[16] Somorjai A *et al.* 1998 *Astron. Astrophys.* **333** 1112
[17] Gyürky Gy *et al.* 2012 *J. Phys.: Conf. Ser.* **337** 012009
[18] Gyürky Gy *et al.* 2012 *Phys. Rev. C Rapid Comm.* **86** 041601
[19] Yalçın C *et al.* 2009 *AIP Conference Proceedings* **1090** 931
[20] Guray R T *et al.* 2009 *Proceedings of Science* **NIC X** 92
[21] Yalçın C *et al.* 2009 *Phys. Rev. C* **79** 065801
[22] Kiss G G *et al.* 2012 *Phys. Rev. C* **86** 035801
[23] Gyürky Gy *et al.* 2011 *Proceedings of Science* **PoS(NIC XI)** 238
[24] Halász Z *et al.* 2012 *Phys. Rev. C* **85** 025804
[25] Gyürky Gy *et al.* 2010 *J. Phys. G* **37** 115201
[26] Gyürky Gy *et al.* 2010 *J. Phys.: Conf. Ser.* **202** 012004
[27] Kiss G G *et al.* 2011 *Phys. Lett B* **695** 419
[28] Kiss G G *et al.* 2011 *Nucl. Phys. A* **867** 52
[29] Rauscher T *et al.* 2012 *Phys. Rev. C* **86** 015804
[30] Rauscher T and Thielemann F-K 2001 *At. Data Nucl. Data Tables* **79** 47, www.nucastro.org
[31] Koning A J *et al.* 2005 *AIP Conf. Proc.* **769** 1154
[32] Rauscher T, code SMARAGD, www.nucastro.org/smaragd.html
[33] Rauscher T 2012 Astrophys. J. Suppl. Ser. **201** 26
[34] Kiss G G *et al.* 2009 *Phys. Rev. C* **80** 045807
[35] Kiss G G *et al.* 2011 *Phys. Rev. C* **83** 065807
[36] Kiss G G *et al.* 2012 *J. Phys.: Conf. Ser.* **337** 012029
[37] Kiss G G *et al.* 2013 *Phys. Rev. C* **88** 045804
[38] Kis G G *et al.* 2013 in this volume
[39] Mohr P *et al.* 2013 *At. Data Nucl. Data Tables* in press, arXiv:1212.2891
[40] Skakun Ye *et al.* 2011 *Proceedings of the 3rd International Conference Current Problems in Nuclear Physics and Atomic Energy (NPAE-Kyiv2010). Kyiv, Ukraine, 7-12 June 2010.* 207
[41] Kiss G G *et al.* 2008 *Phys. Rev. Lett.* **101** 191101
[42] Rauscher T *et al.* 2009 *Phys. Rev. C* **80** 035801
[43] Gyürky Gy *et al.* 2014 *Nucl. Phys. A* **922** 112
[44] Gyürky Gy *et al.* 2009 *Nucl. Phys. A* **828** 1
[45] Farkas J *et al.* 2011 *Eur. Phys. J. A* **47** 7
[46] Gyürky Gy *et al.* 2012 *Appl. Rad. Isot.* **70** 278
[47] Rauscher T *et al.* 2013 *Rep. Prog. Phys.* **76** 066201
[48] Korkulu Z *et al.* 2013 in this volume
[49] Yalçın C *et al.* 2015 *Phys. Rev. C* **91** 034610
[50] Kiss G G *et al.* 2014 *Phys. Lett B* **735** 40
[51] Kiss G G *et al.* 2015 *J. Phys. G* **42** 055103
[52] Szücs T *et al.* 2013 in this volume
[53] Sauerwein A *et al.* 2011 *Phys. Rev. C* **84** 045808
[54] Halász Z *et al.* 2012 *J. Phys.: Conf. Ser.* **337** 012063
[55] Harissopulos S *et al.* 2013 *Phys. Rev. C* **87** 025806
[56] Simon A *et al.* 2013 *Phys. Rev. C* **87** 055802
[57] Zhong Q *et al.* 2010 *J. Phys.: Conf. Ser.* **202** 012011

Measurement of the ^{92,93,94,100}Mo(γ,n) reactions by Coulomb Dissociation

K Göbel[1], P Adrich[2], S Altstadt[1], H Alvarez-Pol[3], F Aksouh[2], T Aumann[2], M Babilon[4], K-H Behr[2], J Benlliure[3], T Berg[5], M Böhmer[6], K Boretzky[2], A Brünle[2], R Beyer[7], E Casarejos[8], M Chartier[9], D Cortina-Gil[2], A Chatillon[3], U Datta Pramanik[10], L Deveaux[11], M Elvers[4,12], T W Elze[1], H Emling[2], M Erhard[7], O Ershova[1,2], B Fernandez-Dominguez[9], H Geissel[2], M Górska[2], T Heftrich[1], M Heil[2], M Hellstroem[2], G Ickert[2], H Johansson[2,13], A R Junghans[7], F Käppeler[14], O Kiselev[5], A Klimkiewicz[2], J V Kratz[5], R Kulessa[15], N Kurz[2], M Labiche[16], C Langer[1,2], T Le Bleis[2,17], R Lemmon[18], K Lindenberg[4], Y A Litvinov[2], P Maierbeck[6], A Movsesyan[2,4], S Müller[4], T Nilsson[13], C Nociforo[2], N Paar[19], R Palit[2], S Paschalis[9], R Plag[1,2], W Prokopowicz[2], R Reifarth[1,2], D M Rossi[2,5], L Schnorrenberger[4], H Simon[2], K Sonnabend[1], K Sümmerer[2], G Surówka[15], D Vretenar[19], A Wagner[7], S Walter[14], W Waluś[15], F Wamers[2], H Weick[2], M Weigand[1], N Winckler[14], M Winkler[2] and A Zilges[4,12]

[1] Goethe-Universität Frankfurt a. M., Germany
[2] GSI Helmholtzzentrum für Schwerionenforschung, Darmstadt, Germany
[3] Universidad de Santiago de Compostela, Spain
[4] Technische Universität Darmstadt, Germany
[5] Johannes Gutenberg-Universität Mainz, Germany
[6] Technische Universität München, Germany
[7] Helmholtz-Zentrum Dresden-Rossendorf, Germany
[8] Universidade de Vigo, Spain
[9] University of Liverpool, United Kingdom
[10] SINP Kolkata, India
[11] Université de Paris Sud, Orsay, France
[12] Institut für Kernphysik, Universität zu Köln, Germany
[13] Chalmers University of Technology, Göteborg, Sweden
[14] FZ Karlsruhe, Germany
[15] Jagellonian University Krakow, Poland
[16] University of Paisley, United Kingdom
[17] University of Strasbourg, France
[18] CCLRC Daresbury Laboratory, United Kingdom
[19] University of Zagreb, Croatia

E-mail: goebel@physik.uni-frankfurt.de

Abstract. The Coulomb Dissociation (CD) cross sections of the stable isotopes ^{92,94,100}Mo and of the unstable isotope ^{93}Mo were measured at the LAND/R^{3}B setup at GSI Helmholtzzentrum für Schwerionenforschung in Darmstadt, Germany. Experimental data on these isotopes may help to explain the problem of the underproduction of ^{92,94}Mo and ^{96,98}Ru in the models of p-process nucleosynthesis. The CD cross sections obtained for the stable Mo isotopes are in good agreement with experiments performed with real photons, thus validating the method of Coulomb Dissociation. The result for the reaction ^{93}Mo(γ,n) is especially important since the corresponding cross section has not been measured before. A preliminary integral Coulomb Dissociation cross section of the ^{94}Mo(γ,n) reaction is presented. Further analysis will complete the experimental database for the (γ,n) production chain of the p-isotopes of molybdenum.

1. Motivation

The p-nuclei between ^{74}Se and ^{196}Hg are produced under explosive conditions in a sequence of photodissociations of s- and r-process seeds and subsequent β-decays [1]. The modeling of the p-process requires a large network with more than 2,000 isotopes linked by more than 20,000 reactions. The necessary nuclear physics input includes masses of the isotopes, half-lives of the nuclei involved, and reaction rates. Only a small fraction of the required reaction rates can be determined experimentally, most of which rely on predictions from statistical model calculations [2]. An experimental validation of the reaction rates included in stellar models is therefore highly desired.

^{92,94}Mo and ^{96,98}Ru are the most abundant, but not sufficiently explained p-nuclei [2]. According to recent stellar model calculations, ^{94}Mo is mainly synthesized via the (γ,n) photodisintegration chain starting from the more neutron-rich, and stable molybdenum isotopes [3] (Fig. 1). In order to understand the abundance ratio of ^{94}Mo to ^{92}Mo, the determination of the cross section of the ^{94}Mo(γ,n) reaction is necessary. If this reaction is possible in a certain stellar environment, the photodisintegration of ^{93}Mo will follow immediately due to the lower neutron separation energy (Fig. 2). The closed neutron shell at ^{92}Mo terminates the photoneutron chain of the molybdenum isotopes.

Figure 1. Portion of the reaction network leading to the production of the p-isotopes ^{92}Mo and ^{94}Mo.

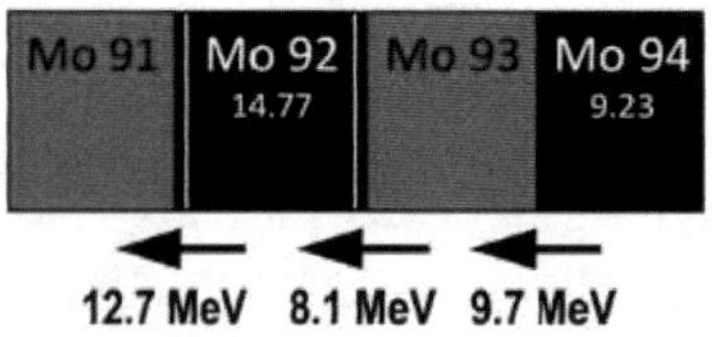

Figure 2. Neutron separation energies of the molybdenum isotopes ^{92}Mo, ^{93}Mo, and ^{94}Mo.

The ^{92,93,94,100}Mo(γ,n) reactions were measured via Coulomb Dissociation at the LAND/R^{3}B setup at GSI Helmholtzzentrum für Schwerionenforschung in Darmstadt, Germany. The (γ,n) reactions are studied in inverse kinematics. The nuclei under investigation are prepared as an ion beam, allowing radioactive nuclei to be investigated. The aim of the present experiment was the validation of the method of Coulomb Dissociation by comparing the results with data from photoactivation measurements [4]. The (γ,n) cross section of the unstable isotope ^{93}Mo could be determined for the first time [5].

2. Experimental method

Most nuclei involved in photodissociation reactions in stellar nucleosynthesis networks are unstable and cannot be prepared as a sample for experiments using real photons. One solution is to study the (γ,n) reaction in inverse kinematics: The nucleus under investigation hits a high-Z target and interacts with the time-varying Coulomb field. This interaction can be interpreted as an absorption of a virtual photon (Fig. 3) [6]. The absorption probability can be translated into a (γ,n) cross section.

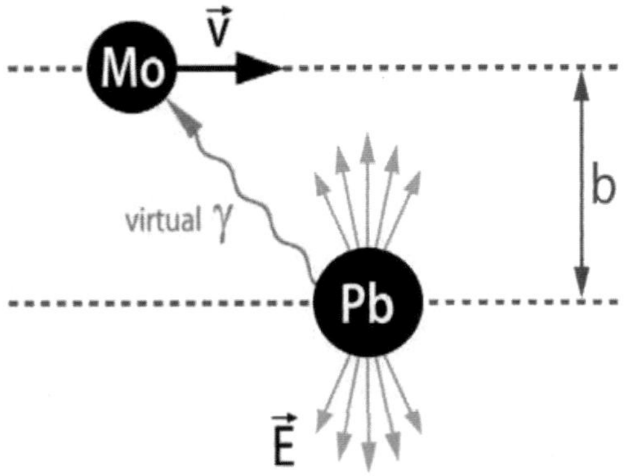

Figure 3. Schematic view of the Coulomb excitation process of a molybdenum nucleus. The Mo projectile impinges on a lead target. The target's electromagnetic field $\vec{E}$ seen by the projectile is Lorentz-contracted in the direction of motion. In a peripheral collision (large impact parameter b) the projectile can be excited by a virtual photon.

The experimental setup is shown in Fig. 4. The ion beam enters the experimental area and is tracked in time by the POS scintillation detector. Position sensitive pin diodes (PSP) [7] detect the position of the ion as well as its charge via energy loss in the detector. The heavy fragment is deflected by the ALADiN magnet (A Large Acceptance Dipole magNet) after the (γ,n) reaction and hits three scintillating fibre detectors (GFI) [8] where the horizontal position is measured. The TFW (Time-of-Flight Wall) at the end of the fragment arm provides position, time and charge information. The emitted neutron remains unaffected by the magnetic field and is detected by the Large Area Neutron Detector (LAND) [9].

3. Analysis and preliminary results

Several steps are needed to extract the cross section from the data of the complex experimental setup.

3.1. Selection of the reaction channel

The reaction channel ^{94}Mo(γ,n)^{93}Mo has to be selected by several conditions applied to the data. A primary beam of ^{94}Mo entered the experimental area. To reject breakup events in the material in front of the target, the incoming projectile ^{94}Mo is identified by the charge (Z) measured by PSP1 and PSP2. In the outgoing channel, the heavy reaction fragment must be identified by its charge and mass. The fragment charge is selected by data obtained by the PSP3 and the TFW. The fragment mass is determined by tracking of the ions through the magnetic field, using position information provided by the PSP and GFI detectors. Furthermore, one neutron has to be detected by LAND.

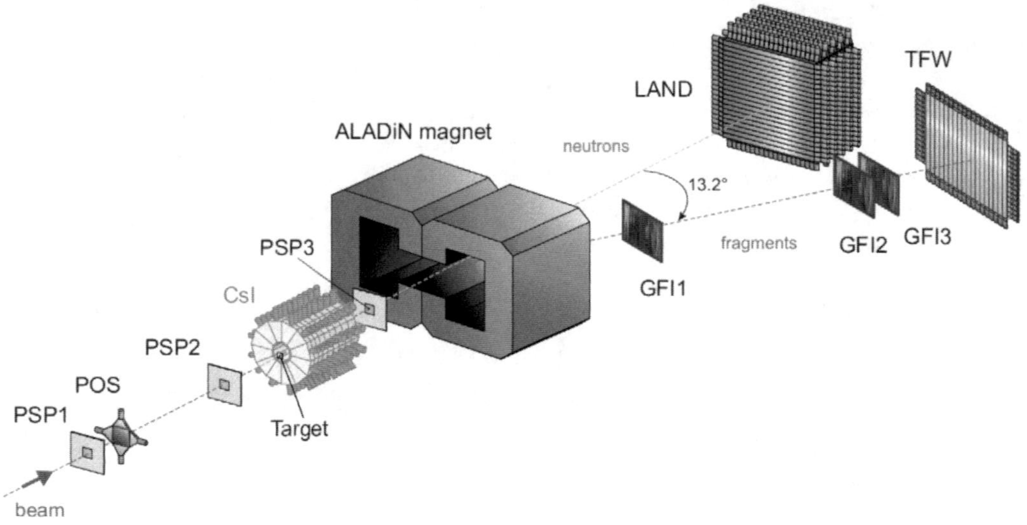

Figure 4. LAND/R^{3}B experimental setup for the measurement of ^{92,93,94,100}Mo(γ,n) by Coulomb Dissociation [10].

3.2. Background subtraction

In addition to Coulomb excitation, the projectile may experience nuclear interactions in the target and reactions with the material outside of the target. In order to determine these background contributions, two additional measurements were performed: one without target ("empty target") and one with a carbon target. Due to the lower atomic number of carbon, the induced electromagnetic excitations are much weaker compared to lead. The cross section of the nuclear interaction is related to the size (radii) of the projectile and of the target nuclei. Therefore, the interaction of the projectile with the carbon target is considered to be purely nuclear [11]. The background distributions are normalized and subtracted from the lead target distributions. The Coulomb Dissociation cross section σ_{CD} is determined by

$$\sigma_{CD} = \left(\frac{M_{Pb}}{d_{Pb}N_A}\right) p_{Pb} - \left(\alpha \cdot \frac{M_C}{d_C N_A}\right) p_C - \left(\frac{M_{Pb}}{d_{Pb}N_A} - \alpha \cdot \frac{M_C}{d_C N_A}\right) p_{Em}, \quad (1)$$

where p is the interaction probability, M the molar mass of the target material [g mol^{-1}], d the target thickness, N_A the Avogadro number [mol^{-1}] and α the nuclear scaling factor. The indices refer to the different targets. In the analysis presented here, α is calculated according to the "black disc model" [12], in which the interacting nuclei are considered fully opaque to each other.

$$\alpha = \frac{A_P^{\frac{1}{3}} + A_{Pb}^{\frac{1}{3}}}{A_P^{\frac{1}{3}} + A_C^{\frac{1}{3}}}. \quad (2)$$

Here, A_P, A_{Pb}, and A_C represent the mass numbers of the projectile, the lead and the carbon target, respectively. The model yields $\alpha = 1.53$ for the ^{94}Mo projectile.

3.3. Preliminary integrated Coulomb Dissociation cross section

For the determination of the integrated Coulomb Dissociation cross section, a distribution of any observable derived according to Eq. 1 can be used. In the preliminary analysis presented here, the neutron kinetic energy in the fragment rest frame E_n was chosen. It is calculated on an event-by-event basis with the kinematics of the incoming ion ^{94}Mo, the outgoing fragment ^{93}Mo and the evaporated neutron considering the Q value of the reaction:

$$E_n = \sqrt{m_{93Mo}^2 + m_n^2 + 2 \cdot \gamma_{93Mo}\gamma_n \cdot m_{93Mo}m_n \cdot (1 - \beta_{93Mo}\beta_n \cos\theta_{93Mo,n})c^2} - m_{94Mo}c^2 - Q. \quad (3)$$

The spectrum was corrected for the LAND efficiency and acceptance. The LAND efficiency and acceptance as a function of the neutron kinetic energy was simulated considering the nominal efficiency of LAND, which was determined in a calibration experiment, the inefficiency resulting from switched-off paddles, and the limited geometrical acceptance of the detector. The data were taken from Ref. [10].

The derived spectrum is shown in Fig. 5. Its integral delivers a preliminary integrated Coulomb Dissociation cross section of 675 mb for the reaction Pb(^{94}Mo,^{93}Mo+n)Pb at a beam energy of 500 MeV.

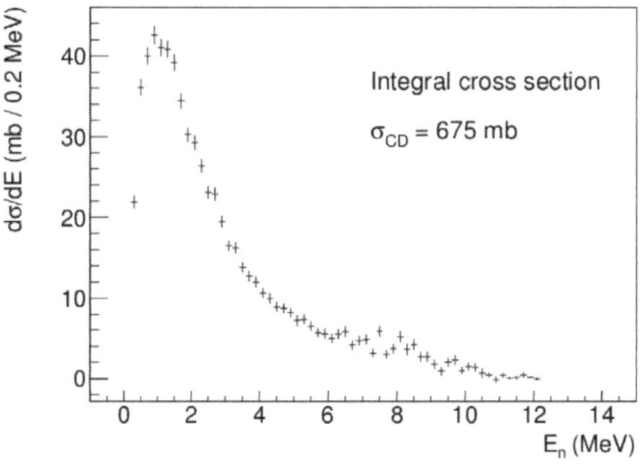

Figure 5. Preliminary Coulomb Dissociation cross section of the reaction Pb(^{94}Mo,^{93}Mo+n)Pb at a beam energy of 500 MeV. The errors indicated are purely statistical.

4. Comparison to results from photoabsorption measurements

The results for the reactions ^{92}Mo(γ,n) and ^{100}Mo(γ,n) are in good agreement with measurements performed with real photons [5]. According to the method described in [10], the photoabsorption data available for the reaction ^{94}Mo(γ,n)^{93}Mo from Beil *et al.* [13] were converted to Coulomb excitation spectra for the E1 and E2 components using the systematics for Giant Quadrupole Resonances. The data are presented in Fig. 6.

From the photoabsorption data, a Coulomb Dissociation cross section of 909 mb is calculated. The ratio of the integral and of the integrated cross section deduced from the experimental data was found to be 0.74. As a first approximation, the obtained value agrees with the normalization factor suggested by Berman *et al.* [14] and Erhard *et al.* [15], that needs to be applied to the photoabsorption data by Beil *et al.* The factor was confirmed in Ref. [10] for ^{92}Mo and ^{100}Mo. The results of further analysis of the data for the ^{94}Mo(γ,n)^{93}Mo reaction, of additional efficiency corrections and of a detailed study of the systematical errors will provide a full picture of the Coulomb Dissociation experiment of the ^{92,93,94,100}Mo(γ,n) reactions at the LAND/R^{3}B setup.

5. Summary and outlook

The Coulomb Dissociation (CD) cross sections of the stable isotopes ^{92,94,100}Mo and of the unstable isotope ^{93}Mo were measured at the LAND/R^{3}B setup at GSI Helmholtzzentrum für Schwerionenforschung in Darmstadt, Germany. The CD cross sections obtained for ^{92}Mo and

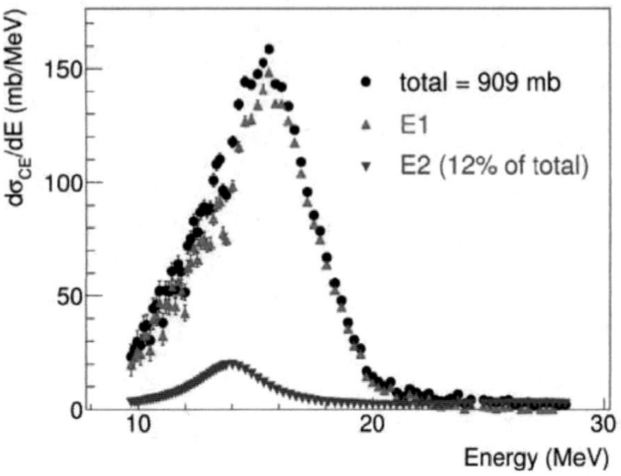

Figure 6. Coulomb excitation cross section of ^{94}Mo derived from the data of Beil *et al.* [13]. The data were obtained by a convolution of the E1 and E2 photoabsorption cross sections with the corresponding calculated virtual photon field. The E2 component was determined from the systematics for Giant Quadrupole Resonances [10].

^{100}Mo and the preliminary value for ^{94}Mo are in good agreement with experiments using real photons, thus validating the method of Coulomb Dissociation. The result of ^{93}Mo(γ,n) is particularly important since the corresponding cross section has not been measured before. The results from the ongoing analysis of ^{94}Mo(γ,n)^{93}Mo will complete the analysis of this series of measurements, hence completing the experimental database for the (γ,n) production chain of the p-isotopes of molybdenum.

Acknowledgments
This project was supported by the Helmholtz International Center for FAIR, the Helmholtz Young Investigator Group VH-NG-327, DFG (SO907/2-1) and HGS-HIRe.

References
[1] Lambert D L 1992 *The Astronomy and Astrophysics Review* **3** 201–256
[2] Arnould M and Goriely S 2003 *Physics Reports* **384** 1–84
[3] Travaglio C, Röpke F K, Gallino R and Hillebrandt W 2011 *The Astrophysical Journal* **739** 93
[4] Sonnabend K, Babilon M, Hasper J, Müller S, Zarza M and Zilges A 2006 *European Physical Journal A* **27** 149–152
[5] Ershova O *et al* 2010 Coulomb dissociation reactions on Mo isotopes for astrophysics applications *Nuclei in the Cosmos*
[6] Bertulani C A and Baur G 1988 *Physics Reports* **163** 299–408
[7] Mahata K *et al* 2009 *Nuclear Instruments and Methods in Physics Research A* **608** 331–335
[8] Cub J *et al* 1998 *Nuclear Instruments and Methods in Physics Research A* **402** 67–74
[9] Blaich *et al* 1992 *Nuclear Instruments and Methods in Physics Research A* **314** 136–154
[10] Ershova O 2012 *Coulomb Dissociation Reactions on Molybdenum Isotopes for Astrophysics Applications* Ph.D. thesis Goethe University Frankfurt, Germany
[11] Bertulani C A and Gade A 2010 *Physics Reports* **485** 195–259
[12] Aumann T, Kratz J V, Stiel E, Sümmerer K, Brüchle W, Schädel M, Wirth G, Fauerbach M and Hill J C 1993 *Physical Review C* **47** 1728–1737
[13] Beil H, Bergère R, Carlos P, Leprêtre A, Miniac A D and Veyssière A 1974 *Nuclear Physics A* **227** 3
[14] Berman B L and Fultz S C 1975 *Reviews of Modern Physics* **47** 3
[15] Erhard M, Junghans A R, Nair C, Schwengner R, Beyer R, Klug J, Kosev K, Wagner A and Grosse E 2010 *Physical Review C* **81** 034319

High precision elastic α scattering on the even-odd ^{115}In nucleus at low energies

G G Kiss[1], T Szücs[1], P Mohr[1,2], Zs Fülöp[1], Gy Gyürky[1], Z Halász,[1,3], R F Soha[4], E Somorjai[1], A Ornelas[5], D. Galaviz[5], C Yalçın[1,6], R T Güray[6], N Özkan[6]

[1] Institute for Nuclear Research (ATOMKI), H-4001 Debrecen, POB.51., Hungary
[2] Diakonie-Klinikum, D-74523 Schwäbisch Hall, Germany
[3] University of Debrecen, Department of Theoretical Physics, H-4001 Debrecen, Hungary
[4] University of Debrecen, Department of Solid State Physics, H-4010 Debrecen, Hungary
[5] Centro de Física Nuclear da Universidade de Lisboa, 1649-003, Lisbon, Portugal
[6] Kocaeli University, Department of Physics, TR-41380 Umuttepe, Kocaeli, Turkey

E-mail: ggkiss@atomki.mta.hu

Abstract. Elastic alpha scattering cross sections on the even-odd ^{115}In nucleus have been measured at energies $E_{lab.}$ = 16.15 MeV and 19.50 MeV. The high precision experimental data are used to derive the parameters of a local α nucleus optical potential.

1. Introduction

Studies in the fields of nuclear structure, nuclear reaction theory, and nuclear astrophysics require the knowledge of α+nucleus optical model potentials (OMP). For example, the OMP plays a role in the determination of the α-decay half-lives of superheavy nuclei [1, 2], and in the unification of the bound and scattering α-particle states [3]. Furthermore, in several astrophysical applications – such as modeling the nucleosynthesis in explosive scenarios like the γ process – the reaction rates are taken from the Hauser-Feshbach (H-F) statistical model [4] using global OMPs [5, 6].

The optical potential combines a Coulomb term with the complex form of the nuclear potential, which consists of a real and an imaginary part. Usually, the parameters of the OMP are derived from the analysis of the angular distributions of elastically scattered α-particles (and are adjusted to experimental α-induced cross sections if they are known).

The variation of the potential parameters of the real part as a function of mass and energy is smooth and relatively well understood [7]. On the contrary, the imaginary part of the optical potential is strongly energy-dependent, especially at energies around the Coulomb barrier. In astrophysical applications the parameters of the OMP have to be known at energies well below the Coulomb barrier. However, at such energies the α+nucleus elastic scattering cross section is non-diffractive and dominated by the Rutherford component. Therefore, the elastic α scattering experiments have to be carried out at slightly higher energies with high precision. From the analysis of the measured angular distributions the parameters of the potential can be derived and have to be extrapolated down to the astrophysically relevant energy region where the relevant α-particle induced-reactions are taking place.

Journal of Physics: Conference Series **665** (2016) 012035 doi:10.1088/1742-6596/665/1/012035

Several α elastic scattering experiments on the target nuclei ^{89}Y, ^{92}Mo, ^{106,110,116}Cd, ^{112,124}Sn, and ^{144}Sm have been performed at ATOMKI in recent years [8-13]. A summary of these works is given in [14, 15]. Previously either semi-magic or even-even target nuclei were investigated. In all of these cases complete angular distributions have been measured at energies close to the Coulomb barrier. The chosen energies were low enough to be close to the region of astrophysical interest and high enough that the scattering cross section differs sufficiently from the Rutherford cross section. This work presents the elastic scattering experiment performed on the ^{115}In non-magic, even-odd nucleus to study further the behavior of the optical potentials at low energies.

2. Experimental technique

The experiment was carried out at the cyclotron laboratory of ATOMKI, Debrecen. A similar experimental setup was used in previous experiments (see e.g. [11]). The following paragraphs provide only a short description of the experimental setup.

The targets for the alpha scattering experiment were made by vacuum evaporation of natural high chemical purity (99.99%) In onto 40 μg/cm^2 thick C foils. During the evaporation, the C backings were fixed in a holder placed 10 cm above the Ta evaporation boat. The target thickness was determined by α energy loss measurment and was found to be $\approx$ 95 μg/cm^2. The energy of the alpha beam provided by the cyclotron accelerator of ATOMKI was E_{lab} = 16.15 and 19.50 MeV (corresponding to $E_{c.m.}$ = 15.59 and 18.83 MeV, after taking into account the energy loss in the target, which is about 15 keV), with a beam current of 150 pnA. The beam enters the scattering chamber through a complex collimator system, necessary to have a point-like beamspot ($\leq$ 4mm^2) at well-defined beam energy (energy spread $\leq$ 100 keV) which is needed for the precise determination of the scattering angle and for fixing the energy dependence of the imaginary potential.

There are two independently rotatable turntables, each of them are holding two (used to scan the forward angular region) or five (used to measure the yield of the scattered alpha particles above $\geq$ 100°) Ion Implanted Silicon Detectors. The thickness of the detectors is 500 μm, and their solid angles vary within $\Delta\Omega$ = 1.0 x 10^{-4} and 1.6 x 10^{-4}. To reach stable operation with higher angular precision and repeatability the mechanical transmission of the scattering chamber was redesigned. The formerly used motors were replaced by 3-phase hybrid stepper motors [17] and closed-loop drives [18] to allow 0.006° stepsize precision and avoid step loss. The system was remote controlled via Labview development environment [19] and all the operations were documented for later analysis.

In addition, two detectors were mounted at a larger distance on the wall of the scattering chamber at fixed angles of 15° left and right to the beam axis. These detectors were used as monitor detectors during the experiment to normalize the measured angular distributions, to determine precisely the beam position on the target and to monitor the target thickness. The solid angle of these detectors are $\Delta\Omega$ = 8.2 x 10^{-6}.

Knowledge on the exact angular position of the detectors is of crucial importance for the precision of a scattering experiment because the Rutherford cross section depends sensitively on the angle. The uncertainty in the angular distribution is dominated by the error of the scattering angles in the forward region. A tiny uncertainty of $\Delta\vartheta$ = 0.3° results in a significant error of approximately 5% in the Rutherford normalized cross sections at very forward angles.

To determine the scattering angle precisely, — similarly to [11] — we measured kinematic coincidences between elastically scattered α particles and the corresponding ^{12}C recoil nuclei using a pure carbon foil target. As the result of this procedure the final angular uncertainty was found to be $\leq$ 0.13° for each detector.

Typical spectra are shown in Fig. 1. The relevant peaks from elastic ^{115}In - α scattering are well separated from elastic and inelastic peaks of target contaminations, and — as expected — peaks from inelastic α scattering on ^{115}In are not visible. The energy of the first excited state

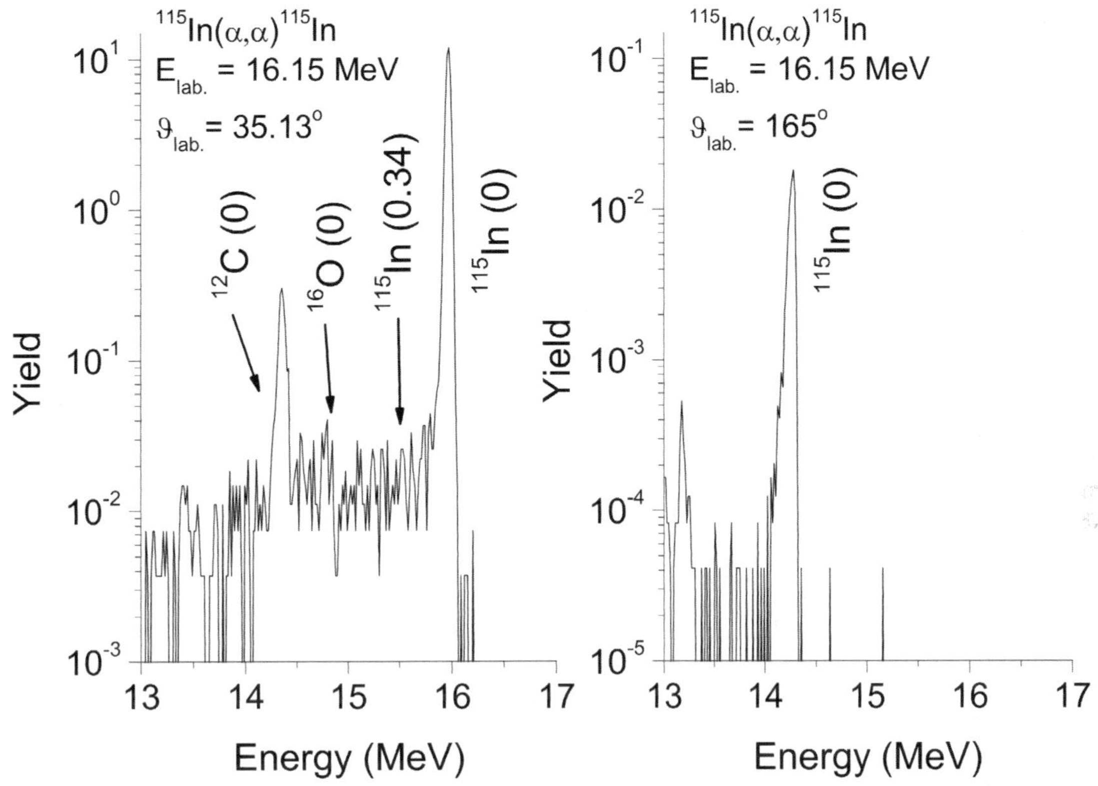

Figure 1. Typical spectra taken at E_{lab} = 16.15 MeV at ϑ = 35.13° and 165.05°. The peak from elastic ^{115}In-α scattering is well resolved from both the ^{12}C-α and ^{16}O-α elastic scattering.

of the ^{115}In nucleus is 336.24 keV. There is a large difference between the spin of the ground and the first excited states ($9/2^+$ and $1/2^-$ respectively) [16], therefore the expected inelastic scattering cross section is very low at the measured energies. This fact explains why events corresponding to inelastic α scattering on ^{115}In are missing from the spectra.

3. Experimental results and the optical potential study

Complete angular distributions between 20° and 175° were measured at energies of E_α = 16.15 and 19.5 MeV in 1° (20° $\leq \vartheta \leq$ 100°), and 2.5° (100° $\leq \vartheta \leq$ 175°) steps. The count rates $N(\vartheta)$ have been normalized to the yield of the monitor detectors $N_{Mon}(\vartheta{=}15°)$

$$\left(\frac{d\sigma}{d\Omega}\right)(\vartheta) = \left(\frac{d\sigma}{d\Omega}\right)_{Mon} \frac{N(\vartheta)}{N_{Mon}} \frac{\Delta\Omega_{Mon}}{\Delta\Omega}, \tag{1}$$

with $\Delta\Omega$ being the solid angles of the detectors. The cross section at the position of the monitor detectors is taken as pure Rutherford. The relative measurement eliminates the typical uncertainties of absolute measurements, coming mainly from changes in target and from the beam current integration.

The measured Rutherford-normalized angular distributions are shown in Fig.2. The experimental absolute cross sections cover five orders of magnitude between the highest (forward angles at E_α=16.15 MeV) and the lowest cross sections (backward angle at E_α=19.5 MeV)

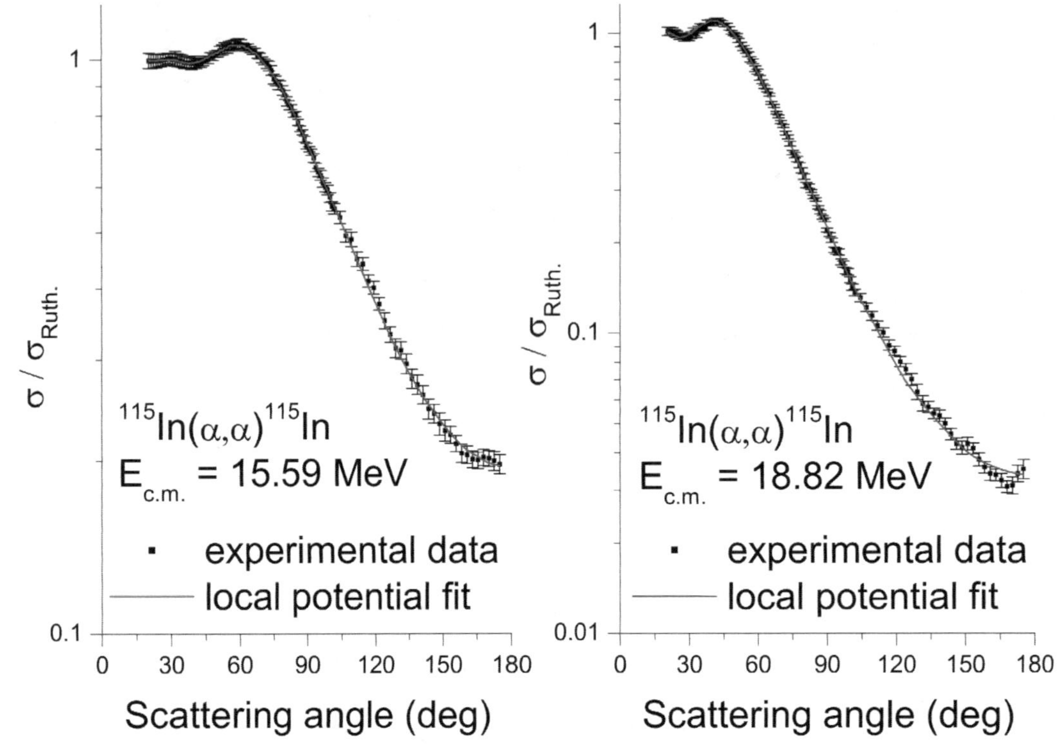

Figure 2. Rutherford normalized elastic scattering cross sections of ^{115}In$(\alpha,\alpha)^{115}$In reaction at $E_{c.m.}$ = 15.59 and 18.83 MeV versus the angle in center-of-mass frame. The lines correspond to the predictions using the local optical potential parameter set derived in the present work (and listed in Tab.2). Note the different scale of the two figures.

with almost the same accuracy (4-5% total uncertainty). This error is mainly caused by the uncertainty of the determination of the scattering angle in the forward region and by the statistical uncertainty in the backward region. The lines correspond to the result of optical model calculations using the local α-nucleus potential, its parameters are listed in Table 2.

The real part $V(r)$ of the nuclear potential is determined by a double-folding procedure using the densities of the α projectile and ^{115}In target (derived from electron scattering [20]) with an effective nucleon-nucleon interaction of the widely used DDM3Y type [21, 22] (for details of the folding procedure see also [23, 14]). The bare folding potential $V_F(r)$ is modified by a strength parameter λ and a width parameter w

$$V(r) = \lambda V_F(r/w) \quad . \tag{2}$$

The strength parameter λ and the width parameter w are adjusted to the experimental ^{115}In$(\alpha,\alpha)^{115}$In elastic scattering angular distributions. The width parameter w should remain close to unity and the strength parameter λ is typically around $1.1 - 1.4$ [7].

The Coulomb potential $V_C(r)$ is taken as usual from a homogeneously charged sphere, with the radius parameter R_C taken from the root-mean-square (rms) radius of the bare folding potential (with $w = 1$).

Table 1. Parameters of the local optical potential derived in the present work.

$E_{c.m.}$ MeV	λ	w	J_R MeV fm^3	r_R fm	W_s MeV	R_s fm	a_s fm	J_I MeV fm^3	r_I fm	χ^2_{red}
15.59	1.3001	0.99781	342.20	5.314	143.77	1.4658	0.3996	80.56	7.310	0.268
18.82	1.4120	0.99193	365.14	5.283	162.59	1.4175	0.4526	96.89	7.133	0.926

The imaginary potential $W(r)$ is parameterized by Woods-Saxon potentials of volume and surface type:

$$W(r) = W_V f(x_V) + W_S \frac{df(x_S)}{dx_S} \quad . \tag{3}$$

The W_i are the depth parameters of the volume and surface imaginary potential, and the Woods-Saxon function $f(x_i)$ is given by

$$f(x_i) = \left[1 + \exp(x_i)\right]^{-1} \tag{4}$$

with $x_i = (r - R_i A_T^{1/3})/a_i$ and $i = V, S$ for the volume and surface part.

In general, at energies far above the Coulomb barrier the volume contribution is dominating whereas at lower energies the surface component becomes more important. The experimental energies of 15.59 MeV and 18.82 MeV are around the Coulomb barrier, therefore it is sufficient to neglect the volume contribution ($W_V = 0$) and to use a pure surface imaginary potential. The parameters of the local potential fits are listed in Table 1. The derived local alpha nucleus optical potential parameters were used to calculate the total reaction cross sections, σ_{reac} [15], which was found to be 350.5 ± 10.6 mb at $E_{c.m.}$=15.59 MeV and 777.0 ± 23.5 mb at $E_{c.m.}$=18.82 MeV, respectively.

4. Conclusions

Elastic ^{115}In(α,α)^{115}In scattering data have been measured at energies around the Coulomb barrier with high precision over the full angular range. A local optical potential has been derived from the new experimental data, and the new data will also be used to obtain an update of the global ATOMKI-V1 potential [14].

Acknowledgments

This work was supported by the EUROGENESIS research program, by the HUNGARIAN PORTUGUESE INTERGOVERNMENTAL S&T COOPERATION PROGRAMME NO. TÉT_10-1-2011-0458, by the European Research Council grant agreement no. 203175, by OTKA (NN83261, K101328, PD104664, K108459), by the Scientific and Technical Research Council of Turkey (TUBITAK, grant number: 109T585). G. G. Kiss acknowledges support from the János Bolyai Research Scholarship of the Hungarian Academy of Sciences. C. Yalçın acknowledges support through The Scientific and Technical Research Council of Turkey (TUBITAK) grant numbers: BIDEP-2219.

References

[1] V.Yu. Denisov, H. Ikezoe, 2005 *Phys. Rev.* C **72** 064613
[2] P. Mohr, 2006 *Phys. Rev.* C **73** 031301(R); 2006 *Phys. Rev.* C **74** 069902
[3] P.E. Hodgson 1990 *Contemp. Phys.* **31** 295
[4] W. Hauser and H. Feshbach, 1952 *Phys. Rev.* **87** 366
[5] T. Rauscher and F.-K. Thielemann 2000 *At. Data Nucl. Data Tables* **75** 1
[6] T. Rauscher and F.-K. Thielemann 2001 *At. Data Nucl. Data Tables* **79** 47
[7] U. Atzrott, P. Mohr, H. Abele, C. Hillenmayer, G. Staudt 1996 *Phys. Rev.* C **53** 1336

[8] G. G. Kiss, P. Mohr, Zs. Fülöp, D. Galaviz, Gy. Gyürky, Z. Elekes, E. Somorjai, A. Kretschmer, K. Sonnabend, A. Zilges and M. Avrigeanu 2009 *Phys. Rev. C* **80** 045807
[9] Zs. Fülöp, Gy. Gyürky, Z. Máté, E. Somorjai, L. Zolnai, D. Galaviz, M. Babilon, P. Mohr, A. Zilges, T. Rauscher, H. Oberhummer and G. Staudt 2001 *Phys. Rev.* C **64** 065805
[10] G. G. Kiss, Zs. Fülöp, Gy. Gyürky, Z. Máté, E. Somorjai, D. Galaviz, A. Kretschmer, K. Sonnabend and A. Zilges 2006 *Eur. Phys. J.* A **27** 197
[11] G. G. Kiss, P. Mohr, Zs. Fülöp, Gy. Gyürky, Z. Elekes, J. Farkas, E. Somorjai, C. Yalçın, D. Galaviz, R. T. Güray, N. Özkan and J. Görres 2011 *Phys. Rev.* C **83** 065807
[12] D. Galaviz, Zs. Fülöp, Gy. Gyürky, Z. Máté, P. Mohr, T. Rauscher, E. Somorjai and A. Zilges 2005 *Phys. Rev.* C **71** 065802
[13] P. Mohr, T. Rauscher, H. Oberhummer, Z. Máté, Zs. Fülöp, E. Somorjai, M. Jaeger and G. Staudt 1997 *Phys. Rev.* C **55** 1523
[14] P. Mohr, G. G. Kiss, Zs. Fülöp, D. Galaviz, Gy. Gyürky and E. Somorjai 2013 *At. Data Nucl. Data Tables* **99** 651
[15] P. Mohr, D. Galaviz, Zs. Fülöp, Gy. Gyürky, G. G. Kiss and E. Somorjai 2010 *Phys. Rev.* C **82** 047601
[16] J. Blanchot 2012 *Nuclear Data Sheets* **113** 2391
[17] 573S20-EC-100, Leadshine Technology Co., Ltd.
[18] HBS57 Stepper Motor Drive with Encoder Feedback, Leadshine Technology Co., Ltd.
[19] http://www.ni.com/labview/
[20] H. de Vries, C. W. de Jager and C. de Vries 1987 *At. Data and Nucl. Data Tables* **36** 495
[21] G. R. Satchler and W. G. Love 1979 *Phys. Rep.* **55** 183
[22] A. M. Kobos, B. A. Brown, R. Lindsay and G. R. Satchler 1984 *Nucl. Phys.* A **425** 205
[23] H. Abele and G. Staudt 1993 *Phys. Rev.* C **47** 742

Investigation of reactions relevant for the γ process using in-beam γ-ray spectroscopy

L Netterdon[1], J Endres[1], J Mayer[1], A Sauerwein[1,2], P Scholz[1], and A Zilges[1]

[1]Institut für Kernphysik, Universität zu Köln, Zülpicher Straße 77, D-50937 Köln, Germany

E-mail: `lnetterdon@ikp.uni-koeln.de`

Abstract. The reaction ^{89}Y(p,γ)^{90}Zr was studied at five proton energies close to the Gamow window. This reaction is of astrophysical importance, since it is located in a mass region, where the p-nuclei abundances are not well reproduced by network calculations. For this purpose, the in-beam technique utilizing the high-efficiency high-purity germanium (HPGe) detector array HORUS at the Tandem ion accelerator at the University of Cologne was used. The excellent agreement of the measured total cross sections with previous data shows, that the setup in Cologne is well suited for such measurements. An additional interesting outcome of this measurement are partial cross sections of the de-excitation of the ^{90}Zr compund nucleus up to the 15th excited state, an observable only accessible in this kind of high-resolution in-beam experiments. The experimental setup and preliminary results of the total and partial cross sections obtained for the ^{89}Y(p,γ) reaction are presented. Additionally, we show results of a first test measurement of the α-capture reaction on the p-nucleus ^{92}Mo using the in-beam technique with HPGe detectors.

1. Introduction

Mostly all of the stable nuclei heavier than iron are synthesized by neutron-capture reactions during the s and r process [1, 2]. However, 30 to 35 neutron-deficient nuclei are bypassed by these processes, referred to as p nuclei [3]. The astrophysical scenario of the p-nuclei production is still under discussion. It is believed, that these nuclei are mainly produced by sequences of photodisintegration reactions in O/Ne shells of core-collapse supernovae [4]. However, type Ia supernovae were also found to be a possible scenario for the synthesis of p nuclei [5]. This production mechanism by photodisintegration reactions is referred to as γ process. In order to account for the huge reaction network of the γ process, reaction rates are usually calculated within the scope of the Hauser-Feshbach (HF) model [6]. Although the model itself is well established, nuclear-physics uncertainties of the models that enter these calculations are significant. In particular, the reproduction of the p-nuclei abundances in the mass region around $90 \lesssim A \lesssim 130$ remains problematic [7]. In order to investigate, whether these deficiencies stem from the nuclear physics or astrophysical side, it is important to experimentally constrain the nuclear physics input parameters entering the HF calculations. These include optical-model potentials, γ-strength functions, and nuclear level densities. Different experimental approaches have been used to study reactions relevant for the γ process, such as the activation technique, see,

[2] Present address: Institut für Angewandte Physik, Goethe-Universität Frankfurt am Main, Germany

e.g., Refs. [8, 9, 10], the 4π summing method [11, 12], and the in-beam method with high-purity germanium (HPGe) detectors [13, 14], which is also used in the present study. This method has the tremendous advantage compared to the activation method, that it enables measurements of astrophysically relevant reactions with a stable reaction product, since the prompt γ-decays of the excited reaction products are detected. Recently, a dedicated setup for in-beam experiments for nuclear astrophysics was developed at the Institute for Nuclear Physics at the University of Cologne. In this article, we report on this experimental setup which has been designed for the measurement of reactions relevant for the γ process at energies far below the Coulomb barrier.

The motivation to measure the ^{89}Y(p,γ)^{90}Zr reaction is twofold. First, this reaction serves as an excellent test case for the recently developed experimental setup in Cologne, since the total cross sections were already measured in two recent experiments, see Ref. [11] and Ref. [14]. Secondly, this reaction is located within a mass region, where the p-nuclei abundances are not well reproduced. Hence, systematic measurements of proton- and α-induced reactions in this mass region are important to reduce these deficiencies from the nuclear physics point of view. This is also the astrophysical motivation of the measurement of the α-capture reaction on the p nucleus ^{92}Mo, besides the fact, that no α-induced measurement of astrophysical interest using the in-beam method with HPGe detectors has been successfully performed so far on target nuclei heavier than ^{70}Ge [15]. Furthermore, an earlier attempt to measure this reaction in Cologne did not succeed due to the enormous beam-induced background [16].

2. Experimental setup and results

The proton and α-particle beam for the respective experiments was delivered by the 10 MV FN TANDEM ion accelerator at the Institute for Nuclear Physics in Cologne. The beam current was determined by the measurement of the charge deposited in a Faraday Cup behind the target, the target itself, and the target chamber. The chamber additionally houses a silicon detector at an angle of 135° to monitor the target stability during the experiment by elastic scattering. The prompt γ decays of the respective reaction products were measured for both reactions using the high-efficiency γ-ray detector array HORUS. This array was equipped with 13 HPGe detectors, five of them had additional BGO shields for an active background suppression. The detectors are mounted at five different angles relative to the beam axis which allows the measurement of angular distributions of the emitted γ rays, which is mandatory to determine absolute cross sections. Additionally, the high granularity of the setup enables the measurement of $\gamma\gamma$ coincidences, which is a powerful tool to suppress beam-induced background.

The absolute full-energy peak efficiency was determined using calibrated ^{152}Eu and ^{226}Ra sources. In order to determine detector efficiencies for γ-ray energies of up to $\approx$10 MeV, the 3674.4- keV resonance of the ^{27}Al(p,γ)^{28}Si reaction was used [17]. Using this resonance, relative full-energy peak efficiencies at energies of E_γ = 4.5 MeV, 6.2 MeV, 8.2 MeV, and 10.5 MeV were obtained. These relative efficiencies were subsequently scaled to fit the slope of the absolute full-energy peak efficiencies that was obtained using the aforementioned calibration sources.

The ^{89}Y(p,γ)^{90}Zr reaction was measured at five different proton energies between E_p = 3.7 MeV and 4.7 MeV. Thus the investigated energy range is very close to the Gamow window, which is located between 1.7 MeV and 3.6 MeV for a temperature of 3 GK [18]. For this experiment, a natural Y target containing 99.9 % of ^{89}Y was used, that was prepared by vacuum evaporation on a 130 $\frac{\text{mg}}{\text{cm}^2}$ thick Ta backing. The target had a thickness of (583 $\pm$ 24) $\frac{\mu\text{g}}{\text{cm}^2}$. The target thickness was measured using the Rutherford Backscattering Spectrometry (RBS) facility at the RUBION Dynamitron-tandem accelerator at the Ruhr-Universität Bochum, Germany. A second measurement after the experiment ensured that no target material was dissipated during the irradiation.

During the experiment, the target was bombarded for several hours with beam currents varying from 1 nA to 60 nA depending on the proton energy. The large span of the beam

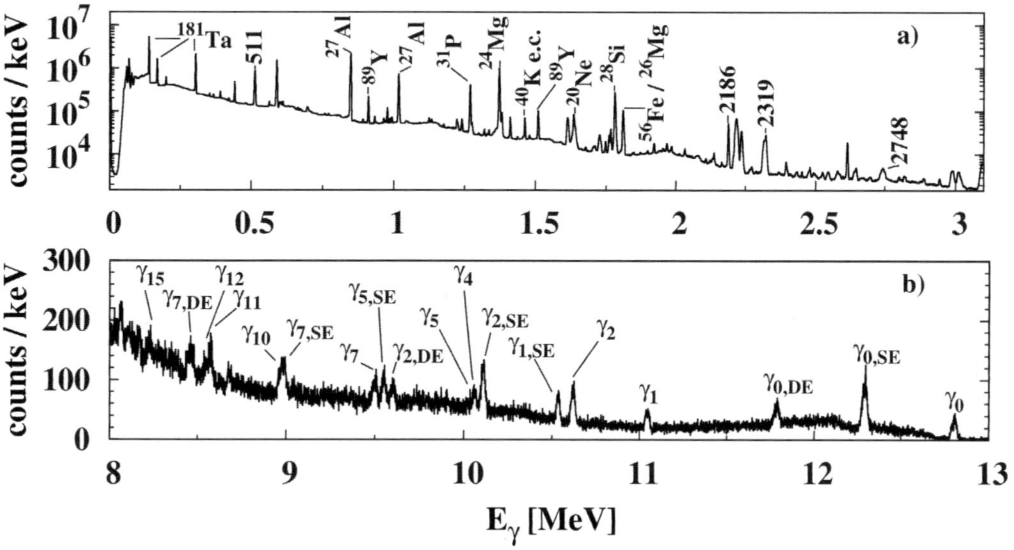

Figure 1. Parts of a γ-ray spectrum of the ^{89}Y(p,γ) reaction. The depicted γ-ray spectrum was recorded by six HPGe detectors at an angle of 90 ° relative to the beam axis. The incident proton energy was $E_p = 4.2$ MeV. In the upper panel a) the low-energy part is shown. Despite strong beam-induced background reactions on, *e.g.*, ^{27}Al, the ground-state transitions at 2186 keV, 2319 keV, and 2748 keV in the reaction product ^{90}Zr are clearly visible. In the lower panel b) the high-energy part of the γ-ray spectrum is shown. One can distinctly identify the de-excitation of the compound nucleus to various excited states (γ_1 up to γ_{15}), as well as to the ground state (denoted as γ_0). Note, that the middle part of the spectrum was omitted, where the remaining ground-state transitions are located. All ground-state transitions are used to determine the total reaction cross section, whereas the de-excitations to the excited states are used to determine the partial cross sections. In total, nine partial cross sections up to γ_{15} could be determined.

current was due to technical limitations regarding the accelerator.

A typical γ-ray spectrum for the (p,γ) reaction that was recorded using a proton energy of $E_p = 4.2$ MeV is shown in Fig. 1 a). This spectrum was obtained by summing the spectra of all HPGe detectors at an angle of 90 ° relative to the beam axis. Besides a lot of beam-induced background transitions, the ground-state transitions in the compound nucleus ^{90}Zr are clearly visible. The total reaction cross section can be determined by summing up the reaction yields of all ground-state transitions. Additionally, from the determination of the reaction yield of the transitions from the compound state to different excited states, it is possible to derive the respective partial cross sections. The de-excitations of the compound nucleus are clearly visible in the high-energy part of the γ-ray spectrum, which is shown in Fig. 1 b). These partial cross sections are an important observable in in-beam experiments, since they may allow additional experimental constraints on γ-ray strength functions upon condition that the γ width is not sensitive to the nuclear level density. One has to emphasize, that the determination of partial cross sections is usually only possible with in-beam experiments using HPGe detectors as in the present case. In the cases of the activation or in-beam 4π-summing method, this is only possible for isomeric states with a sufficiently long half-life.

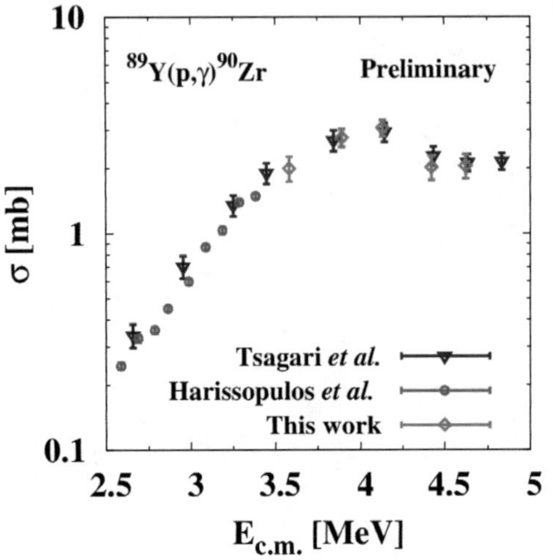

Figure 2. Experimental total cross sections of the ^{89}Y(p,γ) reaction as a function of center-of-mass energies. For comparison, the former results obtained by the 4π-summing method from Tsagari *et al.* [11] and the in-beam measurement with HPGe detectors by Harissopulos *et al.* [14] are shown. The results obtained in this work are in excellent agreement with the previous measurements.

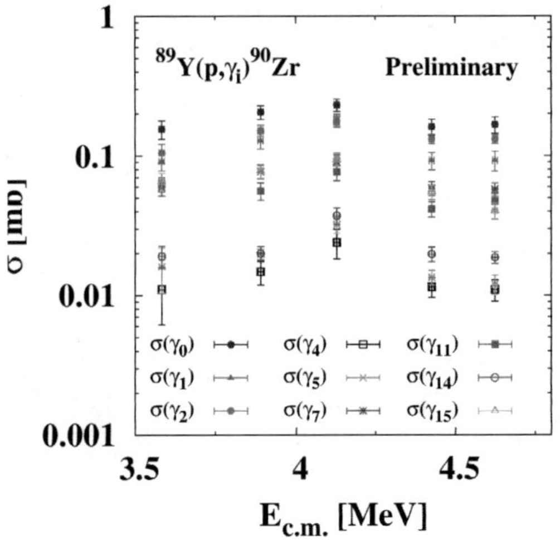

Figure 3. Experimental partial cross sections of the ^{89}Y(p,γ) reaction as a function of center-of-mass energies. In total, nine partial cross sections of the de-excitation of the compound nucleus up to γ_{15} could be observed. The determination of partial cross sections can yield experimental constraints on γ-strength functions.

Figure 2 shows the total cross-section results for the ^{89}Y(p,γ) reaction. These results are compared to previous measurements using the 4π summing method [11] and the in-beam technique with HPGe detectors [14]. The present results are in excellent agreement with the previous experiments. Hence, with respect to this results we can conclude that the recently developed setup at the Institute for Nuclear Physics in Cologne has become fully operational and the deficiencies of the earlier measurements from Ref. [16] are removed. Note, that the uncertainties of the present data are significantly higher than the ones from Ref. [14]. This is mainly due to the fact, that only a relative full-energy peak efficiency calibration for high-energy γ-rays was possible in the present case. This results in a higher systematic uncertainty of the efficiency calibration. Figure 3 shows the measured partial cross sections of the ^{89}Y(p,γ) reaction. In total, nine partial cross sections of de-excitations up to the 15th excited state were determined. The remaining partial cross sections could not be analyzed because of the low transition probabilities from the compound-state to the excited states, caused by the large angular momentum difference.

For the α-induced measurement on ^{92}Mo, a self-supporting 700 $\frac{\mu g}{cm^2}$ thick Mo foil with an isotopic enrichment of 94.1 % in ^{92}Mo was used as a target. It was irradiated with α-particles with an energy of E_α = 9.3 MeV and an average beam current of $\approx$20 nA. It was found in Ref. [16], that the beam-induced background is too high to measure the cross section of the ^{92}Mo(α,γ) reaction. Hence, a test measurement of this reaction using the same α-particle energy was

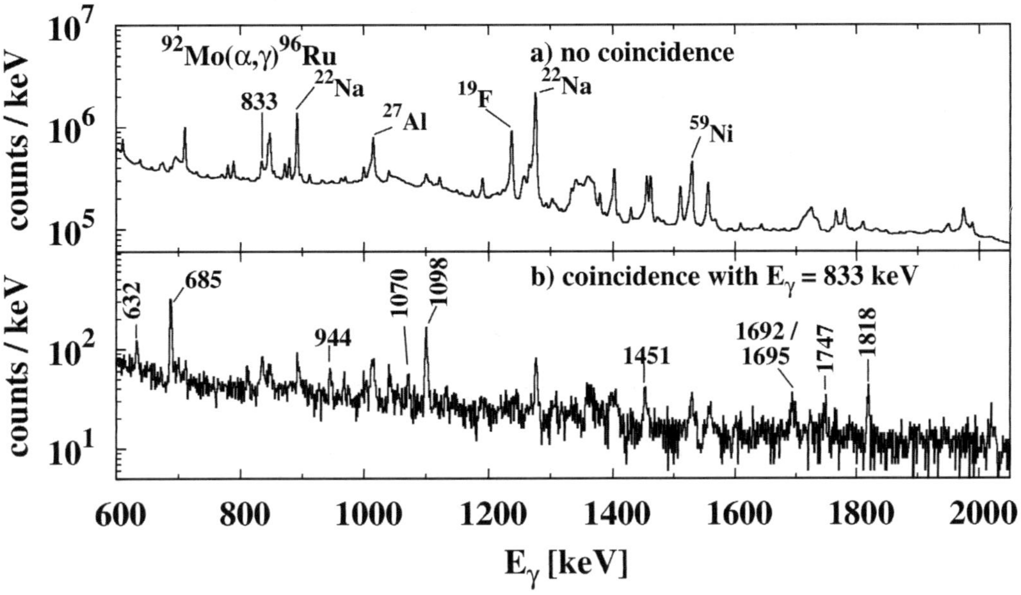

Figure 4. Measured γ-ray spectrum of the ^{92}Mo(α,γ)^{96}Ru reaction. The incident α-particle energy was $E_\alpha = 9.3$ MeV. In the upper panel a) only the transition from the first excited 2^+ state to the 0^+ ground-state in the reaction product ^{96}Ru at $E_\gamma = 833$ keV can be observed due to the enormous beam-induced background. The strongest background transitions are also marked in the spectrum. The lower panel b) shows a γ-ray spectrum obtained by claiming a γ coincidence with the γ-ray transition of 833 keV. The beam-induced background can be strongly suppressed and several feeding transitions of the 833- keV state are revealed.

performed to facilitate a comparison between these two experiments. Figure 4 a) depicts a γ-ray spectrum recorded for the ^{92}Mo(α,γ)^{96}Ru reaction. Contrary to proton-induced reactions, the beam-induced background is by orders of magnitude higher. This is mainly caused by (α,n) and (α,α') reactions on ^{181}Ta, ^{19}F, and ^{56}Fe. Due to the enormous background, only one ground-state transition ($2_1^+ \rightarrow 0_{\mathrm{g.s.}}^+$) in the compound nucleus ^{96}Ru with an energy of $E_\gamma = 833$ keV could be observed. This makes a determination of the reaction cross section impossible. Nevertheless, a powerful tool to suppress the beam-induced background is the requirement of $\gamma\gamma$ coincidences. In Fig. 4 b), a γ-ray spectrum that was obtained by claiming a γ coincidence with the ground-state transition with $E_\gamma = 833$ keV is shown. By this procedure, a lot of γ-ray transitions feeding the 833-keV state are revealed. However, although the beam-induced background is reduced compared to Ref. [16], a determination of the ^{92}Mo(α,γ) reaction cross-section is still challenging. Hence, further improvements of the experimental setup are necessary, including a lead and copper shielding of the apertures and collimators.

3. Conclusion and Outlook

The high-efficiency γ-ray detector array HORUS was used to determine total and partial cross sections of the ^{89}Y(p,γ)^{90}Zr reaction at five different proton energies between $E_p = 3.7$ MeV and 4.7 MeV close to the Gamow window for T = 3 GK. The excellent agreement of the total cross sections with previous measurements of Refs. [11] and [14] shows, that the setup at the Institute for Nuclear Physics in Cologne is fully operational. Moreover, nine partial cross sections up to the 15th excited state were measured, which may allow experimental constraints on the γ-strength function. Since the method of in-beam γ-ray spectroscopy allows to study reactions with stable reaction products, the experimental data needed for γ-process network calculations can be widely extended. Hence, the HORUS array embodies an excellent tool to study charged-particle induced reactions at energies of astrophysical interest.

Additionally, a test experiment on the ^{92}Mo(α,γ)^{96}Ru reaction with an α-particle energy of $E_\alpha = 9.3\,\mathrm{MeV}$ was performed. The dominating beam-induced background makes a determination of the reaction cross section impossible. For this reason, improvements of the target chamber are planned, which include a lead and copper shielding of the tantalum apertures and collimators.

Acknowledgments

The authors gratefully acknowledge the assistance of H.-W. Becker and D. Rogalla of the Ruhr-Universiät Bochum on RBS measurements. Furthermore, we thank the accelerator staff at the Institute for Nuclear Physics in Cologne for providing excellent beams. The authors would also like to thank C. Müller-Gatermann for his magnificent help during the experiments. JM and PS are member of the Bonn-Cologne Graduate School of Physics and Astronomy. This project was partly supported by the Deutsche Forschungsgemeinschaft under contract ZI 510/5-1.

References

[1] F. Käppeler, R. Gallino, S. Bisterzo, and W. Aoki, Rev. Mod. Phys. **83** (2011) 157
[2] M Arnould, S. Goriely, and K. Takahashi, Phys. Rep. **450** (2007) 97
[3] M. Arnould and S. Goriely, Phys. Rep. **384** (2003) 1
[4] M. Rayet, M. Arnould, M. Hashimoti, N. Prantzos, and K. Nomoto, Astron. Astrophys. **298** (1995) 517
[5] C. Travaglio, F. K. Röpke, R. Gallino, and W. Hillebrandt, Astrophys. J. **739** (2011) 93
[6] W. Hauser and H. Feshbach, Phys. Rev. **87** (1952) 366
[7] S. E. Woosley and A. Heger, Phys. Rep. **442** (2007) 269
[8] C. Yalçin, R. T. Güray, N. Özkan, S. Kutlu, Gy. Gyürky, J. Farkas, G. G. Kiss, Zs. Fülöp, A. Simon, E. Somorjai, and T. Rauscher, Phys. Rev. C **79** (2009) 065801
[9] A. Sauerwein *et al.*, Phys. Rev. C **84** (2011) 045808
[10] Z. Halász, Gy. Gyürky, J. Farkas, Zs. Fülöp, T. Szücs, E. Somorjai, and T. Rauscher, Phys. Rev. C **85** (2012) 025804
[11] P. Tsagari, M. Kokkoris, E. Skreti, A. G. Karydas, S. Harissopulos, T. Paradellis, and P. Demetriou, Phys. Rev. C **70** (2004) 015802
[12] A. Spyrou, H.-W. Becker, A. Lagoyannis, S. Harissopulos, and C. Rolfs, Phys. Rev. C **76** (2007) 015802
[13] A. Sauerwein *et al.*, Phys. Rev. C **86** (2012) 035802
[14] S. Harissopulos, A. Spyrou, A. Lagoyannis, M. Axiotis, P. Demetriou, J. W. Hammer, R. Kunz, and H.-W. Becker, Phys. Rev. C **87** (2013) 025806
[15] Zs. Fülöp, Á. Z. Kiss, E. Somorjai, C. E. Rolfs, H. P. Trautvetter, T. Rauscher, and H. Oberhummer, Z. Phys. A **355** (1996) 203
[16] J. Hasper, M. Büssing, M. Elvers, J. Endres, and A. Zilges, J. Phys. Conf. Ser. **202** (2010) 012005
[17] J. Brenneisen, D. Grathwohl, M. Lickert, R. Ott, H. Röpke, J. Schmälzlin, P. Siedle, B. H. Wildenthal, Z. Phys. A **352** (1995) 149
[18] T. Rauscher, Phys. Rev. C **81** (2010) 045807

Electron screening in metals

J Gajević[1] M Lipoglavšek[1] A Likar[1,2] T Petrovič[1,3] P Pelicon[1] A Sánchez Ortiz[1] and A Cvetinović[1]

[1] Jožef Stefan Institute, Jamova cesta 39, Ljubljana, Slovenia
[2] Faculty of Mathematics and Physics, Jadranska ulica 19, Ljubljana, Slovenia
[3] Cosylab d.d, Teslova ulica 30, Ljubljana, Slovenia

E-mail: jelena.gajevic@ijs.si

Abstract. In order to further investigate electron screening effect we studied proton induced nuclear reactions over an energy range from 0.8 to 2.6 MeV for different environments: Mn and Cd metals and MnO and CdO insulators. Shifts in resonance energy for metallic relative to insulator targets were not observed. No large electron screening could be deduced for (p,γ) and (p,n) reactions in Mn. Moreover, no large electron screening can be deduced with certainty for the (p,n) reaction in Cd.

1. Introduction

The fusion reaction can be divided into two parts which are approximately independent of each other, namely the atomic physics of the nuclei approaching each other and the nuclear physics valid if they are close enough to feel the nuclear forces. In order to fuse, two positively charged nuclei must come into contact overcoming the Coulomb repulsion. Due to Coulomb repulsion, the cross section σ for charged-particle-induced nuclear reactions drops rapidly with decreasing beam energy. The astrophysical S-factor is usually introduced to separate the strong energy dependence from effects of pure nuclear interaction and in the case of non-resonant reactions it only slowly varies with energy [1]. The S-factor is defined by writing the cross section as [1]:

$$\sigma(E) = \frac{S(E)}{E} e^{-2\pi\eta}, \tag{1}$$

where E is the center of mass energy, $\eta = Z_1 Z_2 e^2/4\pi\epsilon_0\hbar\sqrt{2E/\mu}$ is the Sommerfeld parameter, Z_1 and Z_2 are the charge numbers of interacting nuclei and μ is their reduced mass. The Gamow factor $e^{-2\pi\eta}$ describes the s-wave penetration through the Coulomb barrier of point-like charges and thus with the geometrical factor $1/E$ accounts for a strong energy dependence of the cross sections at sub-Coulomb energies. The cross section increases at low energies when the interacting nuclei are not bare but surrounded with atomic electron clouds [1]. As a result, the measured cross sections are enhanced compared to cross sections for bare nuclei with an enhancement factor:

$$f(E) = \frac{\sigma(E + U_e)}{\sigma(E)}, \tag{2}$$

where U_e is the electron screening potential. Experimental studies of various nuclear reactions in metallic environments have shown the expected cross section enhancement at low energies

[2-6]. Furthermore, it was observed that the magnitude of the screening effect strongly depends on the host material and the reason of this dependence is still not known. A large electron screening potential of $U_D = 27 \pm 9$ keV [4] (U_D is the difference between screening potentials in metallic and insulating targets) was observed for the (p,n) reaction in V metal relative to the VO_2 insulator, while an electron screening potential of $U_D = 31 \pm 13$ keV was deduced for the (p,n) reaction in Ni metal relative to the NiO insulator [5]. Moreover, Kettner et al. [4] observed a narrow resonance in the ^{176}Lu(p, n)^{176}Hf reaction at proton energy $E_p = 810$ keV and noticed a lowering of this resonance energy by $U_D = 32 \pm 2$ keV for the Lu metal relative to the insulator. The sizeable resonance shifts were interpreted as a demonstration of the acceleration effect by the valence electrons.

2. Experiment

Proton beams with energies between 0.8 and 2.6 MeV were accelerated by the 2 MV Tandetron accelerator at Jožef Stefan Institute. Neutrons and γ rays produced in reactions: ^{55}Mn(p, n)^{55}Fe, ^{55}Mn(p, γ)^{56}Fe, ^{55}Mn(p, p$'\gamma$)^{55}Mn, ^{110,111,112,113,114,116}Cd(p, p$'\gamma$)^{110,111,112,113,114,116}Cd, ^{113}Cd(p, n)^{113}In and ^{112,114}Cd(p, γ)^{113,115}In were detected with a neutron detector and a HPGe detector placed 4.2 cm from the target at an angle of 135° with respect to the beam direction. Neutrons were detected in a liquid organic scintillator detector positioned 4.8 cm from the target at an angle of 45° with respect to the beam direction. This scintillator detector was chosen because of the good neutron-γ separation at low energies. The detector was connected to a fast asynchronous digitizer which digitized the incoming signals. In order to suppress the large γ-ray background, a 6 mm thick lead absorber was placed in front of the neutron detector. The method employed for neutron-γ separation was the digital charge comparison method [7]. In order to properly take into account background neutrons, we measured background neutron activity continuously for 238 h. A SiLi X-ray detector positioned 9.1 cm from the target at an angle of 135° with respect to the beam direction was used for measurement of X rays. Beam current was on the average about 1μA. Only targets with natural isotopic abundance were used. A 125 μm thick $Mn_{88\%}Ni_{12\%}$ foil was obtained from Goodfellow while a 99.99 % pure and 250 μm thick Cd foil was obtained from Chempur. 99 % pure MnO powder and 99.99 % pure CdO powder were obtained from Chempur. Oxide targets were prepared by pressing the powder into a 14 mm diameter and 1 mm deep cylindrical hole of a 2 mm thick Cu backing. The conductive backing helped cooling the insulating targets.

3. Results

We chose to study the (p,n) reaction in Mn due to the reasonably low threshold (E_{th}=1.032 MeV) for this reaction and due to the fact that Mn has only one naturally occurring isotope, ^{55}Mn. We searched for a shift in resonance energy in metallic compared to insulator environment, similarly to the one observed in ^{176}Lu(p, n)^{176}Hf. To search for a shift in resonance energy, we studied two low energy resonances at respective beam energies of 1.38 and 1.54 MeV [8]. Thick target neutron yields from the ^{55}Mn(p, n)^{55}Fe reaction near the $E_p = 1.54$ MeV resonance as a function of laboratory beam energy E_p for Mn metal and MnO insulator are shown in fig. 1. From the observed thick target neutron yields near both studied resonances in Mn, we did not notice any shifts in ^{55}Fe resonance energies in metallic Mn compared to insulating MnO targets. We used a 125 keV Coulomb excitation peak for inferring incident dose. While studying the (p,n) reaction in Mn we also observed the (p,γ) resonances in ^{56}Fe. Thick target 847 keV γ-ray yields from the ^{55}Mn(p, γ)^{56}Fe reaction near E_p=1.54 MeV are shown in fig. 2. Shifts in ^{56}Fe resonance energy in metallic Mn compared to insulating MnO target were not observed. Moreover, the ratios of 847 keV γ-ray (see fig. 3) and neutron yields for Mn and MnO targets do not show any expected enhancement with decreasing beam energy and therefore no large electron screening can be deduced for (p,γ) and (p,n) reactions in Mn. We chose to study the (p,n) reaction in Cd,

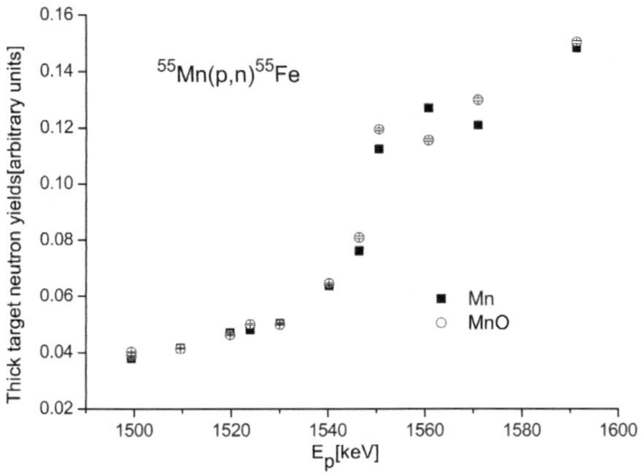

Figure 1. Thick target neutron yields from the $^{55}\mathrm{Mn(p, n)}^{55}\mathrm{Fe}$ reaction near the E_p= 1.54 MeV resonance.

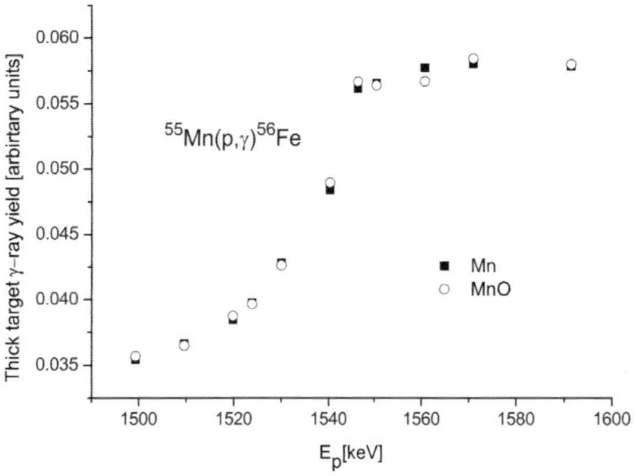

Figure 2. Thick target 847 keV γ-ray yields from the $^{55}\mathrm{Mn(p, \gamma)}^{56}\mathrm{Fe}$ reaction near the E_p= 1.54 MeV resonance.

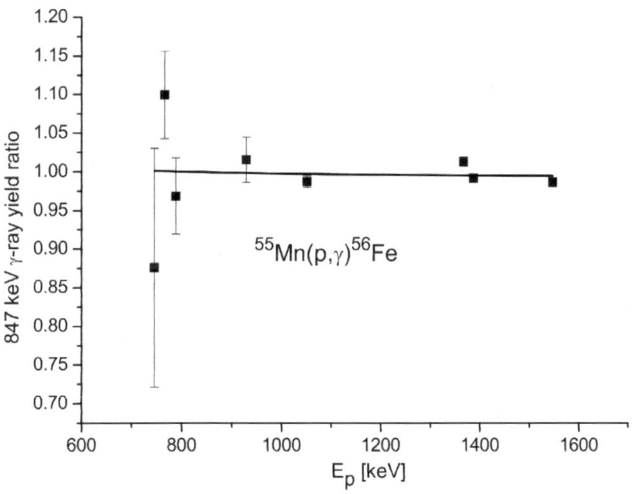

Figure 3. The ratio of 847 keV γ-ray yields from the $^{55}\mathrm{Mn(p, \gamma)}^{56}\mathrm{Fe}$ reaction as a function of laboratory beam energy for Mn metal and MnO insulator.

since it has an even lower threshold than Mn that is for the $^{113}\mathrm{Cd(p, n)}^{113}\mathrm{In}$ reaction E_{th} =0.46 MeV. Cd has a higher atomic number than Ni and V and if we take into account the suggested increment of the screening potential that is roughly proportional to the proton number Z of the target [2-6], we expected a higher electron screening potential in Cd. Thick target neutron yields from the $^{113}\mathrm{Cd(p, n)}^{113}\mathrm{In}$ reaction as a function of laboratory beam energy for Cd metal and CdO insulator are shown in fig. 4. The 299 keV Coulomb excitation peak was used for dose normalization. The neutron yields in Cd and CdO show a trend that is not well understood at the moment. The lowest energy point in fig. 4 is significantly higher in Cd than in CdO. This enchancement could be due to electron screening. However, the increase of the neutron yields at the lowest energy is not presently understood. Improved measurements are certainly needed to clarify the mentioned data point.

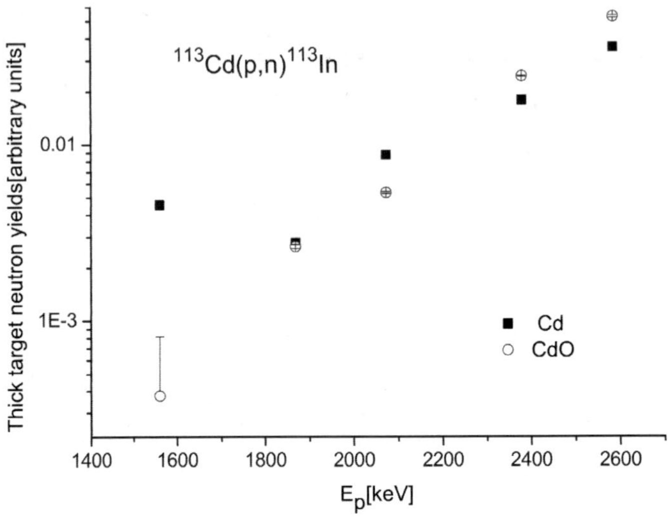

Figure 4. Thick target neutron yields from the $^{113}\mathrm{Cd(p,n)}^{113}\mathrm{In}$ reaction as a function of laboratory beam energy for Cd metal and CdO insulator.

4. Conclusions

Contrary to observed shifts in $^{176}\mathrm{Hf(p,n)}^{176}\mathrm{Lu}$ reaction, shifts in resonance energy for metallic targets relative to insulator ones were not observed in $^{55}\mathrm{Mn(p,n)}^{55}\mathrm{Fe}$ and $^{55}\mathrm{Mn}(\mathrm{p},\gamma)^{56}\mathrm{Fe}$ reactions. Besides, no large electron screening was observed in $^{55}\mathrm{Mn(p,n)}^{55}\mathrm{Fe}$ and $^{55}\mathrm{Mn}(\mathrm{p},\gamma)^{56}\mathrm{Fe}$ reactions, although similar electron screening values to the measured ones in Ni and V were expected. However, there is an indication of a large electron screeing in Cd, although more precise measurements are needed to confirm this indication. The reported results are quite unexpected and certainly motivate further studies of the electron screening effect.

5. References

[1] H. J. Assenbaum, K. Langanke, C. Rolfs, Z. Phys. A, **327**, 461 (1987).
[2] C. Rolfs, Prog. Theor. Phys. Suppl **154**, 373 (2004).
[3] J. Cruz et al., Phys. Lett. B **624**, 181 (2005).
[4] K. U. Kettner et al., J.Phys G **32**, 489 (2006).
[5] J. Gajević et al., Eur. Phys. J., A **49**, 70 (2013).
[6] D. Zahnow, et al., Z Phys. A **359**, 211 (1997).
[7] Chen Xiaohui et al., Radiation Measurements, **49**, 13 (2013).
[8] Y. P. Viyogi et al., Phys. Rev. C18, 1178 (1978).

Latest results of Skyrme-Hartree-Fock-Bogoliubov mass formulas

S Goriely[1], N Chamel[1], J M Pearson[2]

[1] Institut d'Astronomie et d'Astrophysique, ULB, CP226, 1050 Bruxelles, Belgium
[2] Dépt. de Physique, Université de Montréal, Montréal (Québec), H3C 3J7 Canada

E-mail: sgoriely@astro.ulb.ac.be

Abstract. The latest developments made in deriving accurate mass predictions within the microscopic Skyrme Hartree-Fock-Bogolyubov (HFB) approach are described. Using the HFB method, we have refitted our BSk20 and BSk21 models, on which were based the HFB-20 and HFB-21 mass tables, to the 2353 measured masses of nuclei with N and $Z \geq 8$ given in the newly available 2012 Atomic Mass Evaluation. In addition, we have now investigated the sensitivity to the symmetry coefficient J and the stiffness of the neutron matter equation of state. We present here in all 5 new Skyrme interactions, along with their corresponding mass tables. The root-mean square deviation obtained with our preferred model (HFB-24) with respect to essentially all the 2353 available mass data falls to 0.549 MeV, the best value ever found within the mean-field framework. Since our Skyrme force is also constrained by the properties of pure neutron matter this new model is particularly well-suited for application to astrophysical problems involving a neutron-rich environment, such as the elucidation of the r-process of nucleosynthesis, and the description of supernova cores and neutron-star crusts.

1. Introduction

In Ref. [1] we presented a family of three Skyrme-type functionals, BSk19, BSk20 and BSk21, along with their corresponding mass tables, HFB-19, HFB-20 and HFB-21, respectively, that we had constructed with a view to providing a unified approach not only to the structure of nuclei as well as of all the different regions of neutron stars (outer crust, inner crust and core) but also to other phenomena associated with the birth and death of neutron stars, such as supernova-core collapse, the r-process of nucleosynthesis in the neutrino-driven wind or during the decompression of neutron-star matter. These three functionals are all based on effective forces with the 16-parameter generalized Skyrme form [1], which is characterized by unconventional terms that have a simultaneous density and momentum dependence.

The parameters of this form of force were determined primarily by fitting measured nuclear masses, which were calculated with the Hartree-Fock-Bogoliubov (HFB) method. For this it was necessary to supplement the Skyrme forces with a microscopic pairing force, phenomenological Wigner terms and correction terms for the spurious collective energy. However, in fitting the mass data we simultaneously constrained the Skyrme force to fit the zero-temperature equation of state (EOS) of homogeneous neutron matter (NeuM), as determined by many-body calculations with realistic two- and three-nucleon forces. Actually, several realistic calculations of the EOS of NeuM have been made, and while they all agree fairly closely at nuclear and subnuclear densities, at the much higher densities that can be encountered towards the centre of neutron stars they differ greatly in their stiffness, and there are very few data, either observational or experimental, to discriminate between the different possibilities. We therefore considered three different constraining EOSs of NeuM, as follows. The softest is the one that we label FP [2] in Ref. [1], the one of intermediate stiffness is the "A18 + δv + UIX*" EOS [3], which we label as APR, while our stiffest constraining EOS is the one labelled "V18" in Ref. [4], which we refer to as LS2 in Ref. [1].

After the publication of our HFB-19 to HFB-21 paper [1] the neutron stars PSR J1614−2230 and PSR J0348+0432 were shown to have masses of 1.97 ± 0.04 $\mathcal{M}_\odot$ [5] and 2.01 ± 0.04 $\mathcal{M}_\odot$ [6], respectively. This new observation confirmed the plausibility of the BSk20 and BSk21 neutron-star matter EOS, but not the one obtained with BSk19 (the corresponding maximum neutron-star mass being $1.86 M_\odot$ [1]) which is definitely too soft, and it must be discarded. This does not necessarily imply that HFB-19 mass model must also be discarded, nor that the NeuM EOS has to be much stiffer than that of FP, because the core of neutron stars may contain non-nucleonic particles (see, e.g., Ref. [7]).

In addition, at the end of 2012 a new Atomic Mass Evaluation (AME) [8] was published, with 2353 measured masses of nuclei having N and $Z \geq 8$. When we calculate the root-mean square (rms) deviation for HFB-20 and HFB-21 with respect to this new data set, we find that it rises by 17 keV for the former and falls by 5 keV for the latter. The fact that the rms deviation in the case of HFB-21 is actually lower for the larger data base of the 2012 AME than for the 2003 AME [9], to which it was fitted, is an indicator of the reliability with which this model can be extrapolated into unknown regions of the nuclear chart.

For these two reasons, a new series of mass fits was performed. In refitting to the 2353 measured masses of nuclei having N and $Z \geq 8$ in the 2012 AME, we took the opportunity of imposing different values of the symmetry coefficient J on our fits. We generated in all five new parameter sets, BSk22 to BSk26, along with the corresponding mass tables, labelled HFB-22 to HFB-26, respectively [10]. BSk22 to BSk25 were fitted to J = 32, 31, 30 and 29 MeV, respectively, and were all constrained to LS2, while BSk26 was also fitted to J = 30 MeV but under the APR constraint. In Fig. 1 we show the NeuM EOSs of our five new models. These new mass models are described below.

2. The HFB-22 to HFB-26 mass models

The BSk22 to BSk26 interactions are all based on the 16-parameter generalized form of Skyrme force given in Eq. (1) of Ref. [1]. In addition to the above-described constraints on the symmetry energy, we also impose, as for our previous mass models, the following constraints: a) an optimal fit to the charge-radii data [11], b) a value of $0.8M$ for the isoscalar effective mass M_s^* in charge-symmetric infinite nuclear matter (INM) at the appropriate equilibrium density n_0, this being the value indicated by calculations with realistic forces (see, e. g., the discussion in Ref. [12]), c) an incompressibility K_v of charge-symmetric INM falling in the experimental range 240 ± 10 MeV [13], d) the stability of NeuM and of β-equilibrated neutron-star matter (i.e., the homogeneous nucleon-lepton mixture of which neutron-star cores are comprised) against an unphysical polarization at any density relevant to neutron-star cores [14, 15], e) an EOS of charge-symmetric INM that is consistent with measurements in heavy-ion collisions of nuclear-matter flow over the density range 1.5 - $4.5 n_0$ [16, 17], f) a qualitatively acceptable distribution of potential energy among the four different spin-isospin channels in INM.

The form of our functionals is sufficiently flexible to allow all these constraints to be satisfied and at the same time for the 2353 measured masses of nuclei with N and $Z \geq 8$ given in the 2012 AME to be fitted with an rms deviation as low as 0.55 MeV as shown in Table 1 The rms and mean (data - theory) values of the deviations between the measured masses and the predictions for the five new models are given in the first and second lines, respectively, of Table 1. The third line gives the the model error σ_{mod}, as defined by Eqns. (42) and (43) of Möller and Nix [18].

Table 1. Rms (σ) and mean ($\bar{\epsilon}$) deviations between data and predictions for the HFB-22 to HFB-26 models. The first pair of lines refers to all the 2353 measured masses M that were fitted [8], the third line to the model error on all the 2353 masses, and the last pair of lines to the masses M_{nr} of the subset of 257 neutron-rich nuclei ($S_n \leq 5.0$ MeV).

	HFB-22	HFB-23	HFB-24	HFB-25	HFB-26
$\sigma(M)$ [MeV]	0.629	0.569	0.549	0.544	0.564
$\bar{\epsilon}(M)$ [MeV]	-0.043	-0.022	-0.012	0.008	0.006
$\sigma_{\rm mod}(M)$ [MeV]	0.619	0.561	0.542	0.537	0.556
$\sigma(M_{nr})$ [MeV]	0.817	0.721	0.702	0.791	0.749
$\bar{\epsilon}(M_{nr})$ [MeV]	0.221	0.090	0.011	0.023	0.230

The last two lines of this table show the corresponding deviations for the subset consisting of the most neutron-rich measured nuclei, here taken as those with a neutron separation energy $S_n \leq 5.0$ MeV (there are 257 nuclei in this subset). All five models display, not surprisingly, some deterioration as we move into the neutron-rich region. From the first line we see that the parameter sets BSk24 ($J = 30$ MeV) and BSk25 ($J = 29$ MeV) give the best global fits of all the new models, and in fact are better than any of our previous models. However, line 4 of Table 1 shows that the deterioration of BSk25 on moving into the neutron-rich region is much stronger than for BSk24: for all the other models the performance in the neutron-rich region correlates fairly well with the global performance. Thus the apparent high performance of this model in the global fit should be interpreted with caution; other defects of model BSk25 are discussed in Ref. [10]. Looking at BSk22 and BSk23, we see from both lines 1, 3 and 4 that $J = 31$ MeV works less well than either 29 or 30 MeV, while $J = 32$ MeV is still more strongly disfavoured.

Comparing BSk24 and BSk26 shows that for $J = 30$ MeV the high-density LS2 constraint (BSk24) gives slightly better fits than APR (BSk26). It must be stressed, however, that this discrimination in favour of LS2 as the constraining EOS of NeuM relates only to nuclear and subnuclear densities, so that we should not conclude that nuclear masses are telling us something about the EOS of NeuM at the higher densities found in neutron-star cores.

Overall, the clearest conclusion that can be drawn from Table 1 is that model BSk22 is the worst performing of all our models, ruling out $J = 32$ MeV. There are also very strong indications that $J = 29$ or 30 MeV (the latter in both its LS2 and APR forms) are to be preferred to $J = 31$ MeV, although we have expressed some concerns with regards to $J = 29$ MeV, i.e., to BSk25 [10]. In any case, it will be seen that the HFB models favour a value of J significantly different than the value of $J \simeq 32$ deduced from the finite-range droplet model [19] (see Ref. [10] for more details).

As was the case with our models BSk19, BSk20 and BSk21, the new models are well adapted to a unified treatment of all parts of neutron stars: the outer crust [20], the inner crust [21] and the core [1]. The relevance of the models to the core of neutron stars arises not only from their fit to a sufficiently stiff EOS of NeuM but also from their fit to nuclear masses, which implies that they take correct account of the presence of protons. For the same reason these models take account of inhomogeneities, and thus are appropriate for the calculation of the inner crust of neutron stars. As for the outer crust, its properties are determined entirely by the mass tables that we have generated for the appropriate interactions. This is in contrast to the very recent Ref. [22], which does not generate mass tables and thus cannot handle the outer crust.

Apart from the EOS of NeuM, other INM properties, such as the incompressibility coefficient, the isoscalar and isovector effective masses, the pressure in charge-symmetric INM (see Fig. 2)

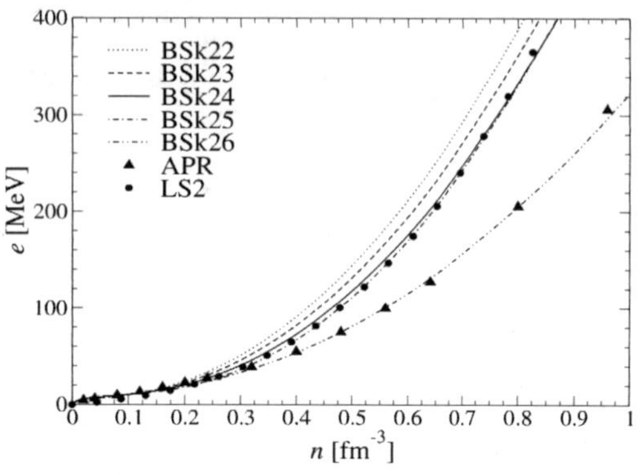

Figure 1. Zero-temperature EOSs for NeuM with BSk22-26. APR is the realistic EOS "A18 + $\delta\,v$ + UIX*" in Ref. [3]. LS2 is the realistic EOS "V18" in Ref. [4].

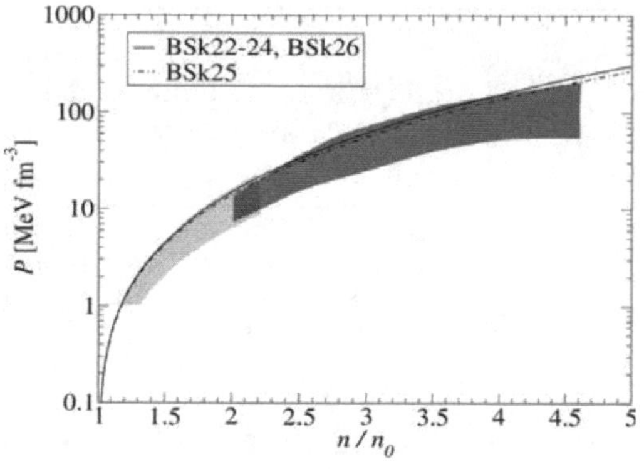

Figure 2. Pressure as a function of density in charge-symmetric INM for our forces. The shaded area represents the analysis of heavy-ion collision experiments obtained in Refs. [16, 17].

are predicted to have values consistent with both experiments and calculations based on realistic nucleon-nucleon potentials. Moreover, INM is found to be stable against spin and spin-isospin fluctuations in agreement with realistic calculations.

Finally, we have found that all the neutron-star EOSs deduced from our new models support the massive neutron stars PSR J1614−2230 and PSR J0348+0432 (see Ref. [10] for more detais).

Acknowledgments

S.G. and N.C. F.R.S.-FNRS research associate and acknowledges the financial support of the "Actions de recherche concertées (ARC)" from the "Communauté française de Belgique", J.M.P. that of NSERC (Canada).

References

[1] Goriely S, Chamel N and Pearson J M 2010 *Phys. Rev.* C **82** 035804
[2] Friedman B and Pandharipande V R 1981 *Nucl. Phys.* A **361** 502
[3] Akmal A, Pandharipande V R and Ravenhall D G 1998 *Phys. Rev.* C **58** 1804
[4] Li Z H and Schulze H-J 2008 *Phys. Rev.* C **78** 028801
[5] Demorest P B, *et al.* 2010 *Nature* **467** 1081
[6] Antoniadis J, *et al.* 2013 *Science* **340** 1233232.
[7] Chamel N, Fantina A F, Pearson J M and Goriely S 2013 *Astron. Astrophys.* **553** A22
[8] Audi G, *et al.* 2012 *Chinese Physics* C **36** 1287
[9] Audi G, Wapstra A H and Thibault C 2003 *Nucl. Phys.* A **729** 337
[10] Goriely S, Chamel N and Pearson J M 2013 *Phys. Rev.* C **88** 024308
[11] Angeli I and Marinova K P 2013 *At. Data Nucl. Data Tables* **99** 69
[12] Goriely S, Samyn M, Bender M and Pearson J M, 2003 *Phys. Rev.* C **68** 054325
[13] Colò G, Giai N V, Meyer J, Bennaceur K and Bonche P 2004 *Phys. Rev.* C **70** 024307
[14] Chamel N, Goriely S and Pearson J M 2009 *Phys. Rev.* C **80** 065804
[15] Chamel N and Goriely S 2010 *Phys. Rev.* C **82** 045804
[16] Danielewicz P, Lacey R and Lynch W G 2002 *Science* **298** 1592
[17] Lynch W G, et al. 2009 *Prog. Part. Nucl. Phys.* **62** 427
[18] Möller P and Nix J R 1988 *At. Data Nucl. Data Tables* **39**, 213
[19] Möller P, Myers W D, Sagawa H and Yoshida S 2012 *Phys. Rev. Lett.* **108** 052501
[20] Pearson J M , Goriely S and N. Chamel N 2011 *Phys. Rev.* C **83** 065810
[21] Pearson J M, Chamel N, Goriely S and Ducoin C 2012 *Phys. Rev.* C **85** 065803
[22] Erler J, Horowitz C J, Nazarewicz W, Rafalski M and Reinhard P G 2013 *Phys. Rev.* C **87** 044320

Investigation of the Photon Strength Function in ^{130}Te

J Isaak[1,2], J Beller[3], E Fiori[1,2], J Glorius[4], M Krtička[5], B Löher[1,2], N Pietralla[3], C Romig[3], G Rusev[6], D Savran[1,2], M Scheck[3,7,8], J Silva[1,2], K Sonnabend[4], A P Tonchev[9], W Tornow[10,11], H R Weller[10,11] and M Zweidinger[3]

[1]ExtreMe Matter Institute EMMI and Research Division, GSI Helmholtzzentrum für Schwerionenforschung GmbH, Darmstadt, Germany
[2]Frankfurt Institute for Advanced Studies FIAS, Frankfurt, Germany
[3]Institut für Kernphysik, Technische Universität Darmstadt, Darmstadt, Germany
[4]Institut für Angewandte Physik, Goethe-Universität Frankfurt, Frankfurt, Germany
[5]Faculty of Mathematics and Physics, Charles University, Prague 8, Czech Republic
[6]Chemistry Division, Los Alamos National Laboratory, Los Alamos, New Mexico 87545, USA
[7]School of Engineering, University of the West of Scotland, Paisley PA1 2BE, UK
[8]SUPA, Scottish Universities Physics Alliance, Glasgow G12 8QQ, UK
[9]Physics Division, Lawrence Livermore National Laboratory, Livermore, CA 94550
[10]Triangle Universities Nuclear Laboratory, Durham, North Carolina 27708, USA
[11]Department of Physics, Duke University, Durham, North Carolina 27708-0308, USA

E-mail: j.isaak@gsi.de

Abstract. The dipole strength distribution of ^{130}Te was investigated with the method of Nuclear Resonance Fluorescence using continuous-energy bremsstrahlung at the Darmstadt High Intensity Photon Setup and quasi-monoenergetic photons at the High Intensity $\vec{\gamma}$-Ray Source. The average decay properties were determined between 5.50 and 8.15 MeV and compared to simulations within the statistical model.

1. Introduction

The properties of highly excited nuclei are important ingredients in astrophysical model calculations of the nucleosynthesis in stellar environments. In particular, Photon Strength Functions (PSF) for electric dipole ($E1$) and magnetic dipole ($M1$) transitions, respectively, are major input parameters for calculations of capture and photodissociation reactions. The PSF can be related to the photoabsorption cross section of atomic nuclei. In the last decades, the photoabsorption cross section has been investigated intensively, especially in the region of the Giant Dipole Resonance (GDR) [1]. Furthermore, for many nuclei additional low-lying $E1$ strength below and in the vicinity of the particle separation energies was observed, which is usually denoted as Pygmy Dipole Resonance (PDR) [2]. So far, the properties of the PDR have been mostly studied in closed-shell nuclei, e.g., in the Z=50 and N=82 mass region [2]. Therefore, the nucleus ^{130}Te was investigated via real photon scattering to extend the studies also to off-shell nuclei.

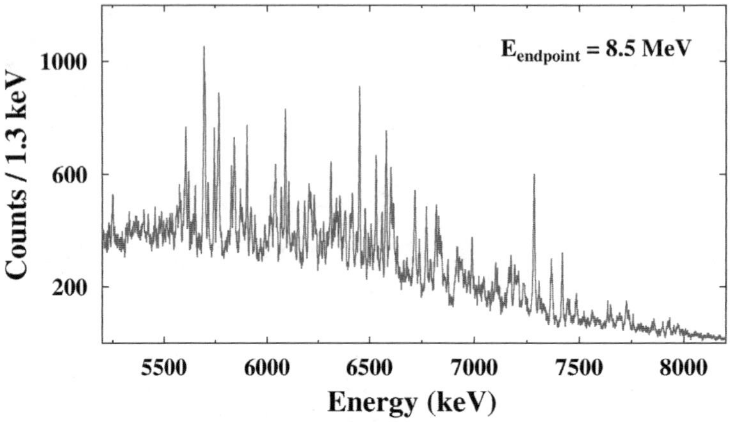

Figure 1. Spectrum of ^{130}Te measured at an endpoint energy of 8.5 MeV.

2. Experiments and preliminary results

Nuclear Resonance Fluorescence (NRF) experiments are well-suited to investigate the photoabsorption cross section below the neutron separation threshold [3]. Experiments with real photons are selective to dipole-excited states, therefore the NRF method was used to investigate the electric dipole response in ^{130}Te. One experiment was performed at the Darmstadt High Intensity Photon Setup (DHIPS) [4] using bremsstrahlung as the photon source. In Fig. 1 the measured spectrum for the endpoint energy of 8.5 MeV is shown. An accumulation of peaks in the spectrum originating from nuclear transitions were observed between 6 and 8 MeV.

To determine the absolute dipole strength, the nucleus of interest was simultaneously measured relative to a known calibration standard ^{11}B (for details see, e.g., Refs. [5, 6]). However, it is not possible to distinguish between $E1$ and $M1$ strength from these data. Therefore, an additional NRF experiment with quasi-monoenergetic and linearly polarized photons has been performed in the energy range from 5.5 to 8.5 MeV at the High Intensity $\vec{\gamma}$-Ray Source (HI$\vec{\gamma}$S) [7] at Triangle Universities Nuclear Laboratory, USA. Due to the linear polarization of the incoming photons it is possible to access parity information of the excited $J = 1$ states and, thus, to separate between $E1$ and $M1$ strength [8, 9]. In the case of ^{130}Te, excited states with negative parity were observed exclusively and therefore the measured dipole strength is assigned to be $E1$ strength.

Another advantage of the HI$\vec{\gamma}$S facility is the quasi-monoenergetic photon beam, which has a FWHM of about 3% of the mean beam energy. This allows for investigating average decay properties of nuclei as a function of the excitation energy [10]. The total photoabsorption cross section σ_{tot} can be determined by the sum of the contribution of photo-excited states that decay directly to the ground state (σ_{elast}) and from cascade transitions ($\sigma_{inelast}$).

This idea was presented, e.g., in Refs [10, 11, 12]. On the right side of Fig. 2 the measured spectrum for a mean beam energy of 7.85 MeV is shown together with the spectral beam distribution. Even though the incoming mono-energetic photon beam exclusively excites states in an energy range of a few hundred keV, decays of the first few excited 2^+ states were observed. This indicates that these excited levels are populated via cascade transitions of the initially excited states. An excitation and de-exciation scheme is shown on the left side of Fig. 2. It is commonly assumed that most of the intensity of cascading transitions (indicated as blue arrows) is collected in the first low-lying excited states (i.e., 2^+ states). Therefore, their decay intensities (indicated as green arrows) are used to estimate $\sigma_{inelast} \approx \sum_i \sigma_{2_i^+}$ (see Ref. [10]).

Preliminary results for σ_{tot} are shown on the left side of Fig. 3 together with data from a

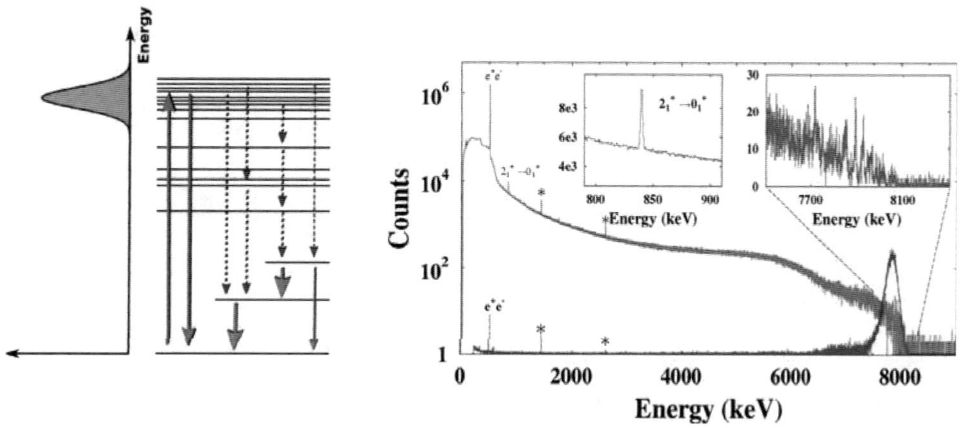

Figure 2. Left panel: Schematic example of the excitation and de-excitation process using a quasi-monoenergetic beam. Right panel: Spectrum of ^{130}Te measured at a mean beam energy of 7.85 MeV using the HI$\vec{\gamma}$S facility.

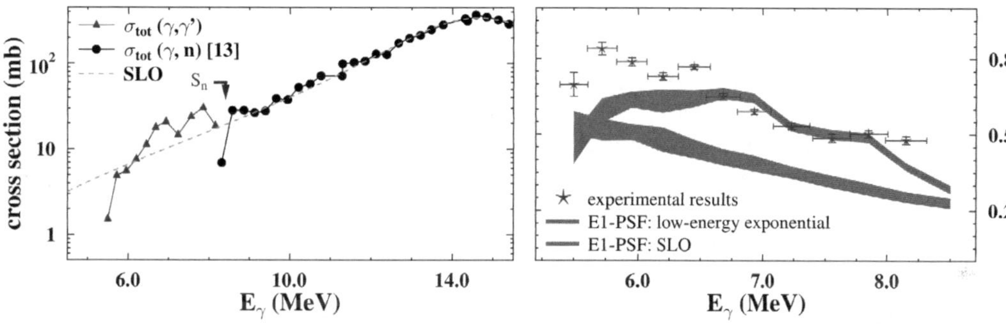

Figure 3. Preliminary results. Left panel: Photoabsorption cross sections for the energy range from 5.5 MeV up to 15 MeV. For details see text. Right panel: Experimental $\langle b_0 \rangle$ (blue stars) compared to DICEBOX simulations.

previous (γ, n) experiment at high energies [13]. An enhancement of the photoabsorption cross section between 6.5 and 8.0 MeV compared to the extrapolation of the GDR is observed. The PDR was observed in several other nuclei so far [2]. Using the extracted cross sections the average ground-state branching ratio $\langle b_0 \rangle = \sigma_{elast}/\sigma_{tot}$ was determined and is shown in the right panel of Fig. 3. The experimental data are compared to results obtained from statistical model simulations using the DICEBOX code [14]. The main input parameters in the simulations are the PSF for the transitions involved and the level density for the nucleus of interest. The two dominant PSFs are the $E1$- and $M1$-PSF. The Single Particle model was used for the $M1$-PSF while for the $E1$-PSF two different approaches were tested. The Standard Lorentzian (SLO) model as well as a model using the experimental data at $E_\gamma > 5.5$ MeV combined with a strong decrease to lower γ-ray energies. The backshifted Fermi-Gas model was used to describe the level density [15]. The SLO model is not able to describe the experimental results, which have higher values for $\langle b_0 \rangle$ at all energies. This observation indicates that the $E1$-PSF at low E_γ (for $E_\gamma \lesssim 3-5$ MeV) must be suppressed strongly with respect to the $E1$-PSF at $E_\gamma \gtrsim 3-5$ MeV than in the SLO model. Therefore, the other model was tested. It shows a better agreement, especially at $E_\gamma > 6.5$ MeV, even though it is not able to fully describe the experimental data at lower energies. It seems that the statistical model is not able to describe the decay properties of

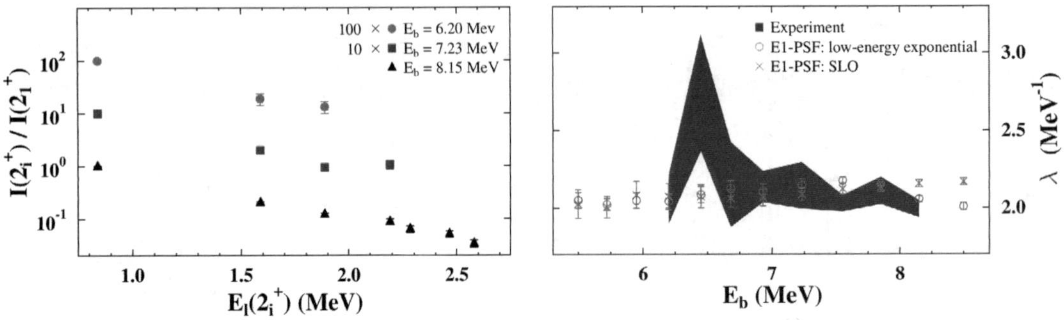

Figure 4. Preliminary results. Left panel: Relative population intensity of low- lying 2^+ states as a function of the level energy $E_l(2_i^+)$. The values for $E_b = 6.20$ MeV and $E_b = 7.23$ MeV are scaled by a factor of 100 and 10, respectively. Right panel: Exponential factor λ as a function of E_b.

^{130}Te at excitation energies below 6.5 MeV. This may indicate that contributions from nuclear structure effects have a significant impact even at rather high excitation energies [12].

As already mentioned, the first excited 2^+ states were populated by the photo-excited states. The population intensities of these states relative to the intensity of the 2_1^+ state as a function of their level energy $E_l(2_i^+)$ are shown for three beam energies E_b in the left panel of Fig. 4. All of them show an exponential behavior. Using an exponential fit function to describe the data ($\propto \exp(-\lambda \cdot E_l(2_i^+))$) the parameter λ can be extracted which serves as a signature for the corresponding function. The values for λ show a constant behavior as a function of E_b (see right side of Fig. 4). The DICEBOX simulations for both $E1$-PSF models show good agreement with the experimental data.

The combined results of both NRF experiments provide an extensive overview of the $E1$ strength distribution in ^{130}Te up to the neutron separation energy. The comparison of the experimental results to simulations in the framework of the statistical model using the DICEBOX code provides a new and deeper insight in the PSFs involved.

*This work was supported by the Alliance Program of the Helmholtz Association (HA216/EMMI), the Deutsche Forschungsgemeinschaft (SFB 634 and SO907/2-1) and by grant No. 13-07117S of the Czech Science Foundation.

3. References

[1] M. N. Harakeh and A. van der Woude, Giant Resonances (Oxford University Press, Oxford, 2001)
[2] D. Savran, T. Aumann and A. Zilges, Prog. Part. Nucl. Phys. **70**, 210 (2013)
[3] U. Kneissl, H. H. Pitz and A. Zilges, Prog. Part. Nucl. Phys. **37**, 349 (1996)
[4] K. Sonnabend *et al.*, Nucl. Instr. and Meth. Phys. Res. A **640**, 6 (2011)
[5] R. Schwengner *et al.*, Phys. Rev C **78**, 064314 (2008)
[6] D. Savran *et al.*, Phys. Rev. C **84**, 024326 (2011)
[7] H.R. Weller *et al.*, Prog. Part. Nucl. Phys. **62**, 257 (2009)
[8] N. Pietralla *et al.*, Phys Rev. Lett. **88**, 012502 (2002)
[9] J. Isaak *et al.*, Phys. Rev. C **83**, 034304 (2011)
[10] A.P. Tonchev *et al.*, Phys. Rev. Lett. **104**, 072501 (2010)
[11] C.T. Angell *et al.*, Phys. Rev. C **86**, 051302 (2012)
[12] M. Scheck *et al.*, Phys. Rev. C **87**, 051304(R) (2013)
[13] A. Leprêtre *et al.*, Nucl. Phys. A **258**, 350 (1976)
[14] F. Bečvář, Nucl. Instr. and Meth. Phys. Res. A **417**, 434 (1998)
[15] T. von Egidy, D. Bucurescu, Phys. Rev. C **80**, 054310 (2009)

Reaction rates of the ^{113}In(γ,n)^{112m}In and ^{115}In(γ,n)^{114m}In

Ye Skakun[1,3], I Semisalov[1], V Kasilov[1], V Popov[1], S Kochetov[1], V Maslyuk[2], V Mazur[2], O Parlag[2] and I Gajnish[2]

[1]NSC KIPT, Institute of High Energy and Nuclear Physics, Kharkiv, Ukraine
[2]Institute of Electron Physics, Uzhgorod, Ukraine

E-mail: skakun@kipt.kharkov.ua

Abstract. The integral yields of the ^{113}In(γ,n)^{112m}In (J^π=9/2$^+$→J^π=4$^+$) and ^{115}In(γ,n)^{114m}In (J^π=9/2$^+$→J^π=5$^+$) photonuclear reactions were measured in the bremsstrahlung end-point energy range from the respective thresholds up to 14 MeV by a conventional activation/decay technique using the ^{197}Au(γ,n)^{196}Au reaction cross sections as the standard for the absolute photon intensity determination. The metallic indium samples of the natural and enriched compositions were irradiated by the bremsstrahlung beams from thin tantalum converters of the electron linear accelerator of NSC KIPT (Kharkiv) and the microtron of IEP (Ughhorod). The integral reaction yields were determined from the activities of the nuclei-products measured by the high resolution γ-ray spectrometry technique with Ge(Li)- and HPGe-detectors. The reaction rates for the Planck spectrum of a thermal photon bath were derived for the ground state target nuclei and compared to the predictions of the statistical model of nuclear reactions.

1. Introduction

Naturally occurred weak-abundant proton-rich isotopes of a group named as *p*-nuclei can not be reached by the stellar slow (*s*) and rapid (*r*) neutron capture processes synthesizing the overwhelming majority of middle and heavy nuclei. Proton capture reactions (*p*-process) and photodisintegration of *s*- and *r*-process seed nuclei (*γ*-process) are considered as most probable mechanisms of the *p*-nuclei synthesis. The large network of low energy nuclear reaction rates (derived from the cross sections) is required for the *p*-nuclei natural abundance simulation and parametrization of the statistical theory of nuclear reactions called to supply non-measurable reaction cross sections.

The *γ*-process temperatures of T_9=2–3 (T_9 is the temperature in 10^9 Kelvin) are provided by the oxygen- and neon-rich layers of type II supernovae and the stellar reaction rates are determined by the product of the Planck spectrum of the black-body radiation characteristic for the thermal interior of a star and the photonuclear reaction cross section [1, 2]. As a result, the significant astrophysical energy range is the region of (1.5–2.0) MeV immediately above the photonuclear reaction threshold. Electron accelerators were earlier used as powerful sources of bremsstrahlung for measurements of photonuclear reaction cross sections. However over the years major portion of photonuclear reaction cross section measurements was fulfilled at larger energies.

The integral yields and derived the (γ,n)-reaction rates for the nuclei of the indium isotopes near the threshold energies would be useful for the problem of the *p*-nuclei stellar synthesis and parametrization of the statistical theory of nuclear reactions.

---

[3] To whom any correspondence should be addressed.

2.Experimental measurements

We used the photoactivation/decay measurement technique with high resolution gamma-spectrometry. The irradiation procedures were performed by using two facilities: one electron linear accelerator of NSC KIPT (Kharkiv) and the other microtron of IEP (Uzhgorod). The electron beam of Kharkiv linac was bent through an angle 15° and passing the 50-μm titanium foil striked to the 100-μm tantalum converter generating bremsstrahlung flux. Having passed the converter the electrons were bent sideward with a sweeping magnet and the Schiff distribution bremsstrahlung flux cut by the 8-mm lead collimator hit the targets under investigation. The Uzhgorod microtron did not have a photon collimator. The Kharkiv and Uzhgorod accelerators were operated with the average beam currents of (20÷30) μA and not more than 10 μA respectively.

The high-purity metallic indium disks of natural isotopic composition with masses of around 500 mg and the ^{113}In enriched (up to 66 %) foil of 60 mg·cm^{-2} surface density and 18 mm diameter were irradiated as targets. The indium targets were sandwiched with gold foils. The ^{197}Au(γ,n)^{196}Au reaction was used as a standard for determination of the photon beam intensity [3]. An ionization chamber was used for monitoring of the beam current fluctuations. The target isotopes, reactions with their thresholds and properties of the radioactive decays of the residual nuclei are summarized in Table.

Table. Target nuclei (spins-parities), reactions and spectroscopic data of the residual indium nuclei used in the data analysis [4].

Target nucleus (Spin-parity)	Reaction, Threshold, MeV	Residual nucleus	Half-life	E_γ, keV	$I_\gamma(\Delta I_\gamma)$, %
^{113}In (9/2$^+$)	(γ,n) 9.597	^{112m}In (4+) ^{112g}In (1+)	20.56 m 14.97 m	157 617	13.2 (3) 4.6
^{115}In (9/2$^+$)	(γ,n) 9.230	^{114m}In (5+) ^{114g}In (1+)	49.51 d 71.9 s	190	15.56 (15)
^{197}Au	(γ,n) 8.071	^{196}Au	6.183 d	333 356	22.9 (5) 87

The spin-parities of the ground state ^{113}In and ^{115}In target nuclei are 9/2$^+$ and the (γ,n)-reactions can result two long-lived isomers in each residual nuclei. However large difference between the spins of the target and residual nuclei ground states (see Table) does small contributions by-passed production cross sections of the 1$^+$ states of the ^{112}In and ^{114}In nuclei. So the production cross sections of the ^{113}In(γ,n)^{112m}In and ^{115}In(γ,n)^{114m}In reactions are practically the total cross sections of the (γ,n)-reactions in the near threshold energy range and it is confirmed by isomeric ratio measurements [5].

3. Results

Figure 1 shows the integral yields of the (γ,n)-reactions calculated by the activation equation for the ^{113}In and ^{115}In targets as the functions of the bombarding end point energy. The data are normalized to the ^{197}Au(γ,n)^{196}Au reaction yields [3]. The dark and light points represent the data from the Kharkiv and Uzhgorod accelerators respectively. The near threshold energy dependence of the (γ,n)-reaction cross section can be expressed [6] as the functional form $\sigma_{(\gamma,n)} = \sigma_0 (\frac{E_\gamma - E_{thr}}{E_{thr}})^{l+1/2}$, where E_γ and E_{thr} are the γ-ray and threshold energies, respectively, and l the angular momentum of the emitted neutron. It gives an opportunity to compare our results with experimental, evaluated or theoretical data of other authors if available. Having collated the integral yields of the reactions under study and the ^{197}Au(γ,n)^{196}Au reaction with parametrization [3] we found the average weighted values of the σ_0 parameter to be equal to 61 mb and 54 mb for the ^{113}In(γ,n)^{112m}In and ^{115}In(γ,n)^{114m}In reactions respectively in the case of the s- wave emission of neutrons. Figure 2 plots the fitted cross sections of these reactions together with the

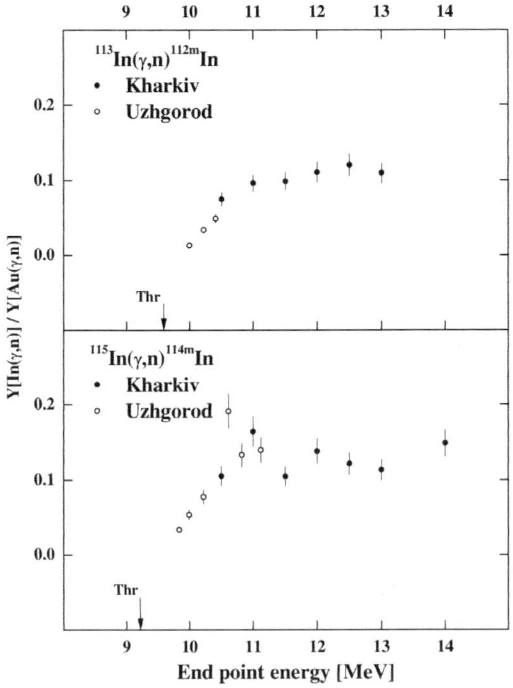

Figure 1. Integral yields of the ^{113}In(γ,n)^{112m}In (upper panel) and ^{115}In(γ,n)^{114m}In (lower panel) reactions relative to the ^{197}Au(γ,n)^{196}Au reaction ones. The dark and light points correspond to the Kharkiv and Uzhgorod measurements respectively The arrows show the reaction thresholds.

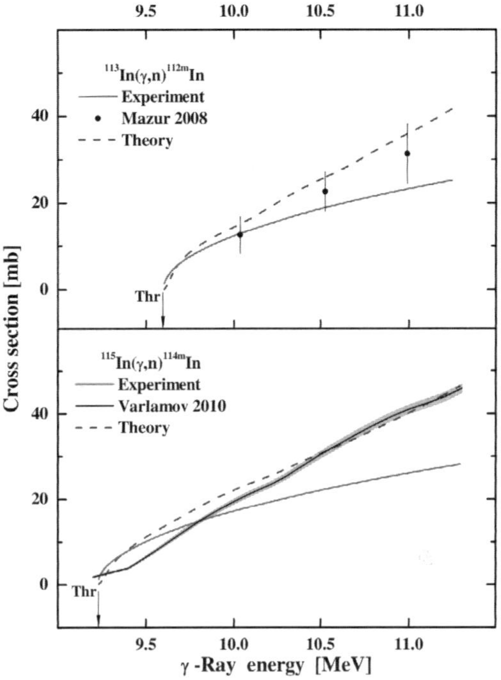

Figure 2. Fitted cross sections (red solid curves) of the ^{113}In(γ,n)^{112m}In (upper panel) and ^{115}In(γ,n)^{114m}In (lower panel) reactions deduced from our measured integral yield data, the literature data (points with error bars in upper panel and shading in lower panel) and theoretical predictions (blue dash curves).

previously evaluated and theoretical data. The red (solid) curves represent our results and the blue (dash) ones correspond to the theoretical values calculated with the NON-SMOKER code [7]. Three points with the error bars in upper panel represent the results of one of our authors (V. M.) with his co-workers [5] for the ^{113}In(γ,n)^{112m}In reaction. The shade band in lower panel depicts the cross section data evaluated by Varlamov et al [8].

From the evaluated in this way the near and above threshold (γ,n)-reaction cross sections the reaction rates $\lambda_{(\gamma,n)}$ for the ground state indium isotope nuclei in a stellar photon bath were calculated using the equation:

$$\lambda_{(\gamma,n)} = \int_0^{\infty} c n_{\gamma}(E)\sigma_{(\gamma,n)}(E)dE,$$

where c is the speed of light and $n_\gamma(E)$ the number of photons per unit volume and energy (the Planck distribution).

In addition to this conventional analysis the method of superposition of several bremsstrahlung spectra [1,2] to approximate the thermal Planck spectrum was used for determination of the reaction rates with no assumption of the (γ,n) cross section energy dependence.

The reaction rates of the ^{113}In(γ,n)^{112m}In and ^{115}In(γ,n)^{114m}In reactions derived from the experimental integral yields by the conventional analysis together with the NON-SMOKER code theoretical predictions are presented in Figure 3. The ratios of the experimental results and theoretical

Journal of Physics: Conference Series **665** (2016) 012040 doi:10.1088/1742-6596/665/1/012040

predictions are shown in the inserts of the graphs. Theory overestimates the reaction rates not more than 20 % for the first reaction and does not go out of the experimental uncertainties for the second one. The superposition method gives similar results.

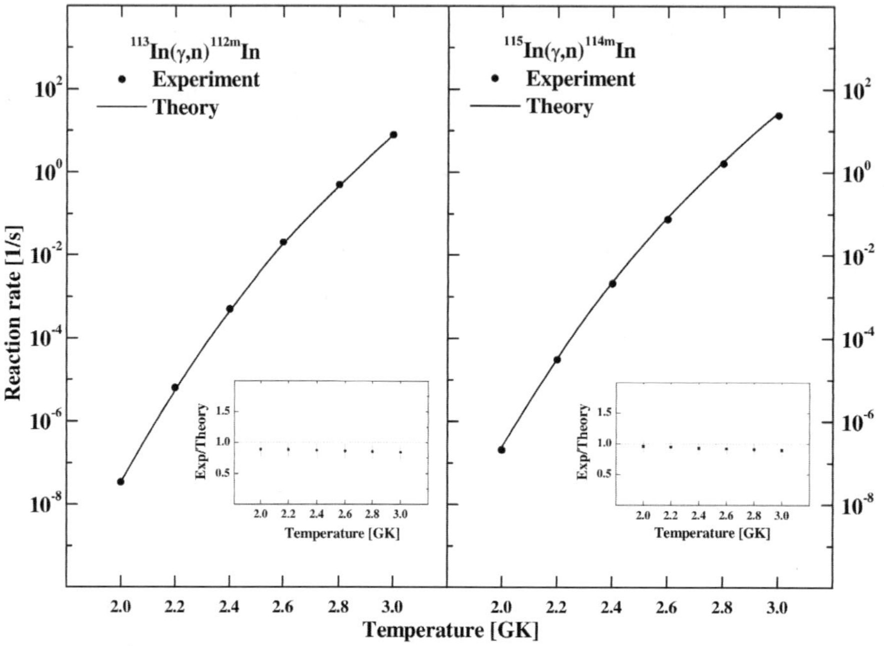

Figure 3. Experimental (points) and the NON-SMOKER code (curves) reaction rates of the ^{113}In(γ,n)^{112m}In (left panel) and ^{115}In(γ,n)^{114m}In (right panel). The inserts present the ratio of the experimental and theoretical values

References

1. Mohr P, Vogt K, Babilon M, Enders J, Hartmann T, Hutter C, Rauscher T, Volz S and Zilges A 2000 *Phys.Lett.*B **488** 127
2. Mohr P, Vogt K, Babilon M, Enders J, Hartmann T, Hutter C, Rauscher T, Volz S and Zilges A 2000 *Phys.Rev.C* **63** 055802
3. Vogt K, Mohr P, Babilon M, Bayer W, Galaviz D, Hartmann T, Hutter C, Rauscher T, Sonnabend K, Volz S and Zilges A 2002 *Nucl. Phys.* A **707** 241.
4. http://nucleardata.nuclear.lu.se/
5. Mazur V M, Bigan Z M and Symochko D M 2008 *Phys. Part. and Nucl. Lett.* **5** 374
6. Wigner E P 1948 *Phys.Rev.* **73** 1002
7. Rauscher T and F.-K. Thielmann F-K 2004 *ADNDT* **80** 1
8. Varlamov V V, Ishkhanov B S, Orlin V N and Sopov A V 2010 *Moscow State Univ.Inst.of Nucl.Phys.Reports* 8/864

Cross section measurements for γ-process studies using a LEPS detector

T Szücs[1], G G Kiss[1], T Rauscher[1,2,3], Zs Török[1], Z Halász[1,4], Zs Fülöp[1], Gy Gyürky[1] and E Somorjai[1]

[1] Institute for Nuclear Research (MTA Atomki), Debrecen, Hungary
[2] University of Hertfordshire, Hatfield, United Kingdom
[3] Department of Physics, University of Basel, Basel, Switzerland
[4] University of Debrecen, Department of Theoretical Physics, Debrecen, Hungary

E-mail: szucs.tamas@atomki.mta.hu

Abstract. In this paper we present the ongoing experiments at ATOMKI related to our systematic γ-process studies. These studies are intended to enlarge the limited experimental database from α-induced reactions on nuclei in the heavier mass range of the γ-process. In all presented cases the activation method was used. The details of the cross section measurements and preliminary results on ^{115}In(α,n)^{118m}Sb, ^{115}In(α,γ)^{119}Sb; ^{162}Er(α,n)^{165}Yb, ^{162}Er(α,γ)^{166}Yb, ^{164}Er(α,n)^{167}Yb, ^{166}Er(α,n)^{169}Yb; ^{191}Ir(α,n)^{194}Au, ^{191}Ir(α,γ)^{195}Au, ^{193}Ir(α,n)^{196m}Au, ^{193}Ir(α,n)^{196}Au reactions are presented.

1. Introduction

The stable proton-rich nuclei with charge number $Z \geq 34$ are the so called p-nuclei [1]. It is generally accepted that the main stellar mechanism synthesizing these nuclei — the so called γ-process — is initiated by (γ,n) photodisintegration reactions on preexisting neutron-rich seed nuclei [2]. As the neutron separation energy increases along the (γ,n) path towards more neutron deficient isotopes, (γ,p) and (γ,α) reactions become stronger and process the material towards lower masses. In order to understand the path of the γ-process in the region of the heavy p-nuclei and to determine precisely the p-isotope abundances, experimental cross section data of the involved reactions are clearly needed. Stellar rates for (γ,α) photodisintegration reactions should always be derived from α-capture to maximize the experimental constraint on the rate [3, 4]. Above the $A \approx 100$ region very few (α,γ) cross sections relevant for the γ-process were measured [5, 6]. This way, the γ-process network calculations rely mostly on theoretical cross sections calculated using the Hauser-Feshbach statistical model. In order to test the cross section predictions of the statistical model in the region of the heavy p-nuclei, radiative alpha-capture and (α,n) cross sections were measured at ATOMKI using a Low Energy Photon Spectrometer (LEPS) [7, 8, 9]. This is a work in progress, we are extending the available database. In this article some details about the latest reactions that are being measured in ATOMKI are presented.

2. Method and equipment

2.1. Activation method

The activation method can be applied only when the reaction product is radioactive. In general the method is the following: the samples are irradiated by an accelerated ion beam resulting in radioactive nuclei. From the produced activity, with known irradiation and sample parameters, the cross section can be derived. The limitations are caused by the decay parameters. The reaction product must have suitable half-life (usually between few minutes and few weeks) and the emitted γ-rays must have suitable branching (i. e. typically more than 10 %). With our setup part of these limitations can be avoided by detecting low energy γ- and X-rays following the decay. The X-ray detection is not able to distinguish between the activity of different isotopes of the same element, however, this problem can be overcame by following the decay curve of the X-ray yield.

2.2. Counting setup

For the measurements a LEPS (Low Energy Photon Spectrometer) detector is used which consists of a thin (15.5 mm) HPGe crystal with relatively large active surface area (20 cm^2) and a 0.5 mm thick beryllium entrance window. The low energy efficiency of this detector is similar to that of a detector with a large HPGe crystal, but it is insensitive to the high energy photons of the background. The resolution is also better than that of a big crystal. The detector is surrounded with a layered shielding consisting of 4 mm copper, 2 mm cadmium and 8 cm lead to suppress the laboratory background. Owing to these facts, the minimum detectable activity of the LEPS detector is less than that of a 100 % relative efficiency HPGe detector, for γ- or X-rays with energy lower than 500 keV.

3. Investigated reactions

α-induced reaction studies are in progress on three γ-process related isotopes, ^{115}In, ^{162}Er and ^{191}Ir. The main goal is to measure the (α,γ) reactions on the given isotopes, but additionally the (α,n) reactions on the same elements can be investigated (see table 1.). Although the latter reactions do not enter directly into the network calculations, they can constrain the statistical models, both reactions have to be described with the same parameter set.

3.1. Indium

Although the ^{115}In is not a p-isotope, the measured reactions can help to construct a global alpha-nucleus optical model potential [11], which enters into the γ-process network calculations. To test the potential was the main reason of this cross section measurement. The element indium has two stable isotopes with natural isotopic ratio of 4.29 % and 95.71 % for ^{113}In and ^{115}In, respectively. For the measurements thin targets produced by vacuum evaporation from natural indium onto thin pure aluminium foils were used. The relatively low abundant ^{113}In did not disturb the X-ray measurement, because of the low cross section of the alpha induced reactions on it, and the relatively short half-lives of the resulting reaction products.

3.2. Erbium

^{162}Er is one of the heavy p-isotopes. The lack of experimental alpha capture cross section data in this mass region calls for experimental investigations. Cross section measurement on this element in or close to the astrophysically relevant energy region is one of the next steps of our systematic studies. The natural isotopic abundance of the measured isotopes on which alpha induced reactions can be investigated with the activation method are 0.139 %, 1.601 %, 33.503 % for ^{162}Er, ^{164}Er, ^{166}Er, respectively. For the measurements thin targets were produced by vacuum evaporation onto thin pure aluminium foils both from natural erbium and from erbium enriched in ^{162}Er up to 25.8 %. From the irradiated natural targets cross sections of the (α,n) reactions on ^{164}Er, ^{166}Er have been determined. The cross section of (α,n) and (α,γ) reactions on the p-isotope have been derived from irradiation of enriched samples.

Nuclear Physics in Astrophysics VI (NPA6) IOP Publishing
Journal of Physics: Conference Series **665** (2016) 012041 doi:10.1088/1742-6596/665/1/012041

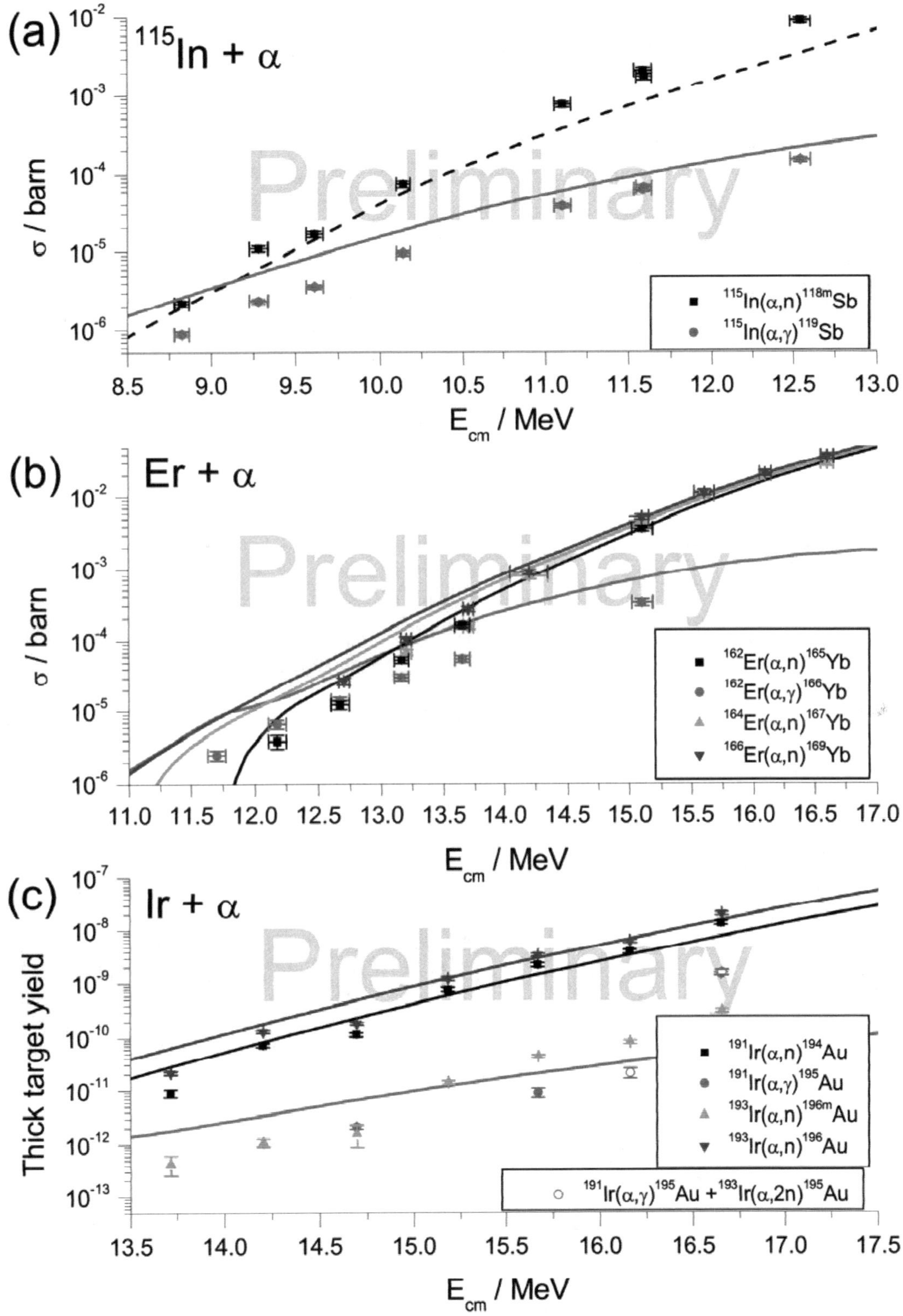

Figure 1. Preliminary results of the investigated reactions. The symbols shows the measured quantities, the curves with the same color are the respective theoretical predictions (see text).

3.3. Iridium

The element has two stable isotopes ^{191}Ir and ^{193}Ir with natural abundance of 37.3 % and 62.7 %, respectively. This measurement is a bit different from the others, because in case of the iridium, thick targets were used in which the alpha particles stops completely. The reactions take place all along the path of the alpha particles with reaction energies between zero and the initial incident energy of the projectiles. In this case from the accumulated activity the so called thick target yield can be directly determined instead of the cross section. There is an integral relation between the cross section and the thick target yield. With precise thick target measurements carried out with small enough energy steps, the two quantities can be derived from each other, but in this paper we present only the preliminary thick target yields.

Table 1. Parameters of the investigated reactions

	Indium		Erbium				Iridium			
Target isotope	^{115}In		^{162}Er		^{164}Er	^{166}Er	^{191}Ir		^{193}Ir	
Studied reaction	(α,n)	(α,γ)	(α,n)	(α,γ)	(α,n)	(α,n)	(α,n)	(α,γ)	(α,n)	
Reaction product	^{118m}Sb	^{119}Sb	^{165}Yb	^{166}Yb	^{167}Yb	^{169}Yb	^{194}Au	^{195}Au	^{196m2}Au	^{196}Au
Half-lives [10]	5.00 h	38.2 h	9.9 m	56.7 h	17.5 m	32 d	38.0 h	186 d	9.6 h	6.2 d
Detected radiation	γ, X	γ, X	γ	γ, X	γ	γ	γ	γ, X	γ	γ

4. Preliminary results

The measured quantities are shown in figure 1. for indium (a), erbium (b) and iridium (c) with the respective Hauser-Feshbach statistical model predictions (solid curves SMARAGD [12], dashed curve TALYS [13] with their default settings). The theoretical discussion of the reasons of the differences are beyond the scope of this paper.

The work is in progress. For each isotopes we are at different stage of the studies. The results will be presented in separate papers devoted to the different elements.

Acknowledgment

This work was supported by the EUROGENESIS research program, by the HUNGARIAN PORTUGUESE INTERGOVERNMENTAL S&T COOPERATION PROGRAMME NO. TÉT_10-1-2011-0458, by the European Research Council grant agreement no. 203175, by OTKA (NN83261, K101328, PD104664, K108459). G. G. Kiss acknowledges support from the János Bolyai Research Scholarship of the Hungarian Academy of Sciences.

References

[1] Burbidge M E, Burbidge G R, Fowler W A and Hoyle F 1957 *Rev. Mod. Phys.* **29** 547-650
[2] Woosley S E and Howard W M 1978 *The Astrophysical Journal Supplement* **36** 285-304
[3] Kiss G G *et al.* 2008 *Phys. Rev. Lett.* **101** 191101
[4] Rauscher T *et al.* 2009 *Phys. Rev. C* **80** 035801
[5] Rauscher T *et al. Rep. Prog. Phys.* **76**, 066201
[6] Szücs T, Dillmann I, Plag R, Fülöp Zs 2012 *J. Phys.: Conf. Series* **337**, 012033
[7] Kiss G G *et al.* 2011 *Phys. Lett. B* **695** 419-23
[8] Kiss G G *et al.* 2011 *Nucl. Phys. A* **867** 52-65
[9] Kiss G G *et al.* 2012 *Phys. Rev. C* **86** 035801
[10] `www.nndc.org`
[11] Mohr P *et al.* 2013 *At. Data Nucl. Data Tables* in press, arXiv:1212.2891
[12] Rauscher T 2011, Code SMARAGD, Version 0.8.4s
[13] Koning A J, Hilaire S and Duijvestijn M C 2008 *Proc. of ND2007*, 211

Alpha capture reaction cross section measurements on Sb isotopes by activation method

Z Korkulu[1,2], N Özkan[1], G G Kiss[2], T Szücs[2], Zs Fülöp[2], R T Güray[1], Gy Gyürky[2], Z Halász[2], E Somorjai[2], Zs Török[2], C Yalçın[1]

[1]Department of Physics, Kocaeli University, Umuttepe Kocaeli, 41380, Turkey

[2]MTA Atomki, P.O. Box. 51, H-4001 Debrecen, Hungary

E-mail: korkulu@atomki.hu

Abstract. Alpha induced reactions on natural and enriched antimony targets were investigated via the activation technique in the energy range from 9.74 MeV to 15.48 MeV, close to the upper end of the Gamow window at a temperature of 3 GK relevant to the γ-process. The experiments were carried out at the Institute for Nuclear Research, the Hungarian Academy of Sciences (MTA Atomki). ^{121}Sb(α,γ)^{125}I, ^{121}Sb(α,n)^{124}I and ^{123}Sb(α,n)^{126}I reactions were measured using a HPGe detector. In this work, the ^{121}Sb(α,n)^{124} cross section results and the comparison with the theoretical predictions (obtained with standard settings of the statistical model codes NON-SMOKER and TALYS) were presented.

1. Introduction

The main products of the last hydrostatic burning stages of massive stars are nuclei in the Fe region. For an efficient production of elements beyond Fe, reactions with neutral particles, i.e. neutrons, are required. Two main neutron capture processes, the s- and r-processes, are involved in the synthesis of nuclei heavier than the Fe group. The s-process (slow neutron capture process) is the mechanism for the formation of about half of the nuclides heavier than Fe up to Bi via neutron capture reactions and following β^--decays. The r-process (rapid neutron capture process) is thought to be responsible for the synthesis of the other half of nuclei heavier than Fe and all nuclei heavier than Bi. Since the s- and r-process build up isotopes at the bottom and on the neutron-rich side of the valley of stability, about 35 proton-rich isotopes cannot be produced in these processes.

The main production mechanism of these so-called p nuclei that are 10 to 100 times less abundant than the s and r nuclei is called the γ-process [1]. The γ-process proceeds by γ-induced reactions starting from pre-existing s- and r-process seed nuclei and requires sufficiently high temperatures (2 - 3 GK) achieved e.g. in pre-explosive or explosive O/Ne burning of massive stars, depending on the their mass [2, 3]. Another

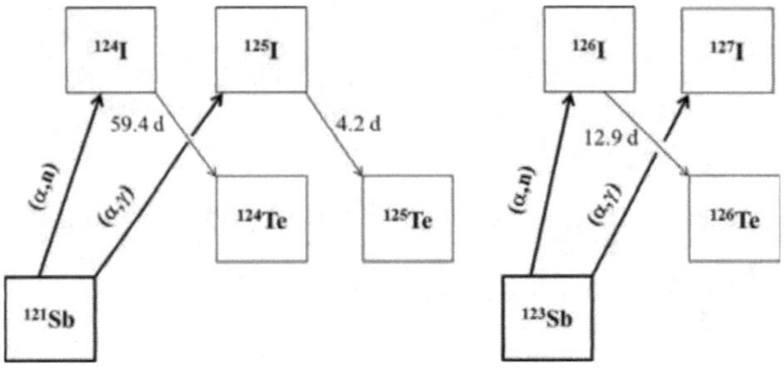

Figure 1. Alpha capture reactions on Sb. The half-lives of the decays of the reaction products are also indicated.

alternative for the astrophysical site is a subclass of type Ia supernovae where the γ-process may occur [4].

This work focuses on the extension of the experimental database for the astrophysical γ-process [5] which has problems in reproducing the p-isotope abundances e.g. in the A < 124 mass range [6], and on testing the reliability of statistical model calculations particularly in the astrophysically relevant energies.

In the following, we present the investigated reactions in Sec. 2, further details of experimental procedure in Sec. 3, and some preliminary results with a comparison to statistical model calculations in Sec. 4.

2. Investigated Reactions

The element antimony has two stable isotopes: ^{121}Sb and ^{123}Sb with the natural abundances of 57.21 % and 42.79 %, respectively. Since the ^{124}I, ^{125}I and ^{126}I isotopes are radioactive, the following reaction channels on the two stable Sb isotopes can be studied with activation method: ^{121}Sb(α,γ)^{125}I, ^{121}Sb(α,n)^{124}I and ^{123}Sb(α,n)^{126}I. These reactions are shown schematically in Fig 1. The cross section of ^{121}Sb(α,γ)^{125}I reaction has not been measured so far. Previous investigations of cross section data for ^{121}Sb(α,n)^{124}I and ^{123}Sb(α,n)^{126}I reactions have been in the literature [11, 12, 13, 14].

The ^{121}Sb(α,γ)^{125}I cross section was measured by the detection of the 35.49 keV γ-ray which follows the ^{125}I decay. At the lowest studied energies, where the yield of this γ-line was not sufficient, the cross section was determined based on the detection of the characteristic X-ray radiation following the electron capture. This is possible since the half-life of ^{125}I is 59.4 days which is much longer than that of the other two produced isotopes (4.2 days for ^{125}I and 12.9 days for ^{126}I).

The cross section of the ^{121}Sb(α,n)^{124}I and ^{123}Sb(α,n)^{126}I reactions were determined based on the detection of the relatively high intensity γ-radiations following the ε decays of the reaction products. The ^{121}Sb(α,n)^{124}I reaction cross sections presented here were measured by measuring the yield of 602.73 keV, 722.78 keV and 1690.96 keV γ-lines.

3. Experimental Method

The targets were produced by evaporating natural Sb and enriched ^{121}Sb (99.59%) in metallic form onto high purity thin aluminum foils with 12 mm in diameter. For the evaporation resistive heating of a tantalum boat was used. The enriched targets were prepared with thicknesses varying between 159 μg/cm^2 and 265 μg/cm^2. The number of target atoms was determined by weighing and Rutherford Backscattering Spectroscopy. During the irradiations the target stability was monitored by detecting the backscattered α-particles from the target. For this purpose an ion implanted silicon detector was built in the chamber at 165° with respect to the beam direction.

The irradiations were carried out at the cyclotron accelerator of the Institute for Nuclear Research, Hungarian Academy of Sciences (MTA Atomki). The length of the irradiations were about 24 hours for each target and the beam current was restricted to 1 μA in order to avoid target deterioration. The current was recorded in every minute by using a multi channel scaler to monitor the changes of the beam intensity. After each irradiation, the target was taken from the irradiation chamber to measure the induced activities. In order to measure the ^{121}Sb(α,γ)^{125}I cross section, a LEPS (Low Energy Photon Spectrometer) detector, which has a thin germanium crystal with large surface and a thin Be entrance window, was used. For (α,n) reactions, γ-rays were detected by a 100 % relative efficiency HPGe detector in an ultra low background (ULB) configuration with a commercial 4π lead shield. The γ-spectra were stored regularly (for each hour) to follow the decay of the reaction products. The absolute full energy peak efficiencies of LEPS and ULB detectors have been measured with several calibrated gamma sources.

4. Experimental Results

Cross section of the ^{121}Sb(α,γ)^{125}I, ^{121}Sb(α,n)^{124}I, and ^{123}Sb(α,n)^{126}I reactions have been measured via γ-ray counting at effective center of mass energies between 9.74 to 15.48 MeV, close to the upper end of the Gamow window (between 6.15 and 8.68 MeV at a temperature of 3 GK). Figure 2 shows the preliminary results of ^{121}Sb(α,n)^{124}I reaction cross sections. The alpha beam energy points of 10.14, 10.46, 12.58, 13.10 and 15.06 MeV were measured with energy degrader foils which eventuate the larger uncertainties. For the cross section of ^{121}Sb(α,n)^{124}I reaction some previous measurements are available [11, 12, 13, 14] (see Figure 2). Statistical model calculations obtained with standard settings of the statistical model codes NON-SMOKER [9] and TALYS [10] gives better agreement with the new results. Further measurements, precise the determination of Al thicknesses and the final analysis of data are in progress.

5. Acknowledgments

This work was supported by the ERASMUS Programme under Lifelong Learning Programme of European Commission, the Scientific and Technological Research Council

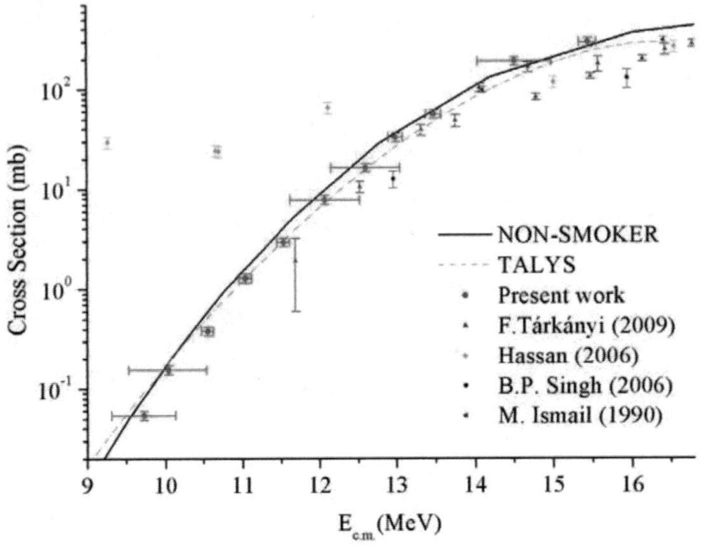

Figure 2. Preliminary experimental cross sections for ^{121}Sb(α,n)^{124}I in comparison with standard NON-SMOKER (solid line), TALYS (dashed line) predictions and with the results of previous experiments.

of Turkey (TUBITAK), Grants No. 109T585 (under the EUROCORES - EuroGENESIS research program), Kocaeli University BAP (grant 2012/026), ERC StG 203175 and OTKA K101328, K108459 and PD104664. C.Y. acknowledges the support through TUBITAK, under the programme of BIDEB-2219.

References

[1] T Rauscher, N Dauphas, I Dillmann, C Fröhlich, Zs Fülöp, Gy Gyürky *et al.* 2013 *Rep. Prog. Phys.* **76** 066201
[2] T Rauscher 2006 *Phys. Rev. C* **73** 015804
[3] W Rapp, J Görres, M Wiesher, H Schatz and F Käppeler 2006 *Astrophys. J.* **653** 474-489
[4] C Travaglio, F Roepke, W. Hillebrndt and R Gallino 2010 *Proc. Int. Symposium on Nuclear Astrophysics-Nuclei in the Cosmos-XI* PoS(NIC-XI) p 065
[5] Szücs T, Dillmann I, Plag R, Fülöp Zs 2012 *J. Phys.: Conf. Series* **337**, 012033
[6] T Rauscher, A Heger, R Hoffman and S Woosley 2002 *Astrophys. J.* **576** 323
[7] G G Kiss, T Rauscher, T Szücs, Zs Kertesz, Zs Fülöp, Gy Gyürky, C Fröhlich, J Farkas, Z Elekes and E Somorjai 2011 *Phys. Lett. B* **695** 419
[8] G G Kiss, T Szücs, Zs Török, Z Korkulu, Gy Gyürky, Z Halász, Zs Fülöp, E Somorjai and T Rauscher 2012 *Phys. Rev. C* **86** 035801
[9] NON-SMOKER http://nucastro.org/nonsmoker.html
[10] A Koning, S Hilaire and M C Duijvestijn 2005 *AIP Conf. Proc.* **769** 1154
[11] F Tarkanyi, S Takacs, B Kiraly, F Szelecsenyi, L Ando, J Bergman, S-J Heselius, O Solin, A Hermanne, Yu N Shubin, A V Ignatyuk 2009 Appl. Radiat. Isot. **67** 10011006
[12] B P Singh, M K Sharma, M M Musthafa, H D Bhardwaj, R Prasad 2006 Nucl. Inst. and Meth. A **562** 717720
[13] K F Hassan, S M Qaim, Z A Saleh, H H Coenen 2006 Appl. Radiat. Isot. **64** 101109
[14] M Ismail 1990 Phys. Rev. C **41** 87

Journal of Physics: Conference Series **665** (2016) 012043 doi:10.1088/1742-6596/665/1/012043

Investigation of Alpha-Induced Reactions on ^{107}Ag at Astrophysical Energies

C Yalçın[1,2], N Özkan[1], R T Güray[1], Gy Gyürky[2], G G Kiss[2], T Szücs[2], Z Halász[2], Zs Fülöp[2], J Farkas[2], E Somorjai[2], Z Korkulu[1] and T Rauscher[2,3,4]

[1] Kocaeli University, Department of Physics, Umuttepe 41380, Kocaeli, Turkey
[2] ATOMKI, H-4001 Debrecen, POB.51., Hungary
[3] Department of Physics, University of Basel, CH-4056 Basel, Switzerland
[4] Centre for Astrophysics Research, University of Hertfordshire, Hatfield, United Kingdom

E-mail: caner.yalcin@kocaeli.edu.tr

Abstract. Cross sections of the ^{107}Ag$(\alpha,\gamma)^{111}$In and ^{107}Ag$(\alpha$,n$)^{110}$In reactions have been measured with the activation method at effective center-of-mass energies between 7.79 MeV and 12.00 MeV close to the astrophysical energy range. The irradiation and counting of the ^{107}Ag targets was carried out at ATOMKI using the cyclotron accelerator and the low background counting facility, respectively. Cross section results are presented and compared with the predictions of Hauser-Feshbach statistical model calculations using the NON-SMOKER and TALYS-1.4 codes. In general, above 10 MeV, the model calculation are able to reproduce reasonably well the experimental data, but below 10 MeV, depending on some input parameters strong deviations are also found.

1. Introduction

The so-called astrophysical p process, which is responsible for production of about 35 rare nuclei along the proton-rich side of the stability line, is one of the least known among those processes which describe the synthesis of elements heavier than iron [1, 2, 3] The p process nucleosynthesis is modeled by using an extended nuclear reaction network, for which reaction rate information of thousands of neutron, proton and α-induced reactions as well as their inverse reactions is needed. Astrophysical reaction rates are mostly derived from theoretical cross sections owing to the lack of experimental data. On the other hand, experiments performed so far have shown that there are considerable differences between predictions and experimental cross sections [4, 5, 6]. In order to extend the experimental database for the astrophysical p process and to test the reliability of statistical model predictions in this mass range, cross sections of ^{107}Ag$(\alpha,\gamma)^{111}$In and ^{107}Ag$(\alpha$,n$)^{110}$In reactions have been measured with the activation method at effective center-of-mass energies between 7.79 MeV and 12.00 MeV, close to the astrophysical energy range (the Gamow window at 3.0 GK is between 5.83 MeV and 8.39 MeV).

2. Experimental Method

Since reaction products are radioactive and their half-lives are convenient, the activation method was used to determine the reaction cross sections. Detailed information about activation method can be found at reference [4].

Table 1. Decay parameters of reaction products taken from [7, 8]. Only the γ-transitions used for the analysis are listed.

Reaction	Half-life	E_γ (keV)	I_γ (%)
$^{107}Ag(\alpha,\gamma)^{111}In$	2.8047 ± 0.0004 d	171.28	90.7 ± 0.9
$^{107}Ag(\alpha,n)^{110g}In$	4.92 ± 0.08 h	641.68	26.0 ± 0.8
		937.16	68.4 ± 1.9
$^{107}Ag(\alpha,n)^{110m}In$	69.1 ± 0.5 min	2129.40	2.15 ± 0.03
		2211.33	1.74 ± 0.03

Natural Ag and isotopically enriched (99.50 %) ^{107}Ag targets were produced by evaporating onto high purity thin aluminum foils. Thickness of the targets were determined with weight measurement and varying between 410 $\mu g/cm^2$ and 1042 $\mu g/cm^2$. The targets were irradiated between 8.16 MeV and 12.50 MeV laboratory energies at the ATOMKI cyclotron accelerator. Because the cyclotron could not produce all beam energies directly, energies of $E_{Lab.}$ = 8.16, 8.51, 10.00, 10.50 MeV were obtained with an energy degrader foil. The thickness of the energy degrader foils was determined by the energy loss of the α particles emitted from ^{241}Am. The typical beam current was between 150 nA and 800 nA. During the irradiation, targets were monitored with a ion implanted Si detector placed into the target chamber at 165° relative to the beam direction and no target deterioration was observed.

After each irradiation, the targets were taken from the reaction chamber and placed into a low-background counting area to measure ^{111}In and ^{110}In activities, which are produced through the $^{107}Ag(\alpha,\gamma)^{111}In$ and $^{107}Ag(\alpha,n)^{110}In$ reactions, respectively. Depending on the count rate of reaction products the target was placed at 10 cm or 1 cm from the end cap of a HPGe detector having 100 % relative efficiency. A typical γ spectrum taken after 8.7 h irradiation with a 10 MeV α beam is shown in Fig. 1. The absolute efficiency calibration of the detection system was done at

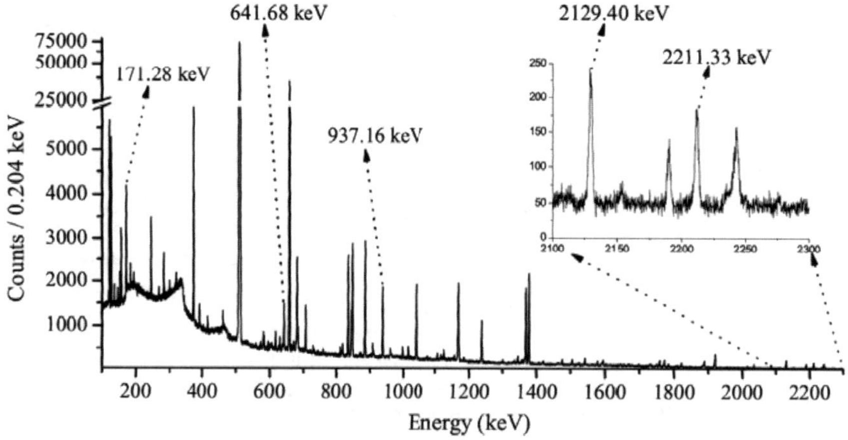

Figure 1. Activation γ spectrum taken after 8.7 h irradiation of a target with a 10 MeV α beam. The γ-lines used for the analysis are indicated in the spectrum. The other peaks are from either laboratory background or other γ-transitions (1 Channel=0.204 keV).

Journal of Physics: Conference Series **665** (2016) 012043 doi:10.1088/1742-6596/665/1/012043

27 cm detector-target distance, where the coincidence-summing effect is negligible. Calibrated multiline ^{57}Co, ^{60}Co, ^{133}Ba, ^{152}Eu, ^{241}Am and single line ^{7}Be, ^{22}Na, ^{54}Mn, ^{60}Co, ^{65}Zn, and ^{137}Cs sources were used for the calibration. The efficiencies at the measurement distances 1 cm and 10 cm were determined by scaling the efficiencies obtained at 27 cm. Experimentally found coincidence correction factors were applied to the efficiency for each γ-line which was used for the analysis (see Table 1) [9].

The (α,n) reactions of ^{107}Ag populated the ground state ($T_{1/2}$ = 4.92 h) and isomeric state ($T_{1/2}$ = 69.1 min) of ^{110}In. The total cross section of the ^{107}Ag(α,n)^{110}In reaction was determined by taking the sum of the partial cross sections of ^{107}Ag(α,n)^{110g}In and ^{107}Ag(α,n)^{110m}In measured independently. Cross section of the ground and isomeric states were found by weighed average of the different gamma lines at Table 1.

3. Results and Discussion

The ^{107}Ag(α,γ)^{111}In and ^{107}Ag(α,n)^{110}In reaction cross sections have been measured at laboratory energies between 8.16 MeV and 12.50 MeV, which include a part of the astrophysically relevant energy range. Laboratory energies have been converted into effective center-of-mass energies ($E^{\rm eff}_{\rm c.m.}$) that correspond to beam energies in the target at which one-half of the yield for the full target thickness is obtained [10]. A comparison of the results with the theoretical calculation using NON-SMOKER [11, 12, 13] and TALYS-1.4 [14, 15] codes is shown in Fig. 2. Previous results from Baglin [16] and Stelson [17] also included in the figure. Two calculations with the NON-SMOKER code are shown, using level densities obtained with microscopic corrections from two different theoretical mass tables (FRDM [18] and ETFSI [19], see [20] for details) but the same optical α+nucleus potential of McFadden-Satchler [21]. In

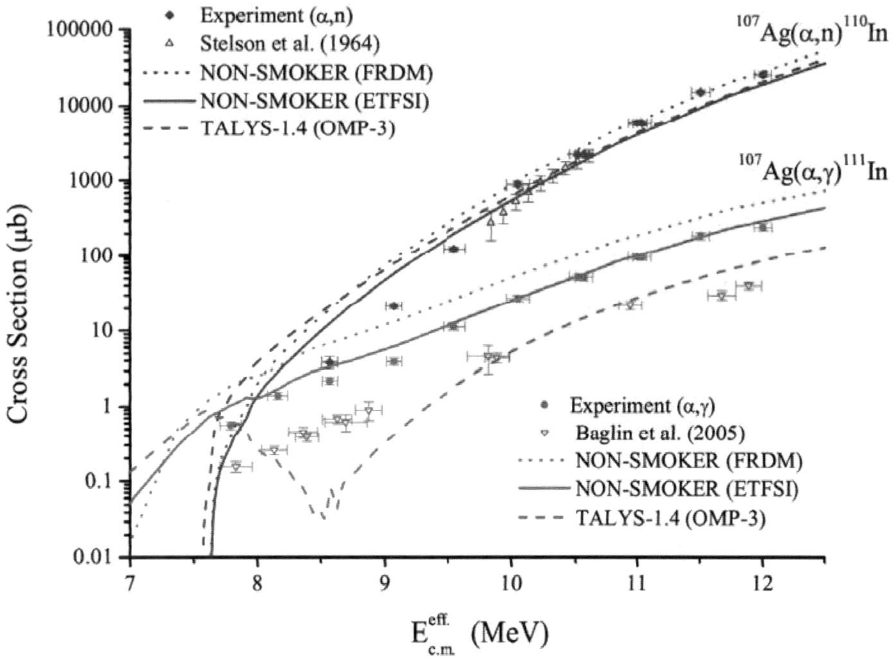

Figure 2. (Color online) Measured cross sections of the ^{107}Ag(α,γ)^{111}In and ^{107}Ag(α,n)^{110}In reactions compared to theory using the NON-SMOKER[11, 12, 13] and TALYS[14, 15] code. Previous results from Baglin [16] and Stelson [17] are also shown.

the TALYS-1.4 calculations, because the phenomenological model for the α break-up reaction is not stable in the low-energy region of the experiment [22], the adjustable parameter for the break-up process (Cbreak) in the (α,n) and (α,p) channels was set to zero and the double folding potential of Demetriou, Grama and Goriely (Table 1 in Ref. [23]) was used.

We conclude that for the ^{107}Ag(α, γ)^{111}In reaction, experimental data are acceptably well described by the NON-SMOKER code when the ETFSI input is used for the calculation, on the other hand the calculation with the FRDM input overestimates the cross sections but the energy dependence is correctly described. These differences in the two calculations stem from the differences in the γ widths obtained with the different level densities. Above 10 MeV, the cross sections predicted with TALYS-1.4 are lower than the experimental results by factors of between 2.8 and 4.5. Below 10 MeV, the energy dependence of the cross section is strongly different and thus around 8.50 MeV the discrepancy between experimental and theoretical results increase up to factors of 60.

For the ^{107}Ag(α,n)^{110}In reaction, all theoretical calculations acceptably reproduce the measured cross sections above 10 MeV but in the same way all calculations overestimate the values below 10 MeV. The deviations from the experimental results are gradually increasing with decreasing energy.

Further details about the experiment and the full analysis will be presented in a forthcoming publication.

Acknowledgments

This work was supported by the Scientific and Technological Research Council of Turkey (TUBITAK), Grants No. 109T585 (under the EUROGENESIS research program) and Grants No. 108T508 and also supported by ERC St.G. 203175 and OTKA (K101328, K108459, PD104664). C Yalçın, acknowledges support through the TUBITAK, under the programme of BIDEB-2219. GG Kiss acknowledges support from the János Bolyai grant.

References

[1] Woosley S E and Howard W M 1978 *Astrophys. J. Suppl. Ser.* **36** 285
[2] Rauscher T, Dauphas N, Dillmann I, Fröhlich C, Fülöp Zs and Gyürky Gy 2013 *Rep. Prog. Phys.* **76** 066201
[3] Arnould M and Goriely S 2003 *Phys. Rep.* **384** 1
[4] Özkan N et al. 2002 *Nucl. Phys. A* **710** 469
[5] Yalçın C et al. 2009 *Phys. Rev. C* **79** 065801
[6] Gyürky Gy et al. 2010 *J. Phys. G.* **37** 115201
[7] Blachot J 2009 *Nuclear Data Sheets* **110** 1239
[8] Gürdal G and Kondev F G 2012 *Nuclear Data Sheets* **113** 1315
[9] Debertin K and Helmer R G 1989 *Gamma- And X-ray Spectrometry With Semiconductor Detectors* (Amsterdam: North-Holland)
[10] Rolfs C E and Rodney W S 1988 *Cauldrons in the Cosmos* (Chicago: The University of Chicago Press)
[11] Rauscher T and Thielemann F K 2000 *At. Data Nucl. Data Tables* **75** 1
[12] Rauscher T and Thielemann F K 2001 *At. Data Nucl. Data Tables* **79** 47
[13] Rauscher T, NON-SMOKER code, http://nucastro.org/nonsmoker.html
[14] Koning A J, Hilaire S and Duijvestijn M C 2008 *Proc. of the Int. Conf. on Nuclear Data for Science and Technology (France)* ed Bersillon O et al. (Les Ulis: EDP Sciences) p 211
[15] Koning A J, Hilaire S and Duijvestijn M C, TALYS code, http://www.talys.eu
[16] Baglin C M et al. 2005 *AIP Conf. Proc.* **769** 1370
[17] Stelson P H and McGowan F K 1964 *Phys. Rev.* **133** B911
[18] Mller P, Nix J R, Myers W D and Swiatecki W 1995 *At. Data Nucl. Data Tables* **59** 185
[19] Aboussir Y, Pearson J M, Dutta A K and Tondeur F 1995 *At. Data Nucl. Data Tables* **61** 127
[20] Rauscher T, Kratz K-L and Thielemann F-K 1997 *Phys. Rev. C* **56** 1613
[21] McFadden L, Satchler G R 1966 *Nucl. Phys.* **84** 177
[22] Koning A J (private communication)
[23] Demetriou P, Grama C and Goriely S 2002 *Nucl. Phys. A* **707** 253

Nuclear astrophysics with radioactive ions at FAIR

R Reifarth[1], S Altstadt[1,2], K Göbel[1], T Heftrich[1,2], M Heil[2], A Koloczek[1,2], C Langer[1,2,60], R Plag[1,2], M Pohl[1], K Sonnabend[1], M Weigand[1], T Adachi[47], F Aksouh[5], J Al-Khalili[18], M AlGarawi[5], S AlGhamdi[5], G Alkhazov[3], N Alkhomashi[4], H Alvarez-Pol[6], R Alvarez-Rodriguez[7], V Andreev[3], B Andrei[8], L Atar[2], T Aumann[9], V Avdeichikov[10], C Bacri[11], S Bagchi[47], C Barbieri[18], S Beceiro[6], C Beck[12], C Beinrucker[1], G Belier[13], D Bemmerer[14], M Bendel[15], J Benlliure[6], G Benzoni[16], R Berjillos[17], D Bertini [2], C Bertulani[19], S Bishop[15], N Blasi[85], T Bloch[9], Y Blumenfeld[21], A Bonaccorso[22], K Boretzky[2], A Botvina[23], A Boudard[24], P Boutachkov[9], I Boztosun[25], A Bracco[85], S Brambilla[85], J Briz Monago[26], M Caamano[6], C Caesar[2], F Camera[85], E Casarejos[27], W Catford[18], J Cederkall[10], B Cederwall[28], M Chartier[31], A Chatillon[13], M Cherciu[32], L Chulkov[33], P Coleman-Smith[34], D Cortina-Gil[6], F Crespi[16], R Crespo[35], J Cresswell[31], M Csatlós[36], F Déchery[24], B Davids[37], T Davinson[38], V Derya[39], P Detistov[40], P Diaz Fernandez[6], D DiJulio[10], S Dmitry[33], D Doré[24], J Dueñas[17], E Dupont[24], P Egelhof[2], I Egorova[8], Z Elekes[14], J Enders[9], J Endres[39], S Ershov[8], O Ershova[1], B Fernandez-Dominguez[6], A Fetisov[3], E Fiori[41], A Fomichev[8], M Fonseca[1], L Fraile[7], M Freer[42], J Friese[15], M G. Borge[26], D Galaviz Redondo[44], S Gannon[31], U Garg[41,84], I Gasparic[9,86], L Gasques[45], B Gastineau[24], H Geissel[2], R Gernhäuser[15], T Ghosh[20], M Gilbert[1], J Glorius[1], P Golubev[10], A Gorshkov[8], A Gourishetty[46], L Grigorenko[8], J Gulyas[36], M Haiduc[32], F Hammache[11], M Harakeh[47], M Hass[48], M Heine[9], A Hennig[39], A Henriques[44], R Herzberg[31], M Holl[9], A Ignatov[9], A Ignatyuk[50], S Ilieva[9], M Ivanov[40], N Iwasa[51], B Jakobsson[10], H Johansson[49], B Jonson[49], P Joshi[52], A Junghans[14], B Jurado[53], G Körner[54], N Kalantar[47], R Kanungo[55], A Kelic-Heil[2], K Kezzar[5], E Khan[11], A Khanzadeev[3], O Kiselev[2], M Kogimtzis[34], D Körper[2], S Kräckmann[1], T Kröll[9], R Krücken[15], A Krasznahorkay[36], J Kratz[56], D Kresan[9], T Krings[57], A Krumbholz[9], S Krupko[8], R Kulessa[58], S Kumar[59], N Kurz[2], E Kuzmin[33], M Labiche[34], K Langanke[2], I Lazarus[34], T Le Bleis[15], C Lederer[1], A Lemasson[60], R Lemmon[34], V Liberati[30], Y Litvinov[2], B Löher[41], J Lopez Herraiz[7], G Münzenberg[2], J Machado[44], E Maev[3], K Mahata[61], D Mancusi[24], J Marganiec[41], M Martinez Perez[7], V Marusov[39], D Mengoni[63], B Million[85], V Morcelle[64], O Moreno[7], A Movsesyan[9], E Nacher[26], M Najafi[47], T Nakamura[65], F Naqvi[66], E Nikolski[33], T Nilsson[49], C Nociforo[2], P Nolan[31], B Novatsky[33], G Nyman[49], A Ornelas[44],

R Palit[67], S Pandit[61], V Panin[9], C Paradela[6], V Parkar[17], S Paschalis[9], P Pawłowski[62], A Perea[26], J Pereira[60], C Petrache[68], M Petri[9], S Pickstone[39], N Pietralla[9], S Pietri[2], Y Pivovarov[69], P Potlog[32], A Prokofiev[70], G Rastrepina[1,2], T Rauscher[72], G Ribeiro[26], M Ricciardi[2], A Richter[9], C Rigollet[47], K Riisager[43], A Rios[18], C Ritter[1], T Rodríguez Frutos[2], J Rodriguez Vignote[26], M Röder[14,71], C Romig[9], D Rossi[2,60], P Roussel-Chomaz[24], P Rout[61], S Roy[67], P Söderström[73], M Saha Sarkar[29], S Sakuta[33], M Salsac[24], J Sampson[31], J Sanchez del Rio Saez[26], J Sanchez Rosado[26], S Sanjari[1], P Sarriguren[26], A Sauerwein[1], D Savran[41], C Scheidenberger[2], H Scheit[9], S Schmidt[1], C Schmitt[74], L Schnorrenberger[9], P Schrock[9], R Schwengner[14], D Seddon[31], B Sherrill[60], A Shrivastava[61], S Sidorchuk[8], J Silva[41], H Simon[2], E Simpson[18], P Singh[2], D Slobodan[75], D Sohler[36], M Spieker[39], D Stach[14], E Stan[32], M Stanoiu[76], S Stepantsov[8], P Stevenson[18], F Strieder[77], L Stuhl[36], T Suda[51], K Sümmerer[2], B Streicher[2], J Taieb[13], M Takechi[2], I Tanihata[78], J Taylor[31], O Tengblad[26], G Ter-Akopian[8], S Terashima[79], P Teubig[44], R Thies[49], M Thoennessen[60], T Thomas[1], J Thornhill[31], G Thungstrom[80], J Timar[36], Y Togano[41], U Tomohiro[73], T Tornyi[36], J Tostevin[18], C Townsley[18], W Trautmann[2], T Trivedi[67], S Typel[2], E Uberseder[84], J Udias[7], T Uesaka[73], L Uvarov[3], Z Vajta[36], P Velho[44], V Vikhrov[3], M Volknandt[1], V Volkov[33], P von Neumann-Cosel[9], M von Schmid[9], A Wagner[14], F Wamers[9], H Weick[2], D Wells[31], L Westerberg[81], O Wieland[16], M Wiescher[84], C Wimmer[1], K Wimmer[60], J S Winfield[2], M Winkel[15], P Woods[38], R Wyss[28], D Yakorev[14], M Yavor[82], J Zamora Cardona[9], I Zartova[83], T Zerguerras[11], I Zgura[32], A Zhdanov[3], M Zhukov[49], M Zieblinski[62], A Zilges[39], K Zuber[71]

[1] University of Frankfurt, Germany
[2] GSI Darmstadt, Germany
[3] PNPI Gatchina, Russia
[4] Atomic Energy Research Institute, Saudi Arabia
[5] King Saudi University, Saudi Arabia
[6] University of Santiago de Compostela, Spain
[7] Universidad Complutense de Madrid, Spain
[8] JINR Dubna, Russia
[9] TU Darmstadt, Germany
[10] Lund University, Sweden
[11] IPN Orsay, France
[12] IPHC - CNRS/UdS Strasbourg, France
[13] CEA Bruyères le Châtel, France
[14] Helmholtz-Zentrum Dresden-Rossendorf, Germany
[15] Technische Universität München, Germany
[16] INFN Milano, Italy
[17] University of Huelva, Spain
[18] University of Surrey, United Kingdom
[19] Texas A&M University-Commerce, USA
[20] VECC Kolkata, India
[21] CERN, Switzerland
[22] INFN Pisa, Italy
[23] INR RAS Moscow, Russia
[24] CEA Saclay, France
[25] Akdeniz University, Turkey

[26] CSIC Madrid, Spain
[27] Universidad de Vigo, Spain
[28] Royal Institute of Technology KTH Stockholm, Sweden
[29] SINP Kolkata, India
[30] University of the West of Scotland, United Kingdom
[31] University of Liverpool, United Kingdom
[32] Institute of Space Sciences, Romania
[33] NRC Kurchatov Institute Moscow, Russia
[34] STFC Daresbury Laboratory, United Kingdom
[35] Instituto Superior Tecnico, Portugal
[36] ATOMKI Debrecen, Hungary
[37] TRIUMF, Canada
[38] University of Edinburgh, United Kingdom
[39] University of Cologne, Germany
[40] INRNE BAS Sofia, Bulgaria
[41] Extreme Matter Institute/GSI Darmstadt, Germany
[42] University of Birmingham, United Kingdom
[43] University of Aarhus, Denmark
[44] University of Lisboa, Portugal
[45] University of Sao Paulo, Brazil
[46] Indian Institute of Technology Roorkee, India
[47] KVI/University of Groningen, The Netherlands
[48] The Weizmann Institute of Science Rehovot, Israel
[49] Chalmers University of Technology, Sweden
[50] IPPE Obninsk, Russia
[51] University of Tohoku, Japan
[52] University of York, United Kingdom
[53] CENBG, France
[54] NuPECC, Europe
[55] Saint Mary University, Canada
[56] University of Mainz, Germany
[57] SEMIKON Detector GmbH, Germany
[58] Jagiellonian University of Krakow, Poland
[59] University of Delhi, India
[60] NSCL/MSU, USA
[61] BARC Mumbai, India
[62] IFJ PAN Krakow, Poland
[63] University of Padova, Italy
[64] Federal Fluminense University, Brazil
[65] Tokyo Institute of Technology, Japan
[66] University of Yale, USA
[67] TIFR Mumbai, India
[68] CSNSM Orsay, France
[69] Polytechnic University of Tomsk, Russia
[70] The Svedberg Laboratory, Sweden
[71] TU Dresden, Germany
[72] University of Basel, Switzerland
[73] RIKEN, Japan
[74] GANIL, France
[75] ESS Bilbao, Spain
[76] IFIN-HH Bucharest, Romania
[77] Ruhr University Bochum, Germany
[78] RCNP Osaka, Japan
[79] Beihang University, China
[80] Mid Sweden University, Sweden
[81] Uppsala University, Sweden
[82] IAI RAS St. Petersburg, Russia
[83] University of Stockholm, Sweden
[84] University of Notre Dame, USA

[85]INFN Rome, Italy
[86]RBI, Zagreb, Croatia

E-mail: reifarth@physik.uni-frankurt.de

Abstract. The nucleosynthesis of elements beyond iron is dominated by neutron captures in the *s* and *r* processes. However, 32 stable, proton-rich isotopes cannot be formed during those processes, because they are shielded from the *s*-process flow and *r*-process β-decay chains. These nuclei are attributed to the *p* and *rp* process.

For all those processes, current research in nuclear astrophysics addresses the need for more precise reaction data involving radioactive isotopes. Depending on the particular reaction, direct or inverse kinematics, forward or time-reversed direction are investigated to determine or at least to constrain the desired reaction cross sections.

The Facility for Antiproton and Ion Research (FAIR) will offer unique, unprecedented opportunities to investigate many of the important reactions. The high yield of radioactive isotopes, even far away from the valley of stability, allows the investigation of isotopes involved in processes as exotic as the *r* or *rp* processes.

1. Introduction

Radioactive beams offer the opportunity to extend the experimentally based knowledge about nuclear structure far beyond the valley of stability. Especially within the planned international Facility for Antiproton and Ion Research (FAIR) at GSI [1], radioactive ions will be produced with highest intensities. It is very often not feasible to collect the respective radioactive ions in order to produce a sample for irradiation with e.g. neutrons, protons or gammas. Experiments in inverse kinematics - irradiating a stable target with the desired radioactive ions - are the solution to that problem. In this article, we will mostly report about performed and upcoming in-beam experiments and not about the wide field of possible ring experiments.

The proposed R^3B setup [2], a universal setup for kinematically complete measurements of Reactions with Relativistic Radioactive Beams will cover experimental reaction studies with exotic nuclei far off stability, with emphasis on nuclear structure and dynamics. Astrophysical aspects and technical applications are also concerned. R^3B is a versatile reaction setup with high efficiency, acceptance, and resolution for reactions with high-energy radioactive beams. The setup will be located at the High Energy Cave which follows the high-energy branch of the new fragment separator (Super-FRS). The experimental configuration is based on a concept similar to the existing LAND setup at GSI introducing substantial improvement with respect to resolution and an extended detection scheme, which comprises the additional detection efficiency of light (target-like) recoil particles and a high-resolution fragment spectrometer. The setup is adapted to the highest beam energies (corresponding to 20 Tm magnetic rigidity) provided by the Super-FRS capitalizing on the highest possible transmission of secondary beams. The experimental setup is suitable for a wide variety of scattering experiments, such as heavy-ion induced electromagnetic excitation, knockout and breakup reactions, or light-ion (in)elastic and quasi-free scattering in inverse kinematics, thus enabling a broad physics program with rare-isotope beams to be performed [2].

Applying the Coulomb dissociation method [3, 4] R^3B contributes already now to almost every astrophysical scenario. With the expected increase in the production of radioactive species at FAIR, even more exotic reactions can be investigated. Both, the current situation and the prospects at FAIR are shown in section 2 for the *s* process, in section 4 for the *r* process and in section 5 for the rp process. Recent experiments and future prospects investigating charge exchange reactions are discussed in section 3 for the *s* process and section 6 for the νp process.

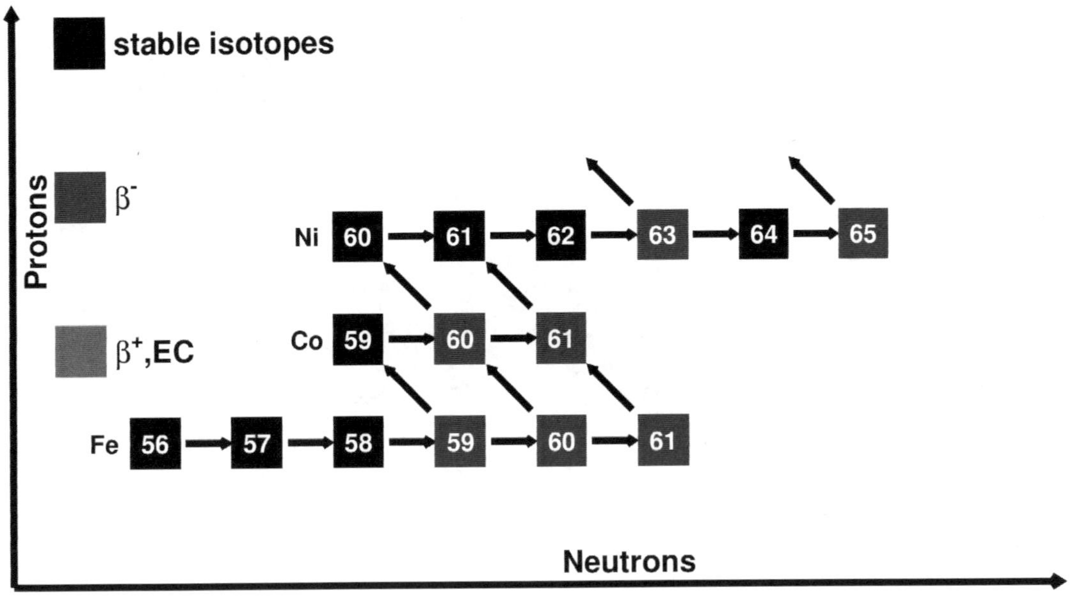

Figure 1. The *s*-process path between Fe and Co. The neutron densities during the *s* process have to be sufficiently high to overcome the rather short-lived isotope ^{59}Fe ($t_{1/2} = 45$ d).

2. ^{60}Fe - a product of the *s* process

A significant contribution to the interstellar abundance of the radiogenic ^{60}Fe is provided by the slow neutron capture (s) process in massive stars, the weak component of the *s* process. The s process in massive stars operates in two major evolutionary stages, first during convective core He-burning and, subsequently, during convective shell C-burning. Neutrons are mainly produced by the ^{22}Ne(α,n) reaction in both cases, but at rather different temperatures and neutron densities [5, 6].

As illustrated in Figure 1, the *s*-process path to ^{60}Fe, which starts from the most abundant seed nucleus ^{56}Fe, is determined by the branching at ^{59}Fe ($t_{1/2} = 44.5$ d). At the low neutron densities during convective core He burning, ^{60}Fe is shielded from the *s*-process chain, because the β^--decay rate of ^{59}Fe dominates over the (n,γ) rate by orders of magnitude. On the other hand, the production of ^{60}Fe becomes efficient during the shell C-burning phase, where higher temperatures of $T = (1.0 - 1.4) \cdot 10^9$ K give rise to the neutron densities in excess of 10^{11} cm^{-3} necessary for bridging the instability gap at ^{59}Fe. The interpretation of all the above observations depends critically on the reliability of the stellar models as well as on the reaction rates for neutron capture relevant to the production and depletion of ^{60}Fe [7]. These rates can only be determined reliably in laboratory experiments, because theoretical calculations are too uncertain. Since the neutron capture cross section of ^{60}Fe(n,γ) has been measured in the astrophysically interesting energy region [8] and a reliable value for the half life of the β^--decay has been provided [9], the most important missing piece to understand the stellar production of ^{60}Fe is the ^{59}Fe(n,γ) cross section under stellar conditions, see Figure 1.

Because the half-life of ^{59}Fe is only 45 d, indirect methods have to be applied to determine the neutron capture cross section. Since ^{60}Fe is unstable too, the method of choice is the determination of the desired A(n,γ)B via the inverse reaction B(γ,n)A applying the Coulomb dissociation (CD) method at the LAND/R^{3}B setup (Figure 2).

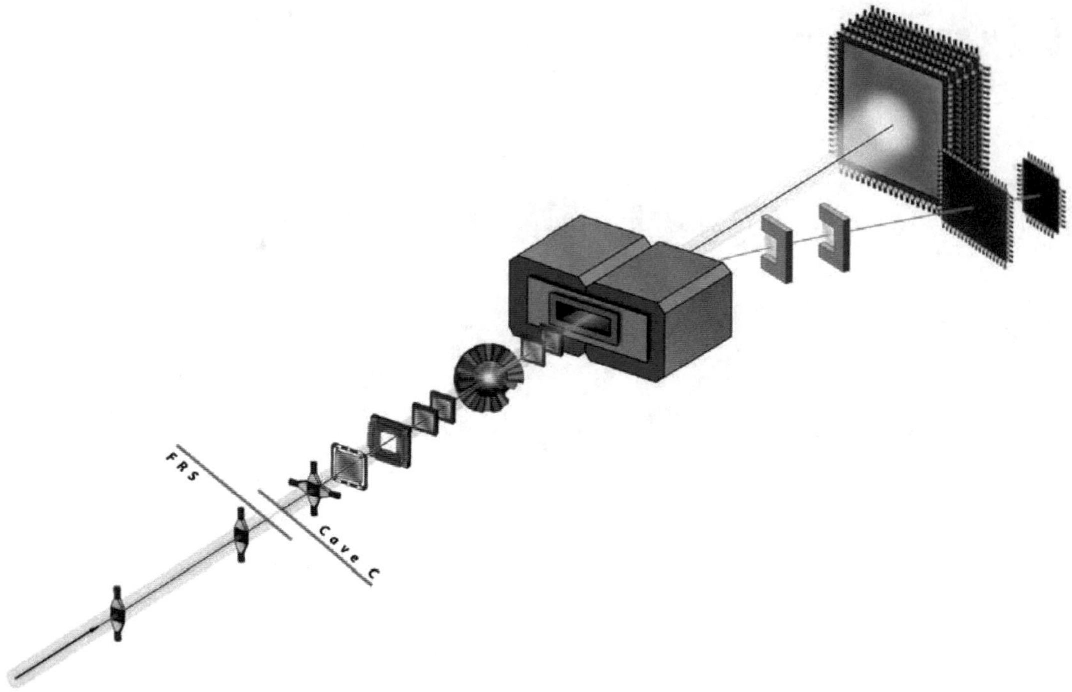

Figure 2. The R^3B setup at GSI optimized for detecting neutrons in the exit channel. The beam enters Cave C after passing the FRS (bottom, left), passes several detectors used for incoming identification and hits the target in the center of th crystal ball (half of the ball is drawn). Afterwards charged fragment and neutrons are separated by the ALADIN magnet. The fragments are bent to the right analysed with a suite of scintillator detectors while the neutrons remain unchanged and are detected with LAND.

^{60}Fe ions have been produced by fragmentation of ^{64}Ni. After passing the fragment separator (FRS, [10]) most of the unwanted species are removed and a beam consisting of almost only ^{60}Fe arrives in at the LAND/R^3B setup, where each ion is identified in charge and mass, Figure 3.

All reaction products are detected and characterised in terms of charge, mass and momentum. This allows the investigation of the neutron removal of ^{60}Fe for different experimetal settings, Figure 4 (left). A lead target was used to determine the Coulomb breakup cross section, while runs with carbon and no sample at all have been performed to determine different background components. After scaling and subtracting the contribution from nuclear interaction (measured with the carbon target) as well as interaction with other components in the beam line (empty), the pure Coulomb breakup can be extracted, Figure 4 (right).

Under certain conditions, stars may experience convective-reactive nucleosynthesis episodes. It has been shown with hydrodynamic simulations that neutron densities in excess of 10^{15} cm^3 can be reached [11, 12], if unprocessed, H-rich material is convectively mixed with an He-burning zone. Under such conditions, which are between the s and r process, the reaction flow occurs a few mass units away from the valley of stability. These conditions are sometimes referred to as the i process (intermediate process). One of the most important rates, but extremely difficult to determine, is the neutron capture on ^{135}I, Figure 5. The half-life time of ^{135}I is about 6 h. Therefore the ^{135}I(n,γ) cross section cannot be measured directly. The much improved

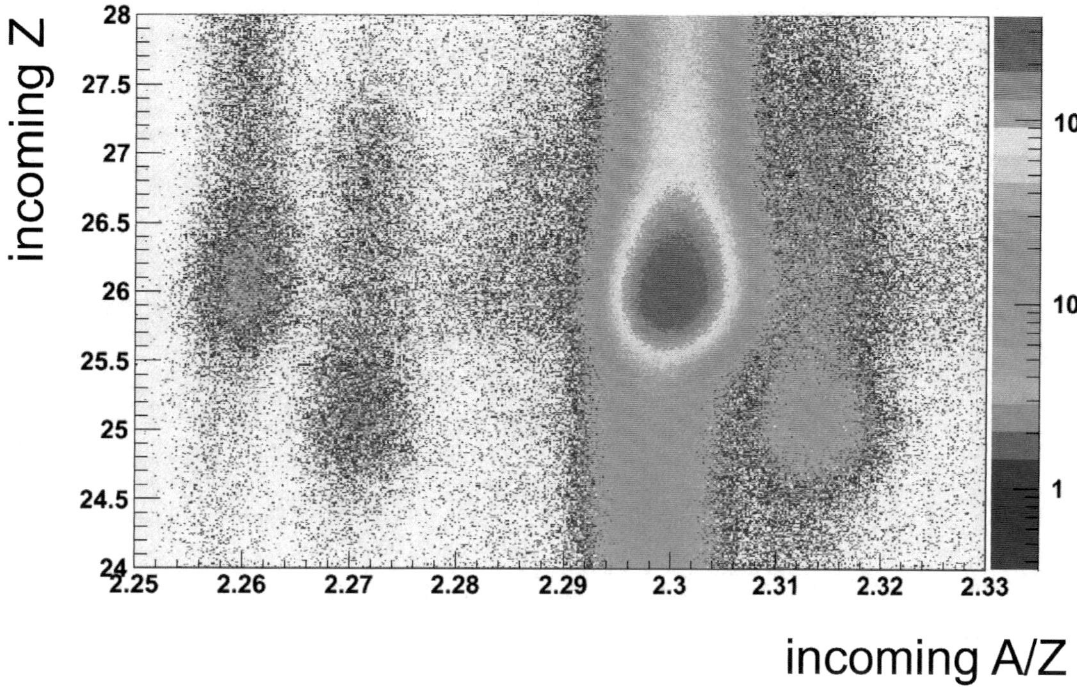

Figure 3. Incoming particle identification at the R^3B setup.

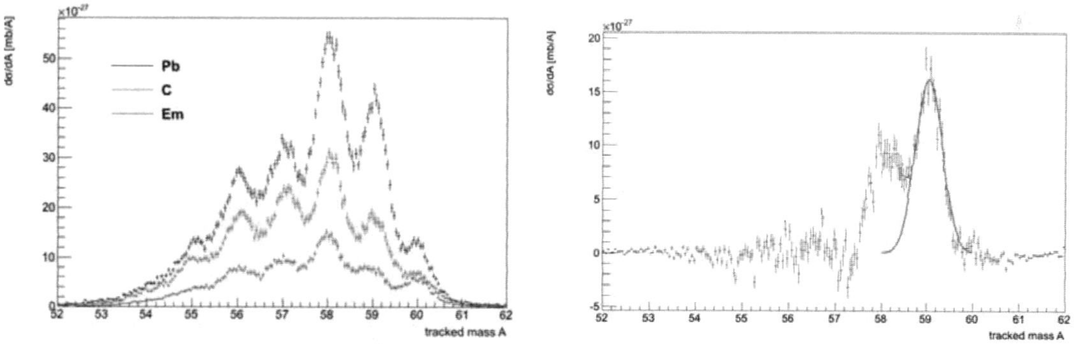

Figure 4. Left: Mass identification of different ion isotopes after requiring an incoming ^{60}Fe ions a a neutron detected by LAND. Right: Spectra shown on the left subtracted such that only the mass distribution of Coulomb breakup events are left.

production rates of radioactive isotopes at FAIR, however, offer the possibility to investigate the Coulomb dissociation of ^{136}I. This reaction can then in turn be used to constrain the $^{135}I(n,\gamma)$ rate.

3. ^{152}Eu - branch point in the *s* process

The laboratory based measurements of beta-decay and electron capture rates can not directly be used in stellar simulations. Electron capture can occur on excited states which are energetically not allowed on earth [13]. Also beta-decays which occur from thermally excited states cannot be

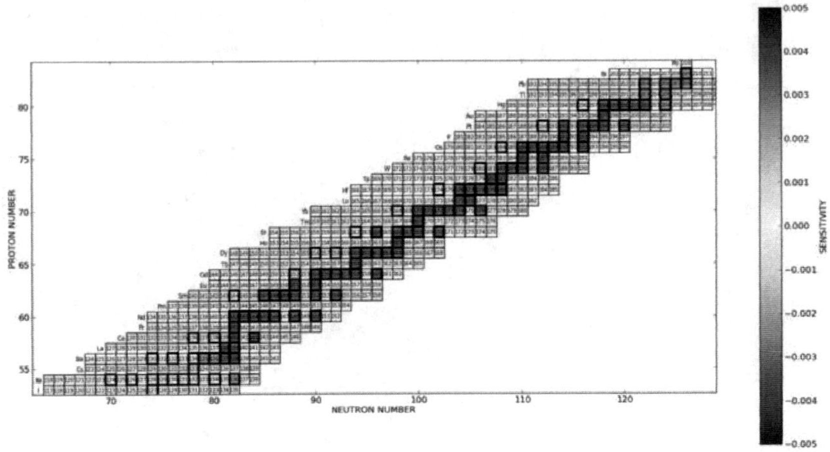

Figure 5. Impact of the ^{135}I(n,γ) rate on the final abundances of the i process. This reaction rate affects most of the abundances beyond ^{135}I and is therefore of global importance.

measured in the laboratory. These effects can sometimes alter the decay rates by a few orders of magnitude [14, 15]. For the theoretical calculations of stellar rates, Gamow-Teller strength distributions B(GT) for low lying states are needed [16, 17, 18]. Charge-exchange reactions, like the (p,n) reaction, allow access to these transitions and can serve as input for rate calculations. In particular, there exists a proportionality between (p,n) cross sections at low momentum transfer (close to 0°) and B(GT) values,

$$\frac{d\sigma^{CE}}{d\Omega}(q = 0) = \hat{\sigma}_{GT}(q = 0)B(GT), \tag{1}$$

where $\hat{\sigma}_{GT}(q = 0)$ is the unit cross section for GT transitions at q=0. [19]. In order to access GT distributions for unstable nuclei experiments have to be carried out in inverse kinematics with radioactive ion beams. This requires the detection of low-energy neutrons at large angles relative to the incoming beam.

An astrophysically interesting test case, ^{152}Sm(p,n), has been investigated at GSI in inverse kinematics. In the case of inverse kinematics, all information about the scattering angle in the center of mass and the excitation of the product nucleus can be determined from the energy and emission angle of the neutron in the laboratory reference system. Therefore a new detector for low-energy neutrons (LENA) has been developed [20] and was used at the LAND/R3B setup, Figure 6. The analysis of this experiment is currently ongoing and first results are very promising.

4. Light elements in the *r* process

The rapid neutron capture process (*r* process) produces half of the elements heavier than iron. However, the nuclear physics properties of the involved nuclei are not well known and its astrophysical site is not yet identified. The neutrino-driven wind model within core-collapse supernovae are currently one of the most promising candidates for a succesful *r* process. These neutrino winds are thought to dissociate all previously formed elements into protons, neutrons

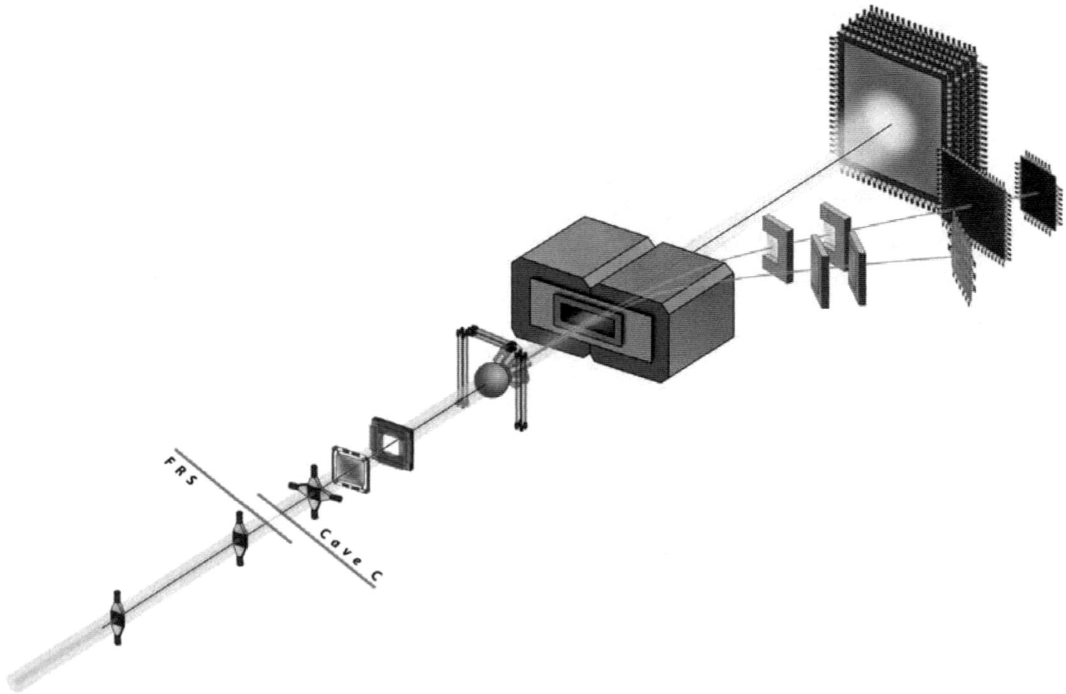

Figure 6. The R^3B setup at GSI optimized for charge-exchange reactions with low-energy neutrons in the exit channel. The LENA detector, optimized for the detection of low-energy neutrons emitted at high angles can be seen surrounding the target area. In forward direction, $LaBr_3$ detectors have been used to detect the decay of excited states.

and α particles before the seed nuclei for the r process are produced. Hence, the neutrino-driven wind model could explain the observational fact that the abundances of r nuclei of old halo-stars are similar to our solar r-process abundances [21]. This also indicates that the r process is a primary process and, thus, independent of the chemical composition of the progenitor star. Therefore, the investigation of the nuclear reactions among light elements forming seed nuclei prior to the r process leads to a better understanding of this process. Model calculations within a neutrino-driven wind scenario find a crucial change in the final r-process abundances by extending the nuclear reaction network towards very light neutron-rich nuclei [22]. Subsequent sensitivity studies point out the most important reactions, which include succesive (n,γ) reactions running through the isotopic chain of the neutron-rich boron isotopes $^{11}B(n,\gamma)^{12}B(n,\gamma)^{13}B(n,\gamma)^{14}B(n,\gamma)^{15}B(\beta^-)^{15}C$ [23]. Almost all reaction rates used in these model calculations are only known theoretically, and their uncertainties were estimated to be at least a factor of two [23]. Since the reaction rates of unstable isotopes are very difficult to determine experimentally, neutron breakup reactions of the neutron-rich beryllium isotopes were investigated in inverse kinematics via Coulomb dissociation. Figure 7 shows the incoming identification as well as the example of Coulomb breakup of ^{11}Be, which served as a benchmark of the measurement. The setup used for this experiment was the same as shown in Figure 2.

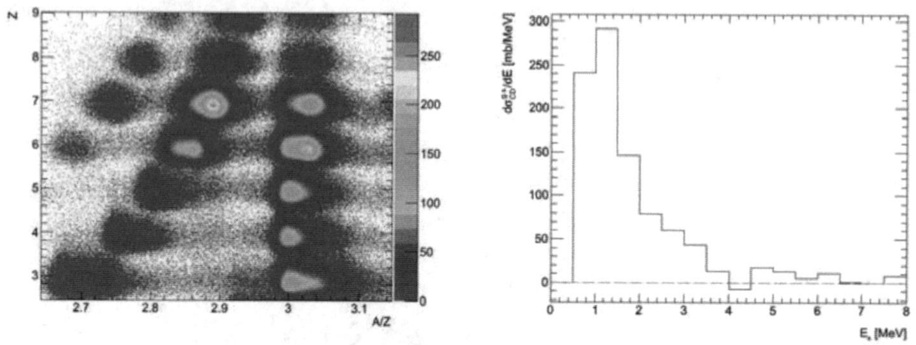

Figure 7. Left: Incoming identification plot. Right: Coulomb breakup cross section of ^{11}Be at almost 500 AMeV.

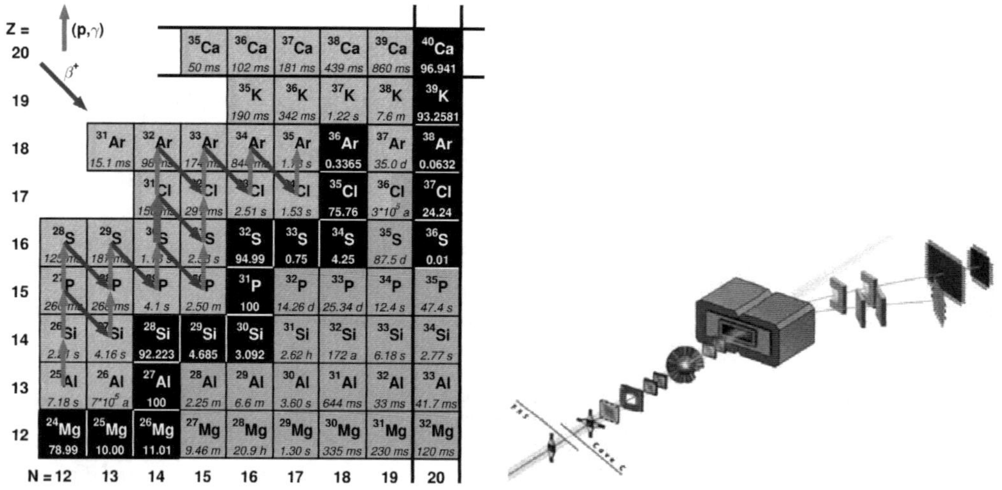

Figure 8. Left: The *rp*-process path at light elements. Right: The R^{3}B setup at GSI optimized for detecting protons in the exit channel. In particular the proton drift chambers are of importance.

5. Break out reactions in the rp process

The most likely astrophysical site of X-ray bursts are a very dense neutron stars, which accrete H/He-rich matter from a close companion [24, 25]. While falling towards the neutron star, the matter is heated up and a thermonuclear runaway is ignited. The exact description of this process is dominated by the properties of a few proton-rich radioactive isotopes, which have a low interaction probability, hence a high abundance (Figure 8, left).

Therefore the short-lived, proton-rich isotopes ^{31}Cl and ^{32}Ar have been investigated applying the Coulomb dissociation method at the GSI. An Ar beam was accelerated to an energy of 825 AMeV and fragmented in a beryllium target. The fragment separator was used to select the desired isotopes with a remaining energy of 650 AMeV. They were subsequently directed onto a ^{208}Pb target. The measurement was performed in inverse kinematics. All reaction products were detected and inclusive and exclusive measurements of the respective Coulomb dissociation

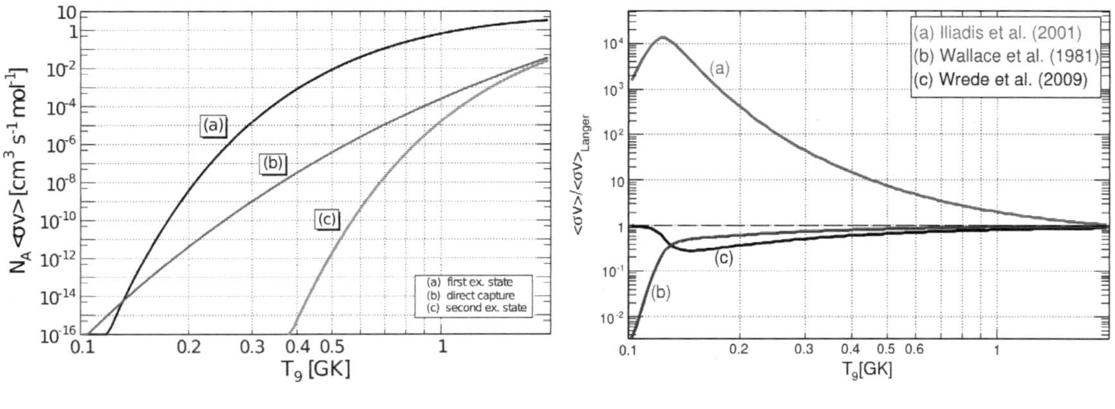

Figure 9. (Preliminary data) Left: The reaction rate for the important ^{30}S(p,γ)^{31}Cl bottleneck reaction in the *rp* process. Maximum peak temperatures in the *rp* process are typically around 2 GK. Three contributions can be seen, from (a) the low-lying resonance, (b) the direct capture, and (c) the second excited state in ^{31}Cl. Right: Comparison of different previous estimations with the reaction rate derived in this work (left). a) Iliadis et al. (red) [26], b) Wallace and Woosley (blue) [27] and c) Wrede et al (black) [28]. Especially in the low-temperature region, a deviation of up to 4 orders of magnitudes is observed.

cross sections were possible, Figure 8, right. Preliminary results for the important ^{30}S(p,γ)^{31}Cl reaction and a comparison with previously known estimates are shown in Figure 9.

6. ^{64}Ge - a waiting point in the νp process

Heavy α-nuclei are typically waiting points in the *rp*-process because of their small (p,γ) cross sections and the long β^+-half lives. Under certain conditions following a core collapse supernova, these waiting points can be overcome via (n,p) reactions in presence of small amount of neutrons. These neutrons stem from reactions like $\bar{\nu} + p \rightarrow n + \beta^+$. This process is therefore called the νp-process, [29]. One important waiting point is ^{64}Ge, which implies the importance of the ^{64}Ge(n,p)^{64}Ga reaction rate [30]. This reaction is very difficult to constrain experimentally. In combination with the planned storage rings, it would be possible to produce ^{64}Ga beam at FAIR and store in one of the rings. In combination with a hydrogen jet target, the inverse reaction ^{64}Ga(p,n)^{64}Ge could be investigated at astrophyscally interesting energies in inverse kinematics. The principle of this approach could be successfully proven with the reaction ^{96}Ru(p,γ), [31]

7. Summary

Nuclear data on radioactive isotopes are extremely important for modern astrophysics. FAIR offers contributions to almost every astrophysical nucleosynthesis process. Important developments are currently ongoing while FAIR is under construction.

Acknowledgments

This project was supported by the HGF Young Investigators Project VH-NG-327, EMMI, H4F, HGS-HIRe, JINA, NAVI, DFG and ATHENA.

References

[1] H H Gutbrod and I Augustin and H Eickhoff and K-D Groß and W F Henning and D Krämer and G Walter 2006 Facility for antiproton and ion research fair baseline technical report, isbn-3-9811298-0-6 Tech. rep. GSI

[2] R^{3}B-collaboration 2005 Technical proposal for the design, construction, commissioning and operation of r^{3}b Tech. rep. GSI http://www-land.gsi.de/r3b/

[3] Baur G and Rebel H 1996 *Annual Review of Nuclear and Particle Science* **46** 321–350

[4] Bertulani C A and Gade A 2010 *Phys. Rep.* **485** 195–259 (*Preprint* 0909.5693)

[5] Pignatari M, Gallino R, Heil M, Wiescher M, Käppeler F, Herwig F and Bisterzo S 2010 *Ap. J.* **710** 1557–1577

[6] Käppeler F, Gallino R, Bisterzo S and Aoki W 2011 *Reviews of Modern Physics* **83** 157–194

[7] Diehl R, Prantzos N and von Ballmoos P 2006 *Nuclear Physics A* **777** 70

[8] Uberseder E, Reifarth R, Schumann D, Dillmann I, Pardo C D, Görres J, Heil M, Käppeler F, Marganiec J, Neuhausen J, Pignatari M, Voss F, Walter S and Wiescher M 2009 *Physical Review Letters* **102** 151101

[9] Rugel G, Faestermann T, Knie K, Korschinek G, Poutivtsev M, Schumann D, Kivel N, Günther-Leopold I, Weinreich R and Wohlmuther M 2009 *Phys. Rev. Lett.* **103** 072502

[10] Geissel H, Armbruster P, Behr K H, Brünle A, Burkard K, Chen M, Folger H, Franczak B, Keller H, Klepper O, Langenbeck B, Nickel F, Pfeng E, Pfützner M, Roeckl E, Rykaczewski K, Schall I, Schardt D, Scheidenberger C, Schmidt K H, Schröter A, Schwab T, Sümmerer K, Weber M, Münzenberg G, Brohm T, Clerc H G, Fauerbach M, Gaimard J J, Grewe A, Hanelt E, Knödler B, Steiner M, Voss B, Weckenmann J, Ziegler C, Magel A, Wollnik H, Dufour J P, Fujita Y, Vieira D J and Sherrill B 1992 *Nuclear Instruments and Methods in Physics Research B* **70** 286–297

[11] Herwig F, Pignatari M, Woodward P R, Porter D H, Rockefeller G, Fryer C L, Bennett M and Hirschi R 2011 *ApJ* **727** 89 (*Preprint* 1002.2241)

[12] García-Hernández D A, Zamora O, Yagüe A, Uttenthaler S, Karakas A I, Lugaro M, Ventura P and Lambert D L 2013 *A&A.* **555** L3 (*Preprint* 1306.2134)

[13] Langanke K and Martinez-Pinedo G 1998 *Physics Letters B* **436** 19–24 (*Preprint* arXiv:nucl-th/9809081)

[14] Takahashi K and Yokoi K 1987 *Atomic Data Nucl. Data Tables* **36** 375

[15] Abbondanno U, Aerts G, Alvarez-Velarde F, Alvarez-Pol H, Andriamonje S *etal* 2004 *Physical Review Letters* **93** 161103

[16] Fuller G M, Fowler W A and Newman M J 1980 *ApJS* **42** 447–473

[17] Langanke K and Martínez-Pinedo G 2000 *Nuclear Physics A* **673** 481–508 (*Preprint* arXiv:nucl-th/0001018)

[18] Langanke K and Martínez-Pinedo G 2003 *Reviews of Modern Physics* **75** 819–862 (*Preprint* arXiv:nucl-th/0203071)

[19] Taddeucci T, Goulding C, Carey T, Byrd R, Goodman C, Gaarde C, Larsen J, Horen D, Rapaport J and Sugarbaker E 1987 *Nucl. Phys.* **A469** 125 – 172

[20] Langer C, Algora A, Couture A, Csatlós M, Gulyás J, Heil M, Krasznahorkay A, O'Donnell J, Plag R, Reifarth R, Stuhl L, Sonnabend K, Tornyi T and Tovesson F 2011 *Nuclear Instruments and Methods A* **659** 411–418 ISSN 0168-9002

[21] Truran J W, Cowan J J, Pilachowski C A and Sneden C 2002 *PASP* **114** 1293–1308

[22] Terasawa M, Sumiyoshi K, Kajino T, Mathews G J and Tanihata I 2001 *ApJ* **562** 470

[23] Sasaqui T, Kajino T, Mathews G J, Otsuki K and Nakamura T 2005 *ApJ* **634** 1173–1189

[24] Schatz H, Aprahamian A, Görres J, Wiescher M, Rauscher T, Rembges J, Thielemann F K, Pfeiffer B, Möller P, Kratz K L, Herndl H, Brown B and Rebel H 1998 *Physics Reports* **294** 167

[25] Schatz H and Rehm K E 2006 *Nuclear Physics A* **777** 601–622 (*Preprint* arXiv:astro-ph/0607624)

[26] Iliadis C, D'Auria J M, Starrfield S, Thompson W J and Wiescher M 2001 *Astrophys. J. Suppl.* **134** 151–171

[27] Wallace R and Woosley S 1981 *Ap. J. Suppl.* **45** 389

[28] Wrede C, Caggiano J A, Clark J A, Deibel C M, Parikh A and Parker P D 2009 *Phys. Rev. C* **79** 045808

[29] Thielemann F K, Dillmann I, Farouqi K, Fischer T, Fröhlich C, Kelic-Heil A, Korneev I, Kratz K L, Langanke K, Liebendörfer M, Panov I V, Martinez-Pinedo G and Rauscher T 2010 *Journal of Physics Conference Series* **202** 012006

[30] Fröhlich C 2012 *Journal of Physics Conference Series* **403** 012034

[31] Zhong Q, Aumann T, Bishop S, Blaum K, Boretzky K, Bosch F, Bräuning H, Brandau C, Davinson T, Dillmann I, Ershova O, Geissel H, Gyürky G, Heil M, Käppeler F, Kelić A, Kozhuharov C, Langer C, Bleis T L, Litvinov Y A, Lotay G, Marganiec J, Petridis N, Plag R, Popp U, Reifarth R, Riese B, Rigollet C, Scheidenberger C, Simon H, Stöhlker T, Szücs T, Weber G, Weick H, Winters D F A, Winters N and Woods P J 2010 *Journal of Physics: Conference Series* **202** 012011

Approaching the precursor nuclei of the third r-process peak with RIBs

C Domingo-Pardo[1,*], R Caballero-Folch[2], J Agramunt[1], A Algora[1,3], A Arcones[4], F Ameil[4], Y Ayyad[5], J Benlliure[5], M Bowry[6], F Calviño[3], D Cano-Ott[7], G Cortés[2], T Davinson[8], I Dillmann[4,9], A Estrade[4,10], A Evdokimov[4,9], T Faestermann[11], F Farinon[4], D Galaviz[12], A García-Rios[7], H Geissel[4,9], W Gelletly[6], R Gernhäuser[11], MB Gómez-Hornillos[3], C Guerrero[13], M Heil[4], C Hinke[11], R Knöbel[4], I Kojouharov[4], J Kurcewicz[4], N Kurz[4], Y Litvinov[4], L Maier[11], J Marganiec[14], M Marta[4,9], T Martínez[7], G Martínez-Pinedo[4], BS Meyer[21], F Montes[15,16], I Mukha[4], DR Napoli[17], Ch Nociforo[4], C Paradela[5], S Pietri[4], Z Podolyák[6], A Prochazka[4], S Rice[6], A Riego[2], B Rubio[1], H Schaffner[4], Ch Scheidenberger[4,9], K Smith[18,19], E Sokol[20], K Steiger[11], B Sun[4], JL Taín[1], M Takechi[4], D Testov[20,22], H Weick[4], E Wilson[6], JS Winfield[4], R Wood[6], P Woods[8] and A Yeremin[20]

[1] IFIC, CSIC-University of Valencia, Valencia, Spain
[2] INTE-DFEN, UPC, Barcelona, Spain
[3] Institute of Nuclear Research of Hungarian Academy of Sciences, Debrecen, Hungary
[4] GSI Helmholtzzentrum für Schweionenforschung GmbH, Darmstadt, Germany
[5] Universidade de Santiago de Compostela, Santiago de Compostela, Spain
[6] Deptartment of Physics, University of Surrey, Guildford, United Kingdom
[7] Centro de Investigaciones Energéticas, Medioambientales y Tecnológicas, Madrid, Spain
[8] School of Physics and Astronomy, University of Edinburgh, United Kingdom
[9] Physikalisches Institut, Justus-Liebig Universität Giessen, Germany
[10] St. Mary's University, Halifax, Nova Scotia, Canada
[11] Physics Department E12, Technische Universität München, Garching, Germany
[12] Centro de Fisica Nuclear da Universidade de Lisboa, Lisboa, Portugal
[13] CERN, Geneva, Switzerland
[14] ExtreMe Mater Institute, Darmstadt, Germany
[15] National Superconducting Cyclotron Laboratory, Michigan State University, EastLansing, USA
[16] Joint Institute for Nuclear Astrophysics, Michigan State University, East Lansing, USA
[17] Instituto Nazionale di Fisica Nucleare, Laboratori Nazionale di Legnaro, Italy
[18] Department of Physics, University of Notre Dame, South Bend, USA
[19] Joint Institute for Nuclear Astrophysics, University of Notre Dame, South Bend, USA
[20] Flerov Laboratory, Joint Institute for Nuclear Research, Dubna, Russia
[21] Clemson University, USA
[22] Institut de Physique Nucléaire d'Orsay, France

E-mail: `domingo@ific.uv.es`

Abstract. The rapid neutron nucleosynthesis process involves an enormous amount of very exotic neutron-rich nuclei, which represent a theoretical and experimental challenge. Two of the main decay properties that affect the final abundance distribution the most are half-lives and neutron branching ratios. Using fragmentation of a primary ^{238}U beam at GSI we were able to

Journal of Physics: Conference Series **665** (2016) 012045 doi:10.1088/1742-6596/665/1/012045

measure such properties for several neutron-rich nuclei from ^{208}Hg to ^{218}Pb. This contribution provides a short update on the status of the data analysis of this experiment, together with a compilation of the latest results published in this mass region, both experimental and theoretical. The impact of the uncertainties connected with the beta-decay rates and with beta-delayed neutron emission is illustrated on the basis of r-process network calculations. In order to obtain a reasonable reproduction of the third r-process peak, it is expected that both half-lives and neutron branching ratios are substantially smaller, than those based on FRDM+QRPA, commonly used in r-process model calculations. Further measurements around $N \sim 126$ are required for a reliable modelling of the underlying nuclear structure, and for performing more realistic r-process abundance calculations.

1. Introduction

The third r-process peak at $A \sim 195$, the platinum peak, seems particularly sensitive to both the nuclear physics input and the conditions of the stellar environment as can be seen in the detailed sensitivity study reported in Ref. [1]. Radioactive Ion Beam (RIB) facilities have enabled the production and measurement of several waiting-point nuclei directly on the r-process path from $N = 50$ to $N = 82$. However all of the nuclei that lead directly to the formation of the third r-process abundance peak still remain in the region of the *terra incognita*. To a large extent this is due to the very low production cross sections, the limited primary beam intensities available at present RIB facilities and also the challenging experimental conditions of large backgrounds and very low production rates. This contribution focuses on the impact of half-lives and beta-delayed neutrons on the formation of the third r-process peak, and how present theoretical models compare with the experimental data available.

2. The r-process paradigm around N=126

The aim of this section is to describe briefly the latest experimental and theoretical results in this mass region, and to explore the impact of the present uncertainties (or discrepancies) on the nucleosynthesis of the r process. On the experimental side, special focus will be made on the latest experiment performed at GSI, thus providing an update on the present status of the data analysis and its future perspective.

As mentioned above, the nuclei which are directly in the path of the r process along $N = 126$, approximately from gadolinium to tantalum, could not be accessed experimentally yet. Nevertheless, one can try to measure neutron-rich nuclei in the higher Z neighbourhood, on both sides of the $N = 126$ shell closure, and use such information as a benchmark for theoretical models. In turn, such models can be applied more reliably to extrapolate the decay properties of the r-path nuclei. Three independent experiments in the neutron-rich region around $N = 126$ have been carried out recently at the GSI facility for heavy ion research (Germany). The FRS fragment separator [2] was used for selection and identification of the ions of interest. The high energy available for primary beams at GSI, of up to 1 GeV/u, has an advantage over other facilities in order to reduce ion identification difficulties related to the charged states of the secondary fragments. Fragmentation of primary ^{238}U and/or ^{208}Pb beams on a beryllium target was used for the production of secondary neutron-rich ion beams. These experiments are briefly described below. For details the reader is referred to the cited publications. Only the main results from each experiment are summarised here.

2.1. *β-decay experiments around N∼126*

The first experiment [3, 4] used relativistic fragmentation of both ^{238}U and ^{208}Pb projectiles for the production of a large number of new neutron-rich nuclei. Several isotopic species were implanted in a stack of four double-sided silicon-strip detectors (DSSSDs) [5], which allowed for the position and time measurement of both implant- and decay-events. β-Decay half-lives were determined after developing a new numerical method [6], which was needed to account for the high and complex background conditions. Half-lives have been published from ^{194}Re ($N = 119$) up to ^{202}Ir ($N = 125$) [7, 8]. From this experiment, another isotopes have been also analysed [9, 10], but they will not be considered in this contribution because they have not been published yet.

New half-lives have been published recently beyond $N = 126$, for ^{219}Bi and ^{211,212,213}Tl [11]. In the latter experiment, in addition to a stack of three DSSDs for measuring ion-implants and β-decays, the RISING array [12] of 105 HPGe detectors was used to detect γ-rays following the β-decays. As is nicely illustrated in Ref. [11], by means of high-resolution γ-ray spectroscopy it was possible to perform a validation of the numerical technique described in Ref. [6], which has been applied in both works [8, 11].

Finally, the latest measurement in this mass region ($N > 126$) used a stack of six SSSDs and three DSSSDs [13] surrounded by a prototype of the BEta deLayEd Neutron (BELEN) detector [14]. The latter allowed for the experimental determination of neutron-branching ratios. Technical details and a few preliminary results have been reported recently in Ref. [15]. In summary, the $N > 126$ nuclei $^{208-211}$Hg, $^{211-215}$Tl and $^{214-218}$Pb were implanted with sufficient statistics for a reliable analysis of their half-lives. As first approach, the numerical method [6] was applied to the measured thallium isotopes. As reported in Ref. [15], the preliminary half-lives thus obtained are compatible, within the error bars, with those reported by Benzoni et al. [11] for the common Tl-nuclei. Alternatively, despite the large and complex background environment, we have investigated the possibility of using the more conventional analytical approach based on determining the half-life from implant-beta time correlations. This seems still feasible in those cases with relatively large implant rates (see Fig. 2 in Ref. [15]). In the case of e.g. ^{212}Tl, an ion implant-rate of $\sim 3\times10^{-5}$ counts/pixel/s was achieved, being the average rate of β-like events of about 4×10^{-4} counts/pixel/s. Using this nuclide as example for illustration purposes, the corresponding implant-beta time correlations are shown in Fig. 1. The correlation area comprises the pixel where the ion was implanted, as well as the eight neighbouring pixels.

In the spectra shown in Fig. 1 each implanted ^{212}Tl ion was correlated with all following β-decay events registered inside the aforementioned correlation area of 9 mm^2. The time-window for correlations spanned several times the expected half-life. The background was estimated in a similar way, by performing backward ion-β time-correlations over the same DSSSD area. After background subtraction the resulting time-spectrum was adjusted to a simple exponential decay (the daughter nucleus ^{212}Pb shows a half-life of 10.64 h[16]), as shown in Fig. 1. The preliminary value obtained for the half-life of ^{212}Tl was $t_{1/2}^{212Tl} = 44(20)$ s. This value, although smaller than the one obtained with the numerical method, is still in perfect agreement (within the quoted error bars) with the half-life reported previously. From the 14 isotopic species implanted, we expect that the analytical analysis method can be applied to derive the half-lives of at least half of them, those showing high implant statistics. The remaining cases will be analysed, most probably, by applying the numerical approach [6].

The results for the half-lives determined in these three experiments are displayed in Fig. 2-left. The solid red circle shows the preliminary result discussed above for the half-life of ^{212}Tl.

2.2. Theory versus experiment

Although still far from the r-process path, these measurements can be used to judge the reliability of theoretical models in the heavy neutron-rich region around $N \sim 126$. Fig. 2-left shows a comparison between the aforementioned experimental data, and the two main theoretical calculations available in this region. The blue up-pointing triangles correspond to the model of P. Möller [18], commonly used in many r-process network calculations. This approach is

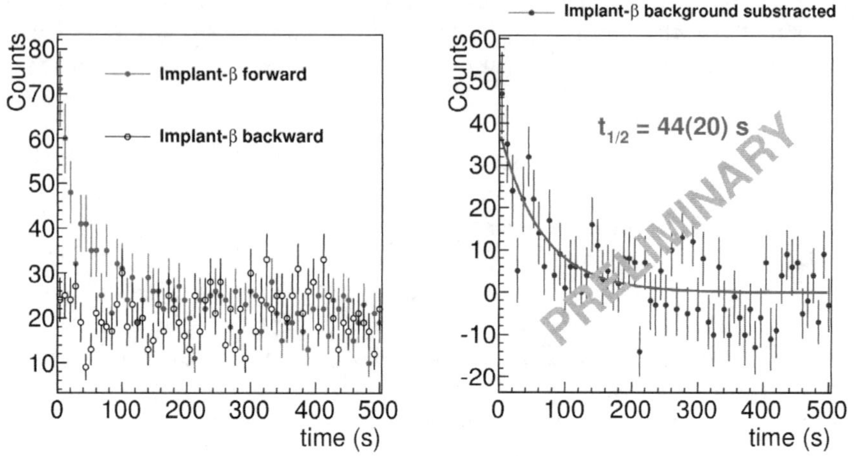

Figure 1. (Left) ^{212}Tl Implant-β time correlations in the forward direction (solid symbols) and in the backward direction (open symbols) for all decay events within the implant pixel and the 8 neighbouring pixels. All decay events over a broad time window of several times the expected half-life were considered. (Right) Background subtracted correlation spectrum. The quoted half-life value is preliminary.

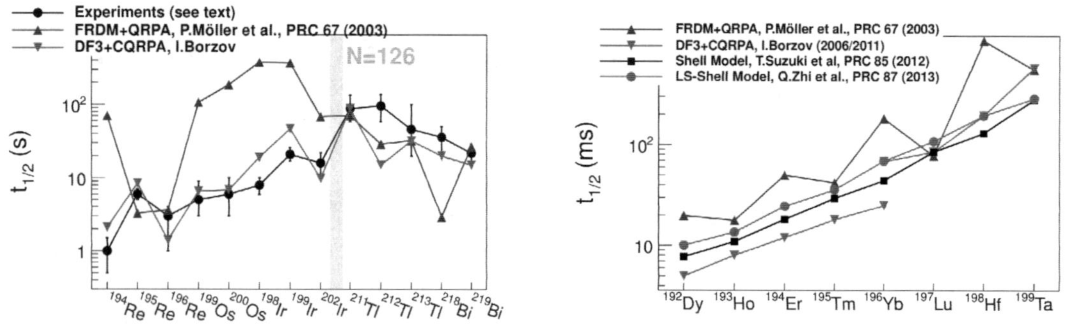

Figure 2. (Left) Comparison between theoretical predictions and measurements in the region around $N = 126$. Experimental values from ^{194}Re to ^{202}Ir are from Ref. [8], values between ^{211}Tl and ^{219}Bi are from Ref. [11]. (Right) Theoretical values published in the literature for the waiting point nuclei along $N = 126$ (adapted from Ref.[17]).

based on the finite-range droplet mass model (FRDM) and uses the quasiparticle random-phase approximation (QRPA) for the Gamow-Teller (GT) part of the β-strength function, whereas the gross-theory is implemented to account for the first-forbidden (FF) part of the β-decay. The pink down-pointing triangles correspond to the calculations by I. Borzov using the Fayans energy-density functional (DF3 version) within a continuum QRPA framework [19, 20]. The latter approach allows for a self-consistent treatment of both allowed GT- and FF-transitions. Although the contribution of FF-transitions can be neglected at the $N = 50$ and $N = 82$ shell closures, it is expected that around $N = 126$ they contribute remarkably to the β-strength distribution and even dominate it above Z$\geq$70, thus providing a remarkable shortening of the half-lives in this mass region [19].

In comparison with the experimental data displayed on Fig.2-left the FRDM+QRPA calculations [18], on average overestimate the measured half-lives by a factor of 24 (average ratio of calculation/experiment) in the region $N \leq 126$, and underestimate them by less than 40% beyond $N = 126$ (the average ratio is 0.62). The CQRPA+DF3 model [20] agrees better with the experimental data in the region $N \leq 126$, showing an average deviation of only a factor of 2 in that region. The discrepancy becomes again $\lesssim$40%, in average, beyond the neutron shell closure. As already noted in Refs. [10, 11], theoretical models seem to overestimate the half-lives before the $N = 126$ neutron shell closure, and to underestimate them beyond it. The question, however, is how well these models will perform when one uses them to extrapolate further off stability, into the region of the waiting-point nuclei. To answer this question one can compare the theoretical models among them. In particular shell-model calculations [21, 17], which became recently available, seem quite helpful given the absence of any experimental information. Such a comparison is shown on the right-hand side of Fig. 2, which has been adapted from Ref. [17]. As expected, the agreement of the different models near the neutron shell closure is much better than toward the valley of stability (see Fig. 2-left), owing to the simplified nature of the shell closure. Indeed, comparing the performance of the FRDM+QRPA model against large-scale shell model calculations [17] one finds discrepancies, which range from ~30% up to a factor of 5, with an average deviation of "only" a factor of 2. In summary, one can conclude two things. First, the uncertainties on the beta-decay half-lives along the shell closure itself are expected to be less severe than in the $N < 126$, $Z < 82$ mass region covered by the r process during freeze-out. Second, the latest measurements on both sides of $N = 126$ provide a very stringent test for theoretical models indeed.

2.3. R-process nucleosynthesis

The impact of the variations in half-lives on the final abundances has been studied in several works (see e.g. [22, 21, 1] and references therein). We have performed new network calculations in order to illustrate the effect of the aforementioned shortening of the half-lives along the $N = 126$ waiting-point nuclei, as well as in the $N < 126$, $Z < 82$ neighbourhood. In these calculations the so-called cold trajectory of Ref. [1] was used in combination with the single-zone code of `NucNet Tools` [23, 24]. Rather standard parameters were used for the electron fraction ($Y_e \approx 0.47$) and the entropy (S$\approx$ $200k_B$/nuc). In order to accelerate the calculation of the r-process phase of the expansions, we used Krylov-space iterative matrix solver routines [25] as implemented in the freely-available software `Sparskit2`. The very sparse network matrix that must be solved during the r-process phase is well-conditioned, so such iterative solvers converge rapidly, and we found roughly a factor of ten speed up over the default sparse matrix solver in `NucNet Tools` when we used these iterative solvers. The resulting abundances as a function of the mass number are shown in Fig. 3-left. The dotted open circles represent the r-process abundances in the solar system. The solid black curve shows a reference calculation performed with the JINA REACLIB library [26]. In the latter database, beta-decay rates are taken from Ref. [18] (FRDM+QRPA) where no data are available. The calculation provides a reasonable description of the solar r-process abundance pattern in the region of the second and third r-process peaks. Discrepancies such as the width of the peak or the region between both shell closures are not relevant for the present discussion, where the aim is to illustrate rather qualitatively the impact of reasonable half-life variations in this region. Reducing by a factor of two the FRDM+QRPA half-lives of the $N = 126$ waiting-point nuclides, from gadolinium up to tantalum, one obtains the abundances represented in Fig. 3-left by the red-dashed line. A small shift towards higher masses is observed, similar to that reported e.g. in Ref. [21]. This seems to be in contradiction with the solar r-process abundances, where the maximum appears at a lower mass number around $A \sim 195$. Furthermore, as discussed above, the uncertainties on the half-lives of the nuclei in the $N < 126$ region seem to be much larger than those at the shell closure itself. Obviously, the uncertainty will depend on the complexity of each nucleus and the difficulties of the model in reproducing the beta-strength distribution and the underlying nuclear structure details. Since it is difficult to assess an uncertainty on each particular nucleus, let us see what is the effect of a variation of the half-lives in that region by a constant factor, which is of the same order-of-magnitude as the discrepancies found between experiment and theory (FRDM+QRPA) discussed above. Assuming an average reduction of "only" a factor 12 (the average ratio theory/experiment found above was $\sim$24) in the half-lives of the nuclei on the left-hand side of $N = 126$, from $N = 116$ up to N$\lesssim$ 125, one obtains a much stronger shift of the third r-process peak, towards higher masses, as shown in Fig. 3-left by the blue dot-dashed line. Additionally, the third r-process peak becomes narrower, an effect that disagrees further with the observed abundance pattern. Thus, one can conclude two things. First, a reduction of the FRDM+QRPA half-lives in the $N \leq 126$ region, which is indeed expected both from experiment and theory, produces a shift of the third r-process peak which seems to go in the wrong direction, i.e. toward higher masses. Secondly, with the data currently available, the impact on the final abundances due to half-life uncertainties for nuclei in the region $N < 126$, may be even more important than the uncertainties for the waiting-point nuclei themselves. From the nuclear

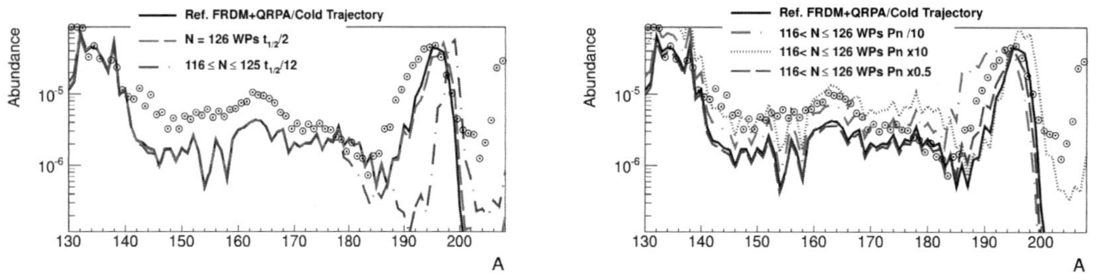

Figure 3. Abundance versus mass number. Dotted open circles represent solar r-process abundances. The solid black curve corresponds to a reference calculation, where the decay rates are essentially based on FRDM+QRPA [18]. (Left panel) The dashed red line shows the same calculation with a factor of 2 reduction on the half-lives of the $N = 126$ waiting point nuclei. The blue dash-dotted line shows the abundances obtained when the half-lives of the nuclei on the left-hand side of the waiting points, from $N = 116$ up to $N = 125$, were shortened by a factor of 12. See text for details. (Right panel) Dashed-dotted line shows the effect of a reduction by a factor of 10 on the neutron emission probabilities for nuclei in the $116 < N \leq 126$ region. The blue dotted line corresponds to an enhancement, by a factor of 10, of the neutron emission probabilities of the same nuclei. The blue dashed line shows the impact of a reduction by a factor of 2 on the neutron branchings.

physics point of view, the discrepancy between shorter half-lives and the position (and width) of the third r-process peak, may be counterbalanced by the effect of β-delayed neutron emission. Indeed, for a given Q_β value, shorter half-lives usually correlate with a small neutron emission probability. The impact of (smaller) neutron emission probabilities on the final abundances is discussed in Sec. 3.

3. β-Delayed neutron emission around N$\sim$126

The emission of β-delayed neutrons plays an important role in the formation of the third r-process peak [18, 1]. However, the situation in terms of experimental information becomes really critical around $N \sim 126$, as demonstrated in Fig. 4-left. In the latter figure, nuclei where any experimental information [16] is known about their beta-delayed neutron emission probability

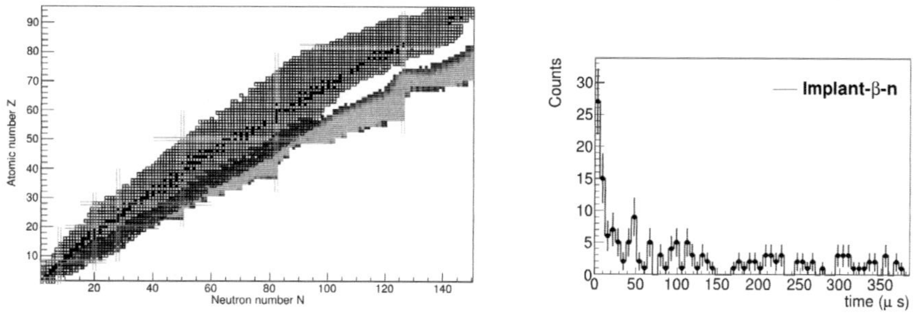

Figure 4. (Left) Nuclear chart showing in red nuclei where experimental information on the neutron branching ratio is available [16]. (Right) Time spectrum showing implant-β-neutron correlations for ^{212}Tl.

are shown in red. For the sake of comparison, the r-process path at some stage of a network calculation is also shown in the same figure. It is worth emphasising that beyond $A \sim 150$ there is practically no experimental information available. The only data available around $N \sim 126$ are on ^{210}Tl [27].

Motivated by this situation we used the BELEN neutron detector in the last experiment performed at GSI around $N \gtrsim 126$. The detector consisted of 30 proportional counters of ^{3}He embedded in a polyethylene matrix which served as moderator. More details about this measurement can be found in Ref. [15]. The analysis of the beta-delayed neutron emission probabilities is still ongoing. Following the example shown before for ^{212}Tl, the spectrum of beta-delayed neutrons registered after the decays of ^{212}Tl is shown in Fig. 4-right. These are implant-β-neutron correlations, which could be observed mostly due to the high efficiency of BELEN, which was of ~40%. Using this kind of correlation analysis, we expect to obtain neutron branching ratios (or upper limits) for all the implanted species.

Since there is no experimental information on neutron-emission probabilities in the mass-region around $N = 126$, it is not possible to study reliably the capability of theoretical models for predicting the effect of neutron emission. However, one can again compare the predictions of different models among themselves. One such comparison is shown in Fig. 17 of Ref. [17] for the three models described above, FRDM+QRPA [18], DF3+CQRPA [20] and LSSM [17]. For the waiting points with Z<69, the neutron emission probabilities predicted by the FRDM+QRPA are, on average, a factor of 3 larger than the values calculated with the LSSM. This ratio becomes ~0.6 for the waiting point nuclei with Z≥69, mostly due to the contribution of FF transitions. It is not straight-forward to predict, what the uncertainties in the calculated neutron-emission probabilities would be, in particular for the neutron-rich nuclei involved during the freeze-out phase. In this case, variations of the neutron-branching ratios are rather arbitrary, but they can still be used for a qualitative interpretation of the large uncertainties in these quantities. In this respect, Fig. 3-right shows the effect of neutron-branching variations in the mass region $116 < N \leq 126$. Final abundances obtained when the neutron emission probabilities are reduced by a factor of 2 are shown by the blue dashed line. The red dash-dotted line shows the effect of a factor of 10 reduction in the neutron branching ratios. The blue dotted line corresponds to a factor of 10 enhancement of the neutron branching ratios.

In summary, the discrepancy between shorter half-lives and the formation of the third peak discussed in Sec. 2 could be compensated, at least to some extent, with a reduction of the neutron emission probabilities. Clearly, new β-delayed neutron emission measurements are needed around $N \sim 126$ in order to confirm such a hypothesis, and to reduce the still very large contribution of the nuclear physics input uncertainties to the calculated r-process abundances.

4. Summary and outlook

In summary, both theory and experiment, indicate that half-lives in the $N \sim 126$ region near the r-process path should be much smaller than those commonly used in r-process model calculations (FRDM+QRPA model [18]). This, however, would imply a shift of the third r-process peak towards higher masses, an effect which is in contradiction with the observed r-process abundances. From the nuclear physics input, such a discrepancy could be removed if neutron emission probabilities were also smaller. This seems to be the case, as it has been shown by recent large-scale shell-model calculations [17]. In the near future, first experimental values for the neutron emission probabilities of several nuclei beyond $N = 126$ will become available from the experiment performed at GSI with the BELEN detector. This will represent a first test of the beta-strength distribution above the neutron separation energy, but clearly new measurements are needed in the neutron-rich heavy-mass region.

5. Acknowledgments

We thank the technical staff of the GSI accelerators, the FRS, and the target laboratory for their support during the S410 experiment. This work has been partially supported by the Spanish Ministry of Economy and Competitivity under grants FPA2011-24553, FPA 2011-28770-C03-03 and AIC-D-2011-0705. I.D., A.Ev., and M.M. are supported by the Helmholtz association via the Young Investigators group LISA (VH-NG-627).

References

[1] Arcones A and Martinez-Pinedo G 2011 *Phys. Rev. C* **83** 045809
[2] Geissel H *et al.* 1992 *Nucl. Instr. and Methods in Phys. Res. B* **70** 286
[3] Alvarez-Pol H *et al.* 2009 *Eur. Phys. J* **42** 485
[4] Alvarez-Pol H *et al.* 2010 *Phys. Rev. C* **82** 041602
[5] Kumar R *et al,* 2009 Nucl. Instrum. and Methods A **598** 754
[6] Kurtukian-Nieto T, Benlliure J and Schmidt K-H 2008 *Nucl. Instr. and Methods in Phys. Res. A* **589** 472
[7] Kurtukian-Nieto T *et al.* 2007 *Phys. Lett. B* (submitted) (arXiv:0711.0101v1)
[8] Kurtukian-Nieto T *et al.* 2009 *Nucl. Phys. A* **827** 587
[9] Morales López AI 2011 *PhD-Thesis* Universiade de Santiago de Compostela
[10] Benlliure J *et al.* 2012 *J. Phys.: Conf. Series* **337** 23
[11] Benzoni G *et al.* 2012 *Phys. Lett. B* **715** 293
[12] Wollersheim HJ *et al.* 2005 *Nucl. Instr. and Methods in Phys. Res. A* **537** 637
[13] Steiger K 2009 *Diploma Thesis* TU-München (Germany).
[14] Gomez-Hornillos MB *et al.* 2011 *J. Phys.: Conf. Series* **312** 052008
[15] Caballero-Folch R *et al.* 2013 *Conf. Proc. Nuc. Data for Science and Technology* (to be published)
[16] `http://www.nndc.bnl.gov/ensdf`
[17] Zhi Q, Caurier E, Cuenca-García JJ, Langanke K, Martinez-Pinedo G and Sieja K 2013 *Phys. Rev. C*
[18] Möller P, Pfeiffer B and Kratz K-L 2003 *Phys. Rev. C* **67** 055802
[19] Borzov IN 2003 *Phys. Rev. C* **67** 025802
[20] Borzov IN 2011 *Physics of Atomic Nuclei* **74** 1435
[21] Suzuki T, Yoshida T, Kajino T and Otsuka T 2012 *Phys. Rev. C* **85** 015802
[22] Borzov IN *et al.* (2008) Nucl. Phys. A **814** 159
[23] Meyer BS 2012 Webnucleo.org *Proc. of Science (NIC XII)* **096**
[24] `http://sourceforge.net/project/nucnet-tools/`
[25] Saad Y 2003 *Iterative Methods for Sparse Linear Systems* SIAM, 2nd Edition
[26] Cyburt RH *et al.* 2010 *The Astrophysical Journal Supplement Series* **189** 240
[27] Kogan AV *et al.* 1957 *Sov. Phys. JETP* **5** 365

Experimental study of the ^{15}O(*2p*,γ)^{17}Ne cross section by Coulomb Dissociation for the *rp* process

J Marganiec[1,5], F Wamers[1,2,3,25], F Aksouh[3,23], Y Aksyutina[3], H Alvarez Pol[6], T Aumann[2,3], S Beceiro[6], C Bertulani[7], K Boretzky[3], M J Borge[8], M Chartier[9], A Chatillon[3], L Chulkov[3], D Cortina-Gil[6], I Egorova[15], H Emling[3], O Ershova[3,4], C Forssén[11], L M Fraile[12,13], H Fynbo[14], D Galaviz[8], H Geissel[3], L Grigorenko[15,27], M Heil[3], D H H Hoffmann[2], J Hoffmann[3], H Johansson[11], B Jonson[11],M Karakoç[24], C Karagiannis[3], O Kiselev[3], J V Kratz[16], R Kulessa[17], N Kurz[3], C Langer[3,4], M Lantz[11,18], K Larsson[3], T Le Bleis[3,19], R Lemmon[9], Yu A Litvinov[3], K Mahata[3,20], C Müntz[4], T Nilsson[11], C Nociforo[3], G Nyman[11], W Ott[3], V Panin[2,3],Yu Parfenova[15,28], S Paschalis[2,9], A Perea[8], R Plag[3,4], R Reifarth[3,4], A Richter[2],K Riisager[14], C Rodríguez Tajes[6,22], D Rossi[3,16,29], G Schrieder[2], N Shulgina[11,21], H Simon[3], J Stroth[4], K Sümmerer[3], J Taylor[9], O Tengblad[8], E Tengborn[11], H Weick[3], M Wiescher[26,5], C Wimmer[3,4], M Zhukov[11]

[1]EMMI, Darmstadt, Germany; [2]TU Darmstadt, Germany; [3]GSI Darmstadt, Germany; [4]Goethe-Universität, Frankfurt am Main, Germany; [5]JINA, Notre Dame, USA; [6]Santiago de Compostela University, Spain; [7]Texas A&M University-Commerce, USA; [8]IEM Madrid, Spain; [9]University of Liverpool, UK; [10]CEA/DAM/DIF Bruyere, France; [11]Chalmers I.T., Sweden; [12]Universidad Complutense de Madrid, Spain; [13]CERN, Geneva, Switzerland; [14]Aarhus University, Denmark; [15]FLNR JINR Dubna, Russia; [16]University of Mainz, Germany; [17]Jagiellonian University, Krakow, Poland; [18]Uppsala University, Uppsala, Sweden; [19]TU München, Germany; [20]BARC Mumbai, India; [21]Kurchatov Institute, Moscow, Russia; [22]GANIL, CEA/DSM-CNRS/IN2P3, France; [23]King Saud University, Kingdom of Saudi Arabia; [24]Akdeniz University, Turkey; [25]FIAS Frankfurt am Main, Germany; [26]University of Notre Dame, USA; [27]RRC KI, Moscow, Russia; [28]INP, Moscow, Russia; [29]NSCL, MSU, East Lansing, USA

E-mail: j.marganiec@gsi.de

Abstract. The time-reversed reaction ^{15}O$(2p,\gamma)^{17}$Ne has been studied by the Coulomb dissociation technique. Secondary ^{17}Ne ion beams at 500 AMeV have been produced by fragmentation reactions of ^{20}Ne in a beryllium production target and dissociated on a secondary Pb target. The incoming beam and the reaction products have been identified with the kinematically complete LAND-R^{3}B experimental setup at GSI. The excitation energy prior to decay has been reconstructed by using the invariant-mass method. The preliminary differential and integral Coulomb Dissociation cross sections (σ_{Coul}) have been calculated, which provide a photoabsorption (σ_{photo}) and a radiative capture cross section (σ_{cap}). Additionally, important information about the nuclear structure of the ^{17}Ne nucleus will be obtained. The analysis is in progress.

Journal of Physics: Conference Series **665** (2016) 012046 doi:10.1088/1742-6596/665/1/012046

1. Introduction

Proton capture reactions play an important role in a nucleosynthesis process, especially in explosive nucleosynthesis such as X-ray bursts [1]. The X-ray binary system (Red Giant - Neutron Star) is characterized by a repeated sudden increase of X-ray emission, which is a consequence of a thermonuclear explosion in the atmosphere of an accreting neutron star. At high temperature and density conditions, the freshly accreted hydrogen and helium ignite. But the slow CNO cycles can be broken out, and the rapid proton capture (rp) process, which is a sequence of proton captures and β^+ decays and which is responsible for the production of proton-rich isotopes up to the mass 100 region, is initiated. The trigger conditions for the burst depend on the efficiency of the breakout reactions from the hot CNO cycle [2].

The possible breakout reactions are still under discussion. At the beginning, only α capture reactions on the waiting-point nuclei of the CNO cycles (^{15}O$(\alpha,\gamma)^{19}$Ne and ^{18}Ne$(\alpha,p)^{21}$Na) have been taken into account [2, 3, 4]. However, at present, the alternative two-proton capture reactions (^{15}O$(2p,\gamma)^{17}$Ne and ^{18}Ne$(2p,\gamma)^{20}$Mg) are also considered [5, 6]. In theoretical predictions, the direct three-particle capture process enhances the reaction rate by a few orders of magnitude [6] compared with a sequential one [5]. In order to verify these calculations, the two-proton radiative capture cross sections of mentioned reactions should be determined. In the present experiment, the ^{15}O$(2p,\gamma)^{17}$Ne reaction has been investigated.

A two-neutron halo structure was already observed for the Borromean nuclei ^{6}He and ^{11}Li [7, 8]. However, an observation of a two-proton halo structure is still a challenge. The Borromean proton dripline nucleus ^{17}Ne is a promising candidate for a two-proton halo, due to a small $2p$ separation energy (S_{2p} = 960 keV) [9]. The mixture of the d^2 and s^2 configurations of the two protons outside the ^{15}O core in the ^{17}Ne ground state is still unknown, and the results of theoretical calculations are controversial. In some papers, the s^2 configuration has been predicted to dominate, while in another, the dominating d^2 component has been expected. It seems that the theoretical conclusion about the properties of ^{17}Ne nucleus is still missing [10]. The solution to this situation is an experimental determination of the ^{17}Ne structure.

2. Experimental technique and setup

The experiment has been performed by means of the Coulomb dissociation method, which is typically used to investigate the nuclear structure of exotic nuclei. It is also an important instrument to study relevant reactions for nuclear astrophysics scenarios using an inverse process [11]. In case of several particles in the entrance channel, the time-reversed process is the only way to measure such a complicated reaction. In this method, the Coulomb field of a heavy nucleus is used as a source of virtual photons. Using the virtual-photon theory, the photoabsorption cross section σ_{photo} can be obtained from the differential Coulomb dissociation cross section σ_{Coul}. It can then be converted into the radiative capture cross section σ_{cap} by using the detailed-balance theorem [12]:

$$\sigma_{cap} = \frac{(2j_a+1)\,2}{(2j_b+1)\,(2j_c+1)}\frac{k_\gamma^2}{k^2}\sigma_{photo}. \tag{1}$$

The LAND-R^{3}B detection setup at GSI was conceived to accommodate such experiments. In the present experiment, the beam was produced by nuclear fragmentation of a ^{20}Ne primary beam on a beryllium production target, situated at the entrance of the fragment separator (FRS), where dipole magnets filter out all species except those with a specific A/Z ratio. The selected secondary beam of ^{17}Ne, with an energy E = 500 AMeV, was transported to the experimental setup (Fig. 1) and identified on an event-by-event basis by means of energy-loss, position, and time-of-flight measurements (Fig. 2a).

After the interaction with secondary targets, position measurements, defining the trajectories of the reaction products, energy loss, and time-of-flight measurements were used to identify

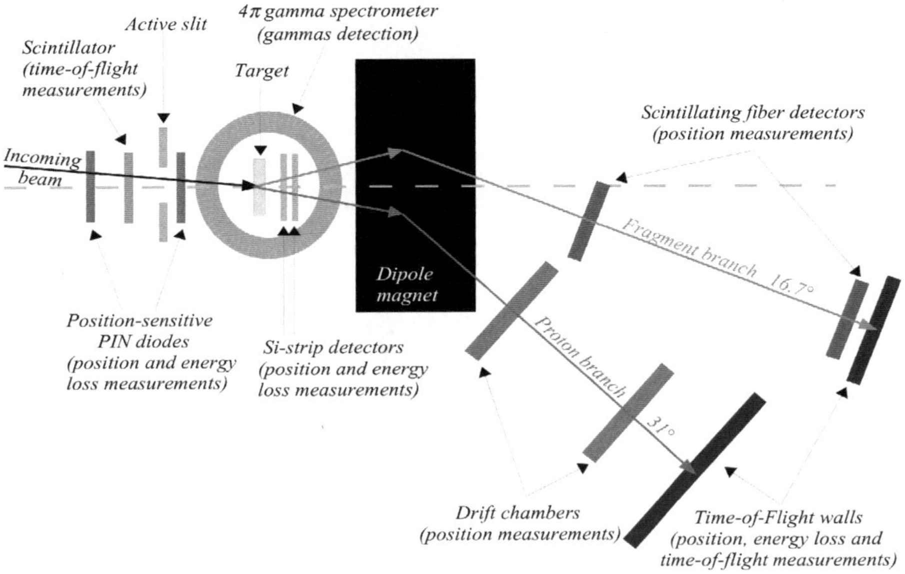

Figure 1. LAND-R^{3}B experimental setup.

outgoing particles (Fig. 2b and 2c), and a 4π gamma spectrometer detected γ-rays emitted by the deexciting fragment.

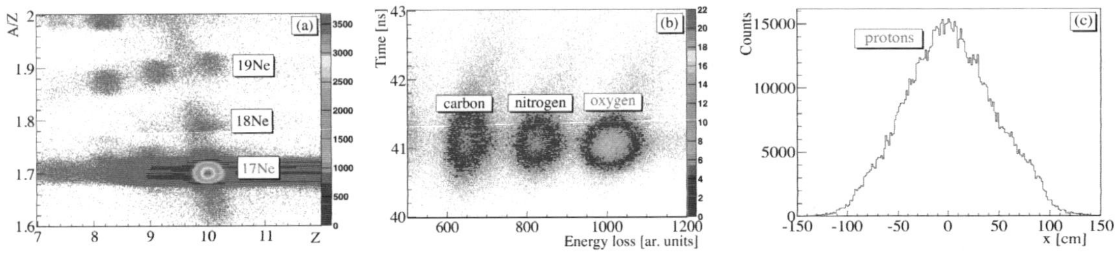

Figure 2. (a) - The identification of incoming beam nuclei; (b) - the identification of outgoing fragments; (c) - detected protons in the Time-of-Flight wall.

To later reconstruct the excitation energy of the desired isotope using the invariant-mass method, the reaction products were tracked on an event-by-event basis. This was done via the magnetic rigidity, time-of-flight, and the known Z (from energy loss) in the outgoing channel:

$$B\rho \propto \frac{A}{Z}\beta\gamma. \tag{2}$$

The measurements were performed using two different targets: a Pb target (200 mg/cm^2) and a C target (370 mg/cm^2). The Pb target was used to investigate the Coulomb dissociation reaction, and the C target to estimate the nuclear contribution. To evaluate the background, the measurements were performed also without any target.

3. Efficiency, acceptance, and γ-rays

The two-proton efficiency of the proton-arm detectors (drift chambers) was estimated to be $55.9 \pm 1.5\%$. Using a simulation within the *R3BRoot* framework, the geometrical acceptance curve of the experimental setup was determined as a function of the relative energy (Fig. 3a).

The γ-rays emitted by the deexciting fragments (^{15}O) were detected in the 4π gamma spectrometer placed around the target. Two groups of excited states of ^{15}O were observed: above 5 MeV and 6 MeV (Fig. 3b). However, only 5% of the events show these excited states, which makes them negligible from the viewpoint of the relative energy spectrum.

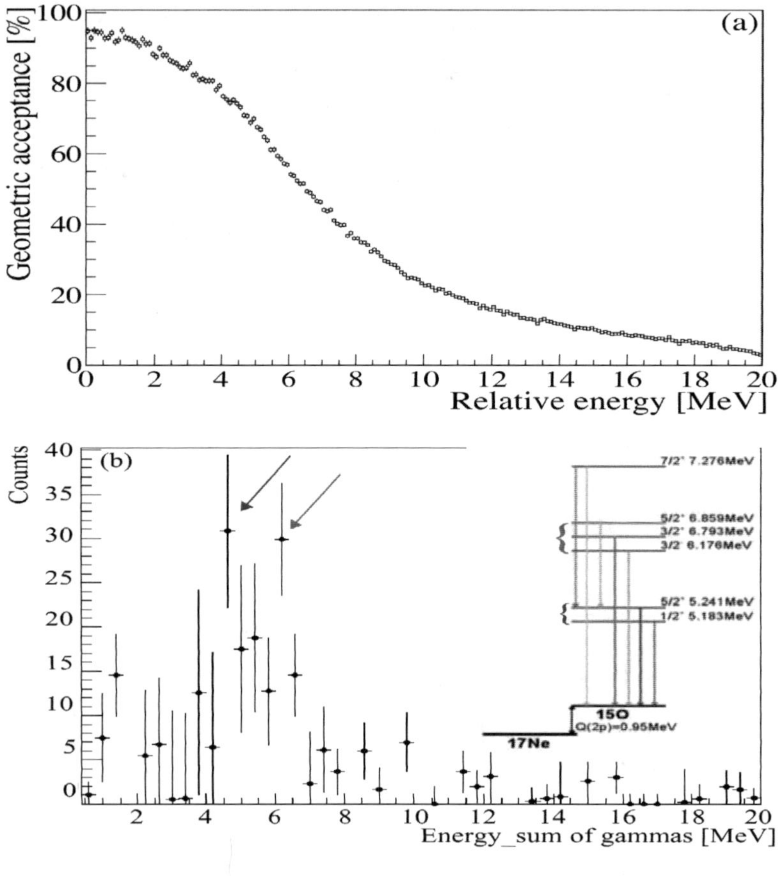

Figure 3. (a) - The geometric acceptance curve; (b) - γ energy sum spectrum.

4. Preliminary results

To calculate a Coulomb dissociation cross section, the following formula was used:

$$\sigma_{Coul} = p_{Pb}\left(\frac{M_{Pb}}{d_{Pb}N_{Av}}\right) - p_C\left(\alpha\frac{M_C}{d_C N_{Av}}\right) - p_{empty}\left(\frac{M_{Pb}}{d_{Pb}N_{Av}} - \alpha\frac{M_C}{d_C N_{Av}}\right), \quad (3)$$

where p is the interaction probability in the target, M is the molar mass of the target material [g/mol], d is the target thickness [g/cm^2], N_{Av} is Avogadro's number [mol^{-1}] and α is a radial scaling factor between Pb and C targets, the value oh which ($\alpha = 1.845$) was estimated from the experimental data. With this formula, the preliminary differential and integral Coulomb dissociation cross section was determined.

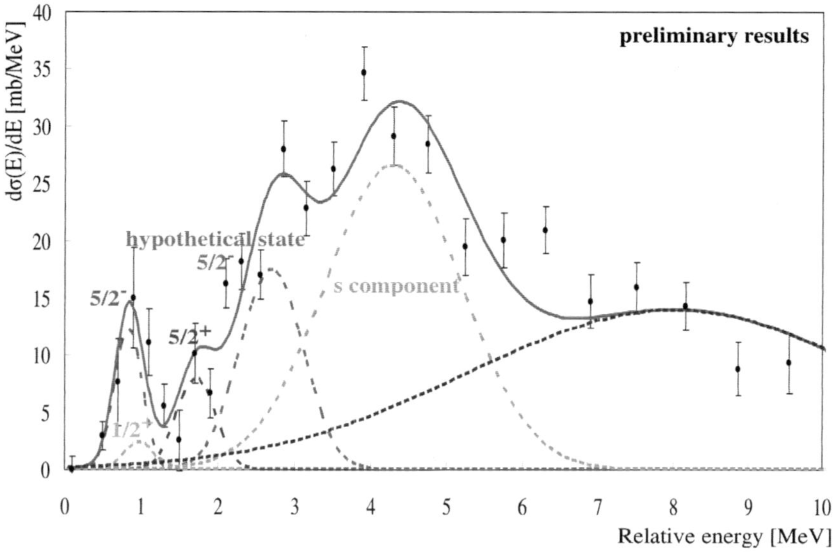

Figure 4. The preliminary fitting of the differential Coulomb dissociation cross section, with existing experimental and theoretical predictions of ^{17}Ne excited states.

For checking the efficiency and the acceptance adjustments, the integral cross section was calculated in two ways, the first using only the ^{15}O data ($\sigma_{Coul_1} = 289 \pm 32(stat.) \pm 35(syst.)$ mb), and the second using the differential cross section spectrum ($\sigma_{Coul_2} = 256 \pm 15(stat.) \pm 18(syst.)$ mb). The reasonable difference ($\Delta = 11\%$) between these two values shows the propriety of these corrections. The shape of the differential cross-section distribution is in agreement with experimental results from Ref. [13] and with the theoretical predictions from Ref. [14]. In the spectrum, the predicted and measured excited states of ^{17}Ne can be distinguished (Fig. 4). The next steps of the analysis will be to calculate the photoabsorption and the radiative capture cross sections.

The $\frac{1}{2}^- \rightarrow \frac{5}{2}^-$ excitation in ^{17}Ne can proceed only via $E2$ photons. Using the virtual photon spectrum (Fig. 5) the $B(E2, ^{17}Ne, \frac{1}{2}^- \rightarrow \frac{5}{2}^-)$ value can be derived to compare with the experimental result from Ref. [13]. The result of present work is $B(E2) = 57.6 \pm 10.6$ e^2fm^4 (only statistical uncertainty has been included), while the result from Ref. [13] is $B(E2) = 124 \pm 18$ e^2fm^4. The source of this discrepancy is not know yet. The analysis is in progress.

The important information about the three-body system ($core + p + p$) can provide $2p$ decays. By analyzing different types of energy and angular correlations between internal clusters ($[core, p+p]$ or $[core+p, p]$) in Jacobi coordinates, the mixture of the d^2 and s^2 configurations can be obtained [15]. This method is also used to analyze the ^{17}Ne case. The experimental data are compared with theoretical predictions provided by Ref. [16] using a simulation in the *R3BRoot* framework. The final conclusion is not obtained yet.

5. Summary

By using a ^{17}Ne secondary beam of an energy $E = 500$ AMeV produced in a fragmentation reaction, the ^{15}O$(2p, \gamma)^{17}$Ne reaction has been investigated. The incoming beam and outgoing

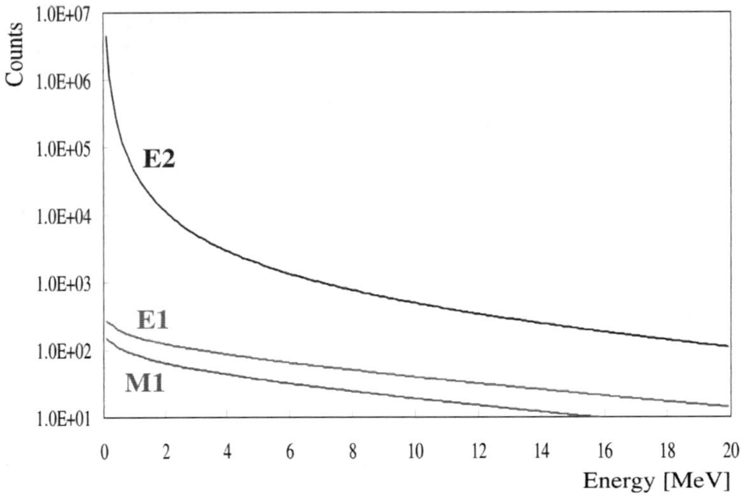

Figure 5. Virtual-photon spectrum for ^{17}Ne @ 500 AMeV on a Pb target.

reaction products have been identified and tracked. The required efficiency and acceptance corrections, as well as the influence of γ-rays emitted by the deexciting fragments have been estimated. The preliminary Coulomb dissociation cross section has been obtained. The photoabsorption and the radiative capture $^{15}O(2p, \gamma)^{17}Ne$ cross section calculations, as well as the mixture of the d^2 and s^2 configurations of ^{17}Ne structure analysis are ongoing.

6. Acknowledgments

This project was supperted by the German Federal Ministry for Education and Research (BMBF), EU(EURONS), ExtreMe Matter Institute EMMI, GSI Darmstadt, FIAS Frankfurt Institute for Advanced Studies, and HIC for FAIR.

References

[1] Lahir Ch and Gangopadhyay G 2012 *Int. J. Mod. Phys. E* **21** 1250074
[2] Wiescher M, Görres J, Uberseder E, Imbriani G, and Pignatari M 2010 *Annu. Rev. Nucl. Part. Sci.* **60** 381
[3] Van Wormer L *et al.* 1994 *Astrophys. J.* **432** 326
[4] Wallace R K, Woosley S E 1981 *Astrophys. J. Suppl.* **45** 389
[5] Görres J, Wiescher M, Thielemann F K 1995 *Phys. Rev. C* **51** 392
[6] Grigorenko L V, Zhukov M V 2005 *Phys. Rev. C* **72** 015803
[7] Tanihata I *et al.* 1985 *Phys. Lett. B* **160** 380
[8] Tanihata I *et al.* 1985 *Phys. Rev. Lett.* **55** 2676
[9] Kanungo R *et al.* 2005 *Eur. Phys. J. A* **25** 327
[10] Grigorenko L V, Parfenova Yu L, Zhukov M V 2005 *Phys. Rev. C* **71** 051604
[11] Aumann T 2005 *Eur. Phys. J. A* **26** 441
[12] Baur G and Bertulani C A 1986 *Nucl. Phys.* **A458** 188
[13] Chromik M J *et al.* 2002 *Phys. Rev. C* **66** 024313
[14] Grigorenko L V *et al.* 2006 *Phys. Lett. B* **641** 254
[15] Grigorenko L V *et al.* 2009 *Phys. Lett. B* **677** 30
[16] Grigorenko L V, Parfenova Yu L, Egorova I, *private comunication.*

CNO cycle: "Soft E1" mode of the ^{17}Ne excitation in the ^{17}Ne+$\gamma \rightarrow$ ^{15}O+$2p$ reaction.

Yu L Parfenova[1,2], **L V Grigorenko**[1,3,4], **I A Egorova**[1], **N B Shulgina**[3], **M V Zhukov**[6]

[1] JINR, Dubna, Russia
[2] INP MSU, Moscow, Russia
[3] RRC KI, Moscow, Russia
[4] GSI Helmholtzzentrum fur Schwerionenforschung, Darmstadt, Germany
[6] Chalmers University of Technology, Göteborg, Sweden

E-mail: parfenova@jinr.ru

Abstract. The ^{15}O(2p,γ)^{17}Ne reaction is studied using the time-reversed reaction of the^{17}Ne E1 Coulomb dissociation on lead target in the context of nuclear astrophysics. Looking for the relation between the data on the Coulomb excitation and the astrophysical 2p-capture rate, one faces problem to extract the Coulomb E1 strength function from the measured Coulex cross section. We use a number of phenomenological approaches to estimate influence of such processes as Coulomb-nuclear interference, populations of states with different J^{π}, etc. We calculate the ^{17}Ne+2p astrophysical capture rate and compare the results with different calculations.

1. Introduction

The last decade, studies of proton-rich nuclei in the vicinity of the drip-lines are a subject of intensive experimental and theoretical investigations. In particular, the study of a candidate to the two-proton Borromean halo nucleus ^{17}Ne (^{15}O+p+p where all binary subsystems are unbound) in dissociation reactions, is of a special interest (see discussions in [1, 2, 3]) In particular, the study of the ^{17}Ne Coulomb break-up can provide a information on ^{15}O(2p,γ)^{17}Ne reaction rate, that is interesting from the astrophysical view point. According to the detailed balance theorem, the 2p-capture reaction can be accessed as the time-reversed one for E1 Coulomb dissociation of ^{17}Ne in heavy target. Knowing the ^{17}Ne excitation spectrum for E1 Coulomb excitation, one can get the E1 strength function, and then make a conclusion on the astrophysical ^{15}O+$2p$ capture rate.

Recently, the ^{17}Ne Coulomb break-up on lead target has been experimentally studied at GSI with LAND setup (experiment S318, 2006) [4]. Analyzing these data and aiming to find the relation between the experimental data on the Coulomb excitation and the astrophysical 2p-capture rate we face a set of problems.

First of all, the "soft E1" mode (SDM) [5] is not well studied for excitations of proton-rich nuclei. For example, theoretically predicted strength functions significantly differs depending on the calculation models (see, for example calculations in cluster model [6] and in collective RPA model [7]).

Another considerable uncertainty concerning the interpretation of SDM excitation is related to the yet unknown mixture of the $[s^2]$ and $[d^2]$ proton configurations in the ^{17}Ne ground state wave function (WF).

Studying the time-reversed reaction of the ^{17}Ne dissociation in the context of nuclear astrophysics, it is necessary to extract the information on the Coulomb-induced E1 dissociation from a wide set of different processes contributing to the ^{15}O production, i.e. nuclear- induced dissociation, proton removal from the ^{15}O core fragment in ^{17}Ne [8] contributing to the total proton removal in ^{17}Ne up to 30% [9], and etc.

When both Coulomb and nuclear mechanisms act, the question unavoidably arise on the Coulomb-nuclear interference which is also poorly studied and may influence the ^{17}Ne excitation spectrum and angular distribution of the ^{17}Ne fragments.

Also, contributions of different Eλ and Nλ transitions in vicinity of SDM peak are not yet understood.

Here, we focus on several of these problems, i.e. (i) the nuclear dipole (N1) excitation of ^{17}Ne in lead and carbon targets, (ii) the Coulomb-nuclear interference for electric dipole (E1) and nuclear dipole (N1) excitations, and (iii) the astrophysical ^{15}O+$2p$ capture rate. For the calculations, a combined approach based on convenient models for description of the Coulomb and nuclear- induced reactions with clusterized systems, is used.

2. Model

This combined approach is based on the model of Bertulani and Baur [10] (BBM) for description of the Coulomb- induced dissociation of projectile, the eikonal approximation of the Glauber model [11] (EAGM), and the Green function method with one final state interaction in the ^{15}O-p subsystem [6] (OFSI model).

For quantitative estimates of the Coulomb-nuclear interference, the total excitation cross section to the ^{15}O+p channel is written as

$$\frac{d\sigma_{dif}(E_T,\theta)}{dE_T\, d\theta} = |A_C(E_T,\theta) + A_N(E_T,\theta)|^2 \,, \tag{1}$$

where $A_C(E_T,\theta)$ and $A_N(E_T,\theta)$ are the Coulomb and nuclear amplitudes depending on the ^{17}Ne energy E_T (calculated from the ^{15}O+p+p threshold) and scattering angle θ in the laboratory system, approximated as

$$A_C \approx \sqrt{\frac{d\sigma^C_{dif}}{d\theta}\frac{dB_C}{B_C dE_T} F_{abs}(\theta)} \;\; ; \;\; A_N \approx e^{i\phi_{rel}} \sqrt{\frac{d\sigma^N_{dif}}{d\theta} \cdot \frac{dB_N}{B_N dE_T}} \tag{2}$$

Here, ϕ_{rel} is the relative Coulomb-nuclear phase-shift, a free parameter (generally depending on E_T), $B_C(E\lambda, E_T)$ and $B_N(E\lambda, E_T)$ are the strength functions of the Coulomb and nuclear-induced excitations with multipolarity λ, depending the ^{17}Ne energy E_T.

The *first modification* we use in calculations of the Coulomb excitation cross section σ^C_{dif} in BBM is that the stepwise function at minimal impact parameter (corresponding to the grazing angle) is replaced by a smooth function $F_{abs}(b)$ describing the absorption of ^{17}Ne by the target and taking into account the diffusive edge of the nucleus. It is defined in EAGM as $F_{abs}(b) = \langle |S_{p_1} S_{p_2} S_C|^2 \rangle$, profile functions of the proton-target (S_{p_i}, $i = 1, 2$) and ^{15}O core-target (S_C) interactions [11] are determined by the corresponding interaction potentials and fragment densities with parameters fitted to reproduce the interaction cross section (see Fig. 1). Fig. 1 shows sensitivity of cross sections calculated in EAGM to the profile functions, in particular, to the size of the ^{15}O fragment. For this, two values of the ^{15}O rms radius are used in the calculations, 2.42 fm and 2.51 fm [12]. For the parameters we refer to [9, 13, 14].

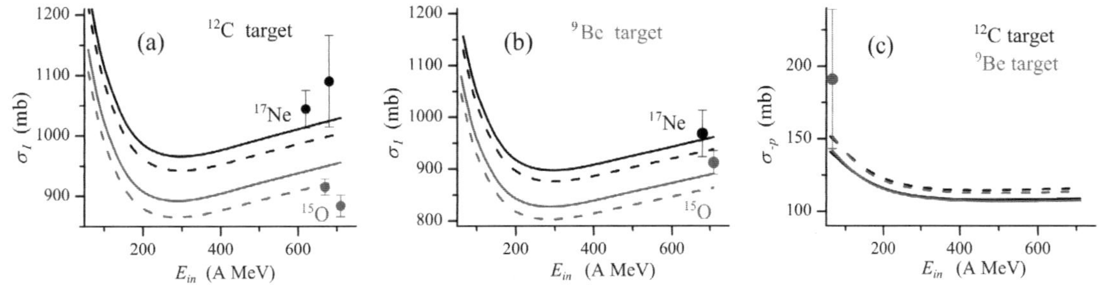

Figure 1. Energy dependence of the ^{15}O (red curves) and ^{17}Ne (black curves) interaction cross sections on the ^{9}Be (a) and ^{12}C (b) targets obtained with the reduced core rms radius (solid lines) and free nucleus rms radius (dashed lines); energy dependence (c) of the two-proton removal cross section in the Be (black curves) and C (red curves) targets. Experimental data are from [15].

To rewrite $F_{abs}(b)$ as a function of the angle θ we make the *second modification*: angular distributions in nuclear-induced dissociation are defined by the Coulomb trajectories and the impact parameter b is related to the angle as θ as $b(\theta) = a \cot \theta/2$ (in non-relativistic limit) where a is a constant [10]. In this case, the free parameter of the BBM, the impact parameter corresponding to the grazing angle, is now fixed in our approach by the profile functions. The nuclear differential diffraction cross section $d\sigma^N_{dif}/d\theta$ for ^{17}Ne takes the form [11]

$$\frac{d\sigma^N_{dif}}{d\theta} = 2\pi \sin\theta \left\{ \left\langle |\, 1 - S_{n_1} S_{n_2} S_C |^2 \right\rangle - |\, \langle 1 - S_{n_1} S_{n_2} S_C \rangle \,|^2 \right\} \tag{3}$$

where $\langle \quad \rangle$ denotes the ground-state expectation value. Note that in the calculations we use the ^{17}Ne ground state WF obtained in [3] with the method of the Hyperspherical Harmonics [16]. This wave function provides 48% weight of s-wave configurations.

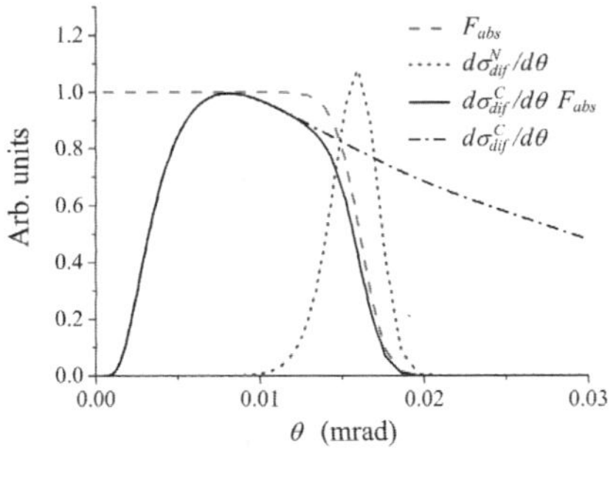

Figure 2. Angular dependence of electromagnetic dissociation cross section $d\sigma^C_{dif}/d\theta$ (dash-dotted black line) and that corrected by nuclear absorption (solid black line), $F_{abs}(b)$ (dashed red line), and nuclear dissociation cross section $d\sigma^N_{dif}/d\theta$ (blue dotted line).

The angular dependence of F_{abs}, the nuclear diffraction cross section obtained in the EAGM and the E1 Coulomb dissociation cross section calculated in BBM are given for lead target in Fig. 2. The nuclear diffraction takes place at the surface of the nucleus leading to larger reflection angles than those corresponding to the Coulomb dissociation. With the diffusive nuclear edge there is a region of angles where both the Coulomb and nuclear mechanisms acts and may interfere. Note, that all calculations were made for the ^{17}Ne incoming energy E_{in}=500 A MeV.

To estimate the Coulomb-nuclear interference using (2) we need the strength function for nuclear-induced excitation. Calculations of the strength function in the OFSI model [6] are rather complicated, and we use the *third simplification* supposing that $B_C(E\lambda, E_T)$ and

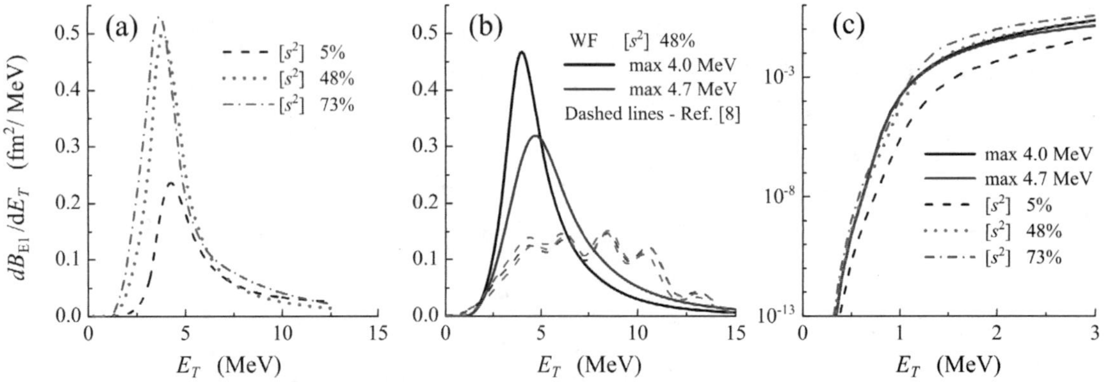

Figure 3. (a) - strength functions obtained in [6] with different ground state wave functions of ^{17}Ne with different weights of s and d-waves; (b) - strength functions obtained within our approach with different FSI (shown by solid lines) in the comparison with results of the RPA model (shown with dashed lines) [7]; (c) - low energy behavior of our strength functions compared with those obtained in [6].

$B_N(E\lambda, E_T)$ are similar. Here, the energy E_T dependence of the strength function is calculated in the OFSI model with the total strengths of the ^{17}Ne Coulomb excitation taken from Refs. [17]. The E1 strength function is found as

$$\frac{dB_{E1}}{dE} = 2\frac{2J_f+1}{2J_i+1}\left(\frac{2}{\pi}\right)^2 E_T \int_0^1 d\varepsilon \frac{|A(\varepsilon)|^2}{v_X v_Y} \tag{4}$$

where energy $E_T = E_X + E_Y$ is the sum of energies in the X- and Y- subsystems (^{15}O+p and ^{16}F+p, respectively), $\varepsilon = E_X/E_T$, v_X and v_Y are corresponding relative velocities of the fragments in the subsystems (for more details of the OFSI model we refer to [6]). The amplitude of excitation $A(\varepsilon)$ is determined by the ^{17}Ne WF and the dipole operator written in the Jacobi coordinates as $\hat{D}_\mu = Z_{eff}\rho\cos\theta Y_{1\mu}(\hat{\mathbf{y}})$ with $Z_{eff} = e^2\frac{98}{255}$.

Calculations within the OFSI model are performed taking into account the antisymmetrization for excited state WFs, the recoil effects for different components, depending on the spin in the subsystems. Also the resonances in the s- and d-waves of the core-p subsystem are taken into account (one final state interaction). In Fig. 3 the calculated strength functions of ^{17}Ne are compared with the results of other authors [6, 7]. In panel (a) strength functions from [6] are shown, obtained with three different ground state wave functions of ^{17}Ne containing respectively $\sim 5\%$, $\sim 48\%$, and $\sim 73\%$ of the $[s^2]$ configuration in the three-body calculations. In panel (b) of this figure the blue and black solid curves are obtained for the ground state WF ($[s^2]$ $\sim$48 %) with different p-wave interaction potentials in the Y-subsystem (^{16}F subsystem and the proton), showing the sensitivity of the results to this final state interaction. Thus, varying the potential in Y-subsystem we change the position of the maximum of the strength function from 4.0 MeV to 4.7 MeV. These results in (b) are compared with those, obtained within RPA collective-type model [7]. In panel (c) our results (solid lines) are compared to those from [6] (dashed lines) at low energies.

Finally, writing the expression for differential cross section $d\sigma^N_{dif}(J, J', P)/d\theta$ of the transition from the ground state J to the excited state J' for the certain multipolarity $\mathbf{P}$ of the nuclear-induced excitation N_P, it is possible to deduce transition rules for change in parity $\pi\pi' = (-1)^P$ and $\mathbf{J}' = \mathbf{P} + \mathbf{J}$, and find the N1 contribution to the cross section of nuclear-induced excitation to the state J' depending on the angle θ and energy E_T.

Table 1. Total cross section, cross sections of the Soft E1 and N1 excitation, cross sections of the Soft E1 and N1 excitation leading to the ^{16}F production in s-wave and d-wave resonant states in C, Si, and Pb targets at the ^{17}Ne energy 500 A MeV. All the cross sections are given in mb.

target	σ^C_{diff} total	σ^C_{diff} E1	σ^C_{diff} E1, s	σ^C_{diff} E1, d	σ^N_{diff} total	σ^N_{diff} N1	σ^N_{diff} N1, s	σ^N_{diff} N1, d
Pb	415	386	368	18	35	1.2	0.1	0.05
Si	17	15	14.3	0.71	13	0.5	0.04	0.02
C	3.2	3.0	2.86	0.14	12	0.4	0.03	0.016

3. Results

The results of the total Coulomb and nuclear cross section calculations are presented in Table 1. For the lead target, the contribution of the N1 nuclear transition is negligibly small compared to that of E1. Depending on the phase ϕ_{rel} in (2) the corresponding deviation from the decoherent value with $\phi_{rel} = \pi/2$ is within 3% for the Pb target, 12% for the Si target, and 15% for the C target. Note, that in the ^{17}Ne Coulomb dissociation, the spectra of the ^{16}F production in the s- and d-wave resonant states have their maxima at $\sim$ 4.5 MeV and $\sim$ 6.5 MeV, respectively. The d-wave ^{16}F production is small in the whole range of the ^{17}Ne energy E_T for both C and Pb targets, and the shape of the SDM peak is mainly determined by the s-wave ^{16}F production.

Using our strength functions $\frac{dB_{E1}}{dE}$ (Fig. 3 (b)) we estimate the non-resonant E1 contribution to the ^{15}O$(2p,\gamma)^{17}$Ne reaction rate as function of the temperature T [6]

$$< \sigma_{2p,\gamma} v >= \left(\frac{17}{15}\right)^{3/2} \left(\frac{2\pi}{mkT}\right)^3 \frac{2J_f+1}{2(2J_i+1)} \int dE \frac{16\pi}{9} e^2 E_\gamma^3 \frac{dB_{E1}(E)}{dE} \exp\left[-\frac{E}{kT}\right] \quad (5)$$

where J_i and J_f are spins of ^{15}O and ^{17}Ne g.s., $k = 8,6173324(78)10^{-5}$ eV/K is the Boltzmann constant.

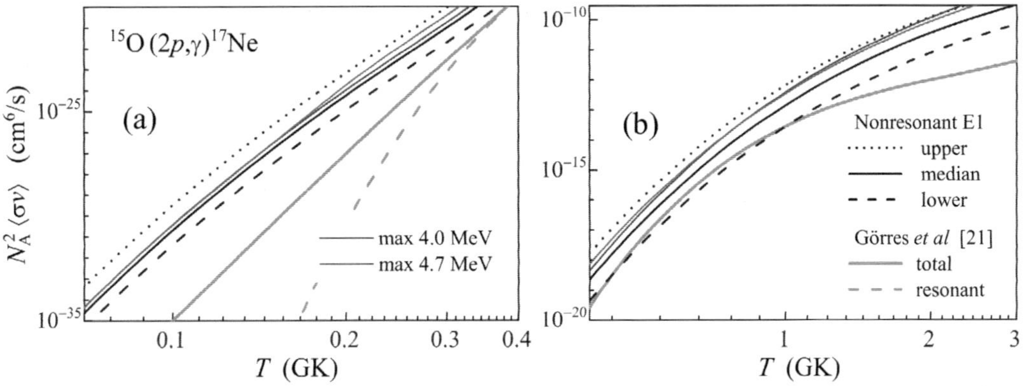

Figure 4. Temperature dependence of the astrophysical rate. Rate 1(2) is shown by blue(red) line. Black curves from [6] are the median (solid line) rate and its upper (dashed line) and lower (dotted line) limits obtained with regard for uncertainties in calculated resonance and non-resonant contributions. Grey curves are total intensity and resonance contribution from Ref. [18].

Figure 4 shows the 2p-capture rate in different temperature regions obtained with the strength functions with maxima at 4.0 and 4.7 MeV. One can see, that the results of our calculations with the WF providing 48% of s-wave are very close to the curves from [6] obtained with the same WF within the same, OFSI, model, but without considering for antisymmetrization of the excited state wave function and recoil effects. From comparison of the red and blue lines, we see, that the final state p-wave interaction potentials in the Y-subsystem, determining the position of the SDM peak, insufficiently affects our results at the energies E_T up to resonance (the difference does not exceed 2 times). At the same time, our results and those from [6] excceeds the results from [18] by few orders of magnitude.

4. Conclusions

The Coulomb-nuclear interference is estimated for the first time for three-body ^{17}Ne dissociation. The applied approach provides a quite realistic description of the E1 Coulomb and N1 nuclear excitation of the ^{17}Ne Borromean nucleus. In the calculations of the 2p-capture rates at astrophysical (low) energies the OFSI model shows to be a reliable substitute for the full three-body model [6], with transparent physical interpretation. In our approach we got, that in the calculations with the same three-body WF, the 2p-capture rate at low temperature (< 0.2 GK) is independent of the position of the maximum of the strength function related with the p-wave potential in the Y-subsystem (^{16}F subsystem and the proton). At highere temperatures, there is a relation between the p-wave potential in the Y-subsystem (^{16}F subsystem and the proton) and the position of the maximum of the SDM strength function. The shape of the calculated ^{17}Ne excitation spectrum is related to the shape of the strength function obtained within the OFSI model, having maximum at about 4 - 5 MeV. Thus, knowing the experimental spectrum of the "Soft E1" mode and its maximum position it will be possible to restore the E1 strength function and to make more precise assessment of the 2p-capture rates.

Besides this, it appeared that in the ^{17}Ne excitation, the Coulomb-nuclear interference in a heavy target like lead target is small due to negligible nuclear contribution compared to the Coulomb one.

It is also important to point out the restrictions of our approach related to simplifications noted above, i.e. the strength function of the nuclear N1 excitation can differ from the E1 Coulomb one, the final state interaction should be taken into account for all fragments.

References

[1] Grigorenko L V 2009, *Physics of Particles and Nuclei* **40** 674
[2] Goldansky V I 1960, *Nucl. Phys.* **19** 482
[3] Grigorenko L V et al. 2003 *Nucl. Phys.* **A713** 372; erratum 2004 *Nucl. Phys.* **A740** 401
[4] Aumann T, et al. 2012 *Journal of Physics: Conference Series* **337** 012011
[5] Ikeda I 1988 INS, JPH-7, 1988 (in Japanese) ; Hansen P and Jonson B 1987 *Europhys. Lett.* **4** 409
[6] Grigorenko L V et al. 2006 *Phys. Lett.* **B641** 254
[7] Oishi T, Hagino K and Sagawa H 2011 *Phys. Rev. C* **84** 057301
[8] Grigorenko L V , Parfenova Yu L and Zhukov M V 2005 *Phys. Rev. C* **71** 051604(R)
[9] Parfenova Y L and Zhukov M V 2005 *AIP Conf. Proc. FINUSTAR (Greece)* **831** 526
[10] Bertulani C A and Baur G 1988 *Phys. Rep.* **163** 299
[11] Bertch G F, Hencken K and Esbensen H 1998 *Phys. Rev. C* **57** 1366
[12] Parfenova Yu L, Zhukov M V and Vaagen J S 2000 *Phys. Rev. C* **62** 044602
[13] Parfenova Y L and Zhukov M V 2006 *Phys. Rev. C* **74** 054607
[14] Grigorenko L V, Parfenova Y L and Zhukov M V 2005 *Phys. Rev. C* **71** 051604R
[15] Ozawa A, Suzuki T and Tanihata I 2001 *Nucl. Phys.* **A693** 32
[16] Zhukov M V *et al.* 1993 *Phys. Rep.* **231** 151 ; Danilin B V *et al.* 1989 *Yad. Fiz.* **49** 351 (*Sov. J. Nucl. Phys.* **49** 351)
[17] Chromik M J *et al.* 1997 *Phys. Rev. C* **55** 1676 ; Chromik M J *et al.* 2002 *Phys. Rev. C* **66** 024313
[18] Görres *et al.* 1995 *Phys. Rev. C* **51** 392

Moments Method for Shell-Model Level Density

V Zelevinsky[1], **M Horoi**[2], **and R A Sen'kov**[2]

[1] Department of Physics and Astronomy and National Superconducting Cyclotron Laboratory, Michigan State University, East Lansing, MI 48824-1321,USA
[2] Department of Physics, Central Michigan University, Mount Pleasant, MI 48859, USA

E-mail: zelevins@nscl.msu.edu

Abstract. The modern form of the Moments Method applied to the calculation of the nuclear shell-model level density is explained and examples of the method at work are given. The calculated level density practically exactly coincides with the result of full diagonalization when the latter is feasible. The method provides the pure level density for given spin and parity with spurious center-of-mass excitations subtracted. The presence and interplay of all correlations leads to the results different from those obtained by the mean-field combinatorics.

The long-standing problem of nuclear level density is important both from a theoretical viewpoint as a powerful instrument for studying nuclear structure and for numerous applications. Nuclear technology requires the detailed knowledge of level density, especially in the region around the neutron separation energy. Astrophysical reactions responsible for the nucleosynthesis in the Universe can be understood only if we know the nuclear level density in the Gamow window. The interest to this problem is also connected with the necessity to develop methods of statistical physics and thermodynamics for small *mesoscopic* systems (complex nuclei, atoms and molecules, nanoscale devices, prototypes of quantum computers).

Application of statistical physics to studies of nuclear properties was started in the early years of nuclear physics by Frenkel [1], Bethe [2, 3] and Landau [4]. Such ideas served as a base for a concept of compound nucleus by Bohr [5]. The review of those ideas and their further development was given in Refs. [6, 7]. At this stage theory of level density and detailed microscopic nuclear theory were growing more or less independently along parallel paths. Their synthesis became possible when the reliable versions of the nuclear shell model were established for several segments of the nuclear chart. Of course, this is related to the modern computational tools.

In this presentation we show the practical way of finding the density of energy levels with fixed exact quantum numbers (spin, parity, isospin projection) for a given shell model Hamiltonian. The method becomes practical because it is avoiding the full diagonalization of prohibitively large matrices. In contrast to popular mean-field approaches [8–10], all many-body correlations are fully accounted for, albeit in a restricted orbital space. The exact classification of states according to the conserved quantities and therefore the absence of artificial standard spin cut-off parameters is the advantage compared to Monte-Carlo calculations although some of them give relatively good results [11, 12]. Similar problems still may arise in approaches based on statistical spectroscopy [13].

Our method was gradually developed in several publications [14–22]. A high-performance code for calculating spin- and parity-dependent level density is available via the Computer

Physics Communication homepage on Science Direct [23]. Technically we use the *Moments method* based on statistical spectroscopy [24, 25], the applications are compared to the results obtained with the full diagonalization in the nuclear shell model with effective interactions when such calculations are practically possible [26]. The method is reinforced by the use of the recurrence relations [17, 27, 28] for eliminating spurious contributions.

We start with the shell-model Hamiltonian that contains mean-field single-particle energies in the restricted space and effective two-body interactions (many-body forces can be added without problems although the equations become more cumbersome). The spherical orbitals $|njm\rangle$ are used as a basis, and the counted many-body levels have definite total spin J and parity. A certain distribution of non-interacting particles over orbitals defines a configuration (*partition*). For each partition p, let $D_{\alpha p}$ be the number of many-body states with exact quantum numbers $\{\alpha\}$ including the number of particles, spin, parity, isospin... The *diagonal* matrix elements of the total (including interaction) Hamiltonian define the energy *centroid* $E_{\alpha p}$ for the states of class α built in the partition p,

$$E_{\alpha p} = \frac{1}{D_{\alpha p}} \mathrm{Tr}^{(\alpha p)}\, H. \tag{1}$$

For each partition we define also the effective *width* (energy dispersion) $\sigma_{\alpha p}$,

$$\sigma^2_{\alpha p} = \frac{1}{D_{\alpha p}} \mathrm{Tr}^{(\alpha p)}\, H^2 - (E_{\alpha p})^2. \tag{2}$$

The contribution of each state to this width, related to the strength function of an unperturbed state fragmented over the exact eigenstates, can be found summing the squares of off-diagonal matrix elements of the full Hamiltonian over a given row of the matrix H. Here the interaction between the states of different partitions is fully accounted for. As known from the shell-model experience [26, 29], the centroids of partitions form a smooth sequence, while the widths of individual unperturbed states fluctuate only weakly which can be considered as one of features of quantum chaos (in particular, *geometric chaoticity* of angular momentum addition [30], that emerges, for a realistic strength of interactions, already at rather low excitation energy). The details of calculating the trace of H^2 and explanation of the computational algorithm can be found in Ref. [23]; a computational method based on group theory but close in spirit to our approach is presented in Ref. [31].

The set of quantities (1) and (2) forms a foundation of the Moments Method [24]. The level density of each partition is known [32] to be close to Gaussian. The first two moments define it with sufficient precision. The total level density is restored by summing the contributions of the partitions,

$$\rho(E;\alpha) = \sum_p D_{\alpha p} G_{\alpha p}(E), \tag{3}$$

where $G_{\alpha p}(E)$ is a *finite range Gaussian* with the centroid at $E_{\alpha p}$ counted from the ground state energy; this Gaussian is cut at a distance $\eta\sigma_{\alpha p}$ from the centroid, where η is an empirical parameter eliminating the unphysical tails of the Gaussian (there are theoretical arguments confirmed by experience that fix η between 2.5 and 3). The ground state energy is to be found independently, for example by the approximate diagonalization in a smaller space and the exponential convergence method [33, 34], where the exponential regime also starts approximately at a distance of 3-4 spreading widths.

The last essential point is exclusion of unphysical states corresponding to the center-of-mass excitation which are present in the shell-model calculations due to cross-shell mixing. If we classify the basis states similarly to the harmonic oscillator field by $N\hbar\omega$ excitations, then for $N = 0$ the spurious excitations are absent, $\rho_{\rm pure}(E, J, N = 0) = \rho_{\rm calc}(E, J, N = 0)$. With $N = 1$

excitations we have to subtract from the shell-model result the states which could be obtained by the action of the center-of-mass vector operator on all states of $N = 0$ class,

$$\rho_{\text{pure}}(E, J, N = 1) = \rho_{\text{calc}}(E, J, N = 1) - \sum_{J'=J,J\pm1} \rho_{\text{calc}}(E, J', N = 0). \tag{4}$$

The final result is expressed by a recurrent relation similar to used in [27, 28]:

$$\rho_{\text{pure}}(E, J, N) = \rho_{\text{calc}}(E, J, N) - \sum_{K=1}^{N} \sum_{J_K=J_{\min}}^{N,\text{step }2} \sum_{J'=|J-J_K|}^{J+J_K} \rho_{\text{calc}}(E, J', N - K). \tag{5}$$

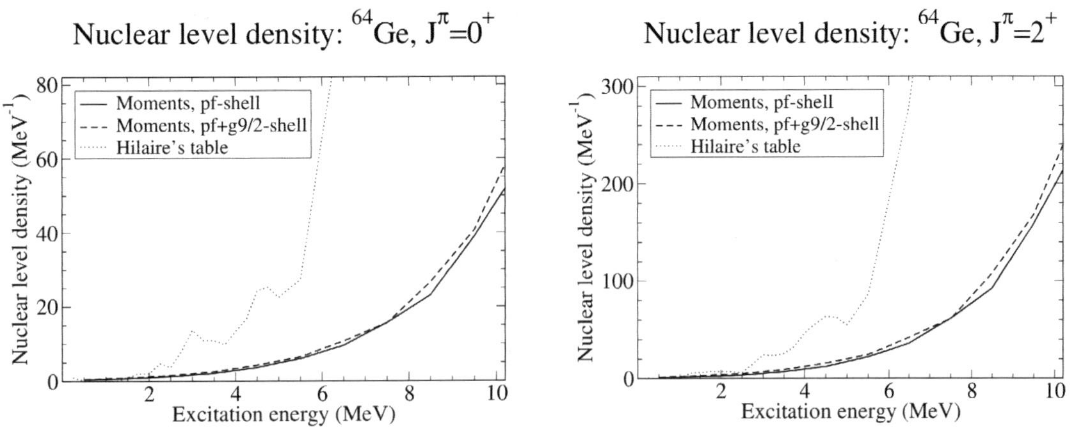

Figure 1. ^{64}Ge nuclear level density for $J = 0, 2$ and positive parity. Solid and dashed lines present calculations by Moments method in pf- (calculated with GXPF1A interaction) and $pf + g_{9/2}$-model spaces correspondingly. Dotted lines present level densities of Ref. [10].

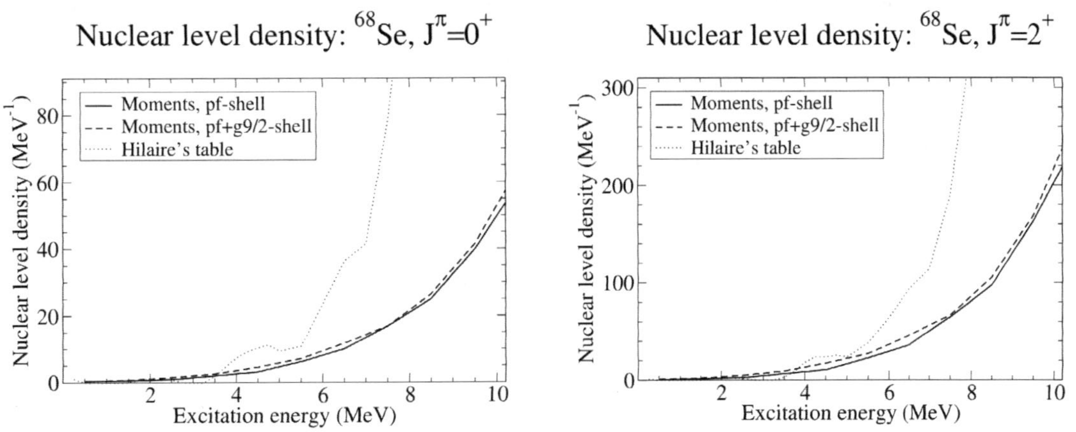

Figure 2. The same as Fig. 1 for ^{68}Se.

We illustrate the results by applications of the method to several nuclei important for astrophysics. Fig. 1 shows the level density for the $N = Z$ nucleus ^{64}Ge, states $J^\pi = 0^+$ and 2^+. Fig. 2 gives the same for ^{68}Se. In all four cases, the results for a larger orbital space, $pf + g_{9/2}$

shells, only slightly exceed those for the pf space. However, as it is typical for all our calculations, the shell-model level density is very smooth as a function of energy and turns out to be much lower than what can be found from the tables [10]. The preliminary conclusion is that the full set of effective shell-model interactions responsible for emergence of quantum chaos is very important smoothing the results based mainly on the mean-field combinatorics.

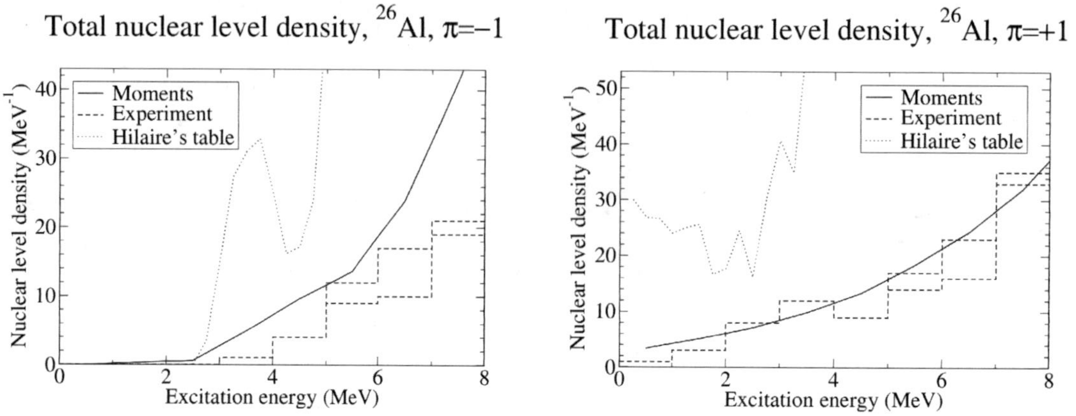

Figure 3. ^{26}Al, *s-p-sd-fp*-model space, all spins, WBT interaction [35]. Nuclear level densities are presented for negative (left figure) and positive (right figure) parities. Solid lines present the results of the Moments method, dashed lines show experimental level density (upper and lower lines correspond to experimental uncertainties in level identification), and dotted lines present the total level densities of Ref. [10].

In Fig. 3 we compare the total level density (for positive and negative parity separately) in ^{26}Al, calculated in the large *s-p-sd-fp*-model space restricted by $1\hbar\omega$ excitations only, with the WBT interaction [35]. For this isotope a good set of data is available and it is possible to juxtapose the results of our approach to those of the full shell-model diagonalization. Again, we obtain much lower and smoother level density than predicted by tables [10]. The two dashed stair-case lines show the data ("optimistic" and "pessimistic" estimates of uncertainties in identification of level parity). The calculations reasonably well describe available data.

Using the Moments method we can study dependence of the level densities on the interaction. Let us scale the shell model two-body interaction with two factors, f and f_p:

$$H = \epsilon + f_p V(\text{pairing}) + fV(\text{non-pairing}), \tag{6}$$

where ϵ is the one-body part of the Hamiltonian, V(pairing) includes only the $J = 0, T = 1$ matrix elements of the two-body interaction, while V(non-pairing) includes all remaining matrix elements. Varying the scale factors f and f_p we can study the influence of different components of interaction onto the level density. Figs. 4 and 5 present examples of such an investigation for ^{28}Si level densities, calculated in the *sd*-model space, with the USDB interaction [36]. Fig. 4 presents results only for $J = 0$ and Fig. 5 presents analogous calculations for all spins. Although dimensions and some details depend on J, the qualitative behavior is universal in all classes of states. The sharp irregularities related to the partition structure are rapidly smoothen even by the pairing interaction when it is accounted for exactly rather than in the BCS or HFB schemes. The remaining interactions broaden the level density and made its final tuning.

The natural next step in development of theory should be an attempt, in application to many nuclei, to extract the global empirical parameters, such as temperature, parameters of phenomenological fits, spin cut-off, and so on in order to bridge the gap between shell-model

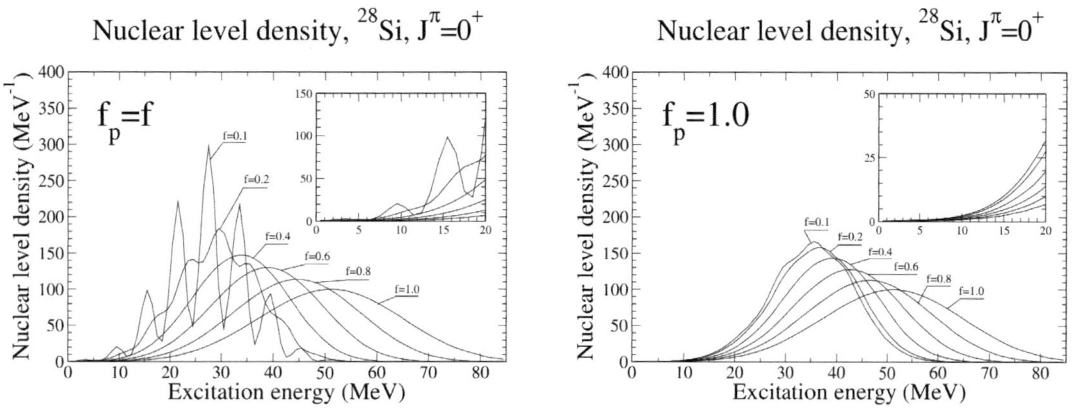

Figure 4. ^{28}Si, $J^{\pi} = 0^{+}$, *sd*-model space, USDB interaction [36]. Different curves correspond to different scale factors f, see Eq. (6). Left panel presents the case when $f_p = f$, i.e. pairing and non-pairing parts of the interaction scales similarly. Right figure presents the case when pairing interaction is always on, $f_p = 1$, while non-pairing interaction scales. The inset shows the low-energy part.

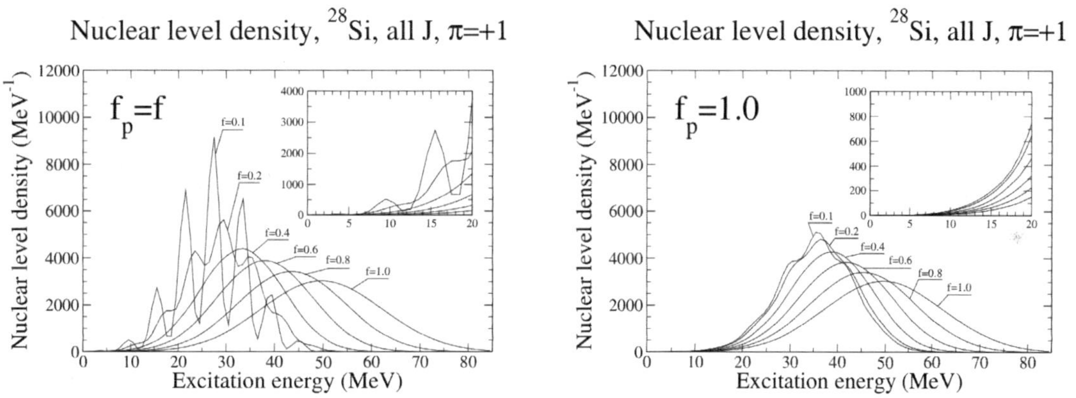

Figure 5. ^{28}Si, all J, $\pi = +1$, *sd*-model space. The description is similar to Fig. 4.

theory and convenient statistical descriptions used by experimentalists. Such studies also could help in selection of more reliable effective interactions for different regions of the periodic table. It might be that the global behavior of the level density is governed by few "coherent" interactions while the remaining abundant matrix elements are responsible for broadening and smoothing the stair-case behavior and can be modeled by an appropriately chosen "noise". With parallelization of calculations [23], they can be extended to larger many-shell spaces. Finally, we have to confess that the problem of continuum effects is still waiting for its proper consideration.

Acknowledgments

V.Z. and M.H. acknowledge support from the NSF grant PHY-1068217, the work of R.A.S and M.H. was supported by the DOE NUCLEI grant DE-SC0008529. V.Z. is grateful to S. Goriely and H. Schatz for constructive discussions during the NPA6 conference.

References

[1] Frenkel Ya I 1936 *Sow. Phys.* **9** 533
[2] Bethe H 1936 *Phys. Rev.* **50** 332

[3] Bethe H 1937 *Rev. Mod. Phys.* **9** 69
[4] Landau L D 1937 *J. Exp. Theor. Phys.* **7** 819
[5] Bohr N 1936 *Nature* **137** 344
[6] Ericson T 1960 *Adv. Phys.* **9** 425
[7] Gilbert A and Cameron A G W 1965 *Can. J. Phys.* **43** 1446
[8] Goriely S, Hilaire S, and Koning A J 2008 *Phys. Rev.* C **78** 064307
[9] Goriely S, Hilaire S, Koning A J, Sin M, and Capote R 2009 *Phys. Rev.* C **79** 024612
[10] http://www.astro.ulb.ac.be/pmwiki/Brusslib/Level
[11] Alhassid Y, Liu S, and Nakada H 2007 *Phys. Rev. Lett.* **99** 162504
[12] Alhassid Y, Fang L, and Nakada H 2008 *Phys. Rev. Lett.* **101** 082501
[13] Teran E and Johnson C W 2006 *Phys. Rev.* C **73** 024303; C **74** 067302
[14] Horoi M, Jora R, Zelevinsky V, Murphy A St J, Boyd R N, and Rauscher T 2002 *Phys. Rev.* C **66** 015801
[15] Horoi M, Kaiser J, and Zelevinsky V 2003 *Phys. Rev.* C **67** 054309
[16] Horoi M, Ghita M, and Zelevinsky V 2004 *Phys. Rev.* C **69**, 041307(R)
[17] Horoi M and Zelevinsky V 2007 *Phys. Rev. Lett.* **98** 263503
[18] Gao Z C and Horoi M 2009 *Phys. Rev.* C **79** 014311
[19] Gao Z C, Horoi M, and Chen Y S 2009 *Phys. Rev.* C **80** 034325
[20] Scott M and Horoi M 2010 *EPL* **91** 52001
[21] Sen'kov R A and Horoi M 2010 *Phys. Rev. C* **82** 024304
[22] Sen'kov R A, Horoi M, and Zelevinsky V 2011 *Phys. Lett.* B **702**, 413
[23] Sen'kov R A, Horoi M, and Zelevinsky V 2013 *Comp. Phys. Comm.* **184** 215
[24] Wong S S M 1986 *Nuclear Statistical Spectroscopy* (Oxford, University Press)
[25] Kota V K B and Haq R U 2010 eds., *Spectral Distributions in Nuclei and Statistical Spectroscopy* (World Scientific, Singapore)
[26] Zelevinsky V, Brown B A, Frazier N, and Horoi M 1996 *Phys. Rep.* **276** 85
[27] Jacquemin C 1981 *Z. Phys.* A **303** 135
[28] Van Isacker P 2002 *Phys. Rev. Lett.* **89** 262502 (2002)
[29] Pillet N, Zelevinsky V, Dupuis M, Berger J-F, and Daugas J M 2012 *Phys. Rev.* C **85** 044315
[30] Zelevinsky V and Volya A 2004 *Phys. Rep.* **391** 311
[31] Launey K D, Sarbadhicary S, Dytrych T and Draayer J P 2013, *Comp. Phys. Comm.* in press
[32] Brody T A, Flores J, French J B, Mello P A, Pandey A, and Wong S S M 1981 *Rev. Mod. Phys.* **53** 385
[33] Horoi M, Volya A, and Zelevinsky V 1999 *Phys. Rev. Lett.* **82** 2064
[34] Horoi M, Brown B A, and Zelevinsky V 2002 *Phys. Rev.* C **65** 027303
[35] Warburton E K and Brown B A 1992 *Phys. Rev.* C **46** 923
[36] Brown B A and Wildenthal B H 1988 *Annu. Rev. Nucl. Part. Sci.* **38** 29

Journal of Physics: Conference Series **665** (2016) 012049 doi:10.1088/1742-6596/665/1/012049

Proton emission from the deformed odd-odd nuclei near drip line

M Patial[1]*, P Arumugam[1], A K Jain[1], E Maglione[2] and L S Ferreira[3]

[1]Department of Physics, Indian Institute of Technology Roorkee, Roorkee - 247667, India.

[2]Dipartimento di Fisica e Astronomia "G. Galilei", Via Marzolo 8, I-35131 Padova, Italy and Istituto Nazionale di Fisica Nucleare, Padova, Italy.

[3]Centro de Física das Interacções Fundamentais, and Departmento de Física, Instituto Superior Técnico, Avenida Rovisco Pais, P1049-001 Lisbon, Portugal.

E-mail: *monikapatial@gmail.com

Abstract. Proton emission from odd-odd nuclei is studied within the two quasiparticle plus rotor model which includes the non-adiabatic effects and the residual interaction between valence proton and neutron. Justification of the formalism is discussed through corroboration of our results with the experimental spectrum of ^{180}Ta. Exact calculations are performed to get the proton emission halflives. Our results for the proton emitter ^{130}Eu leads to the assignment of spin and parity $J^{\pi} = 1^{+}$ for the ground state. The role of Coriolis and residual neutron-proton interactions on the proton emission halflives and their interplay are also discussed.

1. Introduction

Proton decay in the exotic nuclei near proton drip-line mostly takes place from the continuum of energy level spectrum which has single particle character due to Fermi level being closer to the continuum. We can study the physics of these nuclei through the decay width which provides relevant information about the structure and we can infer the details of amplitudes of wave functions of the decaying state. Hence the proton emission decay widths, complemented by the spectra built on the decaying states, is a perfect tool to study the structure and decay of nuclei right on the proton drip line. In a theoretical description within the adiabatic limit, it is assumed that decaying proton moves in the single-particle Nilsson resonances with the unbound core plus neutron system. It has been clearly demonstrated [1] in the case of odd-even nuclei that it is important to consider the nonadiabatic effects, which take care of the rotational excitations of the daughter nucleus, through the Coriolis interaction in a particle plus rotor model approach. A proper treatment of the pairing residual interaction [1] provides a more complete and consistent description of proton emission in agreement with the experimental data. The extension of this model to triaxial nuclei confirmed strong triaxial deformations in ^{145}Tm [2] and ^{141}Ho [3]. These studies show how powerful the study of proton radioactivity can be for the knowledge of nuclear structure.

A proper theoretical framework has been formulated recently [4, 5] to study the structure of odd-odd nuclei which is least explored. An exciting feature that can be studied in these nuclei is the interaction between the odd proton and the odd neutron [6]. Though this residual neutron-proton (np) interaction could be weak, in combination with the zero point rotational energy, it leads to several possible combinations of spin and parity for the ground state of the nucleus.

Here we present our results for the spectrum of ^{180}Ta and for the proton emission halflife of ^{130}Eu. We focus on the role of the Coriolis and the residual np interactions and their interplay while describing the rotational states and the decay widths.

2. The nonadiabatic quasiparticle approach

In this approach, we consider a two quasiparticle plus rotor model [4, 6] based on the mean field of Woods-Saxon potential with deformed spin-orbital potential. The presence of residual pairing interactions are considered by the transformation of single-particle energy to quasiparticle energy through the relation $\tilde{\epsilon}_k = \sqrt{\Delta^2 + (\epsilon_k - \lambda)^2}$, where λ is the Fermi energy, ϵ_k is the single-particle energy and Δ is the pairing gap chosen as $12/\sqrt{A}$. The complete Hamiltonian for the odd-odd nucleus can be expressed as sum of intrinsic part and rotational part where latter consists of sum of various terms including the Coriolis interaction between valence particles and collective rotation, the particle-particle coupling and the recoil term. We have introduced by a suitable constant value, the residual np interaction (GM splitting [7]) and the Newby shift [8]. We consider a variable moment of inertia (VMI) defined as $\Im(I) = \Im_0\sqrt{1 + b\,I(I+1)}$, where b is the VMI parameter and the constant $\Im_0$, is fixed such that $\frac{\hbar^2}{2\Im(I=2)} = \frac{E^{2+}}{6}$, where the E^{2+} is the first excited 2^+ energy of the core. The symmetrized wave function for the parent nucleus takes the form as shown below:

$$\Psi^{I,M} = \sum_{K_p,K_n,K_T=K_n\pm K_p} \sqrt{\frac{2I+1}{16\pi^2(1+\delta_{K_T0})}} a^I_{K_T,K_n,K_p} \left\{ D^I_{M,K_T} \Upsilon^{K_p,K_n}_{K_T} + (-1)^{I+K_T} D^I_{M,-K_T} R_i \Upsilon^{K_p,K_n}_{K_T} \right\}. \quad (1)$$

$a^I_{K_T,K_n,K_p}$ is the mixing coefficient obtained after diagonalising the total Hamiltonian matrix for the parent nucleus. Similarly, the mixing coefficient, $c^{I_d}_{K_n}$ for the daughter nucleus is obtained by solving the Hamiltonian for the rotor plus neutron system. The decay width is expressed in terms of the overlap between the initial (parent nucleus) and final (daughter and proton) states, where the parent is considered to be a two quasiparticle plus rotor system. In this way the partial decay width can be written as

$$\Gamma^{I,I_d}_{l_pj_p} = \frac{\hbar^2\kappa}{\mu} \left| \sum_{K_n,K_p(K_T=K_n\pm K_p)} a^I_{K_T,K_n,K_p} c^{I_d}_{K_n} \times \sqrt{\frac{(2I_d+1)}{(2I+1)}} \langle I_d, K_n, j_p, K_p | I, K_T\rangle \, u_{K_p} N^{K_p}_{l_pj_p} \right|^2 , \quad (2)$$

where $N^{K_p}_{l_pj_p}$ represents asymptotic normalization factor, u_{K_p} is the spectroscopic factor calculated from BCS approach and hence $u^2_{K_p}$ represents the probability of unoccupancy of proton level in daughter nucleus. The CG coefficients define angular momentum coupling between daughter and emitted proton. The total decay width is obtained by taking sum of partial decay widths from every possible combination of the angular momentum values of proton (l_p and j_p values) and hence

$$\Gamma^{I,J_d} = \sum_{j_p=|I-I_d|}^{I+I_d} \Gamma^{I,Jd}_{l_p,j_p}. \quad (3)$$

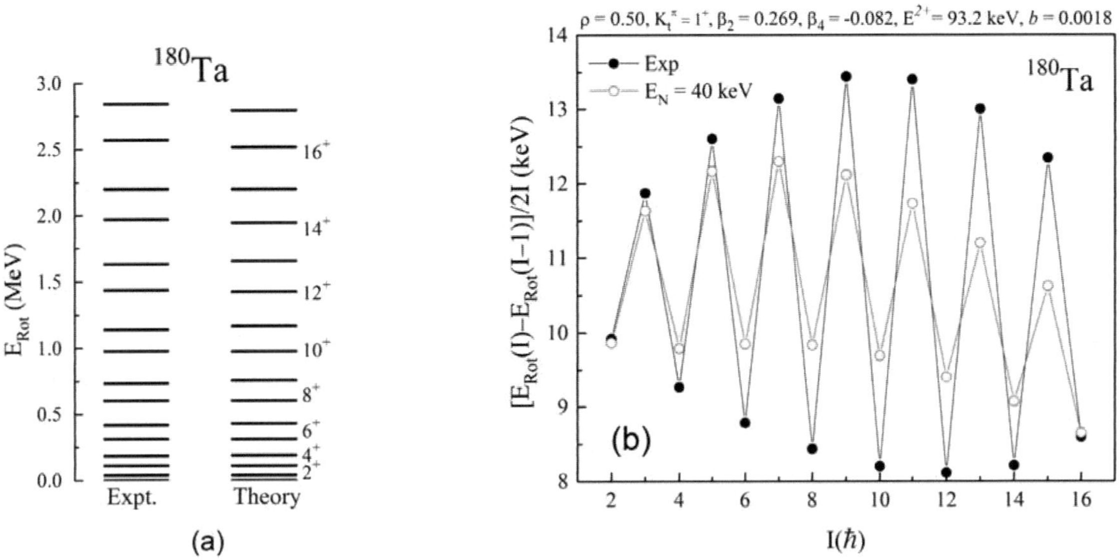

Figure 1. (a) Comparison between the theoretical and experimental [9] level spectrum of the ground state band of ^{180}Ta. (b) Calculated odd-even staggering pattern in the ground state band ($K_t^\pi = 1^+$) of ^{180}Ta. Solid black circles represent the experimental data and the open circles represent the calculated values. The parameters used for the calculations are also presented.

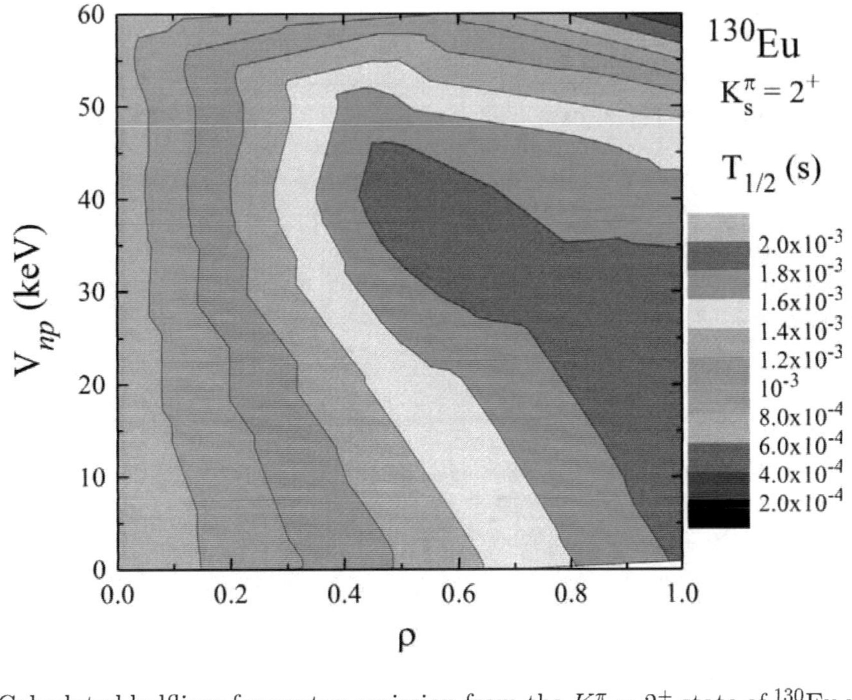

Figure 2. Calculated halflives for proton emission from the $K_s^\pi = 2^+$ state of ^{130}Eu plotted as a contour with the x and y axes representing the Coriolis attenuation coefficient and the residual np interaction.

3. Results

We present the calculated energy levels for ^{180}Ta in Fig. 1, for the band built on the ground state. From Fig. 1(a), it is clear that we have a very good agreement with the experimental results even for very high spins. It has to be noted that the parameters involved in the calculations such as the Coriolis attenuation coefficient (ρ) and VMI parameter (b) are tuned to achieve a good fit for the spectrum. The odd-even (Newby) shift in $K = 0$ band can be transmitted further to high-K bands through Coriolis mixing and such a shift can occur in the ground state ($K^{\pi} = 1^{+}$) band of ^{180}Ta. Introducing a Newby shift leads to a better agreement with the experimental values as shown in the odd-even staggering presented in Fig. 1(b). The Newby shift is tuned to reproduce the first excited state (2^{+}) accurately and the staggering pattern grossly. This leads to the best fit at 40 keV. With our results for ^{180}Ta we demonstrate that our formalism could explain the measured spectrum very well while the Coriolis and residual np interactions are taken in account.

Our results for the highly deformed proton emitter ^{130}Eu are presented in Ref. [4], where it is shown that the decay widths corresponding to both the probable ground states $I^{\pi} = 1^{+}, 2^{+}$ are close to the experimental value. The $I^{\pi} = 1^{+}$ state is a triplet state which is favoured by GM rule as lower in energy than the singlet state. Hence $I^{\pi} = 1^{+}$ is assigned to be the ground state of ^{130}Eu from which the proton is emitted and this conclusion does not depend on the attenuation of the Coriolis interaction. In Fig. 2, we present the half-lives for the decay from the 2^{+} state where we can identify the effect of interplay between the Coriolis and residual np interactions on the proton emission half-lives [5]. For example, at very low Coriolis attenuation $\rho \sim 0.0$, the halflives are almost independent of V_{np} and for very high ρ, the halflives increase for lower V_{np} but start decreasing for higher V_{np}. We note that V_{np} affects only the singlet state because it can mix through Coriolis interaction with a level of same spin built on the triplet state. If such a singlet state or other low-lying state which is sensitive to V_{np}, happens to be an isomeric state, it would be interesting to study proton emission from such states.

4. Conclusion

A proper formalism to study the low lying states of odd-odd nuclei is justified with our results for ^{180}Ta and has been successfully extended to explain the proton emission from ^{130}Eu. We propose that it would be interesting to study proton emission from isomeric states because the corresponding decay widths could be quite sensitive to the residual np interaction.

Acknowledgments

This work was supported by the Fundação para a Ciência e a Tecnologia (Portugal), Project: CERN/FP/123606/2011. Authors (MP and AKJ) thank the CSIR and DST (Govt. of India) for providing the financial support.

References

[1] Fiorin G, Maglione E and Ferreira L S 2003 *Phys. Rev. C* **67** 054302
[2] Arumugam P, Ferreira L S and Maglione E 2008 *Phys. Rev. C* **78** 041305
[3] Arumugam P, Ferreira L S and Maglione E 2009 *Phys. Lett. B* **680** 443
[4] Patial M, Arumugam P, Jain A K, Maglione E and Ferreira L S 2013 *Phys. Rev. C* **88** 054302
[5] Patial M, Arumugam P, Jain A K, Maglione E and Ferreira L S 2013 *Phys. Lett. B* **718** 979
[6] Jain A K, Sheline R K, Headly D M, Sood P C, Burke D G, Hrǐvnácová I, Kvasil J, Nosek D and Hoff R W 1998 *Rev. Mod. Phys.* **70** 843
[7] Gallagher Jr C J and Moszkowski S A 1958 *Phys. Rev.* **111** 1282
[8] Newby N D 1962 *Phys. Rev.* **125** 2063
[9] Wu S C and Niu H 2003 *Nucl. Data Sheets* **100** 483

A newly developed wrapping method for scintillator detectors

L Stuhl[1], **A Krasznahorkay**[1], **M Csatlós**[1], **A Algora**[1,2], **J Gulyás**[1], **G Kalinka**[1], **Zs I Kertész**[1], **J Timár**[1]

[1]MTA Atomki, Institute for Nuclear Research, H. A. S., Debrecen, Hungary
[2]Inst. de Fisica Corp., Univ. Valencia - C.S.I.C. Edificio de Inst. de Paterna, Valencia, Spain

E-mail: stuhl@riken.jp, stuhl@atomki.jp

Abstract. A neutron spectrometer, the European Low-Energy Neutron Spectrometer (ELENS), has been constructed to study exotic nuclei in inverse kinematics experiments. The spectrometer consisting of scintillator bars can be used in the neutron energy range of 100 keV to 10 MeV. To increase the light collection efficiency a special wrapping method was developed for each bars of ELENS. By using the specially heat treated reflector foil 15-20% better light collection is available. The development of wrapping process, the results of the test experiments are also presented.

1. Introduction

The most exotic nuclei are usually the most interesting ones but their production rate decreases exponentially with the increase of the proton-neutron asymmetry. In order to counterbalance this effect, high-efficiency detector setups and thick reaction targets are required. The information of interest is then extracted from the kinematical characteristics of the reaction products such as their scattering angles and energies. Using inverse kinematics, the kinetic energy of the emitted neutrons is relatively low and a short flight path is sufficient for the time-of-flight measurements. The detection of the slow neutrons with good efficiency, as well as the measurement of their energy and angular resolution requires specially designed spectrometers. The detection of a low-energy recoil product is often rather difficult, one needs to detect the neutrons at small angles, while their energies usually go below 1 MeV. Such neutrons interact with the hydrogen content of the detector material predominantly by elastic scattering on the protons. The recoiled protons have low-energy and they produce only a small number of photons in the scintillator material during their slowing down process, due to the quenching effect [1]. Such photons should be collected with high efficiency, so very good light collection and small light attenuation is necessary along the detector bars. An important part of the development was the investigation of different wrappings. Such neutron spectrometers have been built recently by Beyer et al. [2] and by Perdikakis et al. [3] and used successfully to study the strength distribution of the Gamow-Teller resonance [4] and have also been developed at RIKEN [5], [6] and at GSI [7]. In the present work we report the development and the concept of the new wrapping method.

2. Properties of ELENS system

The array consists of 16 plastic scintillator bars, each of them attached to photomultiplier tubes (PMTs) at both ends. The UPS89 fast plastic scintillator material was chosen, which has similar properties to NE102. The size of the detector is: $10 \times 45 \times 1000$ mm^3. A variable detector support was constructed for ELENS. The array is designed to measure neutron energies in the kinetic energy region from a few hundreds keV to a few MeV.

3. The wrapping

The quality of scintillation detectors is determined primarily by the quality of the scintillator material and the detector wrapping. The size and geometrical arrangements of the bars provide high detection efficiency, and at the same time high angular resolution for ELENS. The main problem of similar configurations is the inefficient light transport over long distances in thin ($d/l \ll 1$) tubes. Efficient light transport can only be achieved by specular reflectance materials of high reflectivity. A very good light collection is necessary to pick up the small signals, so a proper wrapping material, and the perfect fitting of the foil onto the plastic was an important criterion. There are several possible candidates, such as wrapping materials for plastic scintillators. The popular materials are: 3M Radiant Mirror Films, Teflon tapes, plastic tapes, gold coated tapes, aluminum foils, white diffuser paint or tape, etc. The main properties of these are the high reflection rate. In addition, these materials must be mechanically stable and flexible within certain limits. The detector paddles in the case of ELENS are wrapped with a specially treated VM2000 multilayer reflector foil (Enhanced Specular reflector), which has been produced by 3M. It is a multilayer reflective foil based on a novel technology [8]. This foil has a good reflection coefficient of $R > 97\%$ for $\lambda \geq 400$ nm, and R = $(98.5 \pm 0.3)\%$ at 430 nm wavelength of light [9]. Because of the thickness and rigidity of the multilayer material, it is very difficult to fold, wrap or create any kind of a crease in VM2000 foil, without decreasing of the light guidance parameters.

4. The novel wrapping method

There are many methods for the optimal wrapping (with VM2000) but all these methods are using some mechanical interactions on the foil. The VM2000 is composed of several different layers, so it is sensitive for the external mechanical influences as cuts or abrasions. Any abrasion or cut may compromise the optical properties of the foil. Along the cut edges the thickness of the material changes (gets thinner) or microscopic cracks

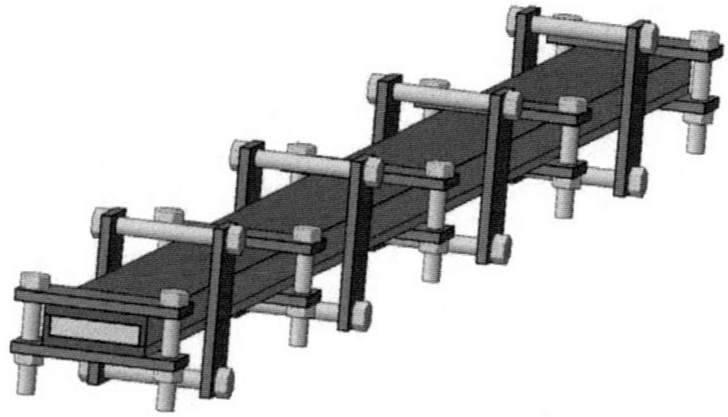

Figure 1. The "baking set" for the heat treating with foil.

are generated and therefore the reflection is reduced. In order to ensure a proper fitting of the reflective wrapping foils to the scintillator bars, the foils were formed by special heat treating prior to the actual wrapping process. A special baking set was made and a heating and cooling cycle was performed, while the foil was placed into the form. In the first step we prepared a special baking set with two bake molds and between them a bake form (see Figure 1). The material of the baking set was aluminum and all surfaces of interest were perfectly flat and smooth. The size of the form was smaller than the plastic scintillator, i.e. $(1000 \times 44.5 \times 9.5)$ mm^3. A sheet of VM2000 foil was cleaned and laid between the molds and the form. This must be a very precise process, as any stretching, stress or shrink of the reflective film can compromise the optical properties of the foil. The use of a clean room was also important. When the foil covered perfectly the aluminum bake form with the help of the exterior aluminum mold fixture, then we fixed it from outside with iron clamps. The distance between the clamps was 10 cm. The wrapping mold with the foil was put into a cylindrical electric oven and it was baked during 2 hours at 115°C. The experimental experiences support this value. The material of the foil is stable and the parameters of reflection is constant till 125°C temperature. After removing the mold from the oven it was cooled over 24 hours. The formed foil was stored for a week before any true wrapping was performed. The optical parameters of the foil did not change after such a procedure. The scintillator paddles have been wrapped with one layer of treated foil, a layer of aluminum foil and finally using black insulating tape to ensure proper light propagation through the bar as well as light-tightness.

5. Test of the ELENS detector with monoenergetic neutrons

The response of the ELENS detectors for monoenergetic neutrons has been investigated at the Physikalisch-Technische Bundesanstalt in Germany. Using the $^7\mathrm{Li}(p,n)^8\mathrm{Be}$ reaction, different quasi-monoenergetic neutrons were produced with kinetic energies of 240, 471, 925 and 2014 keV. Three types of wrapping were compared. In Table 1, the light collection efficiencies of the bars wrapped with Teflon tape as well as with

Table 1. The relative light-collection efficiencies of the detector bars at various energies and with various wrappings. The uncertainties are less than 2%.

	Neutron energy			
Wrapping	210 keV	471 keV	925 keV	2014 keV
Teflon tape	95%	96%	96%	98%
VM2000	100%	100%	100%	100%
VM2000*	116%	115%	118%	120%

specially treated VM2000 foil are compared to bars which were wrapped with VM2000 foil without baking. The values were normalized to data obtained with the simple VM2000 foil. We concluded that using the specially treated VM2000 foil (referred to as VM2000* in Table 1), we can gain about 15-20% in light collection efficiency. The light attenuation length [10] of a plastic scintillator bar is defined as the length reducing the light signal by a factor of e and depending upon bulk transmission of the scintillator, its thickness and shape. A measurement to study the effective attenuation length of the bars of ELENS was performed by moving a ^{60}Co source along the paddle. The obtained light attenuation length is (128.2 ± 1.7) cm, i.e. light can travel relatively long distances in the scintillator material without significant attenuation. The light attenuation of the detector material with the specially treated VM2000 reflecting foil, is less than 50%. The advantage of the wrapping used is also reflected by the fact that our measured speed of the light in the scintillator bar is (17.28 ± 0.3) cm/ns.

6. Acknowledgments

This research was supported by the European Union and the State of Hungary, co-financed by the European Social Fund in the framework of TÁMOP 4.2.4. A/2-11-1-2012-0001 "National Excellence Program" (Nemzeti Kiválóság Program). This work has been supported by the European Community FP7-Capacities, contract ENSAR No.262010 and the Hungarian OTKA Foundation No. K106035.

References

[1] J. B. Birks 1964 The Theory and Practice of Scintillation counting, Pergamon Press., Oxford
[2] R. Beyer *et al.* 2007 *Nucl. Instr. Meth. Phys. Res. A* **575** 449
[3] G. Perdikakis *et al.* 2012 *Nucl. Instr. Meth. Phys. Res. A* **686** 117
[4] M. Sasano *et al.* 2011 *Phys. Rev. Lett.* **107** 202501
[5] T. Uesaka *et al.* 2012 *Progress of Theoretical Physics Supplement* 196
[6] K. Yako *et al.* 2012 *RIKEN Accel. Prog. Rep.* 45,
[7] C. Langer *et al.* 2011 *Nucl. Instr. Meth. Phys. Res. A* **659** 411
[8] M.F. Weber *et al.* 2000 *Science* **287** 2451
[9] D. Motta *et al.* 2005 *Nucl. Inst. and Meth. A* **547** 368
[10] T. Tanimori *et al.* 1983 *Nucl. Inst. and Meth. A* **216** 57

Journal of Physics: Conference Series **665** (2016) 012051 doi:10.1088/1742-6596/665/1/012051

A new r-process star with low abundances of r-process elements

A Frebel, Q Yu, H R Jacobson

Department of Physics, Massachusetts Institute of Technology & Kavli Institute for Astrophysics and Space Research, Cambridge, MA 02139

E-mail: `afrebel@mit.edu`

Abstract. Metal-poor stars with measurable r-process element abundances provide key clues to the production site(s) of the r-process and how its products are mixed with the surrounding medium. While the number of stars exhibiting strong enhancements of r-process elements has grown over the years, the lower "floor" of r-process enrichment in metal-poor stars has yet to be established, largely in part due to the difficulty in detecting weak neutron-capture element absorption lines in stellar spectra. Here we present detailed abundances of 16 neutron-capture elements for a star exhibiting the lowest level of r-process enrichment yet detected and still following the solar system r-process pattern. Taken into consideration with most of the r-process enriched stars currently in the literature, the range of r-process element enrichment spanned by this sample is at least $\sim$1.3 dex or a factor of more than 20. That the r-process abundance pattern is unchanged while the degree of enrichment varies may suggest that the r-process yields are constant while the gas mass into which they are mixed varies. Given that all stars have similar [Fe/H] values then suggests that only one or few previous stellar generations provided the observed chemical abundances, meaning that perhaps only one r-process event occurred prior to their formation. This would be consistent with a (near) constant r-process yield per event. Obtaining detailed element abundances for stars with mild r-process element enhancements is necessary to better constrain the ubiquity of the r-process pattern, the yields of r-process elements, and the site of its production.

1. Introduction

Understanding r-process nucleosynthesis is a central topic of nuclear physics theory and experiments. Much has to still be learned about it, including its astrophysical site and whether it is universal across the full range of neutron-capture elements or just for the heaviest elements. Observations of old metal-poor stars displaying the chemical fingerprint of the r-process in their spectra help to constrain some of these questions. We refer the reader to [1] for a detailed discussion of how metal-poor stars serve to constrain the history of neutron-capture elements. Examining the range of r-process enhancements in metal-poor stars may thus prove to be useful in observationally constraining the levels of r-process yields, how those yields are mixed into star-forming gas, and at what metallicities r-process enhancement becomes dominant.

So far, only stars with large overabundances have been analyzed in detail since they have proved to be the most promising stars to study the details of the stellar r-process pattern [2]. This led to the remarkable discovery that the r-process element abundance pattern appears to be universal, as the scaled solar r-process pattern matches that of these metal-poor stars nearly perfectly (for $Z \geq 56$) [3, 4, 5]. However, discrepancies among lighter neutron-capture elements

Journal of Physics: Conference Series **665** (2016) 012051 doi:10.1088/1742-6596/665/1/012051

were also found, suggesting different contributions or production mechanisms for elements such as Sr, Y and Zr [6, 7]. While many stars with mild r-process enhancements have also been discovered [8], comparably few of those have been analyzed in detail. Accordingly, little is known about the range of the actual r-process yields (as opposed to the pattern, i.e., the abundance ratios). Moreover, it remains to be seen how "diluted" an r-process pattern can become, i.e. what the lowest absolute values for a confirmed r-process pattern in a metal-poor star could be.

In this study, we attempt to address some of these questions with a newly discovered metal-poor star that shows the signature of the r-process, but at an overall level of enrichment that is the lowest one yet studied in detail.

2. A new r-process star with low levels of r-process abundances

The red giant HE 0147−4606 was initially selected from the sample of bright metal-poor stars from the Hamburg/ESO survey [9]. The metallicity determined from its medium- resolution spectrum was [Fe/H] = −2.5, which warranted high-resolution spectroscopic follow-up. We obtained a high-resolution MIKE [10] spectrum with the Magellan-Clay Telescope at Las Campanas Observatory, Chile, in 2013 January 08 and 09, and 2013 May 29, 30 and 31, observing HE 0147−4606 for a total of about 2 h.

Using a 0."7 slit, the spectral resolution is $R \sim 35,000$ in the blue spectral range, and the signal-to-noise is above 300 at the Mg I b triplet. We show a portion of the spectrum around the Eu line at 4129 Å in Figure 1.

Using a standard 1D LTE plane parallel model atmosphere analysis, we derived stellar parameters of $T_{eff} = 4944\,\mathrm{K}$, $\log g = 1.15$, $v_{mic} = 2.15\,\mathrm{km\,s^{-1}}$ and [Fe/H] = −2.72. Furthermore, we found this star to show a mild enhancement of r-process elements, as evidenced by its [Eu/Fe] = 0.7. According to [8] it is thus an r-I star.

Given the relative brightness of this object ($V \sim 10.7$ mag), its high S/N-spectrum allowed us to measure abundances for total of 16 neutron-capture elements to establish a detailed abundance pattern. Comparison with the scaled solar r-process pattern revealed this star to show signs of

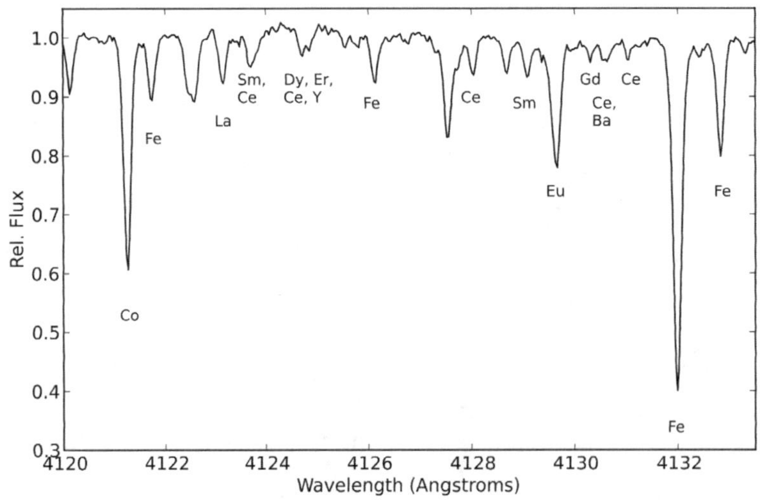

Figure 1. Portion of the spectrum in HE 0147−4606 around the Eu line at 4129 Å. Several elements are labeled.

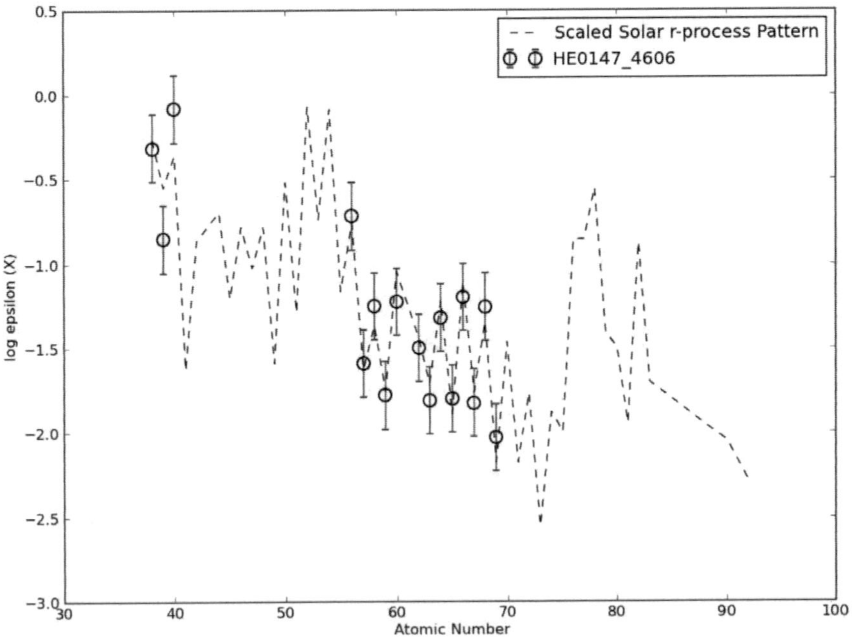

Figure 2. Neutron-capture abundances (circles) in our new r-process metal-poor star HE 0147−4606 compared with the scaled solar r-process pattern [11] denoted with the dashed line. Note the good agreement for elements heavier than Ba ($Z \geq 56$).

r-process nucleosynthesis, i.e. HE 0147−4606 formed from a gas cloud that was enriched by an r-process event, although at a very modest level. We show our abundances compared to the scaled solar pattern [11] in Figure 2.

The r-process pattern in HE 0147−4606 is of the lowest level studied in detail. All other r-process stars with published abundance patterns have larger r-process abundances in terms of their $\log\epsilon(\mathrm{X})$ values. This is mostly due to a bias, in that stars with stronger enhancements show stronger spectral features making them easier to analyze. However, this leaves the actual enrichment level of r-process elements in star forming gas clouds largely unexplored.

As stellar ages can be derived for many r-process stars exhibiting detectable features of unstable isotopes of elements such as Th and U, we attempted to obtain a thorium abundance for HE 0147−4606. Given than the r-process enhancement is not as strong, this has proven difficult. More detailed results will thus be reported in a forthcoming paper.

3. The range of r-process element abundances in metal-poor stars

Figure 3 shows the $\log\epsilon(\mathrm{X})$ values of neutron-capture elements of 12 r-process enhanced stars [12, 13, 14, 15, 13, 16, 17, 4, 13, 5, 18] as well as HE 0147−4606. This is a near-complete sample of (r-II) stars for which detailed neutron-capture elements are available in the literature, with the addition of the r-I star studied here. Rather than scaling them all to a particular element, e.g., Eu, we plot the abundances on top of each other without any scaling. As a visual aid, we scaled the solar r-process pattern to the level of each star (normalized to its Eu abundance) and plotted it together with each star. What is immediately apparent in Figure 3 is the large scatter among patterns of nearly 1.5 dex in $log\epsilon$. This corresponds to range of r-process material of more than

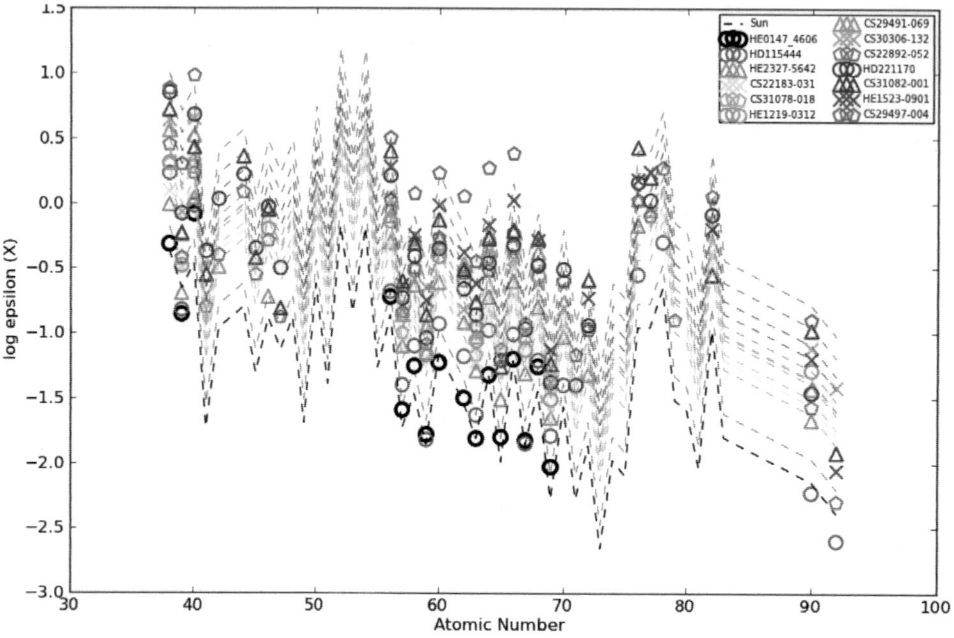

Figure 3. Same as Figure 2, but for 12 out of 14 r-process stars from the literature for which detailed abundance analysis results are available. Rather than scaling them all to a particular element, e.g., Eu, we plot them on top of each other without any scaling. This illustrates the variations in the level of r-process enrichment which are more than 1 dex. HE 0147−4606 is shown with black symbols and the star whose r-process pattern is at lowest levels.

20. For comparison, the metallicity spread of our sample is only ~ 0.5 dex. We have included HE 0147−4606 there, and it has the pattern at the lowest abundance level. Furthermore, there appears to be a continuum between the stars' abundance levels (seen in the variations of the $log\epsilon$ values of each element) over those ~ 1.5 dex.

Given that stars with lower amounts of r-process elements can often not be analyzed in detail due to their weak neutron-capture features, it is likely that many more r-I stars would contain little r-process material that would place them towards the bottom of what is shown in Figure 3, i.e. they would have abundances even lower than what HE 0147−4606 has. This implies that the scatter among the levels of r-process enrichment is likely even larger, perhaps more than a factor of 100, and that we are thus far only seeing the tip of the iceberg.

This could mean that there is either a smooth range of r-process yields, or that – if r-process yields were always of the same amount – the dilution into the star forming gas mass varies a great deal, thus providing the range of observed r-process levels. Either way, investigating the amounts of r-process material in large samples of r-process metal-poor stars should provide constraints on these questions.

4. When did the first r-process event happen?

The two most feasible proposed sites of the r-process are core collapse supernovae [19] and neutron-star mergers [20] (though both scenarios have their problems). A key clue to distinguish between these two scenarios is the frequency of stars exhibiting a clear r-process signature, and establishing when the first r-process events may have occurred e.g., at what [Fe/H] the r-process first appears, and how many stellar generations contributed elements by that time. Observations have already suggested that there is a "main" r-process that appears to run to to completion, and likely also a "failed" r-process that preferentially produces lighter neutron-capture elements.

Only about 5% of metal-poor stars show a strong enhancement in r-process elements (r-II stars), although up to 14% or more may show a mild enhancement (r-I stars) [8]. However, all the strongly-enhanced stars are found in a relatively narrow metallicity range of [Fe/H] ~ -2.5 to ~ -3.0, while r-I stars spread to higher metallicities of [Fe/H] ~ -1.5. With [Fe/H] ~ -2.7, HE 0147−4606 fits right into the r-II metallicity range, though. We show the r-II star metallicity range in Figure 4. For the following discussion we focus on r-II stars, but as Figure 3 demonstrates that at least one r-I star follows the r-process pattern, we will extend this discussion to r-I stars in a forthcoming paper.

Considering only the ~dozen stars known so far with strong r-process enhancement (r-II stars) and their 5% occurance rate among metal-poor stars immediately suggests that the r-process is a rare event [8]. The fact that they have a narrow metallicity range near [Fe/H] ~ -3 furthermore suggests that only few, perhaps just one, supernovae provided all the elements with $Z \leq 30$ observed in a given star (although each star likely formed in a different gas cloud). If the r-process operated in supernova then the r-process elements would have come from these events also.

This conclusion is independent of the neutron-capture element abundances: canonically, only one supernova iron yield of $0.1\,M_\odot$ (instantaneously) mixed into a hydrogen gas mass of $10^5\,M_\odot$ results in a gas, and hence, stellar metallicity of [Fe/H] $= -3.2$. It should be noted that in this example, the iron and gas mass can easily vary by more than a factor of 10. However it illustrates that only few supernovae are required – otherwise the metallicity of the resulting stars would become higher, as in the cases of many r-I stars. If the r-process occurred within supernovae, then the r-process must have operated already in the first few generations of massive stars. But given the rarity of r-II stars, the r-process perhaps only operated in a particular mass range (8 to $10\,M_\odot$) [21].

Additionally, stars with metallicities down to [Fe/H] ~ -5 are also known by now [22, 23, 24, 25]. They have not been found to show clear signs of r-process nucleosynthesis, mostly because no neutron-capture elements could be detected Strontium was detected at the level [Sr/Fe] $= 1.2$ in HE 1327−2326, but its origin remains uncertain since Ba could not be detected in the star [26]. While an r-process origin cannot be excluded when taking the upper limit of Ba into account, the s-process operating in massive near-zero metallicity stars [27] could also provide large amounts of Sr. Overall, the current body of data at [Fe/H] ~ -5 suggests that the r-process did not occur in the very first stars or perhaps the second generation, depending on the mass functions of the respective earliest stellar generations.

Finally, we can also turn the argument around: given that the spread in amounts of r-process material in the r-II stars is extensive, we speculate that it could also be possible that the initial r-process yields are always of the same magnitude, and that it is just the star forming gas mass that is varying. Alternatively, after the r-process event, some inhomogeneous mixing and dilution could occur, especially if the supernova is not spherically symmetric. Both effects would result in variations in the observed abundance levels. Calculating backwards in the same simple way as above suggests that an average Eu r-process yield could be $\sim 10^{-7}\,M_\odot$ per event. Taking into account the pattern spread from Figure 3, the maximum yield would then $6 \times 10^{-6}\,M_\odot$ per event. Assuming a smaller gas mass of $10^5\,M_\odot$, this would result in $6 \times 10^{-4}\,M_\odot$ per event. The numbers of 10^{-4} to 10^{-6} actually correspond to yields calculated for r-process element production in core-collapse supernovae [28].

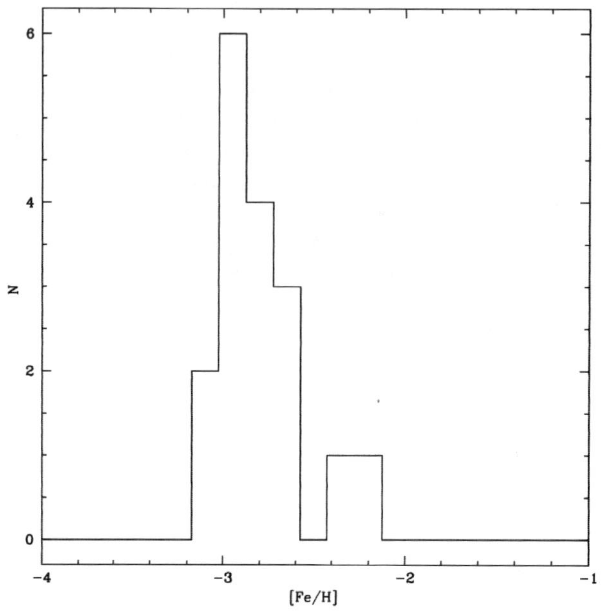

Figure 4. Histogram of [Fe/H] metallicities of the r-II stars shown in Figure 3.

5. Conclusion

Metal-poor stars showing any sign of r-process content are a "cosmic lab" for studying the r-process, its astrophysical site and the chemical evolution of r-process elements at the earliest times. By increasing the sample of stars exhibiting r-process enhancements and for which there are detailed element abundances, we can constrain the level or r-process yields as well as the robustness and universality of the r-process. This, in turn, may shed lights on the astrophysical site or at least the frequency of r-process events in the early universe.

Acknowledgments

A.F. acknowledges support from from NSF grant AST-1255160. Q. Y. thanks the MIT-UROP program for support.

References

[1] Jacobson H R and Frebel A 2013 *ArXiv e-prints* (*Preprint* 1309.0037)
[2] Sneden C, Cowan J J and Gallino R 2008 *ARA&A* **46** 241–288
[3] Sneden C, McWilliam A, Preston G W, Cowan J J, Burris D L and Amorsky B J 1996 *ApJ* **467** 819–840
[4] Hill V, Plez B, Cayrel R, Nordström T B B, Andersen J, Spite M, Spite F, Barbuy B, Bonifacio P, Depagne E, François P and Primas F 2002 *A&A* **387** 560–579
[5] Frebel A, Christlieb N, Norris J E, Thom C, Beers T C and Rhee J 2007 *ApJ* **660** L117–L120 (*Preprint* arXiv:astro-ph/0703414)
[6] Honda S, Aoki W, Ishimaru Y, Wanajo S and Ryan S G 2006 *ApJ* **643** 1180–1189 (*Preprint* arXiv:astro-ph/0602107)
[7] Honda S, Aoki W, Ishimaru Y and Wanajo S 2007 *ApJ* **666** 1189–1197 (*Preprint* 0705.3975)
[8] Barklem P S, Christlieb N, Beers T C, Hill V, Bessell M S, Holmberg J, Marsteller B, Rossi S, Zickgraf F J and Reimers D 2005 *A&A* **439** 129–151

[9] Frebel A, Christlieb N, Norris J E, Beers T C, Bessell M S, Rhee J, Fechner C, Marsteller B, Rossi S, Thom C, Wisotzki L and Reimers D 2006 *ApJ* **652** 1585–1603

[10] Bernstein R, Shectman S A, Gunnels S M, Mochnacki S and Athey A E 2003 MIKE: A Double Echelle Spectrograph for the Magellan Telescopes at Las Campanas Observatory *Society of Photo-Optical Instrumentation Engineers (SPIE) Conference Series* vol 4841 ed Iye M and Moorwood A F M p 1694

[11] Burris D L, Pilachowski C A, Armandroff T E, Sneden C, Cowan J J and Roe H 2000 *ApJ* **544** 302

[12] Mashonkina L, Christlieb N, Barklem P S, Hill V, Beers T C and Velichko A 2010 *A&A* **516** A46+ (*Preprint* 1003.3571)

[13] Honda S, Aoki W, Kajino T, Ando H, Beers T C, Izumiura H, Sadakane K and Takada-Hidai M 2004 *ApJ* **607** 474–498

[14] Lai D K, Bolte M, Johnson J A, Lucatello S, Heger A and Woosley S E 2008 *ApJ* **681** 1524–1556 (*Preprint* 0804.1370)

[15] Hayek W, Wiesendahl U, Christlieb N, Eriksson K, Korn A J, Barklem P S, Hill V, Beers T C, Farouqi K, Pfeiffer B and Kratz K 2009 *A&A* **504** 511–524 (*Preprint* 0910.0707)

[16] Sneden C, Cowan J J, Lawler J E, Ivans I I, Burles S, Beers T C, Primas F, Hill V, Truran J W, Fuller G M, Pfeiffer B and Kratz K L 2003 *ApJ* **591** 936–953 (*Preprint* astro-ph/0303542)

[17] Ivans I I, Simmerer J, Sneden C, Lawler J E, Cowan J J, Gallino R and Bisterzo S 2006 *ApJ* **645** 613–633 (*Preprint* arXiv:astro-ph/0604180)

[18] Christlieb N, Beers T C, Barklem P S, Bessell M, Hill V, Holmberg J, Korn A J, Marsteller B, Mashonkina L, Qian Y Z, Rossi S, Wasserburg G J, Zickgraf F J, Kratz K L, Nordström B, Pfeiffer B, Rhee J and Ryan S G 2004 *A&A* **428** 1027–1037

[19] Wanajo S, Itoh N, Ishimaru Y, Nozawa S and Beers T C 2002 *ApJ* **577** 853–865

[20] Goriely S, Bauswein A and Janka H T 2011 *ApJ* **738** L32 (*Preprint* 1107.0899)

[21] Qian Y Z and Wasserburg G J 2003 *ApJ* **588** 1099–1109

[22] Christlieb N, Bessell M S, Beers T C, Gustafsson B, Korn A, Barklem P S, Karlsson T, Mizuno-Wiedner M and Rossi S 2002 *Nature* **419** 904–906

[23] Norris J E, Christlieb N, Korn A J, Eriksson K, Bessell M S, Beers T C, Wisotzki L and Reimers D 2007 *ApJ* **670** 774–788 (*Preprint* arXiv:0707.2657)

[24] Frebel A, Aoki W, Christlieb N, Ando H, Asplund M, Barklem P S, Beers T C, Eriksson K, Fechner C and Fujimoto M Y 2005 *Nature* **434** 871–873

[25] Caffau E, Bonifacio P, François P, Sbordone L, Monaco L, Spite M, Spite F, Ludwig H G, Cayrel R, Zaggia S, Hammer F, Randich S, Molaro P and Hill V 2011 *Nature* **477** 67–69 (*Preprint* 1203.2612)

[26] Aoki W, Frebel A, Christlieb N, Norris J E, Beers T C, Minezaki T, Barklem P S, Honda S, Takada-Hidai M, Asplund M, Ryan S G, Tsangarides S, Eriksson K, Steinhauer A, Deliyannis C P, Nomoto K, Fujimoto M Y, Ando H, Yoshii Y and Kajino T 2006 *ApJ* **639** 897–917

[27] Chiappini C, Frischknecht U, Meynet G, Hirschi R, Barbuy B, Pignatari M, Decressin T and Maeder A 2011 *Nature* **472** 454–457

[28] Farouqi K, Kratz K L, Pfeiffer B, Rauscher T, Thielemann F K and Truran J W 2010 *ApJ* **712** 1359–1377 (*Preprint* 1002.2346)

The r-process nucleosynthesis during the decompression of neutron star crust material

S Goriely[1], A Bauswein[2], H-T Janka[2], S Panebianco[3], J-L Sida[3], J-F Lemaître[3], S Hilaire[4], N Dubray[4]

[1] Institut d'Astronomie et d'Astrophysique, ULB, CP226, 1050 Bruxelles, Belgium
[2] Max-Planck-Institut für Astrophysik, Postfach 1317, 85741 Garching, Germany
[3] CEA Saclay, Irfu/Service de Physique Nuclaire, 91191 Gif-sur-Yvette, France
[4] CEA, DAM, DIF, F-91297 Arpajon, France

E-mail: sgoriely@astro.ulb.ac.be

Abstract. About half of the nuclei heavier than iron observed in nature are produced by the so-called rapid neutron capture process, or r-process, of nucleosynthesis. The identification of the astrophysics site and the specific conditions in which the r-process takes place remains, however, one of the still-unsolved mysteries of modern astrophysics. Another underlying difficulty associated with our understanding of the r-process concerns the uncertainties in the predictions of nuclear properties for the few thousands exotic neutron-rich nuclei involved, for which essentially no experimental data exist.

The present paper emphasizes some important future challenges faced by nuclear physics in this problem, particularly in the determination of the nuclear structure properties of exotic neutron-rich nuclei as well as their radiative neutron capture rates and their fission probabilities. These quantities are particularly relevant to determine the composition of the matter resulting from the r-process. Both the astrophysics and the nuclear physics difficulties are critically reviewed with special attention paid to the r-process taking place during the decompression of neutron star matter following the merging of two neutron stars.

1. Introduction

The r-process of stellar nucleosynthesis is invoked to explain the production of the stable (and some long-lived radioactive) neutron-rich nuclides that are heavier than iron and observed in stars of various metallicities, as well as in the solar system (for a review, see [1]). In recent years nuclear astrophysicists have developed more and more sophisticated r-process models, trying to explain the solar system composition in a satisfactory way by adding new astrophysical or nuclear physics ingredients. However, the site(s) of the r-process has (have) not been identified yet and for this reason, the r-process nucleosynthesis remains one of the main puzzles of modern astrophysics.

Progress in the modelling of type-II supernovae and γ-ray bursts has raised a lot of excitement about the so-called neutrino-driven wind environment [1, 2, 3]. However, until now a successful r-process has not been obtained *ab initio* without tuning the relevant parameters (neutron excess, entropy, expansion timescale) in a way that is not supported by the most sophisticated existing models. Early in the development of the theory of nucleosynthesis, an alternative to the r-process in high-temperature supernova environments was proposed. It concerns the decompression of cold neutronized matter. While such a connection was suggested decades ago [4, 5] and decompressed neutron star (NS) matter was found to be favorable for strong r-processing [6], only more recent and increasingly sophisticated hydrodynamic simulations could determine the ejecta mass to be about 10^{-3} to $10^{-2} M_\odot$ [7, 8, 9, 10, 11]. This mass, combined with the predicted astrophysical event rate (around $10^{-5}\,\mathrm{yr}^{-1}$ in the Milky Way [12]) can account for the majority of r-material in our Galaxy [10]. Nearly all of the ejecta are converted to r-process nuclei, whose radioactive decay heating leads to potentially observable electromagnetic radiation in the optical and infrared bands [13, 14] with 100–1000 times fainter peak brightnesses than those of typical supernovae and durations of only days [9, 10, 11, 15]. These "macronovae" or "kilonovae" are intensely searched for (with a recent, possible first success [16, 17]) and their unambiguous discovery would constitute the first detection of r-material in situ.

In the present contribution, we describe the r-process nucleosynthesis resulting from the decompression of NS crust material following the merging of two NSs and show that the predicted abundances can be in remarkable agreement with the solar r-abundance distribution for nuclei with $A > 140$ (Sect. 2). Sect 3 is devoted to the study of the sensitivity of the r-abundance prediction to some nuclear models, in particular nuclear masses and fission properties.

2. The r-process from NS mergers

Our NS–NS merger simulations were performed with a general relativistic Smoothed Particle Hydrodynamics scheme [11, 18, 19] representing the fluid by a set of particles with constant rest mass, whose hydrodynamical properties were evolved according to Lagrangian hydrodynamics, keeping the electron fraction of fluid elements fixed. The Einstein field equations were solved assuming a conformally flat spatial metric. It was shown in [11] that the r-abundance distributions resulting from binaries simulations with different mass ratios or different equation of state (EOS) were virtually identical. For this reason, in the present analysis, only the symmetric $1.35M_\odot$–$1.35M_\odot$ system with the DD2 EOS [20], including thermal effects, and a resolution of about 550,000 particles is considered. The mass ejected from the system is about $3\ 10^{-3}M_\odot$(see Ref. [19, 11] for more details on the gross properties of the ejecta, the influence of the EOS and the post-processing nucleosynthesis calculation). Note that the $1.35M_\odot$–$1.35M_\odot$ case is of particular interest since, according to population synthesis studies and pulsar observations, it represents the most abundant systems [21].

The r-process nucleosynthesis is calculated by post-processing the ejected mass elements taking into account their dynamics determined by the hydrodynamics simulation. The nucleosynthesis is followed with a reaction network including all 5000 species from protons up to Z=110 lying between the valley of β-stability and the neutron-drip line. All charged-particle fusion reactions on light and medium-mass elements that play a role when the nuclear statistical equilibrium freezes out are included in addition to radiative neutron captures and photodisintegrations. The reaction rates on light species are taken from the NETGEN library, which includes all the latest compilations of experimentally determined reaction rates [22]. Experimentally unknown reactions are estimated with the TALYS code [23, 24] on the basis of the Skyrme Hartree-Fock-Bogolyubov (HFB) nuclear mass model, HFB-21 [25]. On top of these reactions, β-decays as well as β-delayed neutron emission probabilities are also included, the corresponding rates being taken from the updated version of the Gross Theory [26] based on the same HFB-21 Q-values. Fission processes as well as α-decays are also included for all heavy species, as described in Sect. 3.2. In the present paper, we will restrict ourselves to study the sensitivity of the r-abundance calculation with respect to nuclear masses and fission properties.

In this specific r-process scenario, the number of free neutrons per seed nucleus is so high that the nuclear flow dominated by radiative neutron captures and β-decays follows the neutron-drip line. In tens of ms, the heaviest drip-line nuclei can be synthesized. However, according to microscopic predictions, neutron-induced, β-delayed as well as spontaneous fission become efficient for drip-line nuclei with $Z \gtrsim 103$, prohibiting the formation of super-heavy nuclei and recycling the heavy material into lighter fragments, which will restart capturing the free neutrons. Such fission recycling can take place up to three times before the neutrons are exhausted. When the neutron density drops to some $N_n = 10^{20}$ cm^{-3}, the timescale of neutron capture by the most abundant nuclei with closed-shell neutron number $N = 126$ and $N = 184$ becomes longer than a few seconds, and the nuclear flow is dominated by β-decays back to the stability line, as well as fission and α-decay for the heaviest species.

The final abundance distribution of the matter ejected from a 1.35–1.35 M$_\odot$ NS merger on grounds of such a relativistic hydrodynamical simulation is shown in Fig. 1 and seen to be in remarkable agreement with the solar r-abundance distribution for nuclei with $A > 140$. The final composition of the ejecta is rather insensitive to details of the initial abundances and the astrophysical conditions, in particular the mass ratio of the two NSs, the quantity of matter ejected, and the EOS [10, 11, 27]. It should however be stressed that the distribution shown in Fig. 1 corresponds to the final composition of the ejecta mass-integrated over all mass elements found to be gravitationally unbound (about 1180 particles in this $1.35M_\odot$–$1.35M_\odot$ system). In our simulation, not all the mass elements lead to the same composition after ejection. Indeed, for a given binary system, mass elements can be ejected with relatively different velocities, so that the density evolution may vary significantly from one trajectory to the next. An example is given in Fig. 2 where it can be seen that fast expanding mass elements might not have time to capture all available free neutrons leading to no (trajectory 3) or only one fission cycle (trajectory 2) while the slowly expanding mass element (trajectory 1) allows for all free neutrons to be captured and three fission cycles to take place. If we define an expansion timescale as the time required for the density to drop from the initial drip density below the arbitrary value of 10^8g/cm^3, we find that for timescales shorter than typically ~ 5 ms no fission recycling takes place. In our relativistic simulations [10, 11], the integrated mass associated with a fast expansion remains small relative to the slowly expanding ones (Fig. 2, right panel), but their contributions to the final abundance distribution are found to be non-negligible, in particular around $A \simeq 200$, where their high abundances tend to fill the trough found in slowly expanding trajectories. For this reason, the Newtonian description of the NS merger hydrodynamics which tends to predict rather slow expansions [27] may give rise to different and unsatisfactory nucleosynthesis predictions. Note that the Newtonian simulations also present the drawback of overestimating the ejecta masses in general [8, 11, 27, 28].

Except for the sensitivity to the density history of the various mass elements ejected, the final abundance distribution remains extremely robust with respect to astrophysical conditions [11]. This robustness, which is compatible with the unique, solar-like abundance pattern of the elements heavier than Ba observed in metal-poor stars [29], supports the possible creation of these elements by fission recycling in NS merger ejecta. However, the estimated abundance distribution remains rather sensitive to the adopted nuclear models. The ejected material is composed almost exclusively of $A > 140$ nuclei, and in particular the $A \simeq 195$ third r-process peak appears in proportions similar to those observed in the solar system (Fig. 1), deviations resulting essentially from the still difficult task to predict neutron capture and β-decay rates for exotic neutron-rich nuclei. The situation for the lightest $110 \lesssim A \lesssim 170$ species has been rather unclear up to now and extremely dependent on fission properties, such as fission probabilities and fission fragment distributions (FFD). Indeed, the $110 \lesssim A \lesssim 170$ nuclei originate essentially from the spontaneous and β-delayed fission recycling that takes place in the $A \simeq 278$ region at the time all neutrons have been captured and the β-decays dominate the nuclear flow. The $A \simeq 278$ isobars correspond to the predominant abundance peak in the actinide region resulting from the turn-off point at the $N = 184$ drip-line shell closure, a bottleneck being created by β-decays along the nuclear flow during the irradiation phase. This specific mass region and its fission properties therefore play a key role for the production of the light species, as further

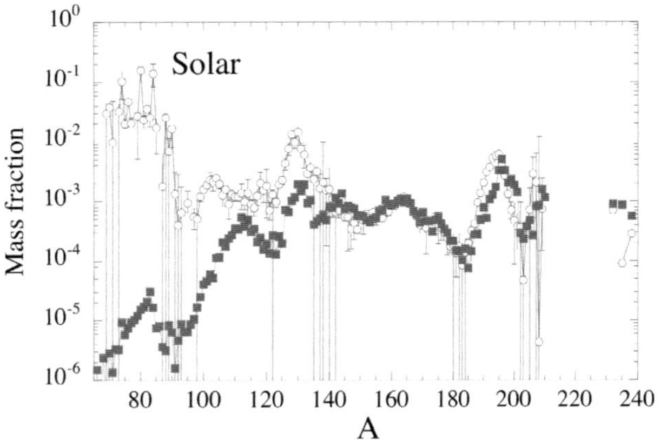

Figure 1. *Left*: Final abundance distribution of the matter ejected from the 1.35–1.35 $M_\odot$ NS merger as a function of the atomic mass. The distribution is normalized to the solar r-abundance distribution (dotted circles).

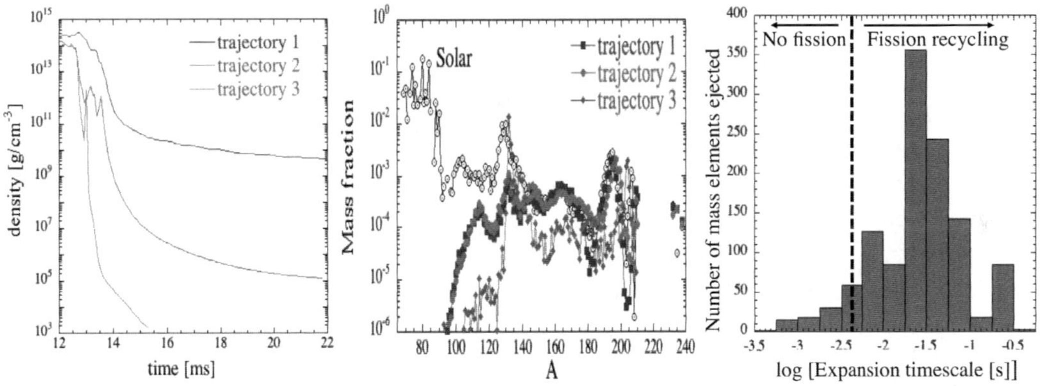

Figure 2. *Left*: Density evolution for three specific mass elements (trajectories) ejected from the 1.35–1.35 $M_\odot$ NS merger. *Middle*: Abundance distribution obtained for the same three trajectories, arbitrarily normalized to the solar system abundances (dotted circles).*Right*: Distribution of the expansion timescales for the 1180 mass elements ejected. The expansion timescale is defined here as the time required for the density to drop below the arbitrary value of 10^8g/cm^3.

discussed below.

3. Sensitivity to nuclear models

3.1. Nuclear mass models

Nuclear masses are known to be a fundamental ingredient for the r-process nucleosynthesis. They define the reaction and β-decay Q-values and consequently strongly affect the corresponding rates. During the decompression of the NS matter, the r-process dominantly takes place at low temperatures so that the radiative neutron capture is in competition with β-decays rather than with the photodisintegrations, as traditionally found in hot environments like in the ν-

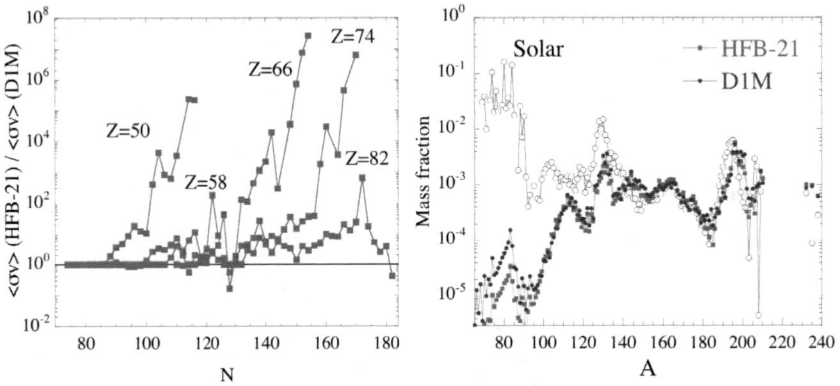

Figure 3. *Left*: Comparison between the neutron capture rates at a temperature $T = 10^9$ K for 5 isotopic chains obtained with the HFB-21 and D1M mass models. *Right*: Abundance distribution of the matter ejected from the 1.35–1.35 $M_{\odot}$ NS merger obtained with reaction rates calculated either with the HFB-21 or the D1M masses.

driven wind [1]. Reaction rates therefore fully enter the reaction network. We have studied the sensitivity of the nucleosynthesis calculation by considering the TALYS neutron capture rates estimated with two different sets of nuclear masses, namely HFB-21 masses based on the Skyrme Hartree-Fock-Bogolyubov (HFB) model [25] and D1M masses based on the Gogny HFB model [30]. The corresponding rates include a compound and a pre-equilibrium component but no direct capture contribution. Both sets of rates are compared in Fig. 3 (left panel) for five illustrative isotopic chains and seen to deviate up to a factor of 10^7 for drip line nuclei, the Skyrme HFB masses predicting essentially faster rates. However, such differences are seen in Fig. 3 (right panel) to have a rather small impact on the calculated r-abundances, deviations being restricted to the $A \simeq 180$ and $A \simeq 200$ regions. Note that the same β-decay rates [26] have been used in both nucleosynthesis calculations. Rather similar abundance distributions can be explained by the mass-averaging due to the large number of contributing trajectories, but also to the fact that the final distribution is shaped by the competition between neutron capture and β-decay for nuclei relatively close to the valley of β-stability at the time of the neutron freeze-out (see Ref. [1] for more details). Despite this similarity, more work on nuclear masses, but also on the determination of neutron capture (and also β-decay) rates is obviously needed, in particular to improve the prediction around the $A \simeq 195$ peak.

3.2. Fission

All fission rates, i.e the neutron-induced, photo-induced, β-delayed and spontaneous fission rates, have been estimated on the basis of the HFB-14 fission paths [31] and the nuclear level densities in the ground state as well as the saddle points within the combinatorial approach [32]. The neutron- and photo-induced fission rates have been estimated on the basis of the TALYS code for all nuclei with $90 \leq Z \leq 110$ [33]. Similarly, the β-delayed and spontaneous fission rates have been estimated with the same TALYS fission barrier penetration calculation. The β-delayed fission rate takes into account the full competition between the fission, neutron and photon channels, weighted by the population probability given by the β-decay strength function [34], i.e

$$\lambda_{\beta df} = \int_{B_f}^{Q_\beta} \frac{\Gamma_f}{\Gamma_f + \Gamma_n + \Gamma_\gamma} S_\beta(E) dE. \tag{1}$$

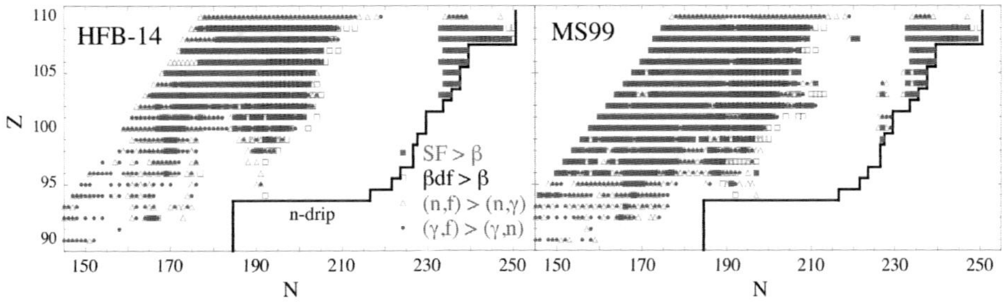

Figure 4. *Left*: Representation of dominant fission regions in the (N, Z) plane. Nuclei for which spontaneous fission (SF) is estimated to be faster than β-decays are shown by full squares, those for which β-delayed fission (βdf) is faster than β-decays by open squares, those for which neutron-induced fission (n,f) is faster than radiative neutron capture (n,γ) at $T = 10^9$ K by open triangles and those for which photofission (γ,f) at $T = 10^9$ K is faster than photoneutron emission (γ,n) by closed circles. For $Z = 110$, β-decay processes are not calculated. All fission probabilities are calculated with the HFB-14 fission paths [31]. *Right*: Same as the left panel when all fission probabilities are calculated with the MS99 fission barriers [35].

where B_f is the fission barrier, Q_β the β-decay Q-value and S_β the β-decay strength function estimated within the Gross Theory [34]. Γ_f, Γ_n, Γ_γ are the fission, neutron emission and electromagnetic de-excitation widths, respectively, at a given energy E which have been calculated consistently with the TALYS reaction code.

The spontaneous fission rate is deduced from the well-known formula $\lambda_{SF} = f_0 P$ where P is the barrier penetrability of the ground state calculated with the TALYS code and $f_0 = \omega_0/2\pi$ the frequency of oscillations in the fission mode for the ground state in the first well at the energy $\hbar\omega_0 \simeq 0.75$ MeV. The main fission regions by one of the four fission processes are illustrated in Fig. 4.

In order to perform sensitivity calculations, the fission probabilities have also been estimated on the basis of the Thomas-Fermi fission barriers compiled in Ref. [35] (hereafter MS99). To do so, the HFB-14 fission paths are renormalized so that the highest barrier corresponds now to the highest MS99 barrier. The latter usually predicts lower barriers than the mean-field approach so that many more nuclei are found to be affected by fission processes when use is made of the MS99 barriers, as shown in Fig. 4 (right panel). Note that MS99 barriers may not always be available for very exotic n-rich nuclei close to the neutron-drip line, especially for $Z \geq 106$. In this case, the HFB-14 barriers were adopted. This explains the relatively similar fissioning region for $Z \geq 103$ at the neutron drip line (Fig. 4). It is seen in Fig. 4 that if we adopt the HFB-14 fission barriers, during the freeze-out phase, fission takes place around $Z \simeq 101 - 102$ along the abundant $A = 278$ isobar, while the lower MS99 barriers [35] lead to fission already around the $Z \simeq 95 - 97$ isotopes.

Another ingredient of prime importance for the r-process nucleosynthesis is the fission FFD. The β-decay nuclei along the $A = 278$ isobar may fission either symmetrically or asymmetrically depending on the adopted model for the FFD. We consider here two models, namely the renewed statistical scission-point model based on microscopic ingredients called SPY [36] and the semi-empirical GEF model [37]. Both models differ not only in the prediction of the FFD for the exotic neutron-rich fissioning nuclei, but also in the estimate of the average number of emitted neutrons.

The final composition of the matter ejected from the 1.35–1.35 $M_\odot$ NS merger is shown

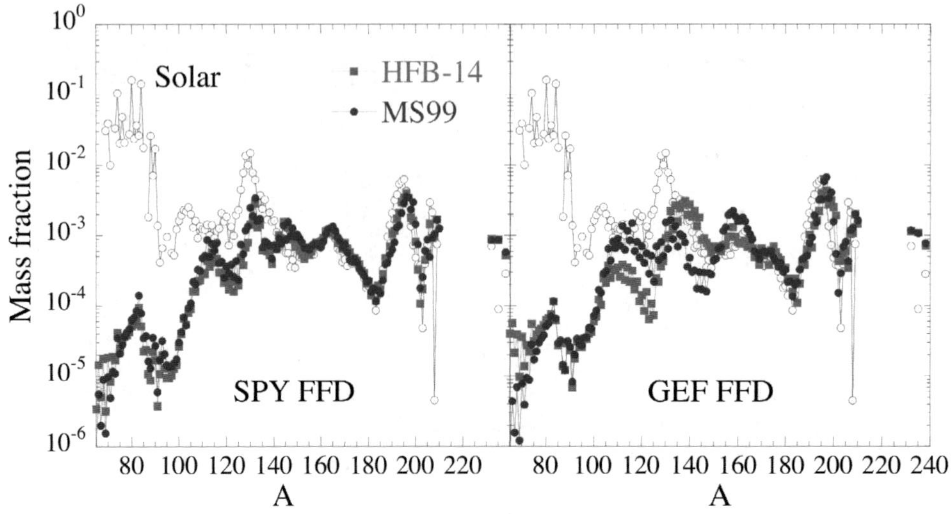

Figure 5. *Left*: Final abundance distribution of the matter ejected from the 1.35–1.35 $M_\odot$ NS merger as a function of the atomic mass. The distributions are normalized to the solar r-abundance distribution (dotted circles). The distributions are obtained with two sets of fission barriers, the HFB-14 [31] and MS99 barriers [35]. The adopted model for the FFD correspond to the SPY model [36]. *Right*: same as left panel when the FFD are estimated with the GEF model [37].

in Fig. 5 for fission rates obtained with both sets of fission barriers, namely the HFB-14 [31] and MS99 barriers [35] and two FFD models, namely the SPY [36] and GEF [37] models. As discussed above, the $A \simeq 278$ isobars that may fission during the freeze-out phase depend on the adopted fission model (see Fig. 4). Depending on the fissioning nuclei, they may fission either symmetrically or asymmetrically following the prediction of the FFD model. For example, considering the GEF FFD model, with the HFB-14 barriers, fission takes place symmetrically around $Z \simeq 101 - 102$ along the abundant $A = 278$ isobar, while the lower MS99 barriers [35] lead to fission already at the $Z \simeq 95 - 97$ isobar which are predicted to fission asymmetrically (see right panel of Fig. 5). In contrast, the SPY model predicts all $A = 278$ isobars to fission asymmetrically, leading to an abundance distribution rather insensitive to the adopted fission barriers (see left panel of Fig. 5). Interestingly, the SPY model can consistently explain the origin of the rare-earth peak by the r-process, in contrast to what is found with more phenomenological models predicting symmetrical mass yields for the $A \simeq 278$ fissioning nuclei.

4. Conclusions

Decompressed NS matter remains a viable site for the r-process, which is extremely robust with respect to many astrophysical uncertainties. This robustness, which is compatible with the unique, solar-like abundance pattern of the elements heavier than Ba observed in metal-poor stars, supports the possible creation of these elements by fission recycling in NS merger ejecta. However, the estimated abundance distribution remains rather sensitive to the adopted nuclear models. The ejected material is composed almost exclusively of $A > 140$ nuclei, and in particular the $A \simeq 195$ third r-process peak appears in proportions similar to those observed in the solar system, deviations resulting essentially from the still difficult task to predict neutron capture and β-decay rates for exotic neutron-rich nuclei. The situation for the lightest $110 \lesssim A \lesssim 170$ species has been rather unclear up to now and extremely dependent on fission properties, including in particular the fission barriers and FFD. The newly derived FFD based on the SPY model can consistently explain the origin of the rare-earth peak by the r-process, in contrast to what is found with more phenomenological models predicting symmetrical mass yields for the $A \simeq 278$ fissioning nuclei. This new finding provides an even stronger hint that NS mergers could be a dominant process in the origin of the $A > 140$ r-nuclei in the Universe.

Acknowledgments

S.G. is F.R.S.-FNRS research associate and acknowledges the financial support of the "Actions de recherche concertées (ARC)" from the "Communauté française de Belgique". A.B. and H.-T.J. acknowledge support by DFG through grants SFB/TR7 and EXC 153.

References

[1] Arnould M, Goriely S and Takahashi K 2007 *Phys. Rep.* **450** 97
[2] Janka H-T 2012 *Ann. Rev. Nuc. Part. Science* **62** 407
[3] Wanajo S, Janka H-T and Müller B 2011 *Astrophys. J. Lett.* **726** L15
[4] Lattimer J M and Schramm D N 1974 *Astrophys. J.* **192** L145
[5] Eichler D, Livio M, Piran T and Schramm D N 1989 *Nature* **340** 126
[6] Meyer B S 1989 *Astrophys. J.* **343** 254
[7] Ruffert M, Janka H-T, Schäfer G 1996 *Astron. Astrophys.* **311** 532
[8] Rosswog S, Liebendörfer M, Thielemann F-K, Davies M B and Piran T 1999 *Astron. Astrophys.* **341** 499
[9] Roberts L F, Kasen D, Lee W H and Ramirez-Ruiz E 2011 *Astrophys. J. Lett.* **736** L2
[10] Goriely S, Bauswein A and Janka H-T 2011 *Astrophys. J. Lett.* **738** L32
[11] Bauswein A, Goriely S and Janka H-T 2013 *Astrophys. J.* **773** 78
[12] Dominik M, Belczynski K, Fryer C, Holz D E, Berti E, Bulik T, Mandel I and O'Shaughnessy R 2012 *Astrophys. J.* **759** 52
[13] Metzger B D, *et al.* 2010 *Month. Not. Roy. Astro. Soc.* **406** 2650
[14] Li L-X and Paczyński B 1998 *Astrophys. J.* **507** L59
[15] Barnes J and Kasen D 2013 *arXiv:1303.5787*
[16] Berger E, Fong W and Chornock R 2013 *arXiv:1306.3960*
[17] Tanvir N R, Levan A J, Fruchter A S, Hjorth J, Hounsell R A, Wiersema K, Tunnicliffe R L 2013 *Nature* 12505
[18] Oechslin R, Janka H-T and Marek A 2007 *Astron. Astrophys.* **467** 395
[19] Bauswein A, Janka H-T and Oechslin R 2010 *Phys. Rev. D* **82** 084043
[20] Typel S, Röpke G, Klähn T, Blaschke D and Wolter H. H. 2010 *Phys. Rev. C* **81** 015803
[21] Belczynski K, O'Shaughnessy R, Kalogera V, Rasio F, Taam R E and Bulik T 2008 *Astrophys. J. Lett.* **680** L129
[22] Xu Y, Goriely S, Jorissen A, Chen G L and Arnould M 2013 *Astron. Astrophys.* **549**, A10
[23] Koning A J, Hilaire S and Duijvestijn M C 2005 *Nuclear Data for Science and Technology* (eds. C. Haight et al.) AIP Conference **769** 1154
[24] Goriely S, Hilaire S and Koning A J 2008 *Astron. Astrophys.* **487** 76
[25] Goriely S, Chamel N and Pearson J M 2010 *Phys. Rev. C* **82** 035804
[26] Tachibana T, Yamada M and Yoshida Y 1990 *Prog. Theor. Phys.* **84** 64
[27] Korobkin O, Rosswog S, Arcones A and Winteler C 2012 *Month. Not. Roy. Astro. Soc.* **426** 1940
[28] Janka H-T, Eberl T, Ruffert M and Fryer C L 1999 *Astrophys. J. Lett.* **527** L39
[29] Sneden C, Cowan J J and Gallino R 2008 *Ann. Rev. Astron. & Astrophys.* **46** 241
[30] Goriely S, Hilaire S, Girod M and Péru S 2009 *Phys. Rev. Lett.* **102** 242501
[31] Goriely S, Samyn M and Pearson J M 2007 *Phys. Rev. C* **75** 064312
[32] Goriely S, Hilaire S and Koning A J 2008 *Phys. Rev. C 78* **064307**
[33] Goriely S, Hilaire S, Koning A J, Sin M and Capote R 2009 *Phys. Rev. C* **79** 024612
[34] Kodoma T and Takahashi K 1975 *Nucl. Phys. A.* **239** 48
[35] Myers W D and Swiatecki W J 1999 *Phys. Rev. C* **60** 014606
[36] Panebianco S, Sida J-L, Goutte H, Lemaître J-F, Dubray N and Hilaire S 2012 *Phys. Rev.* C **86** 064601
[37] Schmidt K-H and Jurado B 2010 *Phys. Rev. Lett.* **104** 212501

Journal of Physics: Conference Series **665** (2016) 012053 doi:10.1088/1742-6596/665/1/012053

Recent results on mass measurements of exotic nuclides in storage rings

X L Yan[1,2,3,4], K Blaum[2,5], Yu A Litvinov[1,2,3,5], X L Tu[1,3], H S Xu[1], Y H Zhang[1], and X H Zhou[1],
for ESR & CSRe Collaborations on In-Ring Mass Measurements

E-mail: y.litvinov@gsi.de
[1] Institute of Modern Physics, Chinese Academy of Sciences (IMP), Lanzhou 730000, China
[2] Max-Planck-Institut für Kernphysik (MPIK), 69117 Heidelberg, Germany
[3] GSI Helmholtzzentrum für Schwerionenforschung (GSI), 64291 Darmstadt, Germany
[4] Graduate University of Chinese Academy of Sciences, Beijing 100049, China
[5] Ruprecht-Karls Universität Heidelberg, 69120 Heidelberg, Germany

Abstract. Storage ring mass spectrometry allows for simultaneous measurements of masses of a large amount of nuclides. Two measurement techniques, isochronous and Schottky mass spectrometry, are applied. The clear advantage of the storage ring mass spectrometry is that only a few particles are required to measure their mass with high accuracy. In this contribution we address the most recent results obtained at the two presently operating facilities, namely the Experimental Storage Ring (ESR) in Darmstadt and the Experimental Cooler-Storage Ring (CSRe) in Lanzhou.

1. Introduction

It is well known in physics that the total energy of a closed system is its fundamental constant which results from all interactions and correlations of the constituent particles. In this respect nuclei are fascinating many-body objects in which the strong, weak, and electromagnetic fundamental interactions take place by acting between two types of nucleons, protons and neutrons. Via the famous equation of A. Einstein, the total energies of the nuclei can directly be connected to their masses.

Nuclear masses are indispensable quantities for investigations of nuclear structure [1, 2] as well as for our understanding of nucleosynthesis processes in stars [1, 3, 4]. For the latter, nuclear masses determine – through nucleon separation energies – the paths of various processes on the nuclidic chart [3, 4]. Today the nuclei of interest with yet unknown masses lie typically far away from the valley of β-stability. We emphasise, that many of nuclei, which for instance are involved in the rapid-neutron capture process (r-process), are not accessible at the present radioactive ion beam facilities and their masses have thus to be obtained from theory. Such nuclides can be characterised by a strong asymmetry of their proton-to-neutron ratio and can reveal unexpected nuclear structure effects. Therefore new experimental nuclear masses are essential to test and improve nuclear models. However, these nuclei are difficult to investigate due to their small production cross-sections and short lifetimes (see, e.g., [5]). Hence, their studies require very efficient and fast experimental techniques [6, 7]. Two highly complementary experimental techniques are routinely used. These are the Penning trap [1] and storage ring mass spectrometry [8].

In this contribution we briefly review recent results obtained with storage ring mass spectrometry.

2. Experimental facilities

Two ion storage ring facilities conducting in-ring mass measurements of exotic nuclides are in operation today [9]. One of these facilities is located at the GSI Helmholtz Center in Darmstadt, Germany (GSI) [10] and the other one is built at the Institute of Modern Physics, Chinese Academy of Sciences in Lanzhou, China (IMP) [11]. The high energy parts of both facilities are based on a similar principle. The driver accelerators are heavy-ion synchrotrons, SIS-18 [12] at GSI and CSRm [13] at IMP, which are respectively connected to the Experimental Storage Ring ESR [14] and the experimental Cooler-Storage Ring CSRe [13] via in-flight fragment separators FRS [15, 16] and RIBLL2 [13]. The storage rings are schematically illustrated in Figure 1.

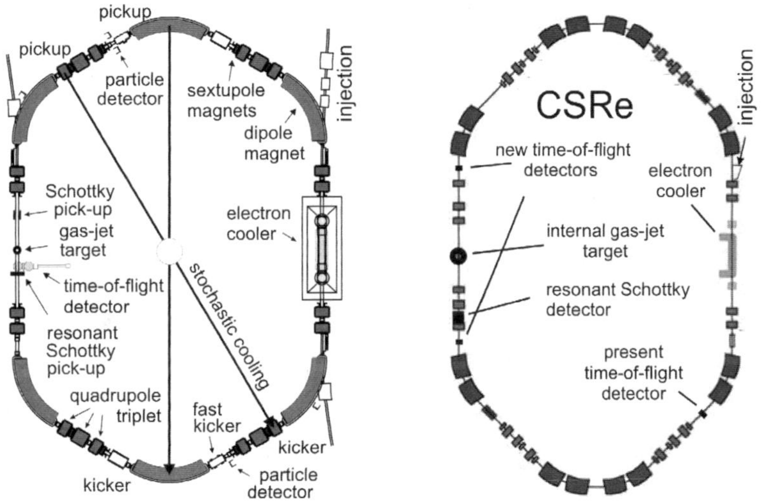

Figure 1. Schematic view of the Experimental Storage Ring (ESR) [14] (left) and the experimental Cooler-Storage Ring (CSRe) [13] (right). The stored beams circulate clockwise in the rings. The rings are equipped with systems for stochastic- [17] and electron [18] cooling, various instrumentations and detection systems [19], internal targets [20], devices for (de)acceleration as well as for bunching of the stored beams [21], etc. Adopted from [9, 22].

The ESR [14] is a storage-cooler ring with two arcs and two straight sections. Its mean circumference is 108.36 m. It consists of 6 dipole magnets (deflection angle is 60°). The ESR can operate up to a maximum magnetic rigidity of 10 Tm and is capable of storing, cooling, accelerating or decelerating stored beams as well as extracting them towards external setups. The CSRe [13] has a mean circumference of 128.8 m and a maximal magnetic rigidity of 8.4 Tm. It has eight 45° bending sections, each of which consists of two dipole magnets. Both storage rings operate under ultra-high vacuum (UHV) conditions of about 10^{-11} mbar, which inevitably requires that in-ring instrumentations are bakeable to about $150-200$ °C and have to be made of UHV materials. In addition to direct mass measurements, which will be discussed in more detail below, the rings are routinely used for a broad range of atomic physics experiments [23, 24], dielectronic recombination on stable and radioactive ions [25, 26, 27, 28], studies of nuclear reactions with internal targets [29, 30, 31, 32, 33]. Owing to the UHV and high kinetic energies, storing and investigations of radioactive decays of highly-charges ions is possible. For more details see, e.g., Refs. [34, 35, 36, 37, 38, 39, 40, 41, 42, 43, 44, 45].

3. Storage ring mass spectrometry

For ions stored in a storage ring holds the following relationship, which connects the relative revolution frequencies (f) or revolution times (t) of the circulating ions to their relative mass-over-charge ratios and velocities [8, 46]:

$$\frac{\Delta f}{f} = -\frac{\Delta t}{t} = -\alpha_p \frac{\Delta(m/q)}{m/q} + (1 - \alpha_p \gamma^2) \frac{\Delta v}{v}, \tag{1}$$

where α_p is the momentum compaction factor, which characterises the relative variation of the orbit length of stored particles per relative variation of their magnetic rigidity. The $\alpha_p = -1/\gamma_t^2$ is nearly constant over the entire revolution frequency acceptance of the storage ring, and γ_t is the so-called transition point of a ring [8].

From Eq. (1) it follows that in order to determine m/q values of the stored ions, one needs to measure revolution frequencies (or revolution times) of the ions. The magnitude of the second term on the left hand side determines the mass resolving power and it has to be made as small as possible.

One way to minimise the second term in Eq. (1) is to reduce the velocity spread $\Delta v/v$, which can be done by applying beam cooling. This allows for the reduction of the energy spread of the ions induced by the production reaction process. Stochastic [17] and/or electron [18, 47] cooling techniques are applied which enable beams with the highest phase-space density. For electron cooled beams, the momentum spread of stored ions $\delta p/p$ is determined by the equilibrium between the Coulomb interactions with cold electrons in the cooler device and the intra-beam scattering [47]. If the electron cooling is applied to low intensity (below about a thousand ions) fragment beams, then the velocity spread as low as a few 10^{-7} can be achieved [48, 49]. Electron cooling requires at least a few seconds which sets a limit on half-lives of the nuclides that can be studied. Applying the stochastic cooling prior to the electron cooling allows for reducing the cooling time [50, 51]. The revolution frequencies are measured with non-destructive Schottky detectors. The details on the Schottky detector and data acquisition can be found in Refs. [8, 43]. This is the so-called *Schottky Mass Spectrometry* (SMS) [52, 53].

The second way to minimise the second term in Eq. (1) is based on a special isochronous ion-optical mode of the ring. The ions are injected into the ring at energies corresponding to $\gamma = \gamma_t$. The γ_t-values are γ_t(ESR) = 1.41 and γ_t(CSRe) = 1.395 [9]. In this mode the velocity spread is compensated by the orbit lengths of the stored ions. This simply means that faster and slower ions of the same nuclear species are stored on the longer or shorter orbits, respectively, such that they have the same mean revolution frequency. This is the prerequisite for the so-called *Isochronous Mass Spectrometry* (IMS) [54, 55, 56]. Since the IMS does not require beam cooling, it is ideally suited for the mass measurements of the shortest-lived nuclides. The revolution frequencies are obtained by using dedicated time-of-flight detectors [57, 58]. Very recently a novel resonant Schottky detector [59, 60] was developed which will allow for an accurate revolution frequency determination within a few tens of milliseconds and which thus can be employed in the IMS [61, 62]. The dramatic improvement in the speed of measurements enabled now applying SMS also in the isochronous mode.

More than 1000 masses have been measured with the IMS and SMS at the ESR and CSRe, which enabled numerous investigations of nuclear structure and astrophysics questions [46, 52, 56, 63, 64, 65, 66, 67, 68, 69, 70, 71, 72, 73, 74, 75, 76, 77, 78, 79].

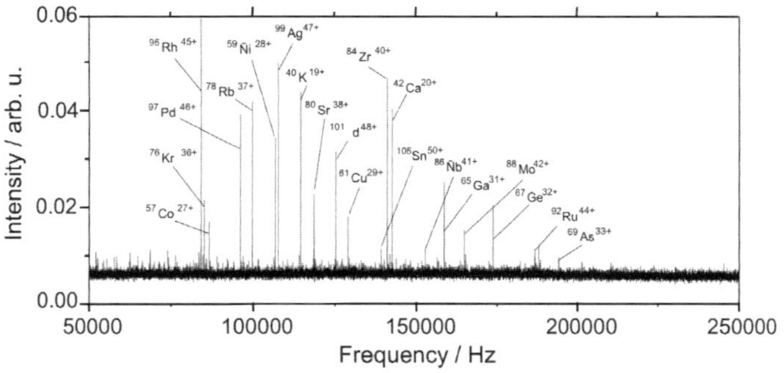

Figure 2. (Colour online) A part of Schottky revolution frequency spectrum of ^{152}Sm projectile fragments. Adopted from Ref. [80].

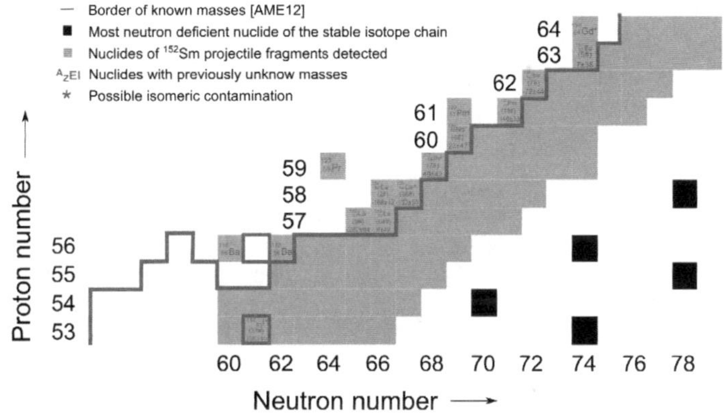

Figure 3. (Colour online) A part of the nuclidic chart above iodine ($Z \geq 53$) with the mass surface measured in this experiment. Fifteen nuclides whose masses were obtained for the first time are indicated. Adopted from Ref. [81].

4. Schottky mass measurements ^{152}Sm projectile fragments in the ESR

To date, the Schottky mass measurements were conducted only in the ESR [8, 62, 82]. For cooled ions the ESR acceptance corresponds to $\Delta(m/q)/(m/q) = \pm1.2\%$ which allows for simultaneous mass measurements of many nuclides [83]. In the last three years, masses of several tens of neutron-rich heavy isotopes were published [73, 78]. Furthermore, several new isotopes [69, 84, 85] and isomers were discovered [66, 68, 70, 75, 76, 79, 86].

Here we report on the mass measurements of neutron-deficient ^{152}Sm projectile fragments, the data analysis of which has been completed. The experiment was performed in 2005 [41, 80]. Primary beam of ^{152}Sm ions was extracted from the SIS at 508 AMeV or 615 AMeV. It was fragmented in 1032 or 4009 mg/cm^2 ^{9}Be production targets, respectively. Two about 12 hours long settings have been done in this experiment. The fragments centred around ^{108}Sb^{51+} or ^{140}Ce^{58+} ions were separated in-flight and injected, stored, and electron-cooled in the ESR. The revolution frequencies were measured with a capacitive Schottky detector. The details of the Schottky data processing and the isotope identification can be found in Refs. [46, 52, 65, 83].

A part of a typical revolution frequency spectrum obtained in this experiment is shown in Figure 2. The mass surface covered in this experiment consists of more than 370 different nuclides. Masses of fifteen nuclides were determined for the first time. A relative mass accuracy

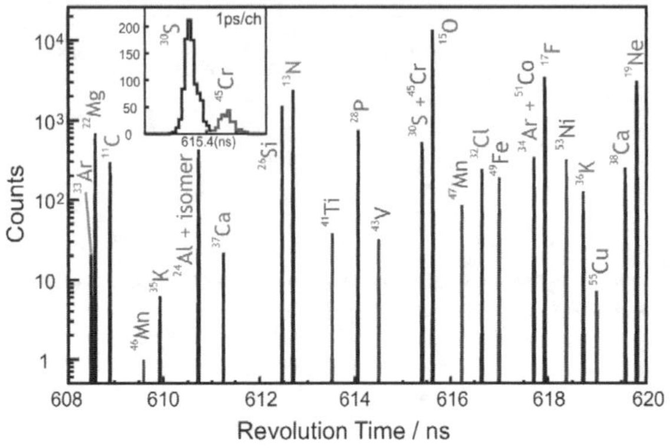

Figure 4. A part of the revolution time spectrum of neutron-deficient ^{58}Ni projectile fragments measured in the CSRe [77]. The insert shows the well-resolved peaks of $^{30}S^{16+}$ and $^{45}Cr^{24+}$ nuclei, which have very similar m/q values. Nuclei with masses determined in this experiment and those used as references are indicated with bold and italic letters, respectively. Taken from [74, 77].

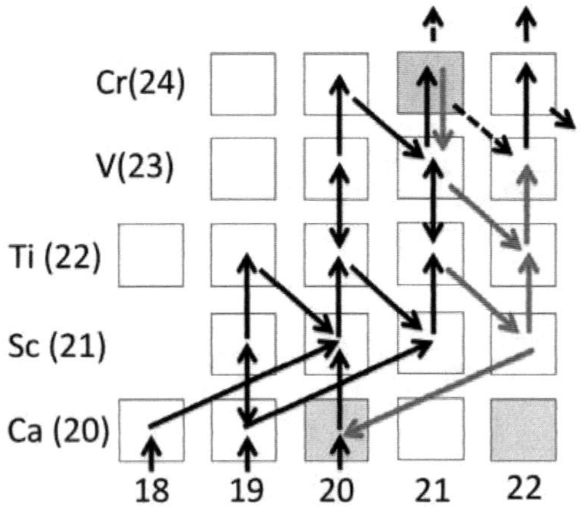

Figure 5. The main time integrated reaction flow in the Ca-Cr region during an X-ray burst. Black arrows show the reaction flow for the new $S_p(^{45}Cr)$ value. The red arrows indicate the reaction flow for the lowest $S_p(^{45}Cr)$ value allowed by the 3σ uncertainties from AME03 [87]. Flows that disappear for the latter case are indicated as black dashed arrows. Taken from [77].

$\Delta m/m$ of a few 10^{-7} was achieved. The new masses are indicated on the nuclidic chart shown in Figure 3. The preparation of the publication is in progress [81].

5. Isochronous mass measurements ^{58}Ni projectile fragments in the CSRe

Isochronous mass measurements are conducted at the ESR as well as at the CSRe [56, 64, 71, 72, 74, 77]. Since no cooling is applied, the measured spectrum is about a six times broader ($\Delta(m/q)/(m/q) \approx 13\%$) than a typical revolution frequency spectrum of cooled ions. In the last three years a series of experiments addressing neutron-deficient nuclides has been performed at the CSRe [71, 72, 74, 77]. Here we address only the latest results on the mass measurement of ^{45}Cr (see [77]).

In this experiment, primary beam of $^{58}Ni^{19+}$ ions accelerated by the CSRm to an energy of 463.36 AMeV was fragmented in about 15 mm thick beryllium target placed at the entrance of the RIBLL2. Neutron-deficent fragments centred around $^{47}Mn^{25+}$ were separated in flight with RIBLL2 and injected into the CSRe. The CSRe was tuned into an isochronous ion-optical mode. Typically, about 10 ions were stored simultaneously after each injection. The revolution times of the stored ions were determined with a dedicated timing detector [58]. The details of the detector operation, data acquisition, and data analysis can be found in Refs. [58, 71, 72, 88]. A typical revolution time spectrum is illustrated in Fig. 4. The inset shows the resolved peaks of $^{30}S^{16+}$ and $^{45}Cr^{24+}$ ions. The mass of the latter was determined in this experiment for the first time and amounts to $ME(^{45}Cr) = -19515(35)$ keV, where a contamination by a recently observed isomeric state at $E^* = 107$ keV [89] had to be taken into account.

The new mass of ^{45}Cr has an effect on the modelling the rapid proton capture process (*rp*-process) in X-ray bursters [90]. The new value of the proton separation energy of ^{45}Cr is $S_p(^{45}Cr = 2684(125)$ keV [77]. Figure 5 shows a part of the nuclidic chart around titanium and the path of the *rp*-process as obtained from a single-zone calculation [91, 92]. The double arrows indicate nuclides which are in $(p,\gamma)-(\gamma,p)$ equilibrium. Important here is the case of ^{44}V and ^{43}Ti nuclides. The flow here is determined by the $^{44}V(p,\gamma)^{45}Cr$ reaction rate, the so-called 2*p*-capture process on ^{43}Ti [91]. The value of $S_p(^{45}Cr)$ effects this rate. For a low $S_p(^{45}Cr)$, the $^{45}Cr(\gamma,p)^{44}V$ reaction becomes effective and thus reduces the proton capture flow at ^{43}Ti and leads to an increased β-decay branch in ^{43}Ti. The proton capture on the β-decay daughter ^{43}Sc nuclei has a large (p,α) branch which leads to ^{40}Ca (see red arrows in Figure 5). This is a so-called Ca-Sc cycle [93, 94] which limits the reaction flow towards heavier elements. The new $S_p(^{45}Cr)$ value constrains the proton capture flow through ^{43}Ti and allows one to exclude a possible formation of a strong Ca-Sc cycle in X-ray bursts.

6. Summary and Outlook

Storage ring mass spectrometry has been applied to neutron-deficient nuclides. Two examples of mass measurements conducted at the ESR and CSRe are presented here.

Schottky mass spectrometry was applied to ^{152}Sm projectile fragments. Masses for fifteen nuclides were obtained for the first time. The publication of these results is in preparation [81].

Since a few years isochronous mass measurements are successfully conducted at the CSRe [95]. In this contribution we addressed the mass measurement of ^{45}Cr and its impact on the *rp*-process of nucleosynthesis.

Concerning the future experiments, it is important to emphasise new technical developments which are ongoing. One of them is the commissioning in the ESR and CSRe of the resonant Schottky detectors which have by a factor of about 100 better signal-to-noise characteristics than the older capacitive pickups [59, 60]. Another one is the planned setup of two time-of-flight detectors which is being done in the CSRe and which will enable an in-ring velocity measurement of every stored ion [95, 96]. The latter is an essential quantity to enable a high resolving power over the entire revolution time spectrum [97].

Storage ring mass measurements are planned also at future radioactive beam facilities. A dedicated isochronous storage ring, the RI-Ring [98], is being commissioned at RIKEN Nishina Center in Japan [99], where the highest beam intensities are presently offered. The ILIMA project [100] aims at mass measurements at the future FAIR facility [101, 102]. The key facilities here are the Super-FRS [103] and the collector ring (CR), which is particularly designed for conducting IMS. The CR will be equipped with two ToF detectors installed in one of the straight sections. Employing the novel resonant Schottky detectors will enable simultaneous broad-band mapping by SMS technique of nuclear masses and lifetimes [61, 100, 104]. In China, a concept for a new project, High-Intensity Accelerator Facility, is being prepared where one of the key experiments will be IMS.

6.1. *Acknowledgments*

We are thankful to all our collaborators working with us over many years on storage ring mass and lifetime measurements. This work is supported by the 973 Program of China (No.

2013CB834401), the NSFC (Grants No. 10925526, 11035007, U1232208, 10675147, 10805059, 11135005, 11075103, 10975008, 11105010, 11128510 and 11205205), the Chinese Academy of Sciences (Grant No. 2009J2-23), the NCET (Grant No. NCET-09-0031), the External Cooperation Program of the Chinese Academy of Sciences (Grant No. GJHZ1305), the BMBF Grant in the framework of the Internationale Zusammenarbeit in Bildung und Forschung (No. 01DO12012), and the Helmholtz-CAS Joint Research Group HCJRG-108. K.B. acknowledges support by the Nuclear Astrophysics Virtual Institute (NAVI) of the Helmholtz Association. K.B. and Y.A.L. thank ESF for support within the EuroGENESIS program.

References

[1] K. Blaum, *Phys. Rep.* 425 (2006) 1
[2] R. B. Cakirli *et al.*, *Phys. Rev. Lett.* 102 (2009) 082501
[3] E. M. Burbidge, G. R. Burbidge, W. A. Fowler, F. Hoyle, *Rev. Mod. Phys.* 29 (1957) 547
[4] K. Langanke, H. Schatz, *Phys. Scripta* T 153 (2013) 014011
[5] J. Kurcewicz *et al.*, *Phys. Lett.* B 717 (2012) 371
[6] K. Blaum *et al.*, *J. Phys. Conf. Series* 312 (2011) 092001
[7] K. Blaum *et al.*, *J. Phys. Conf. Series* 381 (2012) 012013
[8] B. Franzke, H. Geissel and G. Münzenberg, *Mass Spectrom. Rev.* 27 (2008) 428
[9] Yu. A. Litvinov *et al.*, *Acta Phys. Polonica* B 41 (2010) 511
[10] GSI: GSI Helmholtzzentrum für Schwerionenforschung GmbH, `http://www.gsi.de`
[11] IMP: Institute of Modern Physics, Chinese Academy of Sciences, `http://www.impcas.ac.cn`
[12] K. Blasche *et al.*, *IEEE Trans. on Nucl. Sci.* Ns-32 (1985) 2657
[13] J. W. Xia *et al.*, *Nucl. Instr. Meth. Phys. Research* A 488 (2002) 11
[14] B. Franzke, *Nucl. Instr. Meth. Phys. Research* B 24/25 (1987) 18
[15] H. Geissel *et al.*, *Nucl. Instr. Meth. Phys. Research* B 70 (1992) 286
[16] P. Armbruster *et al.*, *AIP Conf. Proc.* 164 (1987) 839.
[17] F. Nolden *et al.*, *Nucl. Instr. Meth. Phys. Research* A 532 (2004) 329
[18] M. Steck *et al.*, *Nucl. Instr. Meth. Phys. Research* A 532 (2004) 357
[19] O. Klepper, C. Kozhuharov, *Nucl. Instr. Meth. Phys. Research* B 204 (2003) 553
[20] R. Grisenti *et al.*, *Europhys. Lett.* 73 (2006) 540
[21] W. Q. Wen *et al.*, *Nucl. Instr. Meth. Phys. Research* A 711 (2013) 90
[22] Yu. A. Litvinov *et al.*, *Phys. Scripta* T 144 (2011) 014001
[23] J. Eichler, T. Stöhlker, *Phys. Rep.* 439 (2007) 1
[24] Th. Stöhlker *et al.*, *AIP Conf. Proc.* 1336 (2011) 132
[25] C. Brandau *et al.*, *Phys. Rev. Lett.* 100 (2008) 073201
[26] C. Brandau *et al.*, *J. Phys. Conf. Series* 194 (2009) 012023
[27] C. Brandau *et al.*, *Hyperfine Interact.* 196 (2010) 115
[28] C. Brandau *et al*, *Phys. Scripta* T (2013) in press
[29] P. Egelhof, *Acta Phys. Polonica* B 24 (1993) 303
[30] Q. Zhong *et al.*, *J. Phys. Conf. Series* 202 (2011) 012011
[31] H. Moeini *et al.*, *Nucl. Instr. Meth. Phys. Research* A 634 (2011) 77
[32] N. Kalantar-Nayestanaki *et al.*, *Int. J. Mod. Phys.* E 18 (2009) 524
[33] M. von Schmid *et al.*, *Nucl. Instr. Meth. Phys. Research* A 629 (2011) 197
[34] H. Irnich *et al.*, *Phys. Rev. Lett.* 75 (1995) 4182
[35] M. Jung *et al.*, *Phys. Rev. Lett.* 69 (1992) 2164
[36] F. Bosch *et al.*, *Phys. Rev. Lett.* 77 (1996) 5190
[37] Yu. A. Litvinov *et al.*, *Phys. Lett.* B 573 (2003) 80
[38] T. Ohtsubo *et al.*, *Phys. Rev. Lett.* 95 (2005) 052501
[39] J. Kurcewicz *et al.*, *Acta Phys. Polonica* B 41 (2010) 525
[40] Yu. A. Litvinov *et al.*, *Phys. Rev. Lett.* 99 (2007) 262501
[41] Yu. A. Litvinov *et al.*, *Phys. Lett.* B 664 (2008) 162
[42] N. Winckler *et al.*, *Phys. Lett.* B 679 (2009) 36
[43] Yu. A. Litvinov, F. Bosch, *Rep. Prog. Phys.* 74 (2011) 016301
[44] D. R. Atanasov *et al.*, *Eur. Phys. J.* A 48 (2012) 22
[45] P. Kienle *et al.*, submitted.
[46] T. Radon *et al.*, *Phys. Rev. Lett.* 78 (1997) 4701
[47] H. Poth, *Phys. Rep.* 196 (1990) 135

[48] M. Steck *et al.*, *Phys. Rev. Lett.* 77 (1996) 3803
[49] M. Steck *et al.*, *J. Phys.* B 36 (2003) 991
[50] H. Geissel *et al.*, *Nucl. Phys.* A 746 (2004) 150c
[51] H. Geissel, Yu. A. Litvinov, *Nucl. Instr. Meth. Phys. Research* B 266 (2008) 4176
[52] T. Radon *et al.*, *Nucl. Phys.* A 677 (2000) 75
[53] Yu.A. Litvinov *et al.*, *Nucl. Phys.* A 734 (2004) 473
[54] M. Hausmann *et al.*, *Nucl. Instr. Meth. Phys. Research* A 446 (2000) 569
[55] M. Hausmann *et al.*, *Hyperfine Interact.* 132 (2001) 291
[56] B. H. Sun *et al.*, *Nucl. Phys.* A 812 (2008) 1
[57] J. Trötscher *et al.*, *Nucl. Instr. Meth. Phys. Research* B 70 (1992) 455
[58] B. Mei *et al.*, *Nucl. Instr. Meth. Phys. Research* A 624 (2010) 109
[59] F. Nolden *et al.*, *Nucl. Instr. Meth. Phys. Research* A 659 (2011) 69
[60] Y. D. Zang *et al.*, *Chin. Phys.* C 35 (2011) 1124
[61] B. H. Sun *et al.*, *GSI Scientific Report* 2011 (2012) PHN-NUSTAR-FRS-21
[62] F. Bosch, Yu. A. Litvinov, T. Stöhlker, *Prog. Part. Nucl. Phys.* (2013) in press
[63] Yu. N. Novikov *et al.*, *Nucl. Phys.* A 697 (2002) 92
[64] J. Stadlmann *et al.*, *Phys. Lett.* B 586 (2004) 27
[65] Yu. A. Litvinov *et al.*, *Nucl. Phys.* A 756 (2005) 3
[66] B. H. Sun *et al.*, *Eur. Phys. J.* A 31 (2007) 393
[67] L. X. Chen *et al.*, *Phys. Rev. Lett.* 102 (2009) 122503
[68] B. H. Sun *et al.*, *Phys. Lett.* B 688 (2010) 294
[69] L. X. Chen *et al.*, *Phys. Lett.* B 691 (2010) 234
[70] M. W. Reed *et al.*, *Phys. Rev. Lett.* 105 (2010) 172501
[71] X. L. Tu *et al.*, *Nucl. Instr. Meth. Phys. Research* A 654 (2011) 213
[72] X. L. Tu *et al*, *Phys. Rev. Lett.* 106 (2011) 112501
[73] L. X. Chen *et al.*, *Nucl. Phys.* A 882 (2012) 71
[74] Y. H. Zhang *et al.*, *Phys. Rev. Lett.* 109 (2012) 102501
[75] M. W. Reed *et al.*, *J. Phys. Conf. Series* 381 (2012) 012058
[76] M. W. Reed *et al.*, *Phys. Rev.* C 86 (2012) 054321
[77] X. L. Yan *et al.*, Astrop. J. Letters 766 (2013) L8
[78] D. Shubina *et al.*, *Phys. Rev.* C 88 (2013) 024310
[79] L. X. Chen *et al.*, *Phys. Rev. Lett.* 110 (2013) 122502
[80] Yu. A. Litvinov *et al.*, *Hyperfine Interact.* 173 (2006) 55
[81] X. L. Yan *et al.*, in preparation.
[82] Yu. A. Litvinov *et al.*, *Nucl. Instr. Meth. Phys. Research* B (2013) in press
[83] Yu. A. Litvinov *et al.*, *Hyperfine Interact.* 132 (2001) 283
[84] F. Bosch *et al.*, *Int. J. Mass Spectrom.* 251 (2006) 212
[85] H. Geissel *et al.*, *AIP Conf. Proc.* 831 (2006) 108
[86] B. H. Sun *et al.*, *Nucl. Phys.* A 834 (2010) 476c
[87] G. Audi *et al.*, *Nucl. Phys.* A 729 (2003) 337
[88] B. H. Sun *et al.*, *Chin. Phys.* C 33 (Suppl. 1) (2009) 161
[89] R. Hoischen *et al.*, *J. Phys.* G 38 (2011) 035104
[90] S. E. Woosley and R. E. Taam, *Nature (London)* 263 (1976) 101
[91] H. Schatz *et al.*, *Phys. Rep.* 294 (1998) 167
[92] H. Schatz *et al.*, *Phys. Rev. Lett.* 86 (2001) 3471
[93] L. Van Wormer *et al.*, *Astroph. J.* 432 (1994) 326
[94] F. Rembges *et al.*, *Astroph. J.* 484 (1997) 412
[95] H. S. Xu, Y. H. Zhang, Yu. A. Litvinov, *Int. J. Mass Spectrom.* 349-350 (2013) 162
[96] H. Geissel and Yu. A. Litvinov, *J. Phys.* G 31 (2005) S1779
[97] H. Geissel *et al.*, *Hyperfine Interact.* 173 (2006) 49
[98] T. Yamaguchi, Y. Yamaguchi, A. Ozawa, *Int. J. Mass Spectrom.* 349-350 (2013) 240
[99] RIKEN Nishina Center for Accelerator-Based Science, `http://www.rarf.riken.go.jp/Eng/`
[100] P. M. Walker, Yu. A. Litvinov, H. Geissel, *Int. J. Mass Spectr.* 349-350 (2013) 247
[101] FAIR: Facility for Antiproton and Ion Research, `http://www.fair-center.de/`
[102] FAIR: The Modul. Start Version, `http://www.gsi.de/documents/DOC-2009-Nov-124-1.pdf`
[103] H. Geissel *et al.*, *Nucl. Instr. Meth. Phys. Research* B 204 (2003) 71
[104] F. Bosch, Yu. A. Litvinov, *Int. J. Mass Spectrom.* 349-350 (2013) 151

The Impact of Fission on R-Process Calculations

M Eichler[1], A Arcones[2], R Käppeli[3], O Korobkin[4], M Liebendörfer[1], G Martinez-Pinedo[2], I V Panov[5], T Rauscher[6,1], S Rosswog[4], F-K Thielemann[1], C Winteler[7]

[1] Department of Physics, Basel University, Klingelbergstrasse 82, 4056 Basel, Switzerland
[2] Institut für Kernphysik, Technische Universität Darmstadt, Schlossgartenstrasse 2, D-64289 Darmstadt, Germany
[3] Seminar for Applied Mathematics, ETH Zürich, Rämistrasse 101, 8092 Zürich, Switzerland
[4] Astronomy and Oskar Klein Centre, Stockholm University, AlbaNova, SE-10691 Stockholm, Sweden
[5] Institute for Theoretical and Experimental Physics, Bol'shaya Cheremushkinskaya ul. 25, Moscow, 117259 Russia
[6] Centre for Astrophysics Research, School of Physics, Astronomy and Mathematics, University of Hertfordshire, Hatfield AL10 9AB, UK
[7] Institut Energie am Bau, Fachhochschule Nordwestschweiz, St. Jakobs-Strasse 84, 4132 Muttenz, Switzerland

Abstract. We have performed r-process calculations in neutron star mergers (NSM) and jets of magneto-hydrodynamically driven (MHD) supernovae. In these very neutron-rich environments the fission model of heavy nuclei has an impact on the shape of the final abundance distribution and the second r-process peak in particular.
We have studied the effect of different fission fragment mass distribution models in calculations of low-Y_e ejecta, ranging from a simple parametrization to extensive statistical treatments (ABLA07).
The r-process path ends when it reaches an area in the nuclear chart where fission dominates over further neutron captures. The position of this point is determined by the fission barriers and the neutron separation energies of the nuclei involved. As these values both depend on the choice of the nuclear mass model, so does the r-process path. Here we present calculations using the FRDM (*Finite Range Droplet Model*) and the ETFSI (*Extended Thomas Fermi with Strutinsky Integral*) mass model with the related TF and ETFSI fission barrier predictions. Utilizing sophisticated fission fragment distribution leads to a highly improved abundance distribution.

1. Introduction

The r-process (*rapid neutron capture process*) is responsible for the production of about half of the heavy elements in our universe. Although the astrophysical site is still unknown, the physical requirements point towards explosive environments. Possible scenarios include core-collapse supernovae and neutron star mergers. The r-process reaction path in the nuclear chart proceeds in the very neutron-rich, unstable region, thus involving nuclei that cannot (yet) be studied in experiments. This means that r-process reaction networks need to rely on theoretically predicted nuclear data.
When the material reaches the top end of the reaction path, fission becomes a dominant decay channel (either via spontaneous fission for very low fission barriers or via delayed or induced fission, if beta-decays or neutron captures produce compound nuclei above their fission barrier energies). Then nuclei split into two lighter ones with roughly half the mass of the parent nucleus, while there is a possibility of several fission neutrons to be emitted. While the number of fission neutrons has been measured to be

2 – 4 for experimentally studied nuclei, it is known to increase with mass number and β–decay energy of the mother nucleus [1]. If the neutron density is still high enough, the freshly made daughter nuclei continue capturing neutrons, thus keeping the r-process alive. In that sense, a fission reaction increases the amount of seed nuclei during the event. Obviously, a realistic description of fission processes are just as important as other nuclear reactions in r-process calculations. However, since fission reactions do not only need a prediction of the reaction rate but also of the fission products, fission treatment is more complex.
The fission probability of a nucleus is directly dependent on the fission barrier, which is a feature predicted by the nuclear mass model. Therefore, the mass model determines the end of the r-process path, having a direct influence on the final abundance distribution.
The mass number and proton number of the daughter nuclei that are produced in a fission reaction are given statistically by fission fragment distribution models. These models are usually fitted to known fission data and vary in complexity.

2. Method

We have tested the impact of two different nuclear mass models and four fission fragment distribution models on the final abundances for both a neutron star merger scenario [2,3] as well as a magneto-hydrodynamically driven supernova [4]. Our nucleosynthesis calculations have been done using the nuclear network WINNET [4], closely following the descriptions in [3] (NSM) and [4] (MHD supernova), respectively. This way, a direct comparison to these works can be made. A more detailed description of the nuclear network and the procedure can be found in the two papers indicated.
The two mass models are the *Finite Range Droplet Model* (FRDM) [5] and the *Extended Thomas Fermi with Strutinsky Integral* (ETFSI) [6] with the related TF [7] and ETFSI [8] fission barrier predictions. These are mass models for which a complete set of data exist in order to perform elaborate r-process

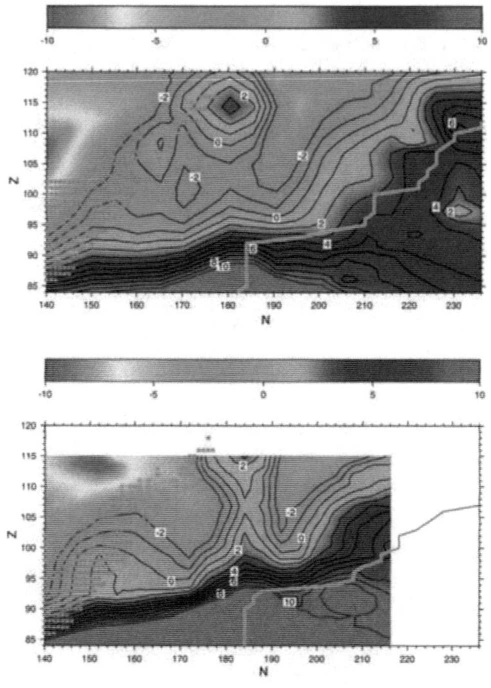

Figure 1: The quantity $B_f - S_n$ at the heavy-mass end of the nuclear chart for the FRDM (top) and the ETFSI (bottom) mass models. See text for discussion. Figure taken from [9].

Journal of Physics: Conference Series **665** (2016) 012054 doi:10.1088/1742-6596/665/1/012054

calculations. Figure 1 shows contour plots of the quantity $B_f - S_n$ (fission barrier minus neutron separation energy) for both mass models. This quantity determines the dominant reaction (or decay) channel in any particular region in the nuclear chart. If along the reaction path this value is close to 0 MeV or negative, fission will dominate over the (n,γ)- or (γ,n)-reactions and the reaction path will end here. It can be seen from Figure 1 that for the FRDM model the quantity $B_f - S_n$ is not larger than 2 MeV even at the drip line for nuclei with $N \geq 184$, preventing the production of heavier nuclei. The ETFSI model, on the other hand, predicts larger fission barriers along the drip line, thus enabling the reaction path to continue beyond $N = 184$. This means that in calculations employing the ETFSI model the fissioning nuclei are heavier, resulting in heavier fission products.

The fission fragment distribution models we use differ in complexity. They predict the fission products and the number of fission neutrons of each fission reaction using a statistical distribution, accounting for nuclear shell effects. The four models are: (a) Panov et al. (2001) [10], (b) Kodama & Takahashi (1975) [11], (c) Panov et al. (2008) [12] and (d) ABLA07 (Kelic et al., 2008) [13]. Figure 2 demonstrates the

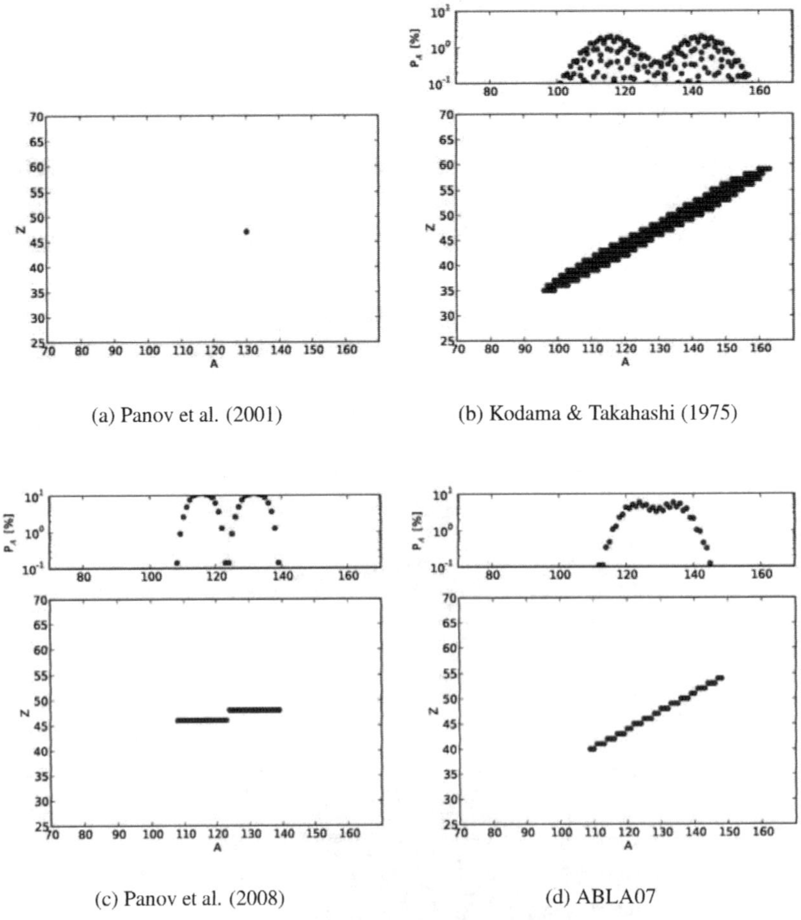

(a) Panov et al. (2001)

(b) Kodama & Takahashi (1975)

(c) Panov et al. (2008)

(d) ABLA07

Figure 2: Comparison of the four different fission fragment distribution models on the example of a fissioning ^{260}Pu nucleus. Note that the Panov et al. (2001) distribution has only one fission channel per reaction, therefore the probability is always 100%. Calculations using all four models are presented in section 3.

differences of the four models on the example of the fission of ^{260}Pu. The lower part of each graph shows the distribution of the fission products in the nuclear chart, while the upper part marks the probability in percent for each fission channel.

3. Results

We have run r-process calculations testing all the models that have been introduced in the previous section and their differences regarding the final abundance distribution. Two different possible r-process scenarios have been employed: (a) a merger of two neutron stars (NSM) with 1.4 $M_\odot$ each [3], and (b) the neutron-rich jets of a magneto-hydrodynamically driven (MHD) supernova [4]. It has been shown that in both models r-process material is synthesized. However, there is a significant difference: While the electron fraction for the neutron star merger is very low (around $Y_e \approx 0.02 - 0.05$), which allows for several fission cycles before the r-process freeze-out, its value is between 0.2 and 0.4 in the MHD jets. This results in only one fission cycle, taking place around the time of the r-process freeze-out.
Figures 3 and 4 demonstrate the effect of the fission fragment distribution model on the final abundances in an NSM and an MHD supernova scenario, respectively.

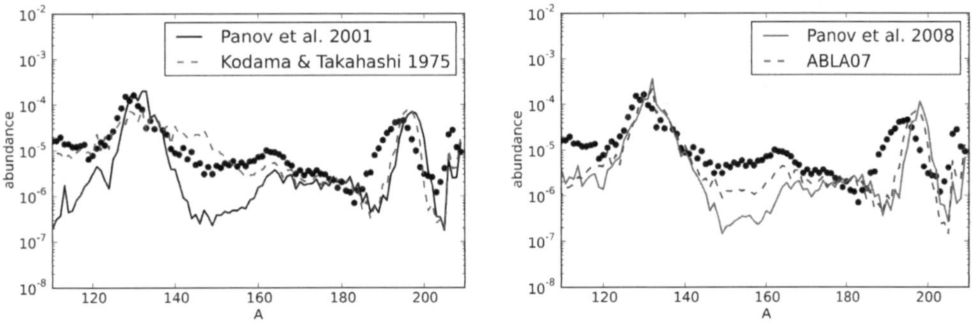

(a) (b)

Figure 3: Final abundances around the second and third peak for a NSM [3] employing four different fission fragment distribution models. For reasons of clarity the results are presented in two graphs. The black dots represent the solar r-process abundances [14]. The mass model is FRDM in all calculations.

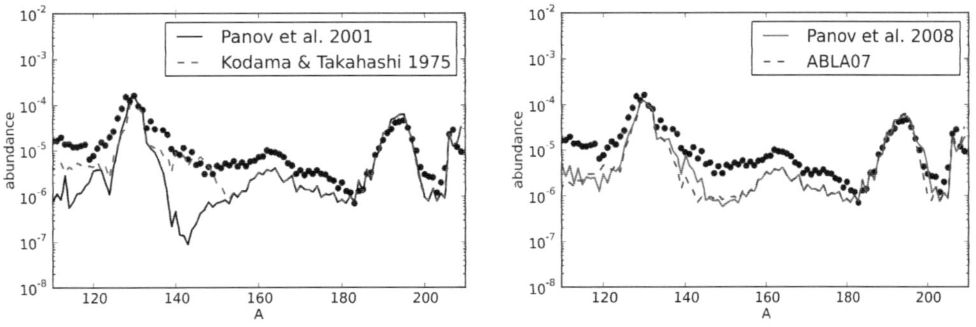

(a) (b)

Figure 4: Same as Figure 3, but for an MHD supernova [4].

The features of the models are clearly visible around the second peak (A $\approx$ 130). The Panov 2001 model produces a very pronounced peak, followed by a sharp drop in abundances. NSM calculations using the Kodama & Takahashi distribution on the other hand do not cause a well-pronounced peak, with the distribution overestimating the production of nuclei with mass number A = 140 − 150, and also the MHD simulations show an unusual abundance pattern. The Panov 2008 and ABLA07 models tend to agree better with the solar abundances, both in terms of shape of the second peak as well as the abundances of the nuclei beyond the peak.

We find that the choice of the nuclear mass model has a smaller impact on the final abundances than the fission fragment distribution model. As already mentioned above, the ETFSI model tends to predict larger fission barriers, allowing the reaction path to run up to heavier nuclei. As the nuclei reach the end of the reaction path and undergo fission, they produce heavier daughter nuclei. Therefore the abundance distribution features heavier nuclei compared to the FRDM model. Figure 5 shows the difference between the two mass models, again for both NSM and MHD supernova scenarios.

A reoccurring feature of NSM calculations is the position of the third peak, which is shifted towards higher masses compared to the solar abundances. This phenomenon has been present in calculations of various authors. We have found that the position of the peak depends on the characteristics of the r-process freeze-out. In particular, neutron captures are still in competition with photodisintegrations after the freeze-out. Therefore, neutron density and temperature are crucial for the further nucleosynthesis processes. A relatively high neutron density and low temperature will result in several neutron captures on heavy nuclei, causing the abundance distribution to shift towards higher masses. This effect can be prevented if the temperature is higher at freeze-out, e.g., if the freeze-out happens earlier. Temperatures of around 1 GK or higher are sufficient for photodisintegrations to effectively compete against the neutron captures, decreasing the average mass number of the heavy nuclei. The second peak material at that time consists to a considerable degree of freshly made fission products. Those are neutron-rich themselves and do not capture additional neutrons. Therefore the second peak does not experience the same shift as the third peak. As a test, we have artificially increased the β−decay rates for all nuclei with charge number Z > 80 by a factor of 10. This includes the β−decays which are followed by spontaneous neutron emission. As a consequence, the reaction flow is accelerated and the r-process freeze-out takes place earlier, i.e., at a higher temperature (Figure 6a). The photodisintegrations act against the neutron captures and prevent the shift in heavy nuclei abundances. The rare earth elements (140 $\leq$ A $\leq$ 185) are also affected.

The above example is only one of several ways of preventing the shift in abundances. Another possibility is to decrease the neutron density after freeze-out, e.g., by increasing the (n,γ) and (γ,n) rates.

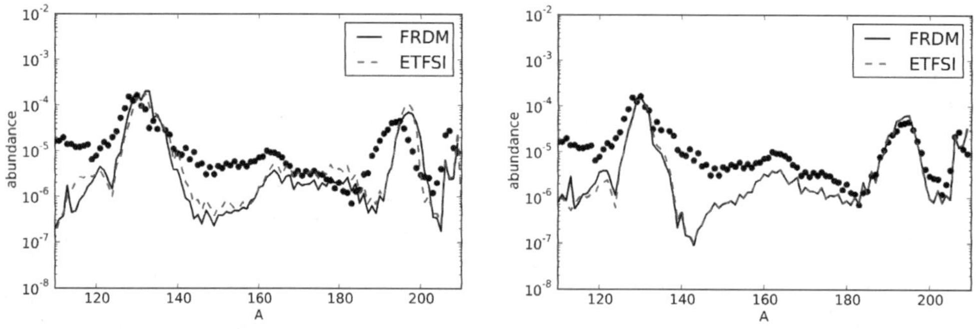

(a) NSM (b) MHD jets

Figure 5: ETFSI mass model (green dashed line) compared to the FRDM model (black solid line) for the NSM scenario (a) and the jets of a MHD supernova (b).

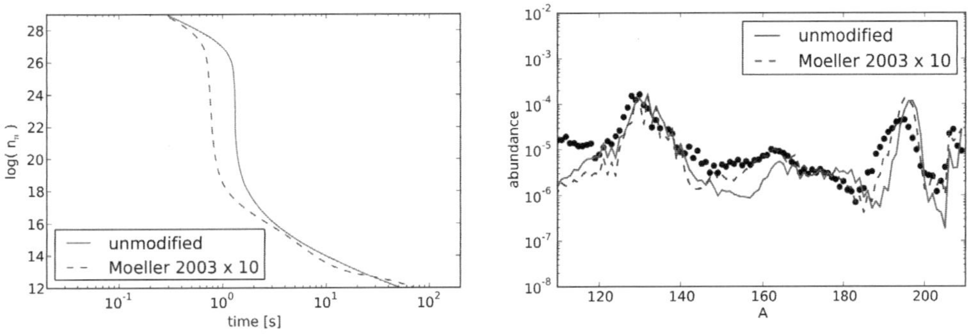

(a) neutron density n_n (b) abundances Y_i

Figure 6: The effect of accelerated β−decays of nuclei with Z > 80 on the neutron density evolution (a) and final abundances (b). The sharp drop in neutron density marks the r-process freeze-out. In this calculation the mass model is FRDM and the fission fragment distribution follows the ABLA07 prescription. See text for further explanations.

4. Conclusion

We have shown that the choice of fission fragment distribution model has a strong effect on the final abundance distribution in r-process calculations. In our case, more elaborate and up-to-date models result in a better agreement with solar r-process abundances. This holds true for both extremely neutron-rich scenarios ($Y_e \approx 0.02 - 0.05$) like neutron star mergers and the moderately neutron-rich ($Y_e \approx 0.2 - 0.4$) jets of an MHD supernova.

Furthermore, we have shown that the nuclear mass model has a smaller impact than the fission distribution model. The ETFSI model tends to predict higher fission barriers compared to the FRDM model, i.e., the reaction path is allowed to run further. As a result, the fission products are heavier, which reflects in the second peak abundances.

We have also addressed the problem of the shifted third peak in NSM calculations. The shift is caused by neutron captures on heavy nuclei after the r-process freeze-out at a relatively low temperature ($T_9 < 1$) and a neutron density of around $n_n \approx 10^{15}$ cm^{-3}. To prevent this effect a higher temperature and/or a lower neutron density after the freeze-out is needed.

References

[1] Möller P, Nix J R, & Kratz K-L, At. Data Nucl. Data Tables **66**, 131 (1997).
[2] Rosswog S, Piran T, & Nakar E, MNRAS **430**, 2585 (2013).
[3] Korobkin O, et al., Mon. Not. R. Astron. Soc. **426**, 1940 (2012).
[4] Winteler C, et al., ApJ **750**, L22 (2012).
[5] Möller P, et al., At. Data Nucl. Data Tables **59**, 185 (1995).
[6] Aboussir Y, et al., At. Data Nucl. Data Tables **61**, 127 (1995).
[7] Myers W D & Swiatecki W J, Phys. Rev. C **60**, 014606 (1999).
[8] Mamdouh A, et al., Nucl. Phys. A **644**, 389 (1998).
[9] Petermann I, et al., Eur. Phys. J. A **48**, 122 (2012).
[10] Panov I V, Freiburghaus C, & Thielemann F-K, NuPhA **688**, 587 (2001).
[11] Kodama T & Takahashi K, NuPhA **239**, 489 (1975).
[12] Panov I V, Korneev I Yu, & Thielemann F-K, Astronomy Letters **34**, 189 (2008).
[13] Kelic A, et al., Proceedings: Dynamical Aspects of Nuclear Fission, World Scientific, p. 203 (2008) [Arxiv:0906.4193].
[14] Sneden C, Cowan J J, & Gallino R, ARA&A **46**, 241 (2008).

EMPIRE: A code for nuclear astrophysics

A Palumbo[1]

[1] *National Nuclear Data Center, Brookhaven National Laboratory, Upton, NY 11973-5000, USA*

E-mail: apalumbo@bnl.gov

Abstract. The nuclear reaction code EMPIRE is presented as a useful tool for nuclear astrophysics. EMPIRE combines a variety of the reaction models with a comprehensive library of input parameters providing a diversity of options for the user. With exclusion of the direct-semidirect capture all reaction mechanisms relevant to the nuclear astrophysics energy range of interest are implemented in the code. Comparison to experimental data show consistent agreement for all relevant channels.

1. Introduction

For the majority of elements heavier than iron, nucleosynthesis occurs mainly via neutron capture reactions in the r- and s- processes. The fission channel needs to be considered in the r-process for heavy nuclei (Z > 80) [1] in which fission barriers are low. Of particular importance are the superheavy elements (SHE) since they may determine the upper end of the nucleosynthesis flow [2] which may only be reached via certain nuclear physics inputs [3]. Although the calculated and measured fission cross sections globally differ by less than a factor of three [4], current extrapolations for r-process simulation models are still unreliable. The neutron spectrum typical of s-process sites can be described by a Maxwell-Boltzmann distribution (MBD). Neutron captures have their relevant energies around the maximum of the Maxwell-Boltzmann distribution (MBD), $E_o \approx kT$ [5]. Stellar neutron capture cross sections in the mass region of the weak s-process are often not available with the required accuracy for abundance determination. The main mechanism of synthesis for the p-nuclei are photodisintegration reactions. A reprocessing of initial s-seed distribution [6] occurs via (γ,n) reactions with subsequent alpha (γ,α) or proton emission (γ,p) branchings. Simulation models have had problems reproducing the abundance distribution for the p-nuclei for Type II supernova [7]. The situation is ameliorated when considering Type 1a supernovae as a site of synthesis [8].

The heavier nuclei are formed via reactions at a high enough level density such that the Hauser Feshbach (HF) formulation applies. EMPIRE [9] is a nuclear reaction model code that accepts nucleons, γ's, and several light ions (including α-particles and deuterons) in the incoming and outgoing channels. In the presence of direct reactions, the Engelbrecht-Weidenmüller transformation can be invoked to account for the interference between compound nucleus (CN) and direct. Resonance treatment (r-matrix formalism) is beyond the scope of the code and will therefore be ignored (neutron resonances [10] can be invoked from the resonance module). The bulk of the incoming flux is treated in terms of the HF formulation with the width fluctuation correction accounted for by the HRTW [11] model. The incident channel can be treated in terms

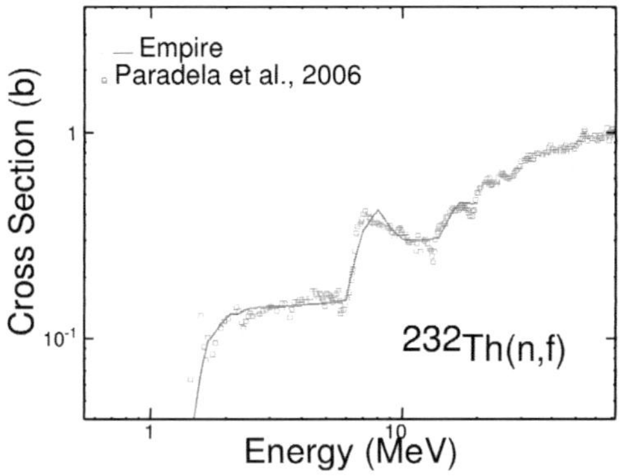

Figure 1. Neutron induced calculated cross section on ^{232}Th using EMPIRE in comparison to n_TOF [21] data

of the spherical optical model, Distorted Wave Born Approximation (DWBA), or Coupled-Channels (CC). Direct contribution to the inelastic scattering can be calculated using either DWBA or CC. The code offers an advanced fission formalism whose treatment is the optical model for fission which allows for a good reproduction of experimental fission cross sections. It offers the choice between an internal library and Hartree-Fock-Bogoliubov (HFB) fission barriers, six models for the γ-ray strength function, and level densities calculated within the Generalized Superfluid Model (EGSM), Gilbert-Cameron, or HFB combined with combinatorial method [12].

Recently, the S-factor formalism [13] [14] for non-resonant reactions (assuming a constant or very weak energy dependence within the Gamow window) has been implemented.

In the calculations below, discrete level schemes were truncated to make them consistent with the level densities. Automatic renormalization of the γ strength function was turned off in all cases [15].

2. ^{232}Th(n, f)

^{232}Th is of unique r-process origin and is a cosmochronometer (the ^{232}Th-^{238}U pair has been used to estimate the age of the r-nuclides). It has been suggested that β-delayed fission has little influence on the actinide chronometer ratios [16] but [17] suggests that spontaneous fission will affect the final abundance pattern of Th. Fig. 1 shows the fission cross section of ^{232}Th. The regional coupled-channel (CC) potential by Capote et al. [18] was used. Seven collective levels were coupled following the prescription in ref. [19]. The parabolic representation of microscopically determined HFB fission barriers were used. The barrier widths and heights were modified by less than 7% from the default values (a variation of up to 1 MeV is considered acceptable [4] [20]). In addition, the default values of the imaginary part in the optical model for the fission potential were increased. EGSM level densities were used in the main potential well and above saddles. For the gamma strength function, the Generalized Fermi Liquid Model was chosen as giving the best results. Default fission inputs were also adjusted via the number of discrete transitional states (increased from four to six), their energies, and the level densities at the saddles. One sees good agreement to the experimental data.

3. ^{186}Os(n,γ)

^{186}Os is produced via the s-process. In order to constrain the s-process contribution for ^{187}Os (Re/Os form a cosmochronometric pair) more accurate MACS are needed for the neutron capture of ^{186}Os and ^{187}Os. The modeling of ^{186}Os used the Koning-Delaroche potential [22], EGSM level densities (the level density a-parameter in the compound was scaled by 7%) and the EGLO gamma strength function (RIPL-3). The flux was treated with the spherical optical model;

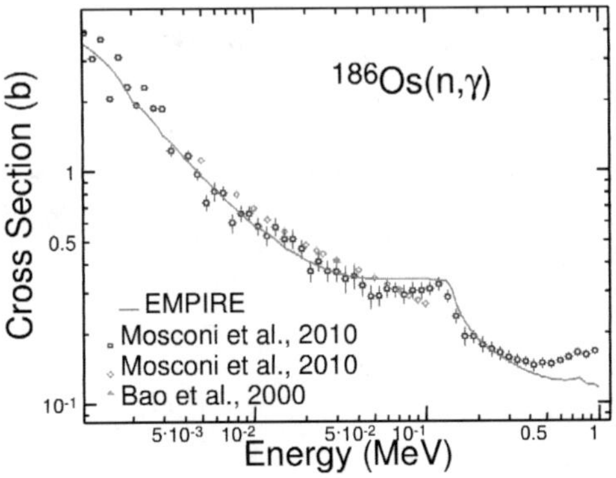

Figure 2. Neutron capture cross section of ^{186}Os using EMPIRE in comparison to the data [23] [24]

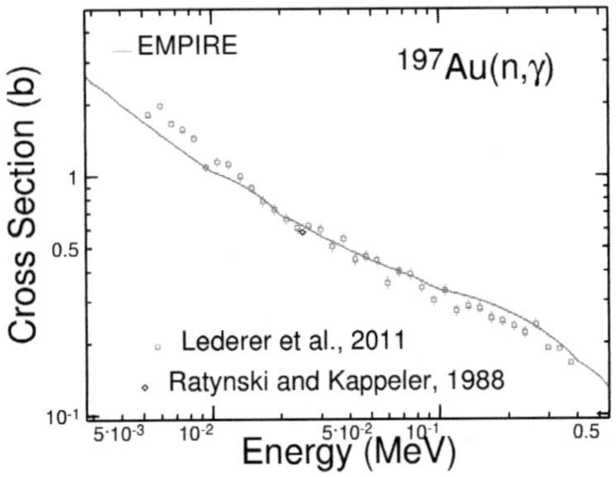

Figure 3. Neutron capture cross section of ^{197}Au using EMPIRE in comparison to the data [25] [27]

however, for those nuclei where DWBA was applied for the inelastic flux, the criterion was that the ground state deformation (β_2) should be ≤ 0.20 (^{186}Os has a β_2 of 0.20). Fig. 2 shows the calculated cross section of ^{186}Os in comparison to the experimental data. One observes a marked drop in the calculated cross section starting around 470 keV in comparison to the experimental data. Such behavior may indicate a competition cusp and/or the overestimation of the inelastic cross section (the second excited state is at 434 keV) due to the lack of a direct component.

4. ^{197}Au(n,γ)

The capture cross section of ^{197}Au [25] is well known and is used as a way to determine the neutron flux in nucleosynthesis measurements. This cross section is considered a standard from 200 keV to 2.8 MeV [26]; however, discrepancies exist with regards to the reference used for nuclear astrophysics experiments. Fig. 3 shows the calculated capture cross section of ^{197}Au in comparison to new TOF data [27]. The calculation used DWBA for the inelastic flux, the potential by Koning-Delaroche, EGSM level densities and the MLO model for the gamma strength function. Shown is the MACS from 5 to 100 keV in addition to the measured ^{197}Au cross sections (the values in the plot are energy averaged). A reasonable agreement between data and theory is observed.

5. ^{74}Se(p,γ)

Although it is photodisintegration reactions that synthesize the p-nuclei in the stellar environment, it is the inverse reaction that is generally measured in the lab. The proton branchings are more dominant at lower masses [28] and higher temperatures. The deviation between data to theory for the p-capture of p-nuclei has generally been within a factor of two (showing sensitivity between the optical potential and the level density). Fig. 4 shows the cross section and S-factor for the proton capture on ^{74}Se. This was modeled using the ESGM level densities (with a 5% adjustment via the a-parameter), the Koning-Delaroche potential, and the MLO gamma strength function. Since the ground state deformation has a value of $\beta_2 = 0.302$, CC were invoked (a default number of four levels were coupled - the conclusions of [19] don't apply for charged particles subbarrier). Although [29] states that a CC approach gives enhancements in the transmission functions for both positive and negative Q-values subbarrier, probably due to the relatively small barrier a minimal overall improvement was observed in comparison to the absense of CC (graph of spherical case not shown).

A cautionary tale should be observed when applying CC to the charged particle reactions of relevance for the γ,α and γ,p branchings in the p-process nucleosynthesis path. In principle, the effect of the coupling alters the transmission functions (the real barrier is replaced by an effective barrier) which may have implications for the nucleosynthesis when the nucleus is α-unbound (it is only the Coulomb barrier which prevents its decay). Depending on the complexity (and strength) of the feedings of a particular p-process isotope, this could potentially (artificially) alter its branch point (γ,α or γ,p). This is in contrast to DWBA where there are no coupled levels (flux is removed to uncoupled levels) and the flux is directly taken into account.

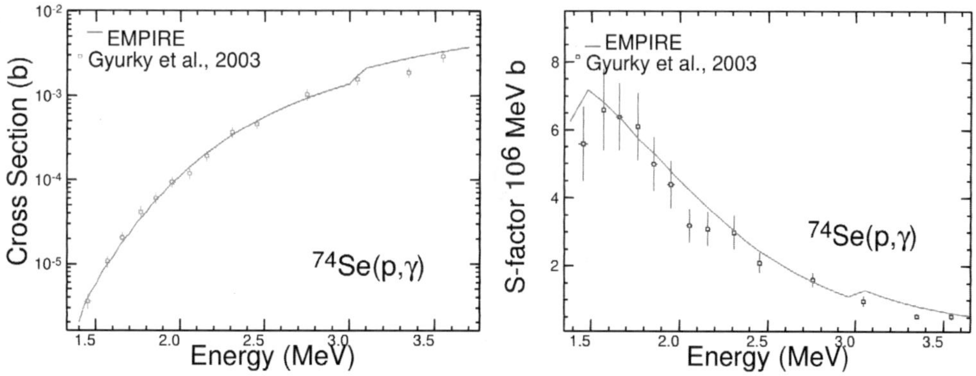

Figure 4. The cross section and S-factor for the p-capture on ^{74}Se [30]

6. ^{130}Ba+α

The agreement for α-capture has been traditionally poor mainly due to problems with the α-optical potential, although it has also been suggested that the deviations could arise due to problems with the γ width [31]. Above the neutron threshold the α-potential can be constrained by (α,n) measurements. This is in part due to the greater (α,n) cross section compared to (α,γ) and since neutron widths are generally broader any possible resonant components can often be excluded. A p-process simulation [7] within the framework of a Type II supernova showed an increased sensitivity of the abundances to a global variation of the (γ,α) and (α,γ) reaction rates in the mass region above A$\approx$ 140.

Fig. 5 shows the α-induced cross section on ^{130}Ba using both the McFadden/Satchler [33] and Avrigeanu potentials [34]. The calculation for the McFadden/Satchler potential uses DWBA for the inelastic treatment while that of the Avrigeanu potential uses the spherical optical model (no inelastic). In addition there is a 15% reduction in the volume diffuseness for the Avrigeanu

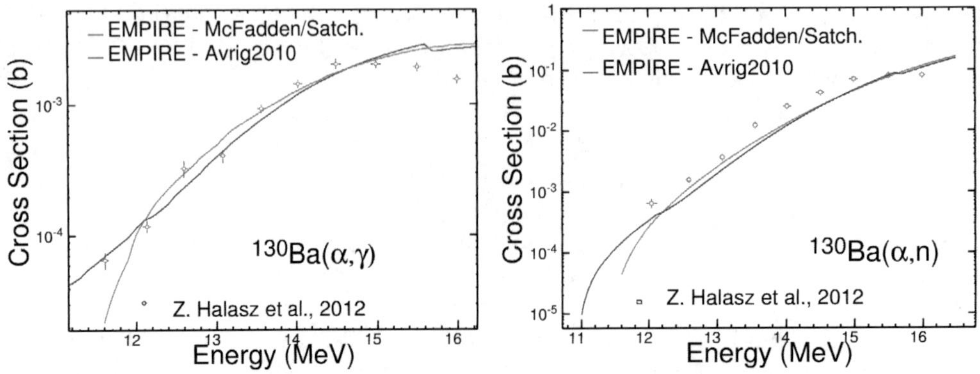

Figure 5. The cross sections for α-capture on ^{130}Ba [32]

potential. The diffuseness is considered to be one of the most sensitive parameters for the real part of the alpha potential [35]. Both use HFB level densities and the MLO model for the gamma strength function.

7. Conclusion

The calculations presented in this paper are of relevance for the r-, s- and p- processes and have been done using the reaction model code EMPIRE. The nuclear inputs are taken from the RIPL-3 database and the calculations have been performed within a nuclear astrophysics framework. The good reproduction of theory to experiment suggests it can be used as a tool for nuclear astrophysics.

Acknowledgments

The author wishes to acknowledge the financial support of the Joint Institute for Nuclear Astrophysics and the Department of Energy, Office of Nuclear Physics, under Contract no. DE-AC02-98CH10886 in presenting this work.

References

[1] M. Arnould, S. Goriely and K. Takahashi 2007 *Phys. Rep.* **450**.
[2] J. Erler *et al.*, 2012 *Phys. Rev. C* **85** 025802.
[3] I. Petermann *et al.*, 2012 *EPJA* **48**.
[4] S. Goriely *et al.*, 2009 *Phys. Rev. C* **79** 024612.
[5] T. Rauscher 2011 *Int. J. of Mod. Phys. E* **20** No. 5, 1071-1169.
[6] M. Wiescher, *private communication*, August 22, 2005.
[7] W. Rapp *et al.*, 2006 *ApJ* **653** 474.
[8] C. Travaglio, F. K. Röpke *et al.*, 2011 *ApJ* **739** 93.
[9] M. Herman *et al.*, 2007 *Nucl. Data Sheets* **108** 2655-2715.
[10] S. Mughabghab 2006 *Atlas of Neutron Resonances* (Elsevier).
[11] H. M. Hofmann *et al.*, 1975 *Ann. Phys.* **90** 403.
[12] R. Capote *et al.*, 2009 *Nucl. Data Sheets* **110** 3107.
[13] C. Rolfs and W. Rodney 1988 *Cauldrons in the Cosmos* (University of Chicago Press).
[14] https://ndclx4.bnl.gov/svn/empire.
[15] S. Goriely *et al.*, 2008 *AA* **487** 767-774.
[16] J.J. Cowan, F.-K. Thielemann, and J. Truran 1991 *Annu. Rev. Astron. Astrophys.* **29** 447-97.
[17] M. Samyn 2004, *Thèse de doctorat*, l'Université Libre de Bruxelles.
[18] R. Capote *et al.*, 2008 *J. Nucl. Sc. Tech. Japan* **45**.
[19] F. Dietrich *et al.*, 2012 *Phys. Rev. C* **85** 044611.

[20] S. Goriely, *private communication*, May 22, 2013.
[21] C. Paradela *et al.*, 2007 *Conf. on Nucl. Data for Sci. and Tech.* **Vol. 1** p. 421.
[22] A.J. Koning and J.P. Delaroche 2003 *Nucl. Phys. A* **713** 231-310.
[23] Z.Y. Bao *et al.*, 2000 *ADNDT* **76** 70-154.
[24] M. Mosconi *et al.*, 2010 *Phys. Rev. C* **82** 015802.
[25] W. Ratynski and F. Käppeler 1988 *Phys. Rev. C* **37** 595.
[26] S.A. Badikov *et al.*, 2007 *International Evaluation of Neutron Cross-Section Standards* **92-0-100807** 227.
[27] C. Lederer *et al.*, 2011 *Phys. Rev. C* **83** 034608.
[28] T. Rauscher 2006 *Phys. Rev. C* **73** 015804.
[29] C.H. Dasso, S. Landowne and A. Winther 1983 *Nucl. Phys. A* **407** 221-232.
[30] Gy. Gyürky *et al.*, 2003 *Phys. Rev. C* **68** 055803.
[31] M. Wiescher, *private communication*, 2006.
[32] Z. Halász *et al.*, 2012 *Phys. Rev. C* **85** 025804.
[33] L. McFadden and G.R. Satchler 1966 *Nucl. Phys.* **84** 177.
[34] M. Avrigeanu and V. Avrigeanu 2010 *Phys. Rev. C* **82** 014606.
[35] H. Abele and G. Staudt 1993 *Phys. Rev. C* **47** 742.

HST/STIS abundances in the uranium rich metal poor star CS 31082-001: Constraints on the r-Process

C Siqueira-Mello[1], M Spite[2], B Barbuy[1], F Spite[2], E Caffau[3,2], V Hill[4], S Wanajo[5], F Primas[6], B Plez[7], R Cayrel[8], J Andersen[9,10], B Nordström[9], C Sneden[11], T C Beers[12,13], P Bonifacio[2], P François[8], and P Molaro[14]

[1]IAG, Universidade de São Paulo, Rua do Matão 1226, Cidade Universitária, 05508-900 São Paulo, Brazil
[2]GEPI, Observatoire de Paris, CNRS, UMR 8111, 92195 Meudon Cedex, France
[3]Zentrum für Astronomie der Universität Heidelberg, Landessternwarte, Königstuhl 12, 69117 Heidelberg, Germany
[4]Laboratoire Lagrange, UMR 7293, Université de Nice Sophia-Antipolis, CNRS, Observatoire de la Côte d'Azur, 06300 Nice, France
[5]National Astronomical Observatory of Japan, 2-21-1 Osawa, Mitaka, 181-8588 Tokyo, Japan
[6]European Southern Observatory, Karl Schwarzschild Strasse 2, 85748 Garching bei München, Germany
[7]LUPM, CNRS, UMR 5299, Université de Montpellier II, 34095 Montpellier Cedex 05, France
[8]GEPI, Observatoire de Paris, CNRS, UMR 8111, 61 Av. de l'Observatoire, 75014 Paris, France
[9]The Niels Bohr Institute, Juliane Maries Vej 30, 2100 Copenhagen, Denmark
[10]Nordic Optical Telescope, Apartado 474, 38700 Santa Cruz de La Palma, Spain
[11]University of Texas at Austin, Department of Astronomy, Austin, TX 78712, USA
[12]National Optical Astronomy Observatory, Tucson, MI 85719, USA
[13]Michigan State University, Department of Physics & Astronomy, and JINA, Joinst Institute for Nuclear Physics, East Lansing, MI 48824, USA
[14]INAF - Osservatorio Astronomico di Trieste, via Tiepolo 11, 34143 Trieste, Italy

E-mail: cesar.mello@usp.br

Abstract.
As a brief revision, the origin of heavy elements and the role of abundances in extremely metal-poor (EMP) stars are presented. Heavy element abundances in the EMP uranium-rich star CS 31082-001 based mainly on near-UV spectra from STIS/HST are presented. These results should be useful for a better characterisation of the neutron exposure(s) that produced the r-process elements in this star, as well as a guide for improving nuclear data and astrophysical site modelling, given that the new element abundances not available in previous works (Ge, Mo, Lu, Ta, W, Re, Pt, Au, and Bi) make CS 31082-001 the most completely well studied r-II object, with a total of 37 detections of n-capture elements.

1. Introduction

The origin of the elements is a fundamental field in modern astrophysics, and the production mechanisms of the neutron-capture elements are still not known with certainty. In their seminal paper B^2FH [1] propose two major mechanisms of neutron capture to explain the origin of the elements beyond iron: the s-process and the r-process. The (slow) s-process occurs with longer rates compared to the half-life of the beta decay of the newly formed nuclei, and consequently the chain of reactions must follow the valley of beta stability, while the (rapid) r-process occurs with shorter rates and it is able to produce neutron-rich nuclei far from the region of stability, which will decay after the action time of the mechanism. The time between the absorption of two neutrons is typically hundreds or thousands of years in the case of the s-process and 0.01 to 0.1 seconds in the case of the r-process. Therefore very different sites are needed to support these mechanisms.

In the case of the r-process, the classical sites are high-entropy neutrino-driven winds of neutron-rich matter in core-collapse supernovae [2,3], but hydrodynamical simulations do not reach the extreme conditions necessary for the r-process and the proton or neutron richness of the wind remains to be investigated in more detail, though the weak r-process [4-6] and the νp-process [7-10] make this scenario an interesting possibility to explain the origin of lighter heavy elements [11], accounting for the light element primary process [12].

Alternative sites have been suggested to explain the so-called "main" r-process, as the merging of two neutron stars or the merging of a neutron star and a black hole [13-19]. Supernova-jet-like explosion is another exciting possibility, where the neutron-rich matter collimated in jets presents the right r-process conditions [20].

Detailed abundances of the elements produced by r-process nucleosynthesis in various circumstances are our best observational clues to the nature of this mechanism. A good picture can be obtained by considering the products of heavy-element production in the first generation(s) of stars, as recorded in the low-mass stars that survive until today, and the extremely metal-poor (EMP) Galactic halo stars have a special role in this problem. As discussed by many authors [21-23], the neutron-capture element abundances in EMP stars should be predominantly due to the r-process, since the main s-process is significant only in later phases of the Galaxy. Consequently, the analysis of these objects provides an insight into the astrophysical site(s) for the r-process.

2. The uranium-rich star CS 31082-001

CS 31082-001 is among the 12 known EMP r-II (following the Beers & Christlieb classification [24]) giant stars. It was observed during the ESO large programme "First Stars" [25-27], showing for the first time a measurable uranium line U II 3859.57 Å, and opening up a new possibility for nucleochronology [28]. This is one of the most extreme r-element enhanced giants, and its abundance pattern was studied in detail in the optical domain [29], showing for the first time in an EMP star a boost of the actinides as compared with the general r-process abundance level. The lead abundance in this star is a puzzle, since in the purely r-process enriched photosphere of CS 31082-001, most of the lead results from the decay of ^{232}Th , ^{235}U, and ^{238}U, which leaves very little space for Pb production during the r-process [30]. Improved transition probabilities and other atomic data were important pieces obtained by several authors [31,32], from which accurate calculations of chemical abundances were possible.

Even if many of the key neutron-capture elements in these star are observable from the ground, observations in space ultraviolet are crucial to obtain abundances of elements that have no measurable lines in the near-UV and visible domain, leading us to observe CS 31082-001 with the Space Telescope Imaging Spectrograph (STIS) in the space near UV. These observations are also important to check the abundances calculated from other lines from the ground spectra. Requiring 45 orbits, the mean spectrum has good S/N $\sim$ 40 in the range 2600 - 3070 Å, with resolution of R = 30 000. In this analysis we also used a new ground UVES spectrum centered at 3400 Å. The abundance determinations are based on OSMARCS 1D LTE atmospheric model [33] and the spectrum synthesis code Turbospectrum [34]. The stellar parameters are adopted from [29]: T_{eff} = 4825$\pm$50 K, log$g$ = 1.5$\pm$0.3 [cgs], [Fe/H] = -2.9$\pm$0.1 dex, and v_t = 1.8$\pm$0.2

km s^{-1}. We also adopted the abundances of the elements from C to Zn determined in previous works. The calculations used the atomic line lists from the VALD2 compilation [35], except if updated oscillator strengths were available in the literature.

The results for the heaviest r-elements were presented in [36], the first determination of all measurable third-peak elements for an EMP r-process enhanced star, including Pt and Au. We were also able to present the first determination of Bi in a r-II star, besides confirming the deficiency of Pb obtained only from the UVES/VLT spectrum. The study of the near-UV spectrum was concluded in [36,37], presenting the analysis of the r-elements, with new abundances for n-elements, 9 of them - Ge, Mo, Lu, Ta, W, Re, Pt, Au, and Bi - not available in previous works. Fig. 1 and fig. 2 show the lines Lu II 2847.505 Å and Mo II 2871.507 Å, as examples of typical fits. When available, the NLTE corrections to these abundances [38-40] have been applied, and in the case of lead we also present a new NLTE+3D corrected abundance.

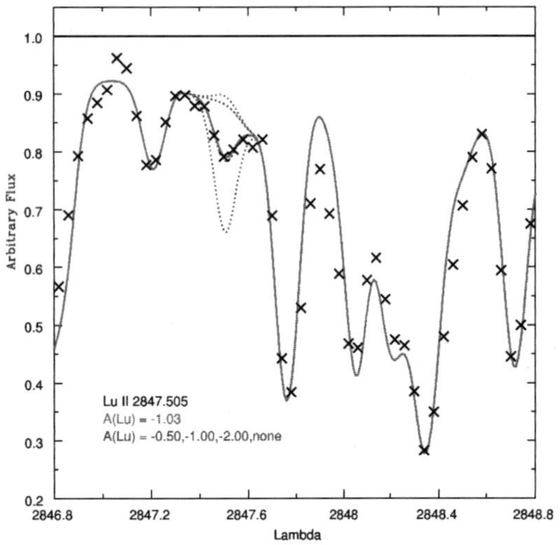

Figure 1. Fit of the observed Lu II 2847.505 Å line in CS 31082-001. Crosses: observations. Dotted lines: synthetic spectra computed for the abundances indicated. Solid line: synthetic spectrum computed with the best abundance, also indicated.

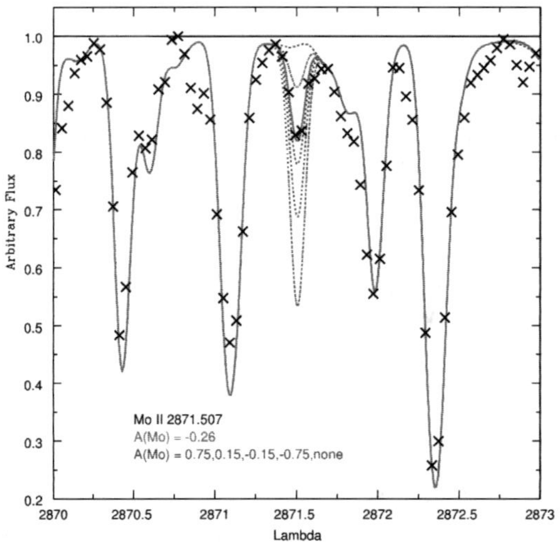

Figure 2. Fit of the observed Mo II 2871.507 Å line in CS 31082-001. Symbols as in Fig. 1.

The result obtained allows us to assess the consistency of the ages obtained from different radioactive chronometer pairs, when combined with theoretical calculations of the production ratios of the third-peak neutron capture elements and actinides. The comparison of the abundance pattern observed in this star with those from different models of r-process production permits to check these calculations. Fig. 3 compares the new complete observed abundances in CS 31082-001 with the predicted abundance patterns in the framework of neutrino-driven winds [41] using different electron abundance (Y_e), showing that the whole mass region can be fitted by using different parameters, in agreement with the need of more than one site for the r-process and/or different conditions into the same enviroment.

3. Conclusions and perspectives

Together with the previous abundances, the new results make CS 31082-001 the most complete r-II object studied, with a total of 37 detections of n-capture elements, and a major template

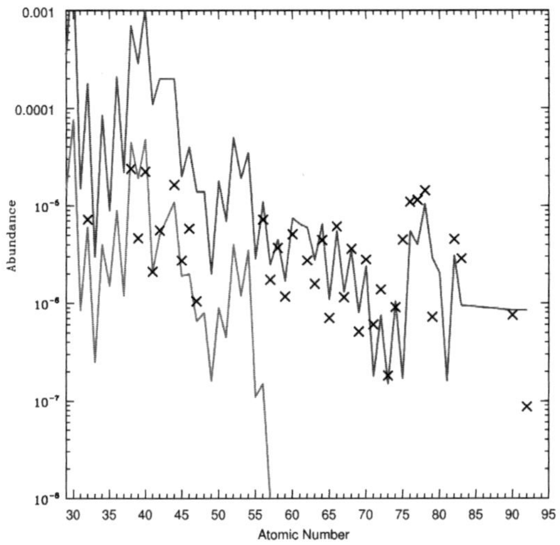

Figure 3. Comparison of the new complete observed abundances in CS 31082-001 (crosses) with theoretical yields [41], using Y_e of 0.498 (magenta solid line) and 0.482 (blue solid line). For each Y_e the superposition of the entropies spans from S = 5 k_B/nucleon to the maximum entropy $S_{final}(Y_e)\sim 300$ k_B/nucleon.

for studies of r-process models in this star, as well as a guide for improving nuclear data and modelling astrophysical site of elements production. In general, the comparisons between calculations and observations do in fact argue for a combination of processes to reproduce the full range of observed stellar abundances, but it is necessary to increase the number of stars with abundance patterns determined to obtain strong conclusions in this field. Forthcoming projects involving the new spectrographs SOAR/STELES and VLT/CUBES are planned.

References

[1] Burbidge E M, Burbidge G R, Fowler W A, and Hoyle F 1957 *Reviews of Modern Physics* **29** 547
[2] Woosley S E, Wilson J R, Mathews G J, et al. 1994 *ApJ* **433** 229
[3] Wanajo S 2007 *ApJ* **666** L77
[4] Wanajo S and Ishimaru Y 2006 *Nuclear Physics A* **777** 676
[5] Farouqi K, Kratz K-L, Pfeiffer B, et al. 2010 *ApJ* **712** 1359
[6] Wanajo S, Janka H-T, and Müller B 2011 *ApJ* **726** L15
[7] Fröhlich C, Martínez-Pinedo G, Liebendörfer M, et al. 2006 *Physical Review Letters* **96** 142502
[8] Pruet J, Hoffman R D, Woosley S E, et al. 2006 *ApJ* **644** 1028
[9] Wanajo S 2006 *ApJ* **647** 1323
[10] Arcones A and Montes F 2011 *ApJ* **731** 5
[11] Arcones A and Thielemann F-K 2013 *Journal of Physics G* **40** 3201
[12] Travaglio C, Gallino R, Arnone E, et al. 2004 *ApJ* **601** 864
[13] Lattimer J M, Mackie F, Ravenhall D G, et al. 1977 *ApJ* **213** 225
[14] Meyer B S 1989 *ApJ* **343** 254
[15] Freiburghaus C, Rosswog S, and Thielemann F-K 1999 *ApJ* **525** L121
[16] Surman R, McLaughlin G C, Ruffert M, et al. 2008 *ApJ* **679** L117
[17] Goriely S, Bauswein A, and Janka H-T 2011 *ApJ* **738** L32
[18] Wanajo S and Janka H-T 2012 *ApJ* **746** 180
[19] Korobkin O, Rosswog S, Arcones A, et al. 2012 *MNRAS* **426** 1940
[20] Winteler C, Käppeli R, Perego A, et al. 2012 *ApJ* **750** 22
[21] François P, Depagne E, Hill V, et al. 2007 *A&A* **476** 935
[22] Sneden C, Cowan J J, and Gallino R 2008 *ARA&A* **46** 241
[23] Siqueira-Mello C, Barbuy B, Spite M, et al. 2012 *A&A* **548** 42
[24] Beers T C and Christlieb N 2005 *ARA&A* **43** 531
[25] Cayrel R, Depagne E, Spite M, et al. 2004 *A&A* **416** 1117
[26] Bonifacio P, Molaro P, Sivarani T, et al. 2007 *A&A* **462** 851
[27] Bonifacio P, Spite M, Cayrel R, et al. 2009 *A&A* **501** 519
[28] Cayrel R, Hill V, Beers T C, et al. 2001 *Nature* **409** 691

[29] Hill V, Plez B, Cayrel R, et al. 2002 *A&A* **387** 560
[30] Plez B, Hill V, Cayrel R, et al. 2004 *A&A* **428** L9
[31] Sneden C, Lawler J E, Cowan J J, et al. 2009 *ApJS* **182** 80
[32] Meléndez J and Barbuy B 2009 *A&A* **497** 611
[33] Gustafsson B, Edvardsson B, Eriksson K, et al. 2008 *A&A* **486** 951
[34] Alvarez R and Plez B 1998 *A&A* **330** 1109
[35] Kupka F, Piskunov N, Ryabchikova T A, et al. 1999 *A&AS* **138** 119
[36] Barbuy B, Spite M, Hill V, et al. 2011 *A&A* **534** A60
[37] Siqueira-Mello C, Spite M, Barbuy B, et al. 2012 *A&A* **550** 122
[38] Andrievsky S M, Spite M, Korotin S A, et al. 2009 *A&A* **494** 1083
[39] Andrievsky S M, Spite F, Korotin S A, et al. 2011 *A&A* **530** A105
[40] Mashonkina L, Ryabtsev A, and Frebel A 2012 *A&A* **540** A98
[41] Farouqi K, Kratz K-L, Pfeiffer B, et al. 2010 *ApJ* **712** 1359

Gamow-Teller strength and beta-decay rate within the self-consistent deformed pnQRPA

M Martini[1,2], S Goriely[1] and S Péru[2]

[1] Institut d'Astronomie et d'Astrophysique, CP-226, Université Libre de Bruxelles, 1050 Brussels, Belgium

[2] CEA, DAM, DIF, F-91297 Arpajon, France

Abstract. In recent years fully consistent quasiparticle random-phase approximation (QRPA) calculations using finite range Gogny force have been performed to study electromagnetic excitations of several axially-symmetric deformed nuclei up to the ^{238}U. Here we present the extension of this approach to the charge-exchange nuclear excitations (pnQRPA). In particular we focus on the Gamow-Teller (GT) excitations which are known to play a crucial role in several fields of physics, in particular in nuclear astrophysics (stellar evolution and nucleosynthesis). A comparison of the predicted GT strength distribution with existing experimental data is presented. The role of nuclear deformation is shown. Special attention is paid to β-decay half-lives calculations for which experimental data exist and for specific isotonic chains of relevance for the r-process nucleosynthesis.

Spin-isospin nuclear excitations, in particular the Gamow-Teller (GT) resonances, play nowadays a crucial role in several fields of physics in particular in astrophysics, since they influence stellar evolution and nucleosynthesis by governing the electroweak processes. Experimentally the spin-isospin nuclear excitations are studied via charge-exchange reactions and β-decay measurements. In spite of the great efforts and interest, most of the exotic nuclei of astrophysical interest cannot be studied experimentally. To study the nuclei experimentally inaccessible one can rely on theoretical models. In this context one of the most employed models is the so called proton-neutron quasiparticle random-phase approximation (pnQRPA). To treat consistently isotopic chains from drip line to drip line two main features of the theoretical model are in order: the possibility to deal with deformed nuclei and the use of an unique effective nuclear force. The term *unique* has here two meanings. First of all, it means that the interaction is the same for all the nuclei; second, that the nuclear interaction used to describe the ground state and the excited states is the same (this is the so-called self-consistency of the calculation). In spite of the relatively large number of pnQRPA calculations, only a few of those include both features. Furthermore, even in such self-consistent calculations, it remains at least one coupling constant, typically in the particle-particle channel, which should be considered as a free parameter usually fitted to half-lives or to the experimental position of the GT excitation energy.

Here we present the fully consistent axially-symmetric-deformed pnQRPA calculation based on the finite range Gogny force. The originality of the present work consists in the use of the Gogny force, since up to now the other self-consistent (spherical or axially deformed) calculations were performed either with a zero-range Skyrme-type force or within the relativistic covariant description. In the present approach, no additional parameters are introduced in the pnQRPA calculation beyond those characterizing the effective nuclear force (namely D1M [1]

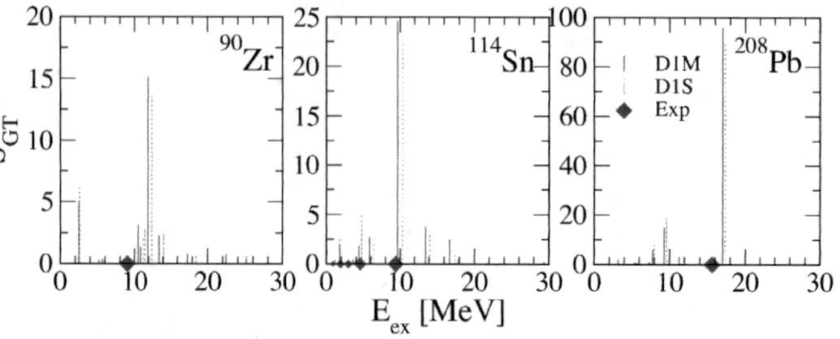

Figure 1. pnQRPA GT strength distributions in ^{90}Zr, ^{114}Sn and in ^{208}Pb calculated the with D1M and D1S forces. The experimental data of the energy peaks position [7, 8, 9] are shown as diamonds on the x-axes.

or D1S [2]). This work represents a transposition to the charge-exchange field of the fully consistent axially-symmetric-deformed QRPA calculations presented in [3] and devoted to the study of the electromagnetic excitation of deformed nuclei [4, 5]. In this approach pairing correlations which play an important role in open shell nuclei are automatically included. The possibility to take into account the nuclear deformation is also fundamental. The β-decay properties of nuclei (including their impact on the r-process nucleosynthesis [6]) as well as the nuclear matrix elements for the double β decay have been shown to depend significantly on the deformation parameter. Furthermore, deformed nuclei present strong fragmentation in their response functions and different nuclear shapes can be experimentally distinguished.

Our approach is based on the pnQRPA on top of axially-symmetric-deformed Hartree-Fock-Bogoliubov (HFB) calculations. The HFB equations are solved in a finite harmonic oscillator (HO) basis. As a consequence, the positive energy continuum is discretized. The number of HO major shells included in the model space depends on the atomic mass number. All HFB quasiparticle states are included to generate the 2-quasiparticle (2-qp) excitations. This means that our calculation can be performed without cut in energy or in occupation probabilities. According to the symmetries imposed in the present axially-symmetric-deformed HFB calculation in even-even nuclei, the projection K of the angular momentum J on the symmetry axis and the parity π are good quantum numbers. Consequently, pnQRPA calculations can be performed separately for each K^π block. To solve the pnQRPA matrix equation we use the same numerical procedure as recently applied to the study of giant resonances of the heavy deformed ^{238}U [4]. This procedure is based on a massive parallel master-slave algorithm. The solution of the pnQRPA matrix equation provides the energies ω_n of the excited states of the parent nucleus and the set of amplitudes describing the wave function of the excited state in terms of the two quasiparticle excitations.

Once the pnQRPA matrix equation is solved, we can calculate the response to the Gamow-Teller operator

$$\hat{O}_{GT} = \sum_{i=1}^{A} \vec{\sigma}(i)\ \tau_-(i) \tag{1}$$

generating a spin-flip ($\Delta S = \Delta J = 1$) response. In an axially-symmetric-deformed nuclear system, the response function of a given J^π contains different $K^\pi = 0^\pi, \pm 1^\pi, ..., \pm J^\pi$ components. In this case the GT $J^\pi = 1^+$ distributions are obtained by adding twice the $K^\pi = 1^+$ result to the $K^\pi = 0^+$ result. Details to go from the intrinsic to the laboratory frame can be found in Ref. [3].

We consider the closed neutron shell ^{90}Zr and the ^{208}Pb, as well as neutron open shell nucleus ^{114}Sn as test cases. Their GT strength distributions calculated with D1M and D1S interactions

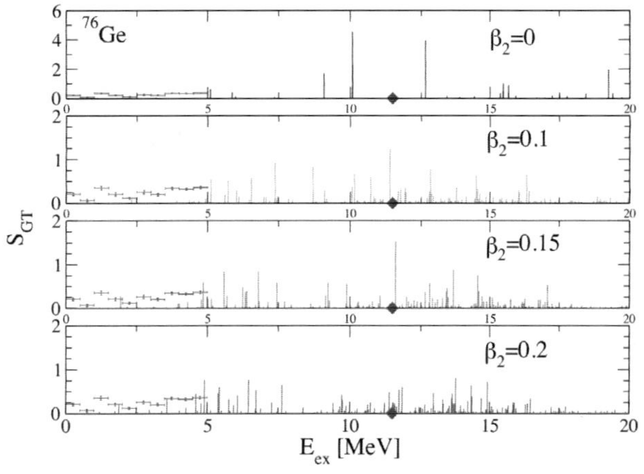

Figure 2. pnQRPA GT strength distributions in ^{76}Ge obtained with the D1M force for several values of the deformation parameter β_2, including the HFB ground state minimum of $\beta_2 = 0.15$. The experimental low energy data [10] as well as the energy position of the main GT peak are also shown.

are shown in Fig. 1. In the same figure, are shown the corresponding experimental values [7, 8, 9] for the major excitation energies obtained from (p, n) scattering data. The results are expressed as a function of the excitation energy E_{ex} referred to the ground state of the daughter nucleus. In our model it is obtained by subtracting the lowest two-quasiparticle energy E_0 from the excitation energy ω_n of the parent nucleus calculated in the pnQRPA, i.e. $E_{ex} = \omega_n - E_0$. Both interactions give quite similar results for the position of the main peak. The agreement between our calculations and experimental data is rather satisfactory. For both interactions we also verified that the Fermi and the Ikeda sum rules are exhausted by our strength distributions.

The above results refer to three spherical nuclei. As already emphasized, our approach described axially symmetric deformed nuclei. As an example of deformed nucleus, we show in in Fig. 2 the ^{76}Ge spin-isospin excitations, more precisely the GT distributions obtained with D1M Gogny interaction for several values of the deformation parameter β_2, including the HFB ground state minimum of $\beta_2 = 0.15$. Experimental data [10] are also included. As expected, the deformation tends to increase the fragmentation of the response. Calculations with different deformations produce peaks that are displaced. Deformation effects also influence the low-energy strength and consequently can be expected to affect β^--decay half-lives.

In the allowed GT decay approximation, the β^--decay half-life $T_{1/2}$ can be expressed in terms of the GT strength function S_{GT} according to

$$\frac{\ln 2}{T_{1/2}} = \frac{(g_A/g_V)^2_{\text{eff}}}{D} \sum_{E_{ex}=0}^{Q_\beta} f_0(Z, A, Q_\beta - E_{ex}) S_{GT}(E_{ex}). \tag{2}$$

For the phase-space volume f_0 as well as the D factor and the vector g_V and axial vector g_A coupling constants (including the quenching factor), we refer to the work in Ref. [11]. For the Q_β mass differences, we consider experimental values, whenever available or the D1M predictions, otherwise.

The pnQRPA calculation provides, as shown in previous figures, a discrete strength distribution. In order to derive a smooth continuous strength function, the pnQRPA GT strength is folded with a Lorentz function, as classically done. We have choosen to parametrize

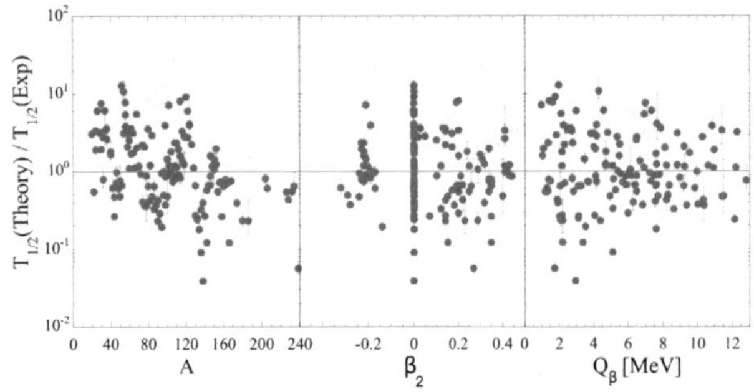

Figure 3. (color online) Ratio between the pnQRPA and experimental β^--decay half-lives as a function of the mass number A, the deformation parameter β_2 and the Q_β value.

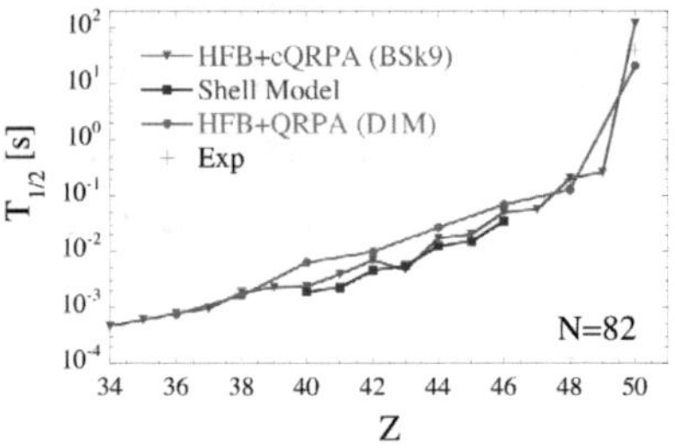

Figure 4. (color online) Comparison between HFB+cQRPA with the BSk9 force [11], shell model [13] and our predictions of the β^--decay half-lives along the $N = 82$ isotonic chain.

the spreading width Γ as shown in Ref. [12] in order to reproduce the experimental GT widths found experimentally in Sn isotopes [9].

To give an idea of the global predictions of our model, we compare in Fig. 3 the pnQRPA β^--decay half-lives with experimental data. The results are plotted as a function of the mass number, the deformation parameter and the Q_β value. They turn to be quite homogeneous with respect to β_2. Larger deviations are found for nuclei close to the valley of β-stability (low Q_β value), as found in most models. Globally, devation with respect to experimental data rarely exceeds one order of magnitude.

A comparison between different theoretical predictions (and with data when available) for the β^--decay half-lives of the $N = 82$ isotones is given in Fig. 4. We choose to focus on this region of the nuclear chart owing to its particular relevance for the $r-$process nucleosynthesis [6]. Our results closely agree with the HFB plus continuum QRPA approach [11] but tend to give rather larger half-lives than the shell-model predictions [13].

In conclusion, we presented here a fully consistent pnQRPA approach using a finite-range Gogny force. We applied our model to the analysis of charge-exchange modes paying a special attention to the GT resonances. The crucial role of deformation, automatically included in our approach, was analyzed. The agreement with experiment is satisfactory both for the strength distribution and the β^--decay half-lives.

References

[1] Goriely S, Hilaire S, Girod M and Péru S 2009 *Phys. Rev. Lett.* **102** 242501
[2] Dechargé J and Gogny D 1980 *Phys. Rev. C* **21** 1568
[3] Péru S and Goutte H 2008 *Phys. Rev. C* **77** 044313
[4] Péru S, Gosselin G, Martini M, Dupuis M, Hilaire S and Devaux J-C 2011 *Phys. Rev. C* **83** 014314
[5] Martini M, Péru S and Dupuis M 2011 *Phys. Rev. C* **83** 034309
[6] Arnould M, Goriely S and Takahashi K 2007 *Phys. Rep.* **450** 97
[7] Wakasa T *et al.* 1997 *Phys. Rev. C* **55** 2909
[8] Akimune H *et al.* 1995 *Phys. Rev. C* **52** 604
[9] Pham K *et al.* 1995 *Phys. Rev. C* **51** 526
[10] Thies J H *et al.* 2012 *Phys. Rev. C* **86** 014304
[11] Borzov I N and Goriely S 2000 *Phys. Rev. C* **62** 035501
[12] Martini M, Péru S and Goriely S 2014 *Phys. Rev. C* **89** 044306
[13] Martinez-Pinedo G and Langanke K 1999 *Phys. Rev. Lett.* **83** 4502

Large-scale deformed QRPA calculations of the gamma-ray strength function based on a Gogny force

M Martini[1,2], S Goriely[1], S Hilaire[2], S Péru[2] and F Minato[3,1]

[1] Institut d'Astronomie et d'Astrophysique, Université Libre de Bruxelles, CP-226, 1050 Brussels, Belgium

[2] CEA, DAM, DIF, F-91297 Arpajon, France

[3] Nuclear Data Center, Japan Atomic Energy Agency, Tokai 319- 1195, Japan

Abstract. The dipole excitations of nuclei play an important role in nuclear astrophysics processes in connection with the photoabsorption and the radiative neutron capture that take place in stellar environment. We present here the results of a large-scale axially-symmetric deformed QRPA calculation of the γ-ray strength function based on the finite-range Gogny force. The newly determined γ-ray strength is compared with experimental photoabsorption data for spherical as well as deformed nuclei. Predictions of γ-ray strength functions and Maxwellian-averaged neutron capture rates for Sn isotopes are also discussed.

1. Introduction

The dipole excitations of nuclei play an important role in nuclear astrophysical processes related to the photoabsorption and the radiative neutron capture that take place in stellar environment. Large scale calculations of E1 γ-ray strength functions are usually performed on the basis of the phenomenological Lorentzian model [1]. To improve the reliability of the γ-ray strength predictions the use of microscopic models is in order. In principle, the more microscopic the underlying theory, the greater the confidence in the extrapolations out towards the experimentally unreachable regions. Microscopic approaches are almost never used for practical applications, because of the numerical difficulty associated with large scale predictions and the fine tuning of the model required to reproduce accurately large experimental data set. A prominent exception is represented by Refs. [2, 3] where a complete set of γ-ray strength functions was derived from microscopic HFB+QRPA calculations. In Refs. [2, 3], zero-range Skyrme forces were considered and deformed nuclei were described through a phenomenological correction.

The present project aims at performing axially-symmetric-deformed HFB+QRPA calculations in a fully consistent way using the finite range Gogny force D1M [4]. This approach is essential in open-shell nuclei, where pairing correlations and deformation affect the γ-ray strength. It has already been applied to study giant resonances in Si and Mg isotopes [5], dipole excitations in Ne isotopes and $N = 16$ isotones [6] as well as electromagnetic excitations of the heavy deformed ^{238}U [7]. We refer to Refs. [5, 6, 7, 8] for further details on this method.

2. Results

The axially-symmetric-deformed HFB+QRPA method is now applied to nuclei for which photoabsorption data exist. The QRPA provides an accurate description of the giant dipole

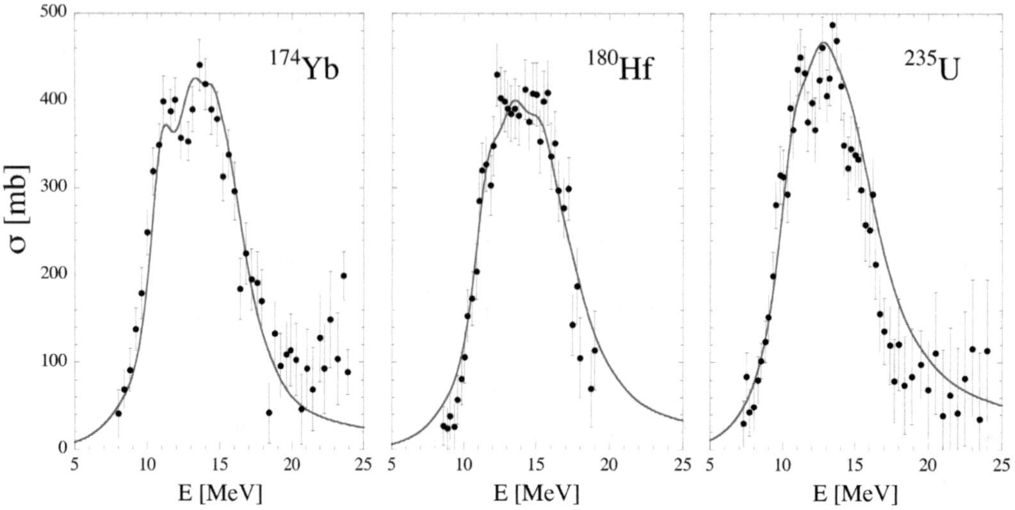

Figure 1. Comparison of the experimental photoabsorption cross sections for ^{174}Yb, ^{180}Hf and ^{235}U nuclei [1] with the axially-symmetric-deformed QRPA predictions using the D1M Gogny force (solid line) obtained with the phenomenological corrections described in the text.

resonance (GDR) centroid and the fraction of the energy-weighted sum rule exhausted by the E1 mode. However, for a proper description of experimental data, it is necessary to go beyond the QRPA scheme by including complex configurations as well as the coupling with phonons. In particular, the GDR is known experimentally to have a relatively large width and therefore a finite lifetime. Different microscopic theories exist to explain the location and width of the GDR. However, for the sake of simplicity and applicability to a large number of nuclei of astrophysical interest, we restrict ourselves to a semi-empirical broadening of the GDR. Such a broadening is obtained by folding the QRPA strength $S_{E_1}(\omega)$ by a normalized Lorentzian function

$$L(E,\omega) = \frac{2}{\pi\Gamma}\frac{\Gamma^2 E^2}{[E^2-(\omega-\Delta)^2]^2+\Gamma^2 E^2} \quad , \tag{1}$$

where the width Γ is adjusted on experimental data and the energy shift is deduced from the GDR centroid assuming the following energy-dependent parametrization: $\Delta(\omega) = \Delta_0 + \Delta_{4qp}(\omega)$, where Δ_0 is a constant shift due to the coupling between quasiparticle (qp)-states and phonons and the quantity $\Delta_{4qp}(\omega)$ is an extra shift taken to be proportional to the number of 4-qp states. The latter correction therefore varies with the excitation energy ω and obviously depends on the nucleus considered. In Fig. 1, QRPA photoabsorption cross sections are compared with experimental data [1] for a sample of 3 deformed nuclei, namely ^{174}Yb, ^{180}Hf and ^{235}U. Similar results can be found in Ref. [9] for ^{76}Ge, ^{156}Gd and ^{192}Os and in Ref. [10] for the ^{238}U. Note that to estimate the strength of odd-A and odd-odd nuclei, we have also derived an interpolation procedure starting from the QRPA calculation of the neighbouring even-even nuclei. This procedure has been tested and found to give satisfactory results as shown in Fig. 1 for ^{235}U. It will be described in a forthcoming paper. A more systematic comparison between our results and experimental data will be provided in a forthcoming paper. We remind here that the split into two components (one for each value of the projection K of the angular momentum $J=1$) of the dipole response in deformed nuclei is naturally taken into account in our approach. Recently a reasonable agreement of our HFB+QRPA calculations with the experimental photoneutron and radiative neutron-capture cross sections for the Mo isotopes has been found [11].

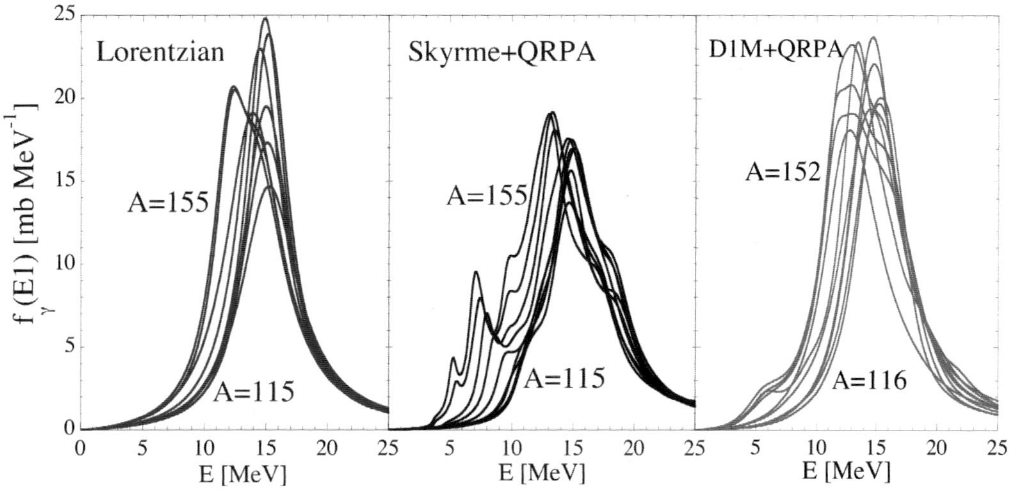

Figure 2. Comparison between the E1 γ-ray strength functions for Sn isotopes obtained with the Lorentzian model [1] (left panel), the microscopic QRPA calculation based on the Skyrme force [3] (central panel) and the present QRPA calculation with the D1M Gogny force (right panel).

As a first application, we have applied the present QRPA approach to the prediction of the γ-ray strength of exotic neutron-rich Sn isotopes, as shown in Fig. 2. If qualitatively speaking, the microscopic E1 strength functions look rather similar to the phenomenological Lorentzian [1] for nuclei close the valley of β-stability, major differences are found for exotic neutron-rich nuclei, in particular in the low-energy region where some extra strength, the so-called pygmy resonance, becomes significant in the Skyrme+QRPA predictions [3]. In our Gogny HFB+QRPA calculation, extra-strength is also found in the low-energy tail of the GDR but the strength is more spread than in the Skyrme calculation.

The differences illustrated in Fig. 2 between analytical and microscopic γ-ray strength predictions for nuclei far from the valley of stability are also known to have a significant impact on the predicted neutron-capture cross section of astrophysical interest. This is illustrated in Fig. 3, where the Maxwellian-averaged neutron capture rate calculated with the TALYS code [12, 13, 14] using the γ-ray strength obtained with the present QRPA approach are compared, for the Sn isotopes, with those obtained using either the Generalized Lorentzian (GLO) model [1] or the Skyrme-HFB+QRPA model of [3]. The ratio between these neutron capture rates is close to one for stable or nearly stable nuclei. However for exotic neutron-rich Sn isotopes our present γ-ray strength is shown to give rise to reaction rates that can be about 20 times larger than those obtained with the GLO model. Rather similar predictions are found when considering either the Skyrme-HFB plus QRPA calculations of Ref. [3] or our new Gogny-HFB plus QRPA predictions. This similarity can be explained by the rather same strenght predicted at low-energy below the GDR, although the Gogny calculation predicts a strength spread on a much wider energy range, as discussed above.

3. Conclusions

The results obtained with the Gogny force for the γ-ray strength functions are quite promising. The encouraging results obtained in parallel to the nuclear masses [4] and the nuclear level densities [15] allow us to plan to include, one by one, the ingredients required to perform microscopic cross sections calculation on the basis of an optimized Gogny interaction. We believe

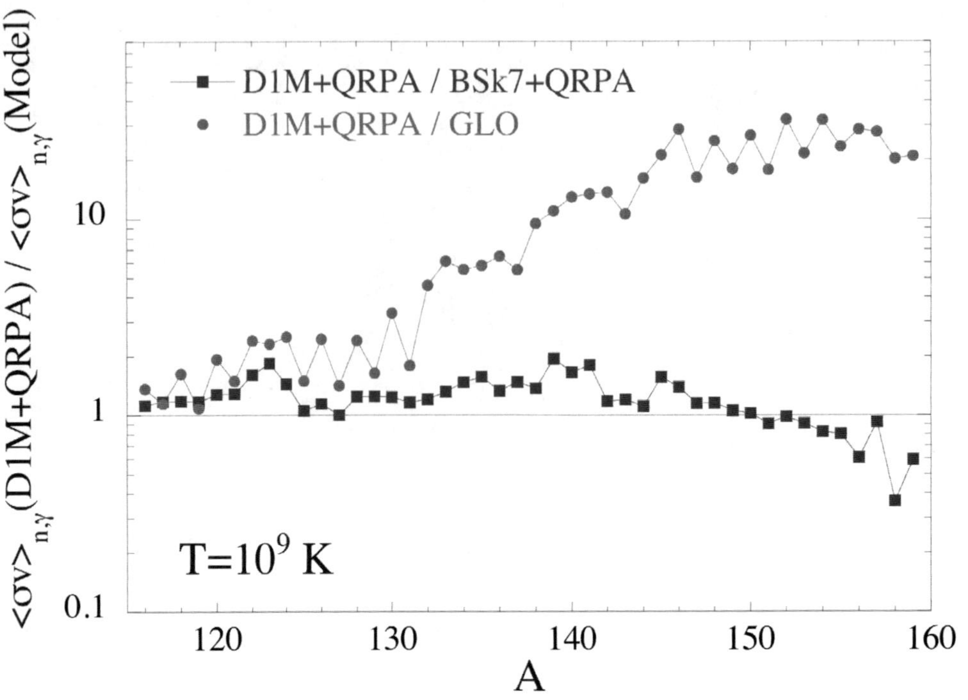

Figure 3. Ratio of the Maxwellian-averaged neutron capture rate (for a temperature of 10^9 K) for the Sn isotopes predicted with the microscopic γ-ray strength taken from the QRPA (with the Gogny D1M force) to the one obtained with the generalized Lorentzian model (circles) [1] and to the one predicted with the Skyrme+QRPA of [3] based on the BSk7 force (squares).

that work along such a path is a way to progress and be able, in the future, to obtain satisfactory cross section evaluations and predictions on the basis of reliable and accurate microscopic inputs.

Acknowledgments

We acknowledge PRACE for awarding us access to resource CURIE based in FRANCE at TGCC-CEA.

References

[1] Capote R *et al.* 2009 *Nuclear Data Sheets* **110** 3107
[2] Goriely S and Khan E 2002 *Nucl. Phys.* **A706** 217-232
[3] Goriely S, Khan E and Samyn M 2004 *Nucl. Phys.* **A739** 331-352
[4] Goriely S, Hilaire S, Girod M and Péru S 2009 *Phys. Rev. Lett.* **102** 242501
[5] Péru S and Goutte H 2008 *Phys. Rev. C* **77** 044313
[6] Martini M, Péru S and Dupuis M 2011 *Phys. Rev. C* **83** 034309
[7] Péru S, Gosselin G, Martini M, Dupuis M, Hilaire S and Devaux J-C 2011 *Phys. Rev. C* **83** 014314
[8] Péru S and Martini M 2014 *Eur. Phys. J. A* **50** 88
[9] Martini M, Goriely S, Hilaire S and Péru S 2012 *AIP Conference Proceedings* **1491** 160
[10] Martini M, Hilaire S, Goriely S, Koning A J and Péru S 2014 *Nuclear Data Sheets* **118** 273-275
[11] Utsunomiya H *et al.* 2013 *Phys. Rev. C* **88** 015805
[12] Hilaire S, Koning A J and Goriely S 2011 *Journal of the Korean Physical Society* **59** 767
[13] Goriely S, Hilaire S and Koning A J 2008 *Astronomy and Astrophysics* **487** 767
[14] Koning A J, Hilaire S and Duijvestijn M 2008 *Nucl. Data for Science and technology* (EDP Sciences, eds Bersillon et al.) 211
[15] Hilaire S, Girod M, Goriely S and Koning A J 2012 *Phys. Rev. C* **86** 064317

The r-process in black hole–neutron star mergers based on a fully general-relativistic simulation

N Nishimura[1], **S Wanajo**[2], **Y Sekiguchi**[3], **K Kiuchi**[4], **K Kyutoku**[2] **and M Shibata**[4]

[1] Astrophysics Group, Keele University, Keele ST5 1BH, UK
[2] iTHES Research Group, RIKEN, Wako 351-0198, Japan
[3] Department of Physics, Toho University, Chiba 274-8510, Japan
[4] YITP, Kyoto University, Kyoto 606-8502, Japan

E-mail: n.nishimura@keele.ac.uk

Abstract. We investigate the black hole–neutron star binary merger in the contest of the r-process nucleosynthesis. Employing a hydrodynamical model simulated in the framework of full general relativity, we perform nuclear reaction network calculations. The extremely neutron-rich matter with the total mass 0.01 $M_{\odot}$ is ejected, in which a strong r-process with fission cycling proceeds due to the high neutron number density. We discuss relevant astrophysical issues such as the origin of r-process elements as well as the r-process powered electromagnetic transients.

1. Introduction

The astrophysical origin of the r-process elements, half the elements heavier than iron including actinides, gold, europium, etc., is still shrouded in mystery despite more than half a century of studies [1]. Although the proto-neutron-star wind scenario has been considered to be the most likely site, recent state-of-the-art hydrodynamic simulations suggest serious problems to achieve the appropriate physical conditions for the r-process. Core-collapse supernovae, therefore, appear to be excluded as the main sources of the r-process elements (but of those with $A \leq 110$ produced by a weak r-process [2, 3]). Besides standard supernova scenarios, some others such as magnetic-jet explosions [4, 5] and those induced via the quark-hadron phase transition [6] have been proposed as possible alternatives.

Mergers of compact object binaries, consisting of double neutron stars (NS-NS) or a black hole and a neutron star (BH-NS), have also been investigated in the context of the r-process origin. Recent nucleosynthetic works [7, 8, 9], however, are based on the hydrodynamical simulations in the framework of particle-based SPH simulations assuming Newtonian or approximative general-relativistic gravity. This paper presents the first r-process study with the use of the hydrodynamical result based on a three-dimensional, full general relativistic treatment. We focus on a BH-NS merger as an initial attempt, which ejects extremely neutron-rich matter via the tidal disruption of the neutron-star material.

2. Neutron-rich ejecta from the BH-NS merger

The hydrodynamical simulation, used in this study, has been performed based on a 3D general relativistic code [10, 11]. This code adopts a tabulated, temperature-dependent equation of

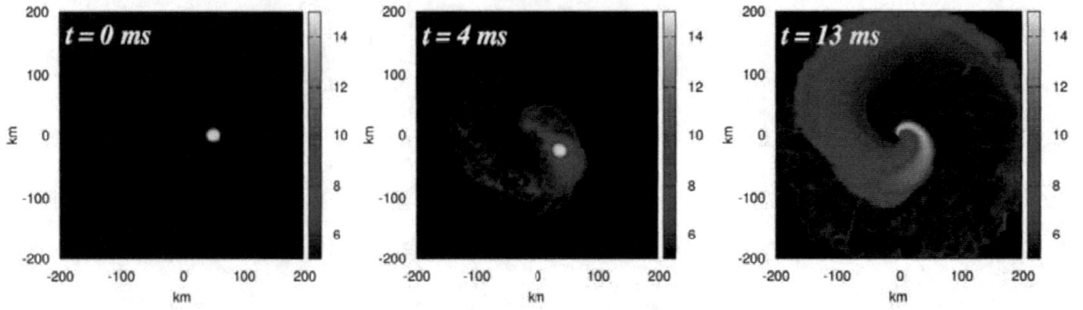

Figure 1. Dynamics of the BH-NS merger. Snapshots of the density distribution are shown in the region of $(400\ \text{km})^2$. The color range (in the log scale) corresponds to $10^5 - 10^{15}\ \text{g cm}^{-3}$.

state of nuclear matter [12] and applies a simplified treatment of neutrino transport including the neutrino cooling via the leakage scheme. We study the BH-NS merger model that initially consists of a non-rotating 4.05 $M_\odot$ black hole and a 1.35 $M_\odot$ neutron star. The amount of ejecta is estimated to be 0.01 $M_\odot$ from the hydrodynamical simulation.

The dynamics of the BH-NS merger is seen in Figure 1, illustrating time slices of the density evolution on the orbital plane. The matter is ejected mainly by the tidal disruption of the neutron-star material. At the coalescence (0 ms), the neutron star is revolving around the black hole with the orbital separation of $\sim$ 60 km. In the next snapshot (4 ms), we observe the neutron star tidally disrupted along with the orbital motion by the strong gravity of the black hole. The disrupted matter falls into the black hole and partially forms an accretion disk except for that having the velocity higher than the escape velocity from the remnant black hole (13 ms).

3. The *r*-process nucleosynthesis

The ejecta of the BH-NS model is extremely neutron-rich as shown in the left panel of Figure 2. The histogram has the peak at $Y_e \sim 0.05$, which qualitatively agrees with previous hydrodynamical studies. Here, the Y_e distribution of the ejecta is evaluated on the two-dimensional orbital plane for simplicity. The ejection is mainly due to the tidal disruption of the neutron-star material without strong shock heating which induces positron-capture on neutrons and increases Y_e. We compute the nuclear abundance evolutions using a reaction network described in [2] applying the thermodynamic trajectories deduced from these ejecta. The network follows the temporal evolutions of density and temperature for each trajectory with the initial Y_e. The network adopts the JINA-REACLIB compilation [13] for experimental rates, the BRUSLIB compilation [14] for theoretical rates, and the HFB-14 fission barriers.

We select two representative thermodynamic trajectories with $Y_e = 0.01$ and 0.04 (a full set of calculations will be presented elsewhere). The nucleosynthetic results are presented in Figure 3 that shows the nuclear abundances with the solar-system r-process distribution. A strong r-process proceeds and produces the elements from the 2nd peak to the 3rd peak and actinides. The two results are almost identical as a consequence of the r-process via strong fission cycling. According to this result, the total ejecta is expected to have a similar abundance distribution to that in Figure 3, in which the light r-process elements ($A < 130$) are severely underproduced.

4. Discussion

We performed r-process calculations with the use of a general relativistic BH-NS merger simulation. Two representative thermodynamic trajectories on the orbital plane were chosen for nucleosynthesis. The r-process proceeded strongly and produced the heavy abundances with

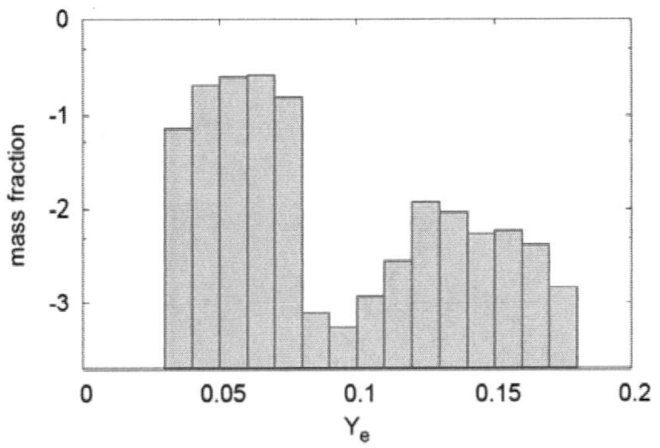

Figure 2. Distribution of Y_e of ejecta. The mass fraction for each Y_e-bin ejecta is plotted in the logarithmic scale.

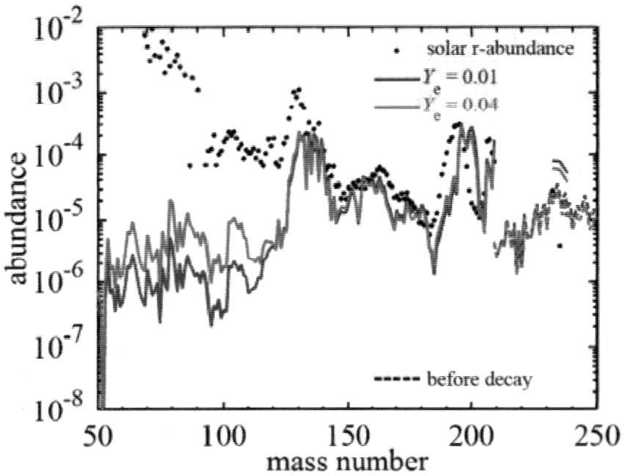

Figure 3. Nuclear abundances for two selected tracer particles of $Y_e = 0.01$ and 0.04.

$A > 130$ only. Therefore, the dynamical ejecta from BH-NS mergers may not be the unique sources of the r-process elements but other contributions from the subsequent accretion tori or from NS-NS mergers to account for the full-range of the solar system r-process pattern [15].

NS-NS and BH-NS mergers are expected to be the sources of the r-process powered electromagnetic transients ("'kilonovae" or 'r-process novae"), being associated with short gamma-ray burst and gravitational-wave events. Detailed nucleosynthetic studies relevant for mergers will be important not only for galactic chemical evolution but also for the gamma-ray and gravitational-wave astronomy. The large ejected mass of 0.01 $M_\odot$ in our BH-NS model may be favorable to account for the brightness of the recently detected (possible) kilonova [16, 17].

This work was supported by EU-FP7-ERC-2012-St Grant 306901 and by the Grant-in-Aid for the Global COE Program "The Next Generation of Physics, Spun from Universality and Emergence". We used computer facilities in CfCA at NAOJ and YITP at Kyoto University.

References

[1] Thielemann F-K, Arcones A, Käppeli R *et al* 2011 *Prog. Part. Nucl. Phys.* **66** 346
[2] Wanajo S, Janka H-T and Müller B 2011 *Astro. Phys. J. Lett.* **726** L15
[3] Wanajo S 2013 *Astro. Phys. J.* **770** L22
[4] Winteler C, Käppeli R, Perego A *et al* 2012 *Astro. Phys. J.* **750** L22
[5] Nishimura N, Takiwaki T and Thielemann 2015 *Astro. Phys. J.* (in press; *Preprint* arXiv:1501.06567)
[6] Nishimura N, Fischer T, Thielemann F-K *et al* 2012 *Astro. Phys. J.* **758** 9
[7] Goriely S, Bauswein A and Janka H-T 2011 *Astro. Phys. J.* **738** L32
[8] Korobkin O, Rosswog S, Arcones A and Winteler C 2012 *MNRAS* **426** 1940
[9] Rosswog S, Korobkin O, Arcones A and Thielemann F-K 2014 *MNRAS* **439** 744
[10] Sekiguchi Y 2012 *Prog. Theor. Phys.* **124** 331
[11] Sekiguchi Y, Kiuchi K, Kyutoku K and Shibata M 2015 *Phys. Rev. D* **91** 064059
[12] Shen H, Toki H, Oyamatsu K and Sumiyoshi K 1998 *Nucl. Phys. A* **637** 435
[13] Cyburt R H, Amthor A M, Ferguson R *et al* 2010 *Astro. Phys. J. Suppl.* **189** 240
[14] Xu Y, Goriely S, Jorissen A, Shen G L and Arnould M 2013 *Astro. Astrophys.* **549** 106
[15] Wanajo S, Sekiguchi S, Nishimura N *et al* 2014 *Astro. Astrophys. lett.* **789** 39
[16] Tanvir N R, Levan A J, Fruchter A S *et al* 2013 *Nature* **500** 547
[17] Berger E, Fong W and Chornock R, 2013 *Astro. Astrophys. J. Lett.* **774** L23

Beta-delayed fission probabilities of transfermium nuclei, involved in the r-process

I Panov[1], Yu Lutostansky[2], F-K Thielemann[3]

[1]Institute for Theoretical and Experimental Physics NRC Kurchatov Institute, Moscow, Russia
[2]NRC Kurchatov Institute, Moscow, Russia
[3]Department of Physics, University of Basel, Basel, Switzerland

E-mail: igor.panov@itep.ru

Abstract. For the nucleosynthesis of heavy and superheavy nuclei fission becomes very important when the r-process runs in a very high neutron density environment. In part, fission is responsible for the formation of heavy nuclei due to the inclusion of fission products as new seed nuclei (fission cycling). More than that, beta-delayed fission, along with spontaneous fission, is responsible in the late stages of the r-process for the suppression of superheavy element yields.

For beta-delayed fission probability calculations a model description of the beta-strength-functions is required. Extended theoretical predictions for astro-physical applications were provided long ago, and new predictions also for superheavy nuclei with uptodate nuclear input are needed. For the further extension of data to heavier transactinides the models of strength-functions should be modified, taking into account more complicated level schemes. In our present calculations the strength-function model is based on the quasi-particle approximation of Finite Fermi Systems Theory.

The probabilities of beta-delayed fission and beta-delayed neutron emission are calculated for some transfermium neutron-rich nuclei, and the influence of beta-delayed fission upon superheavy element formation is discussed.

1. Introduction

In physical scenarios with long r-process duration fission plays an important role [1], [2]. Fission, in part beta-delayed fission, effects strongly on formation of heavy and superheavy elements and nuclei-cosmo-chronometers as well. Nuclei of all chemical elements, participating in the nucleosynthesis under very high neutron density environment during multiple neutron captures achieve an equilibrium among the isotopes of each heavy element and following beta-decay of neutron-rich unstable isotope results in formation of new element. During beta-decay of such neutron-rich nuclei, emission of delayed neutron occurs leading to broading path of the r-process. In transuranium region in addition to beta-delayed neutron emission the beta-delayed fission appeared, leading together with neutron-induced fission to termination of the nucleosynthesis of more heavier nuclei during the r-process and (mainly) after neutron freeze-out. All these beta-delayed processes [3, 4] influence on the formation of heavy (and superheavy) elements yields. For a number of transactinide nuclei, participating in the r-process, neutron separation energy is less than 3 MeV, and total beta-decay energy is about 10 MeV or even more. That is why a real competition between beta-delayed neutron emission and beta-delayed fission exists, depending on nuclear structure, fission barrier and neutron separation energy values and leads to big changes of their values. Though the number of extended calculations of beta-delayed processes probabilities are exist up to Z=100 [2-5], the knowledge of these data also important [6] for superheavy region (for nuclei with Z>100), as well as reevaluation of existed data.

2. Model and calculations

Part of the r-process nuclear stream can escape fission and reach the superheavy region and the surviving of the formed superheavy elements, unstabale against beta-decay, depends in part on beta-delayed fission rates, which strongly depend on the values of beta-delayed fission probabilities $P_{\beta df}$. For beta-delayed fission calculations, except a number of important characteristics, such as nuclear masses and fission barriers, the beta strength-functions are required. The beta-strength function $S_\beta(E)$ defines the probabilities of charge-exchanges processes and beta-delayed processes as well [4,7,8]. The most important of beta-delayed processes for very neutron rich nuclei are: probabilities of one, two and three neutron emission, and for actinide region - beta-delayed fission. For these nuclei values of the probabilities can be high and probably may reach 100% in total.

The main contribution in $S_\beta(E)$ belongs to transitions of Gamov-Teller (GT) type, including Giant GT-resonance, "pigmy"-resonances and other isobaric states (IS). The beta-strength function is usually defined [9] as depending on average energy distribution of matrix elements of excitation states E_i in daughter nucleus:

$$S_\beta = \frac{\overline{d\sum M^2(E_i)}}{dE}\frac{1}{6260}, \tag{1}$$

and is expressed usually as reduced transition probability B_{GT}: $B_{GT} = D\frac{g_V^2}{g_A^2}\cdot S_\beta$, where $\overline{dM^2(E_i)}$ is the averaged addition of matrix elements of all beta-transitions in range dE.

The beta-strength function S_β has resonant structure which is strongly depends on the model used. In present calculations the matrix elements values M_i^2 as well as energies of their states were got in numerical solution of Finite Fermi Systems Theory (FFST) equations [10] when quasi-classical approach to equations for effective nuclear field of GT-type was applied [7, 8].

In constructing of $S_\beta(E)$-function we took into account the fragmentation of the high-lying IS, and their consequent broadening as it was done in [11]. According to [10] the serial expansion of the width is: $\Gamma(E) = \alpha E^2 + \beta E^3 + \ldots$. In calculation of $S_\beta(E)$ it suffices to use only the first term of $\Gamma(E)$ serial expansion, which effectively takes into account three-quasiparticle configurations. The value $\alpha \approx \varepsilon_F^{-1}$ and in present calculations α =0.018 MeV^{-1} , obtained from the averaged experimental widths of GTR, was used.

The probabilities of beta-delayed processes were derived according to formulae

$$P_{\beta df}(Z,A) = \int_{S_n+\delta}^{Q_\beta} \sum_i S_\beta^i(E)\cdot f(Q_\beta - E)\cdot \frac{\Gamma_f}{\Gamma_{tot}} dE \Big/ \int_0^{Q_\beta} \sum_i S_\beta^i(E)\cdot f(Q_\beta - E) dE\,. \tag{2}$$

where f(Z, Q_b - E) is the Fermi function and Γ_j is the fission partial width. When $E - B_f > \delta$ we can let $\Gamma_n \gg \Gamma_f$ and δ is equal $\approx$50 KeV according to [12].

The comparison of calculated probabilities of beta-delayed fission $P_{\beta df}$, based on different strength-function models, show that $P_{\beta df}$-values approximately in factor 2 lower when the proposed model of $S_\beta(E)$ was used instead of the old one [5]. In present calculations as in previous works for calculations of different rates for the r-process [2, 14], the predictions of masses and fission barriers, based on ETFSI model [13], were used.

For the very neutron rich isotopes of Fermium and heavier elements the fission barriers increase when N approaches 184, that is reflected in strong decrease of $P_{\beta df}$ values for the nuclei with N $\approx$ 184 (Fig.1).

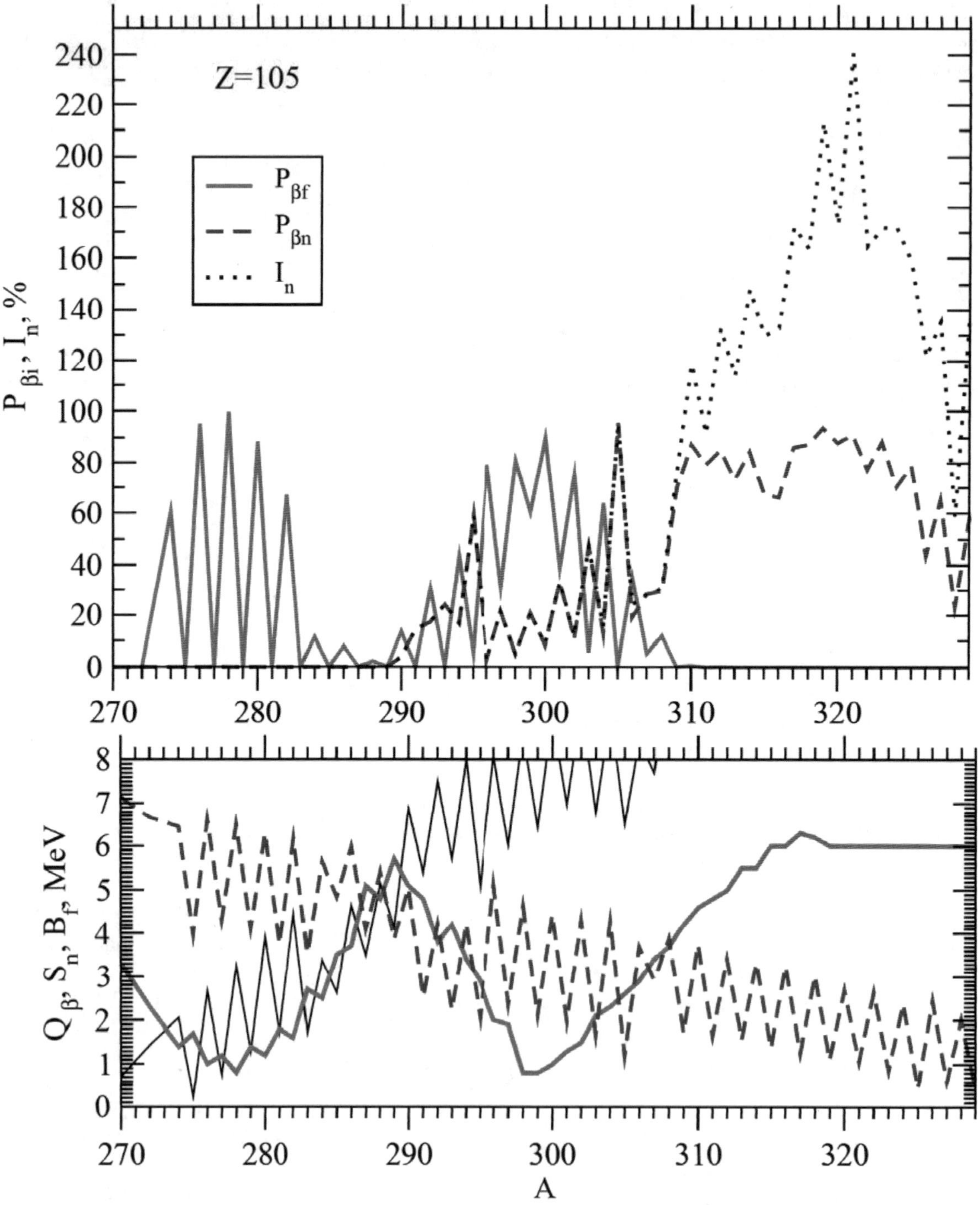

Fig. 1. Upper panel: the neutron beta-delayed emission probabilities $P_{\beta dn}$ (dashed line), beta-delayed fission probabilities $P_{\beta df}$ (line) and number of delayed neutrons per one decay (in percents) I_n (dotted line) for isotopes of Dubnium (Z=105); down panel: total energy of beta-decay Q_β (line), neutron separation energy S_n (dashed line) and fission barriers (bold line) for the same isotopes (in MeV).

At the upper plot of Fig.1 the present calculations of $P_{\beta df}$ for different isotopes of Dubnium are presented, as well as $P_{\beta dn}$-values and number of emitted neutrons during each beta-decay. Strong odd-even effect is explained by the structure of strength function and strong dependence of neutron separation energy on the number of neutrons, which in turn leads to strong variations of the energy window $S_n - B_f$ and changes of "pigmy"-resonances contributions into $P_{\beta df}$ and $P_{\beta dn}$.

3. Conclusions

Calculations of beta-strength function $S_\beta(E)$ for neutron-rich nuclei is complicated task because high-energy charge-exchange excitations in continuum should be taken into account. This task is under consideration and can be solved with arbitrary good accuracy in the framework of approach, based on FFST. In this work the previously developed approach [15] for calculation of beta-delayed processes based on quasi-classical approximation for beta-decay field [6] was used. For the consistency with previously produced nuclear data, beta-strength-functions and probabilities of delayed processes were derived with the same nuclear mass and fission barrier predictions [13] as previously calculated neutron-induced rates [14], used in astrophysical nucleosynthesis modelling.

Preliminary calculations of $P_{\beta df}$, presented in this report, pointed out the less role of delayed fission, and results in that data produced during last decades [2, 3] were overestimated, that is confirm the old predictions of beta-decay fission probabilities made on the basis of simple physical evaluations [16].

The extended calculations of $P_{\beta df}$ and $P_{\beta dn}$ made simultaneously on the basis of the same approach can change the character of the r-process passage in the transactinide region, and may be interesting for experiments of synthesis of superheavy elements [17]. These results are needed either for different modelling of astrophysical nucleosynthsis of heavy and superheavy nuclei and neutron impulse nucleosynthsis at experimental istallations as well [18, 19]. The derived results can be improved after further development of strength function model and utilization of nuclear deformations as well.

This work was supported by SNF grant IZ73Z0_152485 SCOPES, Russian Foundation for Basic Research Grants RFBR 13-02-12106_ofi-m, 14-22-03040_ofi-m

4. References

[1] I. V. Panov, F.-K. Thielemann. // Nuclear Physics A, 718, 647 (2003).
[2] I. V. Panov, E. Kolbe, B. Pfeier, et al. // Nucl. Phys. A. 747, 633 (2005).
[3] F.-K. Thielemann, J. Metzinger, H. V. Klapdor-Kleingrothaus. // Zt. Phys. A309, 301 (1983).
[4] Yu. S. Lyutostansky. // Bull. Acad. Sci. USSR, Phys. Ser. 50, 834 (1986).
[5] P. Moller, J. R. Nix, K.-L. Kratz. // ADNDT 66, 131 (1997).
[6] I. Petermann, K. Langanke, G. Martinez-Pinedo, et al. // Eur. Phys. J. A 48, 122 (2012).
[7] Yu.V. Gaponov and Yu. S. Lutostansky // Phys. At. Nucl. 73, 1360 (2010).
[8] Yu. S. Lutostansky // Physics of Atomic Nuclei, 74, 1176 (2011).
[9] P. G. Hansen. // Adv. Nucl. Phys. 45, 159 (1975).
[10] A. B. Migdal. Theory of Finite Fermi Systems. Wiley, New York, 1967.
[11] Yu. S. Lutostansky and N. B. Shulgina. // Phys. Rev. Lett. 67, 430 (1991).
[12] Yu. S. Lyutostansky, I. V. Panov, V. K. Sirotkin. // Phys. Lett. 161, 9 (1985).
[13] A. Mamdouh, J. M. Pearson, M. Rayet, et al. // Nucl. Phys.A 337, 679 (2001).
[14] I. V. Panov et al. // Astronomy and Astrophysics, 513, id.A61 (2010).
[15] V. G. Aleksankin, Yu. S. Lyutostansky, I. V. Panov // Yad. Fiz. 34, 1451 (1981).
[16] I. V. Panov, Yu. S. Lyutostansky, V. I. Ljashuk. // Bull. Acad. Sci. Ussr, Phys. ser. (USA) 54, 2137 (1990).
[17] Yu. Ts. Oganessian. //J. Phys. G, 34, R165 (2007).
[18] Yu. S. Lutostanskii, V. I. Lyashuk, I. V. Panov. // Bull. Russ. Acad. Sci. Phys. 75. 533 (2011).
[19] I. V. Panov, I. Yu. Korneev, G. Martinez-Pinedo, F.-K. Thielemann. // Astronomy Letters. 39 150 (2013).

Maximum pulsar mass, equation of state and structure of neutron-star cores

P Haensel and JL Zdunik

N. Copernicus Astronomical Center, Polish Academy of Sciences, Bartycka 18, PL-00-716 Warszawa, Poland

E-mail: haensel@camk.edu.pl jlz@camk.edu.pl

Abstract.
The structure of neutron stars is determined by the equation of state of dense matter in their interiors. Brief review of the equation of state from neutron star surface to its center is presented. Recent discovery of two two-solar-mass pulsars puts interesting constraints on the poorly known equation of state of neutron-star cores for densities greater than normal nuclear matter density. Namely, this equation of state has to be stiff enough to yield maximum allowable mass of neutron stars greater than two solar masses. There are many models of neutron stars cores involving exclusively nucleons that satisfy this constraint. However, for neutron-star models based on recent realistic baryon interaction, and allowing for the presence of hyperons, the hyperon softening of the equation of state yields maximum masses significantly lower than two solar masses. Proposed ways out from this "hyperon puzzle" are presented. They require a very fine tuning of parameters of dense hadronic matter and quark matter models. Consequences for the mass-radius relation for neutron stars are illustrated. A summary of the present situation and possible perspectives/challenges, as well as possible observational tests, are given.

1. Introduction

The equation of state (EOS) of neutron stars (NS) is one of their mysteries. For astrophysically interesting masses, $M > \mathrm{M}_\odot$, and for NS which are older than a year, the effects of temperature, magnetic field, and of rotation, on the NS structure are small [1], and will be neglected here. Then, the structure of NS, and in particular their maximum allowable mass, $M_{\rm max}$, is determined by the EOS [1].

To construct models of NS, from the surface (or rather from an atmosphere, which is however typically only a few cm thick) to the star's center, we need the EOS for density from a fraction of g cm^{-3} up to some 5×10^{15} g cm^{-3}. A general overview of theoretical EOSs for NS, from the surface to the center, is given in Sect. 2.

As we stress in Sect. 2, at density larger than 10^6 g cm^{-3}, matter in NS is strongly degenerate, and therefore pressure becomes a function of the density only, $P = P(\rho)$. Up to the normal nuclear density $\rho_0 = 2.7 \times 10^{14}$ g cm^{-3}, corresponding to the baryon (number) density $n_0 = 0.16$ fm^{-3}, the EOS is rather well established [2]. However, for $\rho > \rho_0$ the uncertainty in the EOS increases rapidly with increasing ρ. This uncertainty is reflected in the uncertainty in the theoretical prediction for $M_{\rm max}$, which is the *functional* of $P(\rho)$: $M_{\rm max}[P(\rho)]$. Recent measurements of two pulsar masses yielded values close to 2 $\mathrm{M}_\odot$: 1.97 ± 0.04 $\mathrm{M}_\odot$ for PSR J1614-2230 [3] and 2.01 ± 0.04 $\mathrm{M}_\odot$ for PSR J0348+0432 [4], both masses are quoted with

$\pm 1\sigma$ errors. These measurements imply an important constraint on the EOS at $\rho > \rho_0$ and tell us about crucial importance of strong interactions in dense matter (Sect. 3). Still, the uncertainty in the EOS for $4\rho_0 < \rho < 10\rho_0$ - density range which is crucial for $M_{\rm max}$ - is very large (Sect. 3). It is therefore interesting to consider absolute upper bounds on $M_{\rm max}$ based on very basic assumptions (Sect. 4). While we expect that NS matter at $\rho_0 < \rho < 2\rho_0$ is composed mostly of neutrons, with a few percent admixture of protons, electrons, and muons, at higher density hyperons or even quark gluon plasma might appear, forming a strangeness carrying NS-core. Possible EOS resulting from various theoretical models are reviewed in Sects. 3-5. The "hyperon puzzle" and possible ways out from it are also mentioned there. The impact of the uncertainty in the EOS($\rho > \rho_0$) on the mass vs. central density dependence of NS is described in Sect. 7. The NS radius for $M = 1.0\ {\rm M}_\odot - 1.5\ {\rm M}_\odot$ turns out to be sensitive to the presence of a strange central core. We discuss a potential significance of this feature in the context of recent measurements of radii of NS (Sect. 8). Some (tentative) conclusions are formulated in Sect. 9.

2. Overview of (theoretical) EOS for neutron stars

In general theory of relativity (GTR) the matter density is defined as $\rho = \mathcal{E}/c^2$, where $\mathcal{E}$ is total energy density (including rest energy of particles). Baryon density $n_{\rm b}$ is baryon number (baryon charge) contained in a unit volume. Using elementary thermodynamics we get relation between pressure P and ρ from calculated $\mathcal{E}(n_{\rm b})$

$$P(n_{\rm b}) = n_{\rm b}^2 {\rm d}(\mathcal{E}/n_{\rm b})/{\rm d}n_{\rm b}\ , \quad \rho(n_{\rm b}) = \mathcal{E}(n_{\rm b})/c^2 \implies P = P(\rho)\ . \tag{1}$$

An example of a theoretically calculated EOS in the density range from 1 g cm^{-3} – 10^{16} g cm^{-3} is shown in Fig. 1, based on [5]. The black curve corresponds to dense matter in full equilibrium and at $T = 0$ (cold catalyzed matter, equivalent to the ground state of the matter, see e.g. [1]). Zero pressure is reached at 8.56 g cm^{-3}, and the ground state of the matter is then a ^{56}Fe crystal [1]. Above 10^4 g cm^{-3} atoms are crushed (full "pressure ionisation") and matter consists of a lattice of nuclei immersed in an electron gas: this is the "outer crust" of NS existing down to the "neutron-drip point" $\rho_{\rm ND} \sim 4 \times 10^{11}$ g cm^{-3} (pressure $P_{\rm ND} \sim 10^{30}$ dyn cm^{-2}). Pressure within the outer crust is determined mainly by the contribution of the electron gas. The inner crust of NS, consisting of a crystal lattice of neutron-rich nuclei permeated by a quasi-uniform electron gas and immersed in a neutron gas, extends down to the crust-core (CC) interface, $\rho_{\rm ND} < \rho < \rho_{\rm CC}$, with inner crust edge at $\rho_{\rm CC} \simeq (0.3 - 0.5)\rho_0$ and $P_{\rm CC} \simeq 10^{33}$ dyn cm^{-2}; the relative density jump at the crust-core interface is less than one percent [1].

NS are born as very hot objects, with internal temperature exceeding 10^{11} K. However, due to efficient neutrino emission, they cool to less than 10^9 K in a few months. For observed isolated NS ($10^2 - 10^4$ y old) a typical (measured) surface temperature is $\sim 10^6$ K; it is expected to be significantly larger $\sim 10^8$ K in the deep interior. As seen in Fig. 1, for observed isolated NS the effect of T on the EOS at $\rho > 10^4$ g cm^{-3} can be safely neglected, because matter is strongly degenerate there.

Actually, the structure of the crust accreted by a NS in a close binary is expected to be very different from the ground-state one [2]. However, the difference in the EOS of the ground-state and accreted crust are significant only in the inner crust in the density range $5 \times 10^{11} - 5 \times 10^{12}$ g cm^{-3}, where the accreted crust is stiffer [2]. The effect of this stiffening due to accretion on $M_{\rm max}$ is negligibly small.

For sub-nuclear density ($\rho < \rho_0$) the EOS can be quite reliably calculated, because: (1) matter involves nucleons only, and nuclear forces are tested in terrestrial nuclear physics; (2) in this density range the many-body theories of nuclear matter are well tested in nuclear physics; (3) theoretical calculations of the EOS of pure neutron matter for $\rho < \rho_0$ are quite precise, so that in this limit the EOS is also under control. Unfortunately, the (outermost) sub-nuclear

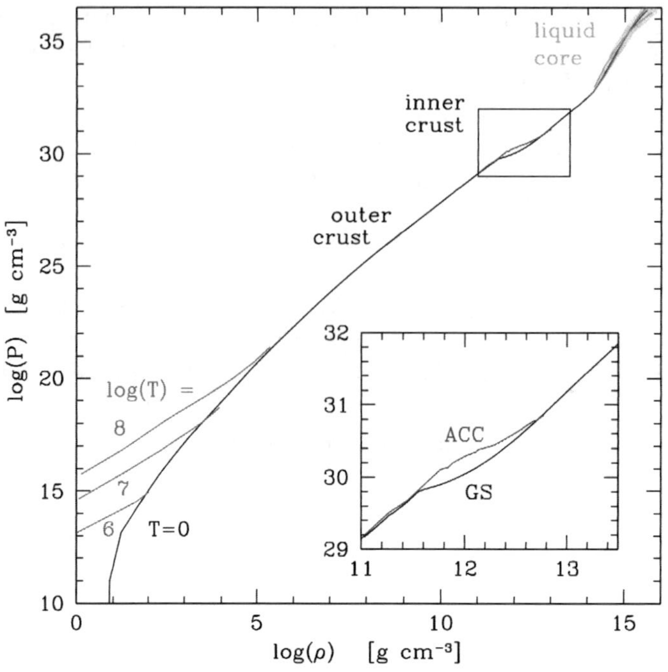

Figure 1. (Color online) An example of the EOS of NS matter from the surface to the center. Based mainly on [5] and [22]. Black line: ground state of dense matter, $T = 0$. Zoomed rectangle: EOS of the inner crust. Blue line - accreted crust, black line - ground state crust [22]. EOS of the liquid core at $\rho > \rho_0$ is uncertain, and the uncertainty grows with increasing ρ. This is visualized by a quasi-triangular colored region. Finite T effect on the EOS of the ^{56}Fe envelope of NS are visible in the bottom-left corner. For $\rho > 10^6$ g cm^{-3} the ^{56}Fe plasma is strongly degenerate even at 10^8 K. Effect of melting of the crust on the EOS for $\rho > 10^6$ g cm^{-3} is so small that it cannot be seen [1].

density layer contains only $(1-3)\%$ of NS mass for $M = 1.4\ M_\odot$ and less than one percent at $M = M_{\rm max} > 2\ M_\odot$.

For $\rho > \rho_0$ uncertainty in the EOS, growing rapidly with increasing ρ, reflects our lack of knowledge of strong interactions in dense matter (e.g., three- and four-body forces) and deficiencies and uncontrolled approximations of the many-body theories. This is visualized by a colored quasi-triangular sector of the high-density EOS in Fig. 1.

3. Uncertainties in the EOS$(\rho > \rho_0)$

First calculation of $M_{\rm max}$ within the general theory of relativity (GTR) was based on an EOS of a free Fermi gas (FFG) of neutrons [6] and yielded $M_{\rm max}[{\rm FFG}] = 0.7\ M_\odot$. The recent measurement of 2 $M_\odot$ masses of two pulsars is an observational (experimental) proof of a dominating rôle of (repulsive) strong interactions in the EOS of NS, because

$$M_{\rm max}^{\rm (obs)}/M_{\rm max}[{\rm FFG}] > 2.8. \qquad (2)$$

A strict condition $M_{\rm max}[{\rm EOS}] > 2\ M_\odot$ restricts very strongly a set of acceptable theoretical EOS for $\rho > \rho_0$. These EOS have to be much stiffer than the FFG one. On the other hand, the EOS should respect a fundamental constraint stemming from the special theory of relativity:

the speed of sound must not exceed c,

$$v_s = (dP/d\rho)^{1/2} \leq c \ . \tag{3}$$

The above condition is usually interpreted as equivalent to a more fundamental condition of *causality* (for a detailed discussion, see [7]). One can easily construct so called *causal limit* (CL) equation of state, matched smoothly to another EOS at the point (ρ_m, P_m) and which is maximally stiff, $v_s = c$ for larger density,

$$P^{(CL)} = P_m + (\rho - \rho_m)c^2 \quad \text{for } \rho > \rho_m \ . \tag{4}$$

A true EOS($\rho > \rho_0$) is a line in the $P - \rho$ plane. Our ignorance implies that there are many proposed theoretical EOS satisfying necessary conditions $\{M_{max}[\text{EOS}] > 2\ M_\odot\} \wedge \{v_s \leq c\}$ and used as "the EOS". They span over a characteristic quasi-triangular region in the $P - \rho$ plane, similar to that shown in the left panel of Fig. 2.

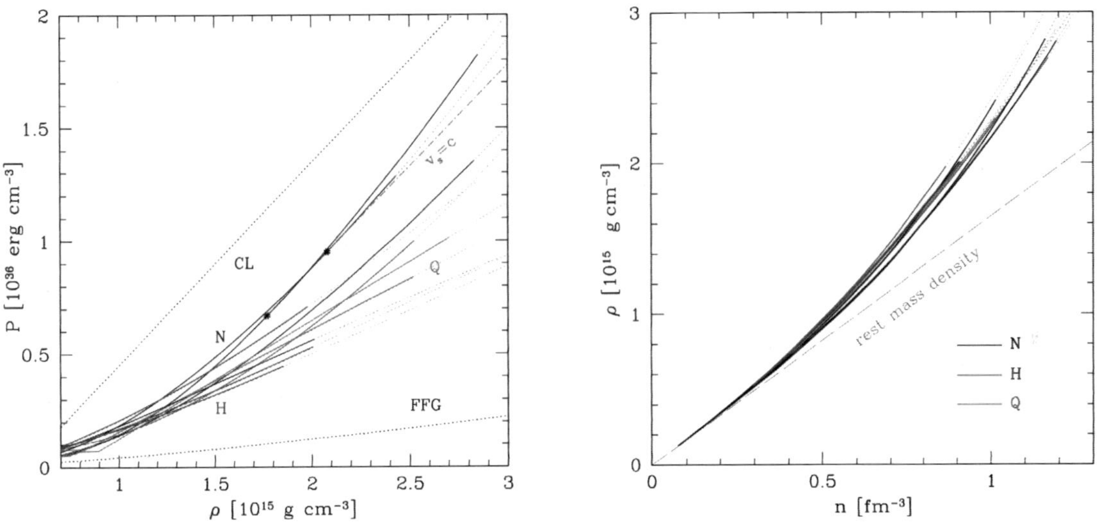

Figure 2. (Color online) **Left panel:** A set of the high-density EOS from the families N (EOS.N), H (EOS.NH), and Q (EOS.BQ), in the $P - \rho$ plane, for $\rho > 7 \times 10^{14}$ g cm^{-3}.(**a**) Dotted line CL: causal limit EOS of the form $P = (\rho - \rho_1)c^2$, where $\rho_1 = 5 \times 10^{14}$ g cm^{-3}. It crosses each of the considered EOS at some point (ρ_m, P_m) such that $\rho_m = \rho_1 + P_m/c^2$; (**b**) Dotted line FFG: EOS of a free Fermi gas of neutrons. (**c**) Asterisk on an EOS indicates the point beyond which EOS is non-causal ($v_s > c$); (**d**) We matched to the APR EOS a CL.EOS starting at the asterisk and denoted by a dot-dash line; the segment of the APR EOS above dot-dash line has to be rejected as non-causal; (**e**) The constant pressure segments in the quark part of the EOS.BQ correspond to the 1st order phase transitions, associated with a density jump.
Right panel: Mass density $\rho = \mathcal{E}/c^2$ vs. baryon density n_b for NS matter. Relation $\rho = \rho(n_b)$ deviates from linearity for $n_b > 0.3$ fm^{-3}. Nonlinearity grows with increasing n_b and is EOS-dependent. For $n_b < 0.2$ fm^{-3} linear approximation $\rho \approx n_b m_n$ (where m_n is neutron mass) is valid.

4. Absolute upper bounds on M_{max}

In view of the uncertainties in the theoretical EOS at supra-nuclear densities, it is of interest to derive an absolute upper bound on M_{max}, based on a fundamental physical argument. It is clear that CL.EOS, Eq. (4), will maximize M_{max} for NS models with "known envelope" $\rho < \rho_m$

and an "unknown core" $\rho > \rho_{\rm m}$". For massive NS with mass close to 2 $M_\odot$, the mass fraction contained in the outer layer with $\rho < \rho_{\rm m}$ is negligibly small. Then, a following inequality holds (see [1] and references therein):

$$M_{\rm max} \leq M_{\rm max}^{\rm (CL)} = 3.0 \left(\frac{5 \times 10^{14}\ {\rm g\ cm^{-3}}}{\rho_{\rm m}} \right)^{\frac{1}{2}} {\rm M}_\odot \ , \tag{5}$$

where $M_{\rm max}^{\rm (CL)}$ is a maximum allowable mass for NS with CL.EOS within the core $\rho > \rho_{\rm m}$. The formula for $M_{\rm max}^{\rm (CL)}$ is quite precise provided $\rho_{\rm m} \leq 5 \times 10^{14}$ g cm^{-3} [1]. The most conservative upper bound is $\rho_{\rm m} = \rho_0 \implies M_{\rm max}^{\rm (CL)} = 4.1\ {\rm M}_\odot$. Therefore the true value of $M_{\rm max}$ fulfills inequality $2.0 < M_{\rm max}/{\rm M}_\odot < 4.1$.

5. Structure of dense matter and fundamental aspects of the EOS

As we will see in the next sections, a suitable unit for the density in NS cores is 10^{15} g cm^{-3}, and the relevant density is written as $\rho_{15} \equiv \rho/(10^{15}\ {\rm g\ cm^{-3}})$. The density range in NS cores is $0.3 < \rho_{15} < 3$, and in order to (theoretically) study NS we need $P(\rho)$ in this interval of matter density.

5.1. Nucleon cores: EOS.N

At $\rho = \rho_0$ NS core consists of nuclear matter (some 95% of neutrons, 5% of protons), and electrons that neutralize proton electric charge [1]. It is expected that for $\rho_0 < \rho \lesssim 2\rho_0$ (outer NS core) nucleons are the only baryons present in NS matter. The "minimal" EOS of NS matter consisting of nucleons and leptons will be denoted as EOS.N. It is based on the 2-body nucleon forces (2BF) fitting a few thousand of data on nucleon-nucleon scattering and the properties of ^{2}H $\boxed{np}$. Moreover, we are forced to consider three-body nucleon forces (3BF), which have to reproduce the properties of ^{3}H $\boxed{nnp}$, ^{3}He $\boxed{ppn}$, ^{4}He $\boxed{nnpp}$, and the semi-empirical parameters of nuclear matter at saturation.

Numerous many-body calculations of EOS.N starting from 2BF+3BF and performed within the quantum many-body theories of dense matter yield $M_{\rm max} > 2\ {\rm M}_\odot$ (see, e.g. [8]).

5.2. Nucleon-hyperon cores: EOS.NH

Since more than fifty years it is believed that above some supranuclear density purely nucleon NS cores become unstable with respect to weak-interaction processes and that hyperons might replace the most energetic nucleons due to the strangeness-changing reactions. Consequently, NS with sufficiently high central density could have strangeness carrying (strange) cores consisting of nucleon-hyperon (hypernuclear) matter; their EOS is denoted by EOS.NH. As nucleons and hyperons are baryons, we will sometimes replace NH by B(baryons). The calculation of EOS.NH necessitates models of nucleon-hyperon and hyperon-hyperon forces. The nucleon-hyperon interaction is obtained from the analyses of hypernuclei and Σ^--atoms, while hyperon-hyperon forces are estimated studying double $\Lambda\Lambda$ hypernuclei. The data on the hyperon interactions are scarce, or even nonexistent in many hyperon-nucleon and hyperon-hyperon channels, and therefore approximate symmetries of strong interactions have to be widely used to get a hypernuclear interaction. Another painful problem is our ignorance concerning the three-body forces involving hyperons.

Appearance of the hyperons leads to a significant softening of the EOS.NH compared to EOS.N where the hyperons are suppressed, because hyperons that replace the most energetic nucleons, are more massive than nucleons and move slowly, because they have small Fermi momenta (densities). Actually, for baryonic force models which are consistent with available nuclear and hypernuclear data, a few existing Brueckner-Bethe-Goldstone theory calculations yield $M_{\rm max}$[EOS.NH] $\lesssim 1.5\ {\rm M}_\odot$ ([9] and references therein). This is in dramatic conflict with observation of 2 $\rm M_\odot$ pulsars: the "hyperon puzzle". One can try to escape from it via a fine tuning of dense matter model (Sect. 6).

Before reviewing proposed "fine tuning" procedures, we will review some fundamental aspects of the structure of dense matter in NS.

5.3. EOS and QCD

According to the modern theory of strong interactions, the Quantum Chromodynamics (QCD), baryons are confined states of three quarks. Only three lightest quarks are involved in terrestrial nuclear and hypernuclear physics. We are dealing with nucleons and nuclei composed of non-strange u and d quarks, and hyperons and hypernuclei, which additionally involve s quark of strangeness -1.

Many-body theory of nuclear and hypernuclear matter tested in low-energy nuclear and hypernuclear physics (i.e., for $\rho \lesssim \rho_0$) does not consider the quark degrees of freedom. Instead, it refers to "effective matter constituents": baryons $\boxed{udd}$, $\boxed{uds}$.... In this "effective theory" nuclear forces result from exchange of (virtual) mesons $\boxed{\overline{q}q}$ The effective theory is very successful in low-energy nuclear physics. However, can it be successfully extrapolated up to $10\rho_0$? We have to ask a basic question: how far this effective theory of hadrons+leptons can be used in dense cold matter?

There are two remarkable features of the QCD: confinement of quarks and asymptotic freedom. QCD predicts also a deconfinement of quarks. Namely, there exists a finite density $\rho_{\rm dec}$ such that for $\rho \gtrsim \rho_{\rm dec}$ cold electrically neutral matter becomes a plasma of quarks interacting via exchange of gluons, maybe with a small admixture of electrons. Unfortunately, both the value of $\rho_{\rm dec}$ and the EOS for $\rho > \rho_{\rm dec}$ are very difficult to calculate: the many-particle system under consideration is a strongly-interacting quark-gluon plasma.

QCD predicts a very important property of the quark-gluon plasma (QGP). For the mean kinetic energy of quarks in QGP (resulting from the Fermi statistics of quarks) $\gg \Lambda_{\rm QCD} \sim$ 1000 MeV, the EOS is simply $P \simeq \frac{1}{3}\rho c^2$ (EOS of an ultra-relativistic Fermi gas)[10]. This is the consequence of the Asymptotic Freedom of the QCD.

The value of $\Lambda_{\rm QCD}$ indicates that the region where the Asymptotic Freedom prevails (the region of simplicity - "Asymptopia") is reached for $\rho > 10^{18}$ g cm^{-3} - far larger than maximum density reached at the center of most massive neutron stars. The maximum density reached in NS does not exceed 5×10^{15} g cm^{-3}. This implies that only the least massive quarks u, d, s are relevant for NS (see [1]).

6. 2$\rm M_\odot$ pulsars with strange neutron-star cores and necessity of a fine tuning of the EOS

It has been shown by many authors that NH cores in 2 $\rm M_\odot$ stars can exist only for very special models of dense hadronic matter. They are all constructed within the relativistic mean-field (RMF) theory of hadronic matter. Additionally, they have specific features leading to a required EOS.NH. First "fine tuned feature" is a strong high-density repulsion acting only between hyperons, and generated by the exchange of the hidden-strangeness ϕ $\boxed{s\overline{s}}$ mesons (see, e.g., [11, 12, 13, 14], and other papers). A second "fine tuned feature" is a sufficiently strong breaking of the SU(6) symmetry in the hyperon-meson coupling constants [15].

Strangeness is associated with s quark, either confined into hyperons or moving in a (deconfined) quark plasma. A stiff superconducting quark core instead of hyperon core (such stars are called "hybrid stars" because they have baryon (B) and quark-matter (Q) components) could allow for $M_{\rm max} > 2\ {\rm M}_\odot$ (e.g., [16, 17, 18, 14, 19]). However, necessary conditions on the baryon - quark-matter phase transition (B $\longrightarrow$ Q), leading to EOS.BQ respecting $M_{\rm max} > 2\ {\rm M}_\odot$,

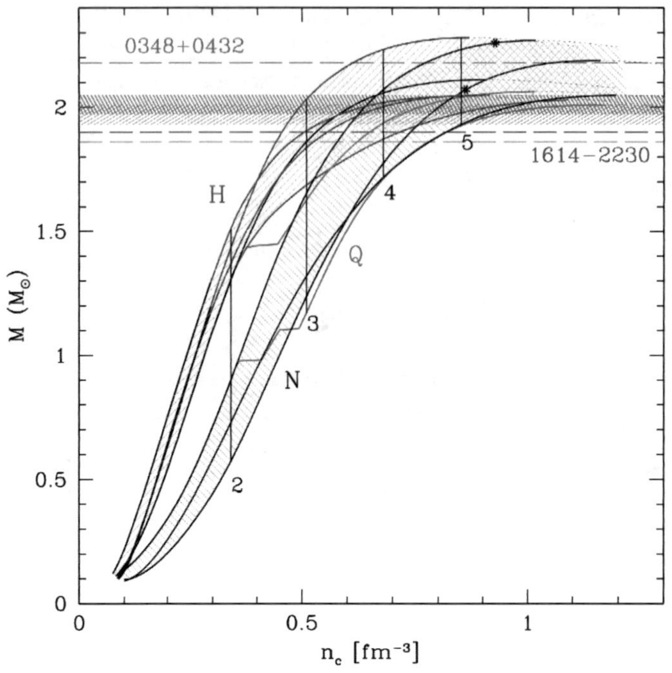

Figure 3. (Color online) Gravitational mass M vs. central baryon density n_c for non-rotating NS models based on the EOS.N (black lines - N family), EOS.NH (blue lines - H family), EOS.BQ (red lines - Q family of hybrid stars). Solid lines: stable NS configurations. Dotted lines: configurations unstable with respect to small radial perturbations. The sign $*$ on a line indicates that $v_s = c$ at the center of this configuration. Configurations to the right of $*$ have $v_s > c$ cores, within which causality is violated. Vertical lines crossing the $M(n_c)$ lines indicate configurations with $n_c/n_0 = 2, 3, \ldots$.

are very strong. First, the density at which first-order phase transition to quark phase occurs should be similar to the threshold density for hyperons, $\sim 2\rho_0 - 3\rho_0$. Second, the relative density jump at the baryon-quark matter phase transition should be below 30%. Third, quark matter should be sufficiently stiff, which can be expressed as a condition on the sound speed in quark plasma, $v_s/c \simeq 0.8 - 0.9$. Fourth, the pairing in the color-superconducting phase should be strong (energy gap $\sim 50 - 100$ MeV), in order to make the Q-phase thermodynamically stable. All in all, this means a "fine tuning" of the quark-matter phase and of the B$\longrightarrow$Q (1-st order) phase transition.

7. Families of neutron stars

7.1. Two densities: n_b and ρ

When investigating the EOS of NS matter, we have to consider two distinct densities, $\rho = \mathcal{E}/c^2$ and n_b (Sect. 2). While ρ is the relevant quantity for the GTR calculations of the NS structure, it is n_b that leads to a correct evaluation of an average distance between baryons (treated as point-like objects), $r_{bb} \propto n_b^{1/3}$. Therefore, knowing n_b, we can compare an actual r_{bb} with average distance between nucleons in nuclear matter at normal nuclear density, r_0, getting $r_{bb}/r_0 = (n_0/n_b)^{1/3}$.

At subnuclear densities, ρ of NS matter can be very well approximated by $n_b m_n$, where m_n is neutron rest mass. However, at supranuclear densities ρ grows nonlinearly with n_b. This

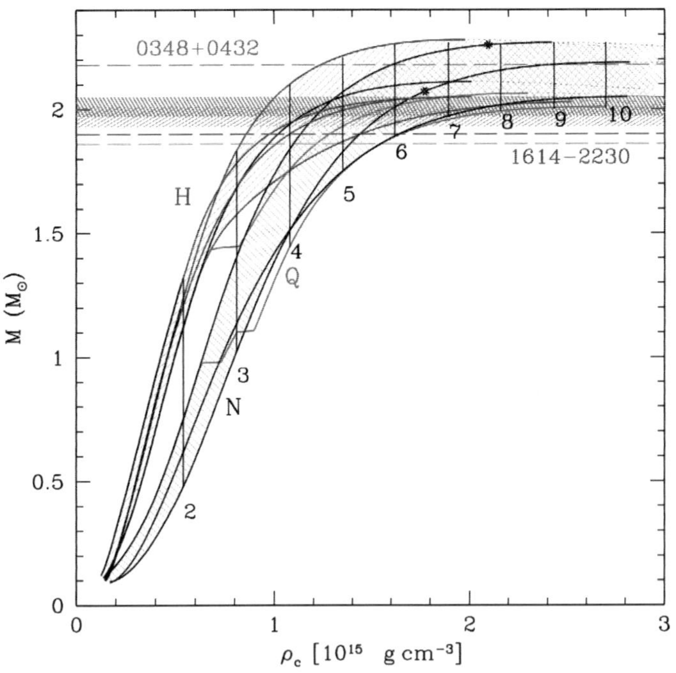

Figure 4. (Color online) Same as for Fig. 3, but with n_c replaced by ρ_c.

nonlinear dependence, as shown in the right panel of Fig. 2, is model dependent.

7.2. Two parametrizations: $M(\rho_c)$ and $M(n_c)$

We restrict to EOS.N, EOS.NH, and EOS.BQ that give $M_{max} > 2.0$ M$_\odot$. There are several lessons stemming from Fig. 3. First, central density in a 2 M$_\odot$ NS is typically $n_c/n_0 = 4 - 5$, so that at the star's center $r_{bb}/r_0 \approx 0.59$. Second, the N segment of EOS.NH corresponding to $n_0 < n_b < 2n_0$ is so much stiffer than the similar segment of EOS.N, that $M^{(NH)}(2n_0) \approx 2\ M^{(N)}(2n_0)$. Therefore, in order to yield $M_{max} > 2.0$ M$_\odot$ in spite of the H-softening, the supranuclear (nucleon) segment $n_0 < n_b < 2n_0$ of EOS.NH has to be very stiff. The fine-tuned H family of $M(n_c)$ curves is, for 0.5 M$_\odot$ $< M <$ 1.5 M$_\odot$, well separated from the N-family and lies above it. This is a well defined result, but n_c is not a measurable quantity, unfortunately.

For the hybrid stars belonging to the Q family the situation is different from that for the N and H families. In view of the n_c and ρ_c jumps associated with 1st order B⟶Q phase transition (or even two jumps, connected with different color-superconducting states, see [18] and the caption to Fig. 3), the $M(n_c)$ curve can cross the gap between the N and H families. This is due to a density jump in a 1st order B⟶Q phase transition combined with an extreme stiffness of the Q-phase, "fine-tuned properties" that still allows the hybrid stars to reach $M_{max} > 2$ M$_\odot$.

The $M(\rho_c)$ curves are displayed in Fig. 4. The gap between N and H families for 0.5 M$_\odot$ $< M <$ 1.5 M$_\odot$ exists, and some other features are also similar to those of the $M(n_c)$ curves. The asterisks $*$ signal the breaking of the $v_s \leq c$ condition.

Vertical lines crossing the $M(\rho_c)$ curves indicate $\rho_c/\rho_0 = 2, 3, \ldots$. One notices that ρ_c/ρ_0 in 2 M$_\odot$ stars can be as high as 6 - 7, significantly larger than the corresponding values of n_c/n_0. For the M_{max} configurations the difference is even larger. However, let us remind that it is n_c and not ρ_c that determines the mean inter-baryon distance at the star's center.

Journal of Physics: Conference Series **665** (2016) 012061 doi:10.1088/1742-6596/665/1/012061

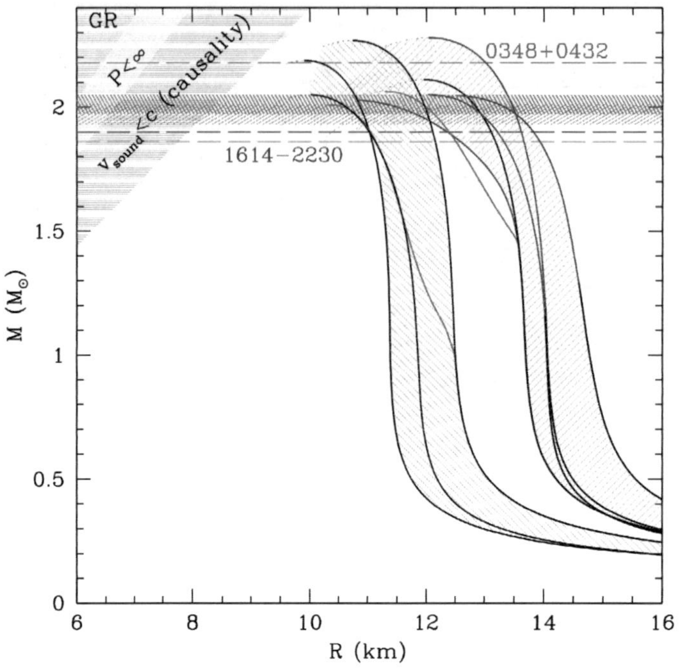

Figure 5. (Color online) Gravitational mass M vs. circumferential radius R for non-rotating NS models. Upper-left corner: (**a**) blue: excluded by the GTR because $R < 2GM/c^2$; (**b**) yellow: excluded due to (GTR)$\wedge(P_c < \infty)$; (**c**) green: excluded due to (GTR)$\wedge(v_s < c)$. Otherwise notation as in Figs. 3,4.

8. Hyperon cores and neutron-star radii

The circumferential radius R is an important measurable global parameter of a NS. It can be extracted in the process of analysis of the spectra of the X-ray radiation from the NS atmosphere, provided we know the distance to the star (for a recent review of attempts to measure R see [20], see also [21]). It should be stressed, that due to the GTR effects (spacetime curvature), the value of M has to be involved in the process of determining R, even in a simplest case of non-rotating and non-magnetized NS. Inclusion of rotation and magnetic field strongly complicates the analysis of the collected X-ray spectra.

In Fig. 5 we show the $M(R)$ curves for the N, H, and Q families of NS. One notices a sizable R-gap, $\sim$ 1.5 km, between the N and H families of NS, for $0.5 < M/M_\odot < 1.5$. This is an important feature, with interesting consequences. Notice that the R-gap does not exist for $M \approx 2\ M_\odot$.

As we mentioned in Sect. 2, the EOS of NS-crust depends on the crust formation scenario. Namely, an accreted crust (ACC) has a significantly stiffer EOS.ACC than the ground-state one, EOS.GS, for 5×10^{11} g cm$^{-3} < \rho < 5 \times 10^{12}$ g cm^{-3}. However, the effects of the difference between EOS.ACC and EOS.ACC for the same NS-core and $M > 1\ M_\odot$ is small. For example, for $M = 1.4\ M_\odot$, $R_{ACC} - R_{GS} \approx 100$ m [22]. This difference decreases with increasing M.

9. Conclusion

In this final section we will focus on the features of the EOS, that stem from the $M_{max} > 2\ M_\odot$ condition. As we have seen, not only an additional repulsion acting between the hyperons has to be included in the hyperon core (due to ϕ-meson echange or a sufficiently strong SU(6) symmetry breaking), but also the EOS *before* the hyperonization has to be very stiff. This gives us a "way out" from the "hyperon puzzle", but as for now, it cannot be called a "solution". For the time being, the "way out" is limited only to a class of the Relativistic Mean Field models of hadronic matter (unsuccessful trials of a sufficient stiffening of EOS.NH in the Brueckner-Bethe-Goldstone theory are reviewed in [23]). However, if we accept the RMF models of the EOS.NH, respecting $M_{\rm max} > 2\ {\rm M}_\odot$, then we find interesting properties of NS that could be confronted with measurements of M and R. Namely, for the M-range $1.0 - 1.5\ {\rm M}_\odot$ the stars with hyperon cores need have $R > 13$ km. Best realistic EOS of purely nucleon NS gives $R < 12$ km. Recent analysis in [20] leads to *lower bound* $R > 14$ km of NS with $M < 2.3\ {\rm M}_\odot$ in 4U 1724-307 X-ray burster. This lower bound of [20] is valid at 90% confidence level, independently of chemical composition of the NS atmosphere. Of course, measuring a radius of NS is a very complicated task (see a critical review of measurements of radii of different NS by other authors, given in [20]). Clearly, measuring R of a NS is a challenge, but as we tried to show it also gives a hope of unveiling the EOS of NS cores.

Acknowledgments
This work was partially supported by the Polish NCN grants no. 2011/01/B/ST9/04838 and 2014/13/B/ST9/02621.

Note Added in Proof

During two years that elapsed since the submission of our article to the NPA6 Conference Proceedings many papers, relevant to the topic of the present article, were published. In this Note we briefly mention some of them. Fourteen RMF EOS.NH allowing for the hyperon cores and $M_{\rm max} > 2\ {\rm M}_\odot$ were considered in [24], where overpressure of these EOS at n_0 compared to the value obtained in [21], correlated with large value of $R_{1.4}$ (radius of neutron star of mass $1.4\ {\rm M}_\odot$) was pointed out. First EOS.NH based on the G-matrix approach yielding $M_{\rm max} > 2\ {\rm M}_\odot$ and $R_{1.4} > 13.5$ km was calculated [25]. An EOS.NH based on the RMF model with hadron masses and coupling constants depending on the scalar field, yielding $M_{\rm max} > 2\ {\rm M}_\odot$ and $R_{1.4} = 12$ km was obtained in [26]. However, even at maximum mass the strangeness per baryon was smaller than one percent. New measurements of the radii of neutron stars were carried out [27, 28, 29, 30, 31]. The lower bound of 14 km found in [20] has been very recently revised and decreased to 10.5 km in a very recent preprint [32]. Generally, controversies and uncertainties associated with measured values of R are still too large to allow for a reliable determination of the true EOS of neutron stars.

References

[1] Haensel, P., Potekhin, A.Y., Yakovlev, D.G. 2007, Neutron Stars 1. Equation of State and Structure (New York, Springer)
[2] Chamel N., Haensel P. 2008, Living Reviews in Relativity, vol.11, no.10
[3] Demorest P.B., et al., 2010, Nature 467, 1081
[4] Antoniadis J. et al. 2013, Science 340, 448
[5] Haensel P., Potekhin A.Y. 2004, A&A 428, 191
[6] Oppenheimer J.R., Volkoff G.M. 1939, Phys. Rev. 54, 540
[7] Fayngold M. 2008, Special Relativity and How it Works, Wiley VCH
[8] Chamel N., Haensel P., Zdunik J.L., Fantina A.F. 2013, Int. J. Mod. Phys. E22, 1330018
[9] Schulze H.-J., Rijken T. 2011, Phys. Rev. C 84, 035801
[10] Collins J.C., Perry M.J. 1975, Phys.Rev.Lett. 34, 1353
[11] Dexheimer V., Schramm, S. 2008, ApJ, 683, 943
[12] Bednarek I., Haensel P., Zdunik J.L., Bejger M., Mańka R., 2012, A&A 543, A157
[13] Weissenborn S., Chatterjee D., Schaeffner-Bielich J. 2012, arXiv:1112.0234
[14] Bonanno, L., Sedrakian, A. 2012, A&A 539, A16
[15] Weissenborn S., Chatterjee D., Schaeffner-Bielich J. 2012, Nucl.Phys.A 881, 62
[16] Blaschke D., Klaehn T., Lastowiecki R., Sandin F. 2010, J. Phys. G: Nucl. Part. Phys 37, 094063

[17] Weissenborn S., Sagert I., Pagliara G., Hempel M., Schaeffner-Bielich J. 2011, ApJ Lett. 740, L14
[18] Zdunik J.L., Haensel P. 2013, A&A 551, A61
[19] Colucci G., Sedrakian, A. 2013, Phys. Rev. C 87, 055806
[20] Suleimanov V., Poutanen J., Revnivtsev M., Werner K. 2011, ApJ, 742, 122
[21] Hebeler K., Lattimer J.M., Pethick C.J., Schwenk A. 2013, Apj 773, 11
[22] Zdunik J.L., Haensel P. 2011, A&A 530, A137
[23] Vidana, I., Logoteta, D., Providencia, C., Bombaci, I., 2011, Europhys. Lett., 94, 11002
[24] Fortin, M., Zdunik, J.L., Haensel, P., Bejger, M., 2015, A& A, 576, A68
[25] Yamamoto, Y., Furumoto, T., Yasutake, N., Rijken, T.A., 2014, Phys. Rev. C, 90, 045805
[26] Maslov, K.A., Kolomeitsev, E.E., Voskresensky, D.N., 2015, Phys. Lett. B, 748, 369
[27] Bogdanov, S., 2013, ApJ, 762, 96
[28] Guver, T., Ozel, F., 2013, ApJ, 765, L1
[29] Poutanen, J., Nattila, J., Kajava, J.J.E., 2014, MNRAS, 442, 3777
[30] Guillot, S., Servillat, M., Webb, N.A., Rutledge, R.E., 2013, ApJ, 772, 7
[31] Lattimer, J.M., Steiner, A., 2014, ApJ, 784, 123
[32] Nattila, J., Steiner, A.W., Kajava, J.J.E., Suleimanov, V.F., Poutanen, J., 2015, arXiv:1509.06561v1[astro-ph.HE] 22 Sep 2015

Nuclear Physics in Astrophysics VI (NPA6) IOP Publishing
Journal of Physics: Conference Series **665** (2016) 012062 doi:10.1088/1742-6596/665/1/012062

The equation of state at finite temperature: Structure and composition of protoneutron stars

G F Burgio[1]*, M Baldo[1], H Chen[2] and H-J Schulze[1]

[1] INFN Sezione di Catania, Via S. Sofia 64, 95123 Catania, Italy
[2] Department of Physics, China University of Geoscience, 430074 Wuhan, People's Republic of China

E-mail: baldo@ct.infn.it, burgio@ct.infn.it, huanchen.phy@gmail.com, schulze@ct.infn.it

Abstract. We study the hadron-quark phase transition at finite temperature in the interior of protoneutron stars, combining the equation of state obtained within the Brueckner-Hartree-Fock approach for hadronic matter with the MIT bag and the Dyson-Schwinger models for quark matter. We discuss the dependence of the results on different nuclear three-body forces and on details of the quark model. We find that a maximum mass exceeding two solar masses can be obtained with a strong three-body force and suitable parameter values in the Dyson-Schwinger model.

1. Introduction

It is generally believed that a neutron star (NS) is formed as a result of the gravitational collapse of a massive star ($M \gtrsim 8M_\odot$) in a type-II supernova [1, 2]. Just after the core bounce, a protoneutron star (PNS) is formed, a very hot and lepton-rich object, where neutrinos are temporarily trapped. The following evolution of the PNS is dominated by neutrino diffusion, which results first in deleptonization and subsequently in cooling. The star stabilizes at practically zero temperature, and no trapped neutrinos are left [3, 4]. The dynamical transformation of a PNS into a NS could be strongly influenced by a hadron-quark phase transition in the central region of the star [5, 6]. Calculations of PNS structure, based on a microscopic nucleonic equation of state (EOS), indicate that for the heaviest PNS, close to the maximum mass (about two solar masses), the central baryon density reaches values larger than $1/\text{fm}^3$. In this density range the nucleon cores (dimension ≈ 0.5 fm) start to touch each other, and it is likely that quark degrees of freedom will play a role.

In previous articles [7, 8, 9] we have studied static properties of PNS using a finite-temperature hadronic EOS including also hyperons [10] derived within the Brueckner-Bethe-Goldstone (BBG) theory of nuclear matter. An eventual hadron-quark phase transition was modeled within an extended MIT bag model [11, 12, 13]. In the present work we consider a more sophisticated quark model, the Dyson-Schwinger model (DSM) [14, 15, 16, 17].

The study of hybrid stars is also important from another point of view: Purely nucleonic EOS are able to accommodate fairly large (P)NS maximum masses [18, 19, 20, 21, 22], but the appearance of hyperons in beta-stable matter could strongly reduce this value [22, 23, 24]. In this case the presence of non-baryonic, i.e., "quark" matter would be a possible manner to stiffen the EOS and reach large NS masses. Heavy NS thus would be hybrid quark stars.

*Speaker

The paper is organized as follows. In section 2 we review the determination of the baryonic EOS in the BHF approach at finite temperature. Section 3 concerns the QM EOS according to the DSM, comparing also with the MIT bag model for reference. In sections 4 and 5 we present the results regarding (P)NS structure, combining the baryonic and QM EOS for beta-stable nuclear matter via a Gibbs phase transition construction. Section 6 contains our conclusions.

2. EOS of nuclear matter at finite temperature in the Brueckner approach

The starting point in the many-body theory of the finite-temperature EOS is the calculation of the grand-canonical potential Ω. In the Bloch – De Dominicis formalism [25] it is given by $\Omega = \Omega_0 + \Delta\Omega$, where

$$\Omega_0 = -\frac{2V}{\pi^2}\int_0^\infty k^2 dk \left[\frac{1}{\beta}\ln\left(1+e^{-\beta(e_k-\mu)}\right) + n(k)U(k)\right] \tag{1}$$

is the grand-canonical potential for independent particles with Hamiltonian

$$H_0 = \sum_k \left(\frac{k^2}{2m} + U(k)\right) a_k^\dagger a_k \,, \tag{2}$$

μ is the chemical potential and U is the single-particle potential. The interaction part $\Delta\Omega$ of the grand-canonical potential is given by

$$\Delta\Omega = \frac{e^{2\beta\mu}}{2}\int_{-\infty}^{\infty}\frac{d\omega}{2\pi}e^{-\beta\omega}\,\mathrm{Tr}_2\left[\arctan\left(\pi\mathscr{K}(\omega)\delta\left(H_0-\omega\right)\right)\right] . \tag{3}$$

The trace in the previous equation is taken in the space of anti-symmetrized two-body states and the two-body scattering matrix $\mathscr{K}$ is defined by

$$\langle 12|\mathscr{K}(\omega)|34\rangle = \langle 12|K(\omega)|34\rangle \prod_{i=1,4}\sqrt{1-n(k_i)} \,, \tag{4}$$

where the scattering matrix K satisfies the integral equation

$$\langle 12|K(\omega)|34\rangle = \langle 12|V|34\rangle + \sum_{5,6}\langle 12|V|56\rangle \frac{[1-n(k_5)][1-n(k_6)]}{\omega - e_5 - e_6}\langle 56|K(\omega)|34\rangle \,. \tag{5}$$

In these equations $n(k)$ is the Fermi distribution function at a given temperature and for the single-particle spectrum $e(k) = k^2/2m + U(k)$. Then Eq. (5) coincides with the Brueckner equation for the G-matrix in the zero-temperature limit. It has to be noted that only the principal part has to be considered in the integration, thus making K a real matrix. The appearance of the arctan in Eq. (5) looks peculiar, but it comes from a ladder summation similar to the one for the zero-temperature G-matrix. More details on the derivation and on the numerical treatment of the equations can be found in Ref. [26].

In this framework, the free energy is $F = \Omega + \mu A$ and the pressure is obtained from the thermodynamical relationship

$$p = \rho^2 \frac{\partial(F/A)}{\partial\rho} \,, \tag{6}$$

where $\rho = A/V$ is the total number density. A typical result is reported for symmetric matter in Fig. 1, for different values of the temperature. The resulting EOS at finite temperature, i.e., pressure versus density, is shown in Fig. 2. One recognizes the familiar van der Waals shape, which entails a liquid-gas phase transition, with a definite critical temperature, i.e., the temperature at which the minimum in the van der Waals isotherm disappears. This is clearly a fundamental property of the nuclear medium: at finite temperature it behaves macroscopically in a way similar to a classical liquid. The critical temperature

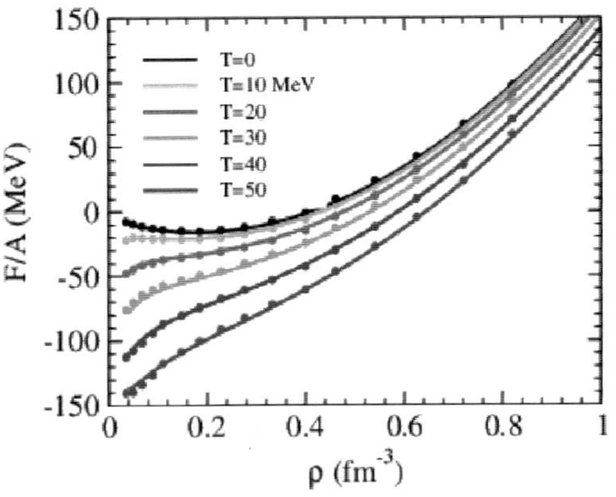

Figure 1. Free energy of symmetric nuclear matter as a function of density at different temperatures. The lines are fits to the calculated points. The curves decrease systematically as the temperature increases.

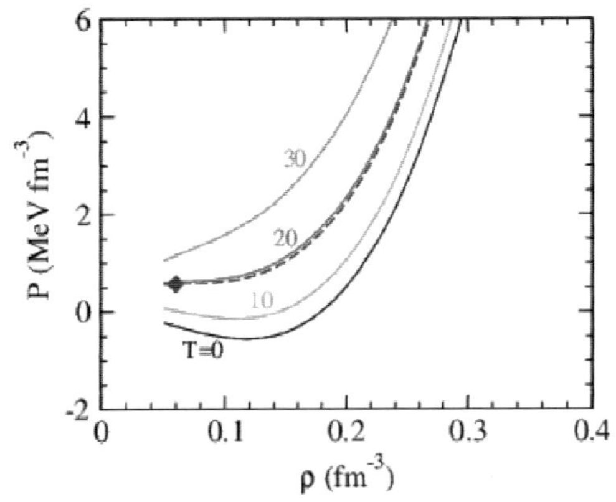

Figure 2. The pressure is displayed vs. baryon density for isotherms of symmetric nuclear matter. The full diamond marks the critical point of the liquid-gas phase transition.

turns out to be around $T_c = 18 - 20$ MeV. This is the result of microscopic calculations, but also of those employing effective forces, e.g., Skyrme forces. The values of the critical temperature, however, depend on the theoretical scheme, and on the particular effective force adopted.

To solve Eq. (5) one needs as input the interactions between hadrons. In the calculations presented in this paper, we have used as nucleon-nucleon two-body force the Argonne V_{18} potential [27], supplemented by three-body forces (TBF) needed in order to reproduce correctly the saturation point of nuclear matter. One should notice that both two- and three-body forces should be consistent, i.e., use the same microscopic parameters in their construction. We discuss results obtained using in the TBF the same meson-exchange parameters as the two-body potential [28]. For completeness, we also present results based on phenomenological models for TBF, such as the Urbana TBF [29], here denoted by UTBF (UIX), which consists of an attractive term due to two-pion exchange with excitation of an intermediate Δ resonance, and a repulsive phenomenological central term.

As far as the hyperonic sector is concerned, we employ the Nijmegen soft-core nucleon-hyperon (NY) potentials NSC89, which are fitted to the available experimental NY scattering data [30, 31]. Those potentials have been widely used in the BBG zero-temperature calculations [22, 23, 32, 33]. We restrict the present study to two hyperon species, Λ and Σ^-, neglecting the appearance of thermal Σ^0 and Σ^+ and heavier hyperons. Moreover we neglect hyperonic two- and three-body forces, due to the lack of experimental data.

3. Quark matter

For the deconfined quark phase, we adopt two different models, the thermodynamical MIT bag model, and one based on the Dyson-Schwinger equations of QCD.

Within the simplest form of the MIT model [34], the quarks are considered to be free inside a bag and the thermodynamic properties are derived from the Fermi distribution,

$$\tilde{f}_q(k;\tilde{\mu}_q,T) = f_q^+(k) - f_q^-(k) \,, \quad f_q^{\pm}(k) = \frac{1}{1+\exp[(E_q(k) \mp \tilde{\mu}_q)/T]} \tag{7}$$

with $E_q(k) = \sqrt{m_q^2 + k^2}$ and $\tilde{\mu}_q$ being the auxiliary quark chemical potentials. The quark number density is obtained from Eq. (14) and the energy density and free energy density for the quark phase are given as

$$\varepsilon_Q = B + g\sum_q \int \frac{d^3k}{(2\pi)^3}\left[f_q^+(k)+f_q^-(k)\right]E_q(k)\,,\quad f_Q = \varepsilon_Q - T\sum_q s_q^{\rm free}\,, \tag{8}$$

where B is the bag constant and $s_q^{\rm free}$ the entropy density of a noninteracting quark gas. The 'true' quark chemical potential, entropy density, and pressure are given by the general relations

$$\mu_q = \frac{\partial f_Q(\rho_u,\rho_d,\rho_s,T)}{\partial \rho_q}\,,\quad s_Q = \frac{\partial f_Q(\rho_u,\rho_d,\rho_s,T)}{\partial T}\,,\quad p_Q = \sum_q \mu_q\rho_q - f_Q\,. \tag{9}$$

We have used massless u and d quarks, and $m_s = 150$ MeV. In the following we present results based on the MIT model using both a constant value of the bag parameter, $B = 90$ MeV fm^{-3} and a Gaussian parametrization for the density dependence,

$$B(\rho) = B_\infty + (B_0 - B_\infty)\exp\left[-\beta\left(\frac{\rho}{\rho_0}\right)^2\right] \tag{10}$$

with $B_\infty = 50$ MeV fm^{-3}, $B_0 = 400$ MeV fm^{-3}, and $\beta = 0.17$, see Ref. [35]. For a more extensive discussion of this topic, the reader is referred to Refs. [35, 36].

The other quark matter model is based on the Dyson-Schwinger equations, which provide a continuum approach to QCD and can simultaneously address both confinement and dynamical chiral symmetry breaking [14, 15]. It has been applied with success to hadron physics in vacuum [37, 38, 39], and to QCD at nonzero chemical potential and temperature [40, 41]. Recently efforts have been made to calculate the EOS for cold quark matter and compact stars [42, 16], and later it has been extended to the finite-temperature case [17]. The starting point is QCD's gap equation for the quark propagator $S(\boldsymbol{p},\omega_n;\mu,T)$ at finite quark chemical potential μ, employing an in-medium gluon propagator

$$D_{\mu\nu}(\boldsymbol{k},\Omega_n) = P^T_{\mu\nu}D_T(\boldsymbol{k}^2,\Omega_n^2) + P^L_{\mu\nu}D_L(\boldsymbol{k}^2,\Omega_n^2) \tag{11}$$

with the gluon Matsubara frequency $\Omega_n = 2n\pi T$,

$$P^T_{\mu\nu} = \begin{cases} \delta_{\mu\nu} - k_\mu k_\nu/\boldsymbol{k}^2 & \mu,\nu = 1,2,3 \\ 0 & \mu,\nu = 4 \end{cases}\,,\quad P^L_{\mu\nu} = P_{\mu\nu} - P^T_{\mu\nu} \tag{12}$$

and

$$D_T(\boldsymbol{k}^2,\Omega_n) = D_L(\boldsymbol{k}^2,\Omega_n) = \frac{4\pi^2 D}{\omega^6}{\rm e}^{-\alpha\mu^2/\omega^2}k^2\,{\rm e}^{-k^2/\omega^2}\,, \tag{13}$$

with only a chemical potential dependence of the coupling strength, quantified by the parameter α, but neglecting any temperature effects. As in Ref. [16], we choose the set of parameters $\omega = 0.5$ GeV, $D = 1$ GeV^2 and use two typical values $\alpha = 2$ and $\alpha = 4$ in our calculation.

The quark propagator is the main input for calculating the single-flavor quark number density as

$$\rho_q(\mu_q,T) = g\int\frac{d^3p}{(2\pi)^3}\tilde{f}_q(|\boldsymbol{p}|;\mu_q,T)\,,\quad \tilde{f}_q(|\boldsymbol{p}|;\mu_q,T) = \frac{T}{2}\sum_{n=-\infty}^{\infty}{\rm tr_D}\left[-\gamma_4 S_q(\boldsymbol{p},\omega_n;\mu_q,T)\right]\,, \tag{14}$$

where $\omega_n = (2n+1)\pi\,T$ is the Matsubara frequency for quarks, $p = (\boldsymbol{p},\omega_n)$, $g = 2N_c = 6$ is the quark degeneracy and the trace is over spinor indices only. Due to the asymptotic freedom at large chemical potential $\mu_{\rm UV}$, which we take as $\mu_{\rm UV} = 1$ GeV, the single-flavor quark thermodynamic pressure at finite chemical potential and temperature can be obtained as

$$P_q(\mu_q,T) = P_q(\mu_{q,0},0) + \int_{\mu_{q,0}}^{\mu_{\rm UV}} d\mu'\rho_q(\mu',0) + \int_0^T dT' s_q^{\rm free}(\mu_{\rm UV},T') + \int_{\mu_{\rm UV}}^{\mu_q} d\mu'\rho_q(\mu',T)\,. \tag{15}$$

The total pressure for the quark phase is given by summing contributions from all flavors. For comparison with the bag model, we write the pressure of the quark phase as

$$p_Q(\mu_u,\mu_d,\mu_s,T) \equiv \sum_{q=u,d,s} \tilde{p}_q(\mu_q,T) - B_{\rm DS} \,, \tag{16}$$

with the bag constant

$$B_{\rm DS} \equiv - \sum_{q=u,d,s} p_q(\mu_{q,0},0) \tag{17}$$

and the reduced pressure $\tilde{p}_q(\mu_q,T) \equiv p_q(\mu_q,T) - p_q(\mu_{q,0},0)$ from the integrals of quark number density and entropy density. As in Ref. [16], we set $B_{\rm DS} = 90$ MeV with $\mu_{u,0} = \mu_{d,0} = 0$ and $\mu_{s,0}$ as the value of the starting point of the deconfined phase of strange quarks. Further details are given in Ref. [17].

4. Phase transition and EOS

In neutrino-trapped beta-stable nuclear matter the chemical potential of any particle $i = n,p,\Lambda,\Sigma^-,u,d,s,l$ is uniquely determined by the conserved quantities baryon number B_i, electric charge Q_i, and weak charges (lepton numbers) $L_i^{(e)}$, $L_i^{(\mu)}$ with corresponding chemical potentials $\mu_B,\mu_Q,\mu_{L_e},\mu_{L_\mu}$:

$$\mu_i = B_i\mu_B + Q_i\mu_Q + L_i^{(e)}\mu_{L_e} + L_i^{(\mu)}\mu_{L_\mu} \,, \tag{18}$$

where $B_i,Q_i,L_i^{(e)},L_i^{(\mu)}$ are the corresponding charges of each particle. The relations of chemical potentials and densities for hadrons and quarks are given in the previous sections, and we treat leptons as free fermions. With such relations, the whole system in each phase can be solved for a given baryon density $\rho_B = \sum_i B_i\rho_i$, imposing the charge neutrality condition $\sum_i Q_i\rho_i = 0$ and lepton number conservation $Y_l\rho_B = \sum_i L_i^{(l)}\rho_i$. We fix the lepton fractions to $Y_e = 0.4$ for neutrino-trapped matter at $T = 40$ MeV, and we neglect muons and muon-neutrinos due to their low fractions, hence $Y_\mu = 0$. When the neutrinos ν_e are untrapped, the lepton number is not conserved any more, the density and the chemical potential of ν_e vanish, and the above equations simplify accordingly.

We study in the following the phase transition at finite temperature with the Gibbs construction [18, 43, 44, 45], which determines a range of baryon densities where both phases coexist, yielding an EOS containing a pure hadronic phase, a mixed phase, and a pure QM region. In the Gibbs construction [43] the stellar matter is treated as a two-component system, and therefore is parametrized by two chemical potentials. Usually the pair (μ_e,μ_n), i.e., electron and baryon chemical potential, is chosen. By imposing mechanical, chemical, and thermal equilibrium, it turns out that the pressure is a monotonically increasing function of the density, at variance with the Maxwell construction, where a plateau in the pressure vs. density plane exists. The Gibbs phase transition has been widely studied in the literature, and the formalism will not be repeated here.

The solid black curves in Fig. 3 represent the hadronic phases obtained with the Argonne V_{18} NN potential supplemented by microscopic TBF (upper panels) or phenomenological Urbana TBF (lower panels). The other curves are the results for beta-stable QM, i.e., the Dyson-Schwinger results with $\alpha = 2$ (dash-dot-dotted blue) and $\alpha = 4$ (dotted red), whereas the other two lines represent the results obtained with the MIT bag model, using either a constant bag parameter $B = 90\ \mathrm{MeV\,fm^{-3}}$ (dashed green) or a density-dependent $B(\rho)$ (dash-dotted pink). The thick lines denote the corresponding mixed phase. The left-hand panels display results for cold beta-stable matter, the central panels for hot ($T = 40$ MeV) and untrapped matter, and the right-hand panels for hot and trapped matter.

We find that the phase transition from hadronic to QM occurs at low values of the baryon density when the MIT bag model is used to describe the quark phase. This holds true irrespective of the EOS adopted for the hadronic matter, since the TBF effects are only important at high densities. On the contrary, with the DSM for QM the phase transition occurs at higher values of the density, since the EOS is generally

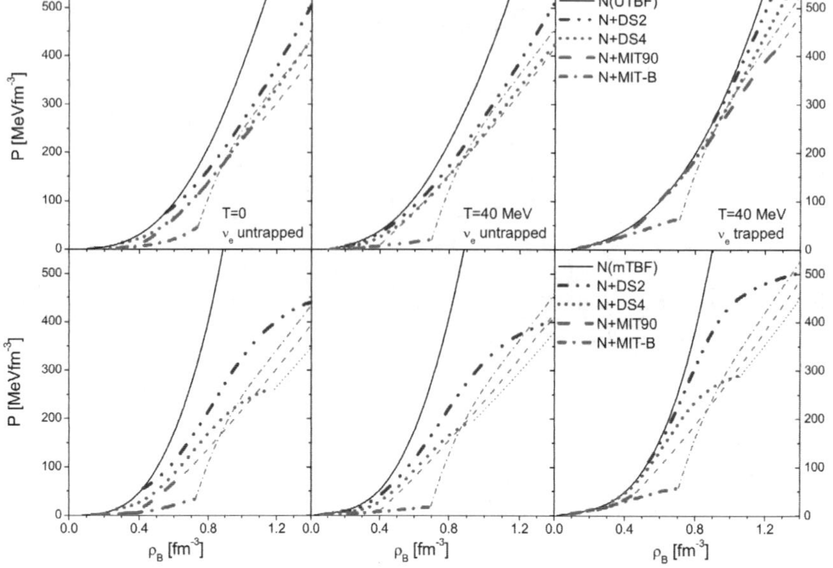

Figure 3. Pressure vs. baryon density of NS matter with the Gibbs phase transition construction for different models. Phenomenological (upper panels) or microscopic (lower panels) TBF are used for the hadronic EOS. The pure hadron or quark phases are marked with thin curves and the mixed phase region with thick curves.

stiffer than the hadronic one. In this case, the onset and the width of the density range characterizing the phase transition depend strongly on the EOS used for the hadronic phase.

Thermal effects play an important role, since they shift the onset of the phase transition to lower values of the baryon density. Neutrino trapping lowers even more the transition density in the MIT case, but increases it for the DSM. This is due to the properties of the nucleonic EOS: At low density, corresponding to the transition density in the MIT case, neutrino trapping increases the pressure and decreases the transition density. At large density, corresponding to the transition density in the DSM case, nuclear matter has a larger symmetry energy, and the decrease of nucleon pressure overwhelms the increase from the leptons. Therefore, neutrino trapping decreases the total pressure and increases the transition density. We have also found that with the DSM no phase transition exists if the hadronic phase contains hyperons and the EOS is very soft, or if the parameter α is too small and the EOS of QM is very stiff, in analogy with the zero-temperature case [16]. Taking all effects into account, the EOS comprising the phase transition turns out to be softer than the hadronic one.

The phase transitions constructed with the DSM and the MIT bag model turn out to be quite different. In the former case, the onset of the phase transition is shifted to a larger baryonic density and the mixed phase is extended to a much larger region. Different nuclear EOSs also affect the onset and the width of the mixed phase, in particular when the DS model is used. In the case of a stiff EOS for the hadronic phase with the microscopic TBF, the onset of the mixed phase is at lower density and the mixed phase has a smaller region, compared to the case of phenomenological TBF. These all have consequences for the internal structure of a PNS, as discussed now.

5. Proto-neutron star structure

The stable configurations of a (P)NS can be obtained from the well-known hydrostatic equilibrium equations of Tolman, Oppenheimer, and Volkov [1], which provide the mass-central density relation. In the low-density range, where nucleonic clustering sets in, we cannot use the BHF approach, and

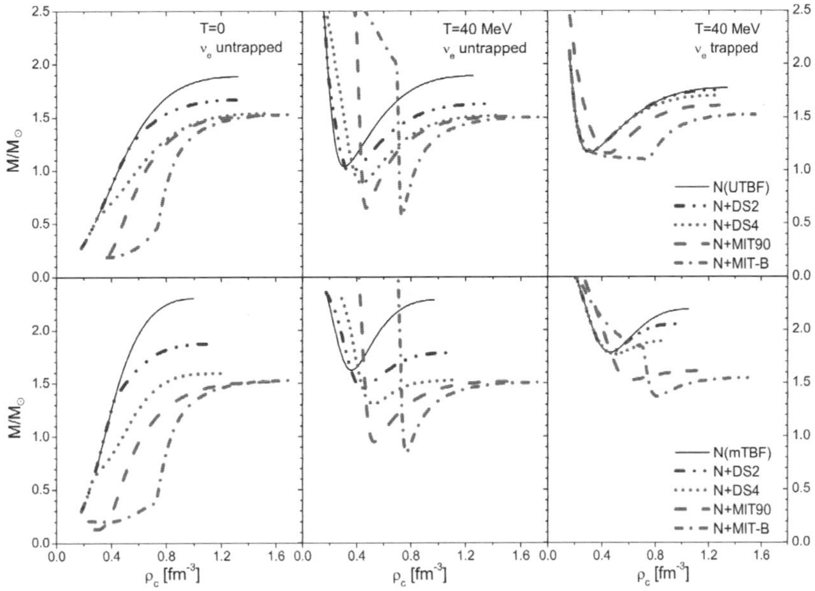

Figure 4. NS gravitational mass vs. central baryon density. Phenomenological (upper panels) or microscopic (lower panels) TBF are used for the hadronic EOS.

therefore we join [9] the BHF EOS to the finite-temperature Shen EOS [46], which is more appropriate at densities $\rho \leq 0.07$ fm^{-3}, since it does include the treatment of finite nuclei.

Our results, using the different EOSs introduced in the previous section, are displayed in Fig. 4, which shows the gravitational mass (in units of the solar mass $M_\odot = 1.98 \times 10^{33}$g) as a function of the central baryon density. We use the same conventions as in the previous figures: The black solid curves represent the calculations performed for purely hadronic matter, using either phenomenological (upper panels) or microscopic (lower panels) TBF. The colored broken curves denote stellar configurations of hybrid (P)NSs with neutrino-free (left and central panels) and neutrino-trapped matter (right panels). We notice that the value of the maximum mass decreases in neutrino-free matter due to thermal effects, both in the purely hadronic case and including the hadron-quark phase transition, where the decrease depends on the EOS used for the quark phase and turns out to be more relevant for the DS model. On the contrary, neutrino trapping decreases further the maximum mass of purely hadronic stars, but increases the value of the maximum mass of hybrid stars, overcoming the thermal effect. Therefore in this case a delayed collapse scenario is possible, just as for hyperon stars [10]. One also notices a dependence on the EOS used for the hadronic phase, which is more important for the DS model, where the QM onset takes place in a range of densities where TBF play an important stiffening role, and this explains the different values of the maximum mass. On the contrary, with the MIT bag model the transition takes place at low baryon densities where the different TBF behave similarly, hence the small influence. We find a maximum mass exceeding 2 $M_\odot$ for the hybrid configurations obtained with the microscopic TBF and suitable values of the parameter α in the DSM.

6. Conclusions

Extending the work of Ref. [16] to finite temperature, we have studied the properties of hybrid PNSs based on a hadronic BHF EOS involving different nuclear TBF joined via a Gibbs construction to the DSM quark EOS. This is only possible if hyperons are excluded from the hadronic phase.

We find a sizeable dependence of the results, in particular for the PNS maximum mass, on the nuclear TBF and on the quark model employed. The DSM model allows to reach larger masses than the MIT model, depending on the value of the interaction strength parameter. In all cases finite temperature reduces slightly the maximum mass, but neutrino trapping increases it sufficiently for hybrid stars, such that a delayed collapse might be possible in principle.

References

[1] Shapiro S L and Teukolsky S A 1983 *Black Holes, White Dwarfs, and Neutron Stars* (New York : John Wiley & Sons)
[2] H. Bethe H 1990 *Rev. Mod. Phys.* **62** 801
[3] Burrows A and Lattimer J M 1986 *Astrophys. J.* **307** 178
[4] Prakash M, Bombaci I, Prakash M, Ellis P J, Lattimer J M and Knorren R 1997 *Phys. Rep.***280** 1
[5] Pons J A, Reddy S, Prakash M, Lattimer J M and Miralles J A 1999 *Astrophys. J.***513** 780
[6] Prakash M, Cooke J R and Lattimer J M 1995 *Phys. Rev.* D **52** 661
Steiner A W, Prakash M and Lattimer J M 2000 *Phys. Lett.* B **486** 239
Pons J A, Steiner A W, Prakash M and Lattimer J M 2001 *Phys. Rev. Lett.* **86** 5223
[7] Nicotra O E, Baldo M, Burgio G F and Schulze H-J 2006 *Astron. Astrophys.***451** 213
[8] Burgio G F and Schulze H-J 2009 *Phys. Atom. Nuc.***72** 1197
[9] Burgio G F and Schulze H-J 2010 *Astron. Astrophys.* **518** A17
[10] Burgio G F, Schulze H-J and Li A 2011 *Phys. Rev.* C **83** 025804
[11] Nicotra O E, Baldo M, Burgio G F and Schulze H-J 2006 *Phys. Rev.* D **74** 123001
[12] Yasutake N, Burgio G F and Schulze H-J 2011 *Phys. Atom. Nucl.* **74** 1502
[13] Chen H, Burgio G F, Schulze H-J and Yasutake N 2013 *Astron. Astrophys.* **551** A13
[14] Roberts C D and Williams A G 1994 *Prog. Part. Nucl. Phys.* **33** 477
[15] Alkofer R and von Smekal L 2001 *Phys. Rep.* **353** 281
[16] Chen H, Baldo M, Burgio G F and Schulze H-J 2011 *Phys. Rev.* D **84** 105023
[17] Chen H, Baldo M, Burgio G F and Schulze H-J 2012 *Phys. Rev.* D **86** 045006
[18] Glendenning N K 2000 *Compact Stars, Nuclear Physics, Particle Physics, and General Relativity* (New York : Springer).
[19] Baldo M, Bombaci I and Burgio G F 1997 *Astron. Astrophys.* **328** 274
[20] Akmal A, Pandharipande V R and Ravenhall D G 1998 *Phys. Rev.* C **58** 1804
[21] Zhou X R, Burgio G F, Lombardo U, Schulze H-J and Zuo W 2004 *Phys. Rev.* C **69** 018801
[22] Li Z H and Schulze H-J 2008 *Phys. Rev.* C **78** 028801
[23] Schulze H-J, Polls A, Ramos A and Vidaña I 2006 *Phys. Rev.* C **73** 058801
[24] Schulze H-J and Rijken T 2011 *Phys. Rev.* C **84** 035801
[25] Bloch C and De Dominicis C 1958 *Nucl. Phys.* **7** 459; Bloch C and De Dominicis C 1959 *Nucl. Phys.* **10** 181
[26] Baldo M and Ferreira L S 1999 *Phys. Rev.* C **59** 682
[27] Wiringa R B, Stoks V G J and Schiavilla R 1995 *Phys. Rev.* C **51** 38
[28] Li Z H, Lombardo U, Schulze H-J and Zuo W 2008 *Phys. Rev.* C **77** 034316
[29] Carlson J, Pandharipande V R and Wiringa R B 1983 *Nucl. Phys.* A **401** 59
[30] Maessen P M M, Rijken Th A and de Swart J J 1989 *Phys. Rev.* C **40** 2226
[31] Rijken Th A, Stoks V G J and Yamamoto Y 1999 *Phys. Rev.* C **59** 21
[32] Schulze H-J, Baldo M, Lombardo U, Cugnon J and Lejeune A 1998 *Phys. Rev.* C **57** 704
Baldo M, Burgio G F and Schulze H-J 1998 *Phys. Rev.* C **58** 3688
Baldo M, Burgio G F and Schulze H-J 2000 *Phys. Rev.* C **61** 055801
[33] Vidaña I, Polls A, Ramos A, Hjorth-Jensen M and Stoks V G J 2000 *Phys. Rev.* C **61** 025802
[34] Chodos A, Jaffe R L, Johnson K, Thorn C B and Weisskopf V F 1974 *Phys. Rev.* D **9** 3471
[35] Burgio G F, Baldo M, Sahu P K, and Schulze H-J 2002 *Phys. Rev.* C **66** 025802
[36] Maieron C, Baldo M, Burgio G F and Schulze H-J 2004 *Phys. Rev.* D **70** 043010
[37] Maris P and Roberts C D 2003 *Int. J. Mod. Phys.* E **12** 297
[38] Eichmann G, Cloet I C, Alkofer R, Krassnigg A and Roberts C D 2009 *Phys. Rev.* C **79** 012202
[39] Chang L and Roberts C D 2009 *Phys. Rev. Lett.* **103** 081601
[40] Chen H, Yuan W, Chang L, Liu Y X, Klähn T and Roberts C D 2008 *Phys. Rev.* D **78** 116015
[41] Qin S X, Chang L, Chen H, Liu Y X and Roberts C D 2011 *Phys. Rev. Lett.* **106** 172301
[42] Klähn T, Roberts C D, Chang L, Chen H and Liu Y X 2010 *Phys. Rev.* C **82** 035801
[43] Glendenning N K 1992 *Phys. Rev.* D **46** 1274
[44] Maruyama T, Chiba S, Schulze H-J and Tatsumi T 2007 *Phys. Rev.* D **76** 123015
[45] Yasutake N, Maruyama T and Tatsumi T 2009 *Phys. Rev.* D **80** 123009
[46] Shen H, Toki H, Oyamatsu K and Sumiyoshi K 1998 *Prog. Theor. Phys.* **100** 1013;
http://user.numazu-ct.ac.jp/~sumi/eos/index.html.

Journal of Physics: Conference Series **665** (2016) 012063 doi:10.1088/1742-6596/665/1/012063

Neutron matter, symmetry energy and neutron stars

S Gandolfi[1] and A W Steiner[2,3]

[1]Theoretical Division, Los Alamos National Laboratory, Los Alamos, New Mexico 87545, USA
[2]Department of Physics and Astronomy, University of Tennessee, Knoxville, Tennessee 37996, USA
[3]Physics Division, Oak Ridge National Laboratory, Oak Ridge, Tennessee 37831, USA

Abstract. Recent progress in quantum Monte Carlo with modern nucleon-nucleon interactions have enabled the successful description of properties of light nuclei and neutron-rich matter. Of particular interest is the nuclear symmetry energy, the energy cost of creating an isospin asymmetry, and its connection to the structure of neutron stars. Combining these advances with recent observations of neutron star masses and radii gives insight into the equation of state of neutron-rich matter near and above the saturation density. In particular, neutron star radius measurements constrain the derivative of the symmetry energy.

1. Introduction

In the last few decades, properties of nuclear systems have been successfully described by nucleon-nucleon potentials like Argonne and Urbana/Illinois forces, that reproduces two-body scattering and properties of light nuclei with very high precision [1, 2]. These nuclear potentials reproduce several properties of nuclear systems extremely well, including binding energies of ground- and excited states, radii, matrix elements, scattering states, and other observables [3, 4, 5, 6]. The Argonne AV18 nucleon-nucleon interaction has small non-local terms and a hard core. The use of correlated wave functions combined with Quantum Monte Carlo (QMC) methods has provided highly accurate solutions of the ground state of many-body nuclear systems [7].

The knowledge of the Equation of State (EoS) of pure neutron matter is an important bridge from the nucleon-nucleon interaction to neutron-rich matter. The symmetry energy $E_{\rm sym}$ is the difference of nuclear matter and neutron matter energy and gives the energy cost of the isospin-asymmetry in the homogeneous nucleonic matter. In the last few years the study of $E_{\rm sym}$ has received considerable attention (see for example Ref. [8] for a recent experimental/theoretical review). The role of the symmetry energy is essential to understand the mechanism of stability of very neutron-rich nuclei, and is also related to many phenomena occurring in neutron stars. The number of protons per baryon, x, is determined by beta-equilibrium and charge neutrality. These imply relationships between the chemical potentials and the symmetry energy, $\mu_e = \mu_n - \mu_p \approx 4E_{\rm sym}(1-2x)$. Matter near the nuclear saturation density is very neutron-rich, because electron degeneracy drives $\mu_n > \mu_p$. Thus neutron star matter is sensitive to $E_{\rm sym}$ and its first derivative. The inner crust of neutron stars, where the density is a fraction of nuclear densities, is mostly composed of neutrons surrounding a matter made of extremely-neutron rich nuclei that, depending on the density, may exhibit very different phases and properties. The extremely rich phase diagram of crustal matter is strongly related to the role of $E_{\rm sym}$. For example, it governs the phase-transition between the crust and the core [9] and the nature of r-mode instabilities [10, 11].

2. The Nuclear Hamiltonian and Quantum Monte Carlo

In our model, neutrons are non-relativistic point-like particles interacting via two- and three-body forces:

$$H = \sum_{i=1}^{A} \frac{p_i^2}{2m} + \sum_{i<j} v_{ij} + \sum_{i<j<k} v_{ijk} \,. \tag{1}$$

The two body-potential that we use is the Argonne AV8′ [12], that is a simplified form of the Argonne AV18 [1]. Although simpler to use in QMC calculations, AV8′ provides almost the same accuracy as AV18 in fitting NN scattering data [13]. The three-body force is not as well constrained as the NN interaction, but its inclusion in realistic nuclear Hamiltonians is important to correctly describe the binding energy of light nuclei [2].

The Urbana IX (UIX) three-body force has been originally proposed in combination with the Argonne AV18 and AV8′ [14]. Although it slightly underbinds the energy of light nuclei, it has been extensively used to study the equation of state of nuclear and neutron matter [15, 16, 17]. The Illinois forces have been introduced to improve the description of both ground- and excited-states of light nuclei, showing an excellent accuracy [2, 3], but it produces an unphysical overbinding in pure neutron systems [18, 19].

Another interesting class of nucleon-nucleon potentials are derived within the chiral effective field theory. Typically, these interactions have strong non-local terms, and as a consequence they cannot be easily included in QMC calculations. Recently it has been showed that these potentials can be designed to be local, and combined with QMC simulations [20]. However, the need to include a cutoff to the nucleon's momentum limits the applicability of chiral forces to study dense neutron matter. The cutoff of these potentials can be controlled in a many-body calculation [20], but the uncertainty is already quite large at saturation density in neutron matter, making the calculation at larger densities unfeasible.

We solve the many-body ground-state with a projection in imaginary-time, i.e.:

$$\Psi(\tau) = \exp[-H\tau]\Psi_v \,, \tag{2}$$

where Ψ_v is a variational ansatz, and H is the Hamiltonian of the system. In the limit of $\tau \to \infty$, Ψ approaches the ground-state of H. The evolution in imaginary-time is performed by sampling configurations of the system using Monte Carlo techniques, and expectation values are evaluated over the sampled configurations. The main difference between GFMC and AFDMC is in the way that spin/isospin states are treated. In GFMC, all the spin/isospin states are explicitly included in the variational wave function. The results obtained are very accurate but limited to the ^{12}C [3] or 16 neutrons [21]. The AFDMC method samples the spin/isospin states using the Hubbard-Stratonovich transformation rather than simpling them explicitly [22]. The calculation can be then extended up to many neutrons, making the simulation of homogeneous matter and heavy nuclear systems possible [23]. The AFDMC has proven to be very accurate when compared to GFMC calculation of energies of neutrons confined in an external potential [21]. We shall present results obtained either using GFMC and AFDMC.

3. The Equation of State of Neutron Matter

In this section we present QMC results for pure neutron matter. There are several reasons to focus on pure neutron matter. First, the three-body interaction is non-zero only in the $T = 3/2$ isospin-channel (T is the total isospin of three-nucleons), while in the presence of protons there are also contributions in $T = 1/2$. The latter term is the dominant one in nuclei, and only

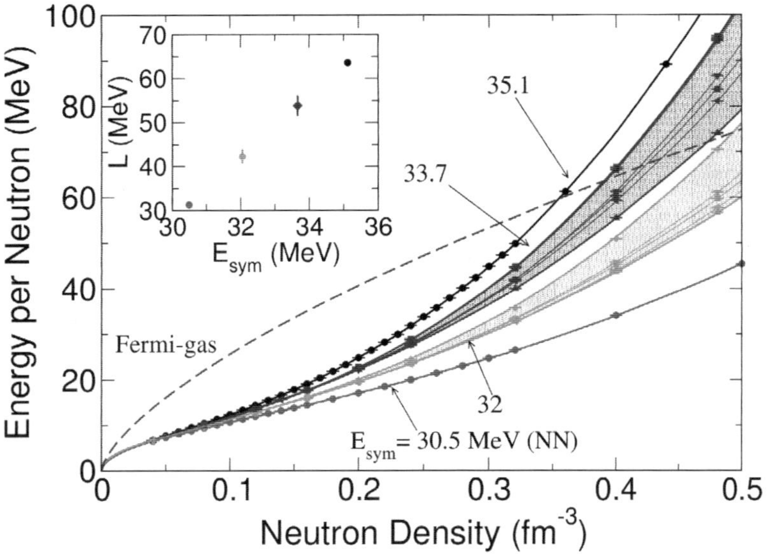

Figure 1. The QMC equation of state of neutron matter for various Hamiltonians. The red (lower) curve is obtained by including the NN (Argonne AV8$'$) alone in the calculation, and the black one is obtained by adding the Urbana IX three-body force. The green and blue bands correspond to EoSs giving the same $E_{\rm sym}$ (32 and 33.7 MeV respectively), and are obtained by using several models of three-neutron force. In the inset we show the value of L as a function of $E_{\rm sym}$ obtained by fitting the EoS. The figure is taken from Ref. [17].

weakly accessible by studying properties of nuclei. Second, the EoS of pure neutron matter is closely related to the structure of neutron stars.

We present several EoSs obtained using different models of three-neutron force in Fig. 1. The two solid lines correspond to the EoSs calculated using the NN potential alone and including the UIX three-body force [14]. The effect of using different models of three-neutron force is clear in the two bands, where the high density behavior is showed up to about $3\rho_0$. At such high density, the various models giving the same symmetry energy at saturation produce an uncertainty in the EoS of about 20 MeV. The EoS obtained using QMC can be conveniently fit using the following functional [16]:

$$E(\rho) = a\left(\frac{\rho}{\rho_0}\right)^{\alpha} + b\left(\frac{\rho}{\rho_0}\right)^{\beta}, \tag{3}$$

where E is the energy per neutron, $\rho_0 = 0.16$ fm^{-3}, and a, b, α and β are free parameters. The parametrizations of the EoS obtained from different nuclear Hamiltonians is given in Ref. [17].

At ρ_0 symmetric nuclear matter saturates, and we can extract the value of $E_{\rm sym}$ and L directly from the pure neutron matter EoS. The result of fitting the pure neutron matter EoS is shown in the inset of Fig. 1. The error bars are obtained by taking the maximum and minimum value of L for a given $E_{\rm sym}$, and the curves obtained with NN and NN+UIX are thus without error bars. From the plot it is clear that within the models we consider, the correlation between L and $E_{\rm sym}$ is linear and quite strong.

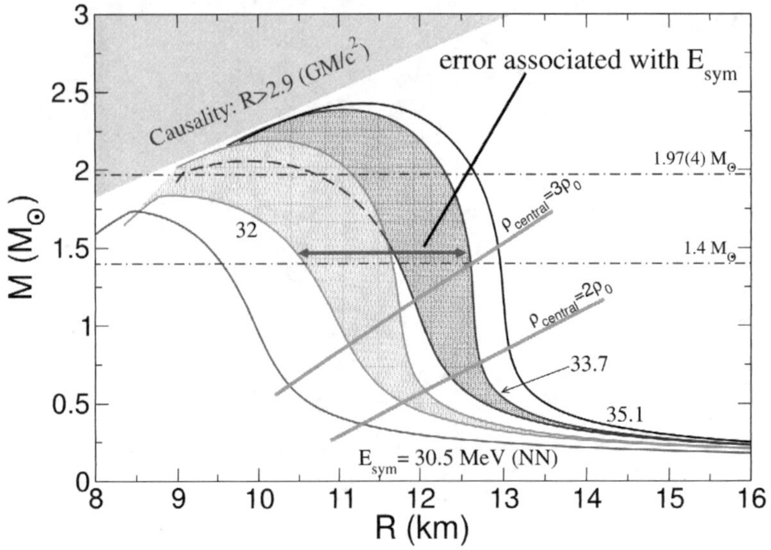

Figure 2. The mass-radius relation of neutron stars obtained from the EoS calculated using QMC. The various colors represent the $M - R$ result obtained from the corresponding EoSs described in Fig. 1. The two horizontal lines show the value of $M = 1.4$ and $1.97(4)M_\odot$ [24]. The figure is adapted from Ref. [17].

4. Connection to Neutron Star Masses and Radii

Neutron stars, unlike planets, are expected to be compositionally uniform, in which case their radius is determined principally by their mass; to a good approximation all neutron stars lie on a universal mass-radius $M-R$ curve. When the EoS of the neutron star matter has been specified, the structure of an idealized spherically-symmetric neutron star model can be calculated by integrating the Tolman-Oppenheimer-Volkoff (TOV) equations.

The neutron star mass measurements which provide the strongest EoS constraints are those which have the highest mass. Recent observations [24, 25] have found two neutron stars with masses near 2 $M_\odot$. These two data points provide some of the strongest constraints on the nature of zero-temperature QCD above the nuclear saturation density. We begin by examining what can be deduced about the M-R relation directly from these mass measurements, without employing a separate model for high-density matter. For lower densities we use the EoS of the crust obtained in Refs. [26] and [27]. For the core, we begin with the parameterization in Eq. 3, employing maximally stiff EoS when the QMC models violate the causality and become superluminal. The mass of a neutron star as a function of its radius is shown in Fig. 2. The two bands correspond to the result obtained using the two sets of EoS giving the same value of $E_{\rm sym}$ indicated in the figure. As in the case of the EoS, it is clear that the main source of uncertainty in the radius of a neutron star with $M = 1.4M_\odot$ is due to the uncertainty of $E_{\rm sym}$ rather than the model of the three-neutron force. The addition of a small proton fraction would change the radius R only slightly [28, 15], smaller than other uncertainties in the EoS that we have discussed. The numbers in the figure indicate the symmetry energy associated with the various equations of state. In the figure we also indicate with the orange lines the density of the neutron matter inside the star. Even at large masses the radius of the neutron star is mainly governed by the equation of state of neutron matter between 1 and 2 ρ_0 [29].

The AV8$'$ Hamiltonian alone does not support the recent observed neutron star with a mass of 1.97(4)$M_\odot$ [24]. However, adding a three-body force to AV8$'$ can provide sufficient repulsion to be consistent with all of the constraints [17]. There is a clear correlation between neutron star radii and the symmetry energy which determines the EoS of neutron matter between 1 and 2 ρ_0. The results in Fig. 2 also show that the most modern neutron matter EoS imply a maximum neutron star radius not larger than about 13 km, unless a drastic repulsion sets in just above the saturation density. This tends to rule out large values of L, typical of Walecka-type mean-field models without higher-order meson couplings which can decrease L.

5. Bayesian Analysis of Neutron Star Masses and Radii

In contrast to the mass measurements described above, neutron star radius measurements have proven more difficult, because they require both a distance measurement and some degree of modeling of the neutron star X-ray spectrum. Low-mass X-ray binaries (LMXBs) are neutron stars accreting matter from a low mass main-sequence or white dwarf companion. There are two types of LMXB observations which have recently provided neutron star radius information. The first type are LMXBs which exhibit photospheric radius expansion (PRE) X-ray bursts, thermonuclear explosions strong enough to temporarily lift the surface (photosphere) of the neutron star outwards [30, 31]. Several neutron stars have exhibited PRE X-ray bursts and four which have have been used to infer the neutron star mass and radius are given in the left panel of Fig. 3, using the methods described in Ref. [32]. The second type are quiescent LMXBs, (QLMXBs), where the accretion from the companion has stopped, allowing observation of the neutron star surface which has been heated by accretion [33]. A recent analysis of five neutron stars [34] including the possibility of both hydrogen and helium atmospheres and distance uncertainties is shown in the right panel of Fig. 3. Note that already from these two figures alone, it is clear that these probability distributions favor neutron star radii near 11 km. Although we will similar (R, M) distributions in our analysis below, it is important to remember that there are several systematic uncertainties which are potentially important. For the QLMXBs, the treatment of the X-ray absorption between the source and the observer, the flux calibration of the observing satellite, and the method used to measure the distance all play important roles. The situation for PRE X-ray bursts is even more challenging: complications such as spherical asymmetry, the time evolution of the spectra, and the location of the photosphere at "touchdown" may all modify the implied masses and radii.

Our goal is to constrain the EoS observational data described above, including the possibility of phase transitions in matter above the nuclear saturation density. In order to do this, we parametrize the EoS of matter at higher densities with a simple expression rich enough to include exotic matter. We perform a Bayesian analysis using data from QLMXBs and neutron stars which exhibit PRE bursts, where our model space is given by the EoS parameters and also one parameter for the mass of each neutron star in the data set. Given an EoS, the TOV equations provide the M-R curve and thus a prediction for the radius of each neutron star from its mass. As described above, we always ensure that our EoS are causal, hydrodynamically stable, and that our M-R curves support a 2 $M_\odot$ neutron stars. For densities near the saturation density, we an EOS parametrized in terms of the symmetry energy and compressibility. At higher densities, we describe matter either in terms of polytropes of the form $P = K_1\epsilon^\Gamma$, line-segments in the $P-\epsilon$ plane, or a simple parameterization of quark matter. The final results for the $M-R$ curve and EoS are given in Fig. 4 from Ref. [35]. The $M - R$ curve obtained is relatively vertical, which naturally implies that almost all neutron stars have approximately the same radius. The EoS obtained from the mass and radius observations is also in concordance with results from quantum Monte Carlo and chiral effective theory described above and constraints obtained from heavy-ion collisions.

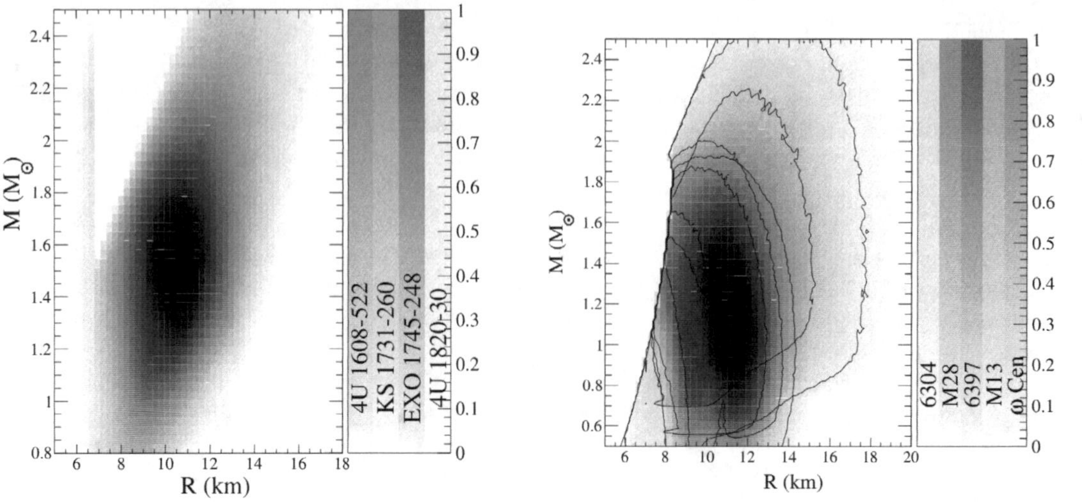

Figure 3. Left panel: Probability distributions in the mass-radius plane for four neutron stars exhibiting PRE X-ray bursts. Colors are added together in RGB color space. Right panel: Probability distributions in the mass-radius plane for five neutron stars in five globular clusters from Ref. [34]. Colors are added together in RGB color space when necessary. The contour lines outline the 90% confidence regions.

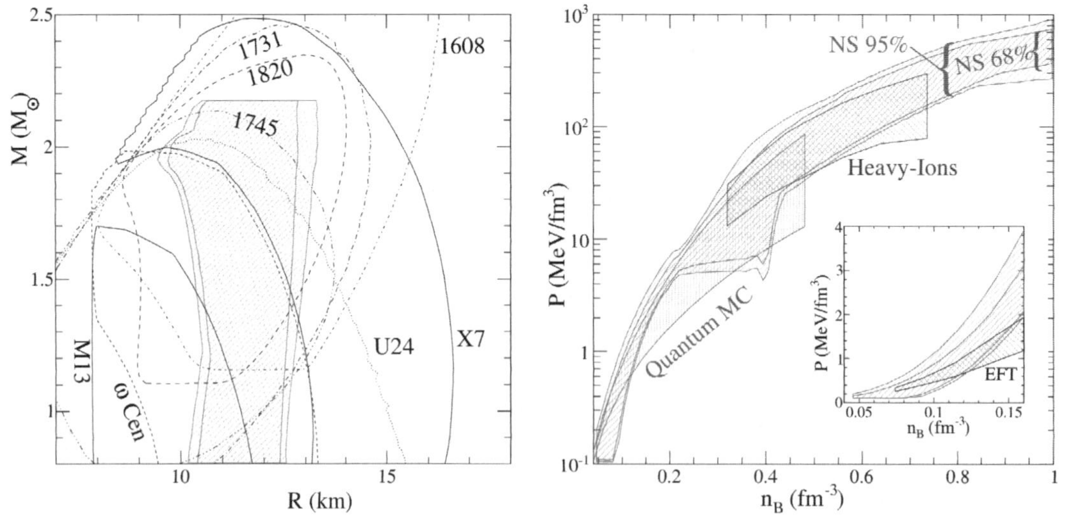

Figure 4. (Left panel) A comparison of the predicted M–R relation with the observations. The shaded regions outline the 68% and 95% confidences for the M–R relation. (Right panel) The predicted pressure as a function of baryon density of neutron-star matter as obtained from astrophysical observations. The region labeled "NS 68%" gives the 68% confidence limits and the region labeled "NS 95 %" gives the 95% confidence limits. Results for neutron-star matter from effective field theory [36] (see inset), from quantum Monte Carlo [17], and from constraints inferred from heavy-ion collisions [37] are also shown for comparison.

6. Determining the Density Dependence of the Symmetry Energy

In order to determine the symmetry energy, we use the parameterization of the neutron matter EoS from the quantum Monte Carlo results in Eq. 3 above. With this parameterization the symmetry energy at the saturation density $E_{\rm sym}$ and the parameter which describes the density dependence of the symmetry energy, L, are given by

$$E_{\rm sym} = a + b + 16\,, \quad L = 3\,(a\alpha + b\beta)\,. \tag{4}$$

Neutron stars contain a small amount of protons, so we multiply the EoS by a small ($\sim 10\%$) and density-dependent correction factor which modifies the pressure. This correction factor is obtained by averaging over Skyrme forces which give similar M-R curves to those suggested by the data.

While we do not obtain significant constraints on a or α, the mass and radius data do constrain the parameters b and β (Fig. 5). While the simple parametrization employed here cannot fully describe the complexities of the nuclear three-body force, it does make it clear that astrophysical data is beginning to rule out some three-body forces which might otherwise be acceptable. We also show constraints on L. From neutron stars we obtain the constraints to the symmetry energy and slope to be $32 < E_{\rm sym} < 34$ MeV and $43 < {\rm L} < 52$ MeV within 68% confidence. The only way to obtain a larger value of L is through a strong phase transition just above the nuclear saturation density which tends to decouple the properties of matter at low- and high-densities and allows for values of L as large as 83 MeV [35]. However, it is not clear that such a strong phase transition at low densities is particularly realistic, as it might have been already ruled out by experimental work in heavy-ion collisions as reviewed in Ref. [8].

7. Acknowledgements

SG is supported by the U.S. Department of Energy, Office of Nuclear Physics, by the NUCLEI SciDAC program and by the LANL LDRD program. AWS is supported by DOE Grant No. DE-FG02-00ER41132.

[1] Wiringa R B, Stoks V G J and Schiavilla R 1995 *Phys. Rev. C* **51** 38
[2] Pieper S C, Pandharipande V R, Wiringa R B and Carlson J 2001 *Phys. Rev. C* **64** 014001
[3] Pieper S C 2008 *AIP Conf. Proc.* **1011** 143
[4] Pieper S C and Wiringa R B 2001 *Annu. Rev. Nucl. Part. Sci.* **51** 53
[5] Nollett K M, Pieper S C, Wiringa R B, Carlson J and Hale G M 2007 *Phys. Rev. Lett.* **99**(2) 022502
[6] Schiavilla R, Wiringa R B, Pieper S C and Carlson J 2007 *Phys. Rev. Lett.* **98**(13) 132501
[7] Pudliner B S, Pandharipande V R, Carlson J, Pieper S C and Wiringa R B 1997 *Phys. Rev. C* **56** 1720
[8] Tsang M B, Stone J R, Camera F, Danielewicz P, Gandolfi S, Hebeler K, Horowitz C J, Lee J, Lynch W G, Kohley Z, Lemmon R, Möller P, Murakami T, Riordan S, Roca-Maza X, Sammarruca F, Steiner A W, Vidaña I and Yennello S J 2012 *Phys. Rev. C* **86** 015803
[9] Newton W G, Gearheart M and Li B A 2011 *arXiv:1110.4043* (*Preprint* 1110.4043)
[10] Wen D H, Newton W G and Li B A 2012 *Phys. Rev. C* **85** 025801
[11] Vidaña I 2012 *Phys. Rev. C* **85** 045808
[12] Wiringa R B and Pieper S C 2002 *Phys. Rev. Lett.* **89** 182501
[13] Gandolfi S, Carlson J, Reddy S, Steiner A W and Wiringa R B 2013 *arXiv:1307.5815*
[14] Pudliner B S, Pandharipande V R, Carlson J and Wiringa R B 1995 *Phys. Rev. Lett.* **74** 4396
[15] Akmal A, Pandharipande V R and Ravenhall D G 1998 *Phys. Rev. C* **58** 1804
[16] Gandolfi S, Illarionov A Y, Schmidt K E, Pederiva F and Fantoni S 2009 *Phys. Rev. C* **79** 054005
[17] Gandolfi S, Carlson J and Reddy S 2012 *Phys. Rev. C* **85** 032801
[18] Sarsa A, Fantoni S, Schmidt K E and Pederiva F 2003 *Phys. Rev. C* **68** 024308
[19] Maris P, Vary J P, Gandolfi S, Carlson J and Pieper S C 2013 *Phys. Rev. C* **87**(5) 054318
[20] Gezerlis A, Tews I, Epelbaum E, Gandolfi S, Hebeler K, Nogga A and Schwenk A 2013 *Phys. Rev. Lett.* **111**(3) 032501
[21] Gandolfi S, Carlson J and Pieper S C 2011 *Phys. Rev. Lett.* **106** 012501
[22] Schmidt K E and Fantoni S 1999 *Phys. Lett. B* **446** 99
[23] Lonardoni D, Gandolfi S and Pederiva F 2013 *Phys. Rev. C* **87**(4) 041303

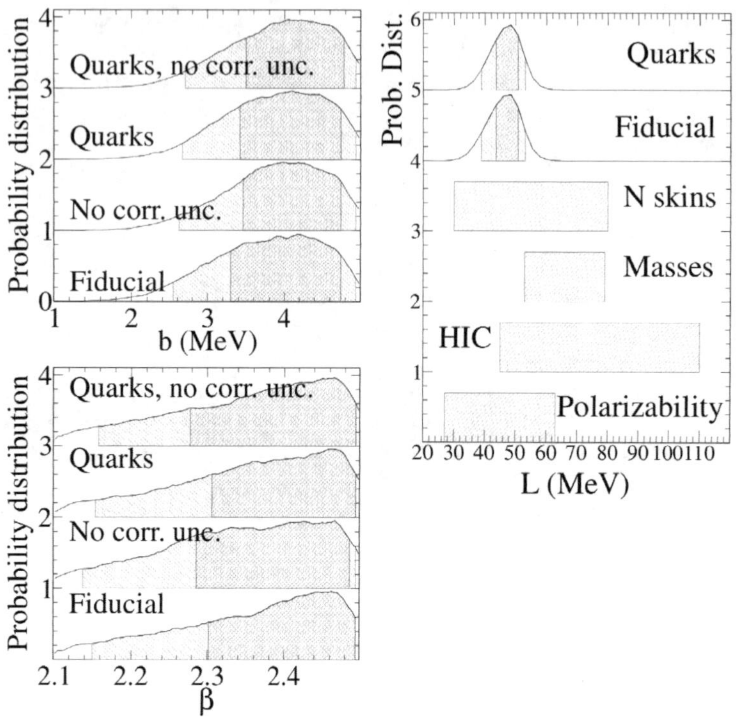

Figure 5. The left panel shows probability distributions of the parameters b and β obtained from the Bayesian analysis. The right panel summarizes constraints on L from observations and experiments. The top two curves show constraints on L as probability distributions assuming either the fiducial model of Ref. [38] or the model containing quarks. The bottom four curves show constraints on L from experiment, from neutron skins [39], nuclear masses [40], heavy-ion collisions [41], and from the electric dipole polarizability [42].

[24] Demorest P B, Pennucci T, Ransom S M, Roberts M S E and Hessels J W T 2010 *Nature* **467** 1081
[25] Antoniadis et al J 2013 *Science* **340** 448
[26] Baym G, Pethick C and Sutherland P 1971 *Astrophys. J.* **170** 299
[27] Negele J W and Vautherin D 1973 *Nucl. Phys. A* **207** 298
[28] Gandolfi S, Illarionov A Y, Fantoni S, Miller J, Pederiva F and Schmidt K 2010 *Mon. Not. R. Astron. Soc.* **404** L35
[29] Lattimer J M and Prakash M 2001 *Astrophys. J.* **550** 426
[30] van Paradijs J 1979 *Astrophys. J.* **234** 609
[31] Özel F, Baym G and Güver T 2010 *Phys. Rev. D* **82** 101301
[32] Steiner A W, Lattimer J M and Brown E F 2010 *Astrophys. J.* **722** 33
[33] Rutledge R, Bildsten L, Brown E, Pavlov G and Zavlin E 1999 *Astrophys. J.* **514** 945
[34] Lattimer J M and Steiner A W 2013 *arXiv:1305.3242*
[35] Steiner A W, Lattimer J M and Brown E F 2013 *Astrophys.J.* **765** L5
[36] Hebeler K and Schwenk A 2010 *Phys. Rev. C* **82** 014314
[37] Danielewicz P, Lacey R and Lynch W G 2002 *Science* **298** 1592
[38] Steiner A W and Gandolfi S 2012 *Phys. Rev. Lett.* **108** 081102
[39] Warda M, Viñas X, Roca-Maza X and Centelles M 2009 *Phys. Rev. C* **80** 024316
[40] Liu M, Wang N, Z-X L and Zhang F S 2010 *Phys. Rev. C* **82** 064306
[41] Tsang M B, Zhang Y, Danielewicz P, Famiano M, Li Z, Lynch W G and Steiner A W 2009 *Phys. Rev. Lett.* **102** 122701
[42] Tamii A, Poltoratska I, von Neumann-Cosel P, Fujita Y, Adachi T *et al.* 2011 *Phys. Rev. Lett.* **107** 062502

Journal of Physics: Conference Series **665** (2016) 012064 doi:10.1088/1742-6596/665/1/012064

Constraints on modern microscopic equations of state

G Taranto[1,2]*, M Baldo[1] and G F Burgio[1]

[1] INFN Sezione di Catania, Via S. Sofia 64, 95123 Catania, Italy

[2] Dipartimento di Fisica e Astronomia, Universitá di Catania, Via S. Sofia 64, 95123 Catania, Italy

E-mail: tarantogabriele@gmail.com, baldo@ct.infn.it, fiorella.burgio@ct.infn.it

Abstract. We compare a set of equations of state derived within microscopic many-body approaches, and study their predictions as far as phenomenological data on nuclei from heavy ion collisions, and astrophysical observations on neutron stars are concerned. All the data, taken together, put strong constraints not easy to be fulfilled accurately. However the results provide an estimate of the uncertainty on the theoretical prediction at a microscopic level of the nuclear equation of state.

1. Introduction

In the past years several efforts have been made in order to constrain the equation of state (EOS). The experimental data on heavy ion collisions and astrophysical observations of compact objects have been used for exploring the density dependence of the EOS, the symmetry energy and its isospin dependence. Recently, a set of equations of state based on the phenomenological Skyrme [1] force and on microscopic many-body methods was tested against the available experimental data [2]. We will present the analysis in this paper, which is based on the results published in ref.[2].

Empirical properties of infinite nuclear matter can be calculated using different microscopic theoretical approaches, the only input being a realistic free nucleon-nucleon (NN) interaction with parameters fitted to NN scattering phase shifts in different partial wave channels, and to properties of the deuteron [3]. In the following we discuss the non relativistic Brueckner-Hartree-Fock (BHF) method [4] and its relativistic counterpart, the Dirac-Brueckner-Hartree-Fock (DBHF) approximation [5] and the variational method [6]. The Brueckner-Bethe-Goldstone (BBG) theory is based on a linked cluster expansion of the energy per nucleon of nuclear matter The basic ingredient in this many-body approach is the Brueckner reaction matrix G, which is the solution of the Bethe-Goldstone equation

$$G[\rho;\omega] = v + \sum_{k_a k_b} v \frac{|k_a k_b\rangle Q \langle k_a k_b|}{\omega - e(k_a) - e(k_b)} G[\rho;\omega], \tag{1}$$

where v is the bare NN interaction, ρ is the nucleon number density, and ω the starting energy. The single-particle energy (assuming $\hbar$=1) $e(k) = e(k;\rho) = \frac{k^2}{2m} + U(k;\rho)$, and the Pauli operator Q determine the propagation of intermediate baryon pairs. $U(k;\rho)$ is the single-particle potential, $U(k;\rho) = \mathrm{Re}\sum_{k' \leq k_F} \langle kk'|G[\rho;e(k)+e(k')]|kk'\rangle_a$, where the subscript "$a$" indicates antisymmetrization of the matrix element. In the BHF approximation the energy per nucleon is

$$\frac{E}{A} = \frac{3}{5}\frac{k_F^2}{2m} + \frac{1}{2\rho} \sum_{k,k' \leq k_F} \langle kk'|G[\rho;e(k)+e(k')]|kk'\rangle_a. \tag{2}$$

*Speaker

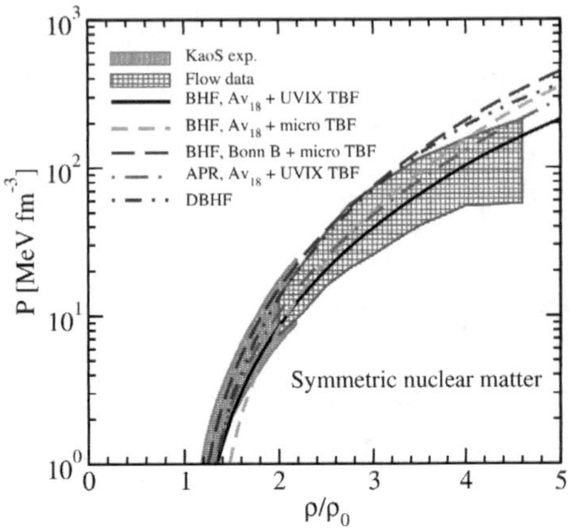

Figure 1. (Color online) Pressure as a function of baryon density for symmetric nuclear matter. See text for details.

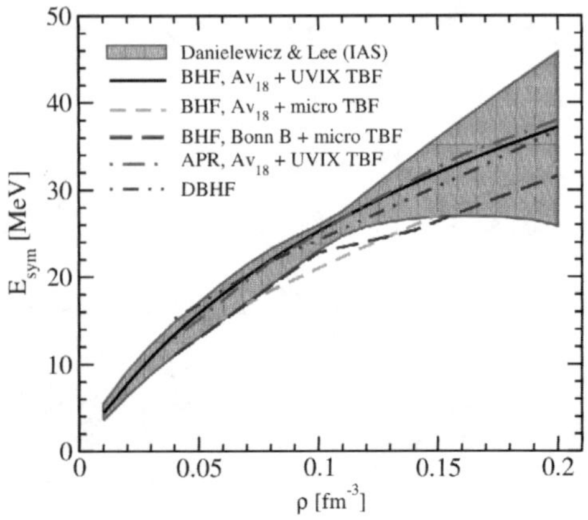

Figure 2. (Color online) The symmetry energy is displayed as a function of the nucleon density for several EOS .

In this scheme, the only input quantity we need is the bare NN interaction v in the Bethe-Goldstone equation (1). It is well known that in order to reproduce the correct point of saturation for nuclear matter we must introduce Three Body Forces (TBF's) in non-relativistic many-body methods. In this work we will illustrate results for two different approaches to the TBF's, i.e., a phenomenological and a microscopic one. The phenomenological approach is based on the so-called Urbana model, which consists of an attractive term due to two-pion exchange with excitation of an intermediate Δ resonance, and a repulsive phenomenological central term [7]. The microscopic approach has been proposed using the same meson-exchange parameters as the underlying NN potential. Results have been obtained with the Argonne v_{18}, the Bonn B, and the Nijmegen 93 potentials [8].

This paper is organized as follows. In section 2 we review the heavy ion phenomenology. In section 3 we present the results regarding astrophysical observation. Section 4 contains our conclusions.

2. Heavy ion phenomenology

In the last two decades intensive studies of heavy ion collisions (HIC) have been performed. The main goal has been the extraction from the data of the gross properties of the nuclear EoS. In principle the dynamics of the collisions should be connected with the properties of the nuclear medium EoS and its viscosity. In the so-called "multifragmentation" regime, the transverse flow, which is strongly affected by the matter compression during the collision, can be measured.

Based on numerical simulations, in Ref.[9] a phenomenological range of densities was proposed where any reasonable EoS for symmetric nuclear matter should pass through in the pressure vs. density plane. The plot is reproduced in Fig.1, where a comparison with the microscopic calculations is made. The green dashed box represents the results of the numerical simulations of the experimental data discussed in Ref.[9], and the brown filled region represents the experimental data on kaon production [10]. We notice that the EoS calculated with the BHF and the variational methods including UVIX three-body (labelled by APR [11]) forces appear to be in agreement with the data in the full density range. On the other hand, the BHF EoS's obtained using microscopic TBF's are only marginally compatible with the experimental data, as well as the DBHF EoS [10], showing that they are too repulsive already at density $\rho \geq 3\rho_0$ if the Bonn potentials are used. Though, it has to be stressed that all EoS are compatible with the data around the saturation density; i.e., their incompressibility is as soft as required by the data. However, the values of the incompressibility do not characterize completely the EoS, since it is density dependent, but in any case the analysis indicates that the EoS at low densities must be soft.

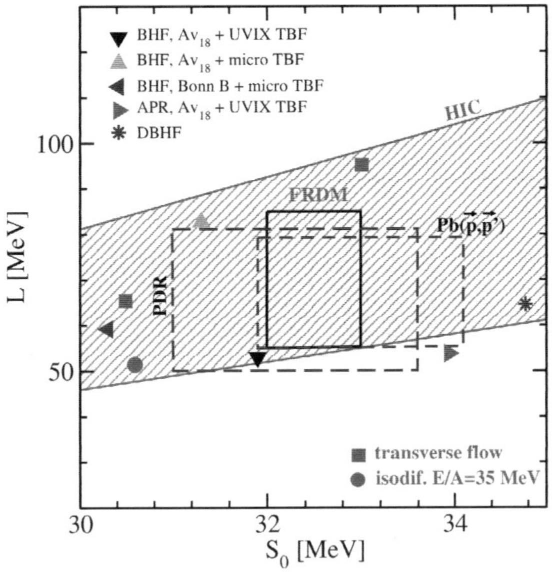

Figure 3. (Color online) The derivative of the symmetry energy L is shown as a function of the symmetry energy at saturation S_0. See text for details.

A further constraint on the EoS is given by the symmetry energy, which has been extensively studied both from the theoretical and experimental points of view in Ref.[12]. The symmetry energy is displayed in Fig.2 as a function of the nucleon density. The green region is the result of a recent analysis performed by Danielewicz and Lee on the isobaric analog states (IAS) in nuclei [13]. This stems from the charge independence of nuclear interactions; i.e., strong interactions between nucleons in the same state do not depend on whether the nucleons are protons or neutrons. Therefore the energy difference between the ground state of a nucleus with $N > Z$ and the isobaric analogs of the ground states of neighboring isobars are given by the symmetry energy, and the Coulomb contributions to the binding energy can be determined using the IAS. Many such states have been identified, and by fitting the available data on the IAS, Danielewicz and Lee obtained the constraint shown as a green region in Fig.2. We observe that all EoS's give results in very good agreement with the experimental data, except the ones of BHF with microscopic TBF at densities below the saturation density.

In Fig.3 we display the slope L of the symmetry energy as a function of the symmetry energy at saturation density, which has been widely discussed in Ref.[12]. Several experimental data sets are displayed. The blue band represents experimental data from HIC, obtained from the neutron and proton spectra from central collisions at 50 MeV/A [14]. At the same incident energy, isospin diffusion was investigated depending on the different N/Z asymmetry of the involved projectiles and targets [15, 16]. The full green circle shows the results from isospin diffusion observables measured for collisions at a lower beam energy of 35 MeV per nucleon [17]. Transverse collective flows of hydrogen and helium isotopes as well as intermediate mass fragments with $Z < 9$ have also been measured at incident energy of 35 MeV/A for ^{70}Zn , ^{64}Zn and ^{64}Ni reactions and compared to transport calculations. The analysis yielded values denoted by the full green squares [18]. The box labeled FRDM (finite-range droplet model) represents a refinement of the droplet model [19], and includes microscopic "shell" effects and the extra binding associated with $N = Z$ nuclei. In Fig.3 the other boxes represent experimental data obtained from measurements of the neutron skin thickness. The measurement of the neutron skin thickness made on the stable nucleus ^{208}Pb reported a value $\delta R_{np} = 0.211^{+0.054}_{-0.063}$ fm. From the experiments constraints on the symmetry energy were derived, and these are plotted in Fig.3 as the short-dashed blue rectangular

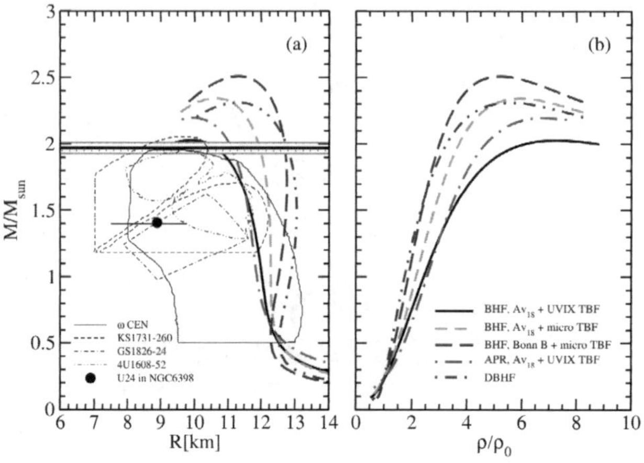

Figure 4. (Color online) The mass-radius (left panel) and the mass-central density (right panel) relations are plotted for the EoS's discussed. Boxes are boundaries extracted from observations, see Ref.[24].

box labelled Pb($\vec{p}, \vec{p}$). Last, we mention the experimental data on the Pygmy Dipole Resonance (PDR) in very neutron-rich nuclei such as ^{68}Ni and ^{132}Sn, which peaks at excitation energies well below the Giant Dipole Resonance (GDR) [20]. Those constraints are shown as a long-dashed rectangle in Fig.3 with the label PDR. The predictions of the different EoS's are also reported in Fig.3 as full symbols. They are distributed within a large region and they span a wide interval in the values of the parameter L. However, the various phenomenological data are at best marginally compatible, and it is difficult to put well definite constraints on the EoS.

3. Astrophysics

A neutron star is bound by gravity, and it is kept in hydrostatic equilibrium only by the pressure produced by the compressed nuclear matter. It is then apparent that the nuclear matter EoS is the main medium property that is relevant in this case, as can be seen in the celebrated Tolman-Oppenheimer-Volkoff [21] equations, valid for spherically symmetric NS. The value of the maximum mass depends on the nuclear EoS, so that the observation of a mass higher than the maximum one allowed by a given EoS simply rules out that EoS. The considered microscopic EoS's are compatible with the largest masses observed up to now, that is the mass of the pulsar PSR J1614- 2230, 1.97 $\pm 0.04\, M_{\odot}$ [22], plotted in the figure, and the mass of the pulsar PSR J0348+0432, 2.01 $\pm 0.04\, M_{\odot}$ [23]. This is clearly shown in Fig.4, where the mass-radius (left panel) and mass-central-density relations (right panel) are plotted for all the considered EoS's as thick lines. It looks unlikely that this value is indeed the largest possible NS mass, and therefore future observational data on NS masses could overcome this limit and strongly constrain the nuclear EoS.

In Fig.4 a sample of observational data taken from Ref.[24] is displayed by closed thin lines for different sources, measured in quiescence and from thermonuclear bursts. It turns out that the current measurements are consistent with radii in the range 8-12 km and disfavor neutron stars with R $\sim$15 km. Those measurements are consistent with the recent observation of the neutron star in SAX J1748.9-2021, which points to the neutron star radius in the 8-11 km range [25].

Additional tentative constraints on the nuclear EoS were obtained in a recent analysis of the data on six NS's based on Bayesian statistical framework [26]. Depending on the hypothesis made on the structure of the NS, the results are slightly different, as shown in Fig.5, where the quantity r_{ph} is the photosphere radius. In the left panel r_{ph} is comparable to the neutron star radius R, whereas in the right panel a substantial expansion of the photosphere during an x-ray burst is assumed to occur. Among the different EoS's, only the one calculated with BHF and phenomenological Urbana model appear to be

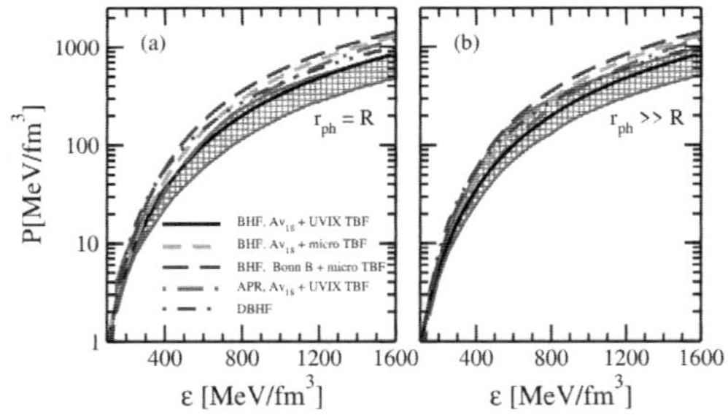

Figure 5. (Color online) Pressure as a function of the mass-energy density in neutron star matter. The shaded areas are taken from ref.[26]. See text for details.

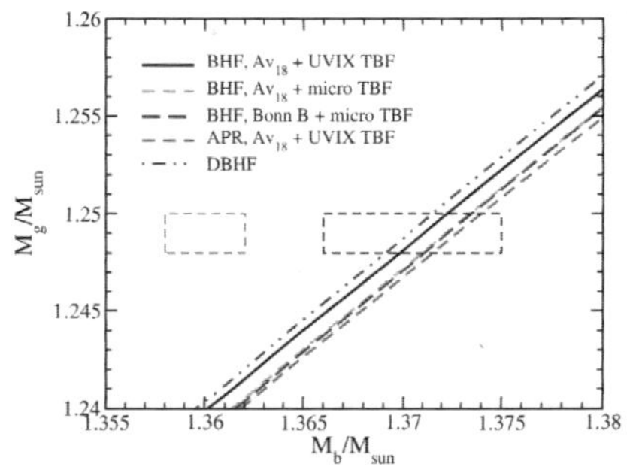

Figure 6. (Color online) The gravitational mass is plotted as a function of the baryon mass for several EoS's. The boxes indicate the boundaries from the simulations.

compatible with the extracted observational constraints over the whole density range. These boundaries obtained from astrophysical data are complementary to the ones obtained from heavy ion reactions. In fact, in heavy ion collisions the tested matter is essentially symmetric, while in a NS the matter is highly asymmetric. Considered together, the two types of constraints probe the density dependence of the symmetry energy.

In relation to the high density region of the nuclear EoS, an additional test is on the speed of sound c_S, which is required to be smaller than the speed of light c (causality condition).The speed of sound is directly connected with the incompressibility and the energy density, and therefore it depends on the matter energy density and asymmetry. We found that at large enough density most of the EoS's show a superluminal speed of sound, except the one calculated with BHF with phenomenological TBF's which doesn't become superluminal at central density corresponding to the maximum mass, unlike the other EoS's.

A further additional constraint on the neutron star EoS is provided by the observation of the double pulsar J0737-3039, and the interpretation given by Podsiadlowski [27]. In fact, the gravitational mass of Pulsar B is very precisely known, $M_G = 1.249 \pm 0.001\,M_\odot$, whereas estimates of the baryonic mass depend upon its detailed mode of formation. As modeled by Podsiadlowski et al., if the pulsar B was formed from a white dwarf with an O-Ne-Mg core in an electron-capture supernova, assuming no or negligible loss of baryonic mass during the collapse, the newly born neutron star will have the same baryonic mass as the precollapse core of the progenitor star, i.e., $M_B \simeq 1.366 - 1.375\,M_\odot$. This result is displayed in Fig.6 as a black box. Though, taking into account the uncertainty in the EoS and the small mass loss during the collapse, Kitaura et al. [28] made another simulation which gave $M_B = 1.360 \pm 0.002\,M_\odot$, which is shown in Fig.6 by the green box. We have calculated for each neutron star matter EoS the relation between the gravitational and baryonic mass, and these are displayed in Fig.6 by the straight curves. We notice that the results of all microscopic EoS's agree very well with the result of Podsiadlowski, at variance with the calculations based on the phenomenological Skyrme forces discussed in Ref.[1], where agreement was found with the result of Kitaura et al.[28], which assumed small mass loss during the collapse.

4. Conclusions

We have presented a systematic confrontation of the nuclear equation of state, obtained within different microscopic many- body methods, with the available constraints coming from phenomenology. Both nuclear structure and heavy ion collisions data were considered, along the same lines of the analysis on the Skyrme forces reported in Ref.[1]. Astrophysical observational data included the measurements of NS masses, some hints on the mass-radius relation from Ref.[24, 25], and the constraints on the EoS presented in Ref [26], obtained from the analysis of transient phenomena in six NS's. If one takes literally all the constraints, among the considered microscopic EoS's only one passes all the tests, namely the BHF with the two-body potential Av18 and the UVIX model for the three-body forces [29]. This shows that the considered phenomenological constraints are well suited in selecting the microscopic EoS. However, before drawing a strong conclusion, it would be appropriate to have a firm estimate of the uncertainties affecting the data. Secondly, from the analysis it appears that one can explain reasonably well all the data with a microscopic EoS that includes only nucleonic degrees of freedom; in particular no exotic components in NS's are needed. Indeed, if a NS with mass much larger than 2 solar masses will be observed, this will rule out most of the considered microscopic EoS's and will introduce a serious and fundamental issue in the physics of NS's. On the other hand, it has been shown that exotic matter such as hyperons or quarks should appear in NS's [30], which will strongly affect the EoS. In the future one can expect that the interplay between theory and observations will continue to play a major role in the worldwide effort of determining the nuclear EoS.

Acknowledgments
We warmly thank P. Danielewicz and J. Lee for providing us the data on the symmetry energy shown in Fig. 2. Moreover we acknowledge H.-J. Schulze for the data concerning the EoS calculated in the BHF approach with Bonn B NN potential and microscopic TBF, and C. Fuchs for the DBHF data set.

References

[1] Dutra M, Laurenço O, Sá Martins J S, Delfino A , Stone J R and Stevenson P D 2012 *Phys. Rev.* C **85** 035201
[2] Taranto G, Baldo M and Burgio G f 2013 *Phys. Rev.* C **87** 045803
[3] Baldo M and Burgio G F 2012 *Rep. Prog. Phys.* **75** 026301
[4] Baldo M 1999 *Nuclear Methods and the Nuclear Equation of State* (Singapore : World Scientific)
[5] Machleidt R 1989 *Advances in Nuclear Physics* **19** 189
[6] Pandharipande V R and Wiringa R B 1979 *Rev. Mod. Phys.* A **51** 821
[7] Carlson J, Pandharipande V R and Wiringa R B 1983 *Nucl. Phys.* C **401** 59
[8] Li Z H, Lombardo U, Schulze H-J and Zuo W 2008 *Phys. Rev.* C **77** 034316
[9] Danielewicz P, Lacey R and Lynch W 2002 *Science* **298** 1592
[10] Fuchs C 2006 *Prog. Part. Nucl. Phys.* **56** 1
[11] Akmal A, Pandharipande V R and Ravenhall D G 1998 *Phys. Rev.* C **58** 1804
[12] Tsang M B *et al* 2012 *Phys. Rev.* C **86** 015803
[13] Danielewicz P and Lee J arXiv:*1307.4130*
[14] Famiano M *et al* 2006 *Phys. Rev. Lett.* **97** 052701
[15] Tsang M B *et al* 2001 *Phys. Rev.* C **64** 054615
[16] Li Bao-An, Chen Lie-Wen and Ko Che Ming 2008 *Phys. Rep.* **464** 113
[17] Sun Z J *et al* 2010 *Phys. Rev.* C **82** 051603(R)
[18] Kohley Z, May L W, Wuenschel S et al. 2011 *Phys. Rev.* C **83** 04460
[19] Möller P, Meyers W D, Sagawa H and Yoshida S 2012 *Phys. Rev. Lett.* **108** 052501
[20] Klimkiewicz A *et al* 2007 *Phys. Rev.* C **76** 051603(R)
[21] Shapiro S L and Teukolsky S A 1983 *Black Holes, White Dwarfs, and Neutron Stars* (New York : John Wiley & Sons)
[22] Demorest P B, Pennucci T, Ransom S M, Roberts M S E and Hessels J W T 2010 *Nature* **467** 1081
[23] Antoniadis J *et al* 2013 *Science* **340** 6131
[24] Özel F 2013 *Rep. Prog. Phys.* **76** 016901
[25] Guver T and Özel F 2013 *Astrophys. J.* **765** L1
[26] Steiner A W, Lattimer J M, Brown E F 2010 *Astrophys. J.* **722** 33
[27] Podsiadlowski Ph, Dewi J, Lesaffre P, Miller J, Newton W and Stone J 2005 *Mon. Not. R. Astron. Soc.* **361** 1243
[28] Kitaura F S, Janka H-Th and Hillebrandt W 2006 *Astron. Astrophys.* **450** 345
[29] Zhou X R, Burgio G F, Lombardo U , Schulze H-J, and Zuo W 2004 *Phys. Rev.* C **69** 018801
[30] Schulze H-J, Baldo M, Lombardo U, Cugnon J and Lejeune A 1998 *Phys. Rev.* C **57** 704; Baldo M, Burgio G F and Schulze H-J 2000 *Phys. Rev.* C **61** 055801

Journal of Physics: Conference Series **665** (2016) 012065 doi:10.1088/1742-6596/665/1/012065

Collective excitations in neutron-star crusts

N Chamel[1], D Page[2] and S Reddy[3]

[1] Institut d'Astronomie et d'Astrophysique, Université Libre de Bruxelles - CP226, 1050 Brussels, Belgium
[2] Instituto de Astronomía, Universidad Nacional Autonóma de México, Mexico D.F. 04510, Mexico
[3] Institute for Nuclear Theory, University of Washington, Seattle, Washington 98195

E-mail: nchamel@ulb.ac.be

Abstract. We explore the spectrum of low-energy collective excitations in the crust of a neutron star, especially in the inner region where neutron-proton clusters are immersed in a sea of superfluid neutrons. The speeds of the different modes are calculated systematically from the nuclear energy density functional theory using a Skyrme functional fitted to essentially all experimental atomic mass data.

1. Introduction

A few meters below the surface of a neutron star at densities above $\sim 10^4$ g·cm^{-3}, atomic nuclei are fully ionized by the pressure thus coexisting with a highly degenerate electron gas. As the density increases, nuclei become more and more neutron-rich due to electron captures [1] until neutrons start to drip out of nuclei at densities above $\sim 4.10^{11}$ g·cm^{-3} [2] thus delimiting the boundary between the outer and inner parts of the crust. Those "free" neutrons are expected to become superfluid below a critical temperature of the order of $\sim 10^{10}$ K. Despite the absence of viscous drag, the neutron superfluid can still be strongly coupled to the crust due to non-dissipative entrainment effects [3].

Although the crust of a neutron star represents only about 1% of the star's mass, it is expected to play a key role in various observed astrophysical phenomena [4]. In particular, the collective excitations in the inner crust of a neutron star can impact the thermal evolution of the star [5]. These excitations have been already studied in the Wigner-Seitz approximation, using self-consistent mean-field methods with Skyrme effective interactions [6, 7]. In this approach, the crust is divided into a set of independent spheres centered around each nucleus. Consequently, only high-energy collective excitations with wavelengths smaller than the size of the Wigner-Seitz cell can be studied in this framework. Moreover, this approximate treatment of the crust does not take into account entrainment effects, which arise from Bragg scattering of unbound neutrons by the Coulomb lattice. For this reason, we have recently followed a different approach by employing the band theory of solids. In this way, we have been able to study low-energy collective modes including entrainment effects [8].

In this paper, we extend our previous study of collective modes by including the coupling of the neutron superfluid to the strain field.

Journal of Physics: Conference Series **665** (2016) 012065 doi:10.1088/1742-6596/665/1/012065

2. Elasto-hydrodynamics of the neutron-star inner crust

The normal modes of oscillations of the neutron superfluid permeating the neutron-star inner crust can be obtained by linearizing the elasto-hydrodynamic equations (see, e.g., Refs. [9, 10]). Treating the crust as an isotropic solid, the normal modes consists of two transverse modes and two longitudinal modes. Using the same notations as in Ref. [8], the speed of the transverse modes is given by $v_t = \sqrt{S/\rho_{\rm I}}$, where S is the shear modulus of the crust and $\rho_{\rm I}$ is the mass density associated with lattice vibrations. Because unbound neutrons are entrained by the crust, the mass density is much larger than the ion mass density, and is given by $\rho_{\rm I} = m(n_p + n_n^{\rm b})$, where m is the nucleon mass, n_p the proton density and $n_n^{\rm b}$ the density of entrained neutrons. In other words, including entrainment significantly reduces the values of v_t. The longitudinal modes satisfy a dispersion relation for two coupled modes, namely

$$(v^2 - v_\phi^2)(v^2 - v_\ell^2) = g_{\rm mix}^2 v^2 + g^4\,, \tag{1}$$

where v_ϕ is the speed of the Bogoliubov-Anderson bosons of the neutron superfluid, and v_ℓ is the speed of longitudinal lattice phonons. The mixing between these two modes is characterized by the parameters $g_{\rm mix}$ and g, given by

$$g_{\rm mix} = \sqrt{\frac{n_n^{\rm b}(2L + E_{nn}n_n^{\rm b})}{\rho_{\rm I}}}\,, \qquad g = \left(\frac{L^2 n_n^{\rm c}}{m\rho_{\rm I}}\right)^{1/4}\,, \tag{2}$$

where $L = n_p\partial\mu_n/\partial n_p$ (μ_n being the neutron chemical potential and n_n the neutron density), $E_{nn} = \partial\mu_n/\partial n_n$, and $n_n^{\rm c} = n_n - n_n^{\rm b}$ is the density of "conduction" neutrons. Note that Eq. (1) is identical to Eq. (34) from Ref. [10] although it is expressed here in a slightly different form.

In our previous study [8], we assumed that the coupling of the superfluid to the strain field embedded in the coefficient L is small, and we thus set $L = 0$. We now include this coefficient. The two solutions of Eq. (1) are given by

$$v_\pm = \frac{V}{\sqrt{2}}\sqrt{1 \pm \sqrt{1 - \frac{4v_\ell^2 v_\phi^2}{V^4} + \frac{4g^4}{V^4}}}\,, \tag{3}$$

where $V = \sqrt{v_\ell^2 + v_\phi^2 + g_{\rm mix}^2}$.

In the non-superfluid phase, any relative motion between the neutron liquid and the crust will be damped by viscosity to the effect that ions, electrons and neutrons will be essentially comoving. Only one longitudinal mode corresponding to ordinary hydrodynamic sound will persist and its speed will be given by $c_s = \sqrt{(K + 4S/3)/\rho} \approx \sqrt{K/\rho}$ where ρ is the total mass density of the crust and K is the total bulk modulus, which is related to the bulk modulus $\widetilde{K}$ of the electron-ion system by $K = \widetilde{K} + 2n_nL + n_n^2E_{nn}$. Assuming that pairing affects neither the composition nor the elastic properties of the crust, the coefficient L can be obtained by inverting the previous equation, thus yielding

$$2n_nL = K - \widetilde{K} - n_n^2E_{nn} \approx \rho c_s^2 - \rho_{\rm I}v_\ell^2 - \frac{mn_n^2v_\phi^2}{n_n^{\rm c}}\,. \tag{4}$$

3. Microscopic model for the inner crust of a neutron star

As in our previous work [8], we use the inner crust composition given in Ref. [11], which was obtained from the fourth-order extended Thomas-Fermi method with proton quantum shell effects added via the Strutinsky-Integral theorem. This so-called ETFSI method is a high-speed approximation to the self-consistent Hartree-Fock equations. The calculations of Ref. [11] were carried out using the BSk14 Skyrme interaction underlying the HFB-14 atomic mass model [12], which yields an excellent fit to essentially all experimental atomic mass data with a root mean square deviation of 0.73 MeV. Moreover, the BSk14 interaction was also constrained to reproduce a realistic neutron-matter equation of state. For these reasons, this interaction is particularly well-suited for describing the inner crust of a neutron star.

Although the neutron quantum shell effects are much smaller than proton ones [13], and therefore are expected to have a small impact on the crustal composition, they drastically affect the entrainment between the neutron superfluid and the crust. Neutron quantum shell effects were systematically studied in Ref. [3] in the framework of the band theory of solids using the mean-field potentials obtained self-consistently from the ETFSI method in Ref. [11]. This allowed the calculation of the conduction neutron density n_n^c. The quantities E_{nn}, v_ϕ, v_ℓ and c_s are calculated as in Ref. [8]. The L coefficient is then estimated from Eq. (4). The numerical results are shown in Table 1. It can be seen that the coupling of the superfluid to the strain field reduces the speed of the highest longitudinal mixed mode, whereas the speed of the lowest longitudinal mixed mode is barely affected. Since the contribution of a collective excitation with velocity v to the heat capacity varies like v^{-3} at low temperatures, taking the coefficient L into account substantially increases the heat capacity of the highest longitudinal mixed mode, which is enhanced by about a factor of 3 at temperature $T = 10^7$ K near the neutron-drip transition, as compared to our previous estimates. However, the contribution of this mode to the crustal heat capacity still remains negligible, as shown in Fig. 1. Likewise, the thermal conductivity of the crust remains almost unchanged compared to our previous estimates. Indeed, the electron-phonon scattering rate, which scales like v^{-3}, is dominated by processes involving transverse phonons.

Table 1. Properties of collective modes in the inner crust of a neutron star for different values of the average baryon number density $\bar{n}$. The speeds are expressed in units of $10^{-2}c$, c being the speed of light. Values in parenthesis were obtained by neglecting the coupling of the superfluid to the strain field.

$\bar{n}$ (fm^{-3})	v_ϕ	v_ℓ	c_s	v_t	v_-	v_+
0.0003	1.11	5.13	5.35	0.578	0.935 (1.02)	3.75 (5.56)
0.001	1.34	3.69	4.46	0.416	1.27 (1.08)	3.50 (4.58)
0.005	1.64	2.60	4.78	0.293	1.01 (0.890)	3.86 (4.80)
0.01	1.77	2.39	5.18	0.251	0.897 (0.814)	4.40 (5.20)
0.02	1.42	2.26	5.84	0.237	0.582 (0.548)	5.35 (5.84)
0.03	1.62	2.31	6.55	0.242	0.595 (0.570)	6.20 (6.56)
0.04	2.18	2.44	7.39	0.256	0.750 (0.720)	6.96 (7.39)
0.05	4.21	2.83	8.30	0.235	1.50 (1.44)	7.68 (8.31)
0.06	5.86	3.31	9.28	0.275	2.16 (2.09)	8.37 (9.29)
0.07	7.76	4.26	10.3	0.354	3.25 (3.20)	9.07 (10.3)
0.08	8.98	4.87	11.4	0.404	3.70 (3.84)	9.82 (11.4)

4. Conclusions

Due to nondissipative entrainment effects, the Bogoliubov-Anderson bosons of the neutron superfluid in the inner crust of a neutron star are strongly mixed with the longitudinal crystal lattice phonons. The coupling of the superfluid to the strain field, which we neglected in our previous study [8], is found to substantially reduce the speed of the highest longitudinal mixed

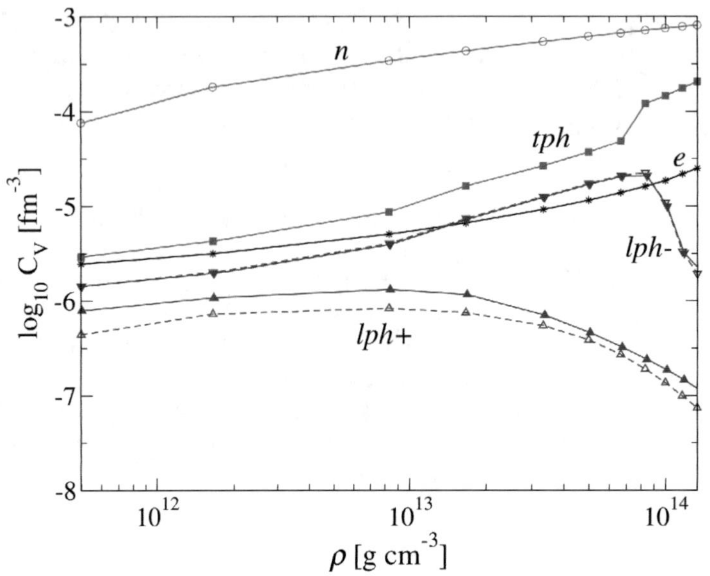

Figure 1. Heat capacity of electrons (e), transverse lattice phonons (tph) and longitudinal excitations ($lph-$ and $lph+$) in the inner crust of neutron stars at temperature $T = 10^9$ K, with the coupling of the superfluid to the strain field (solid lines) and without (dashed lines). For comparison, is also shown the normal neutron contribution (n), but it is strongly suppressed by superfluidity except in the shallowest and densest parts of the inner crust where the neutron $^1S_0$ pairing gap becomes vanishingly small.

mode, especially in the shallowest layers of the inner crust. However, the thermal properties of the crust are almost unchanged.

Acknowledgments
The financial support of F.R.S.-FNRS (Belgium) is gratefully acknowledged. The work of S.R. was supported by DOE Grant No. #DE-FG02-00ER41132 and by the Topical Collaboration to study *neutrinos and nucleosynthesis in hot dense matter*. D.P's work was supported by grants from Conacyt (Grant No. CB-2009/132400) and UNAM-DGAPA (Grant No. PAPIIT IN113211).

5. References

[1] Pearson J M, Goriely S and Chamel N 2011 *Phys. Rev. C* **83** 065810
[2] Pearson J M, Chamel N, Goriely S and Ducoin C 2012 *Phys. Rev. C* **85** 065803
[3] Chamel N 2012, *Phys. Rev. C* **85**, 035801
[4] Chamel N and Haensel P 2008 *Living Rev. Relativity* **11**, 10. http://www.livingreviews.org/lrr-2008-10
[5] Page D and Reddy S, in "Neutron Star Crust, edited by C. A. Bertulani and J. Piekarewicz.
[6] Grasso M, Khan E, Margueron J, van Giai N 2008, *Nucl. Phys.* **A 807**, 1
[7] Baroni S, Pastore A, Raimondi F, Barranco F, Broglia R A and Vigezzi E 2010, *Phys. Rev. C* **82**, 015807
[8] Chamel N, Page D, Reddy S 2013, *Phys. Rev. C* **87**, 035803
[9] Pethick C J , Chamel N and Reddy S 2010, *Prog. Theor. Phys. Supplement* **186**, 9
[10] Kobyakov D, Pethick C J 2013, *Phys. Rev. C***87**, 055803
[11] Onsi M, Dutta A K, Chatri H, Goriely S, Chamel N and Pearson J M 2008, *Phys. Rev. C* **77**, 065805
[12] Goriely S, Samyn M and Pearson J M 2007, *Phys. Rev C***75**, 064312
[13] Oyamatsu K, Yamada M 1994, *Nucl. Phys.* **A578**, 181

Symmetry energy from nuclear masses and neutron-star observations using generalised Skyrme functionals

N Chamel[1], A F Fantina[1], J M Pearson[2] and S Goriely[1]

[1] Institut d'Astronomie et d'Astrophysique, Université Libre de Bruxelles - CP226, 1050 Brussels, Belgium
[2] Département de Physique, Université de Montréal, Montréal, Québec, H3C 3J7 Canada

E-mail: nchamel@ulb.ac.be

Abstract. We study the constraints imposed by nuclear mass measurements and neutron-star observations on the symmetry energy. For this purpose, we use a family of unified equations of state of neutron-star interiors, based on generalised Skyrme functionals that were fitted to essentially all the experimental nuclear mass data while ensuring a realistic neutron-matter equation of state.

1. Introduction

Born from gravitational core-collapse supernova, neutron stars (NSs) are among the most compact objects in the Universe, with a central density which can reach several times nuclear matter density $n_0 \simeq 0.16$ fm^{-3} (see, e.g., Ref. [1]). The interior of a NS is expected to contain at least four different regions: (i) an "outer crust", at densities above $\sim 10^4$ g cm^{-3}, composed of fully ionised atoms, arranged in a Coulomb lattice and neutralised by a degenerate electron gas, (ii) an "inner crust", at densities above $\sim 4 \times 10^{11}$ g cm^{-3}, where neutron-proton clusters and electrons coexist with unbound neutrons, (iii) an outer core, at densities above $\sim 10^{14}$ g cm^{-3}, consisting of a neutron-rich liquid with a small admixture of protons and leptons, and (iv) an inner core, whose composition is still a matter of debate (see, e.g., Ref. [1]). Throughout this paper, we will consider nucleons and leptons only. The composition of dense matter, which can impact various NS phenomena like cooling, is mainly determined by the symmetry energy. The slope of the symmetry energy is related to the pressure of neutron-matter, and is therefore related to the NS structure. More generally, the equation of state (EoS) of dense matter, i.e., the relation between the average baryon number density n, the pressure P and the mass-energy density ρ, is a fundamental ingredient for modeling NSs.

In this paper, we consider three different EoSs of dense matter based on generalised Skyrme functionals, which provide a unified description of all regions of a NS. Although all functionals yield the same value for the symmetry energy at saturation, each of them leads to a different prediction for the high-density behaviour of the symmetry energy. We use these EoSs to compute the mass and radius of NSs and we test our predictions against astrophysical observations. In this way, we can obtain a constraint on the symmetry energy. This constraint is compared to other estimates coming from nuclear physics.

2. Unified dense matter equations of state

Although many EoSs of dense matter have been already constructed, most of them are only applicable to specific parts of a NS. On the other hand, a thermodynamically consistent description of all regions of the star is of utmost importance for dynamical simulations since the ad hoc matching of different EoSs may trigger spurious instabilities. For this reason, a family of three different unified EoSs, referred to as BSk19, BSk20, and BSk21 respectively, and reflecting the current lack of knowledge of high-density matter, has been recently developed [2, 3, 4, 5]. These EoSs are based on the nuclear energy density functional (EDF) theory derived from generalised Skyrme effective interactions [2, 6], supplemented with a microscopic contact pairing interaction [7]. These EDFs, employed in the Hartree-Fock-Bogoliubov (HFB) framework with phenomenological collective corrections, fit the 2149 measured masses of nuclei from the 2003 Atomic Mass Evaluation [8] with a root-mean square deviation as low as 0.58 MeV for all three models. These EDFs are also constrained to reproduce three different realistic neutron matter EoSs: BSk19 (BSk21) was adjusted to a soft (stiff) EoS, while BSk20 was fitted to an EoS with an intermediate stiffness (see Ref. [2] for details). Therefore, these EDFs are very well suited for a unified treatment of NS interiors. The corresponding EoSs of the crust were obtained in Refs. [3, 4, 5]. The EoS of the core can be readily obtained from the EDF [2].

3. Symmetry energy constraints

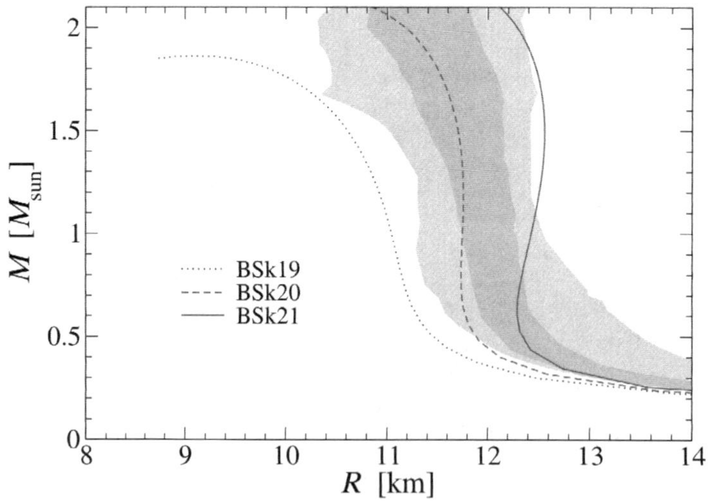

Figure 1. Mass-radius relation of neutron stars for models BSk19, BSk20 and BSk21. The dark (light) shaded regions correspond to the 1-σ (2-σ) probability distributions implied by observations, as shown in the upper right panel of Fig. 9 of Ref. [11].

Using these EoSs, we studied the global properties of NSs. In particular, we found that the maximum masses of non-rotating NSs obtained with our EoSs BSk19, BSk20, and BSk21 are respectively 1.86 $M_\odot$, 2.15 $M_\odot$, and 2.28 $M_\odot$ [9]. The softest EoS BSk19 is incompatible with the recently measured mass of PSR J0348+0432, namely $M = 2.01 \pm 0.04\ M_\odot$ [10]. This conclusion remains unchanged if rotation is taken into account [9], even for the most rapidly spinning pulsar known. The maximum NS mass mainly probes the high-density part of the dense-matter EoS. Constraints on the low-density part of the EoS can be obtained from the measurements of the NS mass and radius. In particular, our EoSs are compared with the constraint obtained from observations of three type I X-ray bursters with photospheric radius expansion, and three transient low-mass X-ray binaries in globular clusters [11]. Once again, as shown in Fi g. 1, the EoS BSk19 seems to be too soft whereas our stiffest EoS BSk21 is marginally consistent. On the other hand, the EDF underlying the soft EoS BSk19 seems to be favoured by the analysis of pion production in heavy-ion collisions. If this analysis is confirmed, it might suggest that the dense inner core of such massive NS may undergo a phase transition to non-nucleonic matter with a stiff EoS [12]. Further constraints on the nuclear-matter EoS at densities around normal density can be obtained from nuclear physics, both from theoretical calculations and from experiments. The energy per nucleon of at density n and charge asymmetry $\eta = (n_n - n_p)/n$ can be approximately written as $e(n,\eta) \approx e(n,\eta=0) + e_{\rm sym}(n)\eta^2$. In this approximation, the symmetry energy $e_{\rm sym}$ is thus simply given by the difference between the energy per nucleon of neutron matter and that of symmetric nuclear matter. However, because of higher order terms in η this identification is generally not exact, and significant deviations can be found with increasing density. Various studies have been carried out on the symmetry energy, because of its importance in understanding both nuclear and astrophysical phenomena (see, e.g., Refs. [13, 14] and Refs. therein). The behaviour of the symmetry energy around normal density can be characterized by the coefficients $J \equiv e_{\rm sym}(n_0)$ and $L \equiv 3n_0 \left.\frac{\partial e_{\rm sym}}{\partial n}\right|_{n=n_0}$. While the values of J obtained from theoretical calculations and experiments lie in a relatively narrow range, the predicted values for L may differ widely. This coefficient is related to the pressure P_0 of neutron matter around the normal density, as $L = 3P_0/n_0$. Therefore, the uncertainties in L reflect the uncertainties in the neutron-matter stiffness around normal density. For our three EDFs, J was set to 30 MeV, a value which was subsequently shown to be optimum for fitting both nuclear masses and realistic neutron-matter EoSs [15]. The values of L corresponding to the EDFs BSk19, BSk20, and BSk21 are respectively 31.9 MeV, 37.4 MeV and 46.6 MeV [2]. Comparing our EoSs with astrophysical observations thus suggests that the slope L of the symmetry energy is most likely to have values around $L \sim 37$ MeV. This constraint should be considered as indicative given our assumptions about NS matter, as well as the uncertainties in the analysis of observations.

Various other constraints on J and L obtained from the analysis of different experiments [13, 14] have been summarised in Fig. 2. It should be pointed out that this required the use of some models so that the uncertainties may be actually larger than those shown. In addition, the constraint may change if high-order terms in the expansion of $e(n,\eta)$ are considered. The figure shows the constraint deduced by Tsang et al. [16] from heavy-ion collisions (HIC), that derived by Chen et al. [17] from measurement of the neutron skin thickness in tin isotopes, and that obtained from the analysis of giant dipole resonance (GDR) [18, 14]. For comparison, the constraint obtained from our HFB nuclear mass models BSk9-26 as well as unpublished ones are also shown. Values of J and L have also been extracted from pygmy dipole resonances [19] and isobaric analogue states [?]. However, we have not included them in the figure because of large experimental and theoretical uncertainties (see, e.g., Refs. [21, 22]). The comprehensive analysis of symmetry energy constraints from Tsang et al. [13] includes data from isobaric analogue states and pygmy dipole resonances. Possibly for this reason, their most preferred values for J and L are higher than those found in Ref. [14], namely $J \sim 32.5$ MeV and $L \sim 70$ MeV.

4. Conclusions

We have found that fitting both nuclear masses and realistic neutron-matter EoSs using generalized Skyrme functionals leads to an optimum value of the symmetry energy coefficient around $J = 30$ MeV. Using unified EoSs of dense matter based on such functionals, we have computed the structure of NSs. Comparing our predictions with astrophysical observations, we have found that the most favoured value for the slope of the symmetry energy is $L \sim 37$ MeV, consistent with other estimates from nuclear physics.

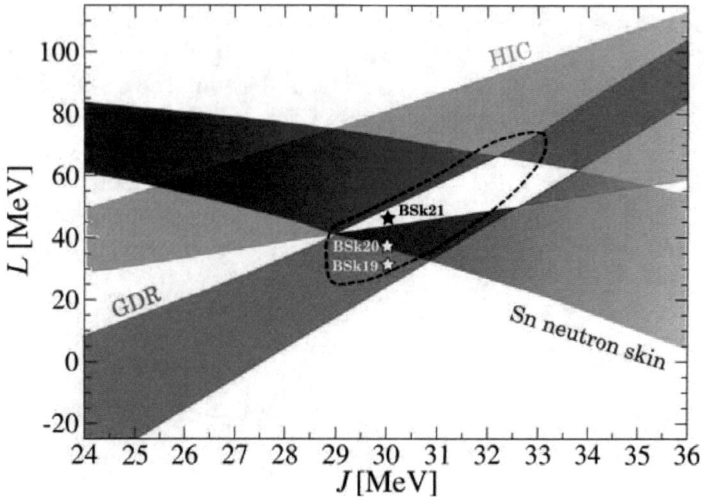

Figure 2. Experimental constraints on the symmetry energy parameters (see text for details), taken from [14]. The dotted line represents the constraint obtained from fitting experimental nuclear masses using the Brussels-Montreal HFB models with a root-mean-square deviation below 0.84 MeV (best fits are for $J = 30$ MeV).

Acknowledgments
The financial support of F.R.S.-FNRS (Belgium) and the NSERC (Canada) is gratefully acknowledged.

5. References

[1] Haensel P, Potekhin A Y and Yakovlev D G 2007 *Neutron Stars 1: Equation of State and Structure* (New York: Springer)
[2] Goriely S, Chamel N and Pearson J M 2010 *Phys. Rev. C* **82** 035804
[3] Pearson J M, Goriely S and Chamel N 2011 *Phys. Rev. C* **83** 065810
[4] Pearson J M, Chamel N, Goriely S and Ducoin C 2012 *Phys. Rev. C* **85** 065803
[5] Chamel N, Pavlov R L, Mihailov L M, et al. 2012 *Phys. Rev. C* **86** 055804
[6] Chamel N, Goriely S and Pearson J M 2009 *Phys. Rev. C* **80** 065804
[7] Chamel N 2010 *Phys. Rev. C* **82** 061307
[8] Audi G, Wapstra A H and Thibault C 2003 *Nucl. Phys. A* **729** 337
[9] Chamel N, Fantina A F, Pearson J M and Goriely S 2011 *Phys. Rev. C* **84** 062802(R)
[10] Antoniadis J, Freire P C C, Wex N, et al. 2013, *Science* **340** 1233232
[11] Steiner A W, Lattimer J M and Brown E F 2010 *Astrophys. J.* **722** 33
[12] Chamel N, Fantina A F, Pearson J M and Goriely S 2013 *Astron. Astrophys.* **553** A22
[13] Tsang M B, Stone J R, Camera F, et al. 2012 *Phys. Rev. C* **86** 015803
[14] Lattimer J M and Lim Y 2013 *Astrophys. J.* **771** 51
[15] Goriely S, Chamel N and Pearson J M 2013 *Phys. Rev. C* **88** 024308
[16] Tsang M B, Zhang Y, Danielewicz P, et al. 2009 *Phys. Rev. Lett.* **102** 122701
[17] Chen L-W, Ko C M, Li B-A and Xu J 2010 *Phys. Rev. C* **82** 024321
[18] Trippa L, Colò G and Vigezzi E 2008 *Phys. Rev. C* **77** 061304
[19] Carbone A, Colò G, Bracco A, et al. 2010 *Phys. Rev. C* **81** 041301
[20] Danielewicz P and Lee J 2009 *International Journal of Modern Physics E* **18** 892
[21] Reinhard P G and Nazarewicz W 2010 *Phys. Rev. C* **81** 051303
[22] Daoutidis I and Goriely S 2011 *Phys. Rev. C* **84** 027301

Neutrino mean free path in neutron matter with Brussels-Montreal Skyrme functionals

A Pastore[1], M Martini[1], D Davesne[2], J Navarro[3], N Chamel[1] and S Goriely[1]

[1] IAA, CP 226, Université Libre de Bruxelles, B-1050 Brussels, Belgium
[2] Université Lyon 1, 43 Bd. du 11 Novembre 1918, F-69622 Villeurbanne cedex, France
CNRS-IN2P3, UMR 5822, Institut de Physique Nucléaire de Lyon
[3] IFIC (CSIC-Universidad de Valencia), Apdo. 22085, E-46.071-Valencia, Spain

Abstract. We calculate the neutrino mean free path in cold neutron matter with some modern Brussels-Montreal functionals. The three typical functionals used in this article produce quite different results implying a possible impact on the cooling mechanism of neutron stars.

1. Introduction

The recent Brussels-Montreal functionals [1] have been very successful in describing both ground state properties of finite nuclei and some astrophysical quantities as the equation of state [2]. These functionals fitted to essentially all the 2149 experimentally known masses are similar to the standard non-relativistic Skyrme effective interaction [3] excepting for the presence of two extra density- and momentum-dependent terms [4]. The latter have been introduced to allow for more flexibilities in reproducing at the same time the properties of nuclei and infinite nuclear matter, including the quality of the functionals against ferromagnetic instabilities in neutron matter [4]. In that respect to avoid such instabilities, the functionals BSk19, BSk20, BSk21 have been fitted to satisfy the Landau inequalities up to very high densities. Furthermore, it has been shown [2] that it was possible to use them for both finite nuclei and astrophysical applications [5, 6].

The neutrino emission is one of the most efficient mechanism to dissipate the energy accumulated during the formation of a neutron star [7]. Since the neutrinos cross different layers of nuclear matter and are scattered by neutrons, they lose energy, making the cooling mechanism less efficient. The neutrino mean free path (NMFP) is a fundamental ingredient for a proper understanding of the cooling process. In the present article, we investigate the NMFP with the 3 above-mentioned functionals.

The article is organized as follows: in Sec.2 we present the linear response formalism necessary for the determination of the response function. Then we show explicitly the influence of the response function on the NMFP in Sec.3 and finally draw our conclusions in Sec.4.

2. Linear response theory

In the following we briefly recall the formalism of the linear response (LR) theory in the case of pure neutron matter at zero temperature. For simplicity we adopt a system of natural units

$\hbar = c = 1$. In order to describe the response of a nuclear system to an external probe $\hat{Q}$, one usually introduces the so-called response function $\chi^{(S,M)}(\omega, \mathbf{q})$ per unit volume defined by :

$$\chi^{(S,M)}(\omega, \mathbf{q}) = \frac{1}{V} \sum_n |\langle n|\hat{Q}^{(S,M)}|0\rangle|^2 \left(\frac{1}{\omega - E_n + i\eta} + \frac{1}{-\omega - E_n + i\eta} \right), \quad (1)$$

where S (M) is the spin (its projection along the z-axis). The variables ω and $\mathbf{q}$ represent the transferred energy and momentum, respectively. We refer to [8, 9, 10, 11] for a more detailed discussion on the LR theory with Skyrme functionals.

The BSk interactions differ from standard Skyrme [3] for the presence of some additional terms, v_{extra}, that are :

$$v_{extra} = \frac{1}{2} t_{1b}(1 + x_{1b}P_\sigma) \left(\mathbf{k}'^2 \delta(\mathbf{r}) \rho(r)^{\beta_1} + \rho(r)^{\beta_1} \delta(\mathbf{r}) \mathbf{k}^2 \right) + t_{2b}(1 + x_{2b}P_\sigma) \mathbf{k}' \rho(r)^{\beta_2} \delta(\mathbf{r}) \mathbf{k}. \quad (2)$$

These terms are the analogous of the terms proportional to t_1, x_1 and t_2, x_2 and thus they follow the same conventions concerning the momenta operators $\mathbf{k}, \mathbf{k}'$. Because of these terms, the expressions for the response functions given in ref. [11] have to be modified. The particular case $\beta_1 = \beta_2 = 1$ was considered in ref.[8]; the complete derivation will be presented in a forthcoming publication.

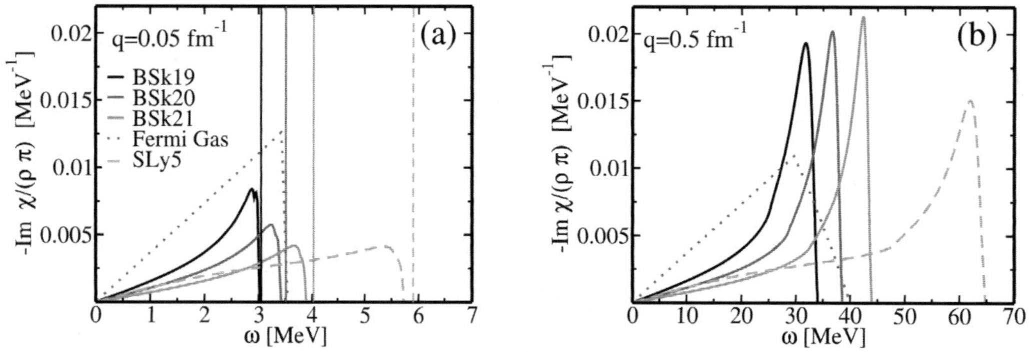

Figure 1. (Color online) Response function in the channel $S = 1, M = \pm 1$ calculated at $\rho = 0.16$ fm^{-3} for two different values of the transferred momentum and for the different functionals adopted in the present article. As a reference, the dotted line represents the response function of the unperturbed Fermi Gas.

In Fig.1, we show the response function at $\rho = 0.16$ fm^{-3} in the $S = 1, M = \pm 1$ channel for the three BSk functionals together with the functional SLy5 [3], commonly used for astrophysical applications, (given here as a reference) , and the Fermi Gas (FG) response function, calculated with the bare mass. We observe that the main effect of the interaction is related to the effective mass, and to a lesser extent to the incompressibility. In the two panels, we compare the case of two different values of the transferred momentum ($q = 0.05$ and 0.5 fm^{-1}). We observe, for the low exchanged momentum of panel (a), the appearance of a collective mode with zero width, the so-called zero-sound mode [12], which strongly suppresses the continuum part of the response function. At a larger transferred momentum (panel (b)), the zero sound mode disappears while the continuum is enhanced.

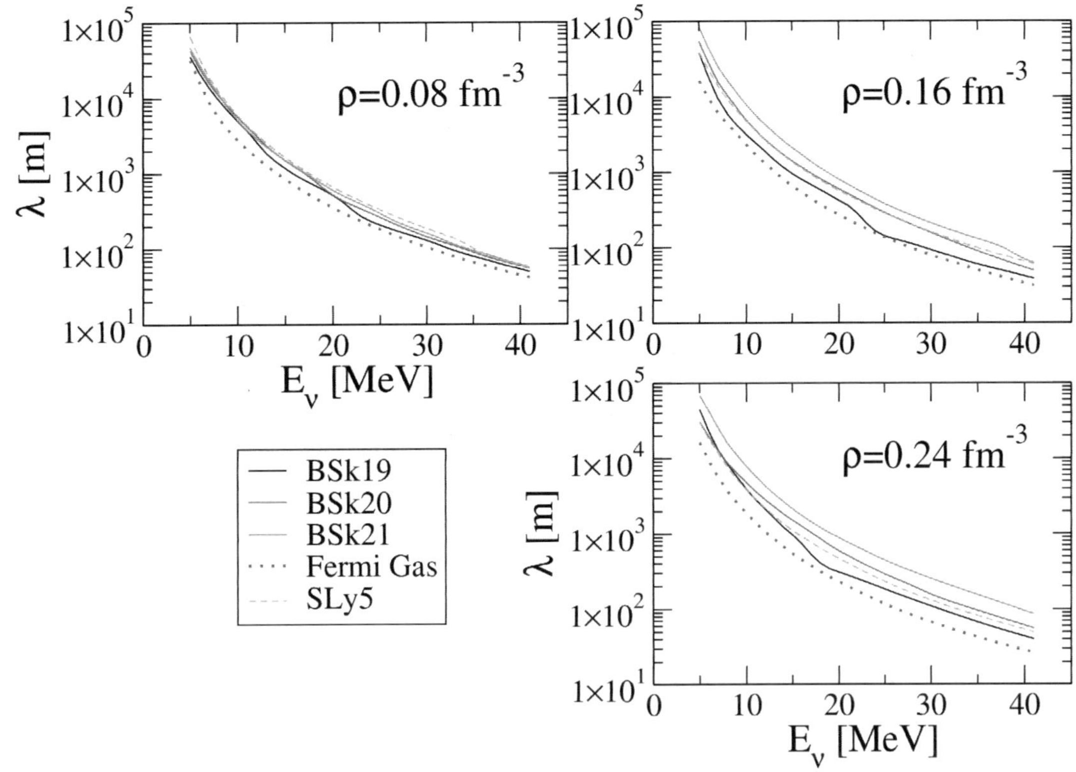

Figure 2. (Color online) Neutrino mean free path calculated for three different densities and for different functionals (see text for details).

3. Neutrino mean free path

The neutrino mean free path is defined as $\lambda = (\sigma\rho)^{-1}$, where σ is the total cross-section for the neutral current reaction $\nu + n \rightarrow \nu' + n'$. The differential cross-section, whose explicit expression is given in Eq.(21) of ref.[11], strongly depends on the response functions of all the different channels (S,M). However, as already illustrated in refs. [11, 13, 14], the cross section is dominated by the spin transverse response $R_T^A \equiv -Im\chi^{(S=1,M=\pm1)}(\omega, \mathbf{q})/(\pi\rho)$.

In Fig.2 we give an example of NMFP for the four functionals adopted here and for three different values of the density that are relevant for astrophysical calculations. We observe that at low density, the four functionals give a comparable value of λ. The differences among them actually start to emerge around the saturation density $\rho = 0.16$ fm^{-3}. In particular, it can be seen that BSk19 always gives the lowest value of λ and BSk21 the highest. These values are always larger that those obtained with the FG, *i.e.* without any residual interaction, thus showing explicitly that the nuclear medium always enhances the NMFP. This is however not a general result, as one can observe in Fig. 9 of ref. [11].

We also observe that SLy5 and BSk20 give very similar results concerning λ although the peaks of their response functions are located at quite different energies (Fig.1). However, when going from the differential cross section to the integrated one, kinematics reduce the integration range to [7] : $0 < \omega < E_\nu$ and $\omega < \mathbf{q} < |2E_\nu - \omega|$. As a consequence, the different response functions have to be compared only in this restricted range. In practice, since the maximum allowed value for the neutrino energy is $E_\nu = 40$ MeV, this corresponds to the low energy domain where SLy5 and BSk20 give actually similar results. Another important feature of Fig.2

is the presence of sudden drops in the NMFP as illustrated for BSk19 at $\rho = 0.16$ fm^{-3} and for $E_\nu \approx 22$ MeV. This rapid change occurs precisely for the values of ρ where the zero sound mode disappears and the response function suddenly increases. Such a phenomenon has been already studied in great details in ref.[15].

4. Conclusions

The calculation of the response functions within the linear response formalism has been extended in order to include the extra terms of the BSk19-21 functionals for the case of pure neutron matter at zero temperature. The resulting response functions have been used to calculate the neutrino mean free path. The preliminary results presented here exhibit some differences due to the parameterization itself (effective mass) and to general characteristic (zero sound). To obtain a more realistic result, it is however mandatory to include the temperature effects. In that case, the maximum transfer momentum between neutrino and nucleon can increase and one expects that it can eventually enter in regions with finite-size instabilities [16, 17, 18, 19]. Work along these lines is in progress.

Acknowledgments

This work was supported by NESQ project (ANR-BLANC 0407, France) and Mineco (Spain), grant FIS2011-28617-C02-02. N.C. and S.G. are F.R.S.-FNRS research associates and acknowledges the financial support of the "Actions de recherche concertées (ARC)" from the "Communauté française de Belgique".

References

[1] Goriely S, Chamel N and Pearson J M 2010 *Phys. Rev.* C **82** 035804
[2] Chamel N, Fantina A F, Pearson J M and Goriely S 2011 *Phys. Rev.* C **84** 062802
[3] Chabanat E, Bonche P, Haensel P, Meyer J and Schaeffer R 1997 *Nucl. Phys.* A **627** 710
[4] Chamel N and Goriely S 2009 *Phys. Rev.* C **82** 065804
[5] Pearson J M, Goriely S and Chamel N 2011 *Phys. Rev.* C **83** 065810
[6] Pearson J M, Chamel N, Goriely S and Ducoin C 2012 *Phys. Rev.* C **85** 065803.
[7] Iwamoto N and Pethick C J 1982 *Phys. Rev.* D **25**, 313
[8] Garcia-Recio C, Navarro J, Van Giai N and Salcedo L L 1992, *Ann. Phys.* (NY) **214** 293
[9] Davesne D, Martini M, Bennaceur K and Meyer J 2009 *Phys. Rev.* C **80**, 024314; 2011 *Phys. Rev. C* **84** 059904(E)
[10] Pastore A, Davesne D, Lallouet Y, Martini M, Bennaceur K and Meyer J 2012 *Phys. Rev.* C **85** 054317
[11] Pastore A, Martini M, Buridon V, Davesne D, Bennaceur K and Meyer J 2012 *Phys. Rev.* C **86** 044308
[12] Fetter A L and Walecka J D 1971 *Quantum Theory of Many-Particle Systems* (McGraw-Hill, New York)
[13] Martini M, Ericson M, Chanfray G and Marteau J 2009 *Phys. Rev.* C **80** 065501
[14] Martini M, Ericson M, Chanfray G and Marteau J 2010 *Phys. Rev.* C **81** 045502
[15] Navarro J, Hernández E S, and Vautherin D 1999 *Phys. Rev.* C **60** 045801
[16] Pastore A, Bennaceur K, Davesne D and Meyer J 2012 *Int. J. Mod. Phys.* E **21** 1250040
[17] Pastore A, Davesne D, Bennaceur K, Meyer J and Hellemans V 2013 *Phys. Scr.* **2013** 014014
[18] Lesinski T, Bennaceur K, Duguet T and Meyer J 2006 *Phys. Rev.* C **74** 044315
[19] Hellemans V, Pastore A, Duguet T, Bennaceur K, Davesne D, Meyer J, Bender M and Heenen P-H 2013 *Phys. Rev.* C **88** 064323

Finite size effects in hadron-quark phase transition by the Dyson-Schwinger method

N Yasutake[1], **H Chen**[2], **T Maruyama**[3], **T Tatsumi**[4]

[1] Department of Physics, Chiba Institute of Technology, 2-1-1 Shibazono, Narashino, Chiba 275-0023, Japan
[2] School of Mathematics and Physics, China University of Geoscience, Wuhan 430074, China
[3] Advanced Science Research Center, Japan Atomic Energy Agency, Tokai, Ibaraki 319-1195, Japan
[4] Department of Physics, Kyoto University, Kyoto 606-8502, Japan

E-mail: nobutoshi.yasutake@p.chibakoudai.jp

Abstract. We study the hadron-quark phase transition, taking into account the finite-size effects for neutron star matter. For the hadron phase, we adopt a realistic equation of state within the framework of the Brueckner-Hartree-Fock theory. For the quark phase, we apply the Dyson-Schwinger method. The properties of the mixed phase are clarified by considering the finite-size effects. We find that, if the surface tension is strong enough, the equation of state becomes to be close the one with the Maxwell condition, though we properly adopt the Gibbs conditions. This result is qualitatively the same with the one by the use of the simple bag model. We also find that the mass-radius relation by the EoS is consistent with the observations of massive neutron stars.

1. Introduction

The equation of sate (EoS) is one of the most important ingredient for the study of compact stars. It is also related to the QCD phase diagram, which has a large uncertainty. The recent observation of Ref. [1] strongly restricts the phenomenological EoS; the EoS must explain $\sim 2M_{\odot}$. On the other hand, the statistical approach from many observations [2] gives us prospect of fixing of the EoS. Theoretically, the study on the baryon-baryon interaction has been recently a hot topic from lattice QCD (LQCD) simulations [3]. Experiments such as JPARC will also provide useful information about the baryon-baryon interaction in the future. Hence, we should not neglect their impacts on the EoS: the EoS in the hadron phase may be provided by using the baryon-baryon interactions thus obtained in near future. In this paper we use the result by the Brueckner-Hartree-Fock (BHF) theory for hadron matter, which is based on the baryon-baryon interactions [4, 5].

For the quark phase, there are a lot of the effective models of QCD, such as the bag model, the (P)NJL model, and so on. However, most of them can not sufficiently describe the realistic features of QCD. For example, the mass functions of quarks should have the momentum dependence as the recent studies by LQCD show [6]. Here, we introduce the Dyson-Schwinger(DS) method which shows the momentum dependence on the mass function. This model is not based on the mean-field approximation. Instead, we need to calculate the quark propagators directly [7].

Journal of Physics: Conference Series **665** (2016) 012068 doi:10.1088/1742-6596/665/1/012068

The main purpose of this study is to know the property of the mixed phase between the quark phase and the hadron phase considering *the finite-size effects.* Assuming the quark deconfinement transition to be of first order, we needs the careful treatment of the effects in the phase transition of the multi-component system. Generally the properties of the mixed phase should strongly depend on the electromagnetic interaction and surface tension, and these effects, sometimes called "*the finite-size effects*", lead to the non-uniform "Pasta" structures [8]. The charge screening is also important for their mechanical instability. In the previous papers these finite-size effects have been properly taken into account to elucidate the properties of the pasta structure and demonstrate the importance of the charge screening adopting a simple bag model [9, 10]. However, the adopted quark model, the bag model, is too simple, and we introduce the DS model for it as mentioned in the last paragraph. Note that the mass-radius relation by our EoS is consistent with the both observations, Ref. [1] and Ref. [2].

This article is organized as follows. In Sec. II, we outline our framework. In Sec. III, we present numerical results. Sec. IV is devoted to the discussion.

2. Equation of State and finite-size effects

2.1. Dyson-Schwinger method

For the deconfined quark phase, we adopt a model based on the DS equation of QCD, which provides a continuum approach to QCD that can simultaneously address both confinement and dynamical chiral symmetry breaking. It has been applied with success to hadron physics in vacuum, and to QCD at nonzero chemical potential and temperature. The details can be found in Ref.[7]. Here, we just show brief points for this article.

Unfortunately, it is not possible to solve the DS equations precisely, and we introduce some approximations; e.g. for the quark-gluon vertex, we use the "rainbow approximation"[11]. For the gluon propagator in the vacuum, we assume a Gaussian-type effective interaction $\mathcal{G}(k^2;\mu)$[12] in the Landau gauge

$$\mathcal{G}(k^2;\mu) = \frac{4\pi^2 D}{\omega^6} e^{-\alpha\mu^2/\omega^2} k^2 e^{-k^2/\omega^2} , \tag{1}$$

with the parameter α we introduced to quantify the asymptotic freedom at large chemical potential μ. At the extreme limit of $\alpha \approx \infty$, the model is reduced to a free quark system at finite chemical potential. Here, we take two typical values $\alpha = 1$ and $\alpha = 2$ in our calculation as in Ref. [7], and choose the other set of parameters $\omega = 0.5$ GeV, $D = 1$ GeV^2 [12]. In the following we will study the parameter dependence of our results, marked as "DSα". We consider three quark flavors $q = u, d, s$, which are independent of each other in our model. We take the current-quark masses $m_{u,d} = 0$ for simplicity and $m_s = 115$ MeV [12]. From the quark propagator, one can calculate the quark number density and correspondingly obtain the physical quantity such as pressure [7].

2.2. Brueckner-Hartree-Fock theory

Our theoretical framework for the hadron phase of matter is the nonrelativistic BHF approximation based on the microscopic nucleon-nucleon potentials. The BHF calculation is a reliable and well-controlled theoretical approach for the study of dense baryonic matter. Detailed procedure can be found in Refs [4, 5]. In this study, we adopt the result with one of the Bonn potentials [13] for the nucleon-nucleon interacti, which is called Bonn-B (BOB) potential. Here we do not consider the degree of freedom of hyperons, since they are reduced by the existence of quarks as we suggested [9]. Of course we should show the behavior including hyperons, and we leave it in near future.

2.3. Mixed phase and finite-size effects

To take into account the finite-size effects, we impose the Gibbs conditions on the mixed phase, which require the pressure balance and the equality of the chemical potentials between two phases besides the thermal equilibrium. We employ the Wigner-Seitz approximation in which the whole space is divided into equivalent cells with given geometrical symmetry, specified by the dimensionality $d = 3$ (droplet or bubble), $d = 2$ (rod or tube), or $d = 1$ (slab). The structures of tube and bubble are characterized by the reversed particle distributions of rod and droplet. A sharp boundary is assumed between two phases and the surface energy is taken into account in terms of a surface-tension parameter σ. The surface tension of the hadron-quark interface is poorly known, but some theoretical estimates (e.g. lattice gauge simulations at finite temperature [14]) suggest a range of $\sigma \approx$ 10–100 MeV fm^{-2}. We show results using $\sigma = 40$ MeV fm^{-2} and $\sigma = 80$ MeV fm^{-2} in this article. The general discussions about the mixed phase in quark-hadron phase transition are given in our recent review [15].

3. Results

In Figure 1, we show the EoS based on the above descriptions with "DS1" as one example. The left panel show the case with $\sigma = 40$ MeV fm^{-2}, and the right one is for $\sigma = 80$ MeVfm^{-2}. Dashed lines show the pure hadron matter, and the solid lines do the pure quark matter. The EoS of the mixed phase becomes similar to the one under the bulk Gibbs calculation for weak surface tension, and to the one given by the Maxwell construction for strong surface tension. This behavior is universal, so that this result is qualitatively the same with the one with simple bag model [9].

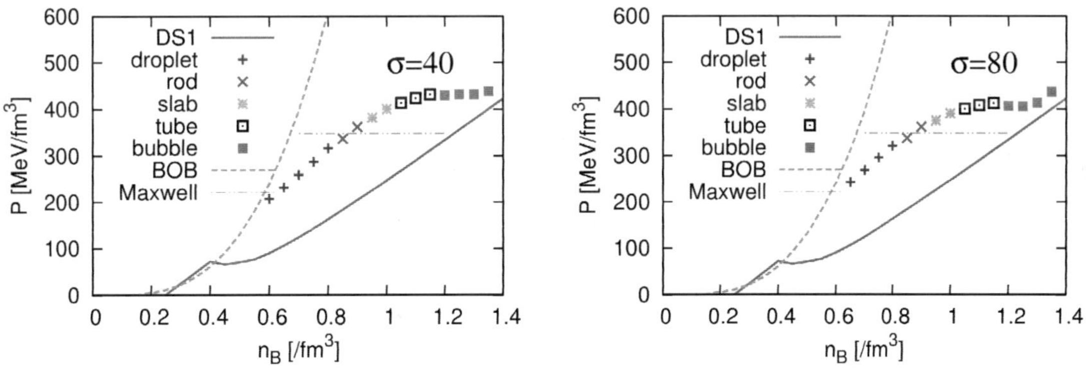

Figure 1. EoS with the quark model " DS1", where the energy of the surface tension σ is $\sigma = 40$ MeV fm^{-2} for the left panel, and $\sigma = 80$ MeV fm^{-2} for the right one. Dashed line shows the one for the hadron EoS, which is labeled with "BOB". Solid curves show the quark model "DS1". Each dot shows the mixed phase with the non-uniform structures. Narrow dashed-solid lines show the phase transition under the Maxwell construction.

At the last part of this section, we briefly discuss some implications of our results on the mass-radius relation of neutron stars shown in Fig. 2, which is obtained by solving the Tolman-Oppenheimer-Volkoff equation. For the crust, we use the BPS EoS which is widely used [16]. Clearly we can see that the maximum masses for all our models are slightly larger than the recent observational data by Demorest et al. [1]. Moreover, most of neutron stars by our models exist around $R \sim 12$ km. They are consistent with the statistical approach based on the observations by Steiner et al.[2], though this constraint is not so strong compared with the one by Demorest et al. [1].

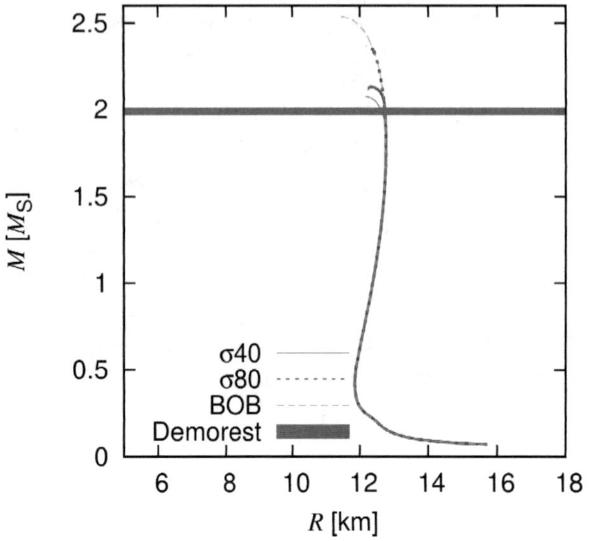

Figure 2. Mass-radius relations for EoS's shown in Figure 1. The solid lines (short-dashed lines) show those with the mixed phase, where the surface tension is $\sigma = 40$ (80) MeV fm^{-2}. Dashed line shows the one for hadron EoS, which is labeled with "BOB". Thick curves show the quark model "DS1", and narrow curves "DS2". Here, we also show the observational result by Demorest et al.; $M = (1.97 \pm 0.04)\ M_\odot$ [1].

4. Discussion

We have studied the hadron-quark phase transition with the finite-size effects for neutron star matter. For the hadron phase, we adopt a realistic equation of state in the framework of the BHF theory, we do the DS method for the quark phase. We find that, if the surface tension is strong, the EoS becomes to be close the one with the Maxwell condition. We also find that the mass-radius relation by the EoS is consistent with the observations of massive neutron stars.

Acknowledgments

NY is grateful to H. J. Schulze, F. Burgio, M. Baldo, and D. Blaschke for their warm hospitality and fruitful discussions. This work was partially supported by the Grant-in-Aid for the Global COE Program "The Next Generation of Physics, Spun from Universality and Emergence" from the Ministry of Education, Culture, Sports, Science and Technology (MEXT) of Japan and the Grant-in-Aid for Scientific Re-search (C) (25105510, 23540325).

References

[1] Demorest P.B., Pennucci T., Ransom S.M., Roberts M.S.E., Hessels J.W.T., 2010 Nature **467**, 1081.
[2] Steiner A.W., Lattimer J.M., Brown E.F., 2010 Astrophys. J. **722**, 33.
[3] Ishii N., Aoki S., Hatsuda T., 2007 Phys. Rev. Lett. **90**, 022001.
[4] Baldo M., 1999 *Nuclear Methods and the Nuclear Equation of State*, World Scientific, Singapore.
[5] Schulze H.-J. et al.,1995 Phys. Lett. B **355**, 21.
[6] Parappilly M.B.,Bowman P.O., Heller U.M., Leinweber D.B., Williams A.G, Zhang J.B., 2006 Phys. Rev. D **73**, 054504.
[7] Chen H., Baldo M., Burgio G.F., Schulze H.-J. 2011 Phys. Rev. D **84**, 105023, and references therein.
[8] Voskresensky D. N., Yasuhira M., Tatsumi T., 2002 Phys. Lett. B **541**, 93.
[9] Maruyama T., Chiba S., Schulze H.-J., Tatsumi T., 2007 Phys. Rev. D **76**, 123015.
[10] Yasutake N., Toshiki M., Tatsumi T., 2009 Phys. Rev. D **80**, 123009.
[11] Bender A., Roberts C.D., Smekal L. von, 1996 Phys. Lett. **B380**, 7.
[12] Alkofer R., Watson P., Weigel H., 2002 Phys. Rev. D **65**, 094026.
[13] Machleidt R., Holinde K., Elster Ch.,1987 Phys. Rep. **149**, 1; Machleidt R. 1989 Adv. Nucl. Phys. **19**, 189.
[14] Kajantie K., Kärkäinen L., Rummukainen K., 1991 Nucl. Phys. B357, 693.
[15] Yasutake N., Noda T., Sotani H., Maruyama T., Tatsumi T., 2013 *Recent Advances in Quarks Research*, Nova, Chap.4, pp.63, ISBN 9781622579709.
[16] Baym G., Pethick C., Sutherland P., 1971, Astrophys. J. **170**, 299.

Journal of Physics: Conference Series **665** (2016) 012069 doi:10.1088/1742-6596/665/1/012069

Early protoneutron star deleptonization - consistent modeling of weak processes and equation of state

T Fischer[1], L Huther[2], A Lohs[2] and G Martínez-Pinedo[2,3]

[1] Institute for Theoretical Physics, University of Wroclaw, pl. M. Borna 9, 50-204 Wroclaw, Poland

[2] Technische Universität Darmstadt, Schlossgartenstraße 2, 64289 Darmstadt, Germany

[3] GSI Helmholtzzentrum für Schwerionenforschung GmbH, Planckstraße 1, 64291 Darmstadt, Germany

E-mail: `fischer@ift.uni.wroc.pl`

Abstract. We discuss the impact of consistent modeling of weak processes and nuclear equation of state (EOS) during the deleptonization phase of protoneutron stars (PNS). They are born in the event of a core-collapse supernova. Being initially hot and lepton rich, PNSs deleptonize via the continuous emission of neutrinos of all flavors. Unlike during the mass accretion phase prior to the onset of the supernova explosion, when the neutrinospheres are located at low densities, they shift to significantly higher densities during the PNS deleptonization phase after the supernova explosion has been launched. Important during this phase is the inclusion of medium modifications to the weak processes, which depend on the nuclear EOS. In particular, these medium modifications depend on the nuclear symmetry energy and its density dependence. We discuss two selected nuclear models and illustrate the early PNS deleptonization phase for a particular EOS. As has been realized recently, including weak processes consistently with the EOS increases the spectral differences of ν_e and $\bar{\nu}_e$ in comparison to simulations that neglect the underlying medium modifications. This has important consequences for the nucleosynthesis relevant conditions of the ejecta as well as for potential neutrino flavor oscillations and hence detection on Earth.

1. Introduction

Stars more massive than about 8 $M_\odot$ end their life's as core-collapse supernovae [1, 2]. They are triggered from the collapse of the stellar core due to electron captures and the photodissociation of heavy nuclei. The collapse continues until normal nuclear matter density is reached. The nucleon pressure from the short-range repulsive nuclear interaction halts the collapse and the supersonically collapsing core bounces back. A sound wave turns into a shock wave which then propagates out of the core. The object which forms at core bounce is the protoneutron star (PNS), being hot and lepton rich in which senses it differs form the final supernova remnant, the neutron star. The moment of core bounce is determined when the maximum central density is reached just before shock break out. The initially outwards propagating bounce shock finally halts due to energy losses from the dissociation of heavy nuclei, which fall onto the shock from the still gravitationally collapsing region above the core, and electron neutrino escapes. The latter are released in the deleptonization burst which is launched from a large number of electron captures on free protons when the bounce shock propagates across the neutrinospheres. The ν_e-luminosity reaches several times 10^{53} erg s^{-1} and lasts only about 10–20 ms after core bounce. These sources of energy loss turn the bounce shock into an accretion front which stalls at about 100–200 km. The revival of this shock wave, the so called supernova problem, is related to the liberation of energy from the central PNS into the layer above the PNS surface behind the standing shock. Several scenarios have been explored in the literature, the magneto-rotational [3], the acoustic [4], the high-density quark-hadron phase transition [5], and the neutrino heating [6]. The moment of explosion onset can be defined when the standing accretion shock starts to propagate continuously to increasingly larger radii and does not retreat back at any later times. It will expel the stellar mantel, ejecting manly α-nuclei up to the iron group.

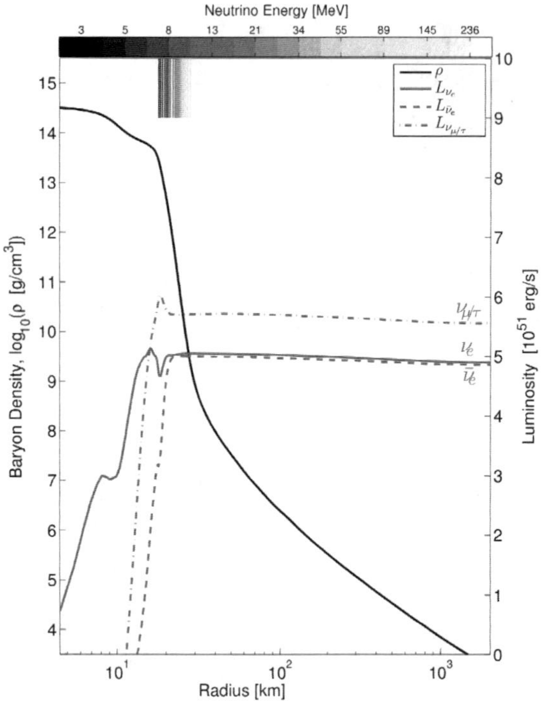

$$\nu_e + n \longrightarrow e^- + p \quad (1a)$$

$$\bar{\nu}_e + p \longrightarrow e^+ + n \quad (1b)$$

$$\nu + N \longrightarrow \nu' + N \quad (1c)$$

$$\nu + e^{\pm} \longrightarrow \nu' + e^{\pm} \quad (1d)$$

$$N + N \longrightarrow \nu + \bar{\nu} + N + N \quad (1e)$$

$$e^- + e^+ \longrightarrow \nu + \bar{\nu} \quad (1f)$$

$$\nu_e + \bar{\nu}_e \longrightarrow \nu_{\mu/\tau} + \bar{\nu}_{\mu/\tau} \quad (1g)$$

$$(N = \{n, p\})$$

$(\nu = \{\nu_e, \nu_{\mu/\tau}\}$ unless stated otherwise)

Figure 1. Left column: Radial profiles of density (solid black line) and luminosities (ν_e: solid red line, $\bar{\nu}_e$: dashed red line, $\nu_{\mu/\tau}$: dash-dotted line) at about 2 s after explosion onset. The region of neutrino decoupling is shown via the energy-dependent transport spheres (for ν_e), where the color-coding is according to the neutrino energies from E_{min} = 3 MeV (black) at about 16.17 km up to E_{max} = 300 MeV (white) at 49.11 km. The energy-averaged ν_e-sphere is shown in the vertical blue solid line. (color online) Right column: List of weak processes.

Independent from the shock-revival mechanism, a layer of net neutrino heating establishes above the PNS surface after the explosion has been launched. It drives a low-mass ($\sim 10^{-4}$ M$_\odot$) outflow, known as neutrino-driven wind. It has been explored in parametric [7, 8, 9, 10] as well as in dynamic studies [11, 12, 13, 14]. It has also long been investigated as nucleosynthesis site for the production of both, light p-nuclei and r-process nuclei [15, 16, 17]. One of the largest uncertainties of neutrino-driven winds is the proton-to-baryon ratio Y_e, which is determined by the weak processes (1a) and (1b) at the neutrinospheres. Addressing this question requires neutrino transport to accurately describe neutrino decoupling from matter at the PNS surface. It determines the neutrino spectra and luminosities and hence the nucleosynthesis relevant conditions in the wind at larger radii [18]. Note that process (1a) will turn matter to the proton-rich side, while process (1b) will turn matter to the neutron-rich side. Moreover, material at neutrino decoupling is extremely neutron rich yielding $Y_e \simeq 0.05 - 0.1$, as a consequence of the deleptonization burst form core bounce. In order to achieve neutron-rich conditions for the material being ejected from the PNS surface in the neutrino-driven wind, it has been demonstrated analytically that for similar ν_e and $\bar{\nu}_e$ luminosities, $\langle E_{\bar{\nu}_e}\rangle - \langle E_{\nu_e}\rangle \sim 5$ MeV [18]. In recent spherically symmetric simulations that are based on general relativistic radiation hydrodynamics and accurate three-flavor Boltztmann neutrino transport [19, 20], it has been found that $\langle E_{\bar{\nu}_e}\rangle - \langle E_{\nu_e}\rangle \sim 2 - 3$ MeV, decreasing towards later times during the ongoing PNS deleptonization. The simulations were carried out for more than 10 s after explosion onset during which generally $Y_e > 0.5$ was found.

In the following section, we will discuss this aspect and demonstrate the complexity of the situation. In particular, we will point to the importance of the consistent implementation of rates for weak processes and nuclear equation of state (EOS). This was not the case in the studies of refs. [19, 20], as has been pointed out very recently in refs. [21, 22].

2. Protoneutron star deleptonization and role of charged-current weak processes

A typical situation during the early PNS deleptonization phase is illustrated in the left panel of Fig. 1 at about 2 s after the explosion has been launched. Note the very steep density profile at the PNS surface, where neutrino decoupling takes place over 5 orders of magnitude in density, which even steepens during the continuous deleptonization. In this region (see the neutrino decoupling in the left panel of Fig. 1) neutrino transport is important to determine the neutrino spectra and luminosities. The magnitude of the luminosity and average energy differences depends on details of the neutrino transport and in particular on the weak processes and their implementation. In addition to the charged-current weak processes, other weak processes are important which are listed in the right panel of Fig. 1. These are neutral current elastic (exchange of momentum) scattering on nucleons (1c), inelastic (exchange of energy and momentum) scattering on electrons/positrons (1d), and pair processes (1e-1g).

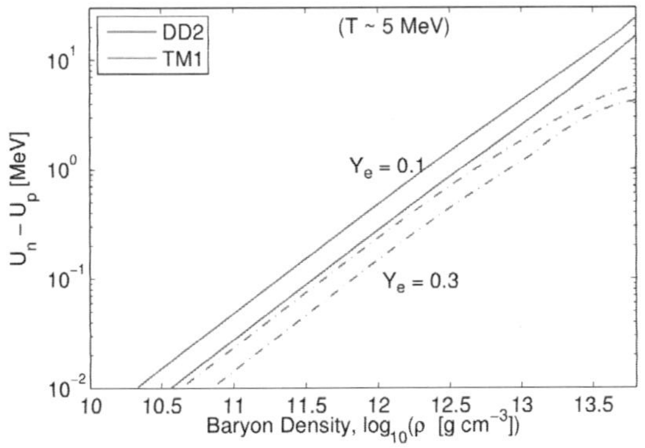

Figure 2. Density dependence of the mean-field potential difference for two selected nuclear interactions DD2 (red lines) and TM1 (blue lines), comparing different proton-to-baryon ratios $Y_e = 0.1$ (solid lines) and $Y_e = 0.3$ (dash-dotted lines) at fixed temperature of about 5 MeV. With increasing isospin asymmetry, i.e. lower Y_e, the differences between U_n and U_p become larger. (color online)

The rate for the charged current process (1a) can be expressed via momentum integrals of the particle occupation numbers, F_i, and the spin-averaged squared matrix element, M, in the zero-momentum transfer approximation ($\mathbf{p}_n \simeq \mathbf{p}_p$) [24, 25]:

$$1/\lambda(E_{\nu_e}) \simeq \frac{G_F^2}{\pi} V_{ud}^2 (g_V^2 + 3g_A^2) \cdot E_e^2 [1 - f_e(E_e)] \frac{n_n - n_p}{1 - e^{\beta(\mu_p^0 - \mu_n^0)}}, \tag{2}$$

with Fermi constant, G_F, vector and axial-vector coupling constants, g_V and g_A, and up-down entry of the Cabibbo-Kobayashi-Maskawa matrix V_{ud}. A similar expression is obtained

Journal of Physics: Conference Series **665** (2016) 012069 doi:10.1088/1742-6596/665/1/012069

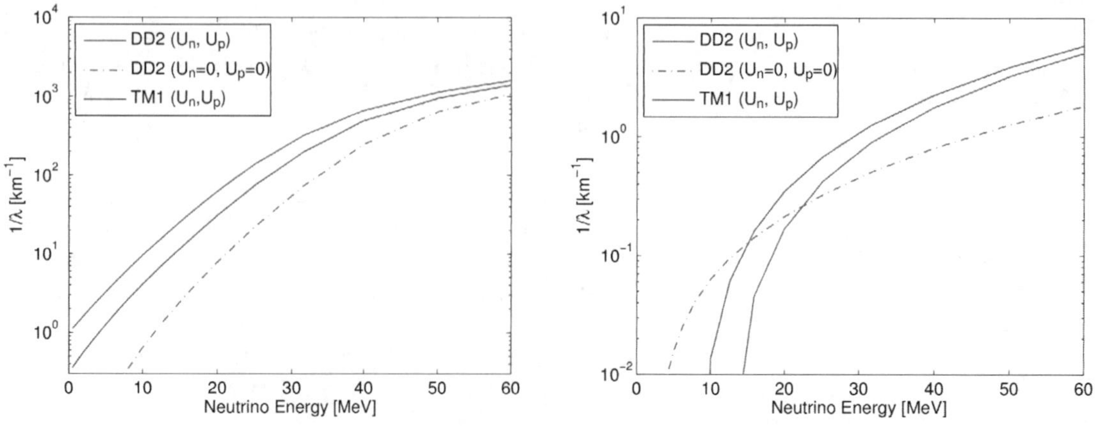

(a) ν_e (b) $\bar{\nu}_e$

Figure 3. Neutrino opacity (Eq. 2) for the two interactions DD2 (solid red lines) and TM1 (solid blue lines) at selected conditions [23], $T = 7.4$ MeV, $\rho = 2.1\times10^{13}$ g cm^{-3}, and $Y_e = 0.035$. It corresponds to $U_n - U_p = 10.5(6.5)$ MeV for DD2(TM1). In addition, the red dashed line shows the opacity for DD2 for which we assume zero potentials. (color online)

for reaction (1b). For the reverse process, i.e. the neutrino emissivity, detailed balance is applied, $j_{\nu_e}(E_{\nu_e}) = \exp\{-\beta(E_{\nu_e} - \mu_{\nu_e}^{\text{eq.}})\}\,1/\lambda(E_{\nu_e})$ with equilibrium chemical potential $\mu_{\nu_e}^{\text{eq.}} = \mu_e - (\mu_n - \mu_p)$ and inverse temperature $\beta = 1/T$. Expression (2) assumes the free Fermi gas of nucleons, i.e. in the limit of low degeneracy the nucleon density of targets, n_N, must be recovered. Hence, the nucleon chemical potentials, μ_N^0, that enter in (2) must be those of the free Fermi gas. In supernova codes, these quantities are typically provided by an EOS based on some nuclear interaction model. Nucleons are treated as quasi-particles that move in a potential, U_N, and obey the following dispersion relation,

$$E_N(\mathbf{p}_N) = \sqrt{\mathbf{p}_N^2 + {m_N^*}^2} + U_N, \qquad (N = \{n, p\}). \tag{3}$$

In the case of uniform matter the single-particle potentials are given by the vector self energy. They and the effective masses, m_N^*, depend on the independent variables density, temperature, and Y_e, such as the entire EOS. Consequently, the nuclear chemical potentials which enter Eq. (2) must be corrected accordingly in order to reproduce the non-interacting case, i.e. $\mu_N^0 = \mu_N - U_N - m_N^*$ where μ_N are the full potentials that contain contributions from interactions. Moreover, since neutrino transport uses charged-current reaction rates with respect to the incoming neutrino energy, E_{ν_e} (see Eq. 2), Eq. (3) can be used to relate electron(positron) energy, $E_{e^-}(E_{e^+})$, and electron (anti)neutrino energy for processes (1a) and (1b) as follows,

$$E_{\nu_e} = E_{e^-} - (m_n - m_p) - (U_n - U_p), \qquad E_{\bar{\nu}_e} = E_{e^+} + (m_n - m_p) + (U_n - U_p). \tag{4, 5}$$

From the expressions (4) and (5) it becomes clear that at low densities, where $U_n - U_p \simeq$ 100 keV (see Fig. 2), the energetics is determined by the vacuum Q-value, $Q_0 = m_n - m_p =$ 1.2935 MeV. The density dependence of the quantity $U_n - U_p$ is shown in Fig. 2 for selected Y_e and at fixed temperature of about 5 MeV. The figure also compares two nuclear interaction models, DD2 [26] and TM1 from the comprehensive supernova catalog of ref. [27] which is based on the statistical model including detailed nuclear composition consistently and have been approved in supernova simulations [28, 29].

The ν_e and $\bar{\nu}_e$-opacity are shown in Figs. 3(a) and 3(b), at selected conditions. From the figures one immediately realizes that neglecting (U_n, U_p) leads to a suppression of low-energy neutrino absorptions (compare the red solid and dash-dotted lines) for ν_e. The magnitude scales almost with $U_n - U_p$. This is attributed mainly to the modified vacuum Q-value and also partly to the chemical potential modifications. For $\bar{\nu}_e$ the opposite holds due to the different sign of the modified vacuum Q-value, including the mean-field potentials leads to a suppression of low $\bar{\nu}_e$ energies. In addition to DD2, Fig. 3 also shows the opacity for TM1 (blue lines), with lower mean-field potentials of about 4 MeV at the same conditions. Hence, also the opacity for ν_e is reduced accordingly, and increased for $\bar{\nu}_e$. Note that the opacity differences between DD2 and TM1 are not entirely due to the different (U_n, U_p) but also related in part to their different neutron and proton abundances.

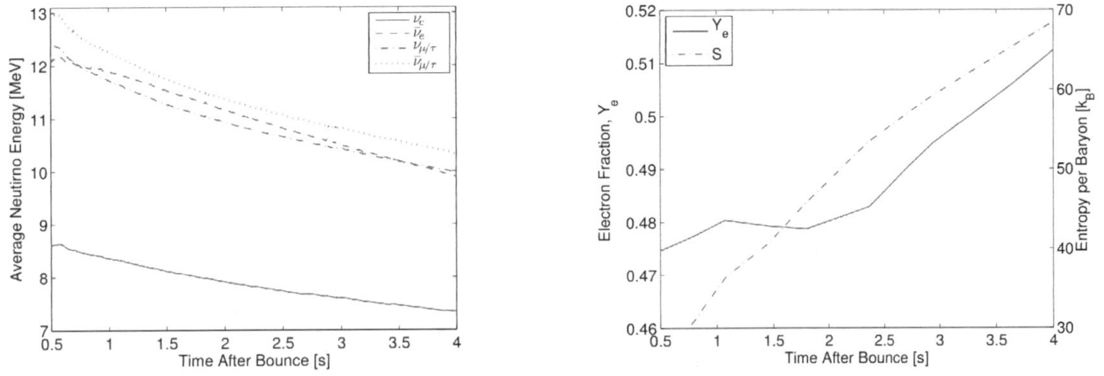

(a) Average neutrino energies.

(b) Electron fraction and entropy per baryon.

Figure 4. Evolution of neutrino energies, Y_e, and entropy per baryon (both obtained in large nucleosynthesis calculations) sampled at a distance of 1000 km. Neutron-rich conditions are obtained between 1.5–2.5 s when the differences between $\langle E_{\bar{\nu}_e} \rangle$ and $\langle E_{\nu_e} \rangle$ are largest.

The impact of the consistent implementation of charged-current weak rates and EOS in simulations of the PNS deleptonization has been shown in refs. [21, 22] for the neutrino spectra and luminosities, as well as the resulting nucleosynthesis relevant conditions. It was found that it leads to generally larger differences of the average energies and luminosities for ν_e and $\bar{\nu}_e$. Here we extend the analysis of ref. [21] and simulate a 11.2 $M_\odot$ progenitor applying our general relativistic radiation hydrodynamics code AGILE-*Boltztran*, which is based on three-flavor Boltzmann neutrino transport [30]. We model a neutrino-driven explosion in our spherically symmetric setup by enhancing the neutrino heating/cooling rates (for details, see ref. [19]). Fig. 4 shows the results obtained for the early PNS deleptonization phase where the charged-current weak rates are implemented consistently with the nuclear EOS DD2. During this early epoch, slightly neutron-rich conditions are found up to about 3 s after bounce. It allows for the production of light neutron-capture elements up to $Z \simeq 45$ but no strong r process. The rising Y_e at later times is related to the reducing opacity of the charged-current reactions, due to the increasing importance of final-state Pauli blocking and the increasing nucleon degeneracy [23]. Instead, elastic scattering on neutrons starts to dominates the opacity, which does not distinguish between different neutrino flavors and hence the spectral differences reduce.

3. Conclusions

We have discussed the consistent implementation of charged-current weak rates and nuclear EOS in radiation hydrodynamics codes and its relevance during the early PNS cooling phase.

The weak rates have now a direct dependence on the nuclear EOS. We have employed rates based on the zero-momentum transfer approximation. It is a valid simplification since inelastic contributions are small for the early deleptonization phase, during which the neutrinospheres are located at intermediate densities $\sim 10^{13}$ g cm^{-3}. Inelastic contributions, as well as nuclear correlations [31], may become of some relevance during the later evolution when the neutrino decoupling region moves to even higher densities ($\sim 10^{14}$ g cm^{-3}). Moreover, corrections due to weak magnetism, which also have to be consistent with the EOS, are not included. Long-term simulations up to 10 s are required to cover the entire PNS deleptonization phase and to study the late neutrino-driven wind, also with particular focus on the nucleosynthesis.

One of the currently largest uncertainties in such studies is the nuclear EOS, in particular at high densities, finite temperatures and low Y_e. Future explorations will also study the impact of different EOS on the PNS deleptonization. E.g., soft EOS that lead typically to a more compact PNS have also a lower symmetry energy. Since the medium modifications to the charged-current rates, $U_n - U_p$, depend sensitively on the symmetry energy also at intermediate densities, a direct impact on the PNS deleptonization can be expected.

TF is supported by the Narodowe Centrum Nauki (NCN) within the "Maestro" program under contract No. DEC-2011/02/A/ST2/00306. GMP, LH, and AL are partly supported by the Deutsche Forschungsgemeinschaft through contract SFB 634, the Helmholtz International Center for FAIR within the framework of the LOEWE program launched by the state of Hesse, the Helmholtz Association through the Nuclear Astrophysics Virtual Institute (VH-VI-417)

References

[1] Janka H T, Langanke K, Marek A, Martínez-Pinedo G and Müller B 2007 *Phys. Rep.* **442** 38–74
[2] Janka H T 2012 *Annual Review of Nuclear and Particle Science* **62** 407–451
[3] LeBlanc J M and Wilson J R 1970 *Astrophys.J.* **161** 541
[4] Burrows A, Livne E, Dessart L, Ott C and Murphy J 2006 *Astrophys.J.* **640** 878–890
[5] Sagert I, Fischer T, Hempel M, Pagliara G, Schaffner-Bielich J *et al.* 2009 *Phys.Rev.Lett.* **102** 081101
[6] Bethe H A and Wilson James R 1985 *Astrophys.J.* **295** 14–23
[7] Duncan R C, Shapiro S L and Wasserman I 1986 *Astrophys. J.* **309** 141–160
[8] Hoffman R, Woosley S and Qian Y 1997 *Astrophys.J.* **482** 951
[9] Thompson T A, Burrows A and Meyer B S 2001 *Astrophys.J.* **562** 887
[10] Thompson T A and Burrows A 2001 *Nucl.Phys.* **A688** 377–381
[11] Woosley S E and Baron E 1992 *Astrophys. J.* **391** 228–235
[12] Woosley S, Wilson J, Mathews G, Hoffman R and Meyer B 1994 *Astrophys.J.* **433** 229–246
[13] Takahashi K, Witti J and Janka H T 1994 *Astron.Astrophys.* **286** 857
[14] Witti J, Janka H T and Takahashi K 1994 *Astronom. Astrophys.* **286** 841–856
[15] Otsuki K, Tagoshi H, Kajino T and Wanajo S 1999 *Astrophys.J.* **533** 424
[16] Wanajo S 2006 *Astrophys.J.* **650** L79–L82
[17] Wanajo S 2006 *Astrophys.J.* **647** 1323–1340
[18] Qian Y and Woosley S 1996 *Astrophys.J.* **471** 331–351
[19] Fischer T, Whitehouse S, Mezzacappa A, Thielemann F K and Liebendörfer M 2010 *Astron.Astrophys.* **517** A80
[20] Hüdepohl L, Müller B, Janka H T, Marek A and Raffelt G G 2010 *Physical Review Letters* **104** 251101
[21] Martínez-Pinedo G, Fischer T, Lohs A and Huther L 2012 *Physical Review Letters* **109** 251104
[22] Roberts L F, Reddy S and Shen G 2012 *Phys. Rev. C* **86** 065803
[23] Fischer T, Martínez-Pinedo G, Hempel M and Liebendörfer M 2012 *Phys. Rev. D* **85** 083003
[24] Bruenn S W 1985 *Astrophys.J.Suppl.* **58** 771–841
[25] Reddy S, Prakash M and Lattimer J M 1998 *Phys. Rev. D* **58** 013009
[26] Typel S 2005 *Phys. Rev. C* **71** 064301
[27] Hempel M and Schaffner-Bielich J 2010 *Nucl.Phys.* **A837** 210–254
[28] Hempel M, Fischer T, Schaffner-Bielich J and Liebendörfer M 2012 *Astrophys. J.* **748** 70
[29] Steiner A W, Hempel M and Fischer T 2013 *Astrophys. J.* **774** 17
[30] Liebendörfer M, Messer O, Mezzacappa A, Bruenn S, Cardall C *et al.* 2004 *Astrophys.J.Suppl.* **150** 263–316
[31] Reddy S, Prakash M, Lattimer J M and Pons J A 1999 *Phys. Rev. C* **59** 2888–2918

Inelastic Neutrino Reactions with Light Nuclei and Standing Accretion Shock Instability in Core-Collapse Supernovae

S Furusawa[1,2], H Nagakura[3], K Sumiyoshi[4], S Yamada[1]

[1] Waseda University, 3-4-1, Okubo, Shinjuku, Tokyo 169-8555, Japan
[2] Center for Computational Astrophysics, National Astronomical Observatory of Japan, Mitaka, Tokyo, 181-8588, Japan
[3] Yukawa Institute for Theoretical Physics, Kyoto University, Oiwake-cho, Kitashirakawa, Sakyo-ku, Kyoto, 606-8502, Japan
[4] Numazu College of Technology, Ooka 3600, Numazu, Shizuoka 410-8501, Japan

E-mail: `furusawa@heap.phys.waseda.ac.jp`

Abstract. We perform numerical experiments to investigate the influence of inelastic neutrino reactions with light nuclei on the standing accretion shock instability. The time evolutions of shock waves are calculated with a simple light-bulb approximation for the neutrino transport and a multi-nuclei equation of state. The neutrino absorptions and inelastic interactions with deuterons, tritons, helions and alpha particles are taken into account in the hydrodynamical simulations in addition to the ordinary charged-current interactions with nucleons. Axial symmetry is assumed but no equatorial symmetry is imposed. We show that the heating rates of deuterons reach as high as $\sim 10\%$ of those of nucleons around the bottom of the gain region. On the other hands, alpha particles heat the matter near the shock wave, which is important when the shock wave expands and density and temperature of matter become low. It is also found that the models with heating by light nuclei have different evolutions from those without it in non-linear evolution phase. The matter in the gain region has various densities and temperatures and there appear regions that are locally rich in deuterons and alpha particles. These results indicate that the inelastic reactions of light nuclei, especially deuterons, should be incorporated in the simulations of core-collapse supernovae.

1. Introduction

The mechanism of core-collapse supernovae is not clearly understood at present because of its intricacy. The neutrino driven mechanism is considered to be the most promising scenario for the core-collapse supernova and it is considered the so-called standing accretion shock instability (SASI) and the convection in multi-dimension are helpful to increase the efficiency of neutrino heating. In addition to these hydrodynamical effects, there are some nuclear physical ingredients that are also supposed to be important for the core-collapse supernovae. The inelastic neutrino interactions and equation of states (EOS) of nuclear matter are also important.

The inelastic interactions between neutrinos and nuclei have been neglected in most hydrodynamical simulations of the neutrino heating phase after the bounce and shock stall. Ohnishi et al. showed that the inelastic neutrino interactions with alpha particles are helpful to revive the shock in 2D simulations if the neutrino luminosity is close to the critical value, which is the threshold for a shock revival [1]. However, the shocked matter is certainly composed not only of nucleons and alpha particles but also of deuterons, tritons and helions [2]. The energy transfer cross sections of deuterons are comparable to those of nucleons and ten times greater than those alpha particles [3].

EOS is another important input physics and its influences on the dynamics of core-collapse supernovae were investigated. There are currently the two EOS's widely used for the simulations of core-collapse supernovae, in which only a single representative nucleus and alpha particles are incorporated to approximate the ensemble of heavy and light nuclei, respectively. We have developed an EOS including a large number of nuclei based on the liquid drop model for heavy nuclei with shell effects and nuclear pasta phases the quantum approach for light nuclei and relativistic mean field theory for free nucleons [4, 5], This EOS can provide the mass fractions of each light nuclei more reliably.

The aim of this article is to investigate effects of these light nuclei in the new EOS on an aspect of supernova dynamics. We explore the impacts of the inelastic neutrino reactions with light nuclei on the SASI. We perform several hydrodynamical simulations of the post-bounce phase in 1D and 2D. In addition to cooling and heating by nucleons, we incorporate the neutrino heating reactions with light nuclei. This article is organized as follows. In section 2, we describe the important ingredients in numerical simulations such as the hydrodynamics code we employ in this study and inelastic reactions with light nuclei. Then the results are shown in section 3, with an emphasis being put on the role of light nuclei in the shock heating. The paper is wrapped up with a summary and in section 4.

2. Models

The basic set-up of dynamical simulations is the same as that of Ohnishi et al. and Nagakura et al. except for the heating reactions of light nuclei [1, 6, 7]. We perform 2D simulations assuming axial symmetry. Spherical coordinates are used and no equatorial symmetry is assumed. We utilize 300 radial mesh points to cover $r_{in} \leq r \leq r_{out} = 500$ km, where r_{in} is the inner boundary and is set to be the radius of the neutrino sphere of ν_e in the initial state. The 60 angular mesh points are adopted to cover the whole meridian section. The mass of a central object is assumed to be constant and set to be $1.4M_{\odot}$. Interactions between neutrinos and nucleons are taken into account in the energy equation and the evolution of electron fraction. The calculations of the contributions of nucleons to the energy and electron fraction are done just in the same way as in Ohnishi et al. [6]. The heating rates of deuterons, tritons, helions and alpha particles are also taken into account in the energy equation and they are the new elements in this work.

The heating rates of light nuclei are calculated from the analytic formula [8]. The energy transfer cross sections of deuterons, tritons, tritons alpha particles are available from some references [3, 8, 9, 10, 11]. The cooling reactions involving light nuclei are ignored since the reaction rates are not available at the moment and we focus only the influences of the heating reactions of light nuclei in this paper. The neutrino transport is solved by the simple light bulb approximation. We assume that the temperatures of ν_e, $\bar{\nu_e}$ and ν_μ emitted from the neutrino spheres are constant and set to be $(T_{\nu_e}, T_{\bar{\nu}_e}, T_{\nu_\mu}) = (4, 5, 10)$ in MeV. The luminosities of ν_e and $\bar{\nu}_e$ are assumed to have the same value: $L_{\nu_e} = L_{\bar{\nu}_e} = L$. The luminosity of ν_μ is set to be $L_{\nu_\mu} = 0.5 \times L$.We employ the multi-nuclei EOS, which gives not only thermodynamical quantities but also the abundance of all nuclei [4, 5].

For the first step of the calculations, we prepare the initial conditions of the stars, which are spherical symmetric steady accretion flows [6, 7]. The neutrino inelastic interactions with light nuclei are included in these calculations for the steady states. We start the dynamical simulations, adding the perturbations of 1 %, which are proportional to $\cos\theta$, to the initial radial velocities. We investigate the influences of light nuclei on the dynamics under different circumstances, varying the luminosity L and mass accretions rate $\dot{M}$. We define the normalized neutrino luminosity $L_{52} \equiv L/(10^{52}\mathrm{erg/s}^{-1})$ and mass accretion rate $\dot{M}_{sun} \equiv -\dot{M}/(M_{\odot}\mathrm{s}^{-1})$. The detail of the model is described in Furusawa et al. [12].

3. Results

3.1. 1D simulation

To obtain the basic features of the heating by light nuclei in the dynamical settings, we perform a spherically symmetric 1D simulation. Although we do not add any perturbation initially, numerical noises induce small radial oscillations that grow gradually. In Fig. 1, we compare the time evolution of the integrated heating rate of each nuclear species along with the shock and gain radii for the $L_{52} = 5.4$ and $\dot{M}_{sun}= 1.0$. The shock radius is defined as the iso-entropic surface of $s = 5.0\ k_B$ where k_B is the Boltzmann constant. The heating rates are integrated over the gain regions as $\int_{\mathrm{gain}} Q_i \mathrm{d}r^3$ and given in normalized unit of 10^{52} erg s^{-1}. We can see that the heating rate of alpha particles changes roughly in step with shock radius, since the larger shock has more alpha particles. It is also found that the heating by alpha particles is more important than that of deuteron after the shock wave revives and goes outward ($t > 400$ ms). On the other hand, the heating rates of deuterons reach a local maximum when the gain and shock radii are small since the matter have high densities and favor deuterons. Furthermore since the deuterons are located closer to the neutrino sphere, the deuterons gain the heating rates as high as which reach $1 - 10\%$ of those of nucleons. These results indicate that the alpha particles and deuterons heat the matter in different phases in the oscillation of the shock wave. Although the tritons and helions are similar to deuterons, they are quite minor.

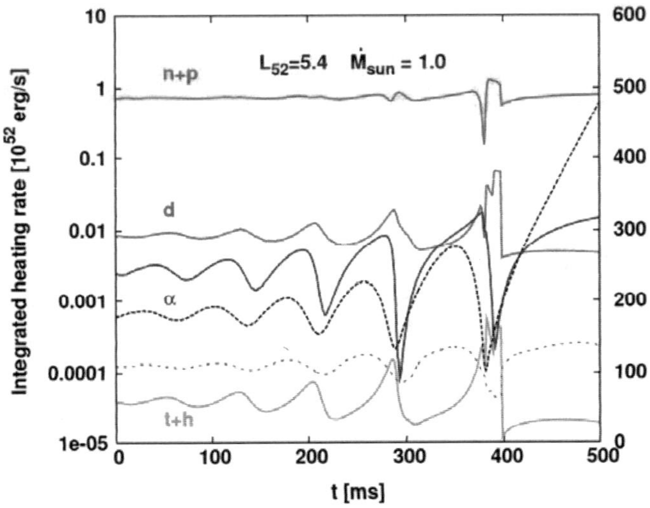

Figure 1. The time evolutions of the shock and gain radii and integrated heating rates of nuclear species. Black dashed and dotted lines denote the shock and gain radii, respectively. Magenta, red, green and blue lines represent the heating rates of A_i=1(nucleons), A_i=2 (deuterons), A_i=3 (tritons and helions) and A_i=4 (alpha particles).

3.2. 2D simulation

Figure 2 displays the time evolutions of average shock radii for four modes, in which all light nuclei, only deuterons, only alpha particles and no light nuclei are taken into account in the heating sources, respectively. The models without deuteron heating do not succeed in the shock revival for $L_{52} = 5.1$, whereas the other two models do though it takes long time. For $L_{52} = 5.2$, on the other hand, all modes produce shock revival. We can see that the heating by deuterons and alpha particles both reduce the time to shock revival. Note that this may be too naive, since the time to shock revival is known to be sensitive to various ingredients such as the initial perturbations when the neutrino luminosity is set to be close to the critical luminosity. We, however, stress that the heating by light nuclei, especially deuterons, brings the clear changes to the evolutions of shock waves in most cases.

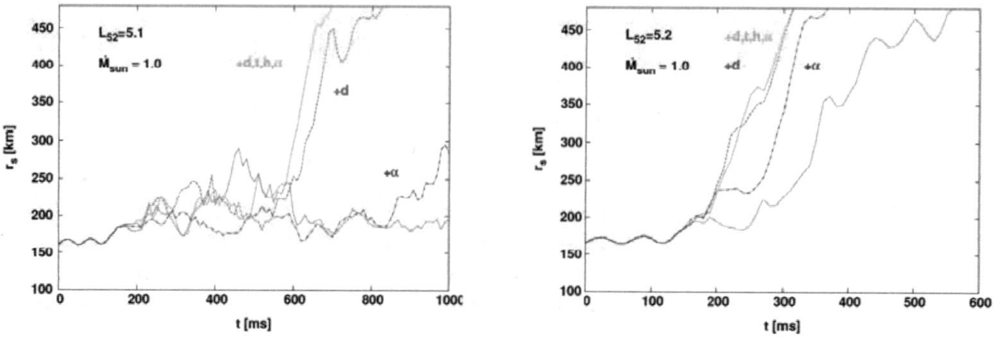

Figure 2. The time evolutions of shock radius of the models with heating by all light nuclei (cyan solid lines), only deuterons (red dashed lines), only alpha particles (blue dashed-dotted lines) and no light nuclei (black dotted lines) for L_{52} = 5.1 and 5.2 with $\dot{M}_{sun}$=1.0.

We now focus on the model including the heating by all light nuclei with $L_{52} = 5.2$ and $\dot{M}_{sun} = 1.0$ to explore the role of light nuclei in the evolution of the shock wave. The shock oscillation grows linearly by $t \sim$150 ms as shown in Fig. 2. The distributions of nucleons and light nuclei are almost spherically symmetric at t=100 ms. The heating rates of light nuclei are large in the narrow region near the quasi-steady shock wave until t=100 ms. At t=200 and 300 ms, however, the shock waves are deformed and have reached the non-linear regime of SASI. In some regions, the light nuclei are abundant indeed and their heating is efficient accordingly. Figure 3 plots the pairs of (ρ, T) obtained along 5 different radial rays. Although the initial distributions in (ρ, T) plane (the black symbols) are not located in the regions that are rich in light nuclei, the turbulence in the non-linear SASI broadens the distributions. Figure 4 shows the mass fractions and the heating rates of nuclear species along the radial ray with $\theta = 180°$ at $t = 200$ ms and the one with $\theta = 0°$ at $t = 300$ ms. The heating rate of deuterons becomes as high as $\sim$ 10% of that of nucleons at t=200 ms around the bottom of the gain region. The shock wave at t = 200 ms moves northwards ($\theta = 0°$) and the matter in the southern ($\theta = 180°$) with low entropies goes down deep into the central regions. The orange in the top panel of Fig. 3 also indicate that the matter in the southern side ($\theta = 180°$) has low entropies at the inner part of the gain regions, resulting in more deuterons than the matter in other parts. At t=300 ms, the shock wave reach at $\sim$ 400 km and the heating by alpha particle is dominant. We can see in the bottom panel of Fig.8 that the matter along the radial ray with $\theta = 0°$ has also lower entropies and as a consequence deuterons and alpha particles are abundant at the regions of high and low densities, respectively. Both at $t = 200$ and 300 ms, the deuterons have the heating rates comparable to those of nucleons near the bottom of the gain regions. The heating rates of alpha particles are $\sim$ 10 % of those of nucleons around the shock wave.

4. Discussion

We have investigated the influences of the inelastic interactions of neutrinos with light nuclei on the dynamics in the post-bounce phase of core-collapse supernovae. We have done numerical simulations of SASI with the assumption of axial symmetry for some representative combinations of the luminosity and mass accretion rate. We have not solved the dynamics of the central part of

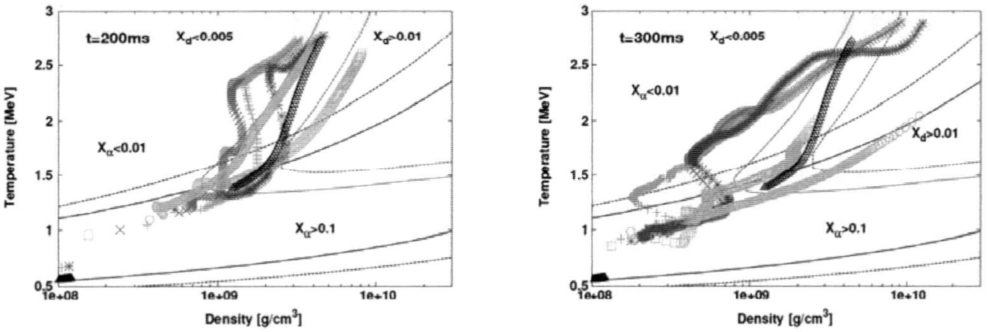

Figure 3. The lines show contours of mass fractions of deuteron X_d and alpha particle X_α (red thin lines for $X_d = 0.005$, red dashed lines for $X_d = 0.01$, blue dashed lines for $X_\alpha = 0.01$ and blue solid lines for $X_\alpha = 0.1$). The symbols show the densities and temperatures in the gain regions at $t = 200$ and 300 ms for the radial rays $\theta = 0°$ (cyan), 45° (magenta), 90° (green), 135° (brown) and 180° (orange). Black symbols are those of the initial spherical symmetric state ($t = 0$ ms).

the core and replaced it with the suitable boundary conditions and have started the simulations from spherically symmetric steady state, adding some perturbations to the radial velocity. The neutrino transport has been handled by the simple light-bulb approximation. In addition to the ordinary heating and cooling reactions with nucleons, we have taken into account the heating reactions with four light nuclei for the first time. The abundance of light nuclei is provided by the multi-nuclei EOS together with other thermodynamical quantities.

We have found that the evolutions of shock waves in 2D are influenced by the light nuclei heating and the deuterons and alpha particles have different roles in dynamical simulations in 1D and 2D. From the results of 1D dynamical simulations, we have found that the integrated heating rates of deuterons and alpha particles become high at different phases in the oscillations of shock radius: the heating rates of deuterons are the highest when the shock radius hits the minimum and the matter compression is the greatest. Whereas the heating by deuterons is constantly helpful for mater to gain the internal energies, the heating by alpha particles has an impact on the shock revival particularly when the shock wave has large radii and the matter becomes low entropies. The dynamics in 2D are more sensitive to the inclusion of the light nuclei heating because SASI in the non-linear regime make the gain regions more inhomogeneous and there appear the regions that have densities and temperatures favorable for the existence of light nuclei. The heating rates of light nuclei reaches about 10 % of that of nucleons locally. As a consequence, the dynamics of the shock revival is influenced by the heating via light nuclei, especially deuterons. Although they are never dominant heating sources, these heating reactions of light nuclei should be included if one were to estimate the critical luminosity and/or explosion energy quantitatively accurately.

The numerical simulations in this paper are admittedly of experimental nature and the numbers we have obtained may be subject to change in more realistic simulations. There is also a room for improvement in our experimental computations. We need more systematic investigations, varying not only the neutrino luminosity and mass accretion rate but also the neutrino temperature, mass of a central object and initial perturbation. The cooling reactions of light nuclei should be incorporated in the calculations. These issues are currently undertaken and will be reported elsewhere.

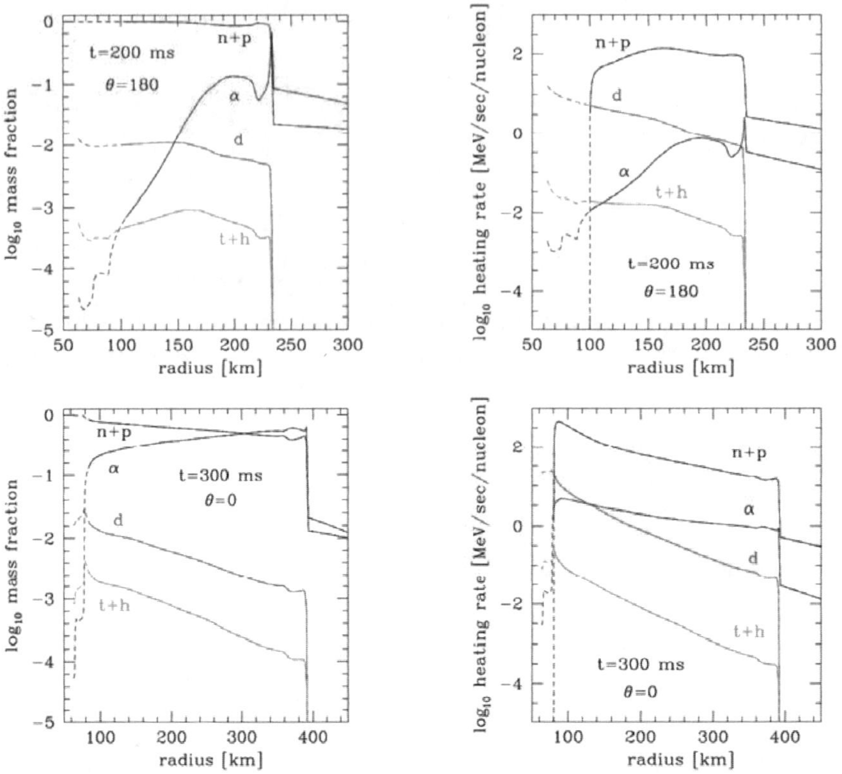

Figure 4. Mass fractions (left panels) and heating rates per baryon (right panels) for protons and neutrons (black), deuterons (red), tritons and helions (green), and alpha particles (blue) for $t = 200$ ms (top panels) and 300 ms (bottom panels). Dashed lines indicate the cooling regions where the cooling reaction of nucleons is dominant.

Acknowledgments
A part of the numerical calculations were carried out on PC cluster at Center for Computational Astrophysics, National Astronomical Observatory of Japan.

References
[1] Ohnishi N, Kotake K and Yamada S 2007, *Astrophys. J.*, **667**, 375
[2] Sumiyoshi K and Röpke G 2008, *Phys. Rev.* C, **77**, 055804
[3] Nakamura S X, Sumiyoshi K and Sato T 2009, *Phys. Rev.* C, **80**, 035802
[4] Furusawa S, Yamada S, Sumiyoshi K and Suzuki H 2011, *Astrophys. J.*, **738**, 178
[5] Furusawa S, Sumiyoshi K, Yamada S and Suzuki H 2013, *Astrophys. J.*, **772**, 95
[6] Ohnishi N, Kotake K and Yamada S 2006, *Astrophys. J.*, **641**, 1018
[7] Nagakura H, Yamamoto Y and Yamada S 2013, *Astrophys. J.*, **765**, 123
[8] Haxton W C 1988, *Phys. Rev. lett.* , **60**, 1999
[9] O'Connor E, Gazit D, Horowitz C J, Schwenk A and Barnea N 2007, *Phys. Rev.* C, **75**, 055803
[10] Arcones A, Martínez-Pinedo G, O'Connor E, Schwenk A, Janka H-T, Horowitz C J and Langanke K 2008, *Phys. Rev.* C, **78**, 015806
[11] Gazit D and Barnea N 2004, *Phys. Rev.* C, **70**, 8801
[12] Furusawa S, Nagakura H, Sumiyoshi K and Yamada S 2013, *Astrophys. J.*, **772**, 95

Journal of Physics: Conference Series **665** (2016) 012071 doi:10.1088/1742-6596/665/1/012071

Power-law distributions in protoneutron stars

G Gervino[1,3], A Lavagno[2,3], D Pigato[2,3]

[1]Dipartimento di Fisica, Università di Torino, I-10126 Torino, Italy
[2]Department of Applied Science and Technology, Politecnico di Torino, I-10129 Torino, Italy
[3]INFN, Sezione di Torino, I-10125 Torino, Italy

Abstract. We investigate the physical properties of the protoneutron stars in the framework of a relativistic mean-field theory based on nonextensive statistical mechanics, characterized by power-law distributions. We study the finite-temperature equation of state in β-stable matter at fixed entropy per baryon, in the absence and in the presence of hyperons and trapped neutrinos. We show that nonextensive power-law effects could play a crucial role in the structure and in the evolution of the protoneutron stars alsc for small deviations from the standard Boltzmann-Gibbs statistics.

1. Introduction

A protoneutron star (PNS) is formed in a stellar remnant after a successful core-collapse supernova explosion of a star with a mass smaller than about 20 solar masses and in the first seconds of its evolution it is a very hot (temperature of up to 50 MeV), lepton rich and β-stable object and a lepton concentration typical of the pre-supernova matter [1].

The essential microphysical ingredients that govern the macrophysical evolution of the PNS in the so-called Kelvin-Helmholtz epoch, during which the remnant changes from a hot and lepton-rich PNS to a cold and deleptonized neutron star, are the equation of state EOS of dense matter and its associated neutrino opacity. The diffused neutrinos are observed by terrestrial detectors and provide unique information on the supernova and the PNS. Nevertheless, the PNS is in quasi-stationary β-equilibrium state during its evolution because the time scale of the weak interaction is much shorter than the time scale of neutrino diffusion.

The knowledge of the nuclear EOS of dense matter at finite temperature plays a crucial role in the determination of the structure and in the macrophysical evolution of the PNS [2, 3]. The processes related to strong interaction should in principle be described by quantum chromodynamics. However, in the energy density range reached in the compact stars, strongly non-perturbative effects in the complex theory of QCD are not negligible [4]. In the absence of a converging method to approach QCD at finite density one often turns to effective and phenomenological model investigations. In the last years there is an increasing evidence that the nonextensive statistical mechanics, originally proposed by C. Tsallis and characterized by power-law quantum equilibrium distributions, can be considered as an appropriate physical and mathematical basis to deal with physical phenomena where strong dynamical correlations, long-range interactions, anomalous diffusion and microscopic memory effects take place [5, 6]. In this framework, several authors have outlined the relevance of nonextensive statistical mechanics effects in high energy physics and astrophysical problems [7, 8, 9, 10].

In this work we limit ourselves to consider only small variation from the standard Boltzmann-Gibbs (BG) statistics (from $q = 0.97$ to $q = 1.03$). In particular, we concentrate our study in the stage immediately after the supernova explosion (called deleptonizazion era), when the PSN assumes the maximum heating and entropy per baryon ($s = 2$) and the presence of nonextensive effects may alter more sensibly the thermodynamical and mechanical proprieties of the PNS.

2. Nonextensive hadronic equation of state

The hadronic EOS is calculated in the framework of the nonextensive statistical mechanics introduced by Tsallis [5, 6]. The nonextensive statistics represents a physical mathematical tool in several physical fields and it is based on the following definition of q-deformed entropy functional

$$S_q[f] = \frac{1}{q-1}\left(1 - \int [f(\mathbf{x})]^q \, d\Omega\right), \left(\int f(\mathbf{x})\, d\Omega = 1\right), \tag{1}$$

where $f(\mathbf{x})$ stands for a normalized probability distribution, $\mathbf{x}$ and $d\Omega$ denoting, respectively, a generic point and the volume element in the corresponding phase space. The nonextensive statistics is, therefore, a generalization of the common BG statistical mechanics and for $q \to 1$ it reduces to the standard BG entropy. Furthermore, nonextensive statistical effects vanishes approaching to zero temperature.

In this context, the nonextensive statistics entails a sensible difference on the power-law particle distribution shape in the high energy region with respect to the standard statistics.

We are going to study the nonextensive hadronic EOS in the framework of a relativistic mean field theory in which baryons interact through the nuclear force mediated by the exchange of virtual isoscalar-scalar (σ), isoscalar-vector (ω) and isovector-vector (ρ) mesons fields [11, 12]. In our analysis we include all the baryon octet in order to reproduce the chemical composition of the PNS at high baryon chemical potential. We also take into account of leptons particle by fixing the lepton fraction $Y_L = Y_e + Y_{\nu_e} = (\rho_e + \rho_{\nu_e})/\rho_B$, where ρ_e, ρ_{ν_e} and ρ_B are the electron, neutrino and baryon number densities, respectively. This is because, in the first stage of PNS evolution, electrons and neutrinos are trapped inside the stellar matter and, therefore, the lepton number must be conserved until neutrinos escape out of the PNS [1].

In the field equations appear the baryon density and the baryon scalar density. In the framework of the nonextensive statistical mechanics, they are given by [9]

$$\rho_B = 2\sum_{i=B}\int \frac{d^3k}{(2\pi)^3}\, n_i(k)\,, \tag{2}$$

$$\rho_S = 2\sum_{i=B}\int \frac{d^3k}{(2\pi)^3}\,\frac{M_i^*}{E_i^*}\, n_i^q(k)\,, \tag{3}$$

where $n_i(k)$ is the q-deformed fermion particle distribution function. For $q > 1$ and $\beta(E_i^* - |\mu_i^*|) > 0$, we have, for example,

$$n_i(k) = \frac{1}{[1 + (q-1)\,\beta(E_i^*(k) - \mu_i^*)]^{1/(q-1)} + 1}\,. \tag{4}$$

The nucleon effective energy is defined as $E_i{}^*(k) = \sqrt{k^2 + M_i{}^{*2}}$, where $M_i{}^* = M_i - g_{\sigma B}\sigma$. The effective chemical potentials μ_i^* are given in terms of the meson fields as follows

$$\mu_i^* = \mu_i - g_{\omega B}\omega - \tau_{3iB}\, g_{\rho B}\rho\,, \tag{5}$$

where μ_i are the thermodynamical chemical potentials ($\mu_i = \partial\epsilon/\partial\rho_i$).

The further conditions, required for the β-stable chemical equilibrium and charge neutrality, can be written as

$$\mu_\Lambda = \mu_{\Sigma^0} = \mu_{\Xi^0} = \mu_n \,, \tag{6}$$
$$\mu_{\Sigma^-} = \mu_{\Xi^-} = \mu_n + \mu_e \,, \tag{7}$$
$$\mu_p = \mu_{\Sigma^+} = \mu_n - \mu_e \,; \tag{8}$$
$$\rho_p + \rho_{\Sigma^+} - \rho_{\Sigma^-} - \rho_{\Xi^-} - \rho_e = 0 \,. \tag{9}$$

In the case of trapped neutrinos, the new equalities are obtained by the replacement of $\mu_e \to \mu_e - \mu_{\nu_e}$. The total entropy per baryon is calculated using $s = (S_B + S_l)/(T\rho_B)$, where $S_B = P_B + \epsilon_B - \sum_{i=B} \mu_i \rho_i$ and $S_l = P_l + \epsilon_l - \sum_{i=l} \mu_i \rho_i$, and the sums are extended over all the baryons and leptons species.

The numerical evaluation of the above thermodynamical quantities can be performed if the meson-nucleon and meson-hyperon coupling constants are known. Concerning the meson-nucleon coupling constants ($g_{\sigma N}$, $g_{\omega N}$, $g_{\rho N}$), they are determined to reproduce properties of equilibrium nuclear matter such as the saturation densities, the binding energy, the symmetric energy coefficient, the compression modulus, and the effective Dirac mass at saturation. Here and in the following, we focus our investigation by considering the so-called GM3 parameter set [11]. The implementation of hyperon degrees of freedom comes from determination of the corresponding meson-hyperon coupling constants that have been fitted to hypernuclear properties and their specific values are taken from Ref.s [13, 14] for the GM3 parameter set.

In the figures, we report some of the most important effects in the structure and in the thermodynamical properties of the PNS in the framework of nonextensive statistical mechanics, during the maximum heating phase ($s = 2$ and $Y_\nu = 0$).

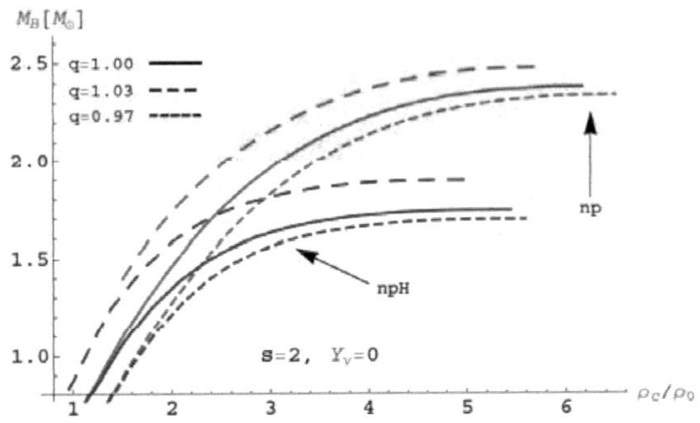

Figure 1. Maximum baryonic mass M_B in units of solar mass $M_\odot$ as a function of the central baryon density ρ_c (in units of the nuclear saturation density ρ_0) for different values of q and for nucleons and hyperons stars.

In Fig. 1, we show the variation of the maximum baryonic mass as a function of the central baryon density ρ_c, for different values of q, both for nucleonic (np) and hyperonic (npH) stars. We observe a remarkable increase (reduction) of the maximum baryonic mass when $q > 1$ ($q < 1$). In Fig. 2, we can see that the corresponding mass-radius relation is also sensibly modified in presence of a nonextensive statistics.

The above behaviors are substantially due to the softening of the EOS in presence sub-extensive statistics ($q < 1$), together with a remarkable reduction of the maximum temperature both for nucleonic and hyperonic PNS, as reported in Fig. 3. Contrariwise, when $q > 1$, we observe an increase of the maximum temperature. These remarkable differences in the stellar temperature have important consequences in the PNS evolution and, consequently, in the cooling of the PNS, making it longer when $q > 1$, and shorter when $q < 1$, with important astrophysical implications. All of these effects are also present when neutrinos are trapped in the PNS ($s = 1$ and $Y_L = 0.4$), however, due to the lower temperature achieved in this phase, they are less pronounced.

Journal of Physics: Conference Series **665** (2016) 012071 doi:10.1088/1742-6596/665/1/012071

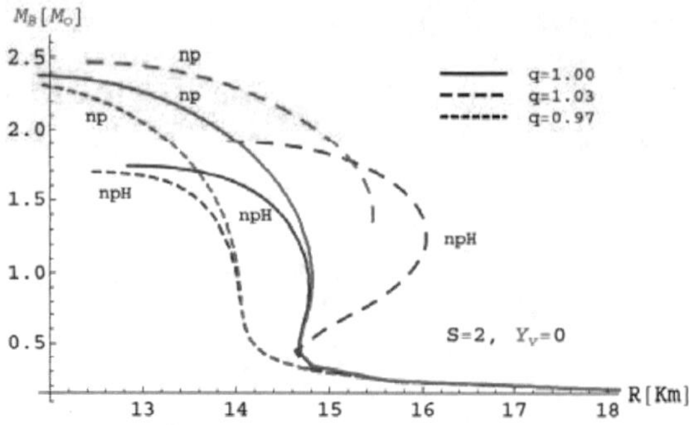

Figure 2. Stellar radius profile, as a function of the central baryon density ρ_c for different values of q and for nucleons and hyperons stars.

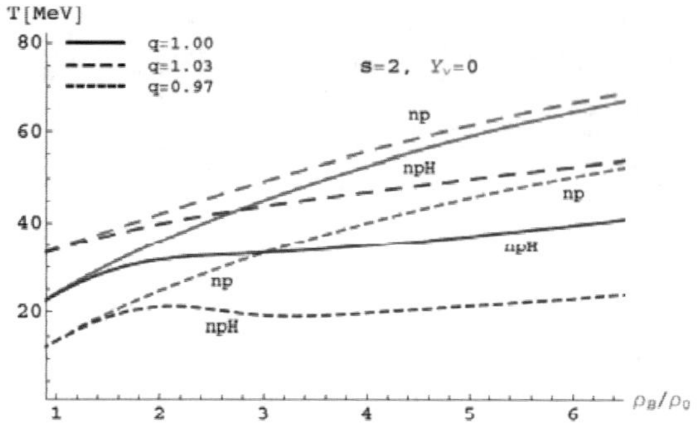

Figure 3. Temperature as a function of the baryon density for nucleonic and hyperonic PNS, for different values of q.

In conclusion, we have investigated the properties of PNS at different stages during its evolution. We have shown that the nonextensive statistical mechanics, characterized by power-law quantum distributions, can play a crucial role on the physical properties of the PNS and can be considered as an effective mathematical basis to investigate the complex structure and evolution of the PNS. From a phenomenological point of view, we have considered the nonextensive index q as a free parameter, even if, in principle, it should depend on the physical conditions inside the PNS, on the fluctuation of the temperature and be related to microscopic quantities (such as, for example, the mean interparticle interaction length).

We have shown that, in the presence of super-extensive statistical effects ($q > 1$) and hyperon degrees of freedom, the realization of a metastable phase is favored, with an enhancement of a possible black-hole formation after the deleptonization era. The variation of the maximum temperature, together with a significant variation of the maximum mass and stellar radius, could be considered as a phenomenological evidence of nonextensive statistical effects in PNS.

References

[1] Prakash M et al. 2001 *Lect. Notes Phys.* **578** 364
[2] Lavagno A and Pigato D 2011 *Eur. Phys. J* A **47** 52
[3] Drago A, Lavagno A and Parenti I 2007 *Astrophys. J.* **659** 1519
[4] Bonanno L, Drago A and Lavagno A 2007 *Phys. Rev. Lett.* **99** 242301
[5] Tsallis C 1988 *J. Stat. Phys.* **52** 479
[6] Tsallis C 2009 *Introduction to Nonextensive Statistical Mechanics* (Springer-Verlag)
[7] Wilk G and Wlodarczyk Z 2009 Eur. Phys. J. A 40 299
[8] Biró T S, Purcsel G and Ürmössy K 2009 *Eur. Phys. J.* A **40** 325
[9] Alberico W M and Lavagno A 2009 *Eur. Phys. J.* A **40** 313
[10] Lavagno A and Pigato D 2012 *J. Phys. G: Nucl. Part. Phys.* **39** 125106
[11] Glendenning N K 1992 *Phys. Rev.* D **46** 1274
[12] Lavagno A and Pigato D 2012 *Phys. Rev.* C **86** 024917
[13] Lavagno A 2010 *Phys. Rev.* C **81** 044909
[14] Lavagno A 2013 *Eur. Phys. J.* A **49** 102

Thermodynamic instabilities in warm and dense asymmetric nuclear matter and in compact stars

A Lavagno[1,3], G Gervino[2,3], D Pigato[1,3]

[1]Department of Applied Science and Technology, Politecnico di Torino, I-10129 Torino, Italy
[2]Dipartimento di Fisica, Università di Torino, I-10126 Torino, Italy
[3]INFN, Sezione di Torino, I-10125 Torino, Italy

Abstract. We investigate the possible thermodynamic instability in a warm and dense nuclear medium where a phase transition from nucleonic matter to resonance-dominated Δ-matter can take place. Such a phase transition is characterized by both mechanical instability (fluctuations on the baryon density) and by chemical-diffusive instability (fluctuations on the isospin concentration) in asymmetric nuclear matter. Similarly to the liquid-gas phase transition, the nucleonic and the Δ-matter phase have a different isospin density in the mixed phase. In the liquid-gas phase transition, the process of producing a larger neutron excess in the gas phase is referred to as isospin fractionation. A similar effects can occur in the nucleon-Δ matter phase transition due essentially to a Δ^- excess in the Δ-matter phase in asymmetric nuclear matter. In this context we also discuss the relevance of Δ-isobar degrees of freedom in the bulk properties and in the maximum mass of compact stars.

1. Introduction

One of the very interesting aspects in nuclear astrophysics and in the heavy-ion collisions experiments is a detailed study of the thermodynamical properties of strongly interacting nuclear matter away from the nuclear ground state.

The new accumulating data from x-ray satellites provide important information on the structure and formation of compact stellar objects. Concerning the structure, these data are at first sight difficult to interpret in a unique and selfconsistent theoretical scenario, since some of the observations indicate rather small radii and other observations indicate large values for the mass of the star.

On the other hand, the information coming from experiments with heavy ions in intermediate- and high-energy collisions is that the EOS depends on the energy beam but also sensibly on the electric charge fraction Z/A of the colliding nuclei, especially at not too high temperature [1, 2, 3, 4]. Moreover, the study of nuclear matter with arbitrary electric charge fraction results to be important in radioactive beam experiments and in the physics of compact stars.

In this article, we study a hadronic equation of state (EOS) at finite temperature and density by means of a relativistic mean-field model with the inclusion Δ(1232)-isobars [5, 6, 7] and by requiring the Gibbs conditions on the global conservation of baryon number and net electric charge. Transport model calculations and experimental results indicate that an excited state of baryonic matter is dominated by the Δ resonance at the energies from the BNL Alternating Gradien Synchrotron (AGS) to RHIC [8]. Moreover, in the framework of the nonlinear Walecka model, it has been predicted that a phase transition from nucleonic matter to Δ-excited nuclear matter can take place and the occurrence of this transition sensibly depends on the Δ-meson coupling constants [9, 10].

The main goal of this paper is to show that, for asymmetric warm and dense nuclear medium, the possible Δ-matter phase transition is characterized by mechanical and chemical-diffusive instabilities. Similarly to the liquid-gas phase transition [11], chemical instabilities play a crucial role in the characterization of the phase transition and can imply a very different electric charge fraction Z/A in the coexisting phases during the phase transition. In this context we also discuss the relevance of Δ-isobar degrees of freedom in the bulk properties of compact stars.

2. Phase transition and stability conditions

We are dealing with the study of a multi-component system at finite temperature and density with two conserved charges: baryon number and electric charge. For such a system, the Helmholtz free energy density F can be written as

$$F(T,\rho_B,\rho_C) = -P(T,\mu_B,\mu_C) + \mu_B\rho_B + \mu_C\rho_C\,, \tag{1}$$

with

$$\mu_B = \left(\frac{\partial F}{\partial \rho_B}\right)_{T,\rho_C}\,, \quad \mu_C = \left(\frac{\partial F}{\partial \rho_C}\right)_{T,\rho_B}\,. \tag{2}$$

In a system with N different particles, the particle chemical potentials are expressed as the linear combination of the two independent chemical potentials μ_B and μ_C and, as a consequence, $\sum_{i=1}^{N}\mu_i\rho_i = \mu_B\rho_B + \mu_C\rho_C$.

Assuming the presence of two phases (denoted as I and II, respectively), the system is stable against the separation in two phases if the free energy of a single phase is lower than the free energy in all two phases configuration. The phase coexistence is given by the Gibbs conditions

$$\mu_B^I = \mu_B^{II}\,, \qquad \mu_C^I = \mu_C^{II}\,, \tag{3}$$

$$P^I(T,\mu_B,\mu_C) = P^{II}(T,\mu_B,\mu_C)\,. \tag{4}$$

Therefore, at a given baryon density ρ_B and at a given net electric charge density $\rho_C = y\,\rho_B$ (with $y = Z/A$), the chemical potentials μ_B are μ_C are univocally determined. An important feature of this conditions is that, unlike the case of a single conserved charge, the pressure in the mixed phase is not constant and, although the total ρ_B and ρ_C are fixed, baryon and charge densities can be different in the two phases. For such a system in thermal equilibrium, the possible phase transition can be characterized by mechanical (fluctuations in the baryon density) and chemical instabilities (fluctuations in the electric charge density). As usual the condition of the mechanical stability implies

$$\rho_B\left(\frac{\partial P}{\partial \rho_B}\right)_{T,\rho_C} > 0\,. \tag{5}$$

By introducing the notation $\mu_{i,j} = (\partial\mu_i/\partial\rho_j)_{T,P}$ (with $i,j = B,C$), the chemical stability for a process at constant P and T can be expressed with the following conditions [10]

$$\rho_B\,\mu_{B,B} + \rho_C\,\mu_{C,B} = 0\,, \tag{6}$$

$$\rho_B\,\mu_{B,C} + \rho_C\,\mu_{C,C} = 0\,. \tag{7}$$

Whenever the above stability conditions are not respected, the system becomes unstable and the phase transition takes place. The coexistence line of a system with one conserved charge becomes in this case a two dimensional surface in (T,P,y) space, enclosing the region where mechanical and diffusive instabilities occur.

By increasing the temperature and the baryon density during the high energy heavy ion collisions ($T \approx 50$ MeV and $\rho_B \geq \rho_0$), a multi-particle system with Δ-isobar and pion degrees of freedom may take place.

In analogy with the liquid-gas case, we are going to investigate the existence of a possible phase transition in the nuclear medium by studying the presence of instabilities (mechanical and/or chemical) in the system. The chemical stability condition is satisfied if [10]

$$\left(\frac{\partial \mu_C}{\partial y}\right)_{T,P} > 0 \quad \text{or} \quad \begin{cases} \left(\frac{\partial \mu_B}{\partial y}\right)_{T,P} < 0\,, & \text{if } y > 0\,, \\ \left(\frac{\partial \mu_B}{\partial y}\right)_{T,P} > 0\,, & \text{if } y < 0\,. \end{cases} \tag{8}$$

In the Fig. 1, we report the baryon and electric charge chemical potential isobars as a function of y, at fixed temperature $T = 50$ MeV and $x_{\sigma\Delta} \equiv g_{\sigma\Delta}/g_{\sigma N} = 1.3$ (the ratio related to the scalar σ meson-Δ coupling constants) in the GM3 parameters set [5].

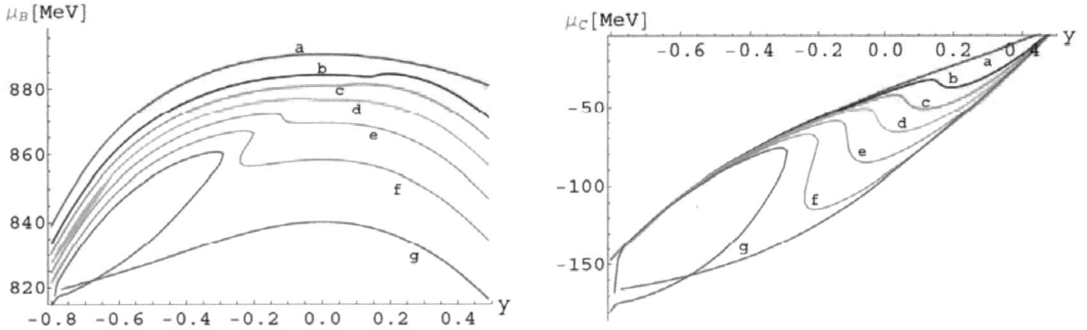

Figure 1. Baryon (right panel) and electric charge (left panel) chemical potential isobars as a function of y at $T = 50$ MeV and $x_{\sigma\Delta} = 1.3$. The curves labeled a through g have pressure P=9,7,6,5,4,3,2 MeV/fm^3, respectively.

From the analysis of the above chemical potential isobars, we are able to construct the binodal surface relative to the nucleon-Δ matter phase transition. In Fig. 2, we show the binodal section at $T = 50$ MeV and $x_{\sigma\Delta} = 1.3$.

The right branch (at lower density) corresponds to the initial phase (I), where the dominant component of the system is given by nucleons. The left branch (II) is related to the final phase at higher densities, where the system is composed primarily by Δ-isobar degrees of freedom (Δ-dominant phase). In presence of Δ-isobars the phase coexistence region results very different from what obtained in the liquid-gas case, in particular it extends up to regions of negative electric charge fraction and the mixed phase region ends in a point of maximum asymmetry with $y = -1$ (corresponding to a system with almost all Δ^--particles, being antiparticles and pions contribution almost negligible in this regime). We analyze the phase evolution of the system during the isothermal compression from an arbitrary initial point A, indicated in Fig. 2. In this point the system becomes unstable and starts to be energetically favorable the separation into two phases, therefore an infinitesimal Δ-dominant phase appears in B, at the same temperature and pressure. Let us observe that, although in B the electric charge fraction is substantially negative, the relative Δ^- abundance must be weighed on the low volume fraction occupied by the phase II near the point B. During the phase transition, each phase evolves towards

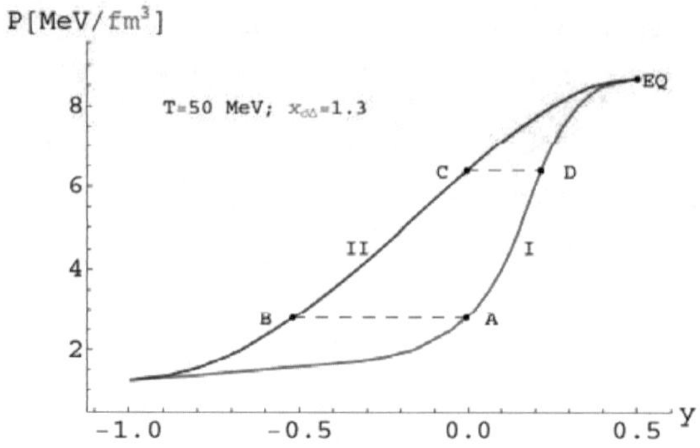

Figure 2. Binodal section at $T = 50$ MeV and $x_{\sigma\Delta} = 1.3$.

a configuration with increasing y, in contrast to the liquid-gas case, where each phase evolves through a configuration with a decreasing value of y (with the exception of the gas phase after the maximum asymmetry point).

Finally, we investigate the relevance Δ-isobar degrees of freedom in the bulk properties of compact star in the framework of the same equation of state discussed above but considering β-stable and electric-charge neutral nuclear matter at $T = 0$.

In Fig. 3, we report the mass-radius relations in absence (no Δ) and in presence of Δ-isobars with different scalar coupling ratios ($x_{\sigma\Delta} = 1.0$ and $x_{\sigma\Delta} = 1.2$). The presence of Δ-isobar degrees of freedom smooths the equation of state and reduces the maximum gravitational mass. On the other hand very compact object with smaller radii can be formed. This matter of fact can be very relevant in the interpretation of recent astrophysical observations [12].

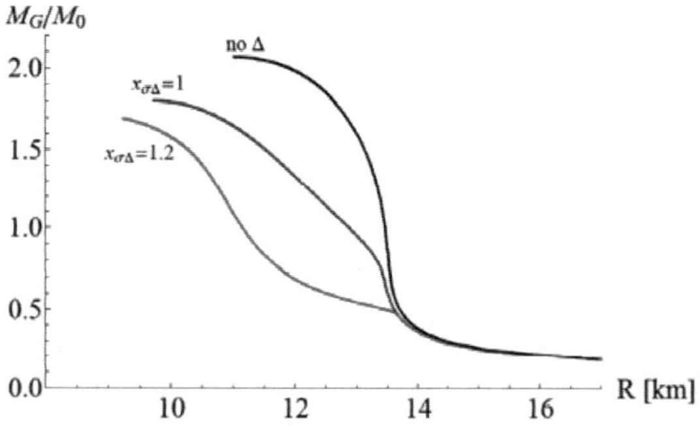

Figure 3. Mass-radius relations in absence (no Δ) and in presence of Δ-isobars with different scalar coupling ratios ($x_{\sigma\Delta} = 1.0$ and $x_{\sigma\Delta} = 1.2$).

References

[1] Di Toro M *et al.* 2006 *Nucl. Phys.* A **775** 102
[2] Bonanno L, Drago A and Lavagno A 2007 *Phys. Rev. Lett.* **99** 242301
[3] Alberico W M and Lavagno A 2009 *Eur. Phys. J.* A **40** 313
[4] Lavagno A 2013 *Eur. Phys. J.* A **49** 102
[5] Glendenning N K 1992 *Phys. Rev.* D **46** 1274
[6] Lavagno A 2010 *Phys. Rev.* C **81** 044909
[7] Lavagno A and Pigato D 2012 *J. Phys. G: Nucl. Part. Phys.* **39** 125106
[8] Hofmann M, Mattiello R, Sorge H, Stocker H and Greiner W 1995 *Phys. Rev.* C **51** 2095
[9] Li Z, Mao G, Zhuo Y and Greiner W 1997 *Phys. Rev.* C **56** 1570
[10] Lavagno A and Pigato D 2012 *Phys. Rev.* C **86** 024917
[11] Müller H and Serot B D 1995 *Phys. Rev.* C **52** 2072
[12] Ozel F, Baym G and Guver T 2010 *Phys. Rev.* D **82** 101301

Query on Accelerating Expansion of the Universe by Error Analysis of Ia Supernova

Q-H Peng[1,2], J Zhang[1], Z-Q Luo[1], Yu Liang [2]

[1]Institute of Theoretical Physics, China West Normal University, Nanchong, 637002, P.R. China

[2]Department of Astronomy, Nanjing University, Nanjing, 210093, P.R. China

E-mail: qhpeng@nju.edu.cn

Abstract. Some serious faults in error analysis of observations for SNIa have been found. Redoing the same error analysis of SNIa, by our idea, it is found that the average total observational error of SNIa is obviously greater than 0.55^m, so we cannot decide whether the universe is accelerating expansion or not.

1. Observation Analysis onto Accelerating Expansion of the Universe

A phenomenon of "Accelerating Expansion of the Universe" is detected by observations of type Ia supernova (SNIa). These analytical methods are based on two assumptions which came from a group of series of experience relationsall of them are named as "Advanced Phillips experience relations": a) Luminosity is related with the width of the light curve of SNIa; b) Luminosity is related with the color index (B-V) in the course of explosion of SNIa.

In the group a), use of type Ia supernovae (SNIa) as lighthouses or "standard candles" for cosmological studies became feasible upon the realization of an empirical correlation of peak luminosity with light-curve shape. The discovery of the relationship between light-curve shape and brightness led to a parameterization by the decline rate,Δm_{15}, in B-band brightness over 15 days after the maximum light [1]. The "multi-color light curve shape" method [2,3] is based on Δ, the difference in peak luminosity between an observed SN and a fiducial template. The "stretch" method [4,5] parameterizes light curves with a factor s which broadens or narrows the rest-frame timescale of a single standard template to match with observed light curves. The stretch factor s can be translated to a corresponding Δm_{15}via the best-fit line $\Delta m_{15} = (1.96\pm0.17)(s^{-1}-1)+1.7$. It was found that stretch describes the observed B and V band light curves well over a month around maximum [5]. Afterwards, most related work in this field are based on the stretch factor parameter, s, including on determining both K-correction and extinction correction of SNeIa [6]. The "Bayesian adapted template match" method introduced by [7] estimates distances by comparing with a large set of well-observed nearby SNeIa rather than a parameterized template. These methods are fundamentally equivalent by making use of light-curve shapes.

In the group b), another distance measurement technique is the CMAGIC method [8], which uses the relationship between light-curve shape and brightness in a more indirect way than the above procedures. It is based on an observed linear relationship in color-magnitude space for a certain period after the maximum light to infer distances. Besides, In a paper the authors address that they have revealed clear evidence for a tight linear correlation between peak luminosities of SNeIa and their B - V colors $\sim$ 12 days after the B maximum denoted by ΔC_{12} [9].

Based on the advanced Philips' experience relations above, SALT2 (Spectral Adaptive Light-curve Template) is a package of the software for Type Ia Supernovae light curve fitting [10]. Following the SALT2, UNION2 [11] deals with the data of 685 SNIa. The system error of the absolute magnitudes of SNIa is found by minimizing χ^2which is the normalized quadratic sum of distance modulus residual [10,11]. with the average total observational error, $\overline{\sigma_{tot}} = 0.31^m$, including intrinsic error $\overline{\sigma_{int}} = 0.18^m$.As a result, they come to a conclusion that universe is accelerating expanding. And Saul Perlmutter, Brian Schmidt and Adam Reiss share the 2011 Nobel Prize in Physics for their observations that type Ia supernovae indicate that the expansion of the universe is accelerating.

Due to advanced progress of researches on SNIa in recent years, all advanced Philips' experience relations have lost their physical basis [12,13]. However, the authors of all related statistical works declare that their statistic errors are rather small. The transferred errors which are caused by transferring from the statistical errors in their cited papers, however, did not be taken in to account. Indeed, they put the transferred errors into an intrinsic error (or system error) which is determined by χ^2- method, $\chi^2 = \sum\limits_{SNe} \frac{(m_B - m_{\text{mod}})^2}{\sigma_{tot}^2}$, $\sigma_{tot}^2 = \sigma_{ext}^2 + \sigma_{tlc}^2 + \sigma_{int}^2$ (σ_{ext} are the errors by the observational method including interstellar extinction, correction by gravitational lens etc., σ_{lc} are errors by fitting both the light curve and color index of SNIa.).

$$m_{\text{mod}} = 5\log_{10} D_L(z_{hel}, z_{cmb}, w, \Omega_m, \Omega_{DE}) - \alpha(s-1) + \beta C + M$$

2. Some serious faults in error analysis

2.1. Error in absolute magnitude

The absolute bolometric magnitudes errors at maximum luminosity of SNIa consist of as follows: a)The intrinsic error (ΔM_{int}) of the absolute magnitude at maximum luminance, in our idea, is just the half width at half-maximum (HWHM) of the statistic distribution curve of the number of SNIa with the maximum luminance, rather than the systematic error founding by using the χ^2 check way [10,11]. b)Some errors of M originates from delivered errors caused by statistical errors of the parameters a and b in original Phillips' relation [11] or α_x, β in the advanced Philips' Relations [2,3]. We call it as the delivered error, $\Delta M(a,b)$. c) Some errors of M, ΔM_{obs} is caused by the errors of some observational quantities of both light-curve and color index in advanced Phillips's way. The total error of absolute bolometric magnitudes at maximum luminosity (ΔM_{total}) is $(\Delta M_{total})^2 = (\Delta M_{int})^2 + (\Delta M_{max}^{(phillips)})^2$, $(\Delta M_{max}^{(phillips)})^2 = (\Delta M(a,b))^2 + (\Delta M_{obs})^2$. SALT2 [10] and UNION2 [11] didn't give the aforementioned errors separately. In fact, they broadly merged them into the system error caused by χ^2 -minimization. Advanced Philips Relation is very complex, but it is sure that the minimum of $\Delta M_{\max}^{(philips)}$ is at least larger than observational apparent magnitude error, $|\Delta M_{max}^{(phillips)}| > |\delta m|$ (the observational error of apparent magnitude. Furthermore, high z SNIa is faint when observed, so its $|\delta m|$ is much larger than nearby SNIa .

2.2. The systematic error by using χ^2 check test is incorrect

However, premise of χ^2 check test is the errors of distance modulus for the set of SNIa obeying a Gaussian distribution. But the set of modeling SNIa in UNION2 [11] including 685 SNeIa is really no Gaussian distribution. Although the average error of UNION2 which contains 685 SNeIa, is 0.16^m, over 10% of total SNeIa are outline of 10σ. If we take a subsample including 217 SNIa with very small observational average error to do the same statistics, it is found that over 10% of total SNeIa are outline of 5σ. Really, the critical permitted outline value for outline of 2.6σ in the standardized normal distribution is 0.805%. Using the χ^2 check test following

SALT2 [10], we find that 3.796% of the data are outline of 2.6σ based on the average total observational error of the distance modulus of SNIa, 0.31^m. Obviously, the distance modulus error deviate Gaussian distribution seriously, and it is not suitable to calculate the systematic error σ_{sys} of SNeIa by χ^2 check test method. In our idea the real intrinsic error of a SNIa compilation should base on statistical distribution diagram of the number of SNIa for their absolute bolometric magnitudes(see Fig. 1). As we don't know the exact luminosity of high z SNeIa, it is the only way to use SALT2 to get the absolute bolometric magnitudes of "modelling SNIa". The intrinsic error (or proper error) of the absolute magnitude at maximum luminosity is just the HMHW of a statistic distribution curve of the number of SNIa with the maximum luminosity. It is $\Delta M_{int} = 0.38^m$, and it is much larger than the systematic error given by χ^2 check test.

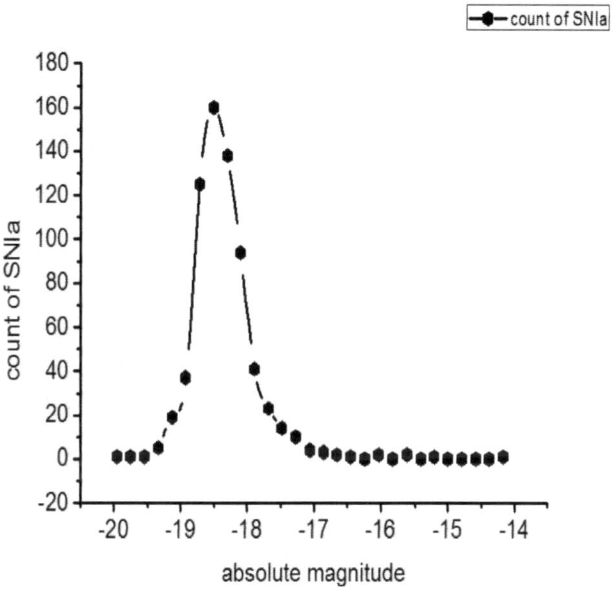

Figure 1. The count of SNIa vs. absolute magnitude.

3. Summary and conclusions

The average total observational errors of distance modulus is $(\Delta\mu)^2 = (\Delta M_{total}^2 + (\delta m)^2$. Using the data of SNIa and observational apparent magnitude error in UNION2, and divided intervals per $\Delta z = 0.1$, we repeat the statistics (by the same χ^2 check test method) to calculate this "modelling SNIa" sample's average total observational errors of distance modulus. It is found that the average total observational error of SNIa is obviously greater than 0.55^m (this is much larger than 0.4^m). So we can't decide whether the universe is accelerating expansion or not (see Fig. 2). And the direct observational evidence of the idea of "dark energy" is also lost by the observational error analysis of SNeIa.

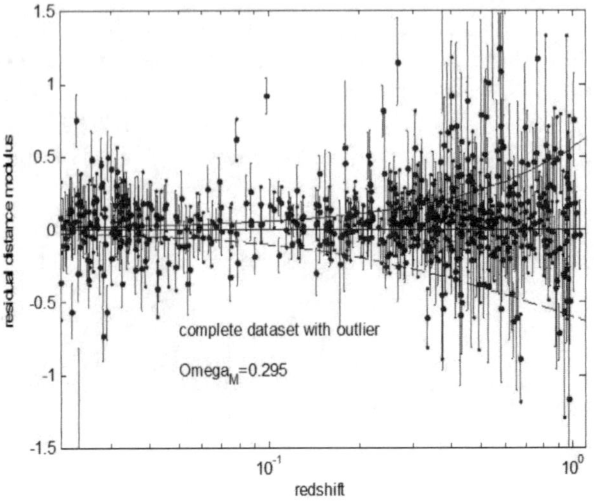

Figure 2. Residual distance modulus as a function of redshift.

Acknowledgment

This research is supported by Chinese National Science Foundation No.11273020.

References

[1] Phillips, M. M. , 1993, *ApJL*, 413, L105
[2] Riess, A. G., Press, W. H., & Kirshner, R. P. 1996, ApJ, 473, 88
[3] Riess, A. G., et al. 1999, AJ, 117, 707
[4] Perlmutter, S., Aldering, G., Goldhaber, G. et al. 1999, *ApJ*, 517, 565
[5] Goldhaber, G., et al. 2001, ApJ, 558, 359
[6] Nugent, P., Kim, A., & Perlmutter, S. 2002, Publications of the Astronomical Society of the Pacific, 114, 803
[7] Tonry, J. L., et al. 2003, ApJ, 594, 1
[8] Wang, L., Goldhaber, G., Aldering, G., & Perlmutter, S. 2003, ApJ, 590, 944
[9] Wang, X., Wang, L., Zhou, X., Lou, Y.-Q., & Li, Z. 2005, ApJL, 620, L87
[10] Guy, J., Astier, P. Baumont, S. et al. 2007, *A&A*, 466, 11 (SALT2)
[11] Amanullah, R., Lidmanet C., Rubin,D. et al. 2010, *ApJ*, 716 712 (UNION2)
[12] Isern J. , 2010, *Type Ia supernovae-Observation and Theory*, In an invited review talk, In 11th International Symposium on Nuclei in the Cosmos, July 19-23, 2010, Heidelberg, Germany.
[13] Wang, X., Wang, L., Filippenko, A. V., Zhang, T., & Zhao, X. 2013, Science, arXiv:1303.260

Nuclear Pasta at Finite Temperature with the Time-Dependent Hartree-Fock Approach

B Schuetrumpf[1], M A Klatt[2], K Iida[3], J A Maruhn[1], K Mecke[2], P-G Reinhard[2]

[1]Institut für Theoretische Physik, Universität Frankfurt, D-60438 Frankfurt, Germany
[2]Institut für Theoretische Physik, Universität Erlangen-Nürnberg, D-91058 Erlangen, Germany
[3]Department of Natural Science, Kochi University, 2-5-1 Akebono-cho, Kochi 780-8520, Japan

Abstract. We present simulations of neutron-rich matter at sub-nuclear densities, like supernova matter. With the time-dependent Hartree-Fock approximation we can study the evolution of the system at temperatures of several MeV employing a full Skyrme interaction in a periodic three-dimensional grid [1].

The initial state consists of α particles randomly distributed in space that have a Maxwell-Boltzmann distribution in momentum space. Adding a neutron background initialized with Fermi distributed plane waves the calculations reflect a reasonable approximation of astrophysical matter.

The matter evolves into spherical, rod-like, connected rod-like and slab-like shapes. Further we observe gyroid-like structures, discussed e.g. in [2], which are formed spontaneously choosing a certain value of the simulation box length. The ρ-T-map of pasta shapes is basically consistent with the phase diagrams obtained from QMD calculations [3]. By an improved topological analysis based on Minkowski functionals [4], all observed pasta shapes can be uniquely identified by only two valuations, namely the Euler characteristic and the integral mean curvature.

In addition we propose the variance in the cell-density distribution as a measure to distinguish pasta matter from uniform matter.

1. Introduction

We observe systems with mean densities of 10% of the nuclear density up to the nuclear density at finite temperatures of an order of 10 MeV. These conditions are present in supernova core collapses and also in neutron stars but with low temperatures. The nuclear matter consists under these conditions no longer of nearly spherical nuclei but rods, slabs and phases where liquid and gas phase are inverted called rod bubbles and spherical bubbles are built up (s. e.g. [5, 6, 7]).

Besides the liquid drop model, Thomas-Fermi and QMD approaches to nuclear pasta (s. e.g. [8, 9, 3]) the time-dependent Hartree-Fock (TDHF) approximation was used to describe matter at zero temperature [10, 11]. Here we describe matter at finite temperature with the TDHF equations with a Skyrme force. For a review see [12].

More details about the calculations of this work, especially about what is presented in the sections 2 and 3, can be found in [13].

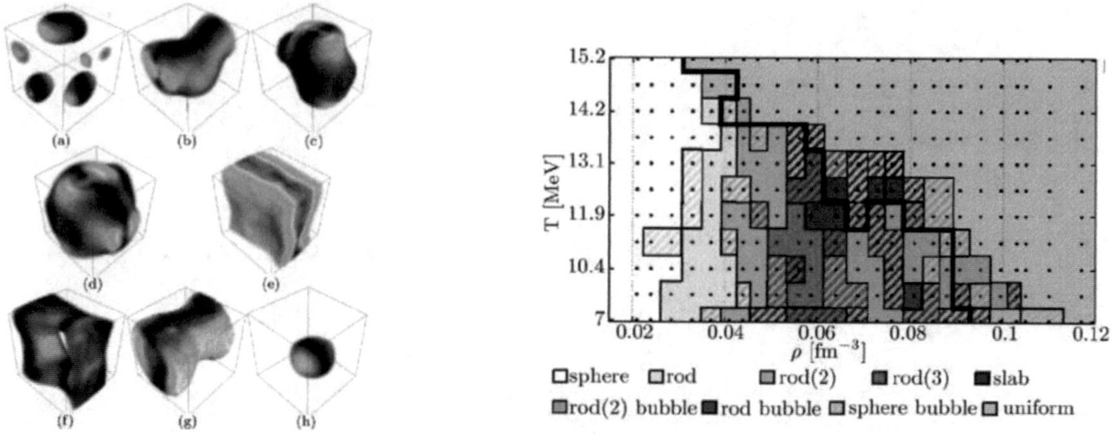

Figure 1. Left: Typical pasta shapes at $T = 7$ MeV. The colors indicate density in a spectral scale from red (high density) to blue (low density). (a) Sphere. (b) Rod. (c) Rod(2). (d) Rod(3). (e) Slab. (f) Rod(2) bubble. (g) Rod bubble. (h) Sphere bubble. Right: Map of pasta shapes achieved in TDHF calculations. Each dot represents two calculations. These pictures were taken from [13].

2. Setup

The calculations are performed on a full 3D grid without any symmetry assumptions with 16 grid points in each direction with a grid spacing of 1 fm and periodic boundary conditions. To vary the density we initialize on this grid a number of α particles with wave functions which were calculated in a static HF calculation, adding plane-wave background neutron wave functions to get a realistic value for the proton fraction of 1/3. The α particles are distributed with the Maxwell Boltzmann distribution in momentum space and randomly in real space. The background neutron wave functions are distributed with the Fermi distribution using the same temperature as for the α particles. The time evolution is performed for 1500 fm/c.

The temperature which was used for the initialization does not stay constant during the fusion of the α particles and background neutrons which takes place in the first few hundred fm/c of the calculation. Therefore the real temperature is measured by observing the excitation energy from the ground state. For a detailed discussion see [13]. Note that even if the α particles are distributed with zero momentum and the background neutrons states are filled up to the Fermi energy, the resulting real temperature is about 7 MeV.

3. The map of pasta shapes

The left side of Fig. 1(a) shows the typical shapes obtained as final states at the lowest temperature. The surfaces of these states are in slight motion because of the high temperature but the topology of the shapes is stable. Additionally to the rod and slab shapes also connected rods, namely rod(2) and rod(3), were observed. The number denotes the connectivity of the rods in the shapes. Similar shapes have been discovered in [7, 10, 14].

On the right side of Fig 1(b) the distribution of the shapes in the density-temperature plane is shown. For the lowest value for the temperature the shapes are well ordered from spherical to rod, rod(2) and rod(3) mixed with the slab phase followed by the corresponding bubble shapes in the reverse order. Note that the slab and rod(3) shapes represent their own bubble shapes because liquid and gas phase are topologically identical. For higher temperature the phases are not ordered anymore and they are overlapping each other. For increasing density smaller temperatures already produce uniform matter.

shape	sph	rod	rod(2)	rod(3)	slab	rod(2) b	rod b	sph b
W_2	>0	>0	>0	- to +	≈ 0	<0	<0	<0
W_3	>0	$=0$	<0	<0	$=0$	<0	$=0$	>0

Table 1. Signs of the integral mean curvature and Euler characteristic as a function of ρ_{th}.

To distinguish the single pasta shapes clearly by a well-defined measure, Minkowski functionals [4] are utilized as already done in [15, 16, 3]. These Minkowski functionals are proportional to the volume $W_0 \propto V$, the surface area $W_1 \propto A$, the integral mean curvature $W_2 \propto \int_{\partial K} \mathrm{d}A(\kappa_1 + \kappa_2)/2$ and the Euler characteristic $W_3 \propto \chi$. The Euler characteristic is a topological constant. It is the number of connected components added to the number of cavities decreased by the number of holes in a structure.

As the densities are given as gray scale data but the Minkowski functionals need a well defined surface, it is useful to calculate the measures in dependence on a threshold density ρ_{th}. If the density on the grid is bigger than ρ_{th} the point is in the liquid phase, otherwise it lies in the gas phase. To reduce voxelization errors, the marching cube algorithm [17] is applied. Then the Minkowski functionals of the polygons are evaluated [18, 19]. It turns out that the different pasta shapes can be identified by the signs of W_2 and W_3 only. The results are shown in Table 1.

The distinction between a pasta phase and the uniform matter phase is done with the cell-density distribution (s. bold line in Fig. 1(a) right side). If the matter tends to be uniform every point on the grid has a density near the mean density. It turned out that the cell-density distribution is approximately a Gaussian. Thus the variance is small. If a pasta phase is the final state, one peak at low densities corresponding to the gas phase and one peak at high densities corresponding to the liquid phase will be built up and therefore the variance is much larger. It came out, that a value which distinguishes pasta from uniform matter is about $1 - 2 \cdot 10^{-4}\mathrm{fm}^{-6}$.

4. The gyroidal structure

In the mid-density range of the map of pasta shapes it emerges, that there are minimal surface-like shapes present. A minimal surface is characterized by a mean curvature equal to zero at each point of the surface. For the slab shape this holds trivially, but also the rod(3) shape shown in Fig. 1(a) left side is topologically equivalent to a minimal surface called primitive (p-) surface. The gyroid is closely connected to this surface via the Bonnet transformation [20] and was discussed as a candidate for pasta matter [21, 2]. In systems with an intrinsic length scale the gyroid should appear at higher box length than the p-surface does. Therefore we made additional calculations with a box length of 22 fm with a mean density of 0.06 fm^{-3} and a temperature of 7 MeV. The initialization was prepared such that the α particles have as large as possible distances between each other but that it is still possible to have random distributions. The experience with the calculations with smaller box length showed that this avoids slab shapes.

Besides shapes which could not be assigned to a certain topological structure, we observed gyroidal shapes as final states of these calculations (s. Fig. 2). These structure were mostly strongly deformed but could be identified to be topologically identical to the gyroid by fitting a gyroid network into the calculated structure.

Although the gyroidal shape appeared only in about 30% of the calculations, the spontaneous ordering of a randomized scenario to a gyroidal structure shows that the gyroid may be at least a meta stable state for pasta matter.

5. Conclusion and Outlook

In this work matter at subnuclear densities was studied by initializing α particles and background neutrons with a certain amount of kinetic energy randomly over a grid with periodic boundary

Figure 2. Dividing surface of a gyroidal pasta structure. The simulated surface is repeated three times in x and y direction. Cyan corresponds to the side facing the liquid phase, magenta to the gas phase. A gyroid network was added (gray/red) to identify the gyroidal structure.

conditions. The time evolution was done with the TDHF method. After about 1000 fm/c the states evolve to topologically constant states. The obtained map of pasta shapes agrees qualitatively with the diagram calculated in QMD [3] calculations. The different shapes are uniquely characterized by the sign of the integral mean curvature and the Euler characteristic and the distinction from uniform matter can be realized by the cell-density distribution.

Additionally gyroidal structures can be identified which are formed from a random initial state at a box length of 22 fm in the mid-density range. This hints that the gyroid could be important at least as a meta stable ground state. This should be investigated further with calculations considering the ground state with varying box lengths and mean densities.

Acknowledgments

This work was supported by the BMBF under contract numbers 06ER9063 and 06FY9086, the DFG under grant ME1361/11 and by Grants-in-Aid for Scientific Research on Innovative Areas through No. 24105008 provided by MEXT. The calculations were performed at the Center for Scientific Computing of Goethe-Universität Frankfurt. B.S., K.I., and J.A.M. acknowledge the hospitality of the Yukawa Institute for Theoretical Physics, where this work was initiated.

6. References

[1] Umar A S, Strayer M R, Reinhard P G, Davies K T R and Lee S J 1989 *Phys. Rev. C* **40**(2) 706–714
[2] Nakazato K, Iida K and Oyamatsu K 2011 *Phys. Rev. C* **83**(6) 065811
[3] Sonoda H *et al.* 2008 *Phys. Rev. C* **77**(3) 035806
[4] Mecke K and Stoyan D 2000 *Statistical Physics and Spatial Statistics - The Art of Analyzing and Modeling Spatial Structures and Pattern Formation* 1st ed (*Lecture Notes in Physics* vol 554) (Springer)
[5] Pethick C J and Ravenhall D G 1995 *Ann. Rev. Nucl. Part. Sci* **45** 429–484
[6] Chamel N and Haensel P 2008 *Living Rev. Relativ.* **11** 10
[7] Newton W G and Stone J R 2009 *Phys. Rev. C* **79**(5) 055801
[8] Ravenhall D G, Pethick C J and Wilson J R 1983 *Phys. Rev. Lett.* **50**(26) 2066–2069
[9] Williams R and Koonin S 1985 *Nucl. Phys.* **435** 844 – 858 ISSN 0375-9474
[10] Sébille F, Figerou S and de la Mota V 2009 *Nucl. Phys. A* **822** 51 – 73 ISSN 0375-9474
[11] Sébille F, de la Mota V and Figerou S 2011 *Phys. Rev. C* **84**(5) 055801
[12] Bender M, Heenen P H and Reinhard P G 2003 *Rev. Mod. Phys.* **75**(1) 121–180
[13] Schuetrumpf B, Klatt M A, Iida K, Maruhn J A, Mecke K and Reinhard P G 2013 *Phys. Rev. C* **87**(5) 055805
[14] Pais H and Stone J R 2012 *Phys. Rev. Lett.* **109**(15) 151101
[15] Watanabe G *et al.* 2003 *Phys. Rev. C* **68**(3) 035806
[16] Watanabe G *et al.* 2004 *Phys. Rev. C* **69**(5) 055805
[17] Lorensen W E and Cline H E 1987 *SIGGRAPH Comput. Graph.* **21** 163–169 ISSN 0097-8930
[18] Schröder-Turk G E *et al.* 2010 *ArXiv e-prints* (*Preprint* 1009.2340)
[19] Schröder-Turk G E *et al.* 2011 *Adv. Mater.* **23** 2535–2553 ISSN 1521-4095
[20] Hyde S, Andersson S, Larsson K, Blum Z, Landh T, Lidin S and Ninham B W 1996 *The Language of Shape* (Elsevier Science; 1 edition)

[21] Nakazato K, Oyamatsu K and Yamada S 2009 *Phys. Rev. Lett.* **103**(13) 132501

NanoSIMS and more: New tools in nuclear astrophysics

P Hoppe

Max Planck Institute for Chemistry, Hahn-Meitner-Weg 1, 55128 Mainz, Germany

E-mail: peter.hoppe@mpic.de

Abstract. Primitive Solar System materials contain nm- to µm-sized presolar grains that formed in the winds of evolved stars and in the ejecta of stellar explosions. These samples of stardust can be analysed in terrestrial laboratories with sophisticated analytical instrumentation in great detail. Of particular importance are coordinated studies of individual grains by Secondary Ion Mass Spectrometry (SIMS), Resonance Ionization Mass Spectrometry (RIMS) and Focused Ion Beam/Transmission Electron Microscopy (FIB/TEM) from which detailed information on isotopic compositions and mineralogies can be obtained. A key tool is the NanoSIMS 50 ion probe which permits to do isotope measurements of light and many intermediate-mass elements with spatial resolutions of <100 nm. A new type of RIMS instrument, "CHILI", is currently under construction and is aimed to provide <100 nm resolution for isotope studies of intermediate-mass and heavy elements. Another promising analysis technique for future studies is Atom Probe Tomography (APT) which might be useful to create 3D-elemental and isotopic maps of presolar grains at the nanometer scale.

1. Introduction

Isotope studies are important tools in cosmochemical research to obtain detailed information on chemical and physical processes involved in the formation and alteration of extraterrestrial materials. One type of extraterrestrial material which is of high relevance for nuclear astrophysics are the so-called presolar grains [1, 2]. Presolar grains are nanometer to micrometer in size and are found in small quantities (up to several 100 ppm) in primitive meteorites, interplanetary dust particles (IDPs), and cometary matter. They are older than our Solar System and formed in the winds of evolved stars and in the ejecta of stellar explosions. Presolar grains exhibit large isotopic anomalies, the nucleosynthetic fingerprints of their parent stars. Among the identified presolar grains are diamonds, graphite, silicon carbide (SiC), silicon nitride (Si_3N_4), refractory oxides, mostly corundum and other forms of Al_2O_3 and spinel ($MgAl_2O_4$), and various silicates. These samples of stardust can be analyzed in the laboratory with sophisticated analytical instrumentation in great detail. Figure 1 illustrates the transport of presolar grains from their stellar sources to terrestrial laboratories.

Of particular importance are coordinated studies of individual grains involving Secondary Ion Mass Spectrometry (SIMS), Resonance Ionization Mass Spectrometry (RIMS), and Focused Ion Beam/Transmission Electron Microscopy (FIB/TEM). Such studies have provided a wealth of astrophysical information, not only on stellar nucleosynthesis and evolution, but also on Galactic chemical evolution, dust formation in stellar environments, chemical and physical processes in the interstellar medium, the types of stars that contributed dust to our Solar System, and on processes in the early Solar System.

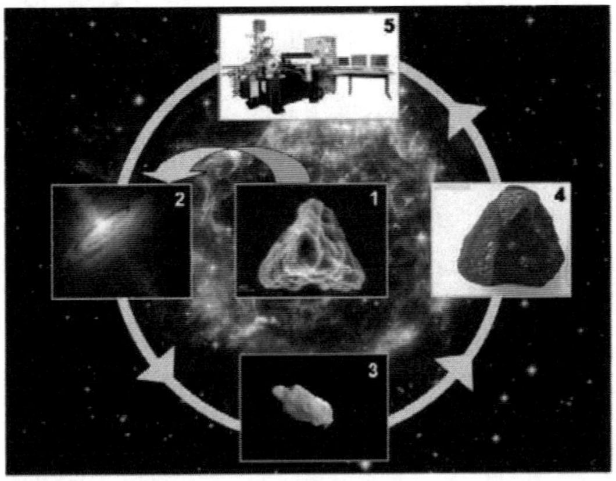

Figure 1. Path of presolar grains from their stellar sources to terrestrial laboratories. After formation in the winds of evolved stars or in the ejecta of stellar explosions (1) and passage through the interstellar medium they became part of the molecular cloud from which our Solar System formed (2). They survived the early history of Solar System formation inside small planetary bodies (3) and are carried to the Earth by meteorites and IDPs (4). Finally they can be studied in detail for isotopic compositions and physical properties in the laboratory (5). Picture credits: NASA/JPL-Caltech (2), NASA/JPL (3), J. Koblitz (4), NASA/JPL-Caltech/STScl/Cxc/SAO (background).

FIB/TEM provides detailed information on the mineralogy and physical properties of individual presolar grains [3, 4]. The latest generation SIMS instrument, the NanoSIMS 50 ion probe, permits to do isotope measurements of the light and intermediate mass elements at the <100 nm scale, making it a powerful tool to study a large number of isotope systems in presolar grains [5]. By NanoSIMS ion imaging it is possible to create isotope maps which can be used (i) to search for presolar grains with specific isotopic fingerprints, e.g., the rare SiC supernova (SN) grains (figure 2), among chemical grain separates and (ii) to identify presolar grains in-situ in thin or FIB sections of meteorites and IDPs. RIMS has been used to measure the isotopic compositions of the heavy elements in presolar grains at the micrometer scale. A new type of RIMS instrument, "CHILI", is currently under construction and is aimed to provide <100 nm resolution and better sensitivity [6]. Another promising analysis technique for future studies is Atom Probe Tomography (APT) which might be useful to create 3D-elemental (and possibly isotopic) maps of presolar grains at the nanometer scale [7].

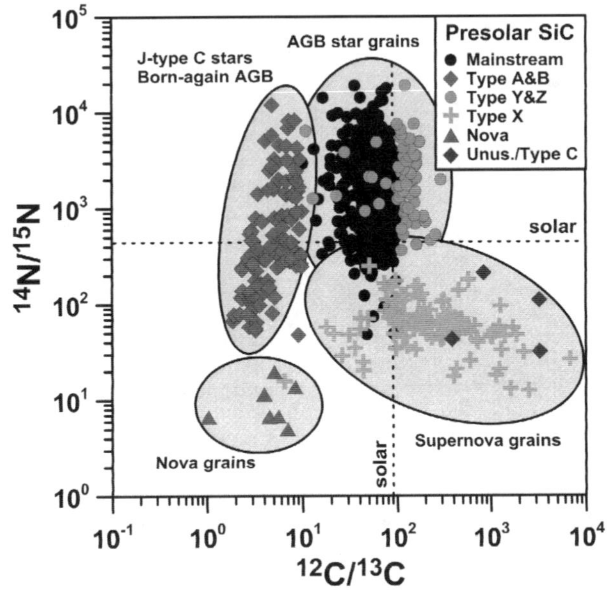

Figure 2. Carbon- and N-isotopic systematics of presolar SiC grains. The vast majority (>90%) of the grains originate from AGB stars. Grains from SNe are the type C and X grains which make up about 1-2 % of all SiC grains. The solar C- and N-isotopic ratios are indicated by the dashed lines. From [8].

In the following I will briefly describe technical aspects of the NanoSIMS (chapter 2), of RIMS (chapter 3), and of APT (chapter 4) as well as their applications in presolar grain research.

2. The NanoSIMS ion probe

In SIMS a solid surface is bombarded by a finely focused primary ion beam. This triggers a collision cascade in the target which causes atoms and small molecules from the upper layers of the sample to be ejected. A small fraction (up to the percent range) of these sputtered particles is ionized and can be analyzed in a mass spectrometer. For a detailed description of the SIMS technique see, e.g., [9].

The Cameca NanoSIMS 50 ion probe is a double-focusing, magnetic sector, multi-collecting SIMS instrument [10, 11]. Primary ions are either oxygen (O^-) or cesium (Cs^+). O^- primary ions favor formation of positive secondary ions (e.g., alkali, alkaline earth, and transition metals) and Cs^+ formation of negative secondary ions (e.g., C, O, CN, Si, S). The ion optics of the NanoSIMS was specifically designed for high spatial resolution (i.e., small primary beam diameter) and high transmission for secondary ions; for a schematic of the ion optics see figure 3. A mass resolution of several 1000 is easily achieved which permits to do isotope measurements of the light and of many intermediate-mass elements.

The NanoSIMS is characterized by four fundamental features, which makes it an extremely useful tool to study presolar grains: (i) High spatial resolution of down to 50 nm for Cs^+ primary ions and down to 200 nm for O^- primary ions. (ii) High useful yields (= detected ions/sputtered atoms) up to the percent level for certain elements. (iii) Simultaneous detection of up to 7 isotopes (depending on the configuration of the instrument). (iv) Ion imaging capabilities (see below).

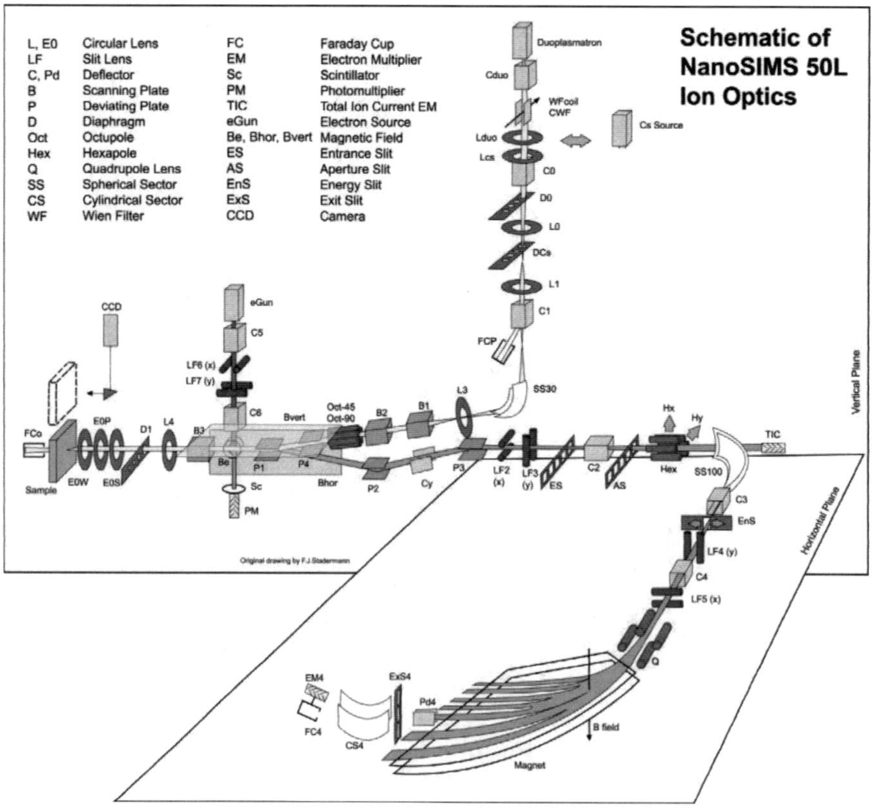

Figure 3. Schematic of the ion optics of the NanoSIMS 50. Picture credit: François Hillion and Frank Stadermann.

Ion imaging is a key feature of the NanoSIMS. Ion images can be acquired by rastering the focused primary ion beam over a region of interest. Samples are typically individual grains dispersed on clean substrates (e.g., Au foils) or meteoritic thin and FIB sections. Specific procedures were developed for grain samples that permit to measure isotopic compositions of a large number of individual grains in a fully automated way in relatively short time [12, 13]. These procedures were used to measure C- and Si- or O-isotopic compositions with Cs^+ primary ions and consist basically of three steps: (i) Ion images of selected isotopes are recorded with a spatial resolution of ~100 nm for a square region with typical side length of 20-30 μm (figure 4). (ii) Individual grains are identified by grain recognition routines in one of the ion images. (iii) Individual grains are measured with much higher precision than in step i in that only small areas with ~2 times the grain size around each identified grain are measured. (iii) Movement of the sample stage to an adjacent area and repetition of steps i to iii. Application of this technique has been particularly useful for the search of rare presolar grain types, e.g., the SiC SN C grains, which exhibit strong enrichments in the heavy Si isotopes [14-16], or oxide grains with extreme O-isotopic compositions [13].

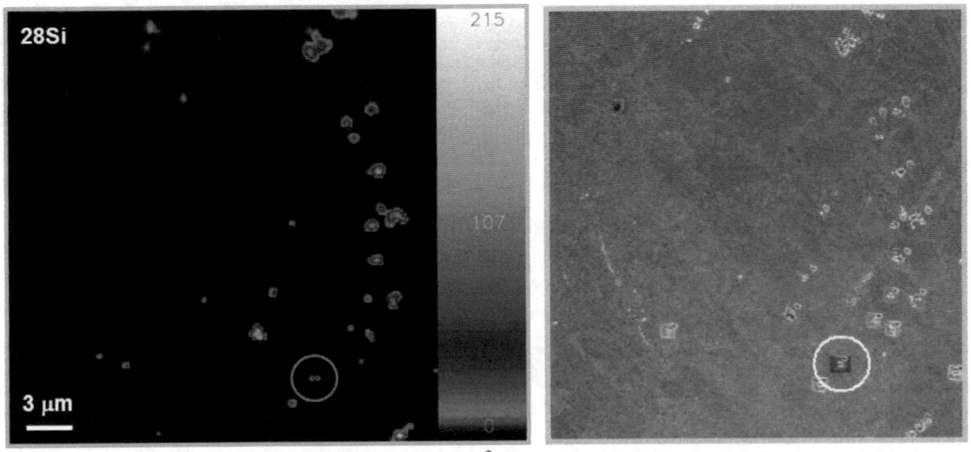

Figure 4. Left: $^{28}Si^-$ ion image (30 x 30 μm^2) of presolar SiC grains from the Murchison CM2 meteorite dispersed on a Au foil. The grain inside the green circle is a submicrometer-sized SN X grain, as indicated by its Si-isotopic signature. Right: Scanning electron microscopy (SEM) image of the same area; several NanoSIMS measurement fields of individual grains (cf. step ii in automated ion imaging) are clearly visible in this picture.

The identification of these rare grains provided new insights into nucleosynthesis and dust formation in explosive environments. One of the oxide grains has massive enrichments in ^{17}O, ^{25}Mg, and ^{26}Mg, which are isotopic signatures indicative of condensation from nova ejecta [13] (figure 5). Another extremely ^{16}O-rich grain shows clear evidence for extinct ^{44}Ti, which can be considered as proof for the SN origin of ^{16}O-rich grains [13] (figure 5). For SiC C grains follow-up measurements showed that many of them have isotopically light S (i.e., ^{32}S enrichments). This is hard to understand in the context of current Type II SN (SNII) models [17] as Si- and S-isotopic signatures (i.e., light, heavy, or solar) follow each other in the different zones. Hoppe et al. [15] invoked preferential incorporation of S from the Si/S zone (which has light S) over S from other zones resulting from molecule chemistry in still unmixed SN ejecta to account for this observatiom. More recently, Pignatari et al. [18] presented an alternative explanation, namely, decay of radioactive ^{32}Si (half life 153 yr). This nuclide is predicted to be produced by explosive He burning in SNeII in large amounts which, if incorporated by the growing SiC grains and after decay, could fully account for the observed ^{32}S enrichments in C grains.

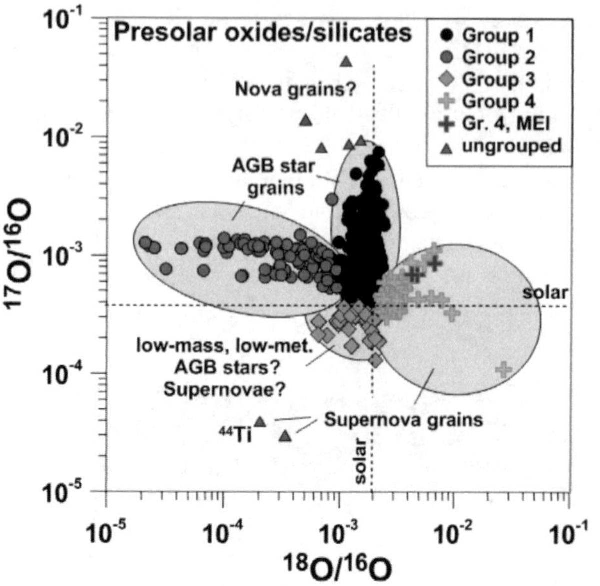

Figure 5. Oxygen-isotopic compositions of different populations of presolar silicate and oxide grains. The solar O-isotopic ratios are indicated by the dashed lines. The ^{16}O-rich SN grain with evidence for extinct ^{44}Ti and the grain with extreme ^{17}O enrichment (possibly a nova grain) were identified by automated ion imaging in the NanoSIMS [13]. From [8].

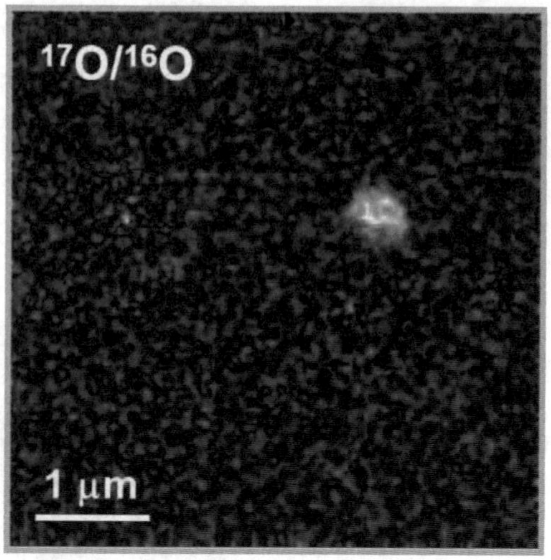

Figure 6. $^{17}O/^{16}O$ ion image of a 5 x 5 μm^2-sized area in the matrix of the Acfer 094 meteorite. This image was acquired with a 50 nm-sized primary Cs^+ ion beam. A presolar silicate grain (490 x 250 nm^2) with an ^{17}O enrichment of a factor of ~3 stands out as bright hotspot. From [19].

Ion imaging of meteoritic thin or FIB sections has been particularly important to identify presolar silicates which have typical sizes of 200-400 nm and which, contrary to refractory presolar oxides and carbonaceous presolar material, can not be separated by chemical treatments from meteorites. They were identified by O ion imaging with the NanoSIMS, first in an IDP [20] and later also in meteorites [21, 22]. Presolar silicates were also identified with a SCAPS detector on a Cameca IMS 1270 ion probe [23]; however, this technique reliably identifies only the largest (>500 nm) presolar grains. NanoSIMS O ion imaging of meteoritic thin and FIB sections is now a well-established technique and hundreds of presolar silicates (along with some presolar oxides which have lower abundances) have been found in the matrix of primitive meteorites and in IDPs (figure 6). Apart from getting information on nucleosynthetic processes in and evolution of their parent stars, observed abundance variations among different primitive Solar System materials provide information on thermal and aqueous alteration on their parent bodies and on possible abundance variations in the Solar nebula [24, 25]. Isotope data other than for O are rare for presolar silicates. This is because the Si^- ion yield is relatively low in O-rich environments, resulting in comparatively large (compared to SiC) uncertainties for Si-isotopic ratios. Other major elements of silicates, e.g., Mg and Fe, must be measured with the O^- primary ion source. Here, the size of the primary ion beam is of comparable size to that of presolar silicates and measured isotope ratios strongly suffer from dilution effects with surrounding solar material. FIB sample preparation techniques have been developed to minimize dilution effects and Mg isotope data were obtained for some grains [26, 27].

3. Resonance Ionization Mass Spectrometry

A fundamental difference between SIMS and RIMS is that RIMS uses the usually >99% of neutral particles released by ion sputtering or laser desorption for mass spectrometric analysis. RIMS can be used to measure the isotopic compositions of most elements in the periodic table; several of the major and minor elements in presolar grains, e.g., C, N, O, and S, however, are not accessible by RIMS. Most of the measurements on presolar grains made by RIMS so far were done with an instrument named CHARISMA (Chicago-Argonne Resonant Ionization Spectrometer for Mass Analysis). In this instrument material from a grain is desorbed by a pulsed laser (Nd-YAG) beam, the element of interest is resonantly ionized with tunable Ti-sapphire lasers, and ions are finally analyzed by time-of-flight (TOF) mass spectrometry [28]. Useful yields of CHARISMA are in the percent range and useful and important results were obtained for the isotopic compositions of intermediate-mass and heavy trace elements, e.g., Sr, Zr, Mo, Ru, and Ba in individual, micrometer-sized presolar grains [29-33].

Currently, a new type of RIMS instrument, "CHILI" (Chicago Instrument for Laser Ionization), is under construction at the University of Chicago [6] (figure 7). This instrument employs a finely focused Ga ion beam with nominal beam diameter of 2.5 nm for sputtering. The sputtered, neutral particles are resonantly ionized by six tunable Ti-sapphire lasers. CHILI is aimed at measuring isotopic compositions of intermediate-mass and heavy elements at sub-micrometer resolution, possibly even with better than 50 nm. Useful yields are much higher than for CHARISMA, reaching values of 40%. This is particularly important as CHILI is intended to carry isotope measurements to much smaller scales than is possible with CHARISMA. CHILI will not only allow to expand isotope measurements to additional trace elements and to access smaller grains but at <50 nm scales even subgrains within presolar grains become accessible for isotope studies. Alltogether this has great potential to get new insights into nucleosynthetic processes and dust formation in stellar environments.

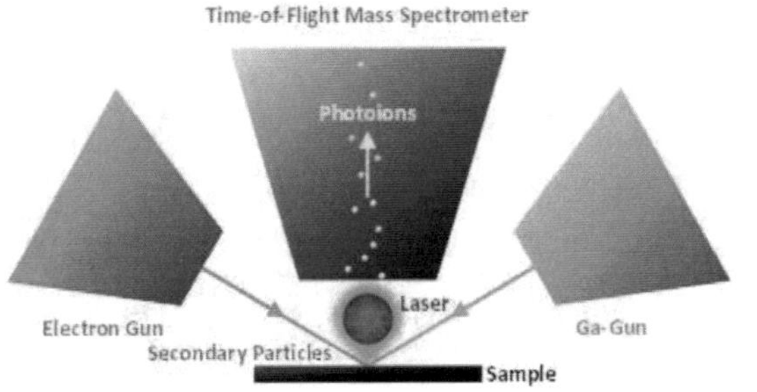

Figure 7. Principle of CHILI. A liquid Ga ion gun is used for sputtering. Photo ions are produced by resonance ionization with tunable Ti-sapphire lasers and analyzed with a time-of-flight mass spectrometer. An electron gun is available to produce secondary electron images of the sample. Picture credit: Thomas Stephan.

4. Atom Probe Tomography

Atom probe tomography is an established technique in the materials science [34]. In recent years, attempts have been made to apply APT to the study of presolar grains [7]. APT uses so-called field evaporation and offers chemical and isotopic analyses with sub-nm spatial resolution. Of particular importance is the preparation of the sample which must be shaped to a very sharp tip (figure 8, left). A high voltage is applied to the sample which results in a very strong electrostatic field close to the tip. Atom removal is triggered by a laser pulse and ionization occurs due to the strong electroststatic field. Ionization is close to 100 % and ions are mass-analyzed (time-of-flight) and monitored with a position-sensitive detector. Repeated application of the laser pulse permits to create 3-dimensional chemical and isotopic maps at the nm- or even sub-nm scale.

A potential application of APT of high interest in presolar grain research are measurements of C-isotopic ratios of individual meteoritic diamonds. Meteoritic diamonds are only 2-3 nm in size, i.e., individual grains are not accessible by currently used isotope measurement techniques. Bulk diamond samples exhibit isotope anomalies in the trace elements Xe and Tc [35, 36], indicative of a presolar origin. But because bulk C- and N-isotopic ratios of meteoritic diamonds are essentially solar [37], it is not clear which fraction of them is true stardust. Attempts to measure C-isotopic ratios in individual diamonds by ATP are currently ongoing [7] (figure 8) and it will be interesting to see whether reliable and useful results can be obtained. If successful, use of APT could open new and interesting applications in presolar grain research.

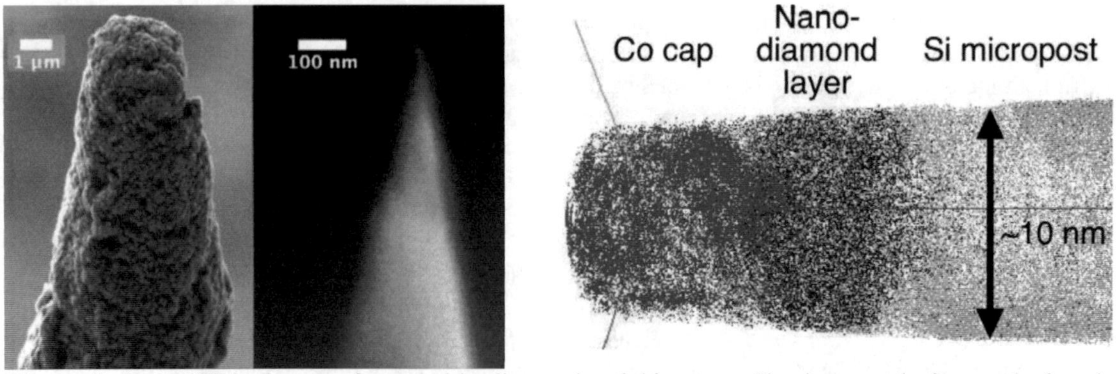

Figure 8. Left: SEM picture of meteoritic diamonds sticking to a Si micropost before and after tip sharpening. Right: 3D reconstruction of the positions of atoms in a diamond layer sandwiched by a Co cap and the Si micropost. From [7].

Acknowledgements
I wish to thank Daniel Galaviz and his team for the organization of the NPA-VI conference and for inviting me to present this paper.

References

[1] Zinner E 2007 Presolar Grains In *Meteorites, Comets, and Planets* Vol. 1.02 (ed. A. M. Davis) pp. 1 Elsevier
[2] Zinner E 2013 *these Proceedings*
[3] Vollmer C, Brenker F, Hoppe P, and Stroud R M 2009 *Astrophys. J.* **700** 774
[4] Zega T J, Nittler L R, Busemann H, Hoppe P, and Stroud R M 2007 *Meteorit. Planet. Sci.* **42** 1373
[5] Hoppe P, Cohen S, and Meibom A 2013 *Geostandards Geoanalytical Res.* **37** 111
[6] Stephan T, Pellin M J, Rost D, Davis A M, Savina M R, Trappitsch R, and Liu N 2013 CHILI The Chicago Instrument for Laser Ionization. In *44th Lunar and Planetary Science Conference*Abstract #2536 Lunar and Planetary Institute.
[7] Heck P R, Pellin M J, Davis A M, Isheim D, Seidman D N, Hiller J, Mane A, Elam J, Savina M R, Stephan T, Stadermann F J, Zhao X, Daulton T L, Floss C, and Amari S 2011 Atom-Probe Tomographic Analyses of Meteoritic Nanodiamond Residue from Allende. In *42nd Lunar and Planetary Science Conference*Abstract #2070 Lunar and Planetary Institute.
[8] Hoppe P 2011 *Proceedings of the 11th Symposium on Nuclei in the Cosmos (NIC XI). July 19 - July 23, 2010 Heidelberg, Germany.* Available online at http://pos.sissa.it/cgi
[9] Benninghoven A, Rüdenauer F G, and Werner H W 1987 *Secondary ion mass spectrometry: Bascic concepts, instrumental aspects, applications, and trends* Wiley
[10] Hillion F, Daigne B, Girard F, and Slodzian G 1993 A new high performance SIMS instrument: The Cameca "Nanosims 50" In *Secondary Ion Mass Spectrometry SIMS IX* (ed. A. Benninghoven, Y. Nihei, R. Shimizu, and H. W. Werner) pp. 254-257 John Wiley & Sons Ltd.
[11] Slodzian G, Daigne B, Girard F, and Hillion F 1993 Ion optics for a high resolution scanning ion microscope and spectrometer : transmission evaluations In *Secondary Ion Mass Spectrometry SIMS IX* (ed. A. Benninghoven, Y. Nihei, R. Shimizu, and H. W. Werner) pp. 294-297 John Wiley & Sons Ltd.
[12] Gröner E and Hoppe P 2006 *Appl. Surf. Sci.* **252** 7148
[13] Gyngard F, Zinner E, Nittler L R, Morgand A, Stadermann F J, and Hynes K M 2010 *Astrophys. J.* **717** 107
[14] Gyngard F, Nittler L R, and Zinner E 2010 *Met. Planet. Sci.* **45** A72

[15] Hoppe P, Fujiya W, and Zinner E 2012 *Astrophys. J.* **745** L26
[16] Xu Y C, Amari S, Gyngard F, Zinner E, and Lin Y 2012 *Meteorit. Planet. Sci.* **47** 5104
[17] Rauscher T, Heger A, Hoffman R D, and Woosley S E 2002 *Astrophys. J.* **576** 323
[18] Pignatari M, Zinner E, Bertolli M G, Trappitsch R, Hoppe P, Rauscher T, Fryer C, Herwig F, Hirschi R, Timmes F X, and Thielemann F-K 2013 *Astrophys. J.* **771** L7 (5pp)
[19] Hoppe P, Leitner J, and Groener E 2013 Search for Small Supernova Silicates in the Acfer 094 Meteorite by High-Resolution NanoSIMS Ion Imaging. In *44th Lunar and Planetary Science Conference*Abstract #1038 Lunar and Planetary Institute.
[20] Messenger S, Keller L P, Stadermann F, Walker R M, and Zinner E 2003 *Science* **300** 105
[21] Mostefaoui S and Hoppe P 2004 *Astrophys. J.* **613** L149
[22] Nguyen A N and Zinner E 2004 *Science* **303** 1496
[23] Nagashima K, Krot A N, and Yurimoto H 2004 *Nature* **428** 921
[24] Floss C and Stadermann F 2009 *Geochim. Cosmochim. Acta* **73** 2415
[25] Leitner J, Vollmer C, Hoppe P, and Zipfel J 2012 *Astrophys. J.* **745** 38
[26] Kodolanyi J and Hoppe P 2011 *Proceedings of the 11th Symposium on Nuclei in the Cosmos (NIC XI). July 19 - July 23, 2010 Heidelberg, Germany.* Available online at http://pos.sissa.it/cgi
[27] Nguyen A, Messenger S, Ito M, and Rahman Z 2010 *Lunar Planet. Sci.* **41** abstract #2413
[28] Savina M R, Pellin M J, Tripa C E, Veryovkin I V, Calaway W F, and Davis A M 2003 *Geochimica et Cosmochimica Acta* **67** 3215
[29] Nicolussi G K, Davis A M, Pellin M J, Lewis R S, Clayton R N, and Amari S 1997 *Science* **277** 1281
[30] Nicolussi G K, Pellin M J, Lewis R S, Davis A M, Amari S, and Clayton R N 1998 *Geochim. Cosmochim. Acta* **62** 1093
[31] Savina M R, Davis A M, Tripa C E, Pellin M J, Clayton R N, Lewis R S, Amari S, Gallino R, and Lugaro M 2003 *Geochim. Cosmochim. Acta* **67** 3201
[32] Savina M R, Davis A M, Tripa C E, Pellin M J, Gallino R, Lewis R S, and Amari S 2004 *Science* **303** 649
[33] Nicolussi G K, Pellin M J, Lewis R S, Davis A M, Clayton R N, and Amari S 1998 *Phys. Rev. Lett.* **81** 3583
[34] Seidman D N 2007 *Ann. Rev. Mater. Res.* **37** 127
[35] Lewis R S, Tang M, Wacker J F, Anders E, and Steel E 1987 *Nature* **326** 160
[36] Richter S, Ott U, and Begemann F 1998 *Nature* **391** 261
[37] Russell S S, Arden J W, and Pillinger C T 1996 *Meteoritics Planet. Sci.* **31** 343

Accelerator Mass Spectrometry in Laboratory Nuclear Astrophysics

O Nusair[1], W. Bauder[2], G Gyürky[3], M Paul[4], P Collon[2], Zs Fülöp[3], J Greene[1], N Kinoshita[5], T Palchan[1], R Pardo[1], K E Rehm[1], R Scott[1], R Vondrasek[1]

[1] Physics Division, Argonne National Laboratory, Argonne, IL 60439
[2] Physics Department, University of Notre Dame, Notre Dame, IN 46556
[3] MTA Atomki, H-4001 Debrecen, Hungary
[4] Racah Institute of Physics, Hebrew University, Jerusalem, Israel 91904
[5] Tandem Accelerator Complex, Research Facility Ctr for Science and Technology, U. of Tsukuba, Tsukuba, Japan

Corresponding author's E-mail: paul@vms.huji.ac.il

Abstract. The extreme sensitivity and discrimination power of accelerator mass spectrometry (AMS) allows for the search and the detection of rare nuclides either in natural samples or produced in the laboratory. At Argonne National Laboratory, we are developing an AMS setup aimed in particular at the detection of medium and heavy nuclides, relying on the high ion energy achievable with the ATLAS superconducting linear accelerator and on gas-filled magnet isobaric separation. The setup was recently used for the detection of the ^{146}Sm *p*-process nuclide and for a new determination of the ^{146}Sm half-life (68.7 My). AMS plays an important role in the measurement of stellar nuclear reaction cross sections by the activation method, extending thus the technique to the study of production of long-lived radionuclides. Preliminary measurements of the ^{147}Sm(γ,*n*)^{146}Sm are described. A measurement of the ^{142}Nd(α,γ)^{146}Sm and ^{142}Nd(α,*n*)^{145}Sm reactions is in preparation.A new laser-ablation method for the feeding of the Electron Cyclotron Resonance (ECR) ion source is described.

1. Introduction

The activation method provides a powerful method of integral cross section measurements for nuclear astrophysics. While the method is mostly used by measuring the radiation emitted in the decay of a reaction product, accelerator mass spectrometry (AMS), through direct counting of the nuclides produced, is able to extend the measurements to cases where the half-life of the product is long ($t_{1/2}$ > 10 y) resulting in unmeasurable decay rates or where the radiation emitted with the decay is hard to detect (*e.g.* electron capture). In recent years, AMS has been widely used in nuclear physics and astrophysics [1]. We review in Section 2 the principle of accelerator mass spectrometry and its use for the determination of nuclear cross sections. In Section 3, we focus on reactions of importance to the understanding of *p*-process nuclide production. Preliminary AMS measurements of the ^{147}Sm(γ,*n*)^{146}Sm reaction cross section are described. (α,γ) reactions in the rare-earth region, shown to be sensitive to calculated abundances of *p*-process nuclides in this region [2], were studied recently [3,4] by γ- and characteristic X-ray decay counting and show systematic deviations from results of statistical calculations. The case of the ^{142}Nd(α,γ)^{146}Sm reaction, in preparation for AMS measurements, is presented and a new technique of sample feeding by laser ablation is described for AMS analysis.

2. Accelerator mass spectrometry and measurements of nuclear reaction cross sections

Radioactive nuclides are detected at ultra-high sensitivity by the technique of accelerator mass spectrometry [5]. AMS is based on individual atom identification and counting, which competes favorably with decay counting (conventional α, β,or γ spectroscopy) for half-lives long compared to

 doi:10.1088/1742-6596/665/1/012076

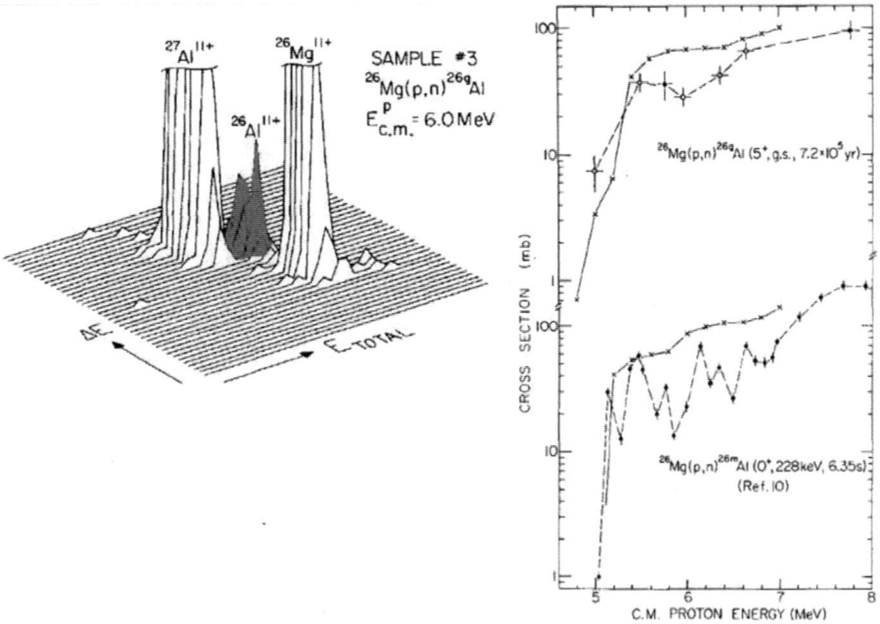

Figure 1: Measurement of the cross section of the ^{26}Mg(p,n)^{26g}Al($t_{1/2}$= 0.72 My) reaction [6], an early application of accelerator mass spectrometry in nuclear astrophysics. The inverse reaction is a destruction process for the ^{26g}Al nuclide. The reaction was studied by bombarding ^{26}Mg targets with a proton beam and, after chemical extraction of the ^{26}Al atoms with a stable Al carrier (Al*) of known amount, the number of ^{26}Al product nuclei and the corresponding cross section was determined through the ratio ^{26}Al/Al* measured by AMS.

typical laboratory measurement times. The quantity measured by AMS is the isotopic abundance of a specific rare nuclide relative to a stable isotope (or in case of actinides where stable isotopes do not exist, a neighboring nuclide taken as reference) and in favorable cases the isotopic abundance measured can be as low as 10^{-15}-10^{-16} with a typical precision of 2-10% (or governed by counting statistics), except for ^{14}C where dedicated facilities achieve subpercent precision [5]. The key in AMS to unambiguous ion identification leading to measurements of these ultra-low abundances is the use of fast accelerated ions: (*i*) the determination of the atomic number Z of a detected ion by nuclear detection methods leads to discrimination (and sometimes physical separation) of isobaric (same A) species. Techniques employed rely on measurement of the differential energy loss *dE/dx*, of the range of an ion in matter or the average ionic charge state (Z-dependent) in a low-pressure gas; (*ii*) the elimination of stable (and possibly overwhelming) molecular ions having same mass number is achieved by stripping in gas or a thin foil (or as described below, in a high charge-state positive ion source), resulting in fragments which are either separated by magnetic and electrostatic analysis or eventually discriminated in the particle detector.

The analysis of these low isotopic abundances in samples with masses typically in the mg range leads to the capability of detecting unambiguously as few as ~10^4 rare atoms present in the sample without relying on their decay, a sensitivity unparalleled by other techniques. It should be emphasized that this capability is mostly limited to unstable species because chemical impurities usually cause unmanageable background in measurements of stable nuclides. The technique proves powerful for the measurement of nuclear reaction cross sections leading to unstable species. Fig. 1 illustrates the first application of AMS in nuclear astrophysics where the cross section of the reaction ^{26}Mg(p,n)^{26g}Al was determined at energies just above threshold [6]; the inverse process is a destruction reaction of the important radioactive nuclide ^{26g}Al ($t_{1/2}$= 0.72 My), observed in the last decade by γ-astronomy.

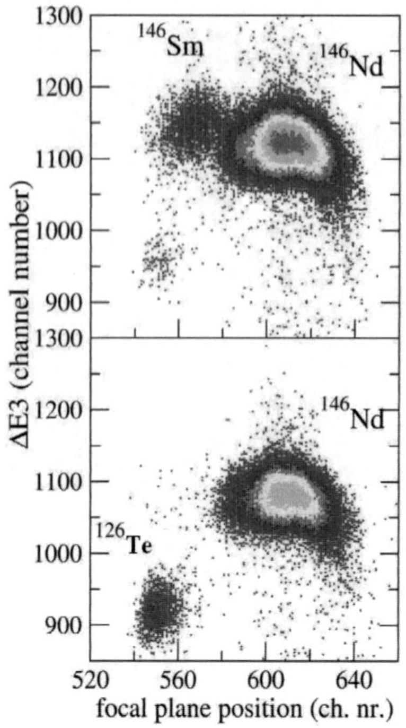

Figure 2: Identification spectrum of ions accelerated from a fast-neutron irradiated ^{147}Sm sample (top) and a natural Sm sample (bottom). The *x*-axis represents distance along the focal plane of the gas-filled spectrograph and the *y*-axis is an energy loss signal. ^{146}Sm produced by the ^{147}Sm(*n*,2*n*) reaction are observed, well separated from stable isobaric ^{146}Nd and chemical impurities, transported by the accelerator due to *m/q* degeneracy. The sensitivity limit determined by the background counts in the ^{146}Sm region for the natural sample corresponds to an isotopic abundance in the range ^{146}Sm/^{147}Sm~10^{-12}.

Measurements of the cross section of nucleosynthesis reactions by AMS have been intensively performed in recent years, especially for *s*-process neutron capture reactions [1]. A few reactions induced by charged-particle have also been performed, e.g. ^{40}Ca(α,γ)^{44}Ti [7], ^{33}S(α,p)^{36}Cl [8]. The method provides an extension of the conventional activation method for products which are long-lived or whose decay radiation is difficult to detect. Importantly, the measurement is independent of the half-life of the reaction product or branching ratios in the product decay.

While most of the reactions studied via AMS were directed to light or medium mass nuclei, we presently use the Argonne ATLAS facility for the detection of heavier nuclides. The higher energies attainable allow us to extend AMS to the rare-earth region, still discriminating efficiently isobaric nuclides. A sample of the material to be measured is fed into the plasma of an Electron Cyclotron Resonance (ECR) ion source which produces highly-charged positive ions. The process eliminates already at this stage any molecular ions. Following magnetic analysis, the ions are accelerated by linear acceleration in ATLAS to energies of about 8 MeV/*u* and analyzed in a gas-filled magnetic spectrograph [9]. Ions are dispersed in the spectrograph according to $mv/\langle q \rangle$ where *v* is the ion velocity (reduced by slowing down in the gas) and $\langle q \rangle$ is an average charge state along the ion trajectory in the gas, separating parasitic ions transported by the accelerator and isobaric nuclides. Fig. 2 shows an identification spectrum showing physical separation of ^{146}Sm from stable isobaric ^{146}Nd and from chemical impurities in the sample. The method was successfully used to detect the nuclide ^{146}Sm and make a new determination of its half-life (68.7 My) [10].

3. p-process studies in the rare-earth region

The ^{146}Sm nuclide, produced like stable ^{144}Sm by the *p*-process, is particularly interesting in view of its presence in the Early Solar System, established by measurements of isotopic anomalies in its ^{142}Nd alpha-daughter abundances in meteorites [11]. Unlike stable nuclides whose Galactic history cannot be established, a live radioactive nuclide must have been synthesized a few half-lives before the Solar system formation and as such, ^{146}Sm had been recognized very early as a cosmochronometer of the *p*-process [12]. The study, both experimental and theoretical, of nuclear reactions responsible for the production of *p*-process nuclei in this region is under intense investigation [13]. It was noted for example that the rate of (γ,α) reactions (or of the inverse (α,γ) reactions) impact sensitively on the calculated abundances of *p*-process nuclides in the region A >140 and N > 82 [2]. Moreover, an experimental study of the ^{144}Sm(α,γ)^{148}Gd reaction showed that the energy dependence of the

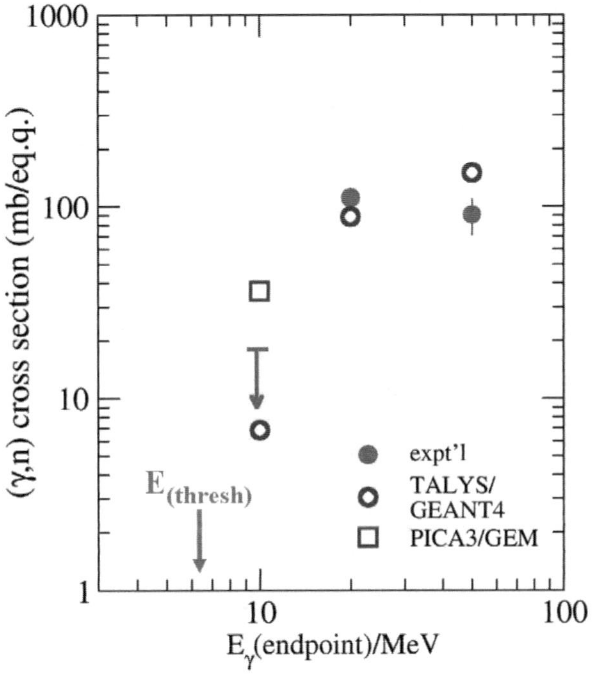

Figure 3: Preliminary measurements of the photonuclear reaction ^{147}Sm(γ,*n*)^{146}Sm performed by electron bremstrahlung irradiation of a ^{147}Sm target and AMS determination of the ^{146}Sm/^{147}Sm ratio.

reaction is hard to understand theoretically [14]. While Somorjai et al. [14] used energy-dependent alpha-nucleus optical potentials to reproduce the data, it was recently shown that taking into account coupled channels involving the Coulomb excitation of the nucleus does not require the use of special optical potentials [15].We are presently applying the ^{146}Sm detection technique described above to the study of nuclear reactions in this region of nuclides. The photonuclear reaction ^{147}Sm(γ,*n*)^{146}Sm was studied by irradiating an enriched ^{147}Sm target by bremsstrahlung photons. The irradiation experiments were performed at Tohoku University and Kyoto University. Preliminary results for endpoint energies of 50 to 10 MeV, demonstrating the feasibility of the study, are shown in fig.3.

The experimental study of the reactions ^{142}Nd(α,γ)^{146}Sm and ^{142}Nd(α,*n*)^{145}Sm is in preparation. Fig. 4 shows the results of theoretical calculations for these reactions. Based on the background conditions shown in fig. 4, we estimate that an AMS measurement of ^{146}Sm production could reach a cross section sensitivity of ~20 μb, corresponding to incident energies down to ~ 11 MeV. Test irradiations of a natural Nd target and an enriched ^{142}Nd target (both as oxide Nd_2O_3, evaporated by Ar sputtering onto a 1.7 μm Al foil) were performed with an alpha particle beam of 13.5 MeV from the ATOMKI facility. Preliminary analysis shows the presence of $^{145}Sm(t_{1/2}= 340\ d)$ through γ- and characteristic X-ray activities of the ^{145}Pm decay product, measured directly from the irradiated targets. After gamma counting, samples were prepared for AMS analysis of ^{146}Sm , using an ad-hoc procedure. The targets are dissolved in a HCl solution together with a known amount (typically 5 mg) of natural Sm carrier and, after transfer to a HNO3 solution, Al is separated in a chromatographic column of TRU resin.

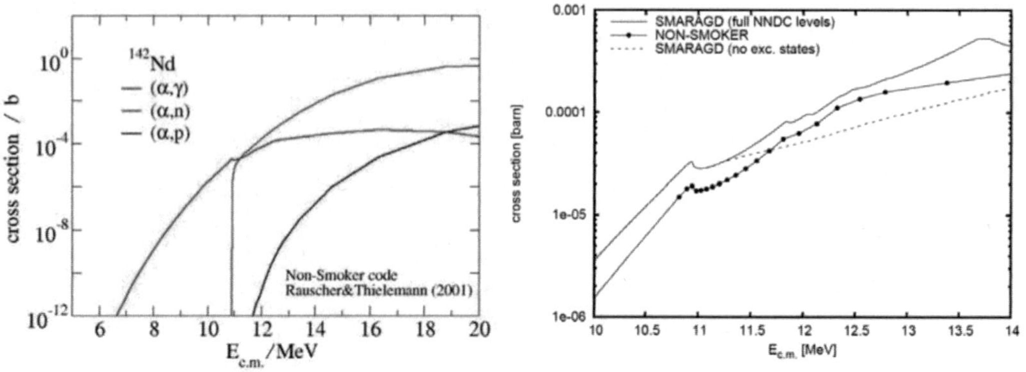

Figure 4: Hauser-Feshbach statistical model calculations of alpha-induced reactions on ^{142}Nd: (left) NON-SMOKER code [16]; (right) SMARAGD code [17]. Calculations were performed (a) taking into account the set of 40 low-lying excited levels in the ^{146}Sm compound nucleus from the NNDC compilation (2010) (labeled full NNDC levels) and (b) with no experimental levels, using a theoretical level density.

The Nd fraction of the residual solution is then carefully separated by a Ln-resin column in order to reduce as much as possible presence of isobaric ^{146}Nd prior to the AMS measurement and the purified Sm fraction (containing $^{145,146}Sm$ activation products) is finally eluted from the Ln column. After precipitation and ignition, the final residues are prepared as Sm_2O_3 oxide samples for AMS. The measurement of the $^{146,145}Sm/Sm$ ratio and the amount of Sm carrier by AMS determine the number of $^{146,145}Sm$ products in the irradiations. It is important to note that this determination based on isotopic ratios is independent of chemical or detection efficiency. These samples are presently awaiting AMS analysis.

4. Laser-ablation accelerator mass spectrometry

In parallel, a new method of material feeding to the ECR ion source is developed using laser ablation of the solid sample [18]. Fig. 5 illustrates the layout of the ion source and the laser system. Laser ablation is performed with a 1064 nm Nd:YAG laser with a 15 ps pulse width, variable repetition rate up to 400 Hz and energy up to 5 mJ per pulse with a maximum power of 3×10^8 W in the pulse. The target material resides inside the ECR plasma chamber in the injection end and the laser pulses enter the ECRIS through the beam extraction hole and are focused to a diameter of 0.5mm at target. Target materials are ejected a few cm into the ECR plasma where they are ionized to a distribution of positive charge states. The ablation by the laser beam, well-focused on the sample surface, is expected to reduce the amount of contaminants entering the plasma compared to conventional methods of sample feeding by sputtering or oven evaporation. In a preliminary experiment using a reagent-grade samarium oxide (Sm_2O_3) sample, charge currents of $^{147,152}Sm^{22,23+}$ in the nA range adequate for AMS analysis were successfully obtained. Figure 6 shows the spectrum of ^{152}Sm ions produced by laser ablation and identified in the focal plane of the gas-filled spectrograph. In the conditions of the experiment, Sm ion rates are ~7-fold higher with laser ablation (laser on) than with laser off. The intensity of an observed ^{126}Xe group (transported by the accelerator as $^{126}Xe^{19+}$ ions due to a *m/q* near degeneracy with $^{152}Sm^{23+}$ ions) attributed to residual atmospheric Xe in the ion source plasma chamber (at a pressure of 10^{-7} Torr) is unchanged by laser irradiation of the Sm_2O_3 sample.

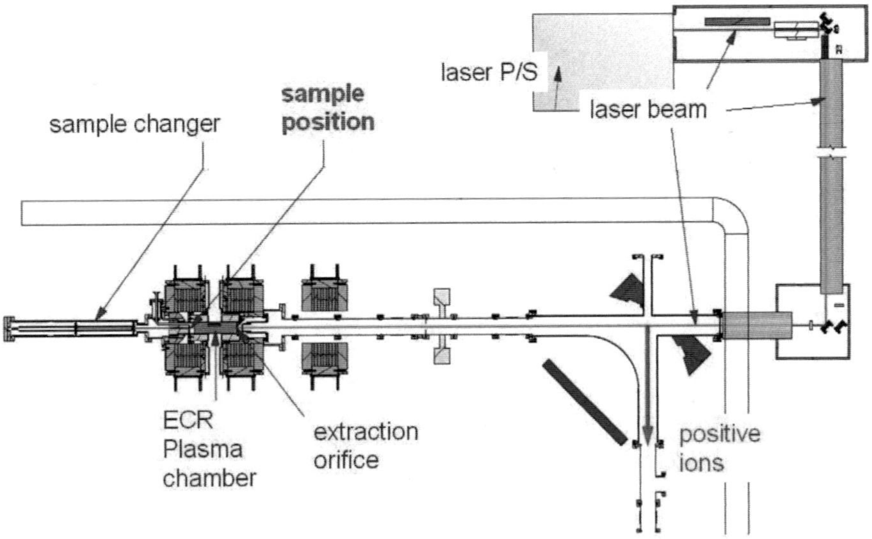

Figure 5: Layout of the laser ablation setup at the ATLAS ECR ion source. The sample is selected and loaded in position at the back edge of the ECR plasma chamber by a multisample changer [19]. The laser beam is directed towards the sample through a zero-degree port of the analysis magnet onto the sample surface co-linearly with the exiting ion beam. Ablation by the fast (15 ps) pulsed laser at a typical energy of 3 mJ/pulse causes a plume of released atoms which penetrates the plasma and gets multiply ionized by the RF accelerated electrons.

5. Conclusion

Accelerator mass spectrometry extends the scope of activation measurements conventionally performed by decay counting to cases of long-lived products or otherwise practically undetectable decay radiations. Although generally more complex, the technique which measures directly the number of nuclides produced by a nuclear reaction, offers also the advantage of being independent of the product half-life or decay branching ratios. The further extension of AMS to heavier elements in the rare-earth region requires higher acceleration energies. The setup at Argonne National Laboratory is being developed for the study of (α,γ) reactions, known to be particularly sensitive to the calculations of abundances of *p*-process nuclides in that region. The new technique of laser ablation feeding of sample material shows good promise for reduction of contaminants from the ion source and for use of smaller-size samples.

6. Acknowledgments

This work was supported by U.S. Department of Energy, Office of Nuclear Physics, under Contract No. DE-AC02-06CH11357 and by OTKA grant K108459.

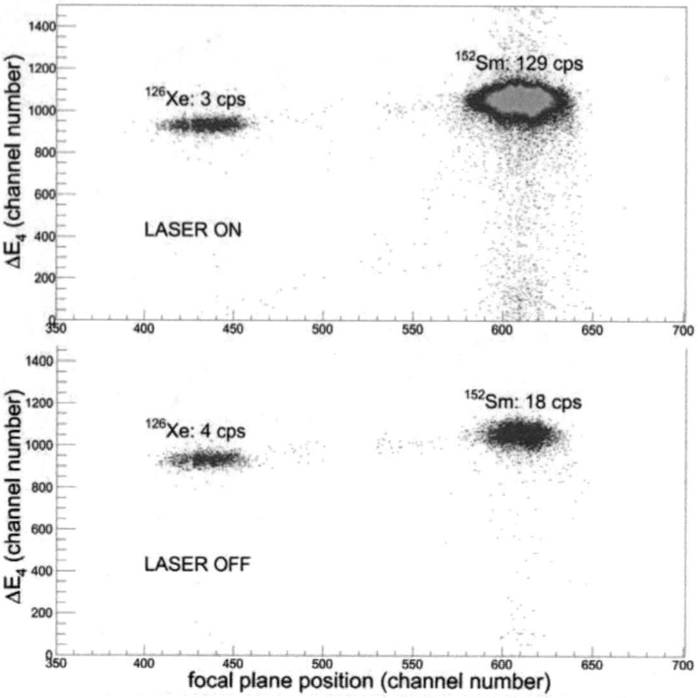

Figure 6: Identification spectra collected in the gas-filled magnetic spectrograph collected with laser ablation on (top) and off (bottom) using a natural Sm_2O_3 sample. The group of ^{126}Xe ions is atributed to residual atmospheric Xe in the ion source plasma and its intensity is unchanged by the laser irradiation while the output of Sm ions is enhanced ~7-fold.

References

[1] A. Wallner, Nucl. Inst. Meth. Nucl. Res. B **268**, 1277 (2010)
[2] W. Rapp *et al.*, Astrophysical J., **653**, 474 (2006)
[3] Gy. Gyürky *et al.*, J. Phys. G: Nucl. Part. Phys. **37,** 115201 (2010)
[4] G.G. Kiss *et al.*, Phys. Lett. B **695**, 419 (2011)
[5] W. Kutschera, Int. J. Mass Spect. **349-350**, 203 (2013)
[6] M. Paul *et a*l., Phys. Lett. **94B,** 303 (1980)
[7] H. Nassar *et al.*, Phys. Rev. Lett. **96**, 041102 (2006)
[8] M. Bowers *et al.*, to be published.
[9] M. Paul *et al.*, Nucl. Instr. and Methods A **277**, 418 (1989)
[10] N. Kinoshita *et al.*, Science **334**, 1614 (2012)
[11] A. Prinzhofer, D.A. Papanastassiou and G.J. Wasserburg, Geochim. Cosmochim. Acta, **56**, 797 (1992)
[12] J. Audouze and D.N. Schramm, Nature **237**, 447 (1972)
[13] T. Rauscher *et al.*, Rep. Prog. Phys. **76**, 066201 (2013)
[14] E. Somorjai *et al.*, Astron. Astrophys. **333**, 1112 (1998)
[15] T. Rauscher, Phys. Rev. Lett. **111**, 061104 (2013)
[16] T. Rauscher and F.-K. Thielemann, At. Data Nucl. Data Tables **79**, 47 (2001)
[17] T. Rauscher, priv. comm.
[18] R. Scott *et al.*, Proc. 15th Int. Conf. on Ion Sources (ICIS2013), 9-13 Sep, 2013, Chiba, Japan; Rev. Sci. Instr., accepted for publication.
[19] R. Vondrasek *et al.*, Proc. 15th Int. Conf. on Ion Sources (ICIS2013), 9-13 Sep, 2013, Chiba, Japan; Rev. Sci. Instr., to be published.

Study of nuclear reactions producing ^{36}Cl by micro-AMS

H Luís[3], A P Jesus[1,2], M Fonseca[1,2], J Cruz[1], D Galaviz[2], N Franco[3], E Alves[3]

[1]Dep. Física, Faculdade de Ciências e Tecnologia, Universidade Nova de Lisboa, 2829-516 Caparica, Portugal
[2]Centro de Física Nuclear da Universidade de Lisboa, 1649-003 Lisboa, Portugal
[3] Campus Tecnológico e Nuclear (CTN), Instituto Superior Técnico, Estrada Nacional 10 2695-066 Bobadela, Portugal

E-mail: heliofluis@gmail.com

Abstract. ^{36}Cl is one of several short to medium lived isotopes (as compared to the earth age) whose abundances at the earlier solar system may help to clarify its formation process. There are two generally accepted possible models for the production of this radionuclide: it originated from the ejecta of a nearby supernova (where ^{36}Cl was most probably produced in the s-process by neutron irradiation of ^{35}Cl) and/or it was produced by in-situ irradiation of nebular dust by energetic particles (mostly, p, a, ^{3}He -X-wind irradiation model).
The objective of the present work is to measure the cross section of the ^{37}Cl(p,d)^{36}Cl and ^{35}Cl(d,p)^{36}Cl nuclear reactions, by measuring the ^{36}Cl content of AgCl samples (previously bombarded with high energy protons and deuterons) with AMS, taking advantage of the very low detection limits of this technique for chlorine measurements.
For that, the micro-AMS system of the LFI/ITN laboratory had to be optimized for chlorine measurements, as to our knowledge this type of measurements had never been performed in such a system (AMS with micro-beam).
Here are presented the first results of these developments, namely the tests in terms of precision and reproducibility that were done by comparing AgCl blanks irradiated at the Portuguese National Reactor with standards produced by the dilution of the NIST SRM 4943 standard material.

1. Introduction

The motivation for measuring cross sections of nuclear reactions producing ^{36}Cl lies in the field of astrophysics that studies the origin and formation of the solar system. ^{36}Cl is one of the short-lived radionuclides (called SLR's for abbreviation), thus named because of their short half-lives (~1 Myrs), whose abundances in the early stages of the formation of the solar system is much higher than expected from the galactic background. These SLR's are ^{7}Be ($t_{1/2}$= 53 days), ^{10}Be ($t_{1/2}$= 1.4 Myrs), ^{26}Al ($t_{1/2}$= 0.74 Myrs), ^{36}Cl ($t_{1/2}$= 0.3 Myrs), ^{41}Ca ($t_{1/2}$= 0.1 Myrs), ^{53}Mn ($t_{1/2}$= 3.5 Myrs) and 60Fe ($t_{1/2}$= 1.5 Myrs), and are known to have been present in the early stages of the solar system through the presence of their daughter isotopes in CAIS (calcium and aluminium rich inclusions) and chondrule present in chondrites, which are the most common form of meteorites found on earth [1]. These structures are thought to be amongst the oldest solid material present in the solar system, and they can provide important information about its formation.

The analysis of Cl-S systematics of sodalite in carbonaceous chondrites yielded a ratio of ^{36}Cl/^{35}Cl around 10^{-6}. Since soladite is in principle a late-stage product of aqueous alteration, this ratio's initial value should have been higher, assuming that the ^{36}Cl supernova injection happened in the prestellar core stage, as it is supposed to have happened with the other SLRs. Following this principle, the expected ^{36}Cl/^{35}Cl ratio at the time of the SLRs injection has been estimated to be around 10^{-4}. This value is already too high for the estimated supernova injection of ^{36}Cl/^{35}Cl~10^{-6}. However, recent analysis of wadalite in the allende carbonaceous chondrite seems to make the SN injection of ^{36}Cl scenario even more implausible, yielding a ratio of ^{36}Cl/^{35}Cl ~= (1.72±0.25) x 10^{-5}, making the initial ratio even higher than 10^{-4} [2]. This seems to definitely invalidate the possibility of ^{36}Cl injection by a supernova. Therefore the irradiation of nebular dust by energetic particles (mostly, p, a, ^{3}He -X-wind irradiation model) becomes a strong candidate to explain the abundance of the ^{36}Cl isotope.

It is in this context that the research group is currently working on measuring the ^{37}Cl(p,d)^{36}Cl and ^{35}Cl(d,p)^{36}Cl nuclear reactions, using the Micro-AMS system at LFI/ITN to measure the amount of ^{36}Cl produced in the AgCl irradiated samples, starting by the optimization of this system for ^{36}Cl measurements.

2. The Micro-AMS system at LFI/ITN

The Micro-AMS system present at LFI/ITN is the former AUSTRALIS system developed by Sie et al. [3] at CSIRO. It is based on a 3 MV tandem accelerator. Its main feature is the ion source, a HICONEX cesium sputter source that was adapted in order to produce a Cs^{+} micro-beam that can be focused to very small diameter (30 um), and a low-energy and high-energy bouncing system that allows quasi-simultaneous measurement of different isotopes.

It was developed with the goal of using the capabilities of a SIMS system (spatial and in-depth analysis of bulk samples) connected to an AMS system that provides the possibility of resolving isobaric and molecular interferences, that in many cases spoil the detection limits of SIMS analysis.

3. Adaptation of the system to the measurement of ^{36}Cl

Measurement of AgCl samples is very common in conventional AMS but has never been tried with Micro-AMS, since its lower intensity primary beam implies much lower detection limits for ^{36}Cl, while in principle not bringing any important advantage. However, since it is possible to produce ^{36}Cl (through the astrophysically relevant ^{37}Cl(p,d)^{36}Cl and ^{35}Cl(d,p)^{36}Cl nuclear reactions) in enough quantity to bring the isotopic ratios present in the sample to the $^{36}Cl/^{35}Cl>10^{-10}$ range, the micro-beam AMS technique becomes a valid technique to measure cross-sections while eventually providing some advantages as, for instance, direct measurement of the irradiated samples without further chemical manipulation, since with the micro-beam it is possible to measure on the exact beam spot of the irradiation in the sample.

The original AUSTRALIS system was not prepared for very low isotopic ratio measurements (below 10^{-5}), and so adaptations had to be made in order for it to be possible to measure the high intensity beams (^{35}Cl and ^{37}Cl) and very low intensity beams (^{36}Cl) quasi-simultaneously, which was done by installing off-axis faraday cups after the Low-energy and high-energy magnets. Another important step was to minimize sulfur contamination in the target chamber.

Sulfur contamination is a problem in conventional Chlorine AMS measurements, since the ^{36}S isotope is an isobaric interference of ^{36}Cl. Up until recently an effective isobar separation of ^{36}S and ^{36}Cl was impossible using 3 MV tandems. However, work developed in ETHZ Zurich and VERA in Vienna [4,5] in the last years has successfully achieved separation for ^{36}Cl and ^{36}S using a 3 MV tandem.

However, these separation procedures for 3 MV tandems were not tried in the case of this work because the current state of the ITN tandem does not permit it to operate much above 2,5 MV. But since the Micro-AMS uses a primary cesium micro-beam, which is on average three orders of magnitude less intense than the conventional AMS primary beam, this means that its detection limits fall by that amount or more when compared to AMS. Our current detection limit, measured by analyzing a blank AgCl sample (produced with sulfur removal chemical treatment [6]) and monitoring the detector background, yielded a background level below $^{36}S/^{35}Cl = 10^{-10}$.

It is in principle possible to keep the ^{36}S reaching the detector on the level of $^{36}S/^{35}Cl = 10^{-12}$, by keeping the sample chamber very clean and by taking special steps in the design of the sample holder, besides using the well-known chemical sample preparation methods for the standards and for the blank AgCl material to be used as blank material and irradiation material.

The contamination of ^{36}S present in the sample holder is a well-known problem in conventional AMS. The micro-beam used in Micro-AMS is an advantage in this case, because it is easier to assure that the whole bombarded area is locate strictly within the target limits, as opposed to conventional AMS where the large diameter beam, maximized for intensity, will easily impact the sample holder as well during operation. However, since there is no possibility to separate ^{36}S from he ^{36}Cl in our system, it is important to assure that no sulfur contamination may be present anywhere close to the sample. This means that the chamber has to be as clean as possible, and the sample holder must be designed so as to minimize any contamination as well.

The original sample holder was covered with a sheet of tantalum. In each of the positions where the samples were to be mounted, a round disk of two centimeters in diameter of compressed AgBr was installed. The AgCl samples (also a compressed powder) were then mounted at the center of these disks. Using AgBr to cover the sample holder in the sample area was shown in the literature [7] to be the most efficient material in minimizing sulfur contamination in the chlorine beam.

Since cross-contamination of samples can happen easily, the maximum number of samples per run was set at 4. This way it was possible to keep the samples at a reasonable distance (> 3 cm) from each other, and still be able to have a blank sample, a standard, and two unknown samples in each run.

4. Measurements

The AgCl samples were irradiated in the RPI nuclear reactor at ITN using the neutron capture nuclear reaction $^{35}Cl(n,\gamma)^{36}Cl$. Neutron flux and irradiation times were set to produce $^{36}Cl/^{35}Cl$ isotopic ratios in the 10^{-6} to 10^{-9} range. The thermal neutron fluxes were controlled with gold monitor samples, by measuring the 411,8 keV gamma emission, emitted by the decay of the ^{198}Au ($t_{1/2}$=2,695 d) produced by the neutron capture reaction $^{197}Au(n,\gamma)^{198}Au$ with a HPGe detector, after each neutron irradiation.

Most of the measurements were performed on irradiated samples with isotopic ratios in the 10^{-6} and 10^{-7} range. The high content of ^{36}Cl in this samples made it easier to tune the system. The standards were prepared to have an isotopic ratio of $^{36}Cl/^{35}Cl = 1 \times 10^{-6}$, by diluting the NIST SRM 4943 standard. Due to the lower intensity of cesium primary beam, the memory effect was not important and it was monitored with a blank sample after each measurement of sample or standard.

5. Results

Figure 1 shows a measurement performed on an irradiated sample. Figure 2 shows three different normalized measurements on the same sample (each of the measurements was succeeded by a standard measurement which was used for normalization); the first two in the same day but with somewhat different beam tunings and the last, two days after the first two. The highest error bar is in the 4% precision range.

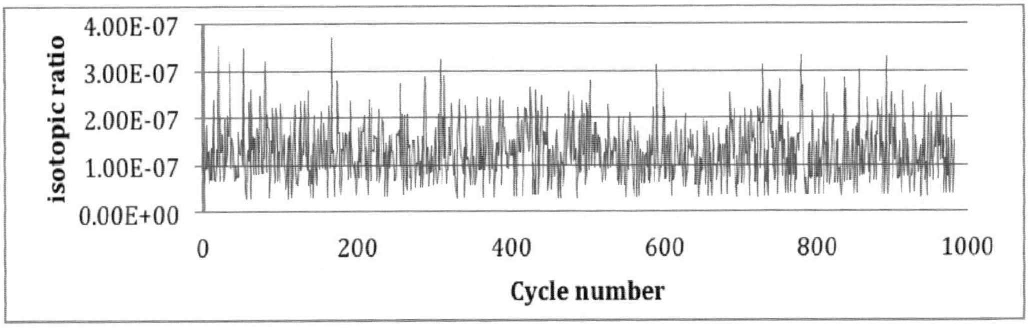

Figure 1: $^{36}Cl/^{35}Cl$ isotopic ratio measured in a irradiated sample vs. cycle number. Each of the cycles represents a 100 ms measurement time for each of the three chlorine isotopes.

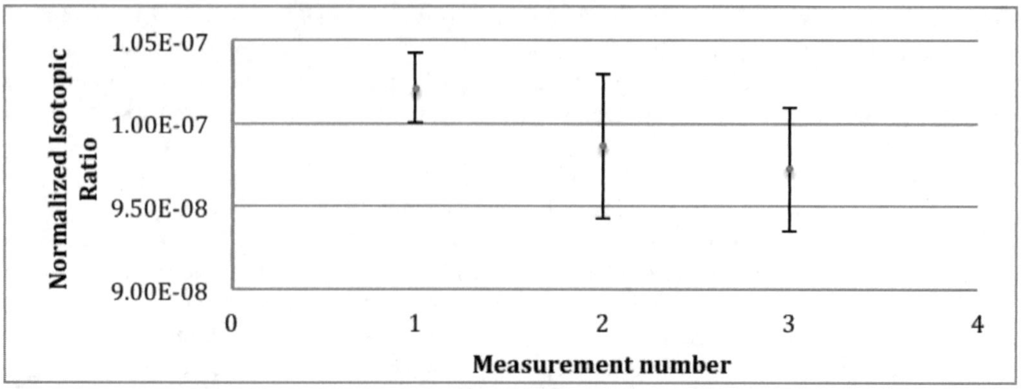

Figure 2: Isotopic ratios of an irradiated sample normalized for the diluted standard. Measurement number 1 and 2 were performed on the same day. Measurement number 3 was performed two days later.

6. Conclusions

The results so far show on the one hand a good agreement in terms of the ^{36}Cl content of the samples between the irradiated samples and the diluted standards and a good reproducibility of the measurements. This validates both the neutron irradiations as a valuable process for the production of standards and the chemical procedures used for the production of standards and blank material.
However, the precision of the measurements is still not as good as expected. This is a problem as the objective of the work is to measure samples with less ^{36}Cl than the ones measured so far, which means an even lower precision, as less ^{36}Cl means less counts per second in the particle detector.

In order to achieve a better precision, steps are being taken to increase the primary beam intensity as well as decreasing the measurement cycle time, especially for the ^{35}Cl and ^{37}Cl beams, by connecting the Faraday cups to a faster current integrator.

References

[1] M. Gounelle, A. Meibom, The Astrophysical Journal, 680 (2008) 781–792
[2] M. Chaussidon, M. Gounelle, C. R. Geoscience 339 (2007) 872–884
[3] S. H. Sie et al., Rev. Sci. Inst. 69, 3 (1998)
[4] Peter Steier et al., Nucl. Instr. and Meth. B B 268 (2010) 744–747
[5] M. Martschini et al. , Nucl. Instr. and Meth. B B 269 (2011) 3188–3191
[6] Y. Nagashima, R. Seki, T. Takahashi, D. Arai, Nucl. Instr. and Meth. B 172 (2000) 129.
[7] Finkel, R. et al , Nucl. Instr. and Meth. B 294 (2013) 121-125.

Nuclear Physics in Astrophysics VI (NPA6) IOP Publishing
Journal of Physics: Conference Series **665** (2016) 012078 doi:10.1088/1742-6596/665/1/012078

The Standard Solar Model and beyond

S Turck-Chièze
SAp/IRFU/CEA, CE Saclay, 91191 Gif sur Yvette, France
E-mail: Sylvaine.Turck-Chieze@cea.fr

Abstract. The Standard Solar Model (SSM) is an important reference in Astrophysics as the Sun stays today the most observed star. This model is used to predict the internal observables like neutrino fluxes and oscillation frequencies and consequently to validate its assumptions for its generalization to other stars. The model outputs result from the resolution of the classical stellar equations and the knowledge of fundamental physics like nuclear reaction rates, screening, photon interaction, plasma physics. The plasma conditions remained unmeasurable in laboratory for long due to the high temperature and high density conditions of the solar interior. Today, neutrino detections and helioseismology aboard SoHO have largely revealed the solar interior, in particular the nuclear solar core so one can estimate the reliability of SSM and also its coherence with the different indicators and between them. This has been possible thanks to a Seismic Solar Model (SeSM) which takes into account in addition the observed sound speed profile. Seismology quantifies also some internal dynamical processes that need to be properly introduced in the description of stars. This review describes the different steps of building of the SSM, its predictions and the comparisons with observations. It discusses the accuracy of such model compared to the accuracy of the SeSM. The noticed differences and observational constraints put some limits on other possible processes like dark matter, magnetic field or waves and determine the directions of progress for the near future that will come from precise emitted neutrino fluxes. High density laser facilities promise also unprecedented checks of energy transfer by photons and nuclear reaction rates.

1. The historical context

Standard Solar Model (SSM) produces the thermodynamical profiles of the present Sun together with all the useful observables. It results from the resolution of stellar evolution equations from the contraction of the Sun (or from the arrival on the main sequence) up to the present age. This model is a reference in Astrophysics, because it is useful for establishing the chemical evolution of the galaxy through helium and oxygen evolution and for validating the physics which relies stellar evolution to cosmology through the age of globular clusters...

SSM has been strongly developed in the eighties, stimulated by the coming observations and detections which were for the first time compared to its detailed predictions. Today it stays a basis on which dynamical processes are slowly introduced.

The first seismic observations appeared in 1962 but have been understood only around 1970, then ground networks have been deployed, the space helioseismology began with the launch of SoHO in December the 2nd 1995. This satellite is still observing the Sun. On board, the resonant spectrometer GOLF [1] is dedicated to the core of the Sun and the MDI instrument looks locally to it, exploring mainly the above layers. The observations of GOLF and of the successor of MDI, HMI on SDO, will continue up to at least 2016. Using both instruments, and in particular the high quality of the GOLF instrument, a precise sound speed profile has been used to extract precise neutrinos fluxes in 2001 [2]. In parallel, the first neutrino limit by Homestake was given in 1968, then followed by Kamiokande, Gallex, SAGE, Super Kamiokande, SNO and BOREXINO. It is well known that the SNO detection has been determinant in the understanding between predictions, detections and neutrinos oscillations, already suspected with Gallex and SAGE results [3]. The originality of SNO is due to its ability to get charge current and neutral current estimates, so to determine the flux of the three different flavours of neutrinos coming initially from the Sun, after transformation of some electronic neutrinos in vacuum and matter in muon and tau neutrinos. The same year 2001, the neutrino puzzle has been definitively solved by the confrontation of seismic predictions of boron flux [2] to SNO neutrino counts [4]. Their remarkable agreement (also with SSM predictions at that time) has been crucial in the attribution of the Nobel Prize to R. Davis and M. Koshiba in 2002, the two precursors of neutrino detections in USA and Japan. For details, quantitative values of these two fields, see the two reviews [5, 6]. The generalization of helioseismology to other stars (asteroseismology) has more recently shown its great potential with the space COROT (2008) and KEPLER (2010) satellites.

Today, 10 years after the Nobel Prize, it is interesting to see what we can say about the SSM, how it compares to the seismic model and to the detailed observations and what we can deduce from the present significant discrepancies. The discussion is justified by the recent progress obtained on the solar photospheric composition and on some reaction rate which has led to poorer neutrino predictions of the SSM.

2. The building of the Standard Solar Model

SSM satisfies the well known structural equations of stellar evolution applied to any star. These equations describe the hydrostatic equilibrium, the conservation of mass and energy and the transport of energy by photons in the radiative zone and by convective motions in the more external layers. The reference [5] summarizes the detailed models and predictions along the last two decades. The resolution of the two first equations (hydrostatic equilibrium and conservation of mass) needs to use a proper equation of state of the plasma that determines P $(\mathrm{T}, \rho, \mathrm{X_i})$, respectively pressure, temperature, density and fraction of mass of each elements. The plasma is mainly formed by hydrogen and helium completely ionized in the central conditions plus about 20 elements for which one needs to know the degree of ionization. The recent progress on the equation of state has been the introduction of the relativistic electronic correction. The other equations justify more caution. SSM assumes that the nuclear energy flux $4\pi r^2\rho\epsilon$ produced per second equilibrates the energy flux $L(r)$ emerging from a sphere of radius r. Of course, the energy loss by neutrinos must be subtracted to this energy flux. This neutrino energy is small during the hydrogen-burning phase. Taking into account quasi-static gravitational readjustment and composition variation, the heat transfer term TdS where S is the total entropy per gram of the gas must be incorporated, and the energy bookkeeping yields:

$$\frac{\mathrm{dL(r)}}{\mathrm{dr}} = 4\pi \mathrm{r}^2 \rho(\mathrm{r}) \left(\epsilon_{\mathrm{nucl}}(\mathrm{r}) - \frac{\mathrm{T(r)}\, \partial \mathrm{S(r)}}{\partial \mathrm{t}} \right) \tag{1}$$

The contraction at the beginning of the pre-main sequence, when the star is formed, allows the increase in the central temperature up to the required level to ignite nuclear reactions. Following this equation, during the main sequence, the star is really equilibrates and the Sun is practically at the middle of this stage. At the end of the hydrogen-burning phase, when the nuclear energy vanishes, a second contraction of the core with dilatation of the envelope leads to a new increase in the central temperature up to the beginning of helium fusion.

A second important equation of stellar evolution describes the transfer of energy from the center to the surface. The star chooses the most favorable type of transport between convection or radiation. In the inner solar radiative region, the diffusion approximation is appropriate, and the relation between temperature gradient and luminosity is:

$$\frac{\mathrm{dT(r)}}{\mathrm{dr}} = \frac{-3}{4\mathrm{ac}}\frac{\kappa(\mathrm{r})\rho(\mathrm{r})}{\mathrm{T^3(r)}}\frac{\mathrm{L(r)}}{4\pi \mathrm{r}^2} \tag{2}$$

The Rosseland mean value $\kappa(\mathrm{r})$ depends on T, ρ and X_i. When it increases strongly, like in the external part of the Sun, the radiative gradient increases and matter becomes convectively unstable. The solar time evolution is governed through the fifth equation which describes the change in composition due to nuclear reaction rates, microscopic diffusion and turbulence [7]:

$$\frac{\partial \mathrm{X_i(r)}}{\partial \mathrm{t}} = -\frac{\partial[(4\pi\rho \mathrm{r}^2)^2(\mathrm{D_i}+\mathrm{D_T})(\partial \mathrm{X_i}/\partial \mathrm{M(r)})]}{\partial \mathrm{M(r)}} + \frac{\partial(4\pi\rho \mathrm{r}^2 \mathrm{v_i X_i})}{\partial \mathrm{M(r)}} + \Sigma_j\Sigma_k \mathrm{R_{ijk}X_jX_k} \tag{3}$$

the last term is the nuclear term and the first one includes both turbulent term (D_T) and microscopic diffusion terms which depends on the concentration gradient, the second term introduces the diffusive velocity. The turbulent term dominates near the transition region between convection and radiation [8]. The five equations are non linear and coupled so they must be solved simultaneously up to the present solar age which is established at 4.6 Gyrs (including premainsequence). At the beginning of the evolution only the mass and the heavy elements relative to hydrogen are given, the initial helium content Y and the coefficient α of convection is adjusted a posteriori to get the proper radius and luminosity at the present time. As appeared in the previous expressions, complex physical processes must be determined to solve these equations that have justified several updated of the SSM along the last decades.

3. The SSM quality and predicted observable accuracy

The SSM delivers radial profiles of all the thermodynamical quantities and of all the different species for the present Sun from which one derives the observables that are detectable (neutrino fluxes) and measurable (mode frequencies or abundances).

Some neutrino fluxes are strongly dependent on the solar central temperature: the whole ^8B flux, associated to the reaction p+^{7}Be, varies as T^{18} and even T^{24} just at the center, ^{7}Be flux, associated to the electronic capture like T^8, so of course, they are strongly depend on the detailed description of the solar core. In contrast, pp neutrino flux appears directly connected to the solar luminosity so very few dependent on such description if equation (1) is correct. But it is very difficult to detect low energy neutrinos (like pp neutrinos) and all the neutrino fluxes independently, so chlorine detector like gallium ones were detecting a range of neutrino energies which favors ^{8}B in the first case and pp in the second case with some contribution of ^{7}Be neutrino flux in both cases, consequently the accuracy of the neutrino flux predictions of the SSM has always been crucial in the comparison with detection.

Three main sources of uncertainties have been quickly identified, see for example table 8 of [9]: the uncertainties coming from reaction rates, from abundances and from the opacity coefficients. So a lot of work has been performed to improve their accuracy. If it is not too difficult to determine an error bar deduced from an experimental cross section, it is more difficult to give an error bar for processes which are only known theoretically like screening effects, Rosseland mean values or microscopic effects.

So the error bars assigned to SSM predictions have always been minimal error bars (introduction of what we know, estimation of what we do not know). Consequently, it is not surprising to note larger variations on the predictions than previously estimated. Nevertheless the predictions have been always improved by time as more physics has been understood and implemented. By chance two groups have dedicated efforts to improve the SSM predictions, they have contributed to these improvements in coordination with physicists of different disciplines. The Bahcall team has predicted values for chlorine experiment from 9 to 7.5 SNU (1SNU=

10^{37}captures/cm^2/s). The Saclay group has given values from about 5.8 to 7 SNU and the two teams have obtained a reasonable agreement around 2000, see figure 1 of [10] with an accuracy of about 12% (of course the uncertainty on pp flux is smaller). These values were far from the mean experimental value of about 2.3 SNU even the fluctuations of the chlorine detection for each run were very large but compatible with the low statistical significance of each run. These fluctuations were largely greater than the 10% accuracy over 20 years which is only justified within the above SSM assumptions, see sections 6 and 7.

The change of the SSM predictions with time was maximal for the boron emitted neutrino flux, due to its extreme sensitivity to the description of the core. The impact of updated physics of SSM on its predictions has been summarized in table 6 of [5]. The predictions have changed by more than 20% and returned practically to the first predictions today as the reduction of the CNO photospheric abundances has compensated the previous effect of the microscopic diffusion. So, the accuracy of 15% attributed to the prediction of that flux could be underestimated.

4. The SSM predictions compared to seismic observations and neutrino detections

Our most recent SSM [11], built using the CESAM code, includes Adelberger reaction rates [13], Mitler screening effects, OPAL opacities, microscopic diffusion and photospheric abundances of Asplund (2009), see all the details on the input physics in [5]. Table 1 recalls the main characteristics of that model. As previously mentioned in both publications, the surface helium is a little low compared to the value determined by helioseismology (around 0.25) and consequently the initial helium also if one follows the general enrichment of the galaxy. Moreover the base of the convective zone at 0.727 is in disagreement with the seismic value of 0.713 $R_\odot$ [14].

Table 1. Main characteristics of our most recent published SSM compared to seismic observations of photospheric helium: 0.25 and BCZ: 0.713 $R_\odot$.

Model	T_c °K	ρ_c g/cm^3	Y_i mass fraction	Y_s mass fraction	Z/X_i	Z/X_s
SSM Saclay	15.54	150.6	0.2645	0.235	0.0181	0.0136

Several publications have also shown that the recent change of the photospheric composition of carbon, nitrogen and oxygen has largely deteriorated the agreement of SSM prediction with the seismic sound speed obtained in 2001, and with the neutrino predictions too, see [16, 17, 18, 19], it is the reason why some authors have been reluctant to use it. Nevertheless the reduced photospheric CNO composition by 20-30 % is justified by the choice of the observed lines, the use of 3D simulation to determine their shape. This result is confirmed by [15], even some small differences exist between the two works. Moreover, as mentioned in [16], it solves a fundamental problem which was the solar oxygen anomaly compared to the oxygen of stars of the neighboring environment as recalled in Figure 1. It is clear also since the beginning of the seismic investigation of the solar interior that the SSM assumptions are not sufficient to reproduce the real Sun. Figure 2 shows the dynamical processes that must be introduced to describe the real Sun, see [20]. They are not considered in the SSM approach even most of them are already seen in the solar interior. I shall come back on this point in Section 6.

Figure 3 shows also the strong sensitivity of the sound speed to the used photospheric solar composition and its time evolution. Consequent central condition changes are observed [5].

Let recall the reactions that produce the different neutrino fluxes: p+p called pp ν, capture electronic on p called pep ν, capture electronic on Be^7 called ^7Be ν, p +Be^7 called ^8B ν, and neutrinos coming from the CNO cycle called CNO ν which are the sum of ^{13}C ν, ^{15}O ν, ^{17}F ν.

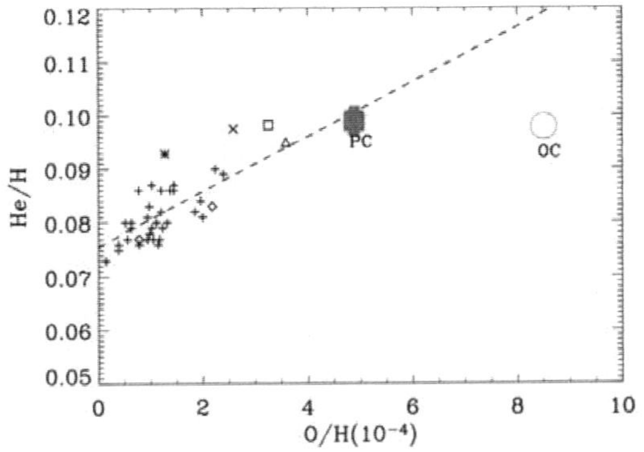

Figure 1. Galactic evolution of helium versus oxygen showing the anomaly of the old value (OC). O/H = 4.90 ±0.06 10^{-4} now (PC red arrow). Extended from [16].

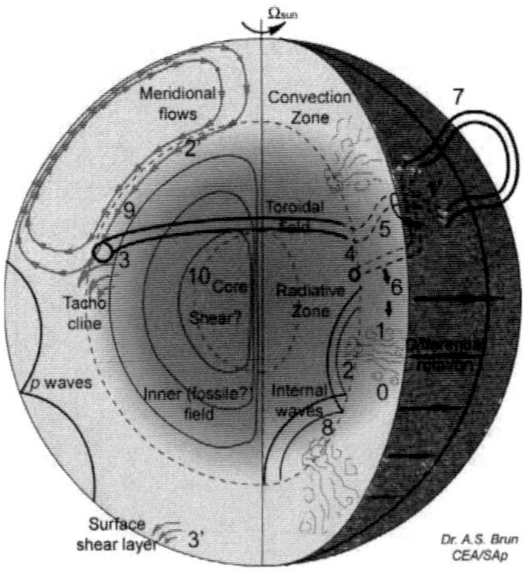

Figure 2. Integrated dynamical view of the solar interior. From [20].

All these individual fluxes differ by the region of emission, the energy and the efficiency of the channels that generate them. ^{8}B ν is the lower flux (weak chain ppIII) but the most energetic, pp ν the higher flux but the less energetic. The pioneer chlorine experiment measures ^{7}Be, ^{8}B and CNO ν fluxes, the gallium detectors have a smaller threshold and are dominated by the pp flux but all the others are present, SNO and SK measure only part of ^{8}B neutrinos. Borexino extracts separately ^{7}Be ν, ^{8}B ν and CNO ν. See [5] for all the details.

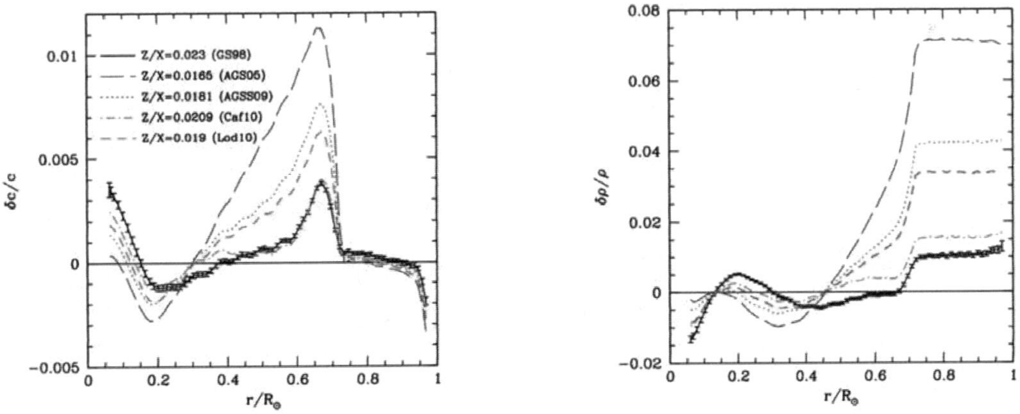

Figure 3. Sound speed and sensitivity profile differences between seismic inversion done by S. Basu and different SSM of the same group using different compositions corresponding to estimates obtained along the last decades. From [21].

Using heavy water, SNO realizes charge current interaction and elastic scattering on electron. The first one is only sensitive to electronic neutrino, but elastic scattering (unique interaction in SK) allows to detect both electronic, tau and muon neutrinos. The elastic scattering detects ν_e + 0.16 (ν_μ + ν_τ). In comparing the two counting rates, one deduces the sum $\nu_e+\nu_\mu+\nu_\tau$ coming from the Sun, of course the last two are a manifestation of the oscillation of neutrinos as the Sun emits only ν_e.

Table 2. Neutrino predictions of our most recent SSM [11] compared to all neutrino experiments. The SNO results come from [24].
1SNU = 10^{-37}interactions/s/cm^2. See our recent reviews for the other references.

	Predictions	Predictions including neutrino oscillation
Chlorine detector		2.56 ± 0.23 SNU
SSM 2011	6.315 SNU	2.24 SNU
Gallium detectors	**GALLEX+GNO+SAGE**	66.1 ± 3. SNU
SSM 2011	120.9 SNU	64.1 SNU
Water detectors	**SNO**	**SNO + SK** in 10^6cm^{-2}s^{-1}
	5.09 ± 0.44 (stat)± 0.45 (syst)	5.27 ± 0.27 (stat)± 0.38 (syst)
SSM 2011	4.21 ± 1.2	4.21 ± 1.2

Table 2 confronts the predictions of the SSM [11] with the detected neutrino fluxes. The direct comparison is only possible for the SNO detector. For the others, the comparison needs to multiply the emitted flux by the probability for a ν_e to stay a ν_e at its arrival on the detector. This probability depends on the values of $\sin^2\theta_{12}$ and Δm^2_{12} that are now well established, see Figure 6 left and [22, 23]. The final results are done in the right column instead the left column corresponds to the emitted flux. One sees that the SSM predictions are systematically below the detections as also noticed in [19]. As the error bars given for these predictions can be wrong, it is only given for boron ν flux where we have a direct comparison, this error bar is quite large. One cannot say that there is a strong disagreement but one sees in Table 2 that the agreement is marginal. If one compares the sound speed and the density profiles (Figures 3 and 4) we note a strong disagreement because in these cases the vertical error bars are very small, We discuss now a better way to estimate the accuracy of the predictions for the different fluxes.

5. Seismic results and Seismic Model predictions of the neutrino fluxes

As already shown on Figure 3, the SoHO satellite has revealed the sound speed and density profiles down to the core, see [5, 6] where detailed final information is given. These two quantities have been compared to SSM predictions all along these two decades and the agreement has varied with time and details of the SSM as neutrino fluxes were varying too. So to stabilize the neutrino emitted fluxes coming from solar models, we have decided to build a model called Seismic Solar Model (SeSM) using the same equations than SSM but using slight modified inputs (reaction rate and opacity coefficients) to micmic as properly as possible the very precise observed sound speed (better than 10^{-4} accuracy), see Figure 4. The modification of the reactions rates has been rather small and the modification of the opacities (never verified experimentally, like screening effect) has been relatively small at the beginning and is now amplified by the effect of recent estimates on the composition [2, 26]. Different checks have been done which lead to thermodynamical quantities which differ from the SSM ones. Of course the physics of the SeSM is not different from SSM but this model is more representative of the solar core and consequently can predict the neutrino emission fluxes with a better accuracy which avoids uncertainties on the most important reaction rates (including screening), on the opacities and detailed composition....

Table 3 shows the SeSM neutrino predictions compared to the detected values discussed previously. One easily notices that the predictions are in better agreement with the detections

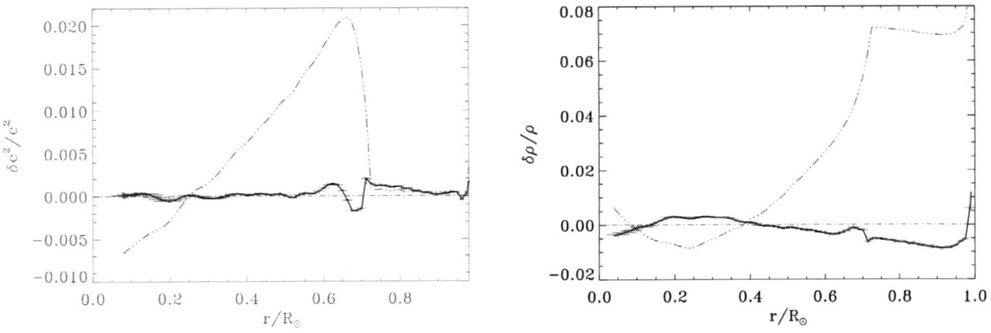

Figure 4. Squared sound speed and density profile differences between seismic inversion from GOLF and MDI [2], SSM (in dashed red line) and SeSM (in black, with error bars). From [5].

Table 3. Neutrino predictions of the SeSM [2, 26] compared to all the neutrino experiments. The Borexino results from [25]. Same references than in Table 2.

	Neutrino Predictions	Predictions including neutrino oscillation
Chlorine detector		2.56 ± 0.23 SNU
Seismic model	7.67 ± 1.1 SNU	2.76 ± 0.4 SNU
Gallium detectors	**GALLEX+GNO+SAGE**	$66.1 \pm 3.$ SNU
Seismic model	123.4 ± 8.2 SNU	67.1 ± 4.4 SNU
Borexino ^{7}Be		$3.36 \pm 0.36\ 10^9 cm^{-2} s^{-1}$
Seismic model	$4.72\ 10^9 cm^{-2} s^{-1}$	$3.05 \pm 0.35\ 10^9 cm^{-2} s^{-1}$
Water detectors	**SNO**	**SNO + SK** in $10^6 cm^{-2} s^{-1}$
	5.09 ± 0.44 (stat)$\pm$ 0.45 (syst)	5.27 ± 0.27 (stat)$\pm$ 0.38 (syst)
Seismic model	5.31 ± 0.6	5.31 ± 0.6

and even in excellent agreement with all of them. This is a great success of the two totally independent approaches of the solar interior to agree with each other through the SeSM. This result does not support the existence of a forth neutrino, the sterile neutrino, except if its existence does not perturbed the detected range of energy. Such model has also shown that it reasonably describes the core because its gravity mode frequency predictions agree with the first observations and rejects the low mass range of WIMP candidates to dark matter as the seismic model predicts a higher temperature of the central region than the SSM, see 7.3, the opposite behavior than the action of WIMPs in the core [27]. We note also that the error bars of the predictions are reduced and easier to determine as the sources of uncertainty are reduced to the weak interaction of neutrinos with matter and some specific cross section like $^7Be + p- >^8 B + \nu_e$ which is not at all constrained by the solar luminosity, no error bar needs to be attributed to deep composition, opacity or plasma effects, which are constrained by the observed sound speed.

As this step has been successful but not sufficient to describe the whole internal dynamics of the Sun, dynamics which could be amplified in more evolved and more massive stars, we can question the future. Can we learn more on the neutrino properties and on the solar interior with the precise solar neutrino fluxes or the gravity mode detections ?

6. Beyond the SSM

The answer is of course yes for two reasons. First one has learned so much from solar neutrinos that one cannot exclude that better statistics will bring other constraints on their properties. The second and important reason is that the Sun is not static and that there are good reasons to learn more in the next two decades.

- We know that at the surface, the Sun appears as a magnetic star, but the dynamo effects inside the Sun are not yet observed except near the surface, so one cannot conclude if there is only one cycle of 22 years with a reversal of the polarity of the poles every 11 year or if there are other longer cycles inside the convective zone (about 100, 200, 400 yrs...).

- We know the rotation profile from the surface to the core but we have not established all the properties of the transition region between radiation and convection, in particular its magnetic topology and its efficiency to generate gravity waves in the radiative zone.

- We have only detected the first gravity modes which suggest a higher rotation profile in the core but it is not established yet if it stays a fossil field in that region.

- We have built a seismic model of the Sun which is able to predict better the neutrino fluxes than SSM but we do not know why the thermodynamical quantities are what we get? is it due to an incorrect microscopic description of the energy transfer, is it due to an incorrect energy balance? is it due to a poor description of the dynamical processes in the radiative zone. Is it due to other fundamental physics like sterile neutrinos acting like dark matter ?

The best is to explore the present and coming data and see how they will help us to interpret the present situation.

7. Precise Nuclear Astrophysics and Future Observational Constraints

The dynamical phenomena in the radiative zone are difficult to catch. Let see how to interpret the present discrepancies and how to progress on the signatures of such phenomena.

7.1. Pure energy transfer problem

The origin of the noticed discrepancies can come from an inaccurate determination of the opacity coefficients used in the SSM calculation in equation (2). Effectively most of the discrepancy or change of the SSM predictions comes from the impact of the deep composition on the sound speed through their Z contribution and in particular its oxygen and iron contribution (but not only). Face to this possibility, we are pushing new opacity calculations OPAS [29] and new measurements of the transfer of radiation through specific elements and mixtures [28]. Internal comparisons have already been made with OPAL tables, some specific differences have been noticed for specific elements, but the mean Rosseland differences on the solar mixture reach only some % mainly at the base of the convective zone. The new calculations OPAS seem in rather good agreement with the first measurement in conditions not too far from the BCZ [30]. New tables will be provided and one cannot exclude some plasma effect which could slightly modify the properties of ionization of the different species at higher temperature and density. To verify these assumptions, experiments are under study on large laser facilities like LMJ+PETAL [31, 32] for specific elements and mixtures equivalent to the solar composition. These experiments are very difficult to perform as one needs to produce density of the order or greater than solid density at Local Thermodynamics Equilibrium or near.

7.2. Interaction between macroscopic and microscopic diffusion in the RZ

The quality of the SSM has been largely improved by the introduction of the microscopic diffusion that depends on both gravitational settling and radiative acceleration built on opacity spectra. These processes, first two terms of equation (3) have not been revised after the observation of the very complex transition region called tachocline (see Figure 2) which delimits the radiative-convective transition. Seismic data show that this region has an extension of about 5% $R_\odot$ and

Journal of Physics: Conference Series **665** (2016) 012078 doi:10.1088/1742-6596/665/1/012078

is not totally spherical [33]. The strong difference of meridional velocity between radiative zone (RZ) and convective zone (CZ) extracted from the modeling of the transport of momentum [34] creates such shear horizontal instability zone. It is a very complex region where the observed differential rotation of the CZ disappears in favor of a solid rotation. As an extension of the convective motions, this layer produces gravity waves in the RZ [36] and contributes probably to regenerate the toroidal field from the poloidal field. One cannot exclude that these dynamical effects can impact more differentially the heavy elements than one gets today. New calculations will estimate such idea but in addition there is some hope to verify their predictions on the central CNO composition if one gets a precise determination of the CNO flux from Borexino or from future neutrino detectors.

7.3. Energetic balance of the Sun and its time evolution

It is interesting to compare the central thermodynamical conditions of the SeSM to those of the SSM given in table 1. For the SeSM, we get:

$$\mathrm{T_C} = 15.74\ 10^6\ {}^\circ\mathrm{K} \quad \rho_\mathrm{C} = 153\,\mathrm{g/cm^3} \quad \mathrm{Y_{init}} = 0.277 \quad \mathrm{Y_{surf}} = 0.250$$

This model agrees with the different helioseismic constraints: photospheric helium, BCZ, sound speed and has a reasonable helium enrichment. If one compares the difference with SSM in term of energetic production, one notices that the difference is about 5% [11]. As the modeling of the Sun assumes that the energy transfer is instantaneous, and that the emergent flux at the surface just compensates the flux emitted through nuclear energy (equation 1), one can question the precise reliability of such equality at the level of accuracy obtained today. Effectively the motions inside the Sun create internal kinetic energy, meridional circulation energy which are partly transformed in magnetic energy. Figure 5 shows the emergence of the energy flux at the surface now measured for about three decades, the total flux increases by about 0.1% near solar maximum [37], although the radius very slightly shrinks simultaneously [35]. We know that this excess of released energy is not totally regular, but we do not know which part of energy stays inside the Sun, part of it could be stored inside the tachocline. Seismology tells us that the energy tank is $< 5\%$. One way to verify if there is a real difference between central production

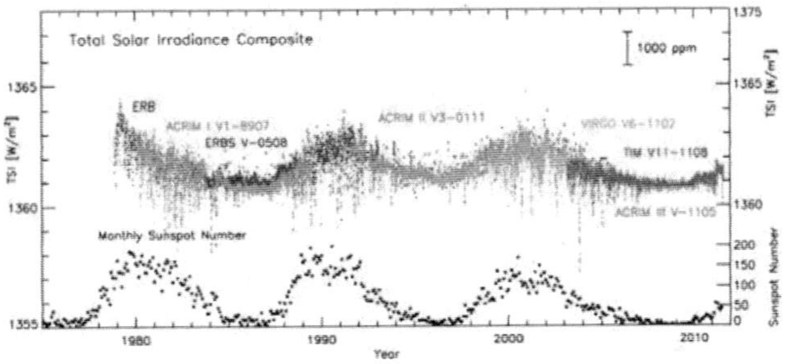

Figure 5. Time variability of the solar irradiance obtained by different satellite measurements. The relative normalization between the data sets has been obtained by Kopp and collaborators.

of energy and external output and how it evolves with time, consists to simulate the dynamical processes in 1D and 3D. This is a difficult work because in 1D one can follow the readjustment of the star but the interplay between processes are better described in 3D which cannot presently mimic sufficient longer series. The other way consists to measure directly the central energetic flux through the simultaneously precise neutrino fluxes coming from ^8B neutrinos (sensitive to central temperature and possible fluctuations) and from pp or pep neutrinos that are directly sensitive to the production of energy. One notes on Figures 6 and 7 that we have already some estimate of the pep flux. Its error bar is today too large to assume really a deviation [38] but this information will be improved during the next decade and we shall be in a position to conclude.

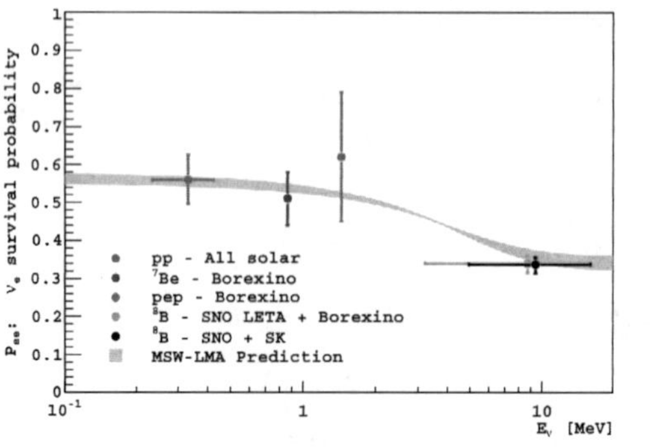

Figure 6. Probability for a ν_e to stay a ν_e. From [38].

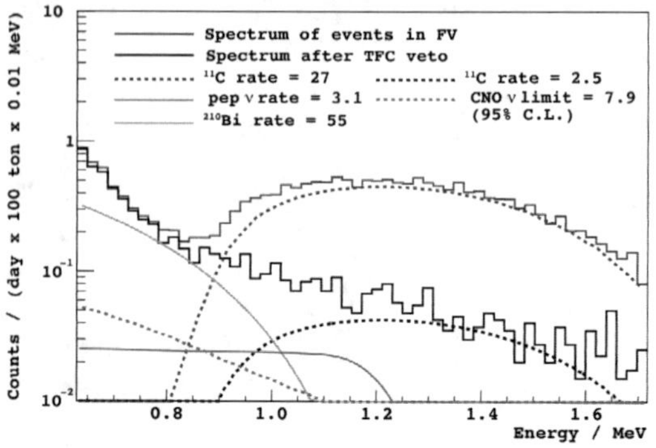

Figure 7. Extraction of the different fluxes of neutrinos from Borexino. From [38].

7.4. Precise nuclear core reaction rates and Fluctuations in the radiative zone

It is important to accompany the coming neutrino projects, SNO+ and HyperKamiokande which will deliver very precise neutrino fluxes. They could see the dynamics of the central core. One can hope to check if they present time variabilities which are compatible with gravity waves in the RZ and if their characteristics agree with improved internal dynamical models of the Sun.

In parallel precise predictions of the fluxes are required. Numerous studies have been dedicated to the knowledge of all the nuclear reactions that act in the Sun [13], but two fundamental problems limit their accuracy: the domain of energy corresponding to very low cross sections and the absence of measurements of the plasma screening effects. Such difficulties can be solved in the near future with the large laser facilities like LMJ+PETAL or the NIF facilities. They provide real opportunities to measure directly the reaction rates that appear in stars which by essence include the electron acceleration of the reaction rates, the maxwellian distribution of the velocity and the role of ion neighbors. This direction is promising and already explored on more modest installation [39]. Such experiments require a lot of expertise and new detectors which are in construction [32].

8. Conclusion

This review shows that the SSM is an excellent basis for stellar evolution, the development of asteroseismology and galactic evolution studies. Nevertheless the impressive progress of the helioseismic probe has confirmed the presence of dynamical processes which are not included in the SSM framework and can partly contribute to the limitation of its predictions of the present neutrino fluxes. The building of a SeSM has stabilized these predictions and has helped to quantify the role of the dynamical effects on different indicators. The very brilliant neutrino program with HyperKamiokande and SNO+ open very important directions of progress. We have shown that precise nuclear astrophysical studies and plasma effects verification need to be performed in parallel. Large laser facilities are well adapted to accompany such progress. Modeling the Sun at timescales which concern the human behavior (hundreds and thousands of years) is a new challenge for the coming decades. Such golden future will contribute also to enrich the modeling of solar-like stars and the relationship between stars and planets.

This review has been enriched by space ESA-NASA SoHO observations, neutrino detections and by numerous collaborations in different fields.

References

[1] Gabriel A H *et al* 1995 *Sol. Phys.* **162** 61
[2] Turck-Chièze S and the GOLF consortium 2001 *ApJ* **555** L69
[3] Abdurashitov J N, Gavrin V N and the SAGE collaboration 2009 *Phys. Rev. C.* **80** 15807
[4] Ahmad Q R and the SNO collaboration 2001 *Phys. Rev. Lett.* **87** 071301
[5] Turck-Chièze S and Couvidat S 2011 *Report on Progr. in Phys.* **74** 086901
[6] Turck-Chièze S and Lopes I 2012 *RAA*, **12** 1107
This reference contains SoHO/GOLF results, density and sound speed profiles.
[7] Proffitt C R and Michaud G 1991 *ApJ* **371** 584
[8] Brun A S, Turck-Chièze S and Zahn J P 1999 *ApJ* **525** 1032
[9] Turck-Chièze S, Cahen S, Cassé M and Doom C 1988 *ApJ* **335** 415
[10] Turck-Chièze S 2004 in Neutrino2004 *Nucl.Phys.B (Proc. Suppl)* **143** 35
[11] Turck-Chièze S, Piau L and Couvidat S 2011*ApJ Lett.* **731** L29
[12] Asplund M, Grevesse N, and Scott P 2009 *ARA&A* **47** 48
[13] Adelberger E *et al* 2011 *Rev. Mod. Phys.* **83** 195
[14] Christensen-Dalsgaard J, Gough D and Thompson M 1991 *ApJ* **378** 413
[15] Caffau E *et al* 2009 *A&A* **498** 877
[16] Turck-Chièze S *et al* 2004 *Phys. Rev. Lett* **93** 211102
[17] Bahcall J N, Serenelli A M and Basu S 2006 *ApJ Supp* **165** 400
[18] Basu S *et al* 2009 *Astrophys. J* **699** 1403
[19] Serenelli A N, Haxton W C and Pena-Garay C 2011 *ApJ* **743** 1
[20] Brun A S and Jouve L *IAU Symposium Waves & Oscillations in the solar atmosphere* **247** 33
[21] Basu S, Grevesse N, Mathis S and Turck-Chièze S 2014 ISSI book on Helioseismology, in press
[22] Bahcall J N and Pena-Garay C 2004 *New J. Phys.***6** 63
[23] Aharmim B *et al*, the SNO Collaboration 2008 *Phys. Rev. Lett.* **101** 111301
[24] Ahmed S N and the SNO collaboration 2004 *Phys. Rev. Lett* **92** 181301
[25] Bellini G*et al*, the Borexino collaboration 2010 *Phys. Rev. D* **82** 033006
[26] Couvidat S, Turck-Chièze S and Kosovichev A 2003 *ApJ* **599** 1434
[27] Turck-Chièze S et al 2003 *ApJ lett* **746**12
[28] Turck-Chièze S et al 2009 *High Energy Density Phys.***5** 132
[29] Blancard C, Cossé P and Faussurier G 2012 *ApJ* **745** 10
[30] Bailey J E *et al* 2009 *Phys Plasma* **16** 058101
[31] Blanchot N *et al* 2008 *Plasma Phys. Control. Fusion* **50** 124045
[32] Ducret J-E *et al* 2013 *NIM Phys Res A* **720** 141
[33] Basu S and Antia H M 2003 *ApJ* **585** 553
[34] Turck-Chièze S, Palacios A, Marques J and Nghiem P 2010 *ApJ* **715** 1539
[35] Piau L et al 2013 *MNRAS* **437** 164
[36] Alvan L, Mathis S and Decressin T 2013 *A&A* **553** 86
[37] Kopp G and Lean J L 2011 *Geo Res Lett* **38** L01706
[38] Bellini G *et al*, the Borexino collaboration 2012 *Phys. Rev. Lett* **108** 051302
[39] Barbui M. *et al* 2013 *Phys. Rev. Lett* **111** 082502

The Sun as a probe of Fundamental Physics and Cosmology

I P Lopes[1]

[1]Centro Multidisciplinar de Astrofísica, Instituto Superior Técnico, Universidade Tecnica de Lisboa , Av. Rovisco Pais, 1049-001 Lisboa, Portugal

E-mail: ilidio.lopes@tecnico.ulisboa.pt

Abstract. The high quality data provided by helioseismology, solar neutrino flux measurements, spectral determination of solar abundances, nuclear reactions rates coefficients among other experimental data, leads to the highly accurate prediction of the internal structure of the present Sun - the standard solar model. In this talk, I have discussed how the standard solar model, the best representation of the real Sun, can be used to study the properties of dark matter, for which two complementary approaches have been developed: - to limit the number of theoretical candidates proposed as the dark matter particles, this analysis complements the experimental search of dark matter, and - as a template for the study of the impact of dark matter in the evolution of stars, which possibly occurs for stellar populations formed in regions of high density of dark matter, such as stars formed in the centre of galaxies and the first generations of stars.

1. Introduction

In this talk, I discussed how the Sun is being used as a probe of fundamental physics and modern cosmology. As a nuclear physics community, if you remember the 30's and 40's, the Sun has played a key role in the progress and development of nuclear physics, namely, by contributing to the understanding of the basic nuclear processes that lead to the discovery of the nuclear chain reactions (PP chains, CNO cycle and 3α reactions) by H. Bethe, C. F. von Weizsäcker and F. Hoyle among others. Therefore, it is no surprise, that in particular the Sun, and in general the other stars, can play an important role to explore the validity of new fundamental laws of physics, such as testing new theories of gravity proposed as alternative to General Gravity [1], to discuss the precision of new measurements of fundamental constants [2], to study the properties of neutrino flavour oscillations [3, 4], and, as discussed in this talk, as a tool to determine the properties of dark matter (DM)) [5–7].

Undoubtedly, the origin of all matter, and which fundamental particles matter is made off, is one of the leading problems of modern physics. Although, our visible Universe is made off baryons, including our own Sun and ourselves, it is by now well-established that most matter (and energy) of the universe is non-baryonic. Its fundamental nature is still a mystery to us. Presently, the unknown non-baryonic matter content of the universe is presented in its two components, dark matter and dark energy – a split based on their two distinct gravitational effects, dark matter accounts for the extra gravity needed to hold baryonic matter together (example: clusters of galaxies, stars in galaxies), and the dark energy must exist to explain the expansion rate of the present universe. Dark matter and dark energy correspond to 24% and

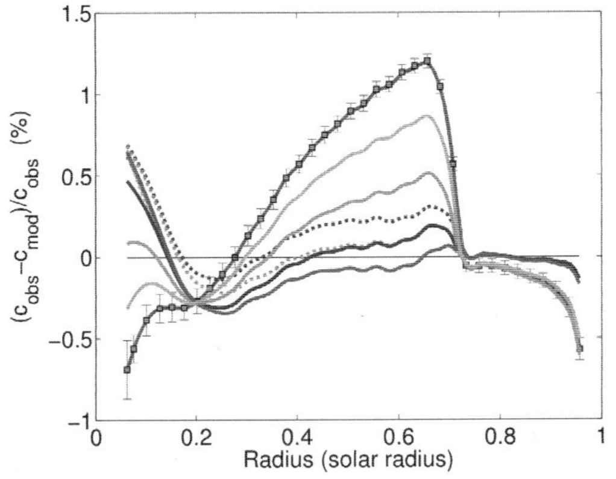

Figure 1. Relative differences between the sound speed inverted using the helioseismic data [8, 9], and the sound speed deduced from the SSM (continuous red curve with error bars). The error bars are multiplied by a factor of 10. This figure also shows a set of solar models with mass-loss, with an anomalous chemical composition. See Ref. [10] for the details.

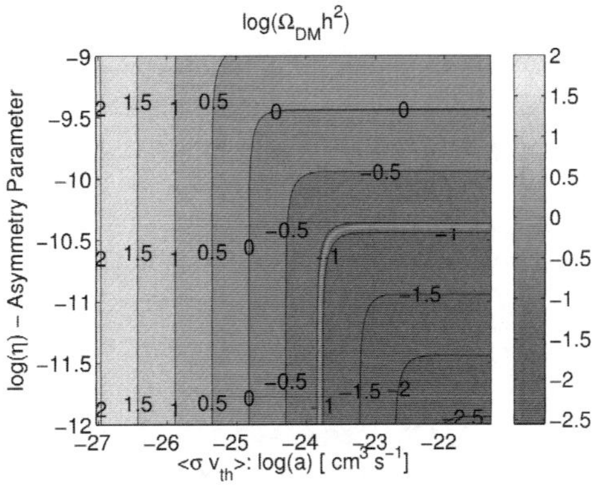

Figure 2. The figure shows iso-curves of the relic dark matter density $\Omega_{DM}h^2$ as a function of the η_{DM} and $\langle\sigma v\rangle$ (s-wave annihilation channel). The model corresponds to a primitive universe made of DM particles with a mass of 10 GeV ($g_\chi = 2$ and $g_\star = 90$). The blue lines define the set of DM models that are compatible with $\Omega_{DM}h^2$ observations. The current measurements $\Omega_{DM}h^2$ and η_B are 0.1109 ± 0.0056 and $(0.88 \pm 0.021) \times 10^{-10}$ [11–13]. The figure shows that η_{DM} and η_B are of the same order of magnitude. See Ref. [14] for the details.

72% of all the matter of the Universe, respectively. It is a quite challenging task to us, trying to understand this new type of matter and energy based on fundamental theories of physics that have been developed to explain the 4% of baryonic matter. In this short review, I will focus on the study of the impact of dark matter in the evolution of the Sun and stars.

The presentation is organized as follows: In the next section, I make a brief summary of the quality of the standard solar model (SSM). In the third section, I make a synopsis of the current knowledge on the origin of dark matter. Finally, in the fourth and fifth sections, I will discuss the impact of dark matter in the Sun and stars and highlight a few DM predictions that are possible to be made using stars.

2. The standard solar model

The Sun is by far the best known star of the Universe. Given the advantage of being our nearest star, we have an exceptional amount of photometric and spectroscopic observational data from the Sun. Moreover, we can probe its interior with very high accuracy, by means of helioseismology and solar neutrino fluxes [10, 15, 16]. Actually, we now know better the core of the Sun than the centre of the Earth. Helioseismology has obtained a large amount of data – more than 7 thousand acoustic modes were measured with a very high precision, better than one part in hundred thousand [17]. During the last three decades, such high quality data has lead to a significant improvement of the description of plasma physics of the Sun's interior, in particular its microscopic physics, such as the nuclear reaction network, the equation of state and the coefficients of radiative transfer (stellar opacities). In particular, Helioseismology played a leading role in the resolution of the solar neutrino problem (see talk of Sylvaine Turck-Chiéze).

The quality of the present solar model is illustrated in Fig. 1 that shows the comparison between the radial sound speed profile obtained from helioseismology data and the one computed from SSM. The difference is smaller than 2%. This difference although small is important. The cause responsible for such difference is not known yet, but very likely it is related with uncertainties on the coefficients of radiative transfer, or an anomalous internal chemical composition related with the formation of the solar system [17, 18]. Presently, the quality of the SSM is simultaneously consistent with helioseismology and solar neutrino flux data. Although solar neutrino predictions show small differences relatively to observations, these are mainly related with the parameters of neutrino flavour oscillations and possibly small changes of the physics of the Sun's core.

I also have highlighted that other fields of observational stellar astrophysics have contributed to this subject. During the recent decades, astronomers have produced high-precision catalogues (including high-precision spectroscopic measurements), that can be used to test the theory of stellar Evolution (and its new variants). An example is the Hipparcos Catalogue [36] with more than 100,000 stars, particularly useful to study stellar populations. Astereoseismology, like the COROT [33] and Kepler [34] missions which have observed more than one hundred pulsating's stars. These new sets of stellar data will significantly improve our knowledge about the physics of stars, but will also open the way to use stars as a tool for fundamental physics and cosmology studies.

In summary, although our understanding of the properties of the solar and stellar plasmas still requires some improvements [17], it is clear that the Sun and sun-like stars can be used as tools to put constraints in fundamental physics, like dark matter particles properties [19–25, 31, 32], as presently is done for other stars, such as neutron stars [26–30].

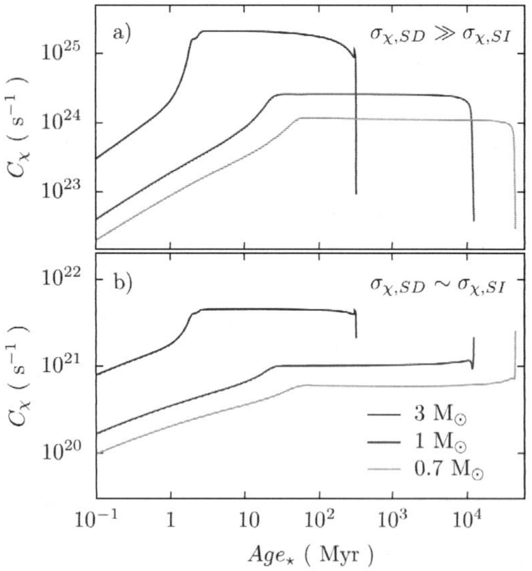

Figure 3. Rate at which DM particles are captured during the life of stars with different masses. The capture rate increases during the pre-MS, is constant through the MS, and varies rapidly in the RGB. We assumed a halo with a density of 0.3 GeV cm^{-3}, constituted by DM particles with a mass of 100 GeV for which the DM-baryon scattering dominated (a) by the SD component $\sigma_{\chi,SD} = 10^{-38}$ cm^2 and (b) by the SI one $\sigma_{\chi,SI} = \sigma_{\chi,SD} = 10^{-44}$ cm^2. See Ref. [35] for the details.

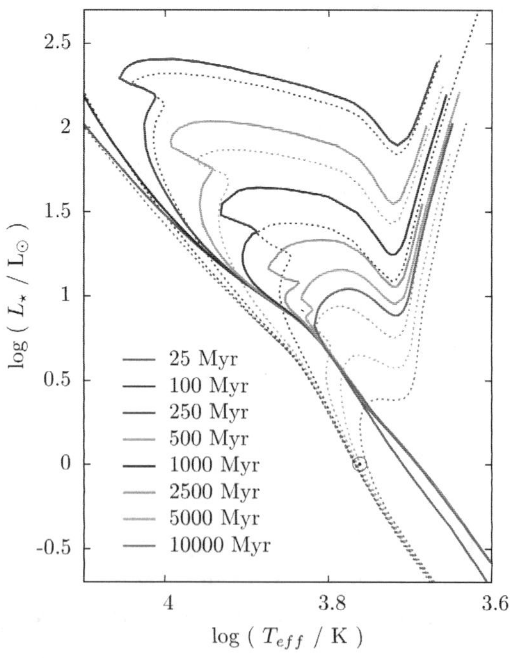

Figure 4. Isochrones for a cluster of stars with masses between 0.7 M$_\odot$ and 3.5 M$_\odot$ that evolved in a halo of DM with a density 10^{10} GeV cm^{-3} (continuous lines) and for the same cluster in the classical scenario without DM (dashed lines). See Ref. [56] for the details.

Journal of Physics: Conference Series **665** (2016) 012079 doi:10.1088/1742-6596/665/1/012079

3. The dark matter in the Universe

Let us remind in a nutshell, what is the current status of the dark matter problem. Significant progress has been achieved in this subject, as a result of the strong interaction between Astrophysics, Cosmology and Particle Physics.

The evidence of the contribution of DM for the formation of the Universe and its structure is very robust. These results come from astrophysical and cosmological observations, as well as from numerical simulations of structure formation [11]. In many observational studies it was found that dark matter is responsible for the existence of external gravitational field sources in different locations of the Universe, leading to quite distinct gravitational effects, such as the velocity of galaxies in clusters, the rotation curves of galaxies, the cosmic microwave background anisotropies, the velocity dispersions of dwarf spheroidal galaxies and the inference of the dark matter by gravitational lensing. All the observational and theoretical results suggest that most of the formation of structure in our Universe can only be explained by the presence of a gravitational field caused by the presence of a new type of particles that must be non-baryonic and non-relativistic [42, 43]. That means that DM particles interact with baryons through gravity and possibly through another unknown mechanism identical to the weak interaction. DM particles form the gravitational web where the first stars and galaxies are created. In summary, these results strongly favour the hypothesis that the dark matter particle is a WIMP, that is a Weak Interacting Massive Particle.

The experimental search of DM by several experimental physics collaborations can be organized into two groups of methods: direct detection of DM particles – infers the properties of the dark matter particles by the type of interaction that the DM particles have with a target-baryon; or an indirect detection of DM particles – by detecting one of its by-products resulting from the annihilation of DM particles, such as high-energy neutrinos. A few of the experiments dedicated to dark matter searches show evidence of positive particle detection, although these results are still very controversial and not universally accepted: Group experiments such as DAMA/LiBRA[44], COGENT[45], CRESST [46] and CDMS II [47] suggest that the DM particle could have been discovered. These experiments show hints of a DM particle with a mass of 10 GeV and a cross section of $10^{-41} cm^2$. But other experiments such as XENON 10-100 [48, 49] exclude the existence of such particles. Nevertheless, several theoretical explanations based on the existence of asymmetric dark matter have been proposed to explain and accommodate all these positive and negative detections [50–55, 58]. Theses hints of DM detection suggest that at best DM-baryon interaction is of the order of the weak interaction.

By now, it is well-established that DM particles, like baryons, are produced in the primitive universe. Currently several theoretical DM models suggested that DM is produced by an identical process to Baryogenesis [37–39] – a physical process during which an asymmetry occurs between baryons and antibaryons, resulting in substantial amounts of residual baryonic matter, i.e. the ordinary matter of the present universe. In a similar manner to Baryogenesis for which the degree of baryonic asymmetry (unbalance between particles and antiparticles) is measured by η_B, the asymmetry between DM particles and antiparticles is measured by η_{DM}. Fig. 2 shows $\Omega_{DM}h^2$ for a primitive Universe composed by DM particles with a mass 10 GeV [20]. As shown in this figure, only a small set of DM models [37, 40] have a $\Omega_{DM}h^2$ compatible with $\Omega_{DM}h^2$ measured by WMAP and Planck experiments [11–13]. Furthermore, the η_{DM} of these DM models are of the same order of magnitude of the observed η_B. Actually, there is no special reason for such coincidence. This suggests that DM and baryonic matter are more closely connected than expected. In the same figure, it is worth noticing that symmetric DM corresponds to low values of η_{DM} and asymmetric DM corresponds to large values of $\langle \sigma v \rangle$.

4. The dark matter inside the Sun and stars

The numerical computation of the impact of dark matter in the evolution of the Sun and stars has progressed considerably in the last two decades [20, 24, 25], however, there are still a few caveats that must be addressed [17].

When a DM particle crosses a star like the Sun, it very occasionally scatters off a proton, or an heavier nucleus including helium, carbon, oxygen and iron. The scattering interaction with hydrogen is predominately spin-dependent (SD) and for the other elements spin-independent (SI), accordingly to the scattering cross-sections $\sigma_{\chi,SD}$ and $\sigma_{\chi,SI}$, respectively. The interaction of DM with baryons depends strongly of the mass and velocity of the DM particle. In certain cases, after collision the DM particle will lose enough energy to become gravitationally bound to the star, and by a sequence of orbital paths around the centre of the star, the DM particle sinks to its core [21, 35, 57]. The total amount of particles (and antiparticles) of DM inside of the star at a given time is regulated by three fundamental processes: capture of DM particles from the DM halo, the evaporation of DM particles from the star (in the Sun only relevant for DM particles with a mass smaller than 5 GeV) and the annihilation of DM particles. It is the balance between these three processes that determines the total amount of DM inside the star and consequently the influence that DM has on the evolution of the host star. Fig. 3 show the variation of the dark matter capture rate for different stars in different phases of their evolution. The capture rate of dark matter depends largely of the radius of the star at each time step evolution. The chemical composition of the star also plays a major role in the capture of DM by the star, for which its content in metals (chemical elements other than hydrogen and helium) determines which is the dominant type of scattering [35].

Once captured, DM particles can change the evolution of the star through two mechanisms: (i) by supplying the star with a new source of energy due to DM annihilation – which complements the PP chain, CNO cycle and 3α nuclear reactions in balancing the self-gravitational contraction of the star, or (ii) by providing a new mechanism for the transport of energy – this complements the radiative and convective energy transport. The two processes can work simultaneously, however, in most of the cases when the evolution of the stars is affected by the presence of a DM energy source, the DM energy transport can be neglected. The first mechanism is important in locations of the Universe with high DM density, such as the regions where the first stars and galaxies are formed, and the center of galaxies like our Milky Way and the spheroidal galaxies. The second mechanism becomes relevant in much less denser DM regions like in the Sun's neighbourhood.

The density of the DM halo where the star is formed is one of the major sources of uncertainty. Usually the DM density is assumed to be 0.38 GeVcm^{-3} [59]. Nevertheless, recently estimations found a DM density of the order of 0.3 GeVcm^{-3} [60] or 0.85 GeVcm^{-3} [61]. This discrepancy has an important impact on our study, nevertheless, researchers in the field use the lower DM density value to make more conservative predictions of DM parameters. However, there are other sources of uncertainty in these computations, which we must resolve to obtain more reliable predictions. Among others, here is a list of a few that I believe must be investigated: (i) Particle Physics and Nuclear Physics: properties of the DM particles, like the mass of DM particle, the scattering cross-section of the DM with baryons, annihilation rate of DM particles and antiparticles, annihilation channel and nuclear factors. (ii) Cosmology: The dynamics and evolution of the dark matter halo, in particular the density of the DM halo where the star is formed, the dynamical properties of the DM particles like their thermal velocity. (iii) Astrophysics and Stellar Physics: The velocity of the star, the detailed internal structure of the star, like its chemical composition. Although, some of these properties are well known in the case of the Sun, it is not yet the case for other stars.

Although, we found in preliminary tests [35] that such caveats have a relative small effect in the accretion of DM by stars, as this research field progresses, a better account of such mechanisms must be included in the models.

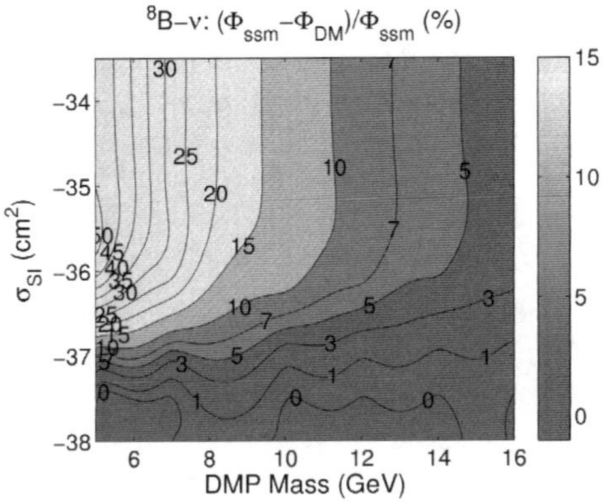

Figure 5. Percentage decrease changes of the solar ^{8}B neutrino flux $\Phi_\nu(^8B)$ for the Sun evolving within different DM halos relatively to the SSM: The properties of the DM particles and the stellar physics can be found in the Ref. [20].

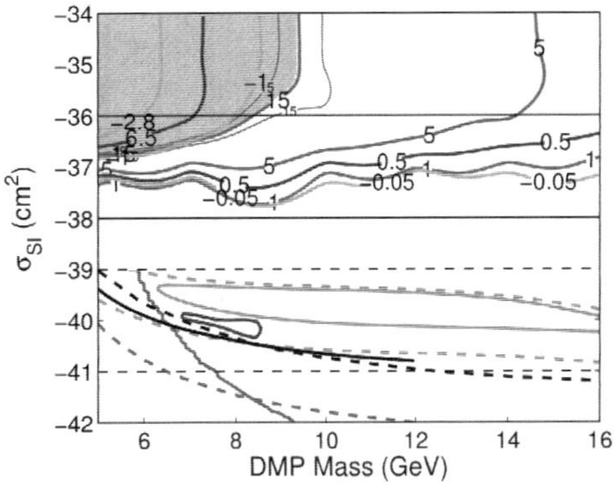

Figure 6. Exclusion Plot of DM searches: **(1)** CoGeNT(blue curve), DAMA/LIBRA a 3σ CL (solid green curve) and 5σ CL (dashed green curve); XENON 10 (dashed magenta curve), XENON100 (solid magenta curve); CDMS II limit (solid black curve) and SIMPLE limit (dashed black curve). **(2)** Solar neutrino flux variations due to DM: $\Phi_\nu(^8B)$ (solid red curve), $\Phi_\nu(^7Be)$ (solid blue curve) and $\Phi_\nu(pep)$ (solid cyan curve). The grey region corresponds to values of $\Phi_\nu(^8B)$ neutrino flux larger than 15% (red curve). See Ref. [20] for the details.

5. Dark matter constraints from Sun and stars

In the following, I provide a few examples for which DM has quite distinct and important effects in the evolution of stars.

The case of a star evolving in a DM of high density: the gravitational contraction of a star during the pre-main sequence phase evolving within DM halo is identical, but not equal to a normal star. In the case of a star forming in a normal molecular cloud the self-gravity of the star leads to a rapid contraction with an increase of its internal temperature, the star follows a well defined path in the H-R diagram [1]. The evolution of this star within a dense DM halo is almost the same, nevertheless due to the extra source of energy provided by DM annihilation, the star gravitational contraction progresses at a much lower time step. This DM effect has a major implication for stellar clusters, namely by changing the location of main-sequence of stars, and the turnoff point [2] at which stars leave the main-sequence. This effect is shown in the isochrones[3] of a stellar population formed in a high density DM halo. In particular, I notice that *the main sequence of low mass stars* has shifted to a high value of Luminosity when compared with a normal stellar cluster (cf. Fig. 4). If such type of isochrones are discovered in a stellar cluster on first generation of stars or at the core of galaxies including in the Milky-Way, then this is a powerful indication that dark matter is contributing for their evolution [62–65].

[1] Hertzsprung-Russell (H-R) Diagram: Luminosity vs. effective temperature of stars - each star of a fix mass describes a unique path in this diagram.

[2] The turnoff point for a star refers to the point on the H-R diagram, where the star leaves the main-sequence after the exhaustion of its main fuel.

[3] Luminosity-temperature plot of a cluster of stars with different masses at a fixed age.

The case of a star evolving in a DM of low density: Another type of impact of dark matter in stars, like the Sun, occurs for stars formed in the disk of galaxies, where the DM density is much smaller than in the case previously discussed. For a star like the Sun, the capture of DM depends strongly of the scattering cross section with baryons: If the DM particle has a relatively large scattering cross section with Hydrogen (spin dependent cross section) or with heavy elements (like Oxygen, Iron among others, spin independent), the number of collisions between DM particles and Baryons is high, consequently in the extreme cases this leads to the formation of an isothermal core [19], otherwise this effect only reduces the central temperature of the star. By using the current probes of the solar interior, such as solar neutrinos, we are able to put constraints on DM parameters. At present, only 3 sources of solar neutrinos (like ^{8}B, ^{7}Be ad PeP) are observed by the current solar neutrino detectors (Borexino, SNO and Kamiokande), for which all are sensitive to the temperature of the Sun's core. Fig. 5 shows the DM impact on ^{8}B neutrino flux for DM solar models, for which the ^{8}B neutrino flux is compared with the ^{8}B flux of the SSM. This example shows DM particles for which the ^{8}B neutrino flux variation is larger than 15% which is larger than the current uncertainty of the SSM, therefore such DM particles can be excluded [22]. These DM constraints are based on the hypothesis that the ^{8}B neutrino flux difference between the observed and predicted SSM is smaller than 15%.

Fig. 6 shows a second interesting example, which uses data of solar neutrino flux measurements: Feng *et al.* [67] have shown the previous experimental data of DM search (including the exclusion limits considering the usual interaction between DM and the target baryons), could be reconciled if one considers that interaction of the DM particle with the target baryon occurs assuming iso-spin violation. As a consequence the spin-independent scattering cross-section of the DM-target Baryon increases from $10^{-40}cm^2$ to $10^{-37}cm^2$. Lopes and Silk [20] have shown that ^{8}B neutrino fluxes can reject the lower masses and higher values of scattering cross section. Further future ^{8}B neutrino flux measurements could lead to much better constraints (cf. Fig. 6).

I would like to notice that this type of analysis could be strongly improved if we could use other sun-like stars of different masses from the Sun for which the impact of dark matter could be more visible. In a recent article Casanellas and Lopes [32] use oscillation spectra of the stars HD 52266 (1.1 $M_\odot$) and alpha Centari B (0.9 $M_\odot$) to put constraints in DM (see poster Casanellas and Lopes). In particular, they found that particles with a mass of $5GeV$ and a spin-dependent scattering cross-section smaller than $3 \times 10^{-36}cm^2$ can be excluded with a 5-σ level.

Finally, I hope that in this talk I have convinced you that the solar and stellar physics and their related fields of nuclear physics can provide fundamental tools to test the different experimental and theoretical DM particle candidates. The author thanks the Astrophysics Portuguese Society (SPA) and the organisation for the invitation to participate in the conference.

References

[1] Casanellas J, Pani P, Lopes I and Cardoso V 2012 *The Astrophysical Journal* **745** 15
[2] Lopes I P and Silk J 2003 *Monthly Notice of the Royal Astronomical Society* **341** 721–728
[3] Lopes I and Turck-Chieze S 2013 *The Astrophysical Journal* **765** 14
[4] Lopes I 2013 *Physical Review D* **88** 45006
[5] Kumar J 2013 *arXiv.org* 4513 (*Preprint* 1308.4513)
[6] Zurek K M 2013 *arXiv.org* 338 (*Preprint* 1308.0338)
[7] Petraki K and Volkas R R 2013 *arXiv.org* 4939 (*Preprint* 1305.4939)
[8] Turck-Chieze S, *et. al.* 1997 *Solar Physics* **175** 247–265
[9] Basu S, Chaplin W J, Elsworth Y, New R and Serenelli A M 2009 *The Astrophysical Journal* **699** 1403–1417
[10] Lopes I and Silk J 2013 *MNRAS* 1–8
[11] Komatsu E, *et. al.* 2011 *The Astrophysical Journal Supplement* **192** 18
[12] Hinshaw G, *et. al.* 2012 *arXiv.org* 5226 (*Preprint* 1212.5226)
[13] Ade P A R,*et. al.* 2013 *arXiv.org* 5062 (*Preprint* 1303.5062v1)

[14] Lopes I and Silk J 2012 *The Astrophysical Journal* **757** 130
[15] Turck-Chieze S and Lopes I 1993 *Astrophysical Journal* **408** 347–367
[16] Serenelli A M, Haxton W C and Pena-Garay C 2011 *The Astrophysical Journal* **743** 24
[17] Turck-Chieze S and Lopes I 2012 *Research in Astronomy and Astrophysics* **12** 1107–1138
[18] Turck-Chieze S and Couvidat S 2011 *Reports on Progress in Physics* **74** 6901
[19] Lopes I P and Silk J 2002 *Physical Review Letters* **88** 151303
[20] Lopes I and Silk J 2012 *The Astrophysical Journal* **752** 129
[21] Lopes I P, Silk J and Hansen S H 2002 *Monthly Notices of the Royal Astronomical Society* **331** 361–368
[22] Lopes I and Silk J 2010 *Science* **330** 462–
[23] Lopes I and Silk J 2010 *The Astrophysical Journal Letters* **722** L95–L99
[24] Taoso M, Iocco F, Meynet G, Bertone G and Eggenberger P 2010 *Physical Review D* **82** 83509
[25] Cumberbatch D T, Guzik J A, Silk J, Watson L S and West S M 2010 *Physical Review D* **82** 103503
[26] Kouvaris C 2012 *Physical Review Letters* **108** 191301
[27] Kouvaris C and Tinyakov P 2011 *arXiv* **83** 83512
[28] Kouvaris C and Tinyakov P 2011 *Physical Review Letters* **107** 91301
[29] Kouvaris C and Tinyakov P 2010 *Physical Review D* **82** 63531
[30] Kouvaris C and Tinyakov P 2012 *arXiv.org* 4075 (*Preprint* `1212.4075`)
[31] Casanellas J and Lopes I 2011 *Monthly Notices of the Royal Astronomical Society* **410** 535–540
[32] Casanellas J and Lopes I 2013 *The Astrophysical Journal Letters* **765** L21
[33] Baglin A,*et. al.* 2007 *Fifty years of romanian astrophysics. AIP Conference Proceedings* pp 201–209
[34] Koch D G, *et. al.* 2010 *The Astrophysical Journal Letters* **713** L79–L86
[35] Lopes I, Casanellas J and Eugénio D 2011 *Physical Review D* **83** 63521
[36] Perryman M A C, *et. al.* 1997 *Astronomy and Astrophysics 323* **323** L49–L52
[37] Drees M, Iminniyaz H and Kakizaki M 2006 *Physical Review D* **73** 123502
[38] Iminniyaz H, Drees M and Chen X 2011 *Journal of Cosmology and Astroparticle Physics* **07** 003
[39] Davoudiasl H, Morrissey D E, Sigurdson K and Tulin S 2011 *Physical Review D* **84** 96008
[40] Dent J B, Dutta S and Scherrer R J 2010 *Physics Letters B* **687** 275–279
[41] Bertone G, Hooper D and Silk J 2005 *Physics Reports* **405** 279–390
[42] Mo H, van den Bosch F C and White S 2010 *Galaxy Formation and Evolution*
[43] Guo Q, *et. al.* 2013 *Monthly Notices of the Royal Astronomical Society* **428** 1351–1365
[44] Belli P,*et. al.* 2011 *Physical Review D* **84** 55014
[45] Aalseth C E, *et. al.* 2011 *Physical Review Letters* **106** 131301
[46] Brown A, Henry S, Kraus H and McCabe C 2012 *Physical Review D* **85** 21301
[47] Agnese R,*et. al.* 2013 *arXiv.org* 4279 (*Preprint* `1304.4279`)
[48] Aprile E,*et. al.* 2011 *Physical Review Letters* **107** 131302
[49] Angle J,*et. al.* 2011 *Physical Review Letters* **107** 51301
[50] Chang S *et. al.* 2010 *Journal of Cosmology and Astroparticle Physics* **08** 018
[51] Farina M *et. al.* 2011 *Journal of Cosmology and Astroparticle Physics* **11** 010
[52] Hooper D and Kelso C 2011 *arXiv.org* 83001 (*Preprint* `1106.1066v1`)
[53] Del Nobile E, Kouvaris C and Sannino F 2011 *arXiv.org* 27301 (*Preprint* `1105.5431v1`)
[54] Fornengo N, Panci P and Regis M 2011 *Physical Review D* **84** 115002
[55] Bell N F, Melatos A and Petraki K 2013 *Physical Review D* **87** 123507
[56] Casanellas J and Lopes I 2011 *The Astrophysical Journal Letters* **733** L51
[57] Lopes I P, Bertone G and Silk J 2002 *Monthly Notice of the Royal Astronomical Society* **337** 1179–1184
[58] Khlopov M Y and Kouvaris C 2008 *Physical Review D* **78** 65040
[59] Catena R and Ullio P 2010 *Journal of Cosmology and Astroparticle Physics* **08** 004
[60] Bovy J and Tremaine S 2012 *The Astrophysical Journal* **756** 89
[61] Garbari S, Liu C, Read J I and Lake G 2012 *Monthly Notices of the Royal Astronomical Society* 3493
[62] Casanellas J and Lopes I 2009 *The Astrophysical Journal* **705** 135–143
[63] Scott P *et. al.* 2011 *The Astrophysical Journal* **742** 129
[64] Sivertsson S and Gondolo P 2011 *The Astrophysical Journal* **729** 51
[65] Ilie C *et. al.* 2012 *Monthly Notices of the Royal Astronomical Society* **422** 2164–2186
[66] Turck-Chieze S *et. al.* 2012 *The Astrophysical Journal Letters* **746** L12
[67] Feng J L, Kumar J, Marfatia D and Sanford D 2011 *Physics Letters B* **703** 124–127

SNO+ status and plans for double beta decay search and other neutrino studies

S Andringa, for the SNO+ Collaboration
LIP, Av. Elias Garcia, 14, 1; 1000-149 Lisboa; Portugal
E-mail: sofia@lip.pt

Abstract. SNO+ is a multi-purpose Neutrino Physics experiment, succeeding to the Sudbury Neutrino Observatory by replacing heavy water with liquid scintillator, which can also be loaded with large quantities of double-beta decaying isotope. The scientific goals of SNO+ are the search for neutrinoless double-beta decay, the study of solar neutrinos and of anti-neutrinos from nuclear reactors and the Earth's natural radioactivity, as well as supernovae neutrinos. The installation of the detector at SNOLAB is being completed and commissioning has already started with a dry run. The detector will soon be filled with water and, later, with scintillator. Here we highlight the main detector developments and address the several Physics analysis being prepared for the several planned SNO+ runs.

1. The SNO+ experiment and its detector

SNO+ re-uses the hardware of the Sudbury Neutrino Observatory detector (see figure 1 and ref. [1] for a global view), located at a depth of 6000 meter water equivalent in SNOLAB, Canada. The active medium changed from heavy water in SNO to a specially designed liquid scintillator in SNO+, allowing for measurements below 1 MeV. The energy and position of each event are reconstructed from the signals of an array of around 9500 PMTs overseeing the isotropic fluorescence light and depend on the precise optical characteristics of the several media.

A new rope system holding the spherical acrylic vessel that separates the active volume from the outside water shielding is already in place, and its capacity to now accommodate a positive buoyancy, will be tested as the detector is filled with water. Also new purification, calibration and monitoring systems are needed to ensure higher levels of radio purity to match the lower energy threshold. The water purification system is an upgrade of the one used in SNO, the scintillator purification is designed to match the goal of 10^{-17} g/g U/Th, as achieved by Borexino; in addition also the loading material for the double beta phases must be purified though to a lower level. A new system is being prepared for the insertion of well known sources in different positions inside the active volume, in order to providing an end-to-end calibration of all the reconstruction and analysis chain. These include alpha, beta, gamma and neutron sources at different energies and also a laser light source for PMT and optical calibration.

In order to allow for an effective monitoring of the detector characteristics with reduced access to the active volume, direct light sources are located in the outer volume and crossing the inner sphere. In this case an array of more than 100 optical fibers carry LED light signals down to the PMT support structure, in short pulses that can be used during normal data-taking. Large aperture fibers allow the calibration and synchronization of the PMTs viewing the same signals,

Figure 1. The SNO+ detector consists of a 12 meter diameter inner sphere, where the active volume is held by a 5 cm thick acrylic vessel; and a 60% optical coverage given by 9500 PMTs placed in a 17 meter diameter geodesical structure, with inside and outside shielding by pure water.

and a partial overlap allows the coverage of the full sphere. Small aperture fibers are used to monitor light attenuation and scattering of the scintillator as a function of time. Part of this system has been installed from the cavity flor and tested in air with the detector empty, the rest will be installed during the water fill. At the same time PMT repairs will also be done.

The experiment will be divided in different phases. Pure liquid scintillator might allow for the measurement of all the solar neutrino lines, depending on the background conditions to be measured in-situ. In order to search for neutrinoless double beta decay, less dependent on low energy backgrounds, loading with specific isotopes is being explored. High intensity Supernovae signals and easily tagged anti-neutrino signals can be identified in all the phases of the experiment.

2. Double Beta Decay

There are only a few isotopes which undergo double beta decays, with simultaneous emission of two electrons and two anti-neutrinos. If the neutrinos are Majorana particles, an extra decay mode can exist in which the two electrons carry all the energy (Q) and no neutrinos are emitted. The neutrinoless double beta decay rate would be proportional to an effective neutrino mass depending on the mass-mixing parameters and Majorana phases. For each isotope it would depend also on phase-space parameters and on the nuclear matrix elements, different from the two-neutrino decay.

We have searched for those double beta decay isotopes which can be dissolved in liquid scintillator, without changing its optical characteristics too much. In fact, liquid scintillator experiments have in general a worse energy resolution than other detector types, but have the advantage of dealing with high quantities in a low background environment. A per-mil loading of a given isotope corresponds to hundreds of kilograms, hard to get in other detector types.

Journal of Physics: Conference Series **665** (2016) 012080 doi:10.1088/1742-6596/665/1/012080

SNO+ has concentrated on two isotopes [1]:

- ^{150}Nd has a high Q value of 3.37 MeV, for which a high energy resolution can be achieved and the main background comes from two-neutrino decay. The two neutrino decay has been measured[3] to be 9×10^{18} years to the ground state, with an extra mode with 10^{20} years. The neutrinoless decay has been calculated as $3 \times 10^{23}/M_{\beta\beta}^2$ years/$(meV)^2$. The natural abundance of the isotope is of only 5.6%, and although enrichment is possible it is nowadays difficult to find in the necessary quantities.
- ^{130}Te has a lower Q of 2.53 MeV, but a much higher natural abundance of 33%, allowing for a more effective loading. The main background is also the two neutrino decay mode, but with a lifetime measured[4] to be large, of 7×10^{20} years. The neutrinoless decay was calculatedto be $4 \times 10^{23}/M_{\beta\beta}^2$ years/$(meV)^2$, under the same framework [5].

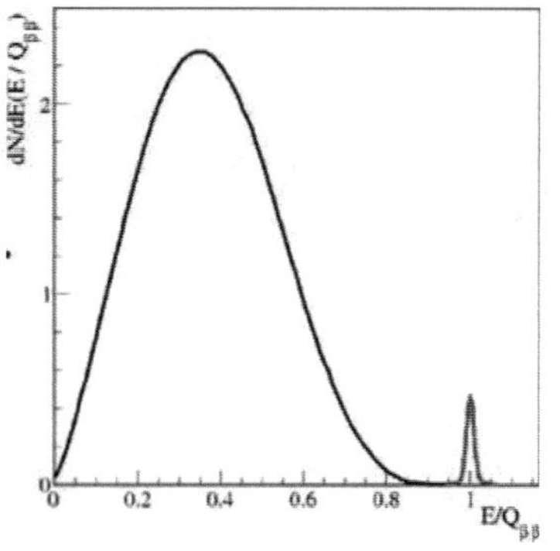

Figure 2. Schematic representation of the expected double beta spectrum. For each isotope, the integral of the continuous curve will be inversely proportional to the $2\nu\beta\beta$ half-life and the peak at Q to the $0\nu\beta\beta$ lifetime.

The two-neutrino decay will be the main background, not only due to the spill-over due to finite energy resolution but also to pile-up events which are reconstructed as a single high energy decay. Both energy and position resolution are fundamental for their treatment. In addition, radiopurity of the loaded scintillator, and namely the cosmogenic activation of some of the isotopes, must be taken into account. ^{214}Bi and ^{208}Tl decays sit in similar energy windows but can be reduced with $\alpha - \beta$ tagging. Solar ^8B neutrinos are an irreducible background, but their energy spectrum is known as is the rate (which can also be measured again as the high energy tail is visible), while all other solar neutrinos will be buried below the beta decay signals.

Preliminary tests indicate the possibility to use ^{130}Te with a 0.3% loading in the first phase of the experiment, which should reach a sensitivity down to 100 meV of effective mass. If the loading can be increased SNO+ could become sensitive to all of the degenerate and even inverted hierarchy mass scenarios.

3. Solar Neutrinos

The oscillation of solar neutrinos was established by SNO which also measured the total ^8B neutrino flux to be consistent with solar models[6]. In fact, the experimental result is more precise than the model predictions, but lays in between two predictions - according to low or high solar metalicity. The goal of the new generation experiments is now to measure the other solar neutrino lines to increase the sensitivity to both oscillation parameters and solar models.

In Table 1 we show the existing measurements and predictions for several solar lines from [7] together with the expected sensitivity of SNO+ after 1 and 3 years of pure liquid scintillator

[1] Another liquid scintillator experiment, KamLAND[2], used ^{136}Xe (with $Q = 2.46$ MeV and enriched from a natural abundance of 10%), and experienced problems with unexpected backgrounds.

Journal of Physics: Conference Series **665** (2016) 012080 doi:10.1088/1742-6596/665/1/012080

Table 1. Summary of Solar Neutrino Measurements, indicating for each line the main uncertainty source in the model, the existing measurements and the expected precision after a SNO+ solar phase of 1 or 3 years. Notice that hep neutrinos, with a very low cross-section are not included, that the pp rate is estimated from a global oscillation fit and that for ^7Be, there are two lines with a fixed ratio, but only one has been measured.

Type	Model	Measurement	SNO+ (1yr)	SNO+ (3yr)
B-8	nuclear(14%)	CC, NC (SNO) and ES	7.5%	6.5%
pep	luminosity(∼1%)	ES by Borexino at 20%	9.1%	6.5% %
Be7	nuclear(7%)	ES by Borexino at 5%	4%	3%
pp	luminosity(≤1%)	radio-chemical	a few %?	a few %
CNO	composition(15% to 30%)	un-measured	∼15%	≤ 15%

running. In comparison to other running experiments, namely Borexino, SNO+ has the advantage of a higher volume but mostly a deeper location. The cosmic ray muon flux at SNOLAB is 100 times smaller, which decreases the cosmogenic production of ^{11}C, whose decay represents an important background for both pep and CNO neutrinos. These are the two main targets of SNO+, for different reasons:

- pep neutrinos have a monochromatic energy of 1.44 MeV, and the solar models predict the flux with very small uncertainty. It's measurements is thus very sensitive to oscillations (and possibly also new physics) effects. The energy lays in the transition region between matter and vacuum dominated oscillations and can be used to measure oscillation parameters with very high precision.
- CNO neutrinos have not been measured yet, they are originated by three beta decays, and the total fluxes depend strongly on the parameters of solar model calculations. This measurement gives a direct probe of the solar metalicity.

The solar neutrino phase of SNO+ might be complicated due to the accumulation of radon daughters on the surface of the AV over the last few years in a significant way. If these leach into the scintillator, the purification system has the capability to remove them. However, depending on the actual leach rate, that removal might be inefficient and the ^{210}Bi levels in the scintillator too high for a pep/CNO solar neutrino measurement without further mitigation. Mitigation could include enhancing online scintillator purification, draining the detector and sanding the AV surface to remove radon daughters, or deploying a bag. $0\nu\beta\beta$ and low-energy ^8B solar neutrino are not affected by these backgrounds.

4. SuperNovae, Geo and Reactor Anti-Neutrinos

Other physics analysis will be pursued during all phases of the experiment. As SNO was, also SNO+ will be one of the neutrino detectors of the SuperNova Early Warning System (SNEWS) network, its size and energy threshold allows the measurement of hundreds of neutrinos and anti-neutrinos of all flavors in case of a Supernova explosion at 10 kiloparsec.

Anti-neutrinos can be also detected due to a characteristic delayed coincidence signal of the inverse beta decay process. In SNO+ the expected rate is of around 100 reactor anti-neutrinos a year. Sensitivity to the squared mass difference parameter becomes competitive with the existing results from KamLAND[8] after five years running because, although the flux is smaller, a lower number of reactors gives rise to very clear oscillation pattern, shown in figure 3.

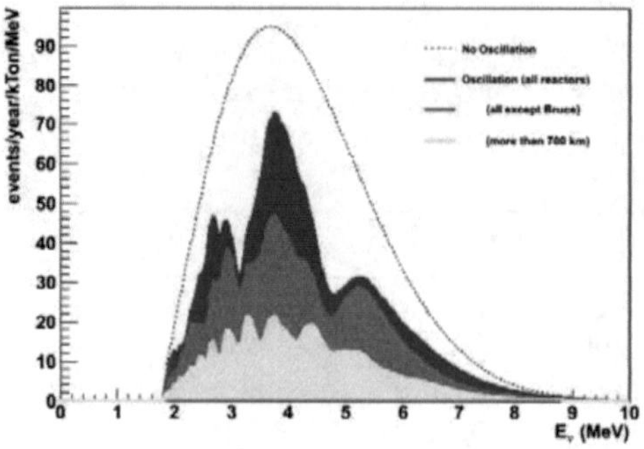

Figure 3. At SNO+, half of the reactor anti-neutrino flux comes from three nuclear power plants, one at 150 km south and two at 250 km south-east in Ontario, and present a very clear oscillation pattern. The rest of the flux is a combination of all other reactors in the world, for which the oscillation is smeared.

For energies below 2.5 MeV, half of the flux is expected to come from geo-neutrinos. Due to its location in a thick continental crust, SNO+ is expected to have an increased geo-neutrino flux compared to the two other experiments that have first measured geo-neutrinos, namely KamLAND[9] and Borexino[10]. While the low reactor rate will decrease the systematics from this background, statistical errors will be smaller because of the detector size. A full combination of the rates in different locations is needed in order to clearly separate the crust from the more interesting mantle contribution. Geo-neutrinos will finally be used to explore Earth models as was done before with Solar neutrinos[11].

5. Summary and Outlook

SNO+ is now completing the installation of the new systems in order to start data-taking in 2014. Some preliminary tests done with an empty detector have allowed early commissioning of PMT and new light calibration systems already. The detector will soon start water fill, allowing to reach the upper part of the detetor and to have first data taking in conditions similar to the latest SNO running. Water will be then replaced by pure liquid scintillator for a small commissioning phase.

In 2014, the scintillator will be loaded with 0.3% of ^{130}Te. In this first phase, neutrinoless double beta decay will be explored, with expected sensitivity to effective masses below 100 meV. Taking into account the results and also background conditions, a second phase will be designed. It can either concentrate on adding extra Te in the detector or to unload it and do a dedicated run for solar neutrinos. During all phases the experiment will look for Supernoave and Anti-neutrinos.

References

[1] SNO Coll. 2000 *Nucl. Inst. Meth.* A **449** 172
[2] KamLAND-Zen Coll. 2013 *Phys. Rev. Lett.* **110** 062502
[3] Argyriades J *et al.* 2009 *Phys. Rev.* C **80** 032501(R)
[4] Arnold R *et al.* 2011 *Phys. Rev. Lett.* **107** 062504
[5] J. Barea, J. Kotila and F. Iachello 2013, *Phys. Rev* C**87**014315;
J. Kotila and F. Iachello 2012, *Phys. Rev.*C**85**034316
[6] SNO Coll. 2010 *Phys. Rev.* C **81** 055504
[7] Borexino Coll. 2012 *Phys. Rev. Lett.* **108** 051302
[8] KamLAND Coll. 2008 *Phys. Rev. Lett.* **100** 221803
[9] KamLAND Coll. 2011 *Nature Geoscience* **4** 647
[10] Borexino Coll. 2013 *Phys. Lett.* B **722** 295
[11] Fogli G.L. *et al* 2010 *Phys. Rev.* D **82** 093006
[12] Perry H.K.C. *etal* 2009 *Earth Plan. Sci. Lett.* **288** 301

Perturbations in the upper layers of α Centauri A

A Brito[1,2] and I Lopes[1,3]

[1] CENTRA, Physics Department, Instituto Superior Técnico, Portugal
[2] Departement of Mathematics, Instituto Superior de Gestão, Portugal
[3] Department of Physics, University of Évora, Portugal

E-mail: ana.brito@ist.utl.pt, ilidio.lopes@ist.utl.pt

Abstract. The emerging field of asteroseismology allows the direct study of stellar interiors with an incredibly high precision. We used a seismic parameter based on the phase shift as a diagnostic tool to infer the presence of a new layer of rapid variation in the external layers of the primary component of the stellar system Alpha Centauri AB. This layer is, apparently, a thin region where the acoustic modes suffer a strong scattering. Our tests indicate that this layer should be located at an acoustical depth of approximately 1400 s (0.939 R), which corresponds to a depth of 6% below the surface of the star. This is somehow unexpected since the internal structure of this sun-like star is predicted to be similar to the Sun.

1. Introduction

Like in the Sun, many sun-like stars have a complex surface structure where it occurs some of the most complex fluid dynamics due to the interaction between magnetic fields, turbulent convection and stellar pulsations. In this work, we present a method to study the structure of the upper layers of a sun-like star as it is possible to be inferred from the most complete observational table of frequencies. In particular we apply the acoustic phase shift method to probe the physics mechanisms operating in these upper layers. We have found that in this sun-like star the existence of a rapid variation (RV) layer in the external region of the star will produce a significant amount of scattering of acoustic waves. This relatively thin layer is very likely produced by a rapid variation in the sound speed similar to variations of sound speed that can occur in transition layers like the known examples in the Sun of the bottom of the convection zone and the He II ionization zone. The diagnostic is done by means of a seismic parameter which allows to infer the scattering of acoustic waves in the surface of the star using the available table of observational frequencies.

2. Determination of the acoustic phase shift

In this work, we determine the seismic observable β, a proxy of the phase shift of the acoustic modes occurring in the envelope of the sun-like stars. We show that a small variation in the structure (typically smaller than $\sim 1\%$ in the sound speed) produces a glitch in the acoustic potential that could explain the origin of the seismic observable β. The fourth-order system of ordinary differential equations describing the linear adiabatic oscillations in the Cowling approximation (the gravitational effects are small in the outer layers) can be reduced to a second order Schrödinger-type equation considering that the trajectories of the acoustic waves in the surface layers are vertical [1]. It follows that the equation of acoustic oscillations reads

$$\frac{\mathrm{d}^2\psi}{\mathrm{d}\tau^2} + (\omega^2 - U^2)\psi = 0 \tag{1}$$

where τ is the acoustic depth, U the acoustic potential, $\omega(\equiv 2\pi\nu)$ the cyclic frequency and ψ the wave function. The acoustic potential U is defined by

$$U^2(\tau) = \frac{g}{c}\left(\frac{g}{c} - \frac{\mathrm{d}\ln h}{\mathrm{d}\tau}\right) + \frac{1}{4}\left(\frac{\mathrm{d}\ln\zeta}{\mathrm{d}\tau}\right)^2 - \frac{1}{2}\frac{\mathrm{d}^2\ln\zeta}{\mathrm{d}\tau^2} \tag{2}$$

with $h(r) = \rho^{-1}\exp\left(-2\int_0^r \frac{g}{c^2}\mathrm{d}r\right)$ and $\zeta(r) = \frac{r^2 h}{c}$, where the variables have their usual meaning [2].

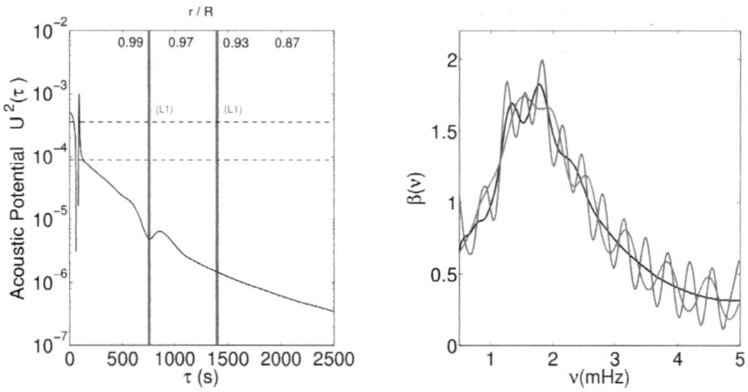

Figure 1. Left panel: The acoustic potential obtained from the stellar envelope of α Centauri A. We can notice the superadiabatic region at $\tau \sim 80$ s ($\sim 0.9931R$), the zone of partial ionization of helium at $\tau \sim 750$ s ($\sim 0.9840R$) and the base of the convective zone at $\tau \sim 2500$ s ($\sim 0.7759R$). The horizontal dashed lines correspond to waves with frequencies $\nu = 1.5$ mHz and $\nu = 3.0$ mHz. The vertical bar L1 located around $\tau = 750$ s (blue bar) indicates the position of one of our hypothesis for the existence of a rapid variation layer whereas the vertical bar L2 located around $\tau = 1400$ s (red bar) indicates the variation layer that better describe the observations. Right panel: The seismic parameter $\beta(\nu)$ in three cases: β obtained from the "unperturbed" acoustic potential (bell shape - dark line); β with a perturbation in the zone of the partial ionization of helium (oscillatory character and longer period - blue line) and finally the β with a perturbation located around the 1400 s (oscillatory character and shorter period - red line).

The outer phase shift of acoustic waves α is computed from the stellar envelope, by numerically solving the following equation

$$\frac{d\theta}{d\tau} = -\omega + \omega^{-1}U^2(\tau)\cos^2\theta \tag{3}$$

where $\theta = \alpha\pi + \pi/4 - \omega\tau$. This equation is solved numerically with the proper initial boundary condition at $\tau = 0$. The details of the numerical procedure can be found in [3]. $\beta(\nu)$ is obtained from numerical differentiation of α/ν as

$$\beta(\omega) = \beta(\nu) = -\nu^2\frac{\mathrm{d}}{\mathrm{d}\nu}\left(\frac{\alpha}{\nu}\right). \tag{4}$$

This method has been applied in the past to determine the outer phase shift by several authors [4], [5], [6] e [7]. The properties of the observable $\beta(\nu)$ were largely discussed among others by [8]. The acoustic potential U used in this calculation was determined from a detailed stellar structure model of α Centauri A specifically computed for this analysis. Fig. 1 shows the acoustic potential and the β calculated for the envelope of the star. This is identical but not equal to the acoustic potential and the calculated for the present standard solar model [10], [11].

The $\beta(\nu)$ seismic parameter is the best proxy of the phase shift of acoustic waves, that occurs as a consequence of the scattering of these waves in the upper layers of the star. $\beta(\nu)$, as previously mentioned, can be computed from a table (theoretical or observational) of frequencies. The algorithm reads

$$\beta_{l,n}(\nu) = \frac{\nu_{l,n} - L\frac{\partial\nu}{\partial L} - n\frac{\partial\nu}{\partial n}}{\frac{\partial\nu}{\partial n}}, \tag{5}$$

where $L = l + 1/2$ and l and n are the degree and radial order of the oscillation mode. The partial derivatives in $\beta_{l,n}(\nu)$ were treated carefully due to the small number of points available in the observational frequency table. This algorithm is based upon the formula proposed by [12], to study the acoustic phase shift in the Sun. $\beta_{l,n}(\nu)$ computed as indicated in equation (5) allows a direct comparison between the star and its stellar model, specifically, highlighting the differences of the physical processes occurring in the upper layers of the star.

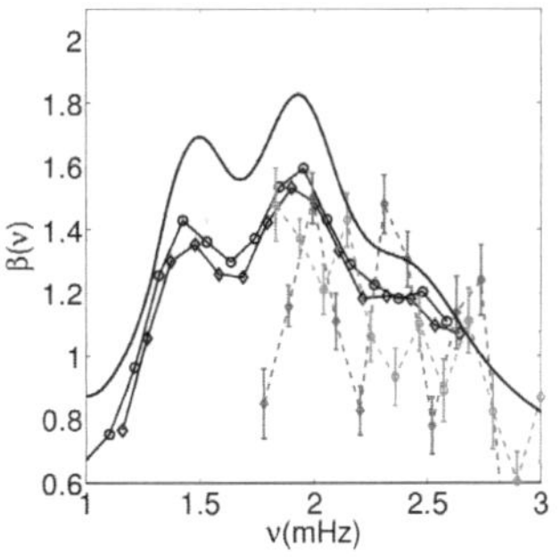

Figure 2. Theoretical β computed from an evolution model of the star α Centauri A: the solid smooth curve corresponds to β obtained from the surface acoustic potential (fig. 1a). We have also calculated β from a theoretical frequency table for the models with degrees l = 1 (diamonds) and l = 2 (circles), these points are connected with solid black lines. With dashed colored lines we represented the observational β for modes also with the degrees l = 1 (magenta) and l = 2 (green).

3. Test case: α Centauri A

In what concerns the observational part, we used the frequency table of [13] to compute β. Several tests were performed to verify the quality of the algorthim to compute β from this observational frequency table. Looking at Figure 2 we can notice a dispersion in the observational modes that is not present in the theoretical modes.

A detailed analysis of the observed β suggests that the real star could have a small rapid varition zone that produces a glitch in the acoustic potential located in the external layers of the star. In Figure 3 we compare the red and the blue (modified; shorter period and larger period respectively) with the black (unmodified) scenarios. These tests with the modified

Journal of Physics: Conference Series **665** (2016) 012081 doi:10.1088/1742-6596/665/1/012081

acoustic potential are able to reproduce the oscillatory behaviour of β better than the unmodified potential.

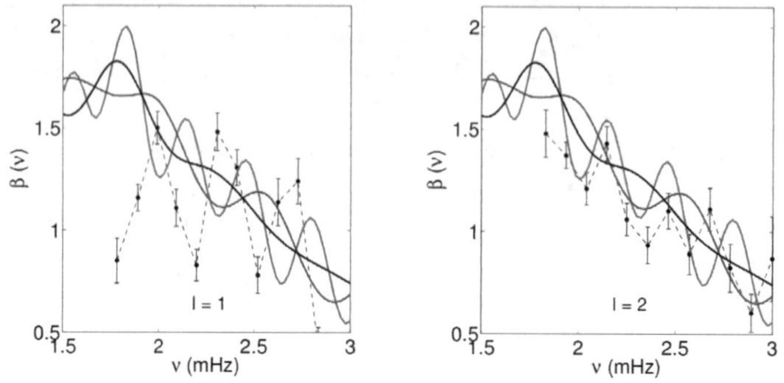

Figure 3. In this figure we compare the two modified scenarios of Fig.1 (blue and red) with the observational signature β (dashed lines) for the modes with the degrees l = 1 and l = 2. We noticed that the modified acoustic potential can reproduce the oscillatory character of the observational β, and the location the better explains the period of the oscillation is arround 1400 s (red line).

4. Summary and conclusion

We study a Sun-like star from the point of view of the acoustic phase shift. Specifically we wanted to know the impact of a small variation of the sound speed in the acoustic potential and consequently in the seismic paramenter β. We derived $\beta(\nu)$ from an observational frequency table, of the star α Centauri A and noticed that there is a disagreement between the theoretical β (obtained numerically by integrating the wave equation through the external layers and from a theoretical frequency table) and the β calculated directly from an observational table. This disagreement is characterized by an oscillatory feature that is present in the obervational β but is absent in the theoretical model. This method allows to determine with precision the location and thickness of a glitch in a star. A preliminary analysis of α Centauri A suggests a glitch located at $\sim 0.94\,R$ ($\sim 1400\,$s) and a thickness $< 0.2\,\%\,R$. This thecnique can also be used together with another different methods of analysis [14] to improve the identification of the location of acoustic glitches in the interios of sun-like stars.

References

[1] Vorontsov S and Zharkov V 1989 *Astrophysics and Space Physics Reviews*, **7**, 1
[2] Unno W, Osaki Y, Ando H, Saio H and Shibahashi H 1989 Nonradial oscillations of stars, Tokyo: University of Tokyo Press
[3] Lopes I and Gough D 2001 *MNRAS*, **322**, 473
[4] Christensen-Dalsgaard J and Perez Hernandez F 1992 *MNRAS*, **257**, 62
[5] Roxburgh I and Vorontsov S 1994 *MNRAS*, **267**, 297
[6] Roxburgh I and Vorontsov S 1994 *MNRAS*, **268**, 143
[7] Roxburgh I and Vorontsov S 1994 *MNRAS*, **268**, 880
[8] Lopes I, Turck-Chièze S, Michel E and Goupil M-J 1997 *ApJ*, **480**, 794
[9] Miglio A, Montalbán J 2005 *A&A*, **441**, 615
[10] Turck-Chieze S and Lopes I 1993 *ApJ*, **408**, 347
[11] Turck-Chieze S and Lopes I 2013 *ApJ*, **765**, 14
[12] Brodskii M and Vorontsov S 1989 *Soviet Astronomy Letters*, **15**, 27
[13] de Meulenaer P, Carrier F, Miglio A, Bedding T, Campante T, Eggenberger P, Kjeldsen H and Montalbán J 2010 *A&A*, **523**, A54
[14] Mazumdar A et al 2012 *Astronomische Nachrichten*, **333**, 1040

Journal of Physics: Conference Series **665** (2016) 012082 doi:10.1088/1742-6596/665/1/012082

Stellar Structure and Evolution With Varying Fundamental Couplings

J P P Vieira, C J A P Martins, M J P F G Monteiro

Centro de Astrofísica da Universidade do Porto
Rua das Estrelas
4150-762 Porto, Portugal

E-mail: Jose.Vieira@astro.up.pt, Carlos.Martins@astro.up.pt, Mario.Monteiro@astro.up.pt

Abstract. We discuss the feasibility of using solar-type main-sequence stars as probes of fundamental physics and unification. We use a simple polytropic stellar structure model and study its sensitivity to variations of the gravitational, strong and electroweak coupling constants in the context of unification scenarios. We quantify the sensitivity of the Sun's interior temperature to these variations, finding $|\Delta\alpha/\alpha| < 1.3\times10-4$ for a 'canonical' choice of unification scenario, and discuss prospects for future improvements. Further details can be found in [1].

1. Introduction

The advent of computational astrophysics has brought about an arsenal of powerful tools to test fundamental Physics in fairly complex astronomical systems. In particular, these new methods can be specially relevant in the context of the discussion of spacetime-dependent cosmological phenomena.

In this work, such a computational approach is presented for the case of varying fundamental couplings in a solar-type star: describing this star by a simple polytropic model, we are able to assess the sensitivity of the structure of the star to variations of the electromagnetic and gravitational fine-structure constants. The formalism used is valid in a broad class of unification models [1].

2. Phenomenology of Unification

We wish to describe phenomenologically a broad class of models which allow for simultaneous variations of several fundamental couplings, such as the electromagnetic fine-structure constant $\alpha=e^2/(\hbar c)$ and its gravitational analog $\alpha_p=Gm_p^2/(\hbar c)$.

The relations between the couplings are model-dependent. Here we will follow [2], considering a class of grand unification models in which the weak scale is determined by dimensional transmutation and further assuming that relative variation of all the Yukawa couplings is the same. Finally we assume that the variation of the couplings is driven by a dilaton-type scalar field. In this case one finds that the variations of α_p and α are related through

$$\frac{\Delta\alpha_p}{\alpha_p} = [0.8R + 0.2(1+S)]\frac{\Delta\alpha}{\alpha} \qquad (1)$$

where R and S can be taken as free phenomenological parameters. Their absolute value can be anything from order unity to several hundreds, but while R can be positive or negative (with the former case being more likely), physically one expects that S > 0. Nevertheless, we can simply treat both as phenomenological parameters to be constrained by astrophysical data.

3. Polytropic Stars

A polytropic star is a very well studied simplified model for the structure of a star in equilibrium, built from the standard equations of continuity and hydrostatic equilibrium as well as a polytropic equation of state, respectively:

$$\frac{dm}{dr} = 4\pi r^2 \rho, \qquad \frac{dp}{dr} = -\frac{Gm\rho}{r^2}, \qquad p = K\rho^{1+1/n}, \tag{2}$$

where r is the radial distance to the centre of the star, m is the mass within the sphere of radius r (assuming spherical symmetry), r is the density, p is the pressure, n is the polytropic index (to be selected), and K is the polytropic constant (to be defined by boundary conditions). The use of this relation avoids the need to include an additional equation for temperature, describing the transport of energy in the interior.

In such a simplified approach the structure of the star is described by the numerical solution of the Lane-Emden Equation, as given by

$$\frac{1}{\xi^2}\frac{d}{d\xi}\left(\xi^2 \frac{d\theta}{d\xi}\right) = -\theta^n \tag{3}$$

by making $\theta^n \equiv \rho_c$ and $\xi \equiv r/a$ where ρ_c is the value of the density at the centre of the star and $a^2 = K(n+1)/(4\pi G \rho_c^{1-1/n})$. The initial conditions that make physical sense are $\theta(\xi=0)=1$ and $\theta'(\xi)=0$. For any $n \in [0,5[$, there is $\xi_s^2 \geq 6$ such that ξ_s is a zero of $\theta(\xi)$ - in fact, we choose it to be the first such zero. We also define θ_s' to be the derivative of the same function in this point. Then, if we consider a star with total mass M_* and radius R_* (not to be confused with the unification parameter R), we have

$$\rho_c = \left(-\frac{\xi_s}{3\theta_s'}\right)\frac{3M_*}{4\pi R_*^3}, \quad K = \frac{G}{n+1}\left[\frac{4\pi}{\xi_s^{n+1}\left(-\theta_s'\right)^{n-1}} \frac{M_*^{n-1}}{R_*^{n-3}}\right]^{1/n}. \tag{4}$$

With this, all of the structural details of our model of a star are defined in terms of n. For a real star, n can be fixed by using its luminosity , L_*, as a boundary condition:

$$L_* = L_n = \frac{M_*}{\xi_s^2\left(-\theta_s'\right)} \int_0^{\xi_s} \varepsilon \xi^2 \theta^n d\xi, \tag{5}$$

where ε is the emissivity, or energy production rate per unit mass and time, inside the star. In general, ε is a function of ξ whose expression will depend on the energy production process considered. The version we adopt here are the global fits for the emissivity of the PP chains (relevant for low mass stars, like the Sun or late-type stars).

4. Impact On The Star From α

For the emissivity, ε_{ij}, of the reaction between species i and j, the following family of expressions can be used;

$$\frac{\varepsilon_{ij}}{m_p^{-1}\hbar^1 c^{-1}} = f\sqrt{\frac{8}{3A_{ij}}}\left(\frac{\pi^2\omega_{ij}}{2}\right)^{1/6} \frac{S_0}{m_p^{-1}\hbar^2} \frac{K^{-2/3}\rho_c^{1-2/3n}}{m_p^4\hbar^{-3}c^{5/3}} \frac{\alpha^{1/3}}{\theta^{n+2/3}} \exp\left[-3\left(\frac{\pi^2\omega_{ij}}{K\rho_c^{1/n}c^{-2}}\frac{\alpha^2}{\theta}\right)^{1/3}\right] \tag{6}$$

where f is the energetic gain of the nuclear reaction in units of m_pc^2, A_{ij} is the reduced atomic mass in units of mp, $w_{ij}=z_iz_jA_{ij}$ where z is the atomic number and S0 the reference value for the S-factor (the effective cross section). This is the expression that will go into equation (5).

A model of a specific star, when we have (M_*,R_*,L_*), can be obtained by solving equation (5) together with equation (6), where we impose that $L_n = L_*$. The solution provides the behavior of ρ and p. Any change in the physics will imply a different solution in the interior for the same global parameters. The model accommodates such a change by having a slightly different value of n that fulfills the equation for the luminosity. This approach allows us to obtain quantitative estimates of the relative change of the stellar structure when the underlying physics changes. Although a polytrope is not the best available model of a real star, it is a close enough model for our purposes. The assumption behind this approach is that the relative change from the real star is of the same magnitude as the relative change of the polytropic model, when every other unknown aspect of the physics is not allowed to change.

5. Results and Constraints

From observations of a specific star, the value of its stellar parameters may be obtained, specifically its mass M_*, radius R_* and luminosity L_*. This allows us to build a model, assuming that the set of equations and physics we use are valid, that fits these boundary conditions and describe the structure of the interior. However, if our physics is changed by a small amount, we must find a new model, adjusted to the same boundary conditions, but with a slightly different structure. This approach can also be implemented for polytropic models, which allow us to quantify with great precision the change that one single aspect of the physics can have on the structure of the interior. Specifically, we can determine by how much specific thermodynamic quantities will change in the interior when we introduce a small change in the physical constants entering the basic equations of stellar structure.

In order to have a useful test, we must refer to a quantity that can be obtained from the observation of real stars. The ideal candidate is the value of the temperature near the centre, since this value has a direct effect on sound speed or neutrino production. The former can be expected to be measured with significant precision with asteroseismology (or helioseismology for the solar case) while the latter may be used for the Sun. We do not include the accumulated effect in time, since that requires a proper modeling of the evolution which is beyond the scope of the present work.

Presently, the precision we have for the temperature (or sound speed) in the interior of the Sun, is well below 1%. While far from this, a very high precision is also expected from the asteroseismology of other stars, as is currently being obtained with data from the NASA space mission Kepler.

We thus use the formalism in the previous sections to estimate the impact that variations of α will have in the internal structure of a star for a given unification model (parametrized by specific values of R and S). We consider a polytropic star similar to our Sun (assumed to have a politropic index n_0), to which will correspond a slightly different value for n. This star will have the same value of M_*/m_p, $R_*/(m_p^{-1}\hbar c^{-1})$, and $L_*/(m_p^2\hbar^{-1}c^4)$ (M_*, R_* and L_* are the Sun's mass, radius and luminosity) but for a range of values where α differs from the standard value. We then calculate by how much its central temperature (in fact the ratio $K_BT_c/(m_pc^2)$, where T_c is the central temperature) must differ from that of our Sun, by solving equation (6) to determine the value of n.

We have fixed specific values of $\Delta\alpha/\alpha$ and S = 0 and let R vary freely. Since we are only considering the PP chains the main sensitivity of the polytropic model will be to changes in α_p (as well as α itself), and therefore we can only hope to constrain a linear combination of R and S. We may then assume S = 0 and get a constraint for R, which from equation (1) will in fact be a constraint on R + S/4. We have also assumed $R_{min}\approx-R_{max}$ in the range we have considered; $10^{-8} < |\Delta\alpha/\alpha| < 10^{-2}$.

This is an excellent approximation except at the top limit where, although $R_{max}/R_{min} \approx -0.7$, it is still adequate in the sense that $R_{max}+R_{min}\approx 0$. This is not critical because variations of this order are not expected in the light of current experimental constraints.

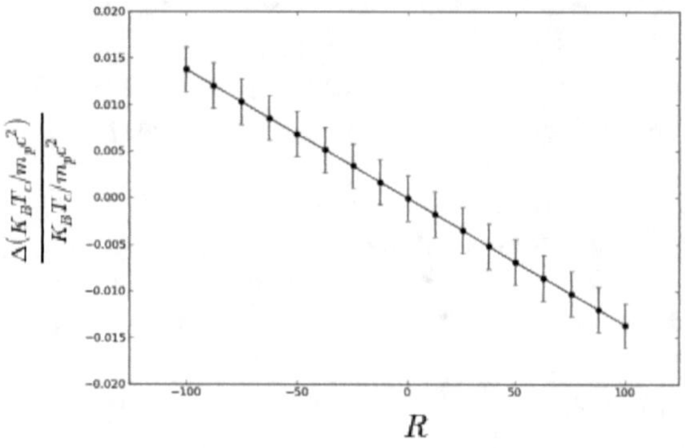

FIG. 1: The relative change of the Sun's central temperature, for $\Delta\alpha/\alpha=-10^{-4}$, as a function of the unification parameter R (with the choice S= 0). Allowing a maximal variation of one percent yields the limit |R|<90. For a description on how the error bars have been estimated, please see the main text.

In Fig. 1 we show an example for $\Delta\alpha/\alpha = -10^{-4}$, in which case Rmax≈90. The error bars are estimates of the upper limit for the uncertainty in our numerical calculation. It is related to both the error incurred when integrating the Lane-Emden equation in search of ξs and θs′ and the uncertainty associated with finding n. The latter is dominant due to the much greater precision that is possible when locating n numerically the surface of the polytrope.

By repeating the above analysis for different values of α it is possible to identify the region in the (R, S, α) parameter-space which is consistent with the currently estimated experimental uncertainty for the central temperature of our Sun. This is summarized in Fig. 2, and yields our final bound

$$|4R+S| \leq 10^{-1.44}\left|\frac{\Delta\alpha}{\alpha}\right|^{-1} \tag{7}$$

As a simple illustration, if we take the typical values suggested in [2] of R≈30 and S≈160, we find |Δα/α|<1.3×10−4. This constraint is slightly weaker than the one obtained in [4] for Population III stars, which is not surprising given the limitations of the polytropic model. Moreover, both our constraint and that of [4] are weaker than those obtained spectroscopically along the line of sight of quasars. However, we emphasise that such measurements are obtained in very low density absorption systems, whereas stars are much denser environments. The two types of contraints therefore complement each other for the purpose of constraining models where nature's fundamental couplings are environment-dependent.

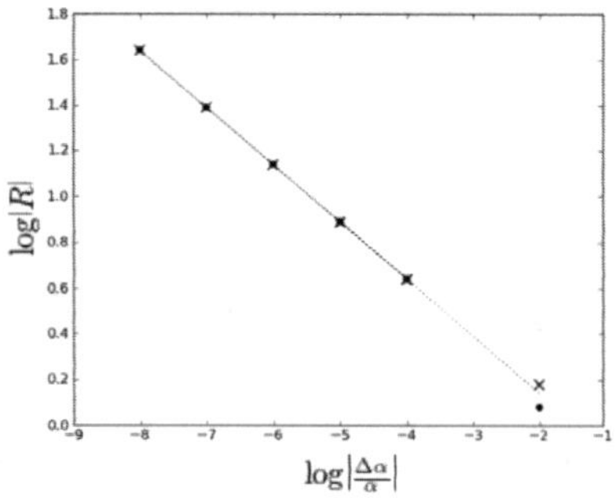

FIG. 2: Bounds on the (α,R) parameter space, for the case S = 0. The region above and to the right of the dashed line is excluded by our analysis. The dots (crosses) show the modulus of the estimated values of the maximum (minimum) allowed R for particular choices of α.

References

[1] J. P. P. Vieira, C. J. A. P. Martins, and M. J. P. F. Monteiro, Phys. Rev. D, 86, 043003_1-043003_6 (2012)
[2] A. Coc, N. J. Nunes, K. A. Olive, J.-P. Uzan, and E. Vangioni, Phys.Rev. D76, 023511 (2007), astro-ph/0610733
[3] F. C. Adams, JCAP 0808, 010 (2008), 0807.3697
[4] S. Ekstrom, A. Coc, P. Descouvemont, G. Meynet, K. A. Olive, et al., A. & A. 514, A62 (2010), 0911.2420